Y0-BTA-888

Nearrings

Geneses and Applications

JAMES R. CLAY

University of Arizona

Oxford · New York · Tokyo
OXFORD UNIVERSITY PRESS
1992

05135102

MATH-STAT.

Oxford University Press, Walton Street, Oxford OX2 6DP

Oxford New York Toronto
Delhi Bombay Calcutta Madras Karachi
Kuala Lumpur Singapore Hong Kong Tokyo
Nairobi Dar es Salaam Cape Town
Melbourne Auckland Madrid
and associated companies in
Berlin Ibadan

Oxford is a trade mark of Oxford University Press

Published in the United States
by Oxford University Press Inc., New York

© *James R. Clay, 1992*

All rights reserved. No part of this publication may be
reproduced, stored in a retrieval system, or transmitted, in any
form or by any means, without the prior permission in writing of Oxford
University Press. Within the UK, exceptions are allowed in respect of any
fair dealing for the purpose of research or private study, or criticism or
review, as permitted under the Copyright, Designs and Patents Act, 1988, or
in the case of reprographic reproduction in accordance with the terms of
licences issued by the Copyright Licensing Agency. Enquiries concerning
reproduction outside those terms and in other countries should be sent to
the Rights Department, Oxford University Press, at the address above.

This book is sold subject to the condition that it shall not,
by way of trade or otherwise, be lent, re-sold, hired out or otherwise
circulated without the publisher's prior consent in any form of binding
or cover other than that in which it is published and without a similar
condition including this condition being imposed
on the subsequent purchaser

A catalogue record for this book is available from the British Library

Library of Congress Cataloging in Publication Data
(data available)

ISBN 0–19–853398–5

Text keyed by the author using TeX
Typeset by the American Mathematical Society
Printed and bound in Great Britain
on acid free paper by
Bookcraft (Bath) Ltd, Midsomer Norton

QA 251
.5
C 53
1992
MATH

To

CAROL CLINE BURGE,

a truly beautiful daughter of Zion.

奇利教授先生雅正

奇士天縱土山解聲説術學

利智笑明四海平生論环環

辛未六月方源撰句杜崇語書

Preface

It has been interesting to organize the material for this book. For one thing, it has given me ample opportunity to reflect over 30 years of activity with nearrings. I must say that it has been a deeply rewarding three decades. There are two very basic reasons for my saying this. First, the subject itself is extremely challenging, offering curiously beautiful results to one who is willing to look for structure where symmetry is not so abundant. Second, the men and women working with this theory represent the finest colleagues with whom one could hope to associate. It has really been a pleasure working with them in developing what theory we have at this time, and I am looking forward to the next 30 years of discovery and association.

The first gathering of those keenly interested in nearrings took place in 1968 in Oberwolfach. Since then, there have been conferences about every two years in various places on three continents. It was very special for me to meet at this first conference men whom I have come to respect highly. Of those in attendance, who had already been very influential in the development of the theory, were H. Wielandt and his student and organizer of the conference, G. Betsch, Ali Fröhlich and his student R. R. Laxton, H. Karzel and his students W. Kerby, H. Wähling, and H. Wefelscheid. W. Nöbauer was absent, but was represented well by two of his students who have made a tremendous positive contribution over the past 25 years: W. Müller and G. Pilz. Also in attendance were others whose impact on the theory would be felt in the quarter century that followed the Oberwolfach Conference. G. Ferrero and K. Magill were there, as was C. J. Maxson. Over the years, these men have continued to make significant contributions. Others have joined the effort, and we have had sporadic visitors. One universally accepted fact is that this party of mathematicians are very dedicated and helpful one to another.

During the latter part of the 1980s, honest applications of nearrings, both to other areas of mathematics and to areas outside mathematics, were beginning to be recognized seriously. In addition, and perhaps even more important, the theory of nearrings is developing a strong personality of its own. Even though a field is a ring, the extra properties of a field give field theory a flavour all its own. The same is true about commutative ring theory, a step between ring theory and field theory. In general, nearrings do not have as much symmetry as rings, and even though every ring is a nearring, if one focuses on the properties that distinguish rings from near-

v

rings, one quickly realizes that the loss of symmetry in nearrings eliminates a lot of *obvious* beauty in these structures.

Further study has shown, however, that nearrings which are not rings, often have very intriguing beauties which completely disappear if one adds one or more of the missing properties of a ring to the nearring. This, of course, is the source of the strong personality of nearring theory. Examples of nearrings which are not rings are very easy to find, and they seem to arise in all branches of mathematics. These various families of nearrings provide the geneses for the theory, and it is then very natural that these families shape the personality of the theory.

In response to these honest applications and the development of a strong personality for nearrings, Dr Martin Gilchrist, Mathematics Editor for Oxford University Press, enquired in December 1988 of my interests in writing a book for OUP which would emphasize the applications of nearrings and provide the foundation for their serious study. Our first meeting took place at the University of Arizona in January 1989, and we felt it might be appropriate to develop a book on nearrings which could be useful for advanced graduate students beginning their serious research, and to all professional mathematicians wishing to broaden their research interests. A large part of the first five sections, together with a proposed table of contents, were sent by OUP to referees for their evaluations and comments. Encouraged and influenced by their report, the Delegates of the Press recommended issuing a contract. They agreed with the basic direction for the book, and recommended (1) emphasis on applications, (2) emphasis on foundations, and (3) suggested minimal overlap with the existing books on nearrings/nearfields by Günter Pilz, J. D. P. Meldrum, and Heinz Wähling.

These guidelines, together with the restrictions of time and page limitations, formed the basis for the choice of topics and the coverage given them. This, of course, leaves much to be included in future works. In particular, I regret not having an introduction to serious structure theory. This, on the one hand, should be included in a work intended for those doing serious research, but, on the other hand, the works of Pilz, Meldrum, and Wähling do provide extensive coverage of the subject. And one could extend Chapter 6 to a whole volume itself!

At this time it seems natural for me to reflect upon those whom I credit with having a strong positive influence upon my mathematical development. This is not to suggest that they would want to claim any responsibility, however. Among my best and most influential teachers are: Golden M. Wood of Burley High School, James R. Bland of the US Naval Academy, E. Allen Davis of the University of Utah, Ross A. Beaumont, R. S. Pierce, and E. Hewitt at the University of Washington, with R. S. Pierce also later at the University of Arizona, Ali Fröhlich at the University of London, and Helmut Karzel at the Technische Universität München. I feel the influence of these men in my life almost daily, and I deeply appreciate what I have learned from them.

It would be a rare event if a book of this type were to appear without the help of many others. I would say that the beginnings occured in 1972–1973 at Eberhard-Karls-Universität Tübingen. During that year, I was enjoying the benefits of a Distinguished US Senior Scientist Award administered by the Alexander von Humboldt Stiftung of Germany. This enabled Gerhardt Betsch and myself to conduct a seminar and to collaborate in research, and results of this seminar and collaboration had a tremendous positive impact for many years. I wish to thank Universität Tübingen for their hospitality during that year, and the people of Germany and the Humboldt Stiftung for this award which made my visit so pleasant.

Conversations with A. Fröhlich over the years have had a profound influence, as I hope can be seen by the content. In 1979–1980, while visiting the Technische Universität München, I further organized material which has influenced the content of this book. Life was considerably more pleasant during this visit because of the financial support from the Deutscher Akademischer Austauschdienst (DAAD) and the Technische Universität München. Also, conversations and collaboration with H. Karzel should be evident throughout these pages. During the 1980s, several students at the University of Arizona have tolerated unorganized lectures and notes while various ideas came together. In recent years, students and colleagues at the University of Stellenbosch, National Cheng Kung University, National Chiao Tung University, Johannes Kepler Universität Linz, Universität der Bundeswehr Hamburg, and the University of Arizona have commented on material in a more organized form.

I would especially like to express my gratitude to the National Science Council of the Republic of China for their support, making possible my visits to Taiwan and the collaboration there with Yuen Fong, Yeong-Nan Yeh, and Tayuan Huang. This collaboration resulted in several of the livlier passages of this work. Likewise, the Foundation for Research Development of the Republic of South Africa made possible my visit to the University of Stellenbosch, and my collaboration with Andries van der Walt during this time will be obvious in Sections 7 and 14. This was a very profitable visit, a visit that is still very much appreciated. During the latter part of 1990, Universität der Bundeswehr Hamburg provided a sanctuary with facilities and atmosphere conducive to much of the writing of these pages. Their hospitality, and especially that of Momme Thomsen, will always be treasured.

Individuals who made numerous positive contributions include Gary Birkenmeier, G. Alan Cannon, Henry Heatherly, Hubert Kiechle, Carl Maxson, Steve Olson, Günter Pilz, Momme Thomsen, and Joel Vaag. To these men I am truly grateful. Yuen Fong's contributions were particularly singular, and for these I am especially appreciative. Wen-Fong Ke, in addition to numerous routine contributions, has left an indelibly positive mark on this work. It should be noted that the contributions of J. D. P. Meldrum were absolutely indispensible. Thank you, John. It is an understatement to note that the staff at OUP has been very professional. Collectively, they did their best to disguise the fact that I was reared in Albion, Idaho. (Being brought up in Albion was *not* sufficiently British.) All remaining errors are the responsibility of the typist and typesetter, Irving P. Schwarz, who has been conditioned to accept such responsibilities. My final bouquet is for the contributions of Roxann Batiste with her copy machine. Her cheerful attitude and willingness to help others make a difference and is hereby permanently recognized.

Tucson   J. R. C.
1992

Le juge: Accusé, vous tâcherez d'être bref.
L'accusé: Je tâcherai d'être clair.

G. Courteline

Contents

CHAPTER 1. Introduction to nearrings   2

§1. Getting acquainted   3
§2. Lots of examples   7
§3. Many cheerful facts about nearrings   20

CHAPTER 2. Planar nearrings   38

§4. Planarity for nearrings   39
    4.1. The Ferrero Planar Nearring Factory   44
    4.2. Output of the Factory   48
§5. Construction of circular planar nearrings   57
    5.1. Basic results   57
    5.2. New structures from old   66
    5.3. Field generated designs   68
    5.4. Ring generated designs   75
    5.5. Other constructions   77
§6. Geometry of circular planar nearrings   84
    6.1. Equivalence classes   85
    6.2. The $(Z_q, \mathcal{B}_p^*, \in)$ 101
    6.3. Compound closed chains   109
§7. Other geometric structures from planar nearrings   115
    7.1. Tactical configurations   115
    7.2. Karzel's observations   126
    7.3. The van der Walt connection   136
    7.4. Karzel's hyperbolae   148
    7.5. The $(N, \mathcal{B}^-, \in)$ 155
    7.6. Partially balanced incomplete block designs   160
    7.7. Double planar nearrings   163
§8. Coding, cryptography, and combinatorics   165
    8.1. Planar nearrings and coding theory   166
    8.2. Planar nearrings and cryptography   178
    8.3. Planar nearrings and combinatorics   183
§9. Sharply transitive groups and nondesarguesian
    geometry from planar nearfields   191

CHAPTER 3. The great unifier   210

    §10. A little category theory   211
    §11. Group and cogroup objects   219
    §12. Examples   233
        12.1. Example 10   233
        12.2. Examples 11   245
        12.3. Examples 12   253

CHAPTER 4. Some first families of nearrings and
               some of their ideals   258

    §13. First, what is a nearring module?   258
    §14. Centralizer and transformation nearrings   270
    §15. Distributively generated nearrings   292
        15.1. Endomorphism nearrings of odd
                 generalized dihedral groups   294
        15.2. Distributively generated nearrings
                 from noncommutative formal group laws   312
    §16. The ideals of abstract affine nearrings   326
    §17. Polynomial nearrings   328
    §18. Power series nearrings   343

CHAPTER 5. Some structure of groups of units   359

    §19. Preliminaries   359
    §20. Direct products in groups of units   366
    §21. Semidirect products and wreath products   368
    §22. Group extensions with factor sets   373
    §23. A mixture of the above   380

CHAPTER 6. *Avant-garde* families of nearrings   383

    §24. Sandwich and laminated nearrings   383
    §25. Syntactic nearrings   397
    §26. The cornucopia   420
        26.1. Distributive nearrings   420
        26.2. Matrix nearrings   427
        26.3. Group nearrings   434
        26.4. Meromorphic products   439

APPENDIX. Various diagrams   445

LIST OF SYMBOLS   456

BIBLIOGRAPHY   459

INDEX   466

Prerequisites

As suggested in the Preface, this book was written with advanced graduate students in mind, those about to begin research for their dissertation. So I have assumed a good foundation in algebra, as might be found in Grove's *Algebra*, Hungerford's *Algebra*, Maclane and Birkhoff's *Algebra*, or Jacobson's *Basic Algebra I*. It will be a rare student who has mastered all the material found in these four excellent works, and, indeed, one will not be called upon to use even the majority of it here. But, it would be very difficult to give a precise description of the material I have used from works such as these.

An honest study of any of the above four works should be adequate preparation provided the reader is willing to supplement his background a little from time to time, when needed.

In addition, just a little knowledge of point set topology is needed in a few places, but these could be eliminated if it is not to the reader's taste.

Below we give a diagram of dependencies among the various sections. This diagram is very conservative. After the first chapter, it is suggested that the reader could study the remaining sections in nearly any order. As he comes to material obviously built upon material from a previous section not yet studied, then he should go back and study that material just enough to continue. As the reader finds material that is particularly interesting to him, then he could study it thoroughly until he is prompted to study something else. Of course, the smoothest path would be as the material is presented.

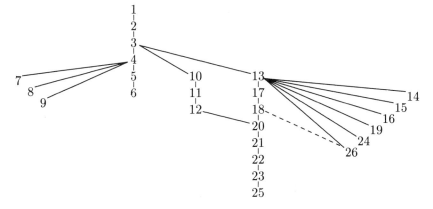

CHAPTER 1
Introduction to nearrings

Even though man has been dealing with nearrings and using their properties since the development of calculus, it was not until about 1905 that steps were taken to formalize the key ideas important to nearrings. Then, Dickson was interested in the independence of the field axioms, and he showed by examples that commutativity of multiplication and a second distributive law were not a consequence of the other field axioms. His example would be called a nearfield today, which is a very important type of nearring.

Its importance was signalled almost immediately when in 1907 Veblen and Wedderburn used Dickson's nearfields to give examples of nondesarguesian planes. Later, in the 1930s, H. Zassenhaus determined all the finite nearfields, a profound contribution, and used these in describing sharply transitive permutation groups.

In the late 1930s, Prof. H. Wielandt commenced the study of nearrings which were not nearfields, but it was not until the 1950s, commencing with Donald Blackett's paper [Bl], that research papers began to appear with any regularity on the subject. Prof. A. Fröhlich's work with distributively generated nearrings germinated considerable interest, and there are still many ideas planted in Fröhlich's papers, awaiting the 'right time' and conditions to develop.

One now recognizes that nearrings abound in all directions of mathematics, and enough study has been made to show that their structure has a beauty all its own, even though they lack the obvious beauty resulting from the extra symmetry provided by rings.

Applications of nearrings vary considerably. In addition to those in geometry and group theory already mentioned, their applications to various other branches of geometry are considerable. There are applications to combinatorics, to the design of statistical experiments, to coding theory, and to cryptography. A nearring is exactly what is needed to describe the structure of the endomorphisms of various mathematical structures adequately.

In this chapter, we introduce one to what a nearring is, where they can be found, and then explore some general properties of nearrings, illustrating these general properties with various influential examples. Throughout, we call attention to various problems and areas worthy of further exploration.

1. Getting acquainted

At about the time one is preparing to study calculus, one learns how to take the composite of two functions, for example, if $f(x) = x^2 + 3$ and $g(x) = \sin(x)$, then $f \circ g(x) = \sin^2(x) + 3$ and $g \circ f(x) = \sin(x^2 + 3)$. Of course, one of the main reasons for doing this is that many functions have their structures defined in terms of composition of simpler functions. If $h(x) = g \circ f(x) = \sin(x^2 + 3)$, one can now apply the chain rule to compute the derivative $h'(x) = [g' \circ f(x)] \cdot f'(x) = [\cos(x^2 + 3)](2x)$.

Sometime later, when being introduced to group theory, one is introduced to the set of all bijections S_X of a nonempty set X onto itself. With respect to composition $\circ$, the pairs $(S_X, \circ)$ provide examples of groups, and it is shown that each group is isomorphic to a subgroup of some $(S_X, \circ)$.

Again, when learning about a ring R and one of its modules M, one is asked to consider $\operatorname{End}_R M$, the set of all R-endomorphisms of M into itself. For $f, g \in \operatorname{End}_R M$, not only is $f \circ g \in \operatorname{End}_R M$, but $f + g \in \operatorname{End}_R M$ where $(f + g)(m) = f(m) + g(m)$ for each $m \in M$. The fact that $f + g \in \operatorname{End}_R M$ is really quite remarkable when considering it in a wider scope. In fact, $(\operatorname{End}_R M, +, \circ)$ is a ring with identity 1_R. These rings are analogous to the $(S_X, \circ)$ in that every ring is isomorphic to a subring of some $(\operatorname{End}_R M, +, \circ)$.

For our purposes, it is important to point out that the two distributive laws for the rings $(\operatorname{End}_R M, +, \circ)$ arise from two very different reasons. To see that $(f + g) \circ h = (f \circ h) + (g \circ h)$ requires the definition of $+$ in $\operatorname{End}_R M$. That is,

$$(f + g) \circ h(x) = (f + g)(h(x)) = f(h(x)) + g(h(x))$$
$$= (f \circ h)(x) + (g \circ h)(x)$$
$$= [(f \circ h) + (g \circ h)](x),$$

for each $x \in M$. Hence,

$$(f + g) \circ h = (f \circ h) + (g \circ h). \tag{1:1}$$

However, the left distributive law $f \circ (g + h) = (f \circ g) + (f \circ h)$ requires that $f \in \operatorname{End}_R M$. To wit,

$$f \circ (g + h)(x) = f((g + h)(x)) = f(g(x) + h(x))$$
$$= f(g(x)) + f(h(x)) = (f \circ g)(x) + (f \circ h)(x)$$
$$= [(f \circ g) + (f \circ h)](x)$$

for each $x \in M$. Hence,

$$f \circ (g + h) = (f \circ g) + (f \circ h). \tag{1:2}$$

Thus, the two distributive laws (1:1) and (1:2) reflect two very different properties. Realizing this is a key to understanding and appreciating near-ring theory. The lack of one of these distributive laws results in a considerable lack of symmetry, but compensation is usually made by a sophisticated beauty in the resulting structures.

Back in calculus, we were very much occupied with differentiable functions, which were necessarily continuous. Continuous functions were more carefully scrutinized later in topology. And examples of differentiable and continuous functions were provided by the already familiar polynomials. It is in this setting that mathematicians get their first introduction to many of the basic ideas of modern algebra, including nearring theory.

For the real numbers $\mathbf{R}$, let $\mathcal{C}(\mathbf{R}, \mathbf{R})$ denote all real valued functions of a real variable, let $\mathcal{D}(\mathbf{R})$ denote all functions in $\mathcal{C}(\mathbf{R}, \mathbf{R})$ that are differentiable on $\mathbf{R}$, and let $\mathbf{R}[x]$ denote all polynomials with coefficients in $\mathbf{R}$ and the single indeterminate x. In addition to $+$, defined by $(f + g)(x) = f(x) + g(x)$, and multiplication $\cdot$, defined by $(f \cdot g)(x) = f(x)g(x)$, we also have composition $\circ$ defined by $(f \circ g)(x) = f(g(x))$.

Whereas it is quite common knowledge that each of the three structures $(\mathcal{C}(\mathbf{R}, \mathbf{R}), +, \cdot)$, $(\mathcal{D}(\mathbf{R}), +, \cdot)$, and $(\mathbf{R}[x], +, \cdot)$ is a ring with identity, it is not so well known that $\circ$ also interacts with $+$, satisfying the right distributive law (1:1), but not, in general, the left distributive law (1:2). In fact, each of the structures $(\mathcal{C}(\mathbf{R}, \mathbf{R}), +, \circ)$, $(\mathcal{D}(\mathbf{R}), +, \circ)$, and $(\mathbf{R}[x], +, \circ)$ satisfies all the requirements for a ring except the left distributive law (1:2). Thus, each of these three structures is a *nearring*. In addition, the function $\iota(x) = x$ belongs to each of these, and is an identity with respect to $\circ$.

The above paragraphs provide the opening defence of the author's thesis that the operation of composition is of paramount importance to mathematics in general, and to algebra, in particular. As we shall continue to see, composition leads the serious student of mathematics to some very elegant results outside the traditional areas of group theory, ring theory, and semigroup theory.

From another point of view, let X again be a nonempty set and let $(\mathcal{F}(X), +)$ denote the free group on X. Even though $+$ denotes the binary operation for $\mathcal{F}(X)$, one should not conclude that $+$ is commutative. Here, for example, $+$ is commutative if and only if $|X| = 1$, where $|X|$ denotes the cardinality of the set X.

The quintessence of a free group $(\mathcal{F}(X), +)$ is that if $f|X \colon X \to G$ is any mapping from X into *any* group G, then there is exactly one group homomorphism $f : \mathcal{F}(X) \to G$ from $\mathcal{F}(X)$ into G which agrees with $f|X$ on the set X. If $f|X$ and $g|X$ are two such mappings, they can be added as usual:

$$(f|X + g|X)(t) = (f|X)(t) + (g|X)(t)$$

for each $t \in X$. (Again, we use $+$ for the operation in G, but we are not suggesting that the reader should infer that $+$ is commutative.) So, now we have maps $f|X$, $g|X$, and $f|X + g|X$ from X into a fixed group G. Since $\mathcal{F}(X)$ is free on X, they can be extended to *homomorphisms* f, g, and $f \oplus g$, respectively, from $\mathcal{F}(X)$ into G. If $t, u \in X$ and $t \neq u$, then $(f \oplus g)(t + u) = (f \oplus g)(t) + (f \oplus g)(u) = f(t) + g(t) + f(u) + g(u)$. For $f + g$ defined, as usual, by $(f + g)(y) = f(y) + g(y)$, then $(f + g)(t + u) = f(t + u) + g(t + u) = f(t) + f(u) + g(t) + g(u)$. So, in order for $f \oplus g = f + g$, we must have $g(t) + f(u) = f(u) + g(t)$ for every $t, u \in X$. It is easy to find a group $(G, +)$ and maps $f|X$, $g|X$ so that $g(t) + f(u) \neq f(u) + g(t)$, if $t \neq u$. Just take $(G, +) = (\mathcal{F}(X), +)$, and let $f = g = \iota$, where $\iota(x) = x$ for each $x \in \mathcal{F}(X)$. Now let $t, u \in X$ with $t \neq u$. Since $t + u \neq u + t$ in $\mathcal{F}(X)$, we have $f \oplus g \neq f + g$. But we do have

(1.1) Theorem. *For a nonempty set X, let $(\mathcal{F}(X), +)$ denote the free group on X. If $(G, +)$ is any group, and if $\mathrm{Hom}(\mathcal{F}(X), G)$ denotes all the homomorphisms from $\mathcal{F}(X)$ into G, then $(\mathrm{Hom}(\mathcal{F}(X), G), \oplus)$ is a group. If G is not abelian then $(\mathrm{Hom}(\mathcal{F}(X), G), \oplus)$ is nonabelian.*

Proof. Let $f, g, h \in \mathrm{Hom}(\mathcal{F}(X), G)$. By the definition of $\oplus$, it is a binary operation on $\mathrm{Hom}(\mathcal{F}(X), G)$. For $t \in X$,

$$
\begin{aligned}
[f \oplus (g \oplus h)](t) &= f(t) + (g \oplus h)(t) \\
&= f(t) + [g(t) + h(t)],
\end{aligned}
$$

and

$$
\begin{aligned}
[(f \oplus g) \oplus h](t) &= (f \oplus g)(t) + h(t) \\
&= [f(t) + g(t)] + h(t).
\end{aligned}
$$

Thus $\oplus$ is associative.

For the identity $0 \in G$, define $\zeta|X : X \to G$ by $\zeta|X(t) = 0$ for each $t \in X$. Extend $\zeta|X$ to $\zeta : \mathcal{F}(X) \to G$. Then $f \oplus \zeta = \zeta \oplus f = f$ for each $f \in \mathrm{Hom}(\mathcal{F}(X), G)$.

For $f \in \mathrm{Hom}(\mathcal{F}(X), G)$, define $-f|X$ by $-f|X(t) = -f(t)$ for each $t \in X$. Then $f|X + (-f|X) = -f|X + f|X = \zeta|X$, so $f \oplus (-f) = (-f) \oplus f = \zeta$.

If $a, b \in G$ and $a + b \neq b + a$, choose $t \in X$ and maps $f|X$, $g|X$ so that $f|X(t) = a$, $g|X(t) = b$. Then $(f \oplus g)(t) = a + b$ and $(g \oplus f)(t) = b + a$, so $f \oplus g \neq g \oplus f$.

(1.2) Corollary. *Let X and $\mathcal{F}(X)$ be as in the theorem, but also with $|X| > 1$. Then $(\mathrm{Hom}(\mathcal{F}(X), \mathcal{F}(X)), \oplus, \circ)$ satisfies all the requirements for a ring except the right distributive law and commutativity of $\oplus$.*

Proof. We have from the theorem that $(\mathrm{Hom}(\mathcal{F}(X), \mathcal{F}(X)), \oplus)$ is a non-abelian group. For $f, g, h \in \mathrm{Hom}(\mathcal{F}(X), \mathcal{F}(X))$, we certainly have $f \circ (g \circ$

$h) = (f \circ g) \circ h$, and for $t \in X$,

$$f \circ [g \oplus h](t) = f([g \oplus h](t)) = f(g(t) + h(t))$$
$$= f(g(t)) + f(h(t)) = (f \circ g)(t) + (f \circ h)(t)$$
$$= \big[(f \circ g) \oplus (f \circ h)\big](t),$$

so $f \circ [g \oplus h] = (f \circ g) \oplus (f \circ h)$. However, if $h(t) = x + y$, where $x, y \in X$ and $x \neq y$, then

$$[f \oplus g] \circ h(t) = [f \oplus g](x + y) = (f \oplus g)(x) + (f \oplus g)(y)$$
$$= f(x) + g(x) + f(y) + g(y),$$

and

$$\big[(f \circ h) \oplus (g \circ h)\big](t) = (f \circ h)(t) + (g \circ h)(t) = f(x + y) + g(x + y)$$
$$= f(x) + f(y) + g(x) + g(y).$$

Take $f = g = \iota$, where $\iota(x) = x$ for all $x \in X$. Then since $x + y \neq y + x$, we have $(f \oplus g) \circ h \neq (f \circ h) \oplus (g \circ h)$, in general.

These examples, $(\mathcal{C}(\mathbf{R}, \mathbf{R}), +, \circ)$, $(\mathcal{D}(\mathbf{R}), +, \circ)$, $(\mathbf{R}[x], +, \circ)$, and for each X with $|X| > 1$, the $(\mathrm{Hom}(\mathcal{F}(X), \mathcal{F}(X)), \oplus, \circ)$ provide part of the geneses for nearring theory.

(1.3) Definition. A *nearring* is a triple $(N, +, \cdot)$ where $+$ and $\cdot$ are binary operations on the set N, where $(N, +)$ is a group (not necessarily abelian), where $(N, \cdot)$ is a semigroup, that is, $\cdot$ is associative, and where $\cdot$ satisfies at least one distributive law (1:1) or (1:2) with respect to $+$.

Of course, there are various modifiers for the various nearrings.

(1.4) Definitions. For a nearring, if the right distributive law (1:1) is valid, then $(N, +, \cdot)$ is often called a *right* nearring. You guessed it, if (1:2) is valid, $(N, +, \cdot)$ is called a *left* nearring. If both (1:1) and (1:2) are valid, $(N, +, \cdot)$ is a *distributive* nearring. If $+$ is commutative and only one distributive law is valid, then $(N, +, \cdot)$ is an *abelian* nearring. If, by chance, $+$ is commutative and both distributive laws are valid, then $(N, +, \cdot)$ is a *ring*.

Now part of the fun (or irritation) of working with nearrings is the *left* or *right* nearring conflict. For each concept, or theorem, about left nearrings, there is an analogous concept, or theorem, about right nearrings, and vice-versa. The novice must take care in formulating the analog, however. There is no sound argument for considering only left nearrings, and, analogously, there is no sound argument for considering only right nearrings. The only solution is to become versatile at both, and to use the one most suitable to the occasion. Putting the element from the domain of a function to the right of a function, such as $f(x)$ or fx, tends to encourage right nearrings. However, (1.2) shows that this is not always the case. Similarly for putting the element from the domain to the left, as in $x\alpha$ for a map α and variable x, tends to force the left distributive law.

In this work, when considering nearrings in general, we shall assume the nearring to be a left nearring. This is based on the following. The author spent many hours as a young man in the US Navy marching to the rhythm of 'left, right, left, right, ...'. In addition, the author wishes to prove that he sides with 'the left' on certain *important* issues.

2. Lots of examples

It has been the author's experience that good examples provide the foundation for some of the best research. Throughout, we shall put an emphasis on examples and what we can learn from them.

Continuous mappings with topological groups

Our first example of a nearring was $(\mathcal{C}(\mathbf{R}, \mathbf{R}), +, \circ)$. Now what went into making this structure a nearring? We used the fact that $(\mathbf{R}, +)$ is a group to define the addition of $f, g \in \mathcal{C}(\mathbf{R}, \mathbf{R})$ and to get that $(\mathcal{C}(\mathbf{R}, \mathbf{R}), +)$ is a group. We used the topology on $\mathbf{R}$ to obtain the elements of $\mathcal{C}(\mathbf{R}, \mathbf{R})$.

(2.1) Definition. Let $(G, +)$ be a group and $(G, \mathcal{T})$ a topological space. The set $G \times G$ inherits the product topology from $\mathcal{T}$. Suppose the map $(x, y) \mapsto x + y$ from $G \times G$ to G is continuous, and that the map $x \mapsto -x$ from G to G is also continuous. Then $G = (G, +, \mathcal{T})$ is a *topological group*.

So we see that $(\mathbf{R}, +, \mathcal{T})$ is a topological group if $\mathcal{T}$ is the usual topology on $\mathbf{R}$.

For any topological group $(G, +, \mathcal{T})$, let $\mathcal{C}(G, G)$ denote all the continuous mappings from G into itself. For $f, g \in \mathcal{C}(G, G)$, define $f + g$ by $(f + g)(x) = f(x) + g(x)$ for each $x \in G$. Then one easily verifies that $(\mathcal{C}(G, G), +, \circ)$ is a right nearring. So, if $(\mathbf{C}, +)$ denotes the additive group of complex numbers and $\mathcal{T}$ is the usual topology, then $(\mathbf{C}, +, \mathcal{T})$ is a topological group. If $(\mathbf{C}^*, \cdot)$ denotes the multiplicative group of nonzero complex numbers, and $\mathcal{T}^*$ is $\mathcal{T}$ restricted to $\mathbf{C}^*$, then $(\mathbf{C}^*, \cdot, \mathcal{T}^*)$ is a topological group. Similarly, let $(T, \cdot)$ denote the multiplicative group of complex numbers of absolute value 1, that is, the unit circle. If $\mathcal{T}_T$ is $\mathcal{T}$ restricted to T, then $(T, \cdot, \mathcal{T}_T)$ is a topological group.

Of course, it is natural to study $\mathcal{C}(G, H)$, where G and H are topological groups, and $\mathcal{C}(G, H)$ consists of the continuous mappings from G to H. One readily gets that $(\mathcal{C}(G, H), +)$ is a group. Kenneth D. Magill showed the world how to make $(\mathcal{C}(G, H), +)$ into a nearring. In fact, one often obtains a very large family of nearrings. Fix $\alpha \in \mathcal{C}(H, G)$ and define $*_\alpha$ by

$$f *_\alpha g = f \circ \alpha \circ g \qquad (2{:}1)$$

for $f, g \in \mathcal{C}(G, H)$. Then one gets that $(\mathcal{C}(G, H), +, *_\alpha)$ is a right nearring.

But Magill's construction is considerably more powerful. Let $\mathcal{C}(X, H)$ denote the continuous maps from a topological space X into a topological group H. If $\alpha : H \to X$ is continuous, then $*_\alpha$ defined by (2:1) makes $(\mathcal{C}(X, H), +, *_\alpha)$ into a right nearring.

Exploratory problems. From time to time we will list some 'problems' and some 'exploratory problems'. What is the difference? A 'problem' is rather well defined, and more restrictive in scope, whereas an 'exploratory problem' can be rather vague and open ended. While both are to stimulate further research, an 'exploratory problem' is intended to open up new branches of research. The researcher is required to discover the interesting and novel features, and to identify specific 'problems'. So 'problems' are a byproduct of exploratory research.

These 'problems' and 'exploratory problems' will be given in context as early as possible. This does not mean that they should be 'solved' or 'explored' at that time, however. The 'exploratory problems', especially, have with them a 'welcome mat' inviting one to return frequently at later dates.

(2.2) Exploratory problem. Consider $\mathcal{C}(T, \mathbf{R})$. Fix an interesting α, for example, $\alpha(t) = \cos(t) + i \sin(t)$, and study the nearring $(\mathcal{C}(T, \mathbf{R}), +, *_\alpha)$. For example, which β make $(\mathcal{C}(T, \mathbf{R}), +, *_\beta)$ isomorphic to $(\mathcal{C}(T, \mathbf{R}), +, *_\alpha)$? What are the ideals of $(\mathcal{C}(T, \mathbf{R}), +, *_\alpha)$? What are other interesting invariants of $(\mathcal{C}(T, \mathbf{R}), +, *_\alpha)$?

(2.3) Exploratory problem. Consider $\mathcal{C}(\mathbf{C}^*, \mathbf{R})$. Fix an interesting α, for example, $\alpha(t) = \cos(t) + i \sin(t)$, and study $(\mathcal{C}(\mathbf{C}^*, \mathbf{R}), +, *_\alpha)$.

(2.4) Exploratory problem. As groups, T and $\mathbf{C}^*$ are isomorphic [Du]. But as topological groups they are not isomorphic. For a common α, compare the nearring $(\mathcal{C}(T, \mathbf{R}), +, *_\alpha)$ with the nearring $(\mathcal{C}(\mathbf{C}^*, \mathbf{R}), +, *_\alpha)$.

(2.5) Exploratory problem. Fix X. Perhaps let X be one of $[-1, 1]$, $(0, 1]$, or $(-\pi/2, \pi/2)$, and choose α to be $\alpha(x) = \sin(x)$, $\alpha(x) = \exp(-x^2)$, or $\alpha(x) = \tan^{-1}(x)$. Explore the properties of the nearring $(\mathcal{C}(X, \mathbf{R}), +, *_\alpha)$.

Group mappings

For a group $(G, +)$, there is always the discrete topology for G so that $\mathcal{C}(G, G) = M(G)$, where $M(G)$ denotes the set of all mappings of G into itself. Certainly, $(M(G), +, \circ)$ is a right nearring.

One usually likes to observe how properties of G are reflected in $M(G)$, or vice-versa. Since the elements of $M(G)$ really have so little restriction on them, it has been more fruitful to look at subnearrings of $M(G)$ whose elements are more closely related to the group properties of $(G, +)$. There have been two main approaches here, each of which has been very fruitful. A third and fourth approach have recently received attention, and promise to bring forth some intriguing results. Let us look at these one at a time.

The first approach, which I identify with Helmut Wielandt, is to fix a semigroup Φ of endomorphisms of our group G. Define

$$M_\Phi(G) = \{f \in M(G) \mid \phi \circ f = f \circ \phi \text{ for each } \phi \in \Phi\}.$$

Then $(M_\Phi(G), +, \circ)$ is a right nearring with identity 1_G. These are *centralizer* nearrings.

(2.6) Definition. A nearring $(M, +, \cdot)$ is a *subnearring* of a nearring $(N, +, \cdot)$ if $M \subseteq N$ and the $+$ and $\cdot$ for M are restrictions of the $+$ and $\cdot$ for N to $M \times M$. Sometimes we shall write $M < N$ to denote that M is a subnearring of N.

(2.7) Proposition. *If $(N, +, \cdot)$ is a nearring and $(M, +)$ is a subgroup of $(N, +)$, then $(M, +, \cdot)$ is a subnearring of $(N, +, \cdot)$ if and only if $a, b \in M$ imply $ab \in M$.*

Proof. If $(M, +, \cdot)$ is a subnearring of $(N, +, \cdot)$, then $a, b \in M$ imply $ab \in M$. If $a, b \in M$ imply $ab \in M$, then $\cdot$ restricted to $M \times M$ defines an associative, left distributive binary operation on M, so $(M, +, \cdot)$ is a subnearring. (Recall that, in general, our nearrings will be considered to be *left* nearrings.)

(2.8) Proposition. *If $\{(M_i, +, \cdot)\}$ is a family of subnearrings of a nearring $(N, +, \cdot)$, then the intersection $(\cap M_i, +, \cdot)$ is also a subnearring of $(N, +, \cdot)$.*

Proof. Certainly, $(\cap M_i, +)$ is a subgroup of $(N, +)$, and if $a, b \in \cap M_i$, then $ab \in \cap M_i$.

The second approach I identify with A. Fröhlich. Again, fix a semigroup Φ of endomorphisms of our group G. Define

$$D(\Phi, G) = \cap\{M_i \mid (M_i, +) \text{ is a subgroup of } (M(G), +) \text{ and } \Phi \subseteq M_i\}.$$

So $f \in D(\Phi, G)$ if and only if f is a finite sum of the form $\sum_{i=1}^n \epsilon_i \phi_i$, where $\epsilon_i \in \{\pm 1\}$ and $\phi_i \in \Phi$. These are the prototypes of *distributively generated* nearrings. To understand the adjective, take $f = \sum_{i=1}^n \epsilon_i \phi_i$,

$g = \sum_{i=1}^{m} \epsilon_i' \phi_i' \in D(\Phi, G)$ and $\lambda \in \Phi$. Then $\lambda \in D(\Phi, G)$ also, and

$$\lambda \circ (f + g) = \lambda \circ \left[\sum_{i=1}^{n} \epsilon_i \phi_i + \sum_{i=1}^{m} \epsilon_i' \phi_i' \right]$$

$$= \sum_{i=1}^{n} \epsilon_i (\lambda \circ \phi_i) + \sum_{i=1}^{m} \epsilon_i' (\lambda \circ \phi_i')$$

$$= \lambda \circ \sum_{i=1}^{n} \epsilon_i \phi_i + \lambda \circ \sum_{i=1}^{m} \epsilon_i' \phi_i'$$

$$= (\lambda \circ f) + (\lambda \circ g).$$

So the element $\lambda \in \Phi$ distributes from the *left*, as well as from the right in $D(\Phi, G)$. This again emphasizes that left and right distributive properties come from different sources.

The third approach is again due to Helmut Wielandt. In 1972, at the 2nd International Conference on Nearrings and Nearfields, held in Oberwolfach, West Germany, Professor Wielandt presented this powerful method to construct subnearrings of $M(G)$.

For a group $(G, +)$ and an index set I, let G^I denote all mappings from I into G. So G^I can be identified with the complete direct sum of $|I|$ copies of G. Let H be a subgroup of G^I. For $\alpha \in M(G)$ and $(g_i)_{i \in I} \in G^I$, define $\overline{\alpha}(g_i)_{i \in I} = (\alpha g_i)_{i \in I}$. Let

$$M(G, I, H) = \{\alpha \in M(G) \mid \overline{\alpha} H \subseteq H\}.$$

As a particularly important application, let $|I| = 1$, and $H = \{0\}$. We let $M_0(G)$ denote this $M(G, I, H)$, and we note that it consists of all the $f \in M(G)$ such that $f(0) = 0$.

The fourth approach comes from M. Holcombe. A *group-semiautomaton* is an ordered triple (G, I, δ), where G is a group and $\delta : G \times I \to G$. For each $a \in I$, one defines $f_a : G \to G$ by $f_a(g) = \delta(g, a)$. If M_S consists of all functions f_a and the identity map 1_G, then M_S generates a subnearring of $M(G)$, denoted by $N(S)$. These are called *syntactic* nearrings.

(2.9) Exploratory problem. For a fixed semigroup Φ of endomorphisms of a group $(G, +)$, Wielandt's path is to study $M_\Phi(G)$, while that for Fröhlich would be to study $D(\Phi, G)$. Each of these nearrings should contain some information about the group G. Take some nice interesting family of groups G and corresponding semigroups Φ. Study in detail the $M_\Phi(G)$ and the $D(\Phi, G)$, and then compare the two nearrings for similarities and differences. For example, let G be any of the dihedral groups and let Φ be the group of inner automorphisms. Which nearring, $M_\Phi(G)$ or $D(\Phi, G)$, has more information about G? Which of these nearrings is more interesting in its own right?

Differentiable functions

Our second example of a nearring was $(\mathcal{D}(\mathbf{R}), +, \circ)$, which is a subnearring of $(\mathcal{C}(\mathbf{R}, \mathbf{R}), +, \circ)$ consisting of the differentiable functions on $\mathbf{R}$. Of course, differentiation has meaning for some of the elements of $(\mathcal{C}(\mathbf{C}, \mathbf{C}), +, \circ)$ also. Let $\mathcal{D}(\mathbf{C})$ be the set of differentiable functions in $\mathcal{C}(\mathbf{C}, \mathbf{C})$. Then $\mathcal{D}(\mathbf{C})$ consists of the *entire* functions. The salient fact about an $f \in \mathcal{D}(\mathbf{C})$ is that it has a power series expansion $f = \sum_{n=0}^{\infty} a_n z^n$ with infinite radius of convergence. Certainly, $(\mathcal{D}(\mathbf{C}), +, \circ)$ is a subnearring of $(\mathcal{C}(\mathbf{C}, \mathbf{C}), +, \circ)$.

Let $(R, +, \cdot)$ be a commutative ring, and let $R[[x]]$ be the formal power series over R. When is $(R[[x]], +, \circ)$ a nearring? If $f = \sum_{n=0}^{\infty} a_n x^n$ and $g = \sum_{n=0}^{\infty} b_n x^n$, we would need, of course, that $f \circ g = \sum_{n=0}^{\infty} a_n (\sum_{k=0}^{\infty} b_k x^k)^n \in R[[x]]$. This means $a_0 + a_1 b_0 + a_2 b_0^2 + a_3 b_0^3 + \cdots = f(b_0)$ would have to have meaning in R, somehow. This could be if b_0 is a nilpotent element of R, or if $b_0 = 0$. If every element of R is nilpotent, then $(R[[x]], +, \circ)$ is a nearring.

If $R_0[[x]]$ denotes all power series with constant term 0, then each $f \in R_0[[x]]$ has form $f = \sum_{n=1}^{\infty} a_n x^n$. If $g = \sum_{n=1}^{\infty} b_n x^n$, then

$$f \circ g = \sum_{n=1}^{\infty} a_n \left(\sum_{k=1}^{\infty} b_k x^k \right)^n = \sum_{n=1}^{\infty} d_n x^n$$

for some well defined $d_n \in R$. Thus, $(R_0[[x]], +, \circ)$ is a right nearring.

In another direction, we can use Magill's trick to modify $\mathcal{D}(\mathbf{R})$ and $\mathcal{D}(\mathbf{C})$. Let $G \in \{\mathbf{R}, \mathbf{C}\}$ and let X be an open subset of G. Define $\mathcal{D}(X, G)$ to be the differentiable functions $f : X \to G$. Fix a differentiable $\alpha : G \to X$. For $f, g \in \mathcal{D}(X, G)$, then $f *_\alpha g = f \circ \alpha \circ g \in \mathcal{D}(X, G)$ and $(\mathcal{D}(X, G), +, *_\alpha)$ is a right nearring.

(2.10) Exploratory problem. Choose some nice X and study the nearrings $(\mathcal{D}(X, G), +, \circ)$ thoroughly. Perhaps this might lead one to discover some powerful or interesting theorems.

Polynomials over a ring

Our third example of a nearring was $(\mathbf{R}[x], +, \circ)$. We can replace $\mathbf{R}$ by any commutative ring with identity. Let $(R, +, \cdot, 1)$ be a commutative ring with identity 1. So $\mathbf{R}[x]$ denotes all polynomials with coefficients from R, and $(\mathbf{R}[x], +, \circ)$ is a right nearring.

(2.11) Exploratory problem. Each $f \in \mathbf{R}[x]$ defines a function, and distinct $f, g \in \mathbf{R}[x]$ define distinct functions. This is certainly not the case when R is a finite field. One could describe an equivalence relation $\sim$ on

$R[X]$ by $f \sim g$ if and only if f and g define the same functions from R to R. (An $f \in R[X]$ defines a function $f^* : R \to R$ by $f^*(r) = f(r)$ for each $r \in R$.) For some interesting classes of suitable rings, investigate the equivalence relation $\sim$ very thoroughly.

What does one do with $R[x, y]$? Certainly, $(R[x, y], +, \cdot)$ is a ring with identity. For $f(x, y), g(x, y) \in R[x, y]$, what should $f(x, y) \circ g(x, y)$ be? If one doesn't think about it very carefully, perhaps one would define

$$f(x, y) \circ g(x, y) = f(g(x, y), g(x, y)).$$

If one were to do this, then $(R[x, y], +, \circ)$ would be a right nearring with at least two left identities.

(2.12) Exploratory problem. The nearrings $(R[x, y], +, \circ)$ have been neglected. So, 'the field is white and ready to harvest'. What are some interesting facts about these $(R[x, y], +, \circ)$? What are some significant problems concerning these nearrings? For a starter, fix a nice ring R and find the left identities with respect to $\circ$. Each of these left identities is a right identity for some subset of $R[x, y]$. Is there a subgroup of $(R[x, y], \circ)$ associated with each left identity? What are some significant substructures? How do the elements $f \in R[x, y]$ relate to functions from $R \times R$ into R?

Something else that one can do with $R[x, y]$ is to take another copy of it. For $(f(x, y), g(x, y)), (m(x, y), n(x, y)) \in R[x, y] \times R[x, y]$, define

$$(f(x, y), g(x, y)) \circ (m(x, y), n(x, y))$$
$$= (f(m(x, y), n(x, y)), \, g(m(x, y), n(x, y))).$$

Then $(R[x, y] \times R[x, y], +, \circ))$ is a right nearring with identity (x, y).

Abstract affine nearrings

In $(R[x], +, \circ)$, let $\mathcal{A}_1(R)$ denote all polynomials of the form $a + bx$. Then $(\mathcal{A}_1(R), +, \circ)$ is a subnearring of $(R[x], +, \circ)$. We identify an $a + bx$ with $(a, b) \in R \times R$. The $+$ and $\circ$ of $(\mathcal{A}_1(R), +, \circ)$ induce the operations $(a, b) + (c, d) = (a+c, b+d)$ and $(a, b) * (c, d) = (a+bc, bd)$ on $R \times R$, so $(R \times R, +, *)$ is a right nearring.

In $(R[x, y] \times R[x, y], +, \circ)$, let $\mathcal{A}_2(R)$ denote all pairs $(\alpha + ax + by, \beta + cx + dy)$. Then $(\mathcal{A}_2(R), +, \circ)$ is a subnearring of $(R[x, y] \times R[x, y], +, \circ)$ We identify

$$(\alpha + ax + by, \beta + cx + dy) \equiv \left[\begin{pmatrix} \alpha \\ \beta \end{pmatrix}, \begin{pmatrix} a & b \\ c & d \end{pmatrix} \right].$$

The $+$ and $\circ$ of $(\mathcal{A}_2(R), +, \circ)$ induce operations

$$\left[\begin{pmatrix} \alpha \\ \beta \end{pmatrix}, \begin{pmatrix} a & b \\ c & d \end{pmatrix} \right] + \left[\begin{pmatrix} \gamma \\ \delta \end{pmatrix}, \begin{pmatrix} m & n \\ s & t \end{pmatrix} \right]$$
$$= \left[\begin{pmatrix} \alpha + \gamma \\ \beta + \delta \end{pmatrix}, \begin{pmatrix} a + m & b + n \\ c + s & d + t \end{pmatrix} \right]$$

and

$$\left[\begin{pmatrix} \alpha \\ \beta \end{pmatrix}, \begin{pmatrix} a & b \\ c & d \end{pmatrix} \right] * \left[\begin{pmatrix} \gamma \\ \delta \end{pmatrix}, \begin{pmatrix} m & n \\ s & t \end{pmatrix} \right]$$
$$= \left[\begin{pmatrix} \alpha \\ \beta \end{pmatrix} + \begin{pmatrix} a & b \\ c & d \end{pmatrix} \begin{pmatrix} \gamma \\ \delta \end{pmatrix}, \begin{pmatrix} a & b \\ c & d \end{pmatrix} \begin{pmatrix} m & n \\ s & t \end{pmatrix} \right].$$

If $\mathcal{M}_2(R) = \left\{ \begin{pmatrix} a & b \\ c & d \end{pmatrix} \,\middle|\, a, b, c, d \in R \right\}$, then $(\mathcal{M}_2(R), +, \cdot)$ is a ring with identity, and $R \oplus R = \left\{ \begin{pmatrix} \alpha \\ \beta \end{pmatrix} \,\middle|\, \alpha, \beta \in R \right\}$ is a unitary $\mathcal{M}_2(R)$-module.

Harry Gonshor noticed that if one has a ring R and an R-module M, then $(M \times R, +, \cdot)$ is a right nearring if $(m_1, r_1) + (m_2, r_2) = (m_1 + m_2, r_1 + r_2)$ and $(m_1, r_1) \cdot (m_2, r_2) = (m_1 + r_1 m_2, r_1 r_2)$. If R has an identity 1, and if M is unitary, then $(0, 1)$ is an identity for $(M \times R, +, \cdot)$.

If R is a field, then $(\mathcal{A}_2(R), +, \circ)$ is essentially the affine transformations of a two-dimensional vector space over R. $(\mathcal{A}_2(R), +, \circ)$ has an easy generalization to $(\mathcal{A}_n(R), +, \circ)$, the affine transformations of an n-dimensional vector space over R. This is why Gonshor called the construction $(M \times R, +, \cdot)$ an *abstract affine nearring*. If $(M \times R, +, \cdot)$ is an abstract affine nearring, we denote this by $M \oplus_A R$.

Endomorphisms of the algebra of polynomials

Actually, $(R[x], +, \cdot)$ and $(R[x, y], +, \cdot)$ are commutative R-algebras with identity. If $(A, +, \cdot)$ is a commutative R-algebra with identity, then an R-endomorphism is a map $\alpha : A \to A$ satisfying (i) $\alpha(a + b) = \alpha a + \alpha b$; (ii) $\alpha(ab) = (\alpha a)(\alpha b)$; (iii) $\alpha(ra) = r\alpha(a)$; and (iv) $\alpha 1_A = 1_A$ for all $a, b \in A$ and for all $r \in R$. If $A = R[x]$, then an R-endomorphism α is completely determined by $\alpha x \in R[x]$, and αx can be any element of $R[x]$. If $A = R[x, y]$, then an R-endomorphism α is completely determined by $\alpha x, \alpha y \in R[x, y]$, and $\alpha x, \alpha y$ can be arbitrary in $R[x, y]$.

For $A = R[x]$, if $\text{End}_R R[x]$ denotes the R-endomorphisms of $R[x]$, then we identify $\text{End}_R R[x]$ with $R[x]$. If $\alpha \leftrightarrow f(x)$ and $\beta \leftrightarrow g(x)$, then $\alpha +$

$\beta \leftrightarrow f(x) + g(x)$, and $\alpha \circ \beta(x) = \alpha(g(x)) = g(f(x)) = g(x) \circ f(x)$. Thus, $(\operatorname{End}_R R[x], +, \circ)$ is a left nearring isomorphic to $(R[x], +, *)$ where $f(x) * g(x) = g(x) \circ f(x) = g(f(x))$. (If we had written $x\alpha$ instead of αx, then we would have obtained $(\operatorname{End}_R R[x], +, \circ)$ as a right nearring isomorphic to $(R[x], +, \circ)$.)

Now take $A = R[x, y]$ and let $\operatorname{End}_R R[x, y]$ denote the R-endomorphisms of $R[x, y]$. We identify $\operatorname{End}_R R[x, y]$ with $R[x, y] \times R[x, y]$. If we let $\alpha \equiv (f(x, y), g(x, y))$ and $\beta \equiv (m(x, y), n(x, y))$, then $\alpha + \beta \equiv (f(x, y) + m(x, y), g(x, y) + n(x, y))$, and $\alpha \circ \beta(x) = \alpha(m(x, y)) = m(f(x, y), g(x, y))$, and $\alpha \circ \beta(y) = \alpha(n(x, y)) = n(f(x, y), g(x, y))$. Hence, $\alpha \circ \beta \equiv (m(x, y), n(x, y)) \circ (f(x, y), g(x, y))$. Thus, $(\operatorname{End}_R R[x, y], +, \circ)$ is a left nearring isomorphic to $(R[x, y] \times R[x, y], +, *)$ where

$$(f(x, y), g(x, y)) * (m(x, y), n(x, y)) = (m(x, y), n(x, y)) \circ (f(x, y), g(x, y)).$$

(Again, had we written $x\alpha$, $y\alpha$ instead of αx, αy, then we would have obtained $(\operatorname{End}_R R[x, y], +, \circ)$, a right nearring, which is isomorphic to $(R[x, y] \times R[x, y], +, \circ)$.)

Note: We have just seen a relatively unknown application of nearring theory. The R-endomorphisms of the R-algebras $R[x]$ or $R[x, y]$ naturally form a nearring. They are essentially the polynomial nearrings $(R[x], +, \circ)$ and $(R[x, y] \times R[x, y], +, \circ)$.

(2.13) Exploratory problem. Each nearring $(R[x], +, \circ)$ and each nearring $(R[x, y] \times R[x, y], +, \circ)$ has an identity, hence a group of units. In light of the above, these units correspond to the R-automorphisms. The group of units for $(R[x], +, \circ)$ has an analogue as a subgroup of units in $(R[x, y] \times R[x, y], +, \circ)$, but here, the group of units will be considerably more complicated. Investigate the group of units for various $(R[x, y] \times R[x, y], +, \circ)$.

Endomorphisms of group algebras

For a commutative ring R with identity and an abelian group G, we let RG denote the corresponding group algebra. An element of RG is a finite sum $r_1 g_1 + r_2 g_2 + \cdots + r_k g_k$ where each $r_i \in R$ and each $g_i \in G$. Addition $+$ and multiplication $\cdot$ for RG make RG a commutative ring with identity, and scalar multiplication by elements of R make RG into an R-algebra. Let $\operatorname{End}_R RG$ denote the R-endomorphisms of RG.

As for $\operatorname{End}_R R[x]$ and $\operatorname{End}_R R[x, y]$, an $\alpha \in \operatorname{End}_R RG$ is completely determined by the values $g\alpha$, $g \in G$. But the $g\alpha$ cannot be arbitrary. They must be units in the R-algebra RG, and $\alpha|G : G \to \mathcal{U}(RG)$ must be a group homomorphism from G into $\mathcal{U}(RG)$, the group of units of the algebra RG. So, we identify $\operatorname{End}_R RG$ with $\operatorname{Hom}(G, \mathcal{U}(RG))$, where $\operatorname{Hom}(G, H)$ denotes the homomorphisms from a group G into a group H. If H is an abelian group, then $\text{Hom}(G, H)$ is an abelian group. Hence, $\text{Hom}(G, \mathcal{U}(RG))$ is an abelian group.

For $\alpha \in \text{End}_R RG, \alpha|G$ is the corresponding element of $\text{Hom}(G, \mathcal{U}(RG))$. Then, if $\beta \in \text{End}_R RG$, we have $\alpha + \beta \equiv \alpha|G + \beta|G$. But, for $g \in G$, if $g\alpha = r_1 g_1 + \cdots + r_s g_s$, then $g(\alpha \circ \beta) = (r_1 g_1 + \cdots + r_s g_s)\beta = r_1(g_1\beta) + \cdots + r_s(g_s\beta)$. We get that $(\text{End}_R RG, +, \circ)$ is a right nearring. (Had we written αg instead of $g\alpha$, then we would have $(\text{End}_R RG, +, \circ)$ as a left nearring.)

Note. It is of interest to point out that in $\text{End}_R R[x]$, $\text{End}_R R[x, y]$, and $\text{End}_R RG$, the additive identity is *not* the zero map, that is, it is not the map $\zeta : b \mapsto 0$ for each b. Each of the R-algebras $\text{End}_R R[x]$, $\text{End}_R R[x, y]$, and $\text{End}_R RG$ has an identity 1. An R-endomorphism must map 1 onto itself, thus precluding it from being the zero map ζ.

(2.14) Exploratory problem. Choose a nice commutative ring R with identity, and a nice abelian group G. Study $\text{End}_R RG$. Find some interesting properties of $\text{End}_R RG$. Now generalize by letting R, G, or both vary. Based on past experience, this is likely to lead to some very nice theorems.

Note. This represents our second application of nearring theory. It is like our first application in that the set of R-endomorphisms of an R-algebra A has the natural structure of a nearring. So, this is part of a bigger picture.

(2.15) Exploratory problem. For a fixed R, which R-algebras A have their corresponding $\text{End}_R A$ resulting in a natural nearring?

Trivial nearrings

Consider the abelian group $(Z_{p^\infty}, +)$ generated by $c_1, c_2, c_3, \cdots$, where $pc_{n+1} = c_n$ and $pc_1 = 0$, and p is a prime. Then this group has the amazing property that if $(Z_{p^\infty}, +, \cdot)$ is a ring, then $ab = 0$ for each $a, b \in Z_{p^\infty}$. Such an abelian group $(A, +)$, where only the *trivial multiplication* $a \cdot b = 0$ makes $(A, +, \cdot)$ a ring, is called a *nil group*.

What would be an analogous concept for nearrings? If $(G, +)$ is any group, and $S \subseteq G \setminus \{0\}$, define $\cdot_S$ by

$$a \cdot_S b = \begin{cases} b, & \text{if } a \in S; \\ 0, & \text{otherwise.} \end{cases}$$

Then $(G, +, \cdot_S)$ is a left nearring. Also, if $a * b = b$ for all $a, b \in G$, then $(G, +, *)$ is also a left nearring. So, as long as G has order greater than 2, then $(G, +)$ can be made into a nearring in numerous ways. The multiplications $*$ and $\cdot_S$ are called *trivial*, as are the resulting nearrings.

(2.16) Problem. Other than the groups of orders 1 and 2, does there exist a group $(G, +)$ having the property that if $(G, +, \cdot)$ is a left nearring, then $\cdot$ is a trivial multiplication?

Nearfields

At the beginning of the twentieth century, Leonard E. Dickson was concerned about the independence of the axioms for a field. For example, if $(F, +, \cdot)$ satisfies all the axioms for a field except the right distributive axiom, then it satisfies that axiom also. Since $\cdot$ is commutative, one has $(a + b) \cdot c = c \cdot (a + b) = (c \cdot a) + (c \cdot b) = (a \cdot c) + (b \cdot c)$. So, the right distributive property is a consequence of the others. Dickson showed in 1905 that if commutativity of $\cdot$ were also eliminated, then one could no longer be assured that the right distributive law would be valid.

(2.17) Definition. A (left) *nearfield* is a nearring $(F, +, \cdot)$ where $(F, +)$ is a group, $(F^* = F \setminus \{0\}, \cdot)$ is a group, and $a \cdot (b + c) = (a \cdot b) + (a \cdot c)$ for all $a, b, c \in F$. (Guess what a right nearfield is!)

Curiously, or unfortunately, if $(Z_2, +)$ is the group of order 2, then $(Z_2, +, *)$ satisfies our definition for a nearfield, where $*$ is the trivial multiplication $a * b = b$ defined above. Many researchers do not recognize $(Z_2, +, *)$ as a nearfield and take steps to disown it. They require $0 \cdot a = a \cdot 0 = 0$ for each $a \in F$.

Let us examine carefully a method for constructing nontrivial nearfields. Let σ be an automorphism of a fixed field $(F, +, \cdot)$, and write x^σ for the image of x by σ. Suppose the multiplicative group $(F^* = F \setminus \{0\}, \cdot)$ has a subgroup $(U, \cdot)$ of index 2. Define a binary operation $\circ$ on F by

$$a \circ b = \begin{cases} a \cdot b, & \text{if } a \in U; \\ a \cdot b^\sigma, & \text{otherwise.} \end{cases} \tag{2:2}$$

First we will investigate associativity. Let $a, b, c \in F$. We want to show that $a \circ (b \circ c) = (a \circ b) \circ c$. There are four cases: (1) $a, b \in U$; (2) $a \in U$, $b \notin U$; (3) $a \notin U$, $b \in U$; and (4) $a, b \notin U$. We consider them in their natural order. First,

$$a \circ (b \circ c) = a \cdot (b \cdot c),$$
$$(a \circ b) \circ c = (a \cdot b) \circ c = (a \cdot b) \cdot c.$$

Second,

$$a \circ (b \circ c) = a \cdot (b \cdot c^\sigma),$$
$$(a \circ b) \circ c = (a \cdot b) \circ c = (a \cdot b) \cdot c^\sigma.$$

Third,

$$a \circ (b \circ c) = a \cdot (b \circ c)^\sigma = a \cdot (b \cdot c)^\sigma = a \cdot (b^\sigma \cdot c^\sigma),$$
$$(a \circ b) \circ c = (a \cdot b^\sigma) \circ c = (a \cdot b^\sigma) \cdot c^\sigma,$$

provided $b^\sigma \in U$. Fourth,

$$a \circ (b \circ c) = a \cdot (b \cdot c^\sigma)^\sigma = a \cdot (b^\sigma \cdot c^{\sigma^2}),$$
$$(a \circ b) \circ c = (a \cdot b^\sigma) \circ c = (a \cdot b^\sigma) \cdot c,$$

provided $b^\sigma \notin U$. Thus, to have associativity, we require that $\sigma|U$ is an automorphism of the group $(U, \cdot)$, and we require that $\sigma^2 = 1$, so σ must have order 2.

Let 1 be the multiplicative identity of the field F. So $1 \in U$. Hence, $1 \circ a = 1 \cdot a = a$, and either $a \circ 1 = a \cdot 1 = a$ or $a \circ 1 = a \cdot 1^\sigma = a \cdot 1 = a$. Thus, 1 is an identity for $\circ$.

What about inverses with respect to $\circ$? Let $a \in U$. Then there is an $a' \in U$ so that $a \cdot a' = 1$. But $a \circ a' = a \cdot a' = 1$. So elements of U have inverses with respect to $\circ$. Now let $b \in F^* \setminus U$. There is an element $b' \in F^* \setminus U$ so that $b \cdot b' = 1$, and there is an element $c \in F^* \setminus U$ so that $c^\sigma = b'$. Thus, $b \circ c = b \cdot c^\sigma = b \cdot b' = 1$. Hence, $(F^*, \circ)$ is a group.

Finally, for the left distributive law, if $a, b, c \in F$, and $a \in U$, then

$$a \circ (b + c) = a \cdot (b + c) = (a \cdot b) + (a \cdot c)$$
$$= (a \cdot b) + (a \cdot c),$$

and if $a \notin U$, then

$$a \circ (b + c) = a \cdot (b + c)^\sigma = a \cdot (b^\sigma + c^\sigma)$$
$$= (a \cdot b^\sigma) + (a \cdot c^\sigma) = (a \circ b) + (a \circ c).$$

In summary we obtain

(2.18) Theorem. *Let $(F, +, \cdot)$ be a field having a subgroup $(U, \cdot)$ of $(F^* = F \setminus \{0\}, \cdot)$ of index 2. Suppose σ is a field automorphism of $(F, +, \cdot)$ of order 2 so that $\sigma|U$ is a group automorphism of U. If $\circ$ is defined by (2:2), then $(F, +, \circ)$ is a left nearfield.*

With (2.18), we can construct an abundance of nontrivial nearfields. Let $(F_{p^k}, +, \cdot)$ denote the finite field with p^k elements, where p is the characteristic of this field. To obtain a subgroup $(U, \cdot)$ of index 2, one only needs that p be an odd prime. Not only is there such a subgroup, but one can describe $U = \{x^2 \mid x \in F_{p^k}^*\}$. So, if $a, b \in U$, then $a \cdot b \in U$. To obtain our automorphism σ, we can take $k = 2n$ and define $x^\sigma = x^{p^n}$, and since $(x^2)^\sigma = (x^\sigma)^2$, we get that $\sigma|U$ is a group automorphism of U. This is all we need, so such $(F_{p^{2n}}, +, \circ)$ are nearfields and not fields.

For infinite examples, we start with a field $(K, +, \cdot)$ of characteristic $\neq 2$, and let $F = K(\sqrt{\nu})$ be a quadratic extension, where $\nu \in K^*$ but ν is not a square. So, the elements of F are of the form $a + b\sqrt{\nu}$ with $a, b \in K$. Define $(a + b\sqrt{\nu})^\sigma = a - b\sqrt{\nu}$. Then $\sigma \neq 1$ but $\sigma^2 = 1$.

We still need a subgroup $(U, \cdot)$ of index 2 where $\sigma|U$ is an automorphism. The only example we give now is where $K = \mathbf{Q}$, the rational numbers, and $\nu = 2$. First note that $q \in \mathbf{Q}^*$ can be written as $q = 2^{\mu(q)} \cdot \frac{r}{s}$, where r, s are odd integers, and where $\mu(q)$ is an integer. Let

$$U = \{a + b \cdot \sqrt{2} \in F^* \mid \mu(a^2 - 2b^2) \text{ is even}\}.$$

We need to show that $(U, \cdot)$ is a subgroup of $(F^*, \cdot)$ of index 2, and that $\sigma|U$ is a group automorphism.

Define $\Psi : F^* \to \mathbf{Q}^*$ by $\Psi(a + b \cdot \sqrt{2}) = a^2 - 2b^2$. Then

$$\begin{aligned}
\Psi[(a + b \cdot \sqrt{2}) \cdot (c + d \cdot \sqrt{2})] &= \Psi[(ac + 2bd) + (ad + bc)\sqrt{2}] \\
&= (ac + 2bd)^2 - 2(ad + bc)^2 \\
&= a^2c^2 - 2a^2d^2 - 2b^2c^2 + 4b^2d^2 \\
&= (a^2 - 2b^2) \cdot (c^2 - 2d^2) \\
&= \Psi(a + b \cdot \sqrt{2}) \cdot \Psi(c + d \cdot \sqrt{2}).
\end{aligned}$$

So Ψ is a group homomorphism.

If $\Psi(F^*)$ denotes the image of Ψ, define $\Gamma : \Psi(F^*) \to \{\pm 1\}$ by

$$\Gamma(a^2 - 2b^2) = \begin{cases} -1, & \text{if } \mu(a^2 - 2b^2) \text{ is odd;} \\ +1, & \text{if } \mu(a^2 - 2b^2) \text{ is even.} \end{cases}$$

Then Γ is also a homomorphism onto $\{\pm 1\}$ and the kernel of $\Gamma \circ \Psi$ is U, that is, $\ker \Gamma \circ \Psi = U$, so U has index 2.

Since $\mu(a^2 - 2b^2)$ is even if and only if $\mu(a^2 - 2(-b)^2)$ is even, we get that $\sigma|U$ is a group automorphism of U. Thus, $(\mathbf{Q}(\sqrt{2}), +, \circ)$ is an infinite nontrivial nearfield.

Distributive nearrings

Honest distributive nearrings exist. The simplest were found by the author while constructing all the nearrings $(S_3, +, \cdot)$ where $(S_3, +)$ is the nonabelian group of order 6. Let $S_3 = \{0, a, b, c, x, y\}$, and consider the addition and multiplication tables below.

+	0	a	b	c	x	y
0	0	a	b	c	x	y
a	a	0	y	x	c	b
b	b	x	0	y	a	c
c	c	y	x	0	b	a
x	x	b	c	a	y	0
y	y	c	a	b	0	x

·	0	a	b	c	x	y
0	0	0	0	0	0	0
a	0	a	a	a	0	0
b	0	a	a	a	0	0
c	0	a	a	a	0	0
x	0	0	0	0	0	0
y	0	0	0	0	0	0

Then $(S_3, +, \cdot)$ is a commutative and distributive nearring. Two additional commutative and distributive nearrings can be constructed from $(S_3, +)$ by altering the multiplication table for $\cdot$ above. Just replace the a by either b or c.

Perhaps the easiest way to construct honest distributive nearrings is to start with a nonabelian group $(G, +)$ and a ring $(R, +, \cdot)$. Let $(N, +)$ be the direct sum of the groups $(G, +)$ and $(R, +)$, so N is nonabelian, and on $N = G \times R$ define $*$ by $(g, r) * (g', r') = (0, rr')$. If R is not commutative, then the resulting distributive nearring is noncommutative.

Certainly, one of the nicest classes of distributive nearrings is the one constructed from the nilpotent groups $(G, +)$ of class 2. Simply define $a \cdot b = [a, b] = -a - b + a + b$. Since $(G, +)$ is nilpotent of class 2, we have $(a \cdot b) \cdot c = [[a, b], c] = 0$ for all $a, b, c \in G$. This means that $[a, b]$ is in the centre of G, $Z(G)$, for all $a, b \in G$. Hence, $a \cdot (b \cdot c) = -a - (b \cdot c) + a + (b \cdot c) = -a - [b, c] + a + [b, c] = 0$. So $\cdot$ is associative.

Now

$$a \cdot (b + c) = -a - c - b + a + b + c$$

and

$$(a \cdot b) + (a \cdot c) = -a - b + a + b - a - c + a + c.$$

So when we show

$$-c - b + a + b = -b + a + b - a - c + a,$$

then we will have shown that $\cdot$ is left distributive over $+$. But

$$-c - b + a + b = -c + a - a - b + a + b = -c + a + [a, b] = [a, b] - c + a$$
$$= a + [a, b] - a - c + a = a - a - b + a + b - a - c + a$$
$$= -b + a + b - a - c + a.$$

Similarly,

$$(a + b) \cdot c = -b - a - c + a + b + c$$

and

$$(a \cdot c) + (b \cdot c) = -a - c + a + c - b - c + b + c.$$

So, to get $\cdot$ being right distributive over $+$, we note that $-b - a - c + a = -b - a - c + a + c - c = -b + [a, c] - c = [a, c] - b - c = -a - c + a + c - b - c$. So, $(G, +, \cdot)$ is a distributive nearring.

(2.19) Exploratory problem. Honest distributive nearrings can be very close to being rings. The examples above for nilpotent groups $(G, +)$ of class 2 are very close since $(G, +)$ is just a step away from being abelian. One would expect numerous important theorems for ring theory to generalize to distributive nearrings. Perhaps just as important are the theorems that cannot be extended. Does the theory of distributive nearrings have a personality of its own? Does the nearring structure for the examples constructed above on the nilpotent groups of class 2 shed any light on the group structure, or conversely?

A lot of nice work on distributive nearrings has been done by H. E. Heatherly [HH], G. F. Birkenmeier [Bi], and C. Ferrero–Cotti [F-C].

3. Many cheerful facts about nearrings

In studying any mathematical structure, there are numerous little facts that one must learn about the structure in order to feel comfortable with that structure. Usually, these facts are rather elementary in nature. It is not unlike becoming acquainted with another person. You want to know the person's name, where she comes from, what his profession is, whether she prefers left nearrings or right nearrings, etc.

One must always remember that the world of nearrings has a dichotomy. There are left nearrings and there are right nearrings. An example of one type can be readily converted to the other, however. If $(N, +, \cdot)$ is a left nearring, define a binary operation $*$ on N by $a * b = b \cdot a$ for each $a, b \in N$. Then $(a + b) * c = c \cdot (a + b) = (c \cdot a) + (c \cdot b) = (a * c) + (b * c)$, and $(a * b) * c = c \cdot (b \cdot a) = (c \cdot b) \cdot a = a * (b * c)$. Hence, $(N, +, *)$ is a right nearring. Of course, right nearrings can be converted to left nearrings in a similar manner.

As mentioned earlier, when we discuss nearrings in general, they are assumed to be left nearrings.

One of the very fundamental sources of symmetry for ring theory is the fact that $0 \cdot a = a \cdot 0 = 0$ for each element a of the ring. Numerous examples in §2 show that one cannot be assured of this for nearrings. Nevertheless, since this little property makes mathematics more comfortable, many researchers restrict themselves to nearrings with this 'zero-symmetric' property.

Consider our first example of a (right) nearring $(\mathcal{C}(\mathbf{R}, \mathbf{R}), +, \circ)$. If $f(x) = \cos(x)$ and $\zeta(x) = 0$, then $f \circ \zeta(x) = f(\zeta(x)) = 1$. So $f \circ \zeta$ is the constant function $\mathbf{1}$ where $\mathbf{1}(x) = 1$ for each $x \in \mathbf{R}$. Thus $f \circ \zeta \neq \zeta$, but ζ is the additive identity or 'zero' of $(\mathcal{C}(\mathbf{R}, \mathbf{R}), +, \circ)$.

(3.1) Proposition. *For a (left) nearring* $(N, +, \cdot)$, *if* 0 *is the additive identity of* $(N, +)$ *and if* $a, b \in N$, *then* $a \cdot 0 = 0$ *and* $a \cdot (-b) = -(a \cdot b)$.

Proof. $a \cdot 0 = a \cdot (0 + 0) = (a \cdot 0) + (a \cdot 0)$. Hence, $a \cdot 0 = 0$. Consequently, $0 = a \cdot 0 = a \cdot (b + (-b)) = (a \cdot b) + (a \cdot (-b))$. Thus $-(a \cdot b) = a \cdot (-b)$.

Note. For a right nearring, one has $0 \cdot a = 0$ and $(-a) \cdot b = -(a \cdot b)$. This is why $a \cdot 0 \neq 0$ is possible in $(\mathcal{C}(\mathbf{R}, \mathbf{R}), +, \circ)$, as illustrated just before

(3.1).

(3.2) Definition. A nearring $(N, +, \cdot)$ for which $a \cdot 0 = 0 \cdot a = 0$ for each $a \in N$ is a *zero-symmetric* nearring.

(3.3) Examples. For a group $(G, +)$, the $M_0(G)$ of §2 is zero-symmetric. If $\zeta(x) = 0$ for each $x \in G$, then ζ is the 'zero' of $(M_0(G), +)$. For $f \in M_0(G)$, $f \circ \zeta(x) = f(0) = 0$, and $\zeta \circ f(x) = \zeta(f(x)) = 0$. Hence, $f \circ \zeta = \zeta \circ f = \zeta$.

(3.4) Proposition. *For a group $(G, +)$ and a subsemigroup Φ of endomorphisms of $(G, +)$ with $\zeta \in \Phi$, the nearring $(M_\Phi(G), +, \circ)$ is zero-symmetric.*

Proof. For $f \in M_\Phi(G)$, since $\zeta \in \Phi$, we have $f \circ \zeta = \zeta \circ f$. But $\zeta \circ f(x) = \zeta(f(x)) = 0$. Thus $\zeta \circ f = \zeta$.

(3.5) Proposition. *Let $(N, +, \cdot)$ be a nearring. Define*

$$N_0 = \{n \in N \mid n \cdot 0 = 0 \cdot n = 0\}.$$

Then $(N_0, +, \cdot)$ is a zero-symmetric subnearring of $(N, +, \cdot)$.

Proof. If $n \in N_0$, then $n \cdot 0 = 0 \cdot n = 0$, so N_0 will be zero-symmetric if all else is satisfactory. Now $0 \in N_0$ since $0 \cdot 0 = 0$. If $m, n \in N_0$, then we already have $(m + n) \cdot 0 = 0$ and $(m \cdot n) \cdot 0 = 0$. Since $0 \cdot (m \cdot n) = (0 \cdot m) \cdot n = 0 \cdot n = 0$, we obtain $m \cdot n \in N_0$. Since $(N, +, \cdot)$ is a *left* nearring, $0 \cdot (m + n) = (0 \cdot m) + (0 \cdot n) = 0 + 0 = 0$. Hence, $m + n \in N_0$. Also, by (3.1), $0 \cdot (-m) = -(0 \cdot m) = -0 = 0$, so since $(-m) \cdot 0 = 0$, we have $-m \in N_0$. Thus, by (2.7), $(N_0, +, \cdot)$ is a subnearring of $(N, +, \cdot)$.

In algebra, there is a strong tradition to write $x\alpha$ for $\alpha(x)$, where α is a function and x is in the domain of α. In §2, we usually wrote $\alpha(x)$, but in the future we shall not hesitate to write $x\alpha$. In doing so, the $M(G)$, $M_0(G)$, and $M_\Phi(G)$ all become *left* nearrings.

(3.6) Corollary. *For any group $(G, +)$, we obtain $M(G)_0 = M_0(G)$.*

Proof. If $f \in M(G)_0$, then $\zeta \circ f = \zeta$, so $0f = 0$ and $f \in M_0(G)$. If $f \in M_0(G)$, then $x(\zeta \circ f) = (x\zeta)f = 0f = 0$, so $\zeta \circ f = \zeta$.

(3.7) Remark. In $(\mathcal{C}(\mathbf{R}, \mathbf{R}), +, \circ)$, $\mathcal{C}(\mathbf{R}, \mathbf{R})_0$ consists of those continuous functions whose graphs pass through $(0,0)$. In $(R[x], +, \circ)$, $R[x]_0$ consists of those polynomials whose constant terms are $0 \in R$, and we will write $R_0[x]$ for $R[x]_0$.

In $(\mathcal{C}(\mathbf{R}, \mathbf{R}), +, \circ)$ and $(M(G), +, \circ)$ there are *constant functions*. For $r \in \mathbf{R}$, the function $\mathbf{r}$ where $\mathbf{r}(x) = r$ for each $x \in \mathbf{R}$, and for $g \in G$ the function $\mathbf{g}$ where $x\mathbf{g} = g$ for each $x \in G$, are *constant functions*. It is not difficult to show that the constant functions form a subnearring in each case. Actually, as indicated in §2, $(\mathcal{C}(\mathbf{R}, \mathbf{R}), +, \circ)$ and $(M(G), +, \circ)$ are special cases of $(\mathcal{C}(G, G), +, \circ)$, and the $\mathbf{g}$ certainly belong to $\mathcal{C}(G, G)$. For $\phi \in \mathcal{C}(G, G)$, we have $\phi \circ \mathbf{g} = \mathbf{g}$ if we write $x\phi$ for $x \in G$. In particular, $\mathbf{0} \circ \mathbf{g} = \mathbf{g}$ when $\mathbf{0} = \zeta$. As a matter of fact, $\mathbf{0} \circ \phi$ is the constant function whose constant value is 0ϕ.

(3.8) Proposition. *Let $(N, +, \cdot)$ be a nearring and define*

$$N_c = \{0 \cdot n \mid n \in N\}.$$

Then $(N_c, +, \cdot)$ is a subnearring of $(N, +, \cdot)$ and if $x, y \in N_c$, then $x \cdot y = y$.

Proof. Take $x, y \in N_c$. Then $x = 0 \cdot n$, $y = 0 \cdot m$ for some $n, m \in N$, so $x + y = (0 \cdot n) + (0 \cdot m) = 0 \cdot (n + m) \in N_c$. Also, $-x = -(0 \cdot n) = 0 \cdot (-n)$, so $-x \in N_c$. Since N_c is not void, $(N_c, +)$ is a subgroup. If $0 \cdot n$, $0 \cdot m \in N_c$, then $(0 \cdot n) \cdot (0 \cdot m) = 0 \cdot [n \cdot (0 \cdot m)]$, an element of N_c, so N_c is a subnearring of $(N, +, \cdot)$ and $x \cdot y = y$ if $x, y \in N_c$.

(3.9) Definition. For a nearring $(N, +, \cdot)$, if $x \cdot y = y$ for all $x, y \in N$, then $(N, +, \cdot)$ is a *constant* nearring.

(3.10) Definitions. For a nearring $(N, +, \cdot)$ the subnearring $(N_0, +, \cdot)$ is the *zero-symmetric part* of $(N, +, \cdot)$ and the subnearring $(N_c, +, \cdot)$ is the *constant part* of $(N, +, \cdot)$.

If $f(x) \in \mathcal{C}(\mathbf{R}, \mathbf{R})$, then $f(x) = (f(x) - f(0)) + f(0) = g(x) + \mathbf{r}(x)$, where $g(x) = f(x) - f(0)$ and $\mathbf{r}(x) = r = f(0)$. Note that $g(0) = 0$ so $g \in \mathcal{C}(\mathbf{R}, \mathbf{R})_0$ and $\mathbf{r} \in \mathcal{C}(\mathbf{R}, \mathbf{R})_c$. Hence, $\mathcal{C}(\mathbf{R}, \mathbf{R}) = \mathcal{C}(\mathbf{R}, \mathbf{R})_0 + \mathcal{C}(\mathbf{R}, \mathbf{R})_c$. This same argument readily extends to obtain $\mathcal{C}(G, G) = \mathcal{C}(G, G)_0 + \mathcal{C}(G, G)_c$. In fact, we get

(3.11) Theorem. *For a nearring $(N, +, \cdot)$, we have $N = N_0 + N_c$.*

Proof. For $n \in N$, $0 \cdot [n - (0 \cdot n)] = 0 \cdot [n + (0 \cdot (-n))] = (0 \cdot n) + [0 \cdot (0 \cdot (-n))] = (0 \cdot n) + (0 \cdot (-n)) = 0$. Hence, $n - (0 \cdot n) \in N_0$, and, certainly, $0 \cdot n \in N_c$. Since $[n - (0 \cdot n)] + (0 \cdot n) = n$, our proof is complete.

Actually, we have more, but first, let us recall that a group $(G, +)$ is a *semidirect product* of a subgroup $(N, +)$ by a subgroup $(K, +)$ if (i) $N \cap K = \{0\}$; (ii) $N + K = G$; and (iii) $(N, +)$ is normal in $(G, +)$. One calls $(K, +)$ the *complement* of $(N, +)$.

(3.12) Corollary. *For a nearring $(N, +, \cdot)$, the group $(N, +)$ is a semidirect product of $(N_0, +)$ by $(N_c, +)$.*

Proof. Let $x \in N_0 \cap N_c$. Then $x = 0n$ for some $n \in N$, and $0x = 0$. Hence, $0 = 0x = 0(0n) = (00)n = 0n = x$, and $N_0 \cap N_c = \{0\}$. From the theorem, $N = N_0 + N_c$. Finally, if $m \in N_0$ and $y \in N$, then $0(y+m-y) = (0y) + (0m) + (0(-y)) = (0y) + (0(-y)) = 0$ since $0m = 0$. Hence, $(N_0, +)$ is normal in $(N, +)$.

There is a much stronger form of (3.12), for which we need only a few more concepts.

(3.13) Definition. An element e of a nearring N is an *idempotent* if $e^2 = e$.

(3.14) Definition. For an element x of a nearring N, the *(right) annihilator* of x is $\text{Ann}(x) = \{n \in N \mid xn = 0\}$. For a subset $Y \neq \varnothing$ of a nearring N, the *annihilator* of Y is $\text{Ann}(Y) = \cap\{\text{Ann}(x) \mid x \in Y\}$.

(3.15) Proposition. *For any element x of a nearring N, $(\text{Ann}(x), +)$ is a normal subgroup of $(N, +)$. So, if $\varnothing \subset Y \subseteq N$, then $(\text{Ann}(Y), +)$ is a normal subgroup of $(N, +)$.*

Proof. Certainly, $0 \in \text{Ann}(x)$. If $m, n \in \text{Ann}(x)$, then $x(m-n) = xm + x(-n) = (xm) - (xn) = 0$, so $m - n \in \text{Ann}(x)$, and $(\text{Ann}(x), +)$ is a subgroup of $(N, +)$. If $y \in N$, then $x(y+m-y) = xy + xm - xy = xy + 0 - xy = 0$, so $y + m - y \in \text{Ann}(x)$, and $(\text{Ann}(x), +)$ is normal in $(N, +)$. By definition of $\text{Ann}(Y)$, and the fact that the intersection of a family of normal subgroups of a group is again a normal subgroup, we have that $(\text{Ann}(Y), +)$ is a normal subgroup of $(N, +)$.

(3.16) Proposition. *If e is any element of a nearring N and, in particular, an idempotent element of N, then $eN = \{en \mid n \in N\}$ is a subgroup of $(N, +)$.*

Proof. Certainly, $eN \neq \varnothing$. For $em, en \in eN$, $em - en = e(m-n) \in eN$, so eN is a subgroup.

(3.17) Theorem. *Let e be an idempotent of a nearring N. Then $(N, +)$ is a semidirect product of $(\text{Ann}(e), +)$ by $(eN, +)$.*

Proof. Let $x \in \text{Ann}(e) \cap eN$. Then $x = en$ for some $n \in N$. So $0 = ex = e(en) = (ee)n = en = x$, hence $eN \cap \text{Ann}(e) = \{0\}$. For $y \in N$, $y = (y - ey) + ey$ and $ey \in eN$. Since $e(y - ey) = ey - e(ey) = 0$, we get $y - ey \in \text{Ann}(e)$. Hence, $\text{Ann}(e) + eN = N$. Now apply (3.15) and (3.16) to complete the proof.

(3.18) Corollary. *For a nearring $(N, +, \cdot)$, the group $(N, +)$ is a semidirect product of $(N_0, +)$ by $(N_c, +)$.*

Proof. Of course, this is exactly (3.12). But here, let $e = 0$, an idempotent. Then $\text{Ann}(0) = N_0$ and $0N = N_c$.

(3.19) Remark. The material (3.10) through (3.18) is a pattern of things to come. It is important to see how nice theorems evolve, and it is equally important to see that good examples are excellent sources for nice theorems.

(3.20) Remark. Our main result from (3.10) through (3.18) is (3.17). More will be said about the eN and the Ann(e) later. The decomposition (3.17) is referred to as the 'Peirce decomposition' because of its analogy to the Peirce decomposition of rings.

Take another look at the definition of a constant nearring as given in (3.9). One could also have said that a nearring $(N, +, \cdot)$ is a *constant* nearring if each element $x \in N$ is a left identity. So then, 0 would be also a left identity. But this is all one really needs!

(3.21) Proposition. *If $0y = y$ for each y in a nearring N, then N is a constant nearring.*

Proof. Take $x \in N$. Then $xy = x(0y) = (x0)y = 0y = y$.

Since each element of a constant nearring is a left identity, it is also an idempotent. So we have

(3.22) Proposition. *For a nearring $(N, +, \cdot)$, if $e \in N_c$ then $e^2 = e$, so e is an idempotent.*

Now (3.17) gives us something to do with idempotents.

(3.23) Exploratory problem. Choose an interesting nearring N where $N_c \neq \{0\}$. By (3.22), each $e \in N_c$ is an idempotent. By (3.17), each idempotent e gives us a decomposition $N = \text{Ann}(e) + eN$. For your chosen N, compute the various decompositions for the various $e \in N_c$. Do some idempotents e provide more interesting decompositions than others? Do they depend upon the nearring N chosen?

With nearrings, as with other mathematical structures, we shall be interested in mappings that preserve some or all of the properties of the nearring. One could summarize a lot of research effort by saying that it is an investigation of properties that are preserved by mappings relevant to the structure.

(3.24) Definitions. Let $(M, \oplus, *)$ and $(N, +, \cdot)$ be nearrings. A mapping $f : M \to N$ such that $f(x \oplus y) = f(x) + f(y)$ and $f(x * y) = f(x) \cdot f(y)$ for all $x, y \in M$ is a *nearring homomorphism*. If $(M, \oplus, *) = (N, +, \cdot)$, then f is an *endomorphism*. If f is injective, that is, one-to-one, then f is a *monomorphism*. If f is surjective, that is, onto, then f is an *epimorphism*. An f that is a monomorphism and an epimorphism is an *isomorphism*. An isomorphism from $(N, +, \cdot)$ onto $(N, +, \cdot)$ is an *automorphism*.

All these nearring homomorphisms are certainly group homomorphisms, which means that we shall use standard facts about group homomorphisms freely.

(3.25) Proposition. *Let $f : M \to N$ be a homomorphism of nearrings. Then the following statements are true.*
(a) *The image $f(M)$ is a subnearring of N.*
(b) *If T is a subnearring of N, then $f^{-1}(T)$ is a subnearring of M.*
(c) $f(M_0) \subseteq N_0$.
(d) $f(M_c) \subseteq N_c$.
(e) *If f is an isomorphism, then so is f^{-1}.*

Proof. Certainly, $f(M)$ and $f^{-1}(T)$ are subgroups. For $a = f(x)$, $b = f(y) \in f(M)$, we also have $ab = f(x)f(y) = f(xy) \in f(M)$. If $f(x), f(y) \in T$, then $f(xy) = f(x)f(y) \in T$. Thus $f(M)$ and $f^{-1}(T)$ are subnearrings by (2.7).

For $m_0 \in M_0$, $0 \cdot f(m_0) = f(0)f(m_0) = f(0m_0) = f(0) = 0$. Hence, $f(M_0) \subseteq N_0$. For $0 \cdot m \in M_c$, $f(0 \cdot m) = f(0)f(m) = 0f(m) \in N_c$. Thus, $f(M_c) \subseteq N_c$.

Let $f : M \to N$ be an isomorphism. Then $f^{-1} : N \to M$ is a group isomorphism. For $n_1, n_2 \in N$. there are unique $m_1, m_2 \in M$ such that $f(m_1) = n_1$ and $f(m_2) = n_2$. Thus $f^{-1}(n_1 n_2) = f^{-1}(f(m_1)f(m_2)) = f^{-1}(f(m_1 m_1)) = m_1 m_2 = f^{-1}(f(m_1))f^{-1}(f(m_2)) = f^{-1}(n_1)f^{-1}(n_2)$.

(3.26) Definitions. Suppose $f : M \to N$ is a nearring monomorphism. We know $f(M)$ is a subnearring of N, and so $f : M \to f(M)$ is an isomorphism. Thus, N has an isomorphic copy of M as a subnearring. We say that M is *embedded* in N and that f is an *embedding*.

Mathematicians enjoy, for one reason or another, embedding a mathematical structure in another structure, especially if the new structure has some desirable property. Workers in ring theory delight in having a multiplicative identity to such an extent that it is quite common for them to ignore rings without identity. Some have even degraded such structures by calling them a 'rng'. (They say that a 'rng' is a 'ring' without identity 'i'.) To some extent they justify themselves by taking comfort in the fact that there are numerous embedding theorems embedding a ring in a ring with identity. There is the general theorem that any ring can be embedded in a ring with identity.

There is also such a theorem for nearrings. In fact, there are several such theorems. The one we give here is due to Heatherly and Malone.

(3.27) Theorem. *Let $(N, +, \cdot)$ be a nearring. Suppose $(G, +)$ is a group with a proper subgroup $(N', +)$ isomorphic to $(N, +)$. Then $(M(G), +, \circ)$ is a nearring with identity and has a subnearring isomorphic to $(N, +, \cdot)$.*

Proof. We identify $(N, +)$ with the subgroup $(N', +)$ of $(G, +)$ where $(N, +)$ and $(N', +)$ are isomorphic. Hence, we think of $(N', +, \cdot)$ as a nearring isomorphic to $(N, +, \cdot)$ with the isomorphism $n \mapsto n'$. For $n \in N$ we associate or identify $n' \in N'$. For each $n \in N$, we define a map $\theta_n \in M(G)$ by

$$g\theta_n = \begin{cases} n', & \text{if } g \notin N'; \\ gn', & \text{if } g \in N'. \end{cases}$$

Consider the map $\Psi : N \to M(G)$ defined by $\Psi(n) = \theta_n$. We will now proceed to show that Ψ is a monomorphism.

For $m, n \in N$, $\Psi(m + n) = \theta_{m+n}$, and

$$g\theta_{m+n} = \begin{cases} m' + n', & \text{if } g \notin N'; \\ g(m' + n') = gm' + gn', & \text{if } g \in N'. \end{cases}$$

Also,

$$g[\theta_m + \theta_n] = g\theta_m + g\theta_n = \begin{cases} m' + n', & \text{if } g \notin N'; \\ gm' + gn', & \text{if } g \in N'. \end{cases}$$

Thus, $\Psi(m + n) = \theta_{m+n} = \theta_m + \theta_n = \Psi(m) + \Psi(n)$. Now $\Psi(mn) = \theta_{mn}$, and

$$g\theta_{mn} = \begin{cases} m'n', & \text{if } g \notin N'; \\ g(m'n') = (gm')n', & \text{if } g \in N'. \end{cases}$$

Also

$$g\theta_m\theta_n = \begin{cases} m'\theta_n = m'n', & \text{if } g \notin N'; \\ (gm')\theta_n = (gm')n', & \text{if } g \in N'. \end{cases}$$

So $\Psi(mn) = \theta_{mn} = \theta_m\theta_n = \Psi(m)\Psi(n)$, and we have a homomorphism Ψ.

It remains to show that Ψ is injective. Take $m, n \in N$, $m \neq n$. We want $\Psi(m) = \theta_m \neq \theta_n = \Psi(n)$. For $g \notin N'$, $g\theta_m = m' \neq n' = g\theta_n$, so we have what we wanted.

Because of (3.25) we have

(3.28) Corollary. *From the embedding of (3.27) we have $\Psi(N_0) \subseteq M_0(G)$ and $\Psi(N_c) \subseteq M(G)_c$.*

(3.29) Corollary. *If $(N, +, \cdot)$ is a zero-symmetric nearring, then N has an embedding into a zero-symmetric nearring with identity.*

(3.30) Note. The nearring $M(G)$ or $M_0(G)$ can be quite large in comparison to N. One can take $G = N^+ \oplus Z_2$, the direct sum of $N^+ = (N, +)$ with Z_2, where Z_2 is the group of order 2, in an effort to minimize the size of $M(G)$ or $M_0(G)$.

What happens when a nearring homomorphism $f : M \to N$ is not a monomorphism? Certainly, $f : M^+ \to N^+$ is a group homomorphism, and so, as groups, $M^+/\ker f \cong f(M)^+$, where $\ker f = \{m \in M \mid f(m) = 0\}$, the *kernel* of f. Let $K = \ker f$. Then M/K consists of all cosets $m + K = K + m$ and $(K, +)$ is a normal subgroup of $(M, +)$. The addition in M defines an addition in M/K by $(m + K) + (n + K) = (m + n) + K$, and this addition is well defined exactly because (1) externally, K is the kernel of a group homomorphism, or (2) internally, $x + K - x \subseteq K$ for each $x \in M$. But our f is more than a group homomorphism. What extra properties does K have? In particular, would it make $(x + K) \cdot (y + K) = xy + K$ a well defined binary operation?

(3.31) First Isomorphism Theorem. *Let K be the kernel of a nearring homomorphism $f : M \to N$. Then $(M/K, +, \cdot)$ is a nearring isomorphic to $f(M)$.*

Proof. It is very easy to see that $(M/K, +, \cdot)$ will be a nearring if $(x + K) \cdot (y + K) = xy + K$ is a well defined binary operation. So this is where we will focus our attention first. Suppose $x' + K = x + K$ and $y' + K = y + K$. Then there are $a, b \in K$ such that $x' = x + a$ and $y' = y + b$. We need to show that $x'y' + K = xy + K$ or, equivalently, $-xy + x'y' \in K$. Now $x'y' = (x + a)(y + b) = (x + a)y + (x + a)b$. Since $f[(x + a)b] = [f(x) + f(a)]f(b) = (f(x) + 0) \cdot 0 = 0$, we know $(x + a)b = d$ for some $d \in K$. We now have $-xy + x'y' = -xy + (x + a)y + d$ with $d \in K$. Since $f[-xy + (x + a)y] = -f(x)f(y) + [f(x) + 0]f(y) = 0$, we have $-xy + (x + a)y = c$ for some $c \in K$. So now we have $-xy + x'y' = c + d \in K$, since $c, d \in K$. This makes $(x + K)(y + K) = xy + K$ a well defined binary operation.

As groups, $(M/K, +) \cong (f(M), +)$ where $F(x + K) = f(x)$ is the isomorphism. Now $F[(x + K)(y + K)] = F(xy + K) = f(xy) = f(x)f(y) = F(x + K)F(y + K)$, so F is a nearring isomorphism.

(3.32) Corollary. *Let $f : M \to N$ be a nearring epimorphism with K denoting the kernel of f. Let $\pi : M \to M/K$ be the natural map $\pi(x) = x + K$. Then π is a nearring epimorphism, the 'natural' epimorphism, and the isomorphism $F : M/K \to N$ defined by $F(x + K) = f(x)$ is the unique mapping such that $F \circ \pi = f$.*

Proof. There is an easy proof that π is a nearring epimorphism and that $F \circ \pi = f$. If $F' \circ \pi = f$, then $f(x) = F' \circ \pi(x) = F'(x + K)$, so $F'(x + K) = F(x + K)$ making $F' = F$.

(3.33) Theorem. *For a nearring $(N, +, \cdot)$ suppose $(K, +)$ is a normal subgroup of $(N, +)$ with the additional property that $xa \in K$ for each $x \in N$ and each $a \in K$. Then the following statements are equivalent:*
(a) K is the kernel of a nearring homomorphism.
(b) $(a + x)y - xy \in K$ for all $x, y \in N$ and all $a \in K$.

(c) $(x + a)y - xy \in K$ *for all* $x, y \in N$ *and all* $a \in K$.
(d) $-xy + (x + a)y \in K$ *for all* $x, y \in N$ *and all* $a \in K$.
(e) $-xy + (a + x)y \in K$ *for all* $x, y \in N$ *and all* $a \in K$.

Proof. (a)→(b). Suppose K is the kernel of a nearring homomorphism f. Then $f[(a + x)y - xy] = [f(a) + f(x)]f(y) - f(x)f(y) = 0$ if $x, y \in N$ and $a \in K$. Hence, $(a + x)y - xy \in K$ and we have (b).

(b)→(c). For suitable a, x, y, that is, for any $a \in K$ and any $x, y \in N$, we have $(x + a)y - xy = (a' + x)y - xy \in K$ where $a' \in K$ and $x + a = a' + x$.

(c)→(d). For suitable a, x, y, we have $-xy + (x + a)y = -[-(x + a)y + xy] = -[(x + a)(-y) - x(-y)] \in K$ since by (c), $(x + a)(-y) - x(-y) \in K$.

(d)→(e). For suitable a, x, y, $-xy + (a + x)y = -xy + (x + \bar{a})y \in K$ since $a + x = x + \bar{a}$ for some $\bar{a} \in K$.

(e)→(a). For the normal subgroup $(K, +)$ of $(N, +)$, there is the quotient group $(N/K, +)$ and the natural group epimorphism $\pi : N \to N/K$ where $\pi(x) = x + K$. Now $K = \ker \pi$ as a group homomorphism. We need only show that $\pi(xy) = \pi(x)\pi(y)$, that is, $xy + K = (x + K)(y + K)$, and to do this, we only need to show that $(x + K)(y + K) = xy + K$ is a well defined binary operation. As in the proof of (3.31), we take $x' + K = x + K$ and $y' + K = y + K$. So there are $a, b \in K$ such that $x' = x + a$ and $y' = y + b$. Hence, $x'y' = (x + a)(y + b) = (x + a)y + (x + a)b$. By hypothesis, $(x + a)b = d \in K$ for some d. Now $-xy + x'y' = -xy + (x + a)y + d$. But $-xy + (x + a)y = -xy + (\hat{a} + x)y$ since $x + a = \hat{a} + x$ for some $\hat{a} \in K$. By (e), $-xy + (\hat{a} + x)y = c \in K$ for some c. We have $-xy + x'y' = c + d \in K$ since $c, d \in K$. This means $x'y' + K = xy + K$, which in turn means that $(x + K)(y + K) = xy + K$ is well defined.

(3.34) Definition. The kernel K of a nearring homomorphism is an *ideal*. Externally, an *ideal* K of a nearring N is the kernel of a nearring homomorphism. Internally, an *ideal* K satisfies (i) $(K, +)$ is a normal subgroup of $(N, +)$; (ii) $xa \in K$ for each $x \in N$ and each $a \in K$; and (iii) $(x + a)y - xy \in K$ for all $x, y \in N$ and all $a \in K$. Certainly, (iii) can be replaced by conditions (b), (d), or (e) of (3.33).

If our nearring $(N, +, \cdot)$ is a ring, then for an ideal I we need not be concerned that $(I, +)$ is normal, since $(N, +)$ is abelian. Also, conditions (b)–(e) of (3.33) are just the condition that $IN \subseteq I$. We do not have this condition, in general, for an ideal of a nearring. If our nearring $(N, +, \cdot)$ is zero-symmetric, we do get it, as we shall see in (3.36).

(3.35) Examples. Let $(G, +)$ be a group with a proper normal subgroup $(K, +)$. Let $*$ be the trivial binary operation $a * b = b$ for all $a, b \in G$. Then $(G, +, *)$ is a nearring with $G_c = G$. So $GK \subseteq K$. But $KG \not\subseteq K$. For $x, y \in G$ and $a \in G$, $(x + a)y - xy = y - y = 0 \in K$, so K is an ideal of the nearring $(G, +, *)$.

(3.36) Proposition. *Let I be an ideal of a nearring N. Then $(I, +, \cdot)$ is a subnearring of $(N, +, \cdot)$. If $N = N_0$, then $NI \subseteq I$ and $IN \subseteq I$.*

Proof. Certainly, $(I, +)$ is a subgroup of $(N, +)$, and if $a, b \in I$, then $ab \in I$, so $(I, +, \cdot)$ is a subnearring. If N is zero-symmetric, $a \in I$, and $n \in N$, then $an = (0 + a)n - 0n \in I$, hence $IN \subseteq I$.

(3.37) Examples. For a nearring $(N, +, \cdot)$, we know that $\{0\}N = \{0\}$ need not be true. But, nonetheless, $\{0\}$ is an ideal of N, as is N itself. They are N's *trivial* ideals. To see that N is an ideal of N is absolutely trivial. It is almost as easy to see that $\{0\}$ is an ideal. Certainly, $(\{0\}, +)$ is a normal subgroup of $(N, +)$, $N\{0\} \subseteq \{0\}$, and $(x + 0)y - xy = 0 \in \{0\}$ for each $x, y \in N$.

(3.38) Definition. A nearring $(N, +, \cdot)$ is *simple* if N has only the two trivial ideals N and $\{0\}$.

(3.39) Examples. Let $(G, +)$ be any simple group and let $(G, +, \cdot)$ be any of the trivial nearrings of §2. Then $(G, +, \cdot)$ is a simple nearring.

Simple nearring are building blocks for more complex nearrings, just as simple groups and simple rings are.

(3.40) Definition. For a group $(G, +)$, we have already mentioned the constant functions. For $g \in G$, g's constant function is **g** where $x\mathbf{g} = g$ for each $x \in G$. But $g \in G$ also has a *punctured constant function* $\mathbf{g}_0$ where

$$x\mathbf{g}_0 = \begin{cases} g, & \text{if } x \in G \text{ and } x \neq 0; \\ 0, & \text{if } x = 0 \in G. \end{cases}$$

So $\mathbf{g}_0 \in M_0(G)$.

(3.41) Proposition. *Let $(S, +)$ be a subgroup of $(M_0(G), +)$ for some group $(G, +)$. Suppose $M_0(G)S \subseteq S$ and $SM_0(G) \subseteq S$ and $S \neq \{0\}$. Then S contains all $\alpha \in M_0(G)$ having a finite range $G\alpha$.*

Proof. Suppose $\alpha \in M_0(G)$ and $G\alpha = \{0 = g_1, g_2, \ldots, g_n\}$. Let $G_i = \{g \in G \mid g\alpha = g_i\} = \alpha^{-1}(g_i)$. The G_i form a partition of G. Define $\gamma_1, \ldots, \gamma_n \in M_0(G)$ by

$$x\gamma_i = \begin{cases} 0, & \text{if } x \notin G_i; \\ g_i, & \text{if } x \in G_i. \end{cases}$$

Then $\alpha = \sum_{i=1}^n \gamma_i \mathbf{g}_{i0}$, for if $x \in G_j \subseteq G$, then $x\alpha = g_j$ and $x \sum_{i=1}^n \gamma_i \mathbf{g}_{i0} = x\gamma_j \mathbf{g}_{j0} = g_j \mathbf{g}_{j0} = g_j$.

Now for $s \in S \setminus \{0\}$, there is a $k \in G$ such that $ks \neq 0$. So $\mathbf{k}_0 s l_0 = l_0 \in S$ for each $l \in G$. This puts each $\mathbf{g}_{i_0} \in S$, hence $\alpha \in S$.

(3.42) Theorem. *If G is a finite group, then $M_0(G)$ is a simple nearring.*

Proof. If $I \neq \{0\}$ is an ideal of $M_0(G)$, then, by (3.36), $IM_0(G) \subseteq I$ and $M_0(G)I \subseteq I$. If $I \neq \{0\}$, $M_0(G) \subseteq I$ follows from (3.41) since G is finite. Hence, $I = M_0(G)$.

It is natural to wonder if $M_0(G)$ is simple for an infinite group G. Well, it is, but for the proof, we shall use a rather curious lemma that divides the group into three parts of equal size. For a set X, we let $|X|$ denote its cardinality.

(3.43) Lemma. *Let G be an infinite group and fix a $g \in G$ where $g \neq 0$. Then there is a subset $T \subset G$ which is maximal with respect to the property that $T \cap (T + g) = \varnothing$. For such a maximal subset T, we have $G = T \cup (T + g) \cup (T - g)$ and $|G| = |T| = |T + g| = |T - g|$.*

Proof. Let $\mathcal{T} = \{T \subset G \mid T \cap (T + g) = \varnothing\}$. Since $g \neq 0$, $\{0\} \in \mathcal{T}$. The family $\mathcal{T}$ is partially ordered by $\subseteq$. Let $\{T_j\}_{j \in J}$ be a chain in $\mathcal{T}$ and let $T = \cup_{j \in J} T_j$. We want to show that $T \in \mathcal{T}$. If $x \in T \cap (T + g)$, then $x = t_m = t_n + g$ for some $t_m \in T_m$ and some $t_n \in T_n$, where $m, n \in J$. We can assume $T_m \subseteq T_n$, so $t_m, t_n \in T_n$. This is exactly what we need to apply Zorn's lemma in order to be assured that $\mathcal{T}$ has at least one maximal element. Now let T also denote such a maximal element.

We already have $T \cap (T + g) = \varnothing$. If $y \in G$ and $y \notin T \cup (T + g)$, we need to see that $y \in T - g$. If $y \notin T - g$, then $y + g \notin T$. Also, $y + g \notin T + g$, for if so, then $y \in T$. But this means $(T \cup \{y\}) \cap (T \cup \{y\} + g) = \varnothing$. Let us look at this assertion carefully. If $x \in T \cup \{y\}$, then either $x \in T$ or $x = y$. If $x \in T$, then $x \notin T + g$. If $x = y + g$, then $y + g \in T$, which, as we have just seen, is not so. If $x = y$, then $y = y + g$ cannot be since $g \neq 0$. Also, $y = t + g \in T + g$ cannot be, since $y \notin T + g$. Thus $(T \cup \{y\}) \cap (T \cup \{y\} + g) = \varnothing$.

So if $y \in G$ and $y \notin T \cup (T + g)$ and $y \notin T - g$, then $T \cup \{y\} \in \mathcal{T}$ and $T \subset T \cup \{y\}$. But $T \in \mathcal{T}$ is maximal. So if we insist that $y \in G$ and $y \notin T \cup (T + g)$, we are forced to conclude that $y \in T - g$.

Obviously, $|T| = |T + g| = |T - g|$. Since G is infinite, we have $|G| \leq 3|T| \leq |G|$, so T is infinite and $|T| = |G|$.

(3.44) Theorem. *Let G be a group. Then $M_0(G)$ is a simple nearring.*

Proof. We have shown this for G finite in (3.42). So we now proceed to the infinite case. Let I be a nonzero ideal of $M_0(G)$. We want to show that I contains all the punctured constant functions $\mathbf{g}_0$. Recall from (3.36) that

$IM_0(G) \subseteq I$ and $M_0(G)I \subseteq I$. Take $s \in I$, with $s \neq 0$, and $k \in G$ so that $ks \neq 0$. Then $\mathbf{k}_0 s l_0 = \mathbf{l}_0 \in I$ for each $l \in G$.

Now let $g \in G$ with $g \neq 0$. Using (3.43), choose a subset $T \subset G$ which is maximal with respect to the property that $T \cap (T + g) = \varnothing$. Define $\iota_T \in M_0(G)$ by

$$x\iota_T = \begin{cases} x, & \text{if } x \in T; \\ 0, & \text{if } x \notin T. \end{cases}$$

Then $\psi = (\iota + \mathbf{g}_0)\iota_T - \iota\iota_T = (\iota + \mathbf{g}_0)\iota_T - \iota_T \in I$, where $x\iota = \dot{x}$ for all $x \in G$. For $x \in T$, $x \neq 0$, $x\psi = (x+g)\iota_T - x = -x$, since $T \cap (T+g) = \varnothing$.

Let $\tau : G \setminus \{0\} \to T \setminus \{0\}$ be a bijection guaranteed by (3.43). Define $\tau_1 \in M_0(G)$ by $x\tau_1 = x\tau$ for $x \in G^* = G \setminus \{0\}$, and $0\tau_1 = 0$, of course. Now define $\tau_2 \in M_0(G)$ by

$$x\tau_2 = \begin{cases} y, & \text{if } x = y\tau, \text{ that is } x \in T \setminus \{0\}; \\ 0, & \text{if } x \notin T; \\ 0, & \text{if } x = 0. \end{cases}$$

We have $\tau_1(-\psi)\tau_2 \in I$. For $x = 0$, $0\tau_1(-\psi)\tau_2 = 0$. For $x \in G^*$, $x\tau_1(-\psi)\tau_2 = (x\tau)(-\psi)\tau_2 = [-(x\tau)\psi]\tau_2 = (x\tau)\tau_2 = x$. Hence, $\tau_1(-\psi)\tau_2 = \iota \in I$. With $\iota \in I$, we obtain $M_0(G) \subseteq I$. This makes $M_0(G)$ simple.

It is now natural to wonder if the $M(G)$ are also simple for a group G. They are, *almost*.

(3.45) Theorem. *Let* $G = Z_2$, *the group of order* 2. *Then* $M(G)$ *has exactly one nontrivial ideal, namely,* $M(G)_c$.

Proof. Here, $M(G) = \{\mathbf{0}, \mathbf{1}, \iota, 1 - \iota\}$ where $x\iota = x$ is the identity function of $M(G)$. We begin with three candidates for nontrivial ideals, namely, $I = M(G)_c = \{\mathbf{0}, \mathbf{1}\}$, $J = \{\mathbf{0}, \iota\}$, and $K = \{\mathbf{0}, 1 - \iota\}$. Since $M(G)J = M(G) \not\subseteq J$, J is eliminated. Since $\mathbf{0} \circ (1 - \iota) = \mathbf{1}$, $M(G)K \not\subseteq K$. This leaves I as the single candidate. For $\mathbf{g} \in I$ and $\alpha, \beta \in M(G)$, we wonder if

$$\gamma = (\alpha + \mathbf{g}) \circ \beta - \alpha \circ \beta$$

is in I. For $\beta \in I$, $\gamma = \mathbf{0}$. For $\beta = \iota$, $\gamma = \mathbf{g}$. For $\beta = 1 - \iota$, $\gamma = -\mathbf{g}$. So $\gamma \in I$ in all cases and thus I is an ideal.

(3.46) Theorem. *If* G *is a group and* G *does not have order* 2, *then* $M(G)$ *is a simple nearring.*

Proof. If $G = \{0\}$, then $M(G) = \{0\}$, which is simple. So assume G has order ≥ 3. By (3.11), $M(G) = M_0(G) + M(G)_c$. Let I be a nontrivial ideal of $M(G)$. We do not have $IM(G) \subseteq I$, but we do have $IM_0(G) \subseteq I$, as we shall now see. Let $\alpha \in M_0(G)$ and take $f \in I$. Then $f\alpha = f\alpha - \mathbf{0} = (f + \mathbf{0})\alpha - \mathbf{0}\alpha \in I$.

Next we show that $M(G)_c \subseteq I$. Take $f \in I \setminus \{\mathbf{0}\}$ and $x \in G$ so that $xf \neq 0$. Then $\mathbf{x}f\mathbf{g}_0 = \mathbf{g}$. But $\mathbf{x}f \in I$ and $\mathbf{x}f\mathbf{g}_0 \in I$, so each $\mathbf{g} \in I$. Thus, $M(G)_c \subseteq I$.

Now we show that $I \cap M_0(G) \neq \{\mathbf{0}\}$. For $k \in G \setminus \{0\}$, define $\theta_k \in M_0(G)$ by

$$x\theta_k = \begin{cases} x, & \text{if } x \notin \{0, k\}; \\ 0, & \text{if } x \in \{0, k\}, \end{cases}$$

and define $\gamma_k = (\iota + \mathbf{k})\theta_k - \iota\theta_k = (\iota + \mathbf{k})\theta_k - \theta_k$, where $x\iota = x$ is the identity function of $M(G)$. First, $0\gamma_k = k\theta_k - 0 = 0$, so $\gamma_k \in M_0(G)$. For $h \in G \setminus \{0, k\}$,

$$h\gamma_k = (h + k)\theta_k - h\theta_k = \begin{cases} h + k - h, & \text{if } h + k \notin \{0, k\}; \\ -h, & \text{otherwise.} \end{cases}$$

Thus $\gamma_k \neq \mathbf{0}$. Certainly, $\gamma_k \in I$. Hence, $\mathbf{0} \neq \gamma_k \in I \cap M_0(G)$.

But $I \cap M_0(G)$ is an ideal of $M_0(G)$. Why is this? Certainly, $(I \cap M_0(G), +)$ is a normal subgroup of $(M_0(G), +)$. Also, $M_0(G)(I \cap M_0(G)) \subseteq I \cap M_0(G)$. For $\alpha, \beta \in M_0(G)$ and $f \in I \cap M_0(G)$, $(\alpha + f)\beta - \alpha\beta \in I \cap M_0(G)$.

Since $I \cap M_0(G) \neq \{\mathbf{0}\}$ and is an ideal of $M_0(G)$, we have $I \cap M_0(G) = M_0(G)$ by (3.44). We have seen that $M(G)_c \subseteq I$; we now have $M_0(G) \subseteq I$, so, since $M(G) = M_0(G) + M(G)_c$, we obtain $I = M(G)$. This makes $M(G)$ simple.

(3.47) Corollary. *Let I be an ideal of a nearring $(N, +, \cdot)$. Then $IN_0 \subseteq I$, and $I \cap N_0$ is an ideal of N_0.*

Proof. The proof can easily be gleaned from the proof of (3.46).

(3.48) Definition. For a commutative ring R and a polynomial $f = a_0 + a_1x + \cdots + a_nx^n \in R[x]$, there is the corresponding *polynomial function* $\bar{f} \in M(R)$, where $\bar{f}(u) = a_0 + a_1u + \cdots + a_nu^n$ for each $u \in R$. Let $\overline{R[x]}$ denote all the polynomial functions $\bar{f}$ for $f \in R[x]$.

(3.49) Theorem. *For a commutative ring R, define the map $\Psi : R[x] \to M(R)$ by $\Psi(f) = \bar{f}$. Then Ψ is a right nearring homomorphism and $\Psi(R[x]) = \overline{R[x]}$.*

Proof. The proof is really easy, but one should study it carefully. Let $f = a_0 + a_1x + \cdots + a_nx^n$, $g = b_0 + b_1x + \cdots + b_nx^n \in R[x]$. (We are not assuming $a_n \neq 0$ or $b_n \neq 0$.) Then $f + g = (a_0 + b_0) + (a_1 + b_1)x + \cdots + (a_n + b_n)x^n$ and $f \circ g = a_0 + a_1(b_0 + b_1x + \cdots + b_nx^n) + \cdots + a_n(b_0 + b_1x + \cdots + b_nx^n)^n$. So $\Psi(f + g) = \overline{f + g}$ where $\overline{f + g}(u) = (a_0 + b_0) + (a_1 + b_1)u + \cdots + (a_n + b_n)u^n =$

$(a_0 + a_1 u + \cdots + a_n u^n) + (b_0 + b_1 u + \cdots + b_n u^n) = \overline{f}(u) + \overline{g}(u)$. Hence, $\Psi(f + g) = \overline{f + g} = \overline{f} + \overline{g} = \Psi(f) + \Psi(g)$.

Also, $\Psi(f \circ g) = \overline{f \circ g}$ and $\overline{f \circ g}(u) = a_0 + a_1(b_0 + b_1 u + \cdots + b_n u^n) + \cdots + a_n(b_0 + b_1 u + \cdots + b_n u^n)^n = \overline{f}(b_0 + b_1 u + \cdots + b_n u^n) = \overline{f}(\overline{g}(u)) = \overline{f} \circ \overline{g}(u)$. So $\Psi(f \circ g) = \overline{f \circ g} = \overline{f} \circ \overline{g} = \Psi(f) \circ \Psi(g)$.

(3.50) Corollary. *For a commutative ring R, $(\overline{R[x]}, +, \circ)$ is a right near-ring.*

For a finite field $(F, +, \cdot)$, $(F[x], +, \circ)$ is infinite and $(M(F), +, \circ)$ is finite. So the kernel of the corresponding Ψ of (3.49) is *not* $\{0\}$, and $\Psi(x) = \iota$, the identity of $M(F)$, so $F[x]$ has a nontrivial ideal. Certainly, $\Psi(f) = \mathbf{0}$ if and only if $\overline{f}(u) = 0$ for each $u \in F$. Thus $f = (x^{p^n} - x) \cdot \phi$ for some $\phi \in F[x]$, where $|F| = p^n$ and $\cdot$ is the usual multiplication from the ring of polynomials $(F[x], +, \cdot)$. Such an f is also in the kernel of Ψ, thus we have

(3.51) Theorem. *For a finite field F where $|F| = p^n$, the kernel of the nearring homomorphism Ψ of (3.49) is*

$$A(p^n) = \{ (x^{p^n} - x) \cdot \phi \mid \phi \in F[x] \}.$$

(3.52) Theorem. *For a finite field F, we have $\overline{F[x]} = M(F)$.*

Proof. Let $|F| = t$ and let $\alpha = \{(a_1, b_1), \ldots, (a_t, b_t)\}$ be an element of $M(F)$. (Here, $\alpha(a_i) = b_i$.) We want to show the existence of an $f \in F[x]$ so that $\overline{f} = \alpha$. Such an f of degree $t-1$ exists as a consequence of Lagrange's interpolation theorem. Such an f has the form

$$f = c_0 + c_1 x + \cdots + c_{t-1} x^{t-1}. \tag{3:1}$$

Thus we need to show the existence of $c_0, c_1, \ldots, c_{t-1} \in F$ so that

$$\begin{cases} c_0 + a_1 c_1 + a_1^2 c_2 + \cdots + a_1^{t-1} c_{t-1} = b_1 \\ c_0 + a_2 c_1 + a_2^2 c_2 + \cdots + a_2^{t-1} c_{t-1} = b_2 \\ \cdots\cdots\cdots\cdots\cdots \\ c_0 + a_t c_1 + a_t^2 c_2 + \cdots + a_t^{t-1} c_{t-1} = b_t. \end{cases} \tag{3:2}$$

These are t linear equations in t unknowns $c_0, c_1, \ldots, c_{t-1}$ with coefficients $a_i^k \in F$, a field, so there will be a unique solution exactly when the van der Monde determinate

$$D = \begin{vmatrix} 1 & a_1 & a_1^2 & \cdots & a_1^{t-1} \\ 1 & a_2 & a_2^2 & \cdots & a_2^{t-1} \\ \cdots & \cdots & \cdots & \cdots & \cdots \\ 1 & a_t & a_t^2 & \cdots & a_t^{t-1} \end{vmatrix} \neq 0.$$

This will be so because $a_1, a_2, \ldots, a_t$ are all distinct. (Certainly, if $a_i = a_j$ for some $i \neq j$, then $D = 0$.) But to see $D \neq 0$, consider

$$
D_x = \begin{vmatrix}
1 & a_1 & a_1^2 & \cdots & a_1^{t-1} \\
1 & a_2 & a_2^2 & \cdots & a_2^{t-1} \\
\cdots & \cdots & \cdots & \cdots & \cdots \\
1 & a_{t-1} & a_{t-1}^2 & \cdots & a_{t-1}^{t-1} \\
1 & x & x^2 & \cdots & x^{t-1}
\end{vmatrix},
$$

a polynomial of degree at most $t-1$ with $t-1$ distinct roots $a_1, a_2, \ldots, a_{t-1}$. Since $a_t \notin \{a_1, a_2, \ldots, a_{t-1}\}$ we must have $\overline{D_x}(a_t) \neq 0$. But $\overline{D_x}(a_t) = D$.

Thus, (3:2) has a unique solution, which means that the f of (3:1) exists with $\overline{f} = \alpha$.

(3.53) Corollary. *For a finite field F, and $A(p^n)$ as in (3.51), we have*

$$
\bigl(F[x]/A(p^n), +, \circ\bigr) \cong \bigl(M(F), +, \circ\bigr).
$$

Proof. Apply (3.52), (3.49), and (3.31).

(3.54) Corollary. *If F is a finite field and $|F| \neq 2$, then the quotient nearring $(F[x]/A(p^n), +, \circ)$ is simple.*

Proof. Apply (3.46) and (3.53).

Corollary (3.54) makes one suspect that $A(p^n)$ is a maximal ideal if $p^n \neq 2$. It also suggests a need for

(3.55) The Correspondence Theorem. *Let $f : M \to N$ be an epimorphism of nearrings. Let I be the kernel of f. So the nearrings M/I and N are isomorphic. This isomorphism defines a bijection between:*
(1) the subnearrings K of N and the subnearrings L of M which contain the ideal I;
(2) the subnearrings K of N having the property that $NK \subseteq K$ and the subnearrings L of M, which contain I, and have the property that $ML \subseteq L$;
(3) the subnearrings K of N having the property that $(x+a)y - xy \in K$ for all $x, y \in N$ and all $a \in K$, and the subnearrings L of M, which contain I, and have the property that $(s + e)t - st \in L$ for all $s, t \in M$ and all $e \in L$;
(4) the ideals K of N and the ideals L of M which contain I.

Proof. The correspondence theorem for group theory gives a bijection between the subgroups $(K, +)$ of $(N, +)$ and the subgroups $(L, +)$ of $(M, +)$ which contain I. This correspondence is given by two maps, f_* and its inverse f^*, where $f_*(L) = f(L) = K$ and where $f^*(K) = f^{-1}(K) = L$. Also, $(L, +)$ is normal if and only if the corresponding $(K, +)$ is normal.

For (1), if L is a subnearring, then so is $f_*(L) = f(L) = K$ by (3.25). Again by (3.25), if K is a subnearring, then $f^*(K) = f^{-1}(K) = L$ is also.

For (2), if $ML \subseteq L$, then $f(ML) \subseteq f(L)$, or $f_*(M)f_*(L) \subseteq f_*(L)$. But $f_*(M) = N$, so $Nf_*(L) \subseteq f_*(L)$. Now suppose $NK \subseteq K$. Is $Mf^*(K) \subseteq f^*(K)$? Take $m \in M$ and $e \in f^*(K) = f^{-1}(K)$. Then $f(me) = f(m)f(e) \in K$, hence $me \in f^{-1}(K) = f^*(K)$.

For (3), if $(x + a)y - xy \in K$ for each $x, y \in N$ and each $a \in K$, consider $e \in f^*(K)$ and $s, t \in M$. Then $f[(s + e)t - st] \in K$ since $f(e) \in K$. Hence, $(s + e)t - st \in f^*(K)$. If $(s + e)t - st \in L$ for each $s, t \in M$ and each $e \in L$, consider $x, y \in N$ and $a \in K = f_*(L)$. Choose $s, t \in M$ so that $f(s) = x$ and $f(t) = y$, and choose $e \in L$ so that $f(e) = a$. We know $(s + e)t - st \in L$, so $(x + a)y - xy = f[(s + a)t - st] \in f(L) = f_*(L) = K$.

(2) and (3), together with the fact that $(L, +)$ is normal if and only if $(K, +)$ is normal, give us (4).

(3.56) Corollary. *For a finite field F with $|F| = p^n$, $A(p^n)$ is a maximal ideal of the nearring $(F[x], +, \circ)$ if $p^n \neq 2$. If $p^n = 2$, then there is exactly one ideal properly between $A(2)$ and $F[x]$. This is $A(2) \cup [1 + A(2)] = B(2)$, the polynomials of $F[x]$ whose polynomial functions are the constant functions.*

Proof. From (3.53), $F[x]/A(p^n) \cong M(F)$. By (3.54) and (3.55), $A(p^n)$ is maximal for $p^n \neq 2$. If $p^n = 2$, then, by (3.53), (3.45) and (3.55), there is exactly one ideal properly between $A(2)$ and $F[x]$, since $M(Z_2)$ has exactly the one nontrivial ideal $M(Z_2)_c$, and it is the image of $B(2)$ by the Ψ of (3.49).

From (3.52), (3.53), and (3.54), we learn that, except for the field of two elements Z_2, a finite field F has a simple nearring of polynomial functions $\overline{F[x]}$. We will soon see that an infinite field F also has a simple nearring of functions $\overline{F[x]}$.

For an infinite field F, we have $M(F)$ simple by (3.46). This next theorem assures us that all subnearrings of $M(F)$ containing $\overline{F[x]}$ are also simple.

(3.57) Theorem. *Let F be an infinite field, and let $(N, +, \circ)$ be a subnearring of $M(F)$ satisfying $\overline{F[x]} \subseteq N \subseteq M(F)$. Then $(N, +, \circ)$ is simple.*

Proof. If Ψ is the map of (3.49), then Ψ is a monomorphism. For if $\Psi(f) = \overline{f} = \mathbf{0}$, then $\overline{f}(u) = 0$ for each $u \in F$. But if $f \neq 0$, then f can have only finitely many roots. Since F is infinite, we are forced to conclude that $f = 0$ if $\Psi(f) = \mathbf{0}$. Thus $F[x] \cong \overline{F[x]}$, and we identify $f \in F[x]$ with $\overline{f} \in \overline{F[x]}$.

Let $I \neq \{0\}$ be an ideal of N, so there is an $f \in I$ and an $r \in F$ so that $a = \overline{f}(r) \neq 0$. Now $f \circ \mathbf{r} = \mathbf{a} \in I$ so $x^2 \circ (x + a) - x^2 \circ x = 2ax + a^2 \in I$ also. Thus $(2ax + a^2) \circ \mathbf{0} = \mathbf{a^2} \in I$, giving us $2ax \in I$. As long as $2a \neq 0$, we obtain $2ax \circ [(2a)^{-1}x] = x \in I$, the identity, so $I = N$, making N simple.

Suppose the characteristic of F is 2, so this is exactly when we can have $2a = 0$. As above, $\mathbf{a} \in I$, so $x^n \circ (x + a) - x^n \circ x \in I$, making $[x^n \circ (x + a) - x^n \circ x] \circ \mathbf{0} = \mathbf{a}^n \in I$ for each positive integer n. Since $g = x^3 \circ (x + a) - x^3 \circ x = ax^2 + a^2x + a^3 \in I$, we obtain $g \circ \mathbf{a}^{-1} = \mathbf{a}^{-1} + \mathbf{a} + \mathbf{a}^3 \in I$, hence $\mathbf{a}^{-1} \in I$. Thus $\mathbf{a}^{-n} \in I$ for each $n = 1, 2, 3, \ldots$. From $g \circ \mathbf{a}^{-2} \in I$ we obtain $\mathbf{1} \in I$. Thus $\mathbf{a}^n \in I$ for each integer n. This gives us $x^2 + x + 1$ and $x^2 + x$ in I, as well as $L \subseteq I$, where $L = \{\mathbf{b}^2 + \mathbf{b} \mid b \in F\}$. Choose $b \in L \setminus \{0, 1\}$ and let $f = b^{-1}x^3$. Then $f \circ (x + b) - f \circ x = x^2 + bx + b^2 \in I$. Since $\mathbf{b}^2 \in I$, we obtain $x^2 + bx \in I$. We have already seen that $x^2 + x \in I$, so we have $(x^2 + bx) + (x^2 + x) = (b + 1)x \in I$. Since $b + 1 \neq 0$, $(b + 1)x \circ [(b + 1)^{-1}x] = x \in I$, and again $I = N$, making N simple.

(3.58) Corollary. *For an infinite field F, the nearring $(F[x], +, \circ)$ is simple.*

(3.59) Corollary. *If $(F, +, \cdot)$ is any subfield of the complex number field $(\mathbf{C}, +, \cdot)$, then $(\mathcal{C}(F, F), +, \circ)$ is simple.*

Proof. Of course, we are assuming that F inherits the usual topology of $\mathbf{C}$. Now $\overline{F[x]} \subseteq \mathcal{C}(F, F) \subseteq M(F)$. So apply (3.57).

In §2, for a group $(G, +)$, we introduced $M(G)$ as $\mathcal{C}(G, G)$, where G has the discrete topology. We saw in (3.46), that if $|G| \neq 2$, then $M(G)$ is simple. Recently, in (3.59), we saw that $\mathcal{C}(G, G)$ is simple for various other topological groups, where G does not have the discrete topology. This motivates

(3.60) Problem. What are necessary and sufficient conditions on a topological group G so that $(\mathcal{C}(G, G), +, \circ)$ is simple?

(3.61) Exploratory problem. Investigate the simplicity of the various $(\mathcal{C}(X, H), +, *_\alpha)$ described in §2.

(3.62) Theorem. *If $(N, +, \cdot)$ is a nearfield and $|N| \neq 2$, then it is zero-symmetric.*

Proof. If x is an element of N such that $0 \cdot x \neq 0$, let $y = 0 \cdot x$. Then $1 = y^{-1}y = y^{-1}(0 \cdot x) = 0 \cdot x = y$. Hence, $1 \in N_c$. For $t \in N$, $t = t \cdot 1 = t \cdot y = t \cdot (0 \cdot x) = 0 \cdot x = 1$. This makes $N = \{0, 1\}$.

For a nearfield $(N, +, \cdot)$, each nonzero element $x \in N$ has the same additive order as 1. If this order is finite, it is a prime p, and we say that N has *characteristic* p, and we write char $N = p$. If this order is infinite, we say that N has *characteristic* 0, and we write char $N = 0$.

(3.63) Lemma. *For a nearfield $(N, +, \cdot)$, if $x \in N$ and $x^2 = 1$, then*

$x \in \{1, -1\}$, *and conversely.*

Proof. Certainly, $x^2 = 1$ if $x \in \{1, -1\}$. Suppose that $x^2 = 1$ and $x \neq 1$. We must single out the case where char $N = 2$, in which case $y + y = 0$ for each $y \in N$, and, consequently, $(N, +)$ is abelian. Also, $x(x+1) = x^2 + x = 1 + x = (1 + x) \cdot 1 = 1 \cdot (x + 1)$. Thus $x = 1$ or $x = -1$.

Now suppose that char $N \neq 2$. Then $2 = 1 + 1 \neq 0$. Also, $(-1) \cdot 2 = (-1) \cdot 1 + (-1) \cdot 1 = (-1) + (-1) = -(1 + 1) = -2 = 2 \cdot (-1)$. From $2 \cdot [(-1) \cdot 2^{-1} + 1] = (-1) + 2 = 1$, we conclude that $(-1) \cdot 2^{-1} + 1 = 2^{-1}$. Let $y = (x-1) \cdot 2^{-1} + 1$. Then $xy = x[(x-1) \cdot 2^{-1} + 1] = x(x-1) \cdot 2^{-1} + x = (x^2 - x) \cdot 2^{-1} + x = (1 - x) \cdot 2^{-1} + x = [-(x - 1)] \cdot 2^{-1} + x = [(x-1)(-1)] \cdot 2^{-1} + x = (x-1)[(-1) \cdot 2^{-1}] + (x - 1) + 1 = (x-1)[(-1) \cdot 2^{-1} + 1] + 1 = (x-1) \cdot 2^{-1} + 1 = y$. From $xy = y = 1 \cdot y$ and $x \neq 1$, we conclude that $y = (x-1) \cdot 2^{-1} + 1 = 0$. So $(x - 1) \cdot 2^{-1} = -1$ and $x + (-1) = x - 1 = (-1) \cdot 2 = -2 = (-1) + (-1)$. Hence, $x = -1$.

(3.64) Lemma. *For a nearfield* $(N, +, \cdot)$, *if* $x, y \in N$, *then* $x(-y) = (-x)y = -(xy)$.

Proof. This is easily true for $y = 0$ and when char $N = 2$. So assume $y \neq 0$ and char $N \neq 2$. Then $[y \cdot (-1) \cdot y^{-1}]^2 = 1$, so $y \cdot (-1) \cdot y^{-1} \in \{1, -1\}$. If $y \cdot (-1) \cdot y^{-1} = 1$, then $-y = y \cdot (-1) = y$, and char $N = 2$. If $y \cdot (-1) \cdot y^{-1} = -1$, then $-y = y \cdot (-1) = (-1) \cdot y$, and so $x(-y) = x[y(-1)] = x[(-1)y]$, and then $-(xy) = x(-y) = (-x)y$.

(3.65) Theorem. *For a nearfield* $(N, +, \cdot)$, *the group* $(N, +)$ *is abelian.*

Proof. From (3.64), $(-1)x = -(1 \cdot x) = -x$. Take $x, y \in N$. Then $x + y = (-1)(-x) + (-1)(-y) = (-1)[(-x) + (-y)] = -[(-x) + (-y)] = -(-y) + [-(-x)] = y + x$.

(3.66) Theorem. *If* $(N, +, \cdot)$ *is a finite nearfield, then* $(N, +, *)$ *is a field for some suitable* $*$, *that is,* $(N, +)$ *is the additive group of some field.*

Proof. Since each $x \in N \setminus \{0\}$ has additive order p, for some prime, and since $(N, +)$ is abelian, then $(N, +)$ is a vector space over the prime field $(Z_p, +, \cdot)$. Hence, $(N, +)$ is the additive group of some field $(N, +, *)$.

Mathematicians do not study objects, but relations among objects; they are indifferent to the replacement of objects by others as long as relations do not change. Matter is not important, only form interests them.

Henri Poincaré

CHAPTER 2
Planar nearrings

In the summer of 1967, the author hoped to discover some geometric interpretations for nearrings. Since planar nearfields had already proven useful to geometry, he hoped that planarity could be extended to nearrings which were not nearfields. After the author developed a suitable concept for planarity for nearrings, Michael Anshel constructed numerous examples in which the algebraic structure of planar nearrings had nice geometric interpretations. Some of these original examples provide the motivation for much of what is being done, and has been done, in planar nearrings.

Almost concurrently, Giovanni Ferrero was concerned with constructing finite integral nearrings. It turned out that his nearrings were planar. Ferrero then gave a new and very important direction to the study of planar nearrings. Parallel to the geometric interpretation of the algebraic structure of Anshel's examples, Ferrero showed that some finite planar nearrings could be used to construct new balanced and partially balanced incomplete block designs. He suggested that these might be useful for coding theory, and his insight has proven to be prophetic.

One of Anshel's examples haunted the author for 20 years. This example described the circles of the usual euclidean plane in terms of the structural parts of a planar nearring $(\mathbf{C}, +, *)$, whose additive group was $(\mathbf{C}, +)$, the additive group of the complex numbers. This description of the circles included a way to identify the radius and the centre of the circles. In 1986, the author was able to extend these ideas to other nearrings. The extension provided a way to obtain new balanced incomplete block designs from *all* finite planar nearrings, and, perhaps more important, provided a way to construct 'planes' with 'circles' with well defined 'radii' and 'centres'. This gave birth to a new branch of geometric investigation. Such planar nearrings are called *circular*.

Circular planar nearrings demand the property that three distinct points lie on *at most* one circle, and that circles have enough 'curvature'. Matthew Modisett used circular properties to apply planar nearrings to cryptology and to coding theory. Peter Fuchs, Gerhard Hofer, and Günter Pilz explored these applications to coding theory further. In addition, they showed that the finite noncircular planar nearrings had positive applications to coding theory as well.

Still another of Anshel's constructions allowed the description of all rays in the complex plane **C** in terms of a planar nearring $(\mathbf{C}, +, *)$. This construction allows the description of the ray from any point P passing through a different point Q in terms of P and Q. With rays, one can describe line segments $\overline{PQ}$ between two distinct points P and Q. These concepts lead to some nice geometric and combinatorial applications of finite planar nearrings, one of which is that the 'segments' sometimes form the blocks of a balanced incomplete block design.

As a guest of the National Cheng-Kung University in Tainan in 1989, the author noticed that it is possible to construct a 'double planar nearring' $(\mathbf{C}, +, *_1, *_2)$. Each $(\mathbf{C}, +, *_i)$ is a planar nearring, and each $*_i$ is also left distributive over the other $*_j$. One $(\mathbf{C}, +, *_1)$ describes the circles of the plane, together with centres and radii, and the other $(\mathbf{C}, +, *_2)$ describes the rays. This construction can have meaning in some finite 'double planar nearrings'. Preliminary investigations indicate that the relationships between the resulting 'circles' and the 'rays', or 'line segments', will lead to some very interesting theorems.

The author feels that there are numerous applications and directions of planar nearrings yet to be explored. As in the first chapter, he will point to some of these with problems and exploratory problems.

4. Planarity for nearrings

Soon after Dickson gave to the world an example of an honest nearfield, somebody came along and found a use for his example. In 1907, Veblen and Wedderburn used Dickson's example to create a plane having unusual properties. (See §9.)

Let F be a nearfield and let $\Pi = F \times F$ denote our set of points. Lines in Π are defined by equations $x = a$ (vertical lines) and $y = mx + b$ (those having slope m). The desire for two nonparallel lines having exactly one point in common leads to require that the system

$$\begin{cases} y = mx + b \\ y = m'x + b' \end{cases} \tag{4:1}$$

has a unique solution if $m \neq m'$. Setting

$$mx + b = m'x + b'$$

yields

$$mx = m'x + (b' - b).$$

This equation has the form

$$ax = bx + c,$$

with $a \neq b$. This leads to

(4.1) Definition. A nearfield $(F, +, \cdot)$ is said to be *planar* if, given constants $a, b, c \in F$, $a \neq b$, the equation

$$ax = bx + c,$$

has exactly one solution for x in F.

Now planar nearfields have served a very useful purpose in geometry. One might hope, therefore, that planarity could be extended to nearrings in such a way as to be a useful concept. Suppose $(N, +, \cdot)$ were a nearring with the property that, given $a, b, c \in N$ with $a \neq b$, the equation $ax = bx + c$ would have exactly one solution for x in N. It is an amusing exercise to show that such a nearring would be a nearfield. Something else is needed.

(4.2) Definitions. For a nearring $(N, +, \cdot)$, define an equivalence relation $=_m$ on N by $a =_m b$ if and only if $ax = bx$ for all $x \in N$. If $a =_m b$, we say that a and b are *equivalent multipliers*.

For a nearring $(N, +, \cdot)$, suppose arbitrary $a, b, c \in N$, $a \neq_m b$, gave us the equation $ax = bx + c$, which had a unique solution for $x \in N$. Such nearrings exist, since each trivial nearring $(G, +, \cdot_S)$ of §2 has this property. Not wanting to be bothered with these $(G, +, \cdot_S)$, we decided to have

(4.3) Definition. A nearring $(N, +, \cdot)$ is *planar* when:
(i) $=_m$ has at least three equivalence classes, that is, $|N/=_m| \geq 3$;
(ii) for constants $a, b, c \in N$ with $a \neq_m b$, the equation

$$ax = bx + c \qquad\qquad (4:2)$$

has a unique solution for x in N.

Of course, it is the right of everyone to expect the mathematician to provide nice, meaningful examples of each concept that he introduces. The three examples we provide next are at the foundation of all that follows concerning planar nearrings.

(4.4) Example 1. Let $(\mathbf{C}, +, \cdot)$ denote the complex number field. For $a, b \in \mathbf{C}$, $a = (a_1, a_2)$, define

$$a *_1 b = \begin{cases} a_1 b, & \text{if } a_1 \neq 0; \\ a_2 b, & \text{if } a_1 = 0. \end{cases}$$

(4.5) Example 2. Let $(\mathbf{C}, +, \cdot)$ denote the complex number field. For $a, b \in \mathbf{C}$, define

$$a *_2 b = |a| \cdot b.$$

(4.6) Example 3. Let $(\mathbf{C}, +, \cdot)$ denote the complex number field. For $a, b \in \mathbf{C}$, define

$$a *_3 b = \begin{cases} 0, & \text{if } a = 0; \\ (a/|a|) \cdot b, & \text{if } a \neq 0. \end{cases}$$

It is direct and easy to show that for each $(\mathbf{C}, +, *_i)$, $i = 1, 2, 3$, is a planar nearring. More examples are also meaningful.

(4.7) Examples 4. Let $(\mathbf{C}, +, \cdot)$ denote the complex number field. For positive numbers a and b, and $u, z \in \mathbf{C}$, define

$$u *_4 z = \left| \frac{x^2}{a^2} - \frac{y^2}{b^2} \right|^{\frac{1}{2}} \cdot z,$$

where $u = (x, y)$.

Examples 1, 2, and 3 are examples of *integral* planar nearrings in that $a * b = 0$ for $b \neq 0$ implies $a = 0$. In examples 4, if $u = (x, (b/a)x)$ and $x \neq 0$, then $u *_4 z = 0$ for each $z \in \mathbf{C}$.

If $a \neq_m 0$, then we may be asked to consider the equation

$$ax = 0x + c.$$

It will be comforting to have

(4.8) Proposition. *For a planar nearring $(N, +, \cdot)$ we have $0 \cdot b = 0$ for each $b \in N$. That is, planar nearrings are zero-symmetric.*

Proof. For $a \neq_m 0$, consider the equation

$$ax = 0x + 0. \tag{4:3}$$

For any $b \in N$, we have $b \cdot 0 = 0$ by (3.1). Let $x = 0 \cdot b$. Then 0 and $0 \cdot b$ are each solutions to (4:3), since $a \cdot 0 = 0 \cdot 0 + 0$, and $a \cdot (0 \cdot b) = (a \cdot 0) \cdot b = 0 \cdot b$ and $0 \cdot (0 \cdot b) + 0 = (0 \cdot 0) \cdot b = 0 \cdot b$.

Examples 4 show that $a =_m 0$ for $a \neq 0$ is meaningful in a planar nearring. For a planar nearring $(N, +, \cdot)$, let

$$A = \{a \in N \mid a =_m 0\}.$$

Throughout §4, we will assume $(N, +, \cdot)$ to be a planar nearring unless otherwise stated.

The equation $ax = 0x + c$ becomes $ax = c$. So one has, for each $a \in N^*$, where $N^* = N \setminus A$, the equation $ax = a$. (Note the more general use of the notation N^*.) Denote the unique solution to this equation by 1_a and call it a's *right identity*. (Of course, we are assuming $a \notin A$.) It may be that a

and b have the same right identity, that is, $1_a = 1_b$ even though $a \neq b$. For $a \notin A$, define

$$B_a = \{b \in N \setminus A \mid b1_a = b\}.$$

Let $L = \{1_a \mid a \in N^*\}$. We are now in a position to prove a theorem of fundamental importance for planar nearrings.

(4.9) MAIN STRUCTURE THEOREM. *For any planar nearring* $(N, +, \cdot)$, *the following statements are true.*
(i) $N = A \cup \left[\bigcup_{a \in N^*} B_a\right]$.
(ii) $N^* B_a = B_a$ *for any* $a \in N^*$.
(iii) $a \in N^*$ *implies that* $(B_a, \cdot)$ *is a group.*
(iv) $\{A\} \cup \{B_a \mid a \in N^*\}$ *is pairwise disjoint.*
(v) *For* $a, c \in N^*$, *the map* $\phi : B_a \to B_c$ *defined by* $\phi(x) = x1_c$
    is a group isomorphism.
(vi) $a \in N^*$ *implies that* 1_a *is a left identity.*

Proof. For $a \in N^*$, $a \in B_a$; hence (i), and the fact that $B_a \neq \varnothing$. For $x \in N^*$ and $b \in B_a$, $(xb)1_a = x(b1_a) = xb$, so $xb \in B_a$. This gives us $N^* B_a \subseteq B_a$ for (ii), and the fact that B_a is closed with respect to $\cdot$, that is, each $(B_a, \cdot)$ is a semigroup.

For $a \in N^*$, $a = a1_a = (a1_a)1_a = a(1_a 1_a)$. The equation $ax = a$ has a unique solution, so $1_a \cdot 1_a = 1_a$. This puts $1_a \in B_a$, and we know that B_a has a right identity 1_a.

Suppose a' is the unique solution to $ax = 1_a$. Then $a(a'1_a) = (aa')1_a = 1_a 1_a = 1_a$. By uniqueness, $a'1_a = a'$, and so $a' \in B_a$. We could justifiably call a' a's *right inverse*. But what about $b \in B_a$? Does it have a right inverse? Certainly, there is a b' such that $bb' = 1_b$ and $b' \in B_b$. Could we have $b \in B_a \cap B_b$ with $B_a \neq B_b$? If $c \in B_a \cap B_b$, then $c1_a = c = c1_b$. So 1_a and 1_b are solutions to $cx = c$. This makes $1_a = 1_b$ and $B_a = B_b$. Thus $bb' = 1_b = 1_a$, which tells us not only that each $b \in B_a$ has a right inverse in B_a making $(B_a, \cdot)$ a group, but also that $B_a \cap B_b \neq \varnothing$ implies $B_a = B_b$. So we obtain (ii), (iii), and (iv).

By (ii), $\phi : B_a \to B_c$, defined by $\phi(x) = x1_c$, is well defined. For $x, y \in B_a$, $\phi(xy) = (xy)1_c = x(y1_c) = x[1_c(y1_c)] = (x1_c)(y1_c) = \phi(x)\phi(y)$, making ϕ a group homomorphism.

For $b \in B_a$, $b1_a = b$. Hence, $b(1_a 1_c) = (b1_a)1_c = b1_c$, and this forces $1_a 1_c = 1_c$, since $bx = (b1_a)1_c$ has a unique solution for x.

If $x1_c = y1_c$ for $x, y \in B_a$, then $(x1_c)1_a = x(1_c 1_a) = x1_a = x = (x1_c)1_a = (y1_c)1_a = y(1_c 1_a) = y1_a = y$. This makes ϕ a monomorphism.

For $y \in B_c$, $y1_a \in B_a$ and $\phi(y1_a) = (y1_a)1_c = y(1_a 1_c) = y1_c = y$. This makes ϕ an epimorphism, and an isomorphism. Thus we have (v).

Consider the equation $1_a x = 1_a d$. It has d as a solution, obviously, but it also has $1_a d$ as a solution, since $1_a(1_a d) = (1_a 1_a)d = 1_a d$. So, we must have $1_a d = d$. Now $d \in N$ represents an arbitrary element of N, hence (vi).

(4.10) Corollary. *Suppose* $(N, +, \cdot)$ *is planar and for* $a, b \in N$, $a =_m b$ *if and only if* $a = b$. *Then* $(N, +, \cdot)$ *is a planar nearfield.*

Proof. For $a, b, c \in N$, $a \neq b$, the equation $ax = bx + c$ has a unique solution. If $a \neq 0$, then $a \neq_m 0$, so $A = \{0\}$. Also, $a, b \in N^* = N \setminus \{0\}$ implies $1_a =_m 1_b$, so $1_a = 1_b$, hence $B_a = B_b = N^*$. Thus $N^* = N \setminus \{0\} = B_a$, and so $(N \setminus \{0\}, \cdot)$ is a group. This makes $(N, +, \cdot)$ a nearfield.

Exercise. Prove that any planar nearring $(N, +, \cdot)$ with identity is a planar nearfield.

Let us now examine our examples relative to the MAIN STRUCTURE THEOREM.

For example 1, $a =_m 0$ exactly when $a = (a_1, a_2) = (0, 0)$. So $A = \{(0, 0)\}$, and $(\mathbf{C}, +, *_1)$ is *integral*. If $a = (a_1, a_2)$ is a left identity, then either $a_1 = 1$ or $a_1 = 0$ and $a_2 = 1$, and conversely. So $L = \{(1, m) \mid m \in \mathbf{R}\} \cup \{(0, 1)\}$, $B_{(0,1)} = \{(0, y) \mid y \in \mathbf{R}^*\}$ and $B_{(1,m)} = \{(x, mx) \mid x \in \mathbf{R}^*\}$ where $\mathbf{R}^* = \mathbf{R} \setminus \{0\}$. So, a B_a is exactly the straight line passing through 0 and a, with the point 0 eliminated.

For example 2, $a =_m 0$ exactly when $a = (0, 0)$. So, again, $A = \{(0, 0)\}$, and $(\mathbf{C}, +, *_2)$ is integral. If $a = (a_1, a_2)$ is a left identity, then $|a| = (a_1^2 + a_2^2)^{1/2} = 1$, or more simply, $a_1^2 + a_2^2 = 1$. So the left identities L are exactly the points on the unit circle. For a point $a \neq 0$, $1_a = (1/|a|) \cdot a$, so $B_a = \{z \in \mathbf{C} \mid (1/|z|) \cdot z = (1/|a|) \cdot a\}$, which consists of all positive scalar multiples of a, that is, the ray from 0 through a, with the point 0 eliminated.

For example 3, $a =_m 0$ exactly when $a = (0, 0)$. As above, $A = \{(0, 0)\}$, and $(\mathbf{C}, +, *_3)$ is integral. Since, for $a \neq 0$, $a/|a|$ is usually *not* a real number, but rather a complex number of absolute value 1, we have $1_a = |a|$. This makes $L = \{r \in \mathbf{R} \mid r > 0\}$. For $r > 0$, what is B_r? Recall that B_r consists of exactly those $a \in \mathbf{C}^*$ such that $a *_3 r = a = (a/|a|) \cdot r$, hence those $a \in \mathbf{C}^*$ with $|a| = r$, that is, the circle with centre 0 and radius r.

Examples 4 will have $A \neq \{(0, 0)\}$. In fact, $u = (x, y) =_m (0, 0)$ exactly when

$$\frac{x^2}{a^2} - \frac{y^2}{b^2} = 0,$$

or, equivalently, $y = \pm|b/a|x$. Thus A consists of the points on the two straight lines through $(0, 0)$ with slopes $\pm|b/a|$. So, planar nearrings need not be integral. Similarly, $u \in L$ exactly when

$$\frac{x^2}{a^2} - \frac{y^2}{b^2} = \pm 1,$$

so L consists of two hyperbolae. To calculate a B_c, just select $c \in \mathbf{C}$ but not in A, that is, c cannot be on the lines $y = \pm |b/a| x$. By (ii) of the MAIN STRUCTURE THEOREM, $\mathbf{C}^* *_4 B_c = B_c$, and by (iii) and (v) of the same THEOREM, $B_c = \mathbf{C}^* *_4 c$. From the definition of $*_4$, $B_c = \{rc \mid r > 0\}$, the ray from 0 through c with 0 eliminated.

Examples 4 come from a curious family of planar nearrings which could possibly provide motivation for interesting research. They are described in

(4.11) Theorem. *Suppose $\phi \neq \zeta$, the zero map, and suppose $\phi : \mathbf{C} \to \mathbf{R}$ satisfies $\phi(tz) = t^\alpha \phi(z)$ for some fixed $\alpha \in \mathbf{R}^*$, for all $t \in \mathbf{R}$, $t \geq 0$, and for all $z \in \mathbf{C}$. Define $*_\phi$ on $\mathbf{C}$ by*

$$u *_\phi z = \left| \phi(u) \right|^{1/\alpha} \cdot z.$$

Then $(\mathbf{C}, +, *_\phi)$ is a planar nearring.

The proof can easily be constructed by the curious reader, and by doing so he may discover something amusing or significant.

Perhaps we should give a hint of things to come by providing a finite example. We will construct a finite integral planar nearring $(Z_5, +, *)$ where $Z_5 = \{0, 1, 2, 3, 4\}$ and $(Z_5, +)$ is the cyclic group of order 5. Now we must define $*$. Let $1 =_m 3$ and define $1 * a = a$. Let $2 =_m 4$ and define $2 * a = -a$. Now, the reader has three choices: (i) just believe that $(Z_5, +, *)$ is a planar nearring; (ii) verify that $(Z_5, +, *)$ is a planar nearring; or (iii) wait until we have explained and verified that the Ferrero Planar Nearring Factory produces planar nearrings.

4.1. The Ferrero Planar Nearring Factory

In Alba, Italy, one finds the headquarters of a factory, with branches throughout the world, which produces chocolate of the highest quality. But in Parma, Italy, home of one of the oldest universities of the world, is the headquarters of another factory which produces planar nearrings. Both factories carry the name Ferrero. As you learn the process of the Ferrero Planar Nearring Factory, you thereby become an authorized branch of this factory with full authority to produce planar nearrings of the highest quality. The president and founder is Professore Giovanni Ferrero.

Raw material for a planar nearring $(N, +, \cdot)$ consists of a pair of groups (N, Φ), where $1 \neq \Phi < \mathrm{Aut}(N, +)$, and Φ is a *regular* group of automorphisms of the group $(N, +)$. This means that if $1 \neq \phi \in \Phi$ and $\phi(x) = x$, then $x = 0$. So the automorphisms of Φ are as *fixed point free* as they can be. We often say that Φ is a nontrivial group of fixed point free automorphisms of the group $(N, +)$. If $(N, +)$ is finite, this is all we need, but if $(N, +)$ is infinite, we require that $-\phi + 1_N$ be a surjective map of

N for each $\phi \in \Phi \setminus \{1_N\}$. In summary, start with (N, Φ), where (i) $(N, +)$ is a group; (ii) Φ is a nontrivial group of fixed point free automorphisms of $(N, +)$, that is, a regular group of automorphisms; and (iii) $-\phi + 1_N$ is a bijection of N for each $\phi \in \Phi \setminus \{1_N\}$. Let us call such a pair (N, Φ) a *Ferrero pair*.* (In our finite example above, $N = Z_5$, and $\Phi = \{\pm 1\}$.) We had better give this a number.

(4.12) Definition. A *Ferrero pair* is a pair of groups (N, Φ) where $1 \neq \Phi < \mathrm{Aut}(N, +)$, and where each $\phi \in \Phi$, $\phi \neq 1$, has the properties that $-\phi + 1_N$ is surjective and $\phi(x) = x$ solely for $x = 0$.

For such a Ferrero pair (N, Φ), an *orbit* of Φ is $\Phi(a) = \{\phi(a) \mid \phi \in \Phi\}$, for some $a \in N$. The orbit $\Phi(0) = \{0\}$ is the *trivial orbit*.

(4.13) Theorem. *Let (N, Φ) be a Ferrero pair. Then:*
(i) *for $0 \neq a \in N$, $|\Phi(a)| = |\Phi|$.*
(ii) *if $b \in \Phi(a)$, then $\Phi(a) = \Phi(b)$, and conversely.*
(iii) *the orbits $\Phi(a)$ form a partition of N.*

Proof. For (i), the map $\phi \mapsto \phi(a)$ is a bijection, for if $\phi(a) = \mu(a)$, then $\mu^{-1}\phi(a) = a$ and so $\mu^{-1}\phi = 1_N$, or $\phi = \mu$.

For (ii), $b \in \Phi(a)$ implies $\phi(a) = b$ for some $\phi \in \Phi$. Hence, $\Phi(\phi(a)) = \Phi(b)$. Since Φ is a group, $\Phi(\phi(a)) = \Phi(a)$. Thus $\Phi(a) = \Phi(b)$. Certainly, $b \in \Phi(b)$, so if $\Phi(a) = \Phi(b)$, then $b \in \Phi(a)$.

For (iii), we have $a \in \Phi(a)$ for each $a \in N$. If $c \in \Phi(a) \cap \Phi(b)$, then by (ii), $\Phi(a) = \Phi(c) = \Phi(b)$. So the orbits $\Phi(a)$ are pairwise disjoint sets whose union is N. This makes a partition of N.

In our finite example above, with $N = Z_5$, and $\Phi = \{\pm 1\}$, we have $\Phi(1) = \{1, 4\}$, and $\Phi(2) = \{2, 3\}$.

So, from a Ferrero pair, we construct the nontrivial orbits $\mathcal{B}_n = \{\Phi(a) \mid a \in N, a \neq 0\}$. The next step is to select any nonempty subset $\mathcal{C} \subseteq \mathcal{B}_n$. For each orbit $B_i \in \mathcal{C}$, select a representative e_i, and then $B_i = \Phi(e_i)$. We are now ready to define $\cdot$ making $(N, +, \cdot)$ a planar nearring.

Let $A = N \setminus \cup \mathcal{C}$, and for $a \in A$, define $a \cdot x = 0$ for each $x \in N$. So $A = \{a \in N \mid a =_m 0\}$, and $\cup \mathcal{C} = N^* = N \setminus A$. For $a \in N^* = \cup \mathcal{C}$, then there is exactly one $\Phi(e_i) \in \mathcal{C}$ with $a \in \Phi(e_i)$. Remember we have chosen and fixed the representatives e_i. By (4.13), there is a unique $\phi_a \in \Phi$ such that $\phi_a(e_i) = a$. Define $a \cdot x = \phi_a(x)$. In summary,

$$a \cdot x = \begin{cases} 0, & \text{if } a \in A = N \setminus \cup \mathcal{C}; \\ \phi_a(x), & \text{if } a \in \Phi(e_i) \in \mathcal{C} \text{ and } \phi_a(e_i) = a. \end{cases} \tag{4:4}$$

---

* When one speaks of *the* Ferrero pair, one is referring to Professore Giovanni Ferrero and his lovely wife Professoressa Celestina Ferrero-Cotti.

Now comes the exciting part, the part where we see that $(N, +, \cdot)$ is actually a planar nearring!

Left distributivity is easy, so we will do that anon. If $a \cdot x = 0$ for all $x \in N$, then $a \cdot (b + c) = 0 = 0 + 0 = (a \cdot b) + (a \cdot c)$. If $a \cdot x = \phi_a(x)$, then $a \cdot (b + c) = \phi_a(b + c) = \phi_a(b) + \phi_a(c) = (a \cdot b) + (a \cdot c)$. There you have it!

For associativity, there are three cases to build up to the climax. Consider

$$a \cdot (b \cdot c) = (a \cdot b) \cdot c. \qquad (4{:}5)$$

The three cases are: (i) $a \in A$; (ii) $a \notin A$, $b \in A$; and (iii) $a, b \notin A$.

Case (i). Since $a \in A$, then $ax = 0$ for each $x \in N$. Hence, $a \cdot (b \cdot c) = 0$ and $(a \cdot b) \cdot c = 0 \cdot c = 0$, since $0 \in A$. Thus (4:5) is valid.

Case (ii). For $a \notin A$ and $b \in A$, $(a \cdot b) \cdot c = \phi_a(b) \cdot c$. But $b \in A$ and $\phi_a \in \Phi$ imply $\phi_a(b) \in A$. Thus, $\phi_a(b) \cdot c = 0$, and $(a \cdot b) \cdot c = 0$. Now $a \cdot (b \cdot c) = a \cdot 0 = \phi_a(0) = 0$, making (4:5) valid.

Case (iii). Since $a, b \notin A$, $(a \cdot b) \cdot c = \phi_a(b) \cdot c$, and $\phi_a(b) \notin A$. Thus, $(a \cdot b) \cdot c = \phi_{\phi_a(b)}(c)$. And $a \cdot (b \cdot c) = a \cdot \phi_b(c) = \phi_a(\phi_b(c)) = \phi_a \circ \phi_b(c)$. Do we really have

$$\phi_{\phi_a(b)} = \phi_a \circ \phi_b?$$

For $b \in \Phi(e_i)$, we have $\phi_a(b) \in \Phi(e_i)$ also. Now $\phi_b(e_i) = b$, and $\phi_{\phi_a(b)}(e_i) = \phi_a(b)$. But $\phi_a \circ \phi_b(e_i) = \phi_a(b)$. So $\phi_a \circ \phi_b(e_i) = \phi_{\phi_a(b)}(e_i)$, making $\phi_{\phi_a(b)} = \phi_a \circ \phi_b$. Thus (4:5) is indeed valid.

This makes $(N, +, \cdot)$ a nearring!

If $a \in A$, then $a \cdot x = 0$ for all $x \in N$. If $\Phi(e_i) \in \mathcal{C}$, then $e_i \cdot x = x$ for each $x \in N$. If $a \in \Phi(e_i) \in \mathcal{C}$ and $a \neq e_i$, then $a \cdot x = \phi_a(x) \neq x$ for each $x \in N$. Hence, $|N/{=_m}| \geq 3$.

Let us now focus on the planarity equation

$$ax = bx + c \qquad (4{:}6)$$

with $a \neq_m b$. Again, there are various cases to consider: (i) $a =_m 0$; (ii) $b =_m 0$; (iii) $a \neq_m 0$ and $b \neq_m 0$.

Case (i). If $a =_m 0$, then (4:6) becomes $0 = bx + c$, or $bx = -c$. Since $b \neq_m 0$, $bx = -c$ is equivalent to $\phi_b(x) = -c$. Since ϕ_b is an automorphism, there is a unique $x \in N$ with $\phi_b(x) = -c$, or, equivalently, with $ax = bx + c$.

Case (ii). If $b =_m 0$, then (4:6) becomes $ax = c$, or $\phi_a(x) = c$. As above, there is a unique x, namely, $\phi_a^{-1}(c)$, and it satisfies $ax = bx + c$.

Case (iii). Equation (4:6) becomes $-bx + ax = c$, or equivalently, $-\phi_b(x) + \phi_a(x) = [-\phi_b + \phi_a](x) = [(-\phi_b) \circ \phi_a^{-1} + 1] \circ \phi_a(x) = [-(\phi_b \circ \phi_a^{-1}) + 1](\phi_a(x)) = c$. Since $\phi_b \circ \phi_a^{-1} \in \Phi \setminus \{1\}$ and since $[-(\phi_b \circ \phi_a^{-1}) + 1]$ is surjective (and injective), there is a unique y, hence a unique x, so that $\phi_a(x) = y$ and $[-(\phi_b \circ \phi_a^{-1}) + 1](y) = c$. (If $|N| < \infty$, then since $-\phi + 1$ is injective for each $\phi \in \Phi \setminus \{1\}$, we obtain that $-\phi + 1$ is surjective automatically, so we need not assume that $-\phi + 1$ is surjective if $|N| < \infty$.) This all means that (4:6) has a unique solution.

But this means $(N, +, \cdot)$ is planar!

Let's summarize the Ferrero Planar Nearring Factory.
(1) Start with a Ferrero pair (N, Φ).
(2) Select some or all of the nontrivial orbits for the set $\mathcal{C}$.
(3) Select a representative e_i for each orbit $\Phi(e_i) \in \mathcal{C}$.
(4) Define $\cdot$ by $(4 : 4)$.
(5) Be assured that $(N, +, \cdot)$ is a planar nearring.

An important thing to remember is that *every* planar nearring $(N, +, \cdot)$ can be constructed from a Ferrero pair (N, Φ). We proceed now to show this.

Using any fixed B_a from $(N, +, \cdot)$, define $\Phi = \{\phi_b \mid b \in B_a\}$ where $\phi_b(x) = bx$. We need to show that $1 \neq \Phi < \mathrm{Aut}(N, +)$, that Φ is regular, and that $-\phi + 1$ is surjective for each $\phi \in \Phi \setminus \{1\}$. From $\phi_b(x + y) = b(x + y) = bx + by = \phi_b(x) + \phi_b(y)$, we know that ϕ_b is a homomorphism. The equation $bx = 0$ only has the solution $x = 0$, so ϕ_b is a monomorphism. For $a \in N$, the equation $bx = a$ has a unique solution for x, so ϕ_b is an epimorphism. Thus, $\phi_b \in \mathrm{Aut}(N, +)$. From $|N/{=_m}| \geq 3$, we have $|\Phi| \geq 2$, so $1 \neq \Phi$. Take $\phi_b \in \Phi \setminus \{1\}$. Then $1_a x = bx + c$ has a unique solution for each $c \in N$, so $-bx + 1_a x = c$ does also, hence $(-\phi_b + 1)(x) = c$ has a unique solution for x, for each $c \in N$. This makes $-\phi_b + 1$ surjective, as well as injective. So (N, Φ) is a Ferrero pair.

If we had started with a different B_c, perhaps we would have got a different Φ? By (4.9), $\phi : B_a \to B_c$ defined by $\phi(b) = b1_c$, is a group isomorphism. Also, 1_c is a left identity. So, take an arbitrary $b1_c \in B_c$. Then $(b1_c)x = b(1_c x) = bx$, for each $x \in N$. So $b \in B_a$ and $b1_c \in B_c$ define the same automorphism. Thus, we get the *same* Φ from each of the B_a's of N.

For each 1_a of N, $\Phi(1_a) = \{\phi_b(1_a) = b \mid b \in B_a\} = B_a$. Thus, $B_a = \Phi(1_a)$, an orbit of Φ with representative 1_a. So $N^* = N \setminus A$ is a union of orbits of Φ. If $a \in A$, then $ax = 0$ for each $x \in N$. Now if $\phi \in \Phi$, then $\phi = \phi_b$ for some $b \in N^*$. So $\phi(a) = \phi_b(a) = ba$. But yet $(ba)x = b(ax) = b0 = 0$, since N is zero symmetric. Hence, $\Phi(a) \subseteq A$, and A is a union of orbits of Φ. This means that $(N, +, \cdot)$ is one of the planar nearrings constructible by the Ferrero Planar Nearring Factory.

We have:

(4.14) Theorem. *Every planar nearring $(N, +, \cdot)$ is constructible from a Ferrero pair (N, Φ) by the Ferrero Planar Nearring Factory.*

(4.15) Corollary. *From a Ferrero pair* (N, Φ), *numerous planar nearrings can be constructed.*

Proof. First note that the family of orbits constituting $\mathcal{C}$ is flexible as long as Φ has more than one nontrivial orbit. Also, for a given $\mathcal{C}$, one can choose a representative of an orbit $\Phi(a) \in \mathcal{C}$ in at least $|\Phi| \geq 2$ ways. Of course, different selections of $\mathcal{C}$ and of representatives of orbits $\Phi(a) \in \mathcal{C}$ lead to different planar nearrings.

4.2. Output of the Factory

So the quest to find planar nearrings can be reduced to finding Ferrero pairs (N, Φ). The easiest place to find them is from a field $(F, +, \cdot)$. Let $(N, +) = (F, +)$, and let Φ' be a subgroup of $(F^* = F \setminus \{0\}, \cdot)$. For $a \in \Phi'$, define $\phi_a : F \to F$ by $\phi_a(x) = ax$. It is easy to show that $\phi_a \in \text{Aut}(N, +)$, and that ϕ_a is fixed point free if $1 \neq a \in N = F$. Let $\Phi = \{\phi_a \mid a \in \Phi'\}$. Then (N, Φ) is a Ferrero pair.

(4.16) Definition. A planar nearring $(N, +, \cdot)$ is *field generated* if it is generated from a Ferrero pair (N, Φ), where $(F, +, \cdot)$ is a field, $(N, +) = (F, +)$, and $\Phi = \{\phi_a \mid a \in \Phi'\}$ where $\Phi' < F^*$. Usually, we simply identify Φ with Φ'.

The same argument above applies to a planar nearfield $(F, +, \cdot)$, so we have

(4.17) Definition. A planar nearring $(N, +, \cdot)$ is *nearfield generated* if it is generated from a Ferrero pair (N, Φ), where $(F, +, \cdot)$ is a planar nearfield, $(N, +) = (F, +)$, and $\Phi = \{\phi_a \mid a \in \Phi'\}$ where $\Phi' < F^*$.

Of course, it is natural now to wonder if one might find a Ferrero pair from a ring $(R, +, \cdot)$ with identity 1. Let $\mathcal{U}(R)$ denote the group of units of R. Certainly, a unit $u \in \mathcal{U}(R)$ defines an automorphism ϕ_u of $(R, +)$ by $\phi_u(x) = ux$. But if $u \neq 1$, then ϕ_u should only have 0 as a fixed point. Thus, if $u \in \mathcal{U}(R)$ and $u \neq 1$ and $x \neq 0$, we want $\phi_u(x) = ux \neq x$. This means $(u - 1)x \neq 0$. As long as $u - 1 \in \mathcal{U}(R)$, we can be assured that $(u - 1)x \neq 0$ for each $x \neq 0$. If $(u - 1)x = 0$ for some $x \neq 0$, then we can be assured that $u - 1 \notin \mathcal{U}(R)$. So, in selecting $1 \neq \Phi' < \mathcal{U}(R)$, if one assures oneself that $u - 1 \in \mathcal{U}(R)$ for each $u \in \Phi' \setminus \{1\}$, then (R, Φ) is a Ferrero pair. But, is this more than is required? For $u \in \mathcal{U}(R)$, suppose $ux \neq x$ for each $x \in R^*$. Then $(u - 1)x \neq 0$ for each $x \in R^*$. That is, $u - 1$ is not a divisor of 0. Conversely, if $u - 1$ is not a divisor of 0, then $(u - 1)x \neq 0$ for each $x \in R^*$, or $ux \neq x$ for each $x \in R^*$, so ϕ_u is fixed point free. Thus, if (R, Φ) is to be a Ferrero pair, then we must be assured that $u - 1$ is not a divisor of 0 for each $u \in \Phi' \setminus \{1\}$. (We are assuming that $\Phi' < \mathcal{U}(R)$.) But remember! We need $-\phi + 1$ to be surjective for each $u \in \Phi' \setminus \{1\}$. This means that $(-u + 1)x = a$ must have a unique solution for each $c \in R$, which forces $-u + 1 \in \mathcal{U}(R)$, or, equivalently, $u - 1 \in \mathcal{U}(R)$.

(4.18) Definition. A planar nearring $(N, +, \cdot)$ is *ring generated* if it is generated from a Ferrero pair (N, Φ), where $(R, +, \cdot)$ is a ring with identity 1, $(N, +) = (R, +)$, $\Phi = \{\phi_u \mid u \in \Phi'\}$, where $\Phi' < \mathcal{U}(R)$, and $u - 1 \in \mathcal{U}(R)$ for each $u \in \Phi' \setminus \{1\}$. Since Φ is so closely identified with Φ', we tend to identify the two, so we essentially say that $\Phi = \Phi'$. And to simplify our writing, we often let $\Phi^* = \Phi \setminus \{1\}$.

Examples 1, 2, 3, and 4 of (4.4)—(4.7) are all field generated planar nearrings from the field of complex numbers. For example 1, $\Phi' = \Phi = \mathbf{R}^*$. For example 2 and examples 4, $\Phi = \mathbf{R}_+ = \{r \in \mathbf{R} \mid r > 0\}$. For example 3, $\Phi = T$, the unit circle.

We shall have occasion later to look at some nearfield generated planar nearrings.

For a family of ring generated planar nearrings, consider the rings of integers modulo p^n, where p is an odd prime. These are $(Z_{p^n}, +, \cdot)$. Now $|\mathcal{U}(Z_{p^n})| = p^{n-1}(p - 1)$ and $\mathcal{U}(Z_{p^n}) \cong Z_{p^{n-1}} \oplus Z_{p-1}$. Let $C(p^{n-1}) = \{1, p+1, 2p+1, \cdots, (p^{n-1}-1)p+1\}$. Since $(kp+1)(lp+1) = klp^2 + (k+l)p+1$, we know that $C(p^{n-1}) \cong Z_{p^{n-1}}$ and $C(p^{n-1}) \triangleleft \mathcal{U}(Z_{p^n})$. These are exactly the $u \in \mathcal{U}(Z_{p^n})$ such that $u - 1 \notin \mathcal{U}(Z_{p^n})$. So, if $\Phi < \mathcal{U}(Z_{p^n})$ and $\Phi \cong Z_{p-1}$, then $u - 1 \in \mathcal{U}(Z_{p^n})$ for each $u \in \Phi^*$. Thus, (Z_{p^n}, Φ) is a Ferrero pair and any planar nearring generated by (Z_{p^n}, Φ) by the Ferrero Planar Nearring Factory will be ring generated.

We shall return to this family of examples to illustrate other important points and applications.

Ferrero pairs (N, Φ) where N is finite come from Frobenius groups. This is an excellent source! If G is a Frobenius group, then it has a normal subgroup $(N, +)$, called its *kernel*, and a subgroup Φ', called its *complement*, so that Φ' is isomorphic to a regular group $\Phi < \text{Aut}(N, +)$, and $G \cong N \times_\theta \Phi$, the natural semidirect product of N by Φ, where $\theta : \Phi \to \text{Aut}(N, +)$ is the natural embedding. Conversely, if $1 \neq \Phi < \text{Aut}(N, +)$ and Φ is a regular group of automorphisms of a finite group $(N, +)$, then $N \times_\theta \Phi$ is a Frobenius group for the natural embedding $\theta : \Phi \to \text{Aut}(N, +)$.

As we shall later see, it is naïve to conclude that the study of finite planar nearrings is equivalent to the study of Frobenius groups. This is because we may have two isomorphic regular groups of automorphisms of $(N, +)$, Φ_1 and Φ_2, so $N \times_{\theta_1} \Phi_1 \cong N \times_{\theta_2} \Phi_2$, but yet the Ferrero pair (N, Φ_1) cannot yield planar nearrings isomorphic to those yielded by the Ferrero pair (N, Φ_2).

What we need now are some examples of Ferrero pairs (N, Φ) which are not derived from fields, planar nearfields, or rings. The following theorem is a major source.

(4.19) Theorem. *Let R be a commutative integral domain with identity 1, and let σ be a ring automorphism of R. If $\mathcal{U}(R)$ denotes the group of units of R, let A be a nontrivial subgroup of $\mathcal{U}(R)$.*

On the set $N = R \times R$, define a binary operation $+_\sigma = +$ by

$$(a, x) + (b, y) = (a + b + xy^\sigma, \, x + y),$$

and for each $\beta \in A$, define a map σ_β by

$$\sigma_\beta(a, x) = (\beta\beta^\sigma a, \beta x).$$

Let $\Phi_A = \{\sigma_\beta \mid \beta \in A\}$. Then:
(i) $(N, +)$ *is a group.*
(ii) $(N, +)$ *is abelian if and only if $\sigma = 1_R$, the identity function on R.*
(iii) Φ_A *is a subgroup of* $\mathrm{Aut}(N, +)$.
(iv) $\Phi_A \cong A$.
(v) *If $\beta \in A^* = A \setminus \{1\}$ implies $\beta^\sigma \neq \beta^{-1}$, then Φ_A is regular.*
(vi) *For $\beta \in A^*$, assume $-1 + \beta$, $-1 + \beta\beta^\sigma \in \mathcal{U}(R)$. Then $-1 + \sigma_\beta$ is a bijection.*
(vii) *For $\beta \in A$, suppose $1 + \beta$, $1 + \beta\beta^\sigma \in \mathcal{U}(R)$. Then $1 + \sigma_\beta$ is a bijection.*

Proof. Let (a, x), (b, y), $(c, z) \in N$. Then $[(a, x) + (b, y)] + (c, z) = (a + b + xy^\sigma + c + (x + y)z^\sigma, \, x + y + z)$ and $(a, x) + [(b, y) + (c, z)] = (a + b + c + yz^\sigma + x(y + z)^\sigma, \, x + y + z)$. So $+$ is associative, since $t \mapsto t^\sigma$ is a ring homomorphism. Certainly, $(a, x) + (0, 0) = (a + 0 + x \cdot 0^\sigma, \, x + 0) = (a, x)$, so $(0, 0)$ is a right identity. Finally, $(a, x) + (-a + xx^\sigma, \, -x) = (0, 0)$, so $-(a, x) = (-a + xx^\sigma, -x)$ is the right inverse of (a, x). So we have that $(N, +)$ is a group.

If $(N, +)$ is abelian, then $(a + b + xy^\sigma, \, x + y) = (a, x) + (b, y) = (b, y) + (a, x) = (b + a + yx^\sigma, \, y + x)$. This requires $yx^\sigma = xy^\sigma$ for all $x, y \in R$. Let $x = 1$. Then $y = y^\sigma$, so $\sigma = 1_R$. Certainly, $\sigma = 1_R$ forces $+$ to be commutative. This gives us (ii).

For (iii), take $\beta \in A$ and proceed to verify that $\sigma_\beta \in \mathrm{Aut}(N, +)$. Thus $\beta \mapsto \sigma_\beta$ will be our map for (iv). For (a, x), $(b, y) \in N$,

$$\begin{aligned}
\sigma_\beta[(a, x) + (b, y)] &= (\beta\beta^\sigma[a + b + xy^\sigma], \, \beta(x + y)) \\
&= (\beta\beta^\sigma a + \beta\beta^\sigma b + (\beta x)(\beta y)^\sigma, \, \beta x + \beta y) \\
&= (\beta\beta^\sigma a, \, \beta x) + (\beta\beta^\sigma b, \, \beta y) \\
&= \sigma_\beta(a, x) + \sigma_\beta(b, y).
\end{aligned}$$

So σ_β is a homomorphism. If $(0, 0) = \sigma_\beta(a, x) = (\beta\beta^\sigma a, \beta x)$, then $\beta x = 0$ and $\beta\beta^\sigma a = 0$. If $(a, x) \neq (0, 0)$, and $x \neq 0$, then $\beta = 0$. But $\beta \in A \triangleleft \mathcal{U}(R)$, so $\beta \neq 0$. Since $\beta \in \mathcal{U}(R)$, then $\beta^\sigma \in \mathcal{U}(R)$, also. This puts $\beta\beta^\sigma \in \mathcal{U}(R)$. So, the consequence of $x = 0$ is $a \neq 0$. But then $\beta\beta^\sigma a \neq 0$. This makes σ_β injective.

To complete (iii), we need σ_β to be surjective. Let $(c, z) \in N$ and let $(a, x) = ((\beta^\sigma)^{-1}\beta^{-1}c, \ \beta^{-1}z)$. Then $\sigma_\beta(a, x) = (c, z)$. So, σ_β is an automorphism, as promised. Hence, $\Phi_A \subseteq \mathrm{Aut}(N, +)$.

Define $\Psi : A \to \mathrm{Aut}(N, +)$ by $\Psi(\beta) = \sigma_\beta$. Then $\Psi(A) = \Phi_A$. If Ψ is a homomorphism, then we will have that Φ_A is a subgroup of $\mathrm{Aut}(N, +)$ and we will be well on our way to our proof of (iv). Now $\Psi(\alpha\beta) = \sigma_{\alpha\beta}$, and $\sigma_{\alpha\beta}(a, x) = ((\alpha\beta)(\alpha\beta)^\sigma a, \ (\alpha\beta)x) = (\alpha\alpha^\sigma \beta\beta^\sigma a, \ \alpha\beta x) = \sigma_\alpha(\sigma_\beta(a, x)) = \sigma_\alpha \circ \sigma_\beta(a, x)$. This makes $\Psi(\alpha\beta) = \sigma_{\alpha\beta} = \sigma_\alpha \circ \sigma_\beta = \Psi(\alpha) \circ \Psi(\beta)$, so Ψ is a group homomorphism. Thus (iii).

If $\sigma_\beta = 1_N$, then $(a, x) = \sigma_\beta(a, x) = (\beta\beta^\sigma a, \ \beta x)$ for each $(a, x) \in N$. Since $\beta x = x$ for all $x \in R$ forces $\beta = 1$, we have $\ker \Psi = \{1\}$, and so Ψ is a monomorphism. Thus $A \cong \Phi_A$, and (iv) is valid.

We want Φ_A to be regular. Suppose $\beta \in A^* = A \setminus \{1\}$ and that (a, x) is a fixed point of σ_β. Then $(a, x) = \sigma_\beta(a, x) = (\beta\beta^\sigma a, \ \beta x)$. From $\beta x = x$ and $\beta \neq 1$, we have $0 = -x + \beta x = (-1 + \beta)x$. But R is an integral domain and $\beta \neq 1$, so $x = 0$. From $a = \beta\beta^\sigma a$, if $a \neq 0$, then $\beta\beta^\sigma = 1$ and $\beta^\sigma = \beta^{-1}$, which we did not allow. So, $a = 0$ and $(a, x) = (0, 0)$. We obtain Φ_A regular, and (v).

For (vi), if $\beta \in A^*$, we have $-1 + \beta\beta^\sigma \in \mathcal{U}(R)$. So, if $\beta^\sigma = \beta^{-1}$, we have $-1 + 1 = 0 \in \mathcal{U}(R)$. Since we cannot tolerate $0 \in \mathcal{U}(R)$, we must conclude that $\beta^\sigma \neq \beta^{-1}$, so by (v), Φ_A is regular. If $\beta \in A^*$, then $-1 + \sigma_\beta$ is injective. Let $(c, z) \in N$ be arbitrary, and $x = (-1 + \beta)^{-1}z$. Then

$$(-1 + \sigma_\beta)\left(\left[\frac{c - (1 - \beta^\sigma)xx^\sigma}{-1 + \beta\beta^\sigma}\right], \ x\right)$$

$$= -\left(\left[\frac{c - (1 - \beta^\sigma)xx^\sigma}{-1 + \beta\beta^\sigma}\right], \ x\right) + \left(\beta\beta^\sigma\left[\frac{c - (1 - \beta^\sigma)xx^\sigma}{-1 + \beta\beta^\sigma}\right], \ \beta x\right)$$

$$= \left(-\left[\frac{c - (1 - \beta^\sigma)xx^\sigma}{-1 + \beta\beta^\sigma}\right] + xx^\sigma, \ -x\right) + \left(\beta\beta^\sigma\left[\frac{c - (1 - \beta^\sigma)xx^\sigma}{-1 + \beta\beta^\sigma}\right], \ \beta x\right)$$

$$= \left(-\left[\frac{c - (1 - \beta^\sigma)xx^\sigma}{-1 + \beta\beta^\sigma}\right] + xx^\sigma + \beta\beta^\sigma\left[\frac{c - (1 - \beta^\sigma)xx^\sigma}{-1 + \beta\beta^\sigma}\right]\right.$$
$$\left. + (-x)(\beta x)^\sigma, \ -x + \beta x\right)$$

$$= \left((-1 + \beta\beta^\sigma)\left[\frac{c - (1 - \beta^\sigma)xx^\sigma}{-1 + \beta\beta^\sigma}\right] + (1 - \beta^\sigma)xx^\sigma,\right.$$
$$\left.(-1 + \beta)(-1 + \beta)^{-1}z\right)$$

$$= (c - (1 - \beta^\sigma)xx^\sigma + (1 - \beta^\sigma)xx^\sigma, \ z) = (c, z).$$

So, $\beta \in A^*$ implies that $-1 + \sigma_\beta$ is surjective and, thus, bijective.

When we have proven (vii), our proof will be complete. Let us first show that $1 + \sigma_\beta$ is surjective for each $\beta \in A$. Let $(c, z) \in N$ be arbitrary, and define $x = (1 + \beta)^{-1}z$. Then

$$(1 + \sigma_\beta)\left(\frac{c - \beta^\sigma xx^\sigma}{1 + \beta\beta^\sigma}, \, x\right)$$

$$= \left(\frac{c - \beta^\sigma xx^\sigma}{1 + \beta\beta^\sigma}, \, x\right) + \left(\beta\beta^\sigma\left[\frac{c - \beta^\sigma xx^\sigma}{1 + \beta\beta^\sigma}\right], \, \beta x\right)$$

$$= \left(\left[\frac{c - \beta^\sigma xx^\sigma}{1 + \beta\beta^\sigma}\right] + \beta\beta^\sigma\left[\frac{c - \beta^\sigma xx^\sigma}{1 + \beta\beta^\sigma}\right] + x(\beta x)^\sigma, \, x + \beta x\right)$$

$$= \left((1 + \beta\beta^\sigma)\left[\frac{c - \beta^\sigma xx^\sigma}{1 + \beta\beta^\sigma}\right] + \beta^\sigma xx^\sigma, \, (1 + \beta)(1 + \beta)^{-1}z\right)$$

$$= (c - \beta^\sigma xx^\sigma + \beta^\sigma xx^\sigma, \, z) = (c, z).$$

Now we want to show that each $1 + \sigma_\beta$ is injective. We use $\Longleftrightarrow$ to denote 'if and only if' or 'is equivalent to'.

$$(1 + \sigma_\beta)(a, x) = (1 + \sigma_\beta)(b, y)$$
$$\Longleftrightarrow (a, x) + (\beta\beta^\sigma a, \beta x) = (b, y) + (\beta\beta^\sigma b, \beta y)$$
$$\Longleftrightarrow (a + \beta\beta^\sigma a + x(\beta x)^\sigma, \, x + \beta x) = (b + \beta\beta^\sigma b + y(\beta y)^\sigma, \, y + \beta y)$$
$$\Longleftrightarrow ((1 + \beta\beta^\sigma)a + \beta^\sigma xx^\sigma, \, (1 + \beta)x) = ((1 + \beta\beta^\sigma)b + \beta^\sigma yy^\sigma, \, (1 + \beta)y)$$
$$\Longleftrightarrow (1 + \beta)x = (1 + \beta)y \qquad \text{and}$$
$$(1 + \beta\beta^\sigma)a + \beta^\sigma xx^\sigma = (1 + \beta\beta^\sigma)b + \beta^\sigma yy^\sigma$$
$$\Longleftrightarrow x = y \text{ and } a = b.$$

This gives us (vii), as well as the theorem.

(4.20) Examples. (a) Let F be a finite field of characteristic p. Now the mapping $x \mapsto x^p$ is an automorphism of the field F. If $x^p = x^{-1}$, Then $x^{p+1} = 1$, so the order of x divides $p + 1$. If we select $1 \neq A \triangleleft F^*$ so that $(|A|, p + 1) = 1$, then for $x \in A^*$, the order of x, denoted by $|x|$, cannot divide the integer $p + 1$.

(b) Let $\sigma = 1$ and select A so that A is a nontrivial multiplicative subgroup of F^* and with $-1 \notin A$.

(c) Let $F = \mathbf{C}$, the complex number field, and let $z^\sigma = \bar{z}$, complex conjugation. Take $A = \mathbf{R}_+$, the multiplicative group of positive real numbers.

(d) Let $F = \mathbf{R}$, $\sigma = 1$, and $A = \mathbf{R}_+$.

(e) Let F be a field strictly between the rational number field $\mathbf{Q}$ and the field of real numbers $\mathbf{R}$, let σ be a field automorphism of the field F, and let and $A = \mathbf{Q}_+ = \{q \in \mathbf{Q} \mid q > 0\}$, the multiplicative group of positive rational numbers.

Another source for Ferrero pairs will now be developed. This source produces some interesting Ferrero pairs as well as nonabelian groups. The examples in (4.23) infra have not been studied, so perhaps they could lead to the discovery of some interesting mathematics.

(4.21) Proposition. *Let A, B, C be abelian groups with $\pi : A \oplus B \to C$ a bilinear map. Let $N = C \times (A \times B)$ and define $+_\pi = +$ on N by*

$$[c, (a, b)] + [z, (x, y)] = [c + z + \pi(x, b), (a + x, b + y)].$$

Then $(N, +)$ is a group, and is abelian exactly when $\pi(x, b) = 0$ for all $(x, b) \in A \times B$.

Proof. For $[c, (a, b)], [z, (x, y)], [p, (m, n)] \in N$,

$$
\begin{aligned}
[c,&(a, b)] + \{[z, (x, y)] + [p, (m, n)]\} \\
&= [c, (a, b)] + [z + p + \pi(m, y), (x + m, y + n)] \\
&= [c + z + p + \pi(m, y) + \pi(x + m, b), (a + x + m, b + y + n)],
\end{aligned}
$$

and

$$
\begin{aligned}
\{[c,&(a, b)] + [z, (x, y)]\} + [p, (m, n)] \\
&= [c + z + \pi(x, b), (a + x, b + y)] + [p, (m, n)] \\
&= [c + z + \pi(x, b) + p + \pi(m, b + y), (a + x + m, b + y + n)].
\end{aligned}
$$

Since $\pi(m, y) + \pi(x + m, b) = \pi(m, y) + \pi(x, b) + \pi(m, b) = \pi(x, b) + \pi(m, b) + \pi(m, y) = \pi(x, b) + \pi(m, b + y)$, we have that $+$ is an associative binary operation .

From $[c, (a, b)] + [0, (0, 0)] = [c + 0 + \pi(0, b), (a + 0, b + 0)] = [c, (a, b)]$, we know that $[0, (0, 0)]$ is a right identity. From $[0, (0, 0)] = [c, (a, b)] + [z, (x, y)] = [c + z + \pi(x, b), (a + x, b + y)]$, we obtain $x = -a$, $y = -b$, and $z = -c - \pi(-a, b) = -c + \pi(a, b)$. Checking, $[c, (a, b)] + [-c + \pi(a, b), (-a, -b)] = [c - c + \pi(a, b) + \pi(-a, b), (a - a, b - b)] = [0, (0, 0)]$. So $-[c, (a, b)] = [-c + \pi(a, b), (-a, -b)]$ is the right inverse for $[c, (a, b)]$. This makes $(N, +)$ a group.

For $[c, (a, b)] + [z, (x, y)] = [z, (x, y)] + [c, (a, b)]$, we need $\pi(x, b) = \pi(a, y)$ for all $a, x \in A$ and for all $b, y \in B$. In particular, $\pi(x, b) = \pi(0, y) = 0$. Conversely, if $\pi(x, b) = 0$ for all $(x, b) \in A \times B$, then $(N, +)$ is an abelian group.

Let $\mathcal{C}(X, \mathbf{R}) = \{f : X \to \mathbf{R} \mid f$ is continuous $\}$. Some subrings of the ring $(\mathcal{C}(X, \mathbf{R}), +, \cdot)$ are useful for constructing infinite planar nearrings, as we shall see in (4.23).

(4.22) Theorem. *Let A, B, C be subrings of $\mathcal{C}(X, \mathbf{R})$ where $X = [0, 1]$. Suppose each of these subrings contains all the constant maps, that is, each contains a copy of $\mathbf{R}$. For a bilinear map $\pi : A^+ \oplus B^+ \to C^+$ of $\mathbf{R}$-modules, let $N = C \times (A \times B)$ and $(N, +)$ be the group of (4.21). Let $\Phi' = \mathbf{R}_+$, the positive real numbers, and for each $\alpha \in \Phi'$, define $F_\alpha : N \to N$ by $F_\alpha[c, (a, b)] = [\alpha^2 c, (\alpha a, \alpha b)]$. If $\Phi = \{F_\alpha \mid \alpha \in \Phi'\}$, then (N, Φ) is a Ferrero pair.*

Proof. The first thing to do is to show that each F_α is a group automorphism of N. Now $(N, +)$ is the group guaranteed by (4.21). Let $[c, (a, b)]$ and $[z, (x, y)]$ be arbitrary elements of the group N. Then, upon calculating, we have

$$
\begin{aligned}
F_\alpha\{[c, (a, b)] + [z, (x, y)]\} &= F_\alpha[c + z + \pi(x, b), (a + x, b + y)] \\
&= [\alpha^2(c + z + \pi(x, b)), (\alpha(a + x), \alpha(b + y)] \\
&= [\alpha^2 c + \alpha^2 z + \alpha^2 \pi(x, b), (\alpha a + \alpha x, \alpha b + \alpha y)] \\
&= [\alpha^2 c + \alpha^2 z + \pi(\alpha x, \alpha b), (\alpha a + \alpha x, \alpha b + \alpha y)] \\
&= [\alpha^2 c, (\alpha a, \alpha b)] + [\alpha^2 z, (\alpha x, \alpha y)] \\
&= F_\alpha[c, (a, b)] + F_\alpha[z, (x, y)].
\end{aligned}
$$

This makes F_α an endomorphism.

If $[0, (0, 0)] = [\alpha^2 c, (\alpha a, \alpha b)] = F_\alpha[c, (a, b)]$, then since $\alpha \neq 0$, we have $a = 0$, $b = 0$, and $c = 0$. So F_α is a monomorphism. For $[c, (a, b)] \in N$, $F_\alpha[\alpha^{-2} c, (\alpha^{-1} a, \alpha^{-1} b)] = [c, (a, b)]$, making F_α an epimorphism, hence an automorphism.

Since $F_\alpha \circ F_\beta[c, (a, b)] = F_\alpha[\beta^2 c, (\beta a, \beta b)] = [\alpha^2 \beta^2 c, (\alpha \beta a, \alpha \beta b)] = F_{\alpha\beta}[c, (a, b)]$, we have that the mapping $\alpha \mapsto F_\alpha$ is a homomorphism. In fact, it is a monomorphism, since $F_\alpha[1, (1, 1)] = F_\beta[1, (1, 1)]$ implies that $[\alpha^2 1, (\alpha 1, \alpha 1)] = [\beta^2 1, (\beta 1, \beta 1)]$ from which we may conclude that $\alpha = \beta$. Thus the groups Φ' and Φ are isomorphic.

Let us now show that Φ is a regular group of automorphisms. Take any $\alpha \neq 1$, and suppose that $[c, (a, b)] = F_\alpha[c, (a, b)] = [\alpha^2 c, (\alpha a, \alpha b)]$. If we assume that $[c, (a, b)] \neq [0, (0, 0)]$, then with $a \neq 0$ we can conclude that $\alpha = 1$. Also if $b \neq 0$ we obtain $\alpha = 1$. If $c \neq 0$, then $\alpha^2 = 1$, which, of course, means that $\alpha = 1$, since $\alpha > 0$. This makes Φ a regular group of automorphisms.

To have (N, Φ) a Ferrero pair, we have yet to show that $\alpha \in \Phi'$ and $\alpha \neq 1$ imply $-F_\alpha + 1$ is surjective. For $[c, (a, b)] \in N$, we will show that there exists a $[z, (x, y)]$ such that $(-F_\alpha + 1)[z, (x, y)] = [c, (a, b)]$. Let $x = (1 - \alpha)^{-1} a$, $y = (1 - \alpha)^{-1} b$, and

$$
z = \frac{\alpha \pi(a, b) + (1 - \alpha) c}{(1 - \alpha)(1 - \alpha^2)}.
$$

Now for the calculations:

$$
\begin{aligned}
(-F_\alpha + 1)[z, (x, y)] &= -F_\alpha[z, (x, y)] + [z, (x, y)] \\
&= -[\alpha^2 z, (\alpha x, \alpha y)] + [z, (x, y)] \\
&= [-\alpha^2 z + \pi(\alpha x, \alpha y), \ (-\alpha x, -\alpha y)] + [z, (x, y)] \\
&= [-\alpha^2 z + \pi(\alpha x, \alpha y) + z + \pi(x, -\alpha y), \ (-\alpha x + x, \ -\alpha y + y)] \\
&= [(1 - \alpha^2)z + \alpha^2 \pi(x, y) - \alpha\pi(x, y), \ ((1-\alpha)x, \ (1-\alpha)y)] \\
&= \left[(1 - \alpha^2)\left[\frac{\alpha\pi(a, b) + (1 - \alpha)c}{(1 - \alpha)(1 - \alpha^2)}\right] + (\alpha^2 - \alpha)\pi(x, y), \ (a, b)\right] \\
&= \left[\frac{\alpha\pi(a, b) + (1 - \alpha)c}{(1 - \alpha)} + \alpha(\alpha - 1)\pi\big((1 - \alpha)^{-1}a, \ (1-\alpha)^{-1}b\big), \ (a, b)\right] \\
&= \left[\frac{\alpha\pi(a, b)}{(1 - \alpha)} + c + \alpha(\alpha - 1)(1 - \alpha)^{-2}\pi(a, b), \ (a, b)\right] \\
&= \left[\frac{\alpha\pi(a, b)}{(1 - \alpha)} + c - \alpha(1 - \alpha)(1 - \alpha)^{-2}\pi(a, b), \ (a, b)\right] \\
&= [c, (a, b)],
\end{aligned}
$$

as promised.

(4.23) Examples. (a) Fix $\phi \in C(X, \mathbf{R})$ where $X = [0, 1] \times [0, 1]$ and define $\pi(a, b) = \int_0^1 \int_0^1 a(x)\phi(x, y)b(y)dx\,dy$. Let A, B, C be various subrings of $C([0, 1], \mathbf{R})$ containing $\mathbf{R}$, for example, $C([0, 1], \mathbf{R})$, $\mathbf{R}[x]$, and $\mathbf{R}$.

(b) Fix $\phi \in C(X, \mathbf{R})$ with $X = [0, 1]$ and define $\pi(a, b) = a(x)\phi(x)b(x)$. Let A, B be various subrings of $C(X, \mathbf{R})$ containing $\mathbf{R}$, but make sure that C contains the set $A\phi(x)B$.

(c) Let $A = B = \mathbf{R}[x]$ and C as in the theorem. Define for $f = \sum_{i=1}^n f_i x^i$, and $g = \sum_{i=1}^n g_i x^i$ the map $\pi(f, g) = \sum_{i=1}^n f_i g_i$.

(d) Let A, B, C be as in the theorem. Define $\pi(a, b) = \int_0^1 a(x)b(x)dx$, or, more generally, $\pi(a, b) = \int_0^1 a(x)\phi(x)b(x)dx$, for some fixed $\phi \in C([0, 1], \mathbf{R})$.

It has been remarked that the examples of (4.4), (4.5), and (4.6) are of the foundation for what we do with planar nearrings. Later, we examined these three examples in light of the MAIN STRUCTURE THEOREM.

For example 1, B_a is the straight line through 0 and a with 0 eliminated. Now $N^* B_a = B_a$, so $NB_a = B_a \cup \{0\}$ is that straight line *with* 0. Also, $NB_a = Na$ since B_a is a group, $a \in B_a$, and $B_a \subseteq N$. So Na is the straight line through 0 and a. This makes $Na + b$ the straight line parallel to Na but passing through b. So the straight lines of the plane are exactly the $Na + b$ with $a, b \in N$ and $a \neq 0$. (Here, we have used N for $\mathbf{C}$.) This suggests that we look at $\mathcal{B} = \{Na + b \mid a, b \in N, \ a \neq 0\}$ for geometric meaning for an arbitrary planar nearring.

Similarly, for example 2, B_a is the ray from 0 through a with 0 removed. Thus $B_a \cup B_{-a}$ is the line through a and 0 with 0 removed. Thus, $B_a \cup \{0\} \cup B_{-a}$ is the entire line through 0 and a, and $B_a \cup \{0\}$ is the ray from 0 through a with 0. As in the preceding paragraph, we have $Na = B_a \cup \{0\}$ and $N\{a, -a\} = B_a \cup \{0\} \cup B_{-a}$. So $N\{a, -a\} + b$ is the line parallel to the line $N\{a, -a\}$ and passing through b and $Na + b$ is the ray from b having the same direction as Na. This is more reason to look at $\mathcal{B}$ as defined in the preceding paragraph, and provides reason to look at $\mathcal{B}^- = \{N\{a, -a\} + b \mid a, b \in N, \ a \neq 0\}$ for geometric meaning in an arbitrary planar nearring.

Finally, for example 3, B_a is the circle with centre 0 and radius $|a|$. From $N^*a = N^*B_a = B_a$, we have that N^*a is this circle, and that $N^*a + b$ is the circle with centre b and radius $|a|$. (We are letting N stand for $\mathbf{C}$ again.) This suggests that we look at $\mathcal{B}^* = \{N^*a + b \mid a, b \in N, \ a \neq 0\}$ for geometric meaning in an arbitrary planar nearring.

So, for an arbitrary planar nearring $(N, +, \cdot)$, we will be looking at the three incidence structures $(N, \mathcal{B}, \in)$, $(N, \mathcal{B}^-, \in)$, and $(N, \mathcal{B}^*, \in)$.

The following is of fundamental importance.

(4.24) Theorem. *For a Ferrero pair (N, Φ), if $(N, +, \cdot)$ and $(N, +, \circ)$ are two planar nearrings constructed from (N, Φ) by the Ferrero Planar Nearring Factory, then for any $a \in N$, $a \neq 0$, we have $N^* \cdot a = N^* \circ a = \Phi(a)$.*

Proof. Let $(N, +, *)$ be a planar nearring constructed from a Ferrero pair (N, Φ). Then $(N, +, *)$ has an A, and if $b \in N^* = N \setminus A$, then $b * a = \phi_b(a)$ for some $\phi_b \in \Phi$. So, for each B_c, we get $B_c * a = \Phi(a)$, and since $N^*B_c = B_c$, we have $N^* * a = \Phi(a)$.

(4.25) Corollary. *If $(N, +, \cdot)$ and $(N, +, \circ)$, are any two distinct planar nearrings constructed from the same Ferrero pair (N, Φ) by the Ferrero Planar Nearring Factory, then they yield identical $(N, \mathcal{B}^*, \in)$, as well as identical $(N, \mathcal{B}, \in)$ and $(N, \mathcal{B}^-, \in)$.*

Proof. $N^* \cdot a + b = \Phi(a) + b = N^* \circ a + b$, $N \cdot a + b = \Phi(a) \cup \{0\} + b = N \circ a + b$, and $N \cdot \{a, -a\} + b = \Phi(a) \cup \{0\} \cup \Phi(-a) + b = N \circ \{a, -a\} + b$.

(4.26) Theorem. *If $(F, +, \cdot)$ is a finite nearfield, then it is a planar nearring as long as $|F| > 2$.*

Proof. If $a, b \in F$ and $a \neq b$, then $a \neq_m b$. This means $|F/{=_m}| = |F| \geq 3$. Certainly, $(F, +, \cdot)$ is a nearring. For $a, b, c \in F$, $a \neq_m b$, does the equation $ax = bx + c$ have a unique solution? We answer this with 'yes', but do so indirectly. If $a \notin \{0, 1\}$, then $x \mapsto ax$ is an automorphism of $(F, +)$, and $ax = x$ for $x \neq 0$ implies $a = 1$, since $(F^*, \cdot)$ is a group. Let $\phi_a(x) = ax$. Then $\Phi = \{\phi_a \mid a \in F^*\}$ is a regular group of automorphisms of $(F, +)$.

Since $|F| < \infty$, we have that $-\phi_a + 1$ is surjective for each $a \in F^* \setminus \{1\}$.

Back to the equation $ax = bx + c$. This is equivalent to $x = (a^{-1}b)x + (a^{-1}c)$, and to $-(a^{-1}b)x + x = (a^{-1}c)$. Does there exist a unique x so that $(-\phi_{a^{-1}b} + 1)(x) = (a^{-1}c)$? The answer is 'yes', since $|F| < \infty$ and $\Phi = \{\phi_c \mid c \in F^*\}$ is regular.

5. Construction of circular planar nearrings

This section gets its motivation from example 3 of §4, namely $(\mathbf{C}, +, *_3)$. Recall that for $0 \neq a \in \mathbf{C}$, $\mathbf{C}^* *_3 a$ is the circle with centre 0 and radius $|a|$. Translate $\mathbf{C}^* *_3 a$ by $b \in \mathbf{C}$ to get $\mathbf{C}^* *_3 a + b$, which is the circle with centre b and radius $|a|$. For a planar nearring $(N, +, \cdot)$ we study here the triple $(N, \mathcal{B}^*, \in)$, where

$$\mathcal{B}^* = \{N^*a + b \mid a, b \in N, \ a \neq 0\}.$$

Of course, $(\mathbf{C}, \mathcal{B}^*, \in)$ is our main example.

5.1. Basic results

(5.1) Definitions. A planar nearring $(N, +, \cdot)$ is *circular*, if every three distinct points $x, y, z \in N$ belong to at most one $N^*a + b \in \mathcal{B}^*$, and if every two distinct points $s, t \in N$ belong to at least two distinct $N^*a_1 + b_1$, $N^*a_2 + b_2 \in \mathcal{B}^*$. In this case, we also say that $(N, \mathcal{B}^*, \in)$ is *circular*. We call $N^*a + b \in \mathcal{B}^*$ the *circle* with *centre* b and *radius* a, and sometimes denote $N^*a + b$ by $C(a; b)$.

As suggested in (5.1), we may want to apply the adjective *circular* to other incidence structures $(N, \mathcal{B}, \in)$. Using $(\mathbf{C}, +, *_1)$ of §4, one attains $(N, \mathcal{B}, \in)$ where every two distinct points $s, t \in N = \mathbf{C}$ belong to exactly one $Na + b \in \mathcal{B}$. In an attempt to put some 'curvature' in our 'circles', we required two distinct points to belong to at least two distinct $N^*a_1 + b_1$, $N^*a_2 + b_2 \in \mathcal{B}^*$.

Perhaps a few words are in order concerning notation such as $(N, \mathcal{B}^*, \in)$. This is a special case of a more general notation $(X, \mathcal{S}, I)$. Here, X and $\mathcal{S}$ are each sets and I is a relation, an *incidence* relation, from X to $\mathcal{S}$, that is, $I \subset X \times \mathcal{S}$. If $(x, S) \in I$, one writes xIS and says that x is *incident* to S, or x is *on* S, or x is *in* S, but also S is *on* x. If we write $(X, \mathcal{S}, \in)$, then $\mathcal{S}$ is a family of subsets of X and $x \in S$ has the usual meaning. One sometimes calls $(X, \mathcal{S}, I)$ an *incidence structure*. So notation such as $(N, \mathcal{B}^*, \in)$ is used to emphasize the geometric interpretations of our planar nearring $(N, +, \cdot)$. We are interested in a rather general incidence space, that of a *tactical configuration*, but of more importance, a special case, that of a *balanced incomplete block design*, or BIBD for short.

(5.2) Definitions. Consider an incidence structure $(X, \mathcal{B}, \in)$. Suppose there are positive integers k and r so that $|B| = k$ if $B \in \mathcal{B}$, and if $x \in X$, then there are exactly r distinct *blocks* or *lines* $B_1, B_2, \ldots, B_r \in \mathcal{B}$ such that $x \in B_i$ for each i, $1 \leq i \leq r$. If so, then $(X, \mathcal{B}, \in)$ is a *tactical configuration*. A tactical configuration $(X, \mathcal{B}, \in)$ is a *balanced incomplete block design*, if there is a positive integer λ so that if $x, y \in X$ are any two distinct *elements* or *varieties* of X, then there are exactly λ distinct blocks $B_1, B_2, \ldots, B_\lambda \in \mathcal{B}$ such that $x, y \in B_i$ for each i, $1 \leq i \leq \lambda$.

For a tactical configuration $(X, \mathcal{B}, \in)$ and for $x_1, x_2, \ldots, x_t$ being t distinct points of X, we let $[x_1, x_2, \ldots, x_t]$ denote the number of distinct blocks of $\mathcal{B}$ which contain $x_1, x_2, \ldots, x_t$. So $[x] = r$ for each $x \in X$, and if $(X, \mathcal{B}, \in)$ is a BIBD, then $[x, y] = \lambda$, if $x, y \in X$ and $x \neq y$.

There are other parameters for a BIBD $(X, \mathcal{B}, \in)$, namely $v = |X|$ and $b = |\mathcal{B}|$. The parameters v, b, k, r, λ are related, as the following popular theorem tells us.

(5.3) Theorem. *Let $(X, \mathcal{B}, \in)$ be a BIBD. Then $vr = bk$ and $\lambda(v - 1) = r(k - 1)$.*

Proof. There are v points in X and each is in r blocks. So there are vr points in the b blocks. But each of the b blocks has k points in it, so there are bk points in the b blocks. This makes $vr = bk$.

There are $v(v - 1)/2$ pairs of distinct points in X, and each pair is in λ blocks. So there are $\lambda v(v - 1)/2$ pairs of distinct points in the b blocks. But each of the b blocks has $k(k - 1)/2$ distinct pairs of points in it, so there are $bk(k - 1)/2$ distinct pairs of points in the b blocks. Thus

$$\frac{\lambda v(v - 1)}{2} = \frac{bk(k - 1)}{2}.$$

Since $bk = vr$, we obtain $\lambda(v - 1) = r(k - 1)$.

For a finite planar nearring $(N, +, \cdot)$ the $(N, \mathcal{B}^*, \in)$ always yield a BIBD. To prepare for the proof of this, we need to know when $N^*a + b = N^*c + d$.

(5.4) Proposition. *For a finite planar nearring $(N, +, \cdot)$, we have $N^*a + b = N^*c + d$ if and only if $b = d$ and $N^*a = N^*c$.*

Proof. From (4.25) we learn that there is no loss in assuming N to be integral. If $N^*a = N^*c$ and $b = d$, then certainly $N^*a + b = N^*c + d$. So the interesting part of the proof is in proving the converse. We shall begin with $N^*a + b = N^*c + d$, and deduce that $N^*a = N^*c + (d - b)$. If we can prove that this forces $d - b = 0$, then we will have $N^*a = N^*c$ and $b = d$, our desired conclusion.

So suppose $N^*a = N^*b + c$ where $a, b, c \in N$ and $0 \notin \{a, b, c\}$. For $u \in N^*$, $N^*a = N^*b + uc$. Let $u_1 = 1_c, u_2, \ldots, u_k$ be representatives of $N^*/=_m$.

Since $k \geq 2$, we obtain $N^*b + c = N^*b + u_2c$ and $N^*b = N^*b + (u_2c - c)$ with $u_2c - c \neq 0$. Let $t = u_2c - c$. From $N^*b = N^*b + t$, we obtain $N^*b = N^*b + t = N^*b + 2t = \cdots = N^*b + (h-1)t$, where h is the order of the subgroup $\langle t \rangle$ of $(N, +)$ generated by t. This makes $N^*b = N^*b + \langle t \rangle$ a union of cosets of $\langle t \rangle$, hence $k = |N^*b| = wh$ for some integer w. If $v = |N|$, then $k|(v-1)$, so $h|(v-1)$ also. Certainly, $h|v$ since $h = |\langle t \rangle|$. So we have h dividing $(v, v-1) = 1$, forcing $h = 1$, and then $t = 0$.

(5.5) Theorem. *If $(N, +, \cdot)$ is a finite planar nearring, then $(N, \mathcal{B}^*, \in)$ is a BIBD with parameters $v = |N|$, $k = |N^*/{=}_m|$, $b = v(v-1)/k$, $r = v-1$, and $\lambda = k - 1$.*

Proof. For a planar nearring $(N, +, \cdot)$, there is a Ferrero pair (N, Φ) from which $(N, +, \cdot)$ could be constructed. Since $N^*a = \Phi(a)$, which is independent of the planar nearring constructed from (N, Φ), there is no loss in assuming that $(N, +, \cdot)$ is integral, and we can conclude that $k = |N^*a + b| = |N^*a| = |\Phi| = |N/{=}_m| - 1 = |N^*/{=}_m|$.

Of course, we have $v = |N|$, and because of (5.4), we obtain $b = [(v-1)/k] \cdot v = v(v-1)/k$. Each of the $(v-1)/k$ blocks N^*a, $a \in N^*$ has k elements. Suppose $N^*a = \{a_1, a_2, \ldots, a_k\}$. For an arbitrary fixed $x \in N$, there are exactly k elements $y_1, y_2, \ldots, y_k \in N$ so that $x = a_i + y_i \in N^*a + y_i$. This means x belongs in exactly the k blocks $N^*a + y_1$, $N^*a + y_2, \ldots, N^*a + y_k$. Repeating this for each of the $(v-1)/k$ blocks N^*a, $a \in N^*$, yields exactly $r = [(v-1)/k] \cdot k = v - 1$ blocks $N^*a + b$ containing the fixed, but arbitrary $x \in N$. It remains to show that $\lambda = k - 1$.

Since $x, y \in N^*a + t$ if and only if $0, y - x \in N^*a + (t - x)$, we need only show that $0, z \in N$ belong to exactly $k - 1$ blocks if $z \neq 0$. Let $1_e = u_1, u_2, \ldots, u_k$ be representatives of the equivalence classes of $N^*/{=}_m$, where $e \in N^*$. Certainly, $0 \in N^*a - u_ia$ for each $a \in N^*$ and each u_i, $1 \leq i \leq k$. If $0 \in N^*a + t$, then $0 = u_ia + t$ for some u_i, and so $t = -u_ia$ and $N^*a + t = N^*a - u_ia$. Thus, the blocks containing 0 are exactly the $N^*a - u_ia$, $a \in N^*$, and $1 \leq i \leq k$.

For $1 < i \leq k$, the equation $1_ex = u_ix + z$ has a unique solution for x, so call it a_i. If $a_i = 0$, then $z = 0$, so we are assured of $a_i \neq 0$. From $1_ea_i = u_ia_i + z$ we get $z = -(u_ia_i) + 1_ea_i = u_i(-a_i) + a_i$. So $z \in N^*(-a_i) + a_i$, which also contains 0. Hence, 0 and z belong to each of $N^*(-a_2) + a_2$, $N^*(-a_3) + a_3, \ldots, N^*(-a_k) + a_k$. Could it be that we do not have $k - 1$ distinct blocks here containing 0 and z? If $i \neq j$ and $a_i = a_j$, then from $u_i(-a_i) + a_i = z = u_j(-a_j) + a_j$, we obtain, since $a_i = a_j$, that $u_i(-a_i) = u_j(-a_j)$, or $u_ia_i = u_ja_j$. Consequently, the equation $u_ix = u_jx + 0$ has a_i as a solution, as well as 0. But the $a_i \neq 0$, so we had better conclude that $a_i \neq a_j$ if $i \neq j$. This gives us at least $k - 1$ distinct blocks containing 0 and z, and by the statement of the theorem, the reader can know that our next step will be to show that there are no more.

If $0, z \in N^*a + b$, then $N^*a + b = N^*a - u_ia$ for some i, $1 \le i \le k$, since these are exactly the blocks containing 0. This means $z = u_ja - u_ia$ for some u_j, where $i \ne j$ since $z \ne 0$. From this we have $-u_ja + z = -u_ia$, $u_j(-a) + z = u_i(-a)$, and $u_i(-a) = u_j(-a) + z$. There is a $u'_i \in N$ so that $(u'_iu_i)y = y$ for all $y \in N$, so $1_e[u_i(-a)] = (u_ju'_i)[u_i(-a)] + z$ and $u_ju'_i =_m u_l$ for some l, $1 < l \le k$. Hence, $1_ex = u_lx + z$ has the unique solution $a_l = u_i(-a) = -(u_ia)$. So $0, z \in N^*a + b = N^*a - u_ia = N^*(u_ia) + u_i(-a) = N^*(-a_l) + a_l$. So, an $N^*a + b$ containing 0 and z must be one of the $N^*(-a_i) + a_i$, that is, one of the $k - 1$ blocks already accounted for. So $\lambda = k - 1$ is the number of distinct blocks containing an arbitrary pair $x, y \in N$, $x \ne y$.

For a given group $(N, +)$, there may be numerous Ferrero pairs (N, Φ) with Φ of various size. If a resulting planar nearring $(N, +, \cdot)$ is to be circular, then the order of Φ cannot be too big.

(5.6) Theorem. *For a finite planar nearring $(N, +, \cdot)$ to be circular, we must have $k \le (3 + \sqrt{4v - 7})/2$, where $v = |N|$, and $k = |N^*a|$ for any $a \ne 0$.*

Proof. Each of N's $v(v - 1)/k$ circles has $k(k - 1)(k - 2)/6$ triples x, y, z of distinct points belonging to it. This makes

$$\frac{v(v - 1)}{k} \cdot \frac{k(k - 1)(k - 2)}{6} = \frac{v(v - 1)(k - 1)(k - 2)}{6},$$

such triples belonging to circles. Each triple x, y, z of distinct points belongs to at most one circle. There are $v(v - 1)(v - 2)/6$ such triples in N. So we cannot have

$$\frac{v(v - 1)(k - 1)(k - 2)}{6} > \frac{v(v - 1)(v - 2)}{6}.$$

This means that we cannot have $(k - 1)(k - 2) > (v - 2)$, which means that we must have $(k - 1)(k - 2) \le (v - 2)$. This means that $k^2 - 3k \le v - 4$, and, consequently, that $k \le (3 + \sqrt{4v - 7})/2$.

There is a lack of examples of circular planar nearrings at this point of our development. Since the theory has till now provided but the one, $(\mathbf{C}, +, *_3)$ from §4, it would be nice for others to be on hand. Hopefully, the following table provides a way to construct all the nontrivial circular planar nearrings from the prime fields $(Z_p, +, \cdot)$, where $13 \le p < 1000$. The table provides data to construct Ferrero pairs (Z_p, Φ) whose planar nearrings $(Z_p, +, *)$ are circular. Certainly, if $|\Phi| = 3$, then the resulting

$(Z_p, +, *)$ would be circular. For $13 \leq p < 1000$, p a prime, we list those divisors d of $p - 1$, $d > 3$, whose cyclic subgroups Φ_d of order d of Z_p^* yield BIBDs $(Z_p, \mathcal{B}_d^*, \in)$ which are circular. We also list a generator for Φ_d as well as for Z_p^*. If there is no entry for such a prime, for example, $p = 23, 47$, and 587, then no $(Z_p, \mathcal{B}_d^*, \in)$ is circular for $d \neq 3$.

A typical entry is the cell for the prime $p = 337$.

337	4	6	7	8	12
10	148	129	8	25	72

It has the structure

p	d_1	d_2	$\cdots$	d_k
g_{p-1}	g_{d_1}	g_{d_2}	$\cdots$	g_{d_k}

where p is a prime, g_{p-1} is a generator of Z_p^*, each d_i divides $p - 1$, each g_{d_i} is a generator of a subgroup Φ_{d_i} of order d_i, and the $(Z_p, \mathcal{B}_{d_i}^*, \in)$ are circular. The $d_1, d_2, \ldots, d_k$ are all those divisors of $p - 1$, $d_i > 3$ for which $(Z_p, \mathcal{B}_{d_i}^*, \in)$ is circular.

Notice in our table that our upper bound of $(3 + \sqrt{4v - 7})/2$, given in (5.6), seems to be very conservative. Perhaps one can do better? Well, perhaps for the Z_p, but not in general. Modisett [MM] has shown that this upper bound is attained infinitely often, and this is the content of the next proposition.

(5.7) Proposition. *If $v = n^2$ and $n > 1$, then $[(3 + \sqrt{4v - 7})/2] = n + 1$, where $[x]$ is the greatest integer $\leq x$.*

Proof. Certainly, $\sqrt{n^2 - 7/4} < n$, and since $(n-1)^2 = n^2 - 2n + 1 < n^2 - 7/4$ for $n > 1$, we obtain $n - 1 < \sqrt{n^2 - 7/4}$. So $n - 1 < \sqrt{n^2 - 7/4} < n$. Adding $3/2$, we have $n + 1/2 < (3 + \sqrt{4n^2 - 7})/2 < n + 3/2$. This puts our upper bound $[(3 + \sqrt{4n^2 - 7}/2] \in \{n, n+1\}$. If $[(3 + \sqrt{4n^2 - 7})/2] = n$, then $n \leq (3 + \sqrt{4n^2 - 7})/2 < n + 1$. Multiplying by 2 and subtracting 3 would yield $0 < 2n - 3 \leq \sqrt{4n^2 - 7} < 2n - 1$. Upon squaring and subtracting the common term of $4n^2$ we get $-12n + 9 \leq -7 < -4n + 1$. From $-7 < -4n + 1$ we learn that $4n < 8$, or $n < 2$. But $n \geq 2$, so we must have the other alternative, namely, $[(3 + \sqrt{4n^2 - 7}/2] = n + 1$.

For a more useful table, one can get a program in C language to create a table for at least the primes p, $13 \leq p \leq 2^{16} - 1$. See [Ke]. Such a table also provides generators for each of the multiplicative subgroups of the prime fields considered, and so one can more easily construct the corresponding planar nearrings using the Ferrero Factory, and then the various geometric structures from these prime fields Z_p.

13	4	17	4	29	4
2	5	3	4	2	12
31	5 6	37	4 6	41	4 5
3	2 6	2	6 11	6	9 10
43	6	53	4	61	4 5 6
3	7	2	23	2	11 9 14
67	6	71	5 7	73	4 6 8
2	30	7	5 20	5	27 9 10
79	6	89	4 8	97	4 6 8
3	24	3	34 12	5	22 36 33
101	4 5	103	6	109	4 6
2	10 36	5	47	6	33 46
113	4 7 8	127	6 7	131	5 10
3	15 16 18	3	20 2	2	53 42
137	4 8	139	6	149	4
3	37 10	2	43	2	44
151	5 6 10	157	4 6	163	6 9
6	8 33 87	5	28 13	2	59 38
173	4	181	4 5 6 9 10	191	5 10
2	80	2	19 42 49 39 46	19	39 7
193	4 6 8	197	4 7	199	6 9
5	81 85 9	2	14 36	3	93 43
211	5 6 7 10	223	6	229	4 6 12
2	55 15 58 23	3	40	6	107 95 18
233	4 8	239	7	241	4 5 6 8 10 12
3	89 12	7	10	7	64 87 16 8 36 4
251	5 10	257	4 8	269	4
6	20 32	3	16 4	2	82
271	5 6 10	277	4 6 12	281	4 5 7 8 10
6	10 29 27	5	60 117 35	3	53 86 59 60 49
283	6	293	4	307	6 9
3	45	2	138	5	18 46
311	5 10	313	4 6 8 12	317	4
17	6 305	10	25 99 5 29	2	114
331	5 6 10 11	337	4 6 7 8 12	349	4 6 12
3	64 32 8 74	10	148 129 8 25 72	2	136 123 24
353	4 8	367	6	373	4 6 12
3	42 70	6	84	2	104 89 69
379	6 7 9	389	4	397	4 6 9 12
2	52 86 84	2	115	5	63 35 14 157

401	4 5 8 10	409	4 6 8 12	419	11
3	20 39 45 29	21	143 54 31 49	2	334
421	4 5 6 7 10 12 14	431	5 10	433	4 6 8 9 12
2	29 252 21 133 44 159 36	7	95 25	5	179 199 354 150 64
439	6	443	13	449	4 7 8 14
15	172	2	35	3	67 18 92 5
457	4 6 8 12	461	4 5 10	463	6 7 11 14
13	109 134 170 18	2	48 88 93	3	22 34 15 51
487	6 9	491	5 7 10	499	6
3	233 41	2	101 138 110	7	140
509	4	521	4 5 8 10	523	6 9
2	208	3	235 25 43 5	2	61 19
541	4 5 6 9 10 12	547	6 7 13	557	4
2	52 48 130 15 313 216	2	41 9 46	2	118
569	4 8	571	5 6 10	577	4 6 8 9 12
3	86 76	3	106 110 90	5	24 214 152 287 57
593	4 8	601	4 5 6 8 10 12 15	607	6
3	77 59	7	125 32 25 59 169 5 18	3	211
613	4 6 9 12	617	4 7 8 11 14	619	6
2	35 66 160 142	3	194 142 139 31 62	2	253
631	5 6 7 9 10 14 18	641	4 5 8 10	643	6
3	288 44 21 32 403 30 138	3	154 354 256 79	11	466
653	4	659	7	661	4 5 6 10 11 12 15
2	149	2	12	2	106 197 297 190 9 246 12
673	4 6 7 8 12 14 16	677	4 13	691	5 6 10 15
5	58 256 117 64 16 23 8	2	26 40	3	89 243 371 113
701	4 5 7 10	709	4 6 12	727	6 11
2	135 89 19 63	2	96 228 91	5	282 46
733	4 6 12	739	6 9	743	7 14
6	353 308 113	3	321 197	5	111 151
751	5 6 10	757	4 6 7 9 12 14	761	4 5 8 10
3	80 73 182	2	87 28 59 3 78 127	6	39 67 62 77
769	4 6 8 12 16	773	4	787	6
11	62 361 40 19 27	2	317	2	380
797	4	809	4 8	811	5 6 9 10 15
2	215	3	318 44	3	212 681 796 311 276
821	4 5 10	823	6	827	7 14
2	295 470 660	3	175	2	124 20
829	4 6 9 12 18	853	4 6 12	857	4 8
2	246 126 5 77 191	2	333 221 98	3	207 188

859	6	11				877	4	6	12		881	4	5	8	10	11
2	261	13				2	151	283	240		3	387	268	177	137	32
883	6	7	9	14	18	907	6				911	5	7	10	13	14
2	339	71	135	134	242	2	385				17	19	49	429	30	7
919	6	9	18			929	4	8	16		937	4 6 8	9	12 13 18		
7	53	440	95			3	324	18	40		5	196 323 14	72	333 36 13		
941	4	5	10			947	11				953	4	7	8	14	
2	97	349	185			2	133				3	442	431	156	79	
967	6	7	14			971	5	10			977	4	8	16		
5	143	97	175			6	341	168			3	252	227	52		
991	5	6	9	10	11 15 18	997	4	6	12							
6	160	114	18	166	42 71 55	7	161	305	91							

Table of Circular Planar Nearrings.

(5.8) Proposition. *Let K and L be finite fields of characteristic p with $L \subset K$, K of order p^{2n}, and L of order p^n, so K is of dimension 2 over L. Let the automorphism group of K over L be $\mathrm{Aut}_L K = \{1, \sigma\}$. The norm of K over L is the function $N : K^* \to L^*$ defined by $N(x) = x\sigma(x)$. Then N is a group homomorphism whose kernel has order $p^n + 1$.*

Proof. For $x, y \in K^*$, $N(xy) = xy\sigma(xy) = xy\sigma(x)\sigma(y) = N(x)N(y)$. Now $|L^*| = p^n - 1$, so if the image of N is L^*, we have, since $|K^*| = p^{2n} - 1$, that $|\ker N| = p^n + 1$. So we proceed to show that the image of N is L^*.

For an $x\sigma(x) = N(x)$, $\sigma(x\sigma(x)) = \sigma(x)\sigma^2(x) = x\sigma(x)$, so $x\sigma(x) \in L^*$, the fixed field of σ. Hence, $N(K^*) \subseteq L^*$. For $H = \{x^2 \mid x \in L^*\}$, we have $|L^*/H| \leq 2$, and if $x^2 \in H$, $x \in L$, then $x^2 = x \cdot \sigma(x)$, so H is in the image of N. We have $H \subseteq N(K^*) \subseteq L^*$, and since $|L^*/H| \leq 2$, either $N(K^*) = H$ or $N(K^*) = L^*$. If $p = 2$, then $H = L^*$, and so $N(K^*) = L^*$, and we have our desired result.

Assume $p \neq 2$. Then $K = L(\omega) = L[X]/(X^2 + \omega^2)$, where $\omega^2 \in L^* \setminus H$ and $\omega \notin L$. So $z \in K$ if and only if $z = x + \omega y$ for some $x, y \in L$. Applying N, we obtain $N(z) = z\sigma(z) = (x + \omega y)(x - \omega y) = x^2 - \omega^2 y^2$. So $N(K^*)$ contains all $x^2 - \omega^2 y^2 \neq 0$ and, in particular, all $-\omega^2 y^2 \neq 0$ and, specifically, $-\omega^2$. If $N(K^*)$ contains one more element than H, then $N(K^*) = L^*$. So, if $-\omega^2 \notin H$, then $N(K^*) = L^*$.

If $-\omega^2 \in H$, and $H = \{h_1, h_2, \ldots, h_t\}$, then $H = \{h_1, h_2, \ldots, h_t\} = \{-\omega^2 h_1, -\omega^2 h_2, \ldots, -\omega^2 h_t\}$. If $N(K^*) = H$, then each $x^2 - \omega^2 y^2 \in H$ for any $x, y \in L$, as long as $\{x, y\} \neq \{0\}$. But $H = \{x^2 \mid x \in L^*\} = \{h_1, h_2, \ldots, h_t\}$, so $\{x^2 - \omega^2 y^2 \mid x, y \in L, \{x, y\} \neq \{0\}\} = \{h_i + h_j \mid h_i, h_j \in H\}$. In particular, this means that $N(K^*) = H$ is closed with respect to $+$, making it a subgroup of L^+, putting $0 \in H$, which cannot be. So the only honourable thing to do is to conclude that $N(K^*) = L^*$. As we said earlier, this forces $|\ker N| = p^n + 1$.

(5.9) Theorem. *For K, L, N, and p as in (5.8), let $\Phi = \ker N$. If $(K, +, *)$ is any planar nearring constructed from the Ferrero pair (K, Φ), then $(K, \mathcal{B}^*, \in)$ is circular.*

Proof. We want to show that two distinct circles can have *at most* two points in common. Take two arbitrary distinct circles $K^*a + b$ and $K^*c + d$. Let $x \in (K^*a + b) \cap (K^*c + d)$. Our goal is to prove that x can have at most two values.

The proof offered here is in terms of field theory. Now $K^*a + b = \Phi a + b$ and $K^*c + d = \Phi c + d$, where $\Phi \triangleleft K^*$. So we are letting $x \in (\Phi a + b) \cap (\Phi c + d)$. This makes $x = ma + b$, and $x = nc + d$ for some $m, n \in \Phi$. We can then write $x - d = ma + (b - d)$ and $x - d = nc$. Being in a field, we have $(x - d)c^{-1} = mac^{-1} + (b - d)c^{-1}$ and $(x - d)c^{-1} = n$, where $m, n \in \Phi = \ker N$. This puts $(x - d)c^{-1} \in [\Phi ac^{-1} + (b - d)c^{-1}] \cap \Phi$.

Now we will show $x \in [\Phi a + b] \cap \Phi$, where $\Phi a + b \neq \Phi$, can have at most two values. Such an x has $x = ma + b$, and $x = n$ for some $m, n \in \Phi$, or, equivalently, $x - b = ma$, and $x = n$. Applying the homomorphism N we obtain

$$\begin{cases} N(x - b) = N(a) \\ N(x) = 1. \end{cases}$$

From the proof of (5.8), we know that $x = x_1 + \omega x_2$ and $b = b_1 + \omega b_2$, with the x_i, $b_i \in L$, and $N(x - b) = (x_1 - b_1)^2 - \omega^2(x_2 - b_2)^2 = N(a) \in H = \{r^2 \mid r \in L^*\}$. (If $\omega^2 = -1$, then this is the traditional equation for a circle of radius r, where $r^2 = N(a)$, and centre (b_1, b_2).) Also, $N(x) = x_1^2 - \omega^2 x_2^2 = 1$. Expanding, we obtain

$$\begin{cases} x_1^2 - 2b_1 x_1 + b_1^2 - \omega^2 x_2^2 + 2\omega^2 b_2 x_2 - \omega^2 b_2^2 = N(a) \\ x_1^2 - \omega^2 x_2^2 = 1. \end{cases}$$

So $x = x_1 + \omega x_2$ must satisfy these equations simultaneously. Hence

$$-2b_1 x_1 + b_1^2 + 2\omega^2 b_2 x_2 - \omega^2 b_2^2 = N(a) - 1,$$

which can be written in the form

$$x_2 = A x_1 + B. \tag{5:1}$$

Substituting into $x_1^2 - \omega^2 x_2^2 = 1$ yields $1 = x_1^2 - \omega^2(A x_1 + B)^2 = x_1^2 - \omega^2(A^2 x_1^2 + 2AB x_1 + B^2) = (1 - \omega^2 A^2)x_1^2 - 2AB\omega^2 x_1 - \omega^2 B^2$, a polynomial of degree at most 2 with coefficients in a field. So, there are at most two solutions for x_1. For each solution, $(5:1)$ gives exactly one solution for x_2. Thus, there are at most two solutions for x.

5.2. New structures from old

For a group $(N, +)$ of order v, we have, in constructing our tables of circular $(Z_p, +, *)$, considered the following question. For which divisors k of $v - 1$ is there a Ferrero pair (N, Φ), yielding a circular planar nearring $(N, +, *)$? Modisett [MM] looked at this from a different point of view. Given a k, for which v with k dividing $v - 1$, does there exist a Ferrero pair (N, Φ) with $|N| = v$, yielding a circular planar nearring? Instead of $k \leq (3 + \sqrt{4v - 7})/2$, we obtain the sequence of inequalities $2k \leq 3 + \sqrt{4v - 7}$, $(2k - 3)^2 \leq 4v - 7$, $4k^2 - 12k + 16 \leq 4v$, $k^2 - 3k + 4 \leq v$, $k^2 - 3k + 2 \leq v - 2$, and finally, $(k - 1)(k - 2) + 2 \leq v$, as a necessary condition. Again, taking a cue from the usual euclidean plane, we define for a fixed x and y of a planar nearring $(N, +, \cdot)$

$$A_{x,y} = \{n \in N \setminus \{x, y\} \mid x, y, n \in B \text{ for some } B \in \mathcal{B}^*\}.$$

One easily obtains

$$A_{x,y} = \bigcup_{\substack{x, y \in B \\ B \in \mathcal{B}^*}} B \setminus \{x, y\}.$$

(Note that $A_{x,y}$ has meaning for any BIBD.) In the usual euclidean plane, with $\mathcal{B}^*$ denoting all circles in the plane, an $A_{x,y}$ is the complement of the line containing x and y. So $A_{x,y}$ is an effort to define 'lines', or at least something else with geometric meaning from a circular planar nearring, or for any planar nearring.

(5.10) Definitions. Let $(N, \mathcal{B}^*, \in)$ and $(M, \mathcal{C}^*, \in)$ be two BIBDs. An *isomorphism* $\phi : (N, \mathcal{B}^*, \in) \to (M, \mathcal{C}^*, \in)$ is a bijective map $\phi : N \to M$ for which $B \mapsto \phi(B)$ is a bijection from $\mathcal{B}^*$ onto $\mathcal{C}^*$. If $(M, \mathcal{C}^*, \in) = (N, \mathcal{B}^*, \in)$, then ϕ is an *automorphism*. The set of all such automorphisms is denoted by $\mathrm{Aut}(N, \mathcal{B}^*, \in)$.

(5.11) Proposition. *For any planar nearring $(N, +, \cdot)$ and an arbitrary $\theta \in \mathrm{Aut}\,(N, \mathcal{B}^*, \in)$,*

$$\theta(A_{x,y}) = A_{\theta(x),\theta(y)}.$$

Proof. For $t \in A_{x,y}$, $x, y, t \in B$ for some $B \in \mathcal{B}^*$. So $\theta(x), \theta(y) \in \theta(B) \in \mathcal{B}^*$. So $\theta(t) \in A_{\theta(x),\theta(y)}$. For $u \in A_{\theta(x),\theta(y)}$, then $u, \theta(x), \theta(y) \in C$ for some $C \in \mathcal{B}^*$. Now $C = \theta(B)$ for some $B \in \mathcal{B}^*$, and so $u = \theta(t)$ for some $t \in B$. Hence $\theta(t), \theta(x), \theta(y) \in \theta(A_{x,y})$.

(5.12) Theorem. *Let $(S, \mathcal{B}, \in)$ be a BIBD. If $x, y \in S$ and $x \neq y$, then $|A_{x,y}| \leq \lambda(k - 2)$, and $(S, \mathcal{B}, \in)$ is circular if and only if $|A_{x,y}| = \lambda(k - 2)$ for each $x, y \in S$, $x \neq y$.*

Proof. The λ blocks containing x and y are B_i, $1 \le i \le \lambda$, and

$$B_i = \{x, y, a_{i1}, a_{i2}, \ldots, a_{i,k-2}\}.$$

Now $A_{x,y} = \{a_{ij} \mid 1 \le i \le \lambda, 1 \le j \le k-2\}$. It may be that $(i,j) \ne (m,n)$ but yet $a_{ij} = a_{mn}$, so $|A_{x,y}| \le \lambda(k-2)$.

However, if $(S, \mathcal{B}, \in)$ is circular, then three points cannot belong to $B_i \cap B_m$ if $i \ne m$. Thus $|A_{x,y}| = \lambda(k-2)$. Conversely, if $|A_{x,y}| = \lambda(k-2)$ for each $x, y \in S$, $x \ne y$, then one cannot have $x, y, a_{ij} \in B_i$ and $x, y, a_{ij} = a_{mn} \in B_m$, if $(i,j) \ne (m,n)$. Thus, $(S, \mathcal{B}, \in)$ is circular.

(5.13) Proposition. *If $(S, \mathcal{B}, \in)$ is a BIBD, then the group of BIBD automorphisms $\Gamma = \mathrm{Aut}\,(S, \mathcal{B}, \in)$ acts on the family $\mathcal{A}(S, \mathcal{B}, \in) = \{A_{x,y} \mid x, y \in S, x \ne y\}$, and so the orbits of Γ will have representatives*

$$A_{x_1, y_1}, \; A_{x_2, y_2}, \ldots, \; A_{x_n, y_n}.$$

(5.14) Lemma. *Let a group Γ act on a set X. Then the orbits of Γ partition X.*

Proof. $x \in X$ implies $x \in \Gamma(x)$, therefore $X = \cup_{x \in X} \Gamma(x)$.

If $b \in \Gamma(x) \cap \Gamma(y)$, then $b = \mu(x)$ and $b = \lambda(y)$ for some $\mu, \lambda \in \Gamma$. Therefore, $\mu^{-1}(b) = x = \mu^{-1}\lambda(y)$, and so $x \in \Gamma(y)$. Similarly, $y \in \Gamma(x)$. From this it follows that $\Gamma(x) \subseteq \Gamma(y)$, $\Gamma(y) \subseteq \Gamma(x)$, and $\Gamma(x) = \Gamma(y)$. Thus $\Gamma(x) \ne \Gamma(y)$ implies $\Gamma(x) \cap \Gamma(y) = \varnothing$.

Now (5.13) is an application of (5.14).

(5.15) Theorem. *Let $(S, \mathcal{B}, \in)$ be a BIBD. Then $(S, \mathcal{B}, \in)$ is circular if and only if $|A_{x_i, y_i}| = \lambda(k-2)$ for each i, $1 \le i \le n$, and where the A_{x_i, y_i} and the n are as in (5.13).*

Proof. If $(S, \mathcal{B}, \in)$ is circular, then $|A_{x_i, y_i}| = \lambda(k-2)$ for each i, $1 \le i \le n$, by (5.12). Conversely, suppose $A_{x,y} \in \mathcal{A}(S, \mathcal{B}, \in)$ and $A_{x,y} \in \Gamma(A_{x_i, y_i})$. Thus $A_{x,y} = A_{\theta(x_i), \theta(y_i)}$ for some $\theta \in \Gamma$. But $|A_{\theta(x_i), \theta(y_i)}| = |A_{x_i, y_i}| = \lambda(k-2)$, so $|A_{x,y}| = \lambda(k-2)$. By (5.12), $(S, \mathcal{B}, \in)$ is circular.

(5.16) Theorem. *Let $\Gamma = \mathrm{Aut}(S, \mathcal{B}, \in)$ act doubly transitively on S, where $(S, \mathcal{B}, \in)$ is a BIBD. Then $|A_{x,y}| = |A_{u,v}|$ for any pairs $\{x, y\}$, $\{u, v\}$ from S, where $x \ne y$ and $u \ne v$.*

Proof. There is a $\theta \in \Gamma$ such that $\theta(x) = u$ and $\theta(y) = v$. Now apply (5.13) to realize that $\Gamma(A_{x,y})$ is the only orbit. Since $\theta \in \Gamma$, $|\theta(A_{x,y})| = |A_{u,v}| = |A_{x,y}|$.

(5.17) Corollary. *For a BIBD $(S, \mathcal{B}, \in)$ with a doubly transitive $\Gamma = \mathrm{Aut}(S, \mathcal{B}, \in)$, we obtain that $(S, \mathcal{B}, \in)$ is circular if and only if $|A_{x,y}| = \lambda(k-2)$ for some $x, y \in S$ with $x \ne y$.*

Proof. Apply (5.16) and (5.15).

5.3. Field generated designs

By a *field generated design*, we mean any BIBD generated by a finite field generated planar nearring. We will now see that field generated nearrings are excellent sources for circular planar nearrings.

(5.18) Theorem. *Let* $(F, +, *)$ *be a finite planar nearring generated from a field* $(F, +, \cdot)$ *and a* $\Phi \lhd F^*$. *Then* $\Gamma = \mathrm{Aut}(F, \mathcal{B}^*, \in)$ *is doubly transitive on* F.

Proof. For $b \in F$, $b^+(x) = x + b$ defines $b^+ \in \Gamma$. Certainly, b^+ is a bijection. If $F^*a + c \in \mathcal{B}^*$, then $b^+(F^*a + c) = F^*a + c + b \in \mathcal{B}^*$, and $x \in F^*a + c$ implies $x = ma + c$ and $b^+(ma + c) = b^+(x) = ma + c + b \in F^*a + c + b$. Thus, $x \in B$ implies $b^+(x) \in b^+(B)$. This makes $b^+ \in \Gamma$.

For $m \in F^*$, $m^\bullet(x) = mx$ defines $m^\bullet \in \Gamma$. Certainly, $m^\bullet$ is a bijection. If $F^*a + b \in \mathcal{B}^*$, and $\phi a + b \in F^*a + b$, $\phi \in \Phi$, then $m^\bullet(\phi a + b) = \phi(ma) + mb \in F^*(ma) + mb \in \mathcal{B}^*$. Hence, $m^\bullet(F^*a + b) = F^*(ma) + mb \in \mathcal{B}^*$, and $x \in F^*a + b$ implies $m^\bullet(x) \in m^\bullet(F^*a + b)$. So $m^\bullet \in \Gamma$.

Let $a, b \in F$, $a \neq b$, and $s, t \in F$, $s \neq t$. We want a $\phi \in \Gamma$ so that $\phi(a) = s$ and $\phi(b) = t$. Now the $d^+ \circ m^\bullet \in \Gamma$, and $d^+ \circ m^\bullet(x) = d^+(mx) = mx + d$. Do there exist suitable d and m so that

$$\begin{cases} ma + d = s \\ mb + d = t \end{cases}$$

in F? For an affirmative answer, we need $\det \begin{vmatrix} a & 1 \\ b & 1 \end{vmatrix} \neq 0$, and this will be the case since $a \neq b$. But what if $m = 0$? Then $s = t$, which cannot be. So let $\phi = d^+ \circ m^\bullet \in \Gamma$, and Γ is doubly transitive on F.

(5.19) Theorem. *For a planar nearring* $(F, +, *)$ *from a finite field* $(F, +, \cdot)$ *and* $\Phi \lhd F^*$, $(F, \mathcal{B}^*, \in)$ *is circular if and only if* $|A_{0,1}| = (k-1)(k-2)$.

Proof. By (5.17), $(F, \mathcal{B}^*, \in)$ is circular if and only if $|A_{0,1}| = \lambda(k - 2)$. But our $\lambda = k - 1$.

(5.20) Theorem. *For a Ferrero pair* (F, Φ) *where* F *is a finite field and* $\Phi \lhd F^*$, *then*

$$A_{0,1} = \{(\mu - 1)(\phi - 1)^{-1} \mid \mu, \phi \in \Phi \setminus \{1\}, \ \mu \neq \phi\}.$$

Proof. We have seen that 0 belongs to the $F^*a - a$, $a \neq 0$. If $1 \in F^*a - a$, then $1 = \phi a - a = (\phi - 1)a$. Hence, $a = (\phi - 1)^{-1}$. Conversely, $\phi(\phi - 1)^{-1} - (\phi - 1)^{-1} = (\phi - 1)(\phi - 1)^{-1} = 1$. So $0, 1$ belong to the $F^*(\phi - 1)^{-1} - (\phi - 1)^{-1}$. Now $x \in F^*(\phi - 1)^{-1} - (\phi - 1)^{-1}$ implies

$x = \mu(\phi - 1)^{-1} - (\phi - 1)^{-1} = (\mu - 1)(\phi - 1)^{-1}$. If $x \neq 0$, then $\mu \neq 1$, and
if $x \neq 1$, then $\mu \neq \phi$. Hence,

$$A_{0,1} = \{(\mu - 1)(\phi - 1)^{-1} \mid \mu, \phi \in \Phi \setminus \{1\}, \ \mu \neq \phi\}$$

since such a $(\mu - 1)(\phi - 1)^{-1} = \mu(\phi - 1)^{-1} - (\phi - 1)^{-1} \in F^*(\phi - 1)^{-1} - (\phi - 1)^{-1}$.

(5.21) Corollary. *Let F and L be finite fields with $F \subset L$, and suppose
that $k|(|F| - 1)$. Then $k|(|L| - 1)$ and $(F, \mathcal{B}_k^*, \in)$ is circular if and only if
$(L, \mathcal{B}_k^*, \in)$ is circular, where $\mathcal{B}_k^*$ denotes the blocks made from Φ_k of order
k in the usual Ferrero way for the corresponding field.*

Proof. $\Phi_k \triangleleft F^* \triangleleft L^*$ implies $\Phi_k \triangleleft L^*$. In each case $A_{0,1}$ is exactly as in (5.20). ∎

(5.22) Theorem. *If $(S, \mathcal{B}, \in)$ is a BIBD and the group $\Gamma = \mathrm{Aut}(S, \mathcal{B}, \in)$
is doubly transitive on S, then the structure $(S, \mathcal{A}(S, \mathcal{B}, \in), \in)$ is a BIBD
and $\mathrm{Aut}(S, \mathcal{A}(S, \mathcal{B}, \in), \in) \supseteq \mathrm{Aut}(S, \mathcal{B}, \in)$.*

Proof. If $\{x, y\} \neq \{w, z\}$, $x \neq y$, and $w \neq z$, then there is a $\theta \in \Gamma = \mathrm{Aut}(S, \mathcal{B}, \in)$ so that $\theta(x) = w$ and $\theta(y) = z$, so $\theta(A_{x,y}) = A_{\theta(x),\theta(y)} = A_{w,z}$.
Therefore, $|A_{x,y}| = |A_{w,z}|$.
  So all the new blocks in $\mathcal{A}(S, \mathcal{B}, \in)$ have the same size.
  Let x belong to $A_{x_1,y_1}, A_{x_2,y_2}, \ldots, A_{x_r,y_r}$, exactly. Then if $y \in S$ and
$\theta(x) = y$, we have $\theta(x) \in A_{\theta(x_1),\theta(y_1)} \cap A_{\theta(x_2),\theta(y_2)} \cap \cdots \cap A_{\theta(x_r),\theta(y_r)}$, so
$[x] \leq [y]$. The same argument shows $[y] \leq [x]$, so $[x] = [y]$.
  Let x and y belong to the $A_{x_1,y_1}, A_{x_2,y_2}, \ldots, A_{x_\lambda,y_\lambda}$, exactly. For
$w, z \in S$, There is a $\theta \in \Gamma$ with $\theta(x) = w$ and $\theta(y) = z$, so $w, z \in A_{\theta(x_1),\theta(y_1)} \cap A_{\theta(x_2),\theta(y_2)} \cap \cdots \cap A_{\theta(x_\lambda),\theta(y_\lambda)}$, hence $[x, y] \leq [w, z]$. Again,
the same argument shows $[w, z] \leq [x, y]$, so $[x, y] = [w, z]$.
  If $\theta \in \Gamma$ and $x \in A_{u,v}$, then $\theta(x) \in A_{\theta(u),\theta(v)} = \theta(A_{u,v})$. So θ is an
automorphism of the new design $(S, \mathcal{A}(S, \mathcal{B}, \in), \in)$. ∎

(5.23) Theorem. *Let F be a finite field. Let $1 \neq \Psi \triangleleft \Phi \triangleleft F^*$. Then
$(F, \mathcal{B}_\Phi^*, \in)$ being circular implies $(F, \mathcal{B}_\Psi^*, \in)$ is circular, where for $\Gamma \triangleleft F^*$,
$\mathcal{B}_\Gamma^* = \{\Gamma a + b \mid a, b \in F, a \neq 0\}$.*

Proof. Assume $(F, \mathcal{B}_\Phi^*, \in)$ is circular. Let $\Psi a + b \neq \Psi c + d$. If $\Phi a \neq \Phi c$ or
$b \neq d$, then $\Phi a + b \neq \Phi c + d$, so $|(\Psi a + b) \cap (\Psi c + d)| \leq |(\Phi a + b) \cap (\Phi c + d)| \leq 2$.
  If $\Phi a = \Phi c$ and $b = d$, then $\Phi a + b = \Phi c + d$. But $\Psi a + b \neq \Psi c + d = \Psi c + b$.
Hence, $\Psi a \neq \Psi c$. But $\Psi a \cap \Psi c = \varnothing$, hence $(\Psi a + b) \cap (\Psi c + b) = \varnothing$. Then
$(\Psi a + b) \cap (\Psi c + d) = \varnothing$. ∎

(5.24) Proposition. *Let L be a field of order $p^n = v$, with p a prime.
Suppose $k|(p^n - 1)$, that is, $k|(v - 1)$. Suppose there exists an $l > 1$ and
an r so that $l|v$, $l \equiv 1 \bmod r$, and $(r - 1)(r - 2) + 2 > l$. Then $(L, \mathcal{B}_k^*, \in)$*

is not circular, where $\mathcal{B}_k^*$ is the family of blocks made from $\Phi_k \lhd L^*$ with $|\Phi_k| = k$. In particular, if $p \equiv 1 \bmod k$, and $(k-1)(k-2) > p-2$, then $(L, \mathcal{B}_k^*, \in)$ is not circular.

Proof. Let F be the subfield of order $l = p^m$ of the field L of order $v = p^n$. If $l \equiv 1 \bmod r$, then $r|(l-1)$, so there is the BIBD $(F, \mathcal{B}_r^*, \in)$, where $B \in \mathcal{B}_r^*$ implies $|B| = r$. Since $(r-1)(r-2) + 2 > l$, then $(F, \mathcal{B}_r^*, \in)$ cannot be circular. Hence, by (5.21), $(L, \mathcal{B}_r^*, \in)$ cannot be circular. (Here the $\mathcal{B}_r^*$ is relative to L in $(L, \mathcal{B}_r^*, \in)$, but relative to F in $(F, \mathcal{B}_r^*, \in)$.) $\cdot$ If $(L, \mathcal{B}_r^*, \in)$ were circular, then so would $(L, \mathcal{B}_r^*, \in)$ be circular.

For the second statement, let $l = p$ and $r = k$.

Let us consider another development due to Modisett [MM]. Fix a k. For each prime p and each positive integer n, consider the sequence of fields, F_{p^n} of order p^n, so $F_{p^n} \subset F_{p^{mn}}$. There may be a smallest n so that $k|(p^n - 1)$, or there may not be. If there is not, put $p \in \mathcal{P}_k'$. If there is, then it may be that $(F_{p^n}, \mathcal{B}_k^*, \in)$ is circular, or it may be that it is not circular. If it is circular, then $(F_{p^{mn}}, \mathcal{B}_k^*, \in)$ is circular for each $m \geq 0$. If $(F_{p^n}, \mathcal{B}_k^*, \in)$ is not circular, then $(F_{p^{mn}}, \mathcal{B}_k^*, \in)$ is not circular for each $m \geq 0$. If $(F_{p^n}, \mathcal{B}_k^*, \in)$ is *not* circular, put $p \in \mathcal{P}_k$.

So $p \in \mathcal{P}_k$ if and only if $k|(p^n - 1)$ for some integer $n \geq 1$, and the resulting $(F_{p^n}, \mathcal{B}_k^*, \in)$ is not circular. That is, there is a finite field of characteristic p that 'has a chance' of having circles of size k, but it fails.

What can we say about $\mathcal{P}_k$ and $\mathcal{P}_k'$?

(5.25) Theorem. *There are circles of every size to be found among the finite fields.*

Proof. Fix k, the size of a desired circle. In $k-1$, $2k-1$, $\ldots$, there is a prime q, by Dirichlet's theorem on primes in an arithmetic progression. Let $q = mk - 1$, so $q + 1 = mk$. But $q + 1$ is the size of the circles in $(F_{q^2}, \mathcal{B}_{q+1}^*, \in)$ where F_{q^2} is the field of order q^2. Since $k|(q+1)$, $(F_{q^2}, \mathcal{B}_k^*, \in)$ is circular also.

(5.26) Corollary. *For any $k \geq 3$, neither $\mathcal{P}_k$ nor $\mathcal{P}_k'$ can be the entire set of primes.*

What we want to do is to show that $\mathcal{P}_k$ is a finite set, and to do this we need to resurrect an idea and some techniques from the past. Given two polynomials $f, g \in F[x]$, where F is a field, the *resultant* of f and g, denoted by $R(f, g)$, is a function that tells us if f and g have a common root. Of course, we may assume that the roots of f and g belong to F. Let $f = a_n x^n + a_{n-1} x^{n-1} + \cdots + a_1 x + a_0$ and $g = b_m x^m + b_{m-1} x^{m-1} + \cdots + b_1 x + b_0$, where the a_i, $b_i \in F$ and $a_n b_m \neq 0$. Let $\alpha_1, \alpha_2, \ldots, \alpha_n$ be the roots of f and $\beta_1, \beta_2, \ldots, \beta_m$ be the roots of g. Then the *resultant* or *eliminant* of f

and g is

$$R(f,g) = a_n^m g(\alpha_1)g(\alpha_2)\cdots g(\alpha_n).$$

We see immediately that f and g have a common root if and only if $R(f,g) = 0$. From f and g we also define *Sylvester's determinant* by

$$S(f,g) = \begin{vmatrix} a_n & a_{n-1} & \cdots & \cdots & a_0 & 0 & 0 & \cdots & 0 \\ 0 & a_n & \cdots & \cdots & a_1 & a_0 & 0 & \cdots & 0 \\ \cdots & \cdots & \cdots & \cdots & \cdots & \cdots & \cdots & \cdots & \cdots \\ 0 & \cdots & 0 & a_n & a_{n-1} & \cdots & \cdots & a_1 & a_0 \\ b_m & b_{m-1} & \cdots & \cdots & b_0 & 0 & 0 & \cdots & 0 \\ 0 & b_m & \cdots & \cdots & b_1 & b_0 & 0 & \cdots & 0 \\ \cdots & \cdots & \cdots & \cdots & \cdots & \cdots & \cdots & \cdots & \cdots \\ 0 & \cdots & 0 & b_m & b_{m-1} & \cdots & \cdots & b_1 & b_0 \end{vmatrix},$$

an $m+n$ by $m+n$ determinant, with the coefficients a_i on m rows and the coefficients b_j on n rows.

The following theorem, whose proof can be found in [J], is remarkable, and will be very useful.

(5.27) Theorem. $R(f,g) = S(f,g)$.

For a fixed $k \geq 4$, we consider the various primes p, and if $p \notin \mathcal{P}'_k$, then we want to know what allows $p \in \mathcal{P}_k$. For this, the kth cyclotomic polynomial will be useful. The kth cyclotomic polynomial has different meanings for the various prime fields, but they are related, and it is crucial that we understand this relationship. For now, let $g_k(x)$ be the kth cyclotomic polynomial for the prime field $\mathbf{Q}$, and let $g_{k,p}(x)$ be the kth cyclotomic polynomial for the prime field Z_p. If $g_k(x) = \sum_{i=0}^r a_i x^i$, then $g_{k,p}(x) = \sum_{i=0}^r b_i x^i$, where the $a_i \in Z$, the $b_i \in Z_p$, and $a_i \equiv b_i \pmod{p}$. This allows us to let $g_k(x)$ denote the kth cyclotomic polynomial for any prime field, except for the Z_p where p divides k.

One may see that the above remarks are so by recalling the definitions of $g_k(x)$ and $g_{k,p}(x)$, verifying a few facts, and making an observation. First, both $g_k(x)$ and $g_{k,p}(x)$ are defined as

$$(x - \xi_1)(x - \xi_2)\cdots(x - \xi_r),$$

where the $\xi_1, \xi_2, \ldots, \xi_r$ are all the distinct primitive kth roots of unity. So, if k divides $p^n - 1$, then each ξ_i generates a multiplicative subgroup of order k. This is why $g_{k,p}(x)$ is undefined if p divides k. The facts to be verified are:

1. $g_1(x) = x - 1$.
1a. $g_{1,p} = x - 1$.
2. For a prime q, $g_q(x) = x^{q-1} + x^{q-2} + \cdots + x + 1$.

2a. For a prime $q \neq p$, $g_{q,p}(x) = x^{q-1} + x^{q-2} + \cdots + x + 1$.
3. $x^n - 1 = \prod_{d|n} g_d(x)$, where the product is over all divisors of n, including 1 and n.
3a. $x^n - 1 = \prod_{d|n} g_{d,p}(x)$, where p does not divide n, and where the product is over all the divisors of n, including 1 and n.
4. $g_n(x) = (x^n - 1)/\prod_{d|n, d\neq n} g_d(x)$.
4a. $g_{n,p}(x) = (x^n - 1)/\prod_{d|n, d\neq n} g_{d,p}(x)$, if p does not divide n.

The observation to be made is the following. $Z[x]$ and the $Z_p[x]$ are Z-algebras, and the $Z_p[x]$ are Z-homomorphic images of $Z[x]$ where $1 \mapsto 1$ and $x \mapsto x$.

So, for $g_k(x) = \sum_{i=0}^r a_i x^i$, if a prime p is larger than each $|a_i|$, then we can say that $g_k(x) = g_{k,p}(x)$, and in all cases, there is no loss if one uses $g_k(x)$ for $g_{k,p}(x)$, as long as p does not divide k. For these reasons, we will speak of *the kth cyclotomic polynomial $g_k(x)$*, and consider it to be in $Z[x]$ or $Z_p[x]$, whichever context is needed.

Getting back to determining if a $p \notin \mathcal{P}'_k$ is in $\mathcal{P}_k$ or not, we will find it useful to consider $R(g_k, f)$ as f varies over some finite set of polynomials $\mathcal{F}(k)$. Let

$$\mathcal{F}(k) = \{\, f_{i,j,s,t} \mid i, j, s, t \in \{1, 2, \ldots, k-1\}, i \neq j, s \neq t, \text{ and } (i,j) \neq (s,t)\,\},$$

where $f_{i,j,s,t} = (x^{t+i} - x^t - x^i) - (x^{s+j} - x^s - x^j)$.

The coefficients of g_k are in $Z_p = \{0, 1, \ldots, p-1\}$, so we can consider them to be in Z also. Likewise, the coefficients of each $f_{i,j,s,t}$ are in Z_p and can be considered to be in Z. We can consider $S(g_k, f_{i,j,s,t})$ as being in Z_p, or in Z, and in doing so, $S(g_k, f_{i,j,s,t}) = 0$ in Z_p if and only if $S(g_k, f_{i,j,s,t})$ is a multiple of p, when considered in Z.

(5.28) Theorem. *If a prime $p \notin \mathcal{P}'_k$, then $p \in \mathcal{P}_k$ if and only if p divides $S(g_k, f_{i,j,s,t})$ in Z for some $f_{i,j,s,t} \in \mathcal{F}(k)$.*

Proof. Let $p \in \mathcal{P}_k$, Then $(F_{p^n}, \mathcal{B}^*_k, \in)$ exists for some n. We know that it is not circular. This means, by (5.19), that $|A_{0,1}| < (k-1)(k-2)$, and by (5.20), that there are $i, j, s, t \in \{1, 2, \ldots, k-1\}$, where $i \neq j$, $s \neq t$, and $(i, j) \neq (s, t)$ and

$$\frac{\xi^i - 1}{\xi^j - 1} = \frac{\xi^s - 1}{\xi^t - 1},$$

where ξ is an element of multiplicative order k in $F^*_{p^n}$.

For the i, j, s, and t, this means $f_{i,j,s,t}(\xi) = 0$, and so $f_{i,j,s,t}$ and g_k have a common root. Thus, by (5.27), $S(g_k, f_{i,j,s,t}) = 0$ in Z_p and so p divides $S(g_k, f_{i,j,s,t})$ when considered in Z.

For the converse, for a $p \notin \mathcal{P}'_k$, if p divides $S(g_k, f_{i,j,s,t})$ in Z for some $f_{i,j,s,t} \in \mathcal{F}(k)$, then $S(g_k, f_{i,j,s,t}) = 0$, when considered in Z_p. So, there is a

$\xi \in F_{p^n}^*$ satisfying $g_k(\xi) = f_{i,j,s,t}(\xi) = 0$. Of course, this means that ξ has multiplicative order k in $F_{p^n}^*$, and that $(\xi^{t+i} - \xi^t - \xi^i) - (\xi^{s+j} - \xi^s - \xi^j) = 0$. From this follows $\xi^{t+i} - \xi^t - \xi^i + 1 = \xi^{s+j} - \xi^s - \xi^j + 1$ and $(\xi^t - 1)(\xi^i - 1) = (\xi^s - 1)(\xi^j - 1)$, and finally

$$\frac{\xi^i - 1}{\xi^j - 1} = \frac{\xi^s - 1}{\xi^t - 1},$$

where $i \neq j$, $s \neq t$, and $(i,j) \neq (s,t)$. This means, of course, that $|A_{0,1}| < (k-1)(k-2)$, and consequently that $(F_{p^n}, \mathcal{B}_k^*, \in)$ is not circular.

(5.29) Theorem. $\mathcal{P}_k' = \{p_1, p_2, \ldots, p_u\}$ *where* $p_1, p_2, \ldots, p_u$ *are exactly the primes which divide* k.

Proof. If p divides k, then $p \notin \mathcal{P}_k'$ implies k divides $p^n - 1$ for some n, so p divides 1. Hence, the primes which divide k are in $\mathcal{P}_k'$.

Suppose p is a prime which does not divide k. Then there is a field F_{p^n} which contains all the roots of $x^k - 1 = 0$. This means that k divides $p^n - 1$, so $p \notin \mathcal{P}_k'$.

(5.30) Theorem. *If* $p \in \mathcal{P}_k$ *implies that* p *does not divide* l, *then* $\mathcal{P}_k \subseteq \mathcal{P}_{kl}$.

Proof. For $p \in \mathcal{P}_k$, then p divides neither k nor l, so p cannot divide kl. Since $p \notin \mathcal{P}_{kl}'$, there is an m such that $(F_{p^m}, \mathcal{B}_{kl}^*, \in)$ exists, that is, kl divides $p^m - 1$. If $(F_{p^m}, \mathcal{B}_{kl}^*, \in)$ were circular, then so would $(F_{p^m}, \mathcal{B}_k^*, \in)$ be circular, by (5.23). But then, $p \notin \mathcal{P}_k$, so $(F_{p^m}, \mathcal{B}_{kl}^*, \in)$ is not circular, and $p \in \mathcal{P}_{kl}$.

(5.31) Theorem. $\mathcal{P}_k$ *is finite.*

Proof. The set $\mathcal{F}(k)$ is finite. By (5.28), $p \in \mathcal{P}_k$ if and only if p divides $S(g_k, f)$ in Z for each $f \in \mathcal{F}(k)$. So, if we show that $S(g_k, f) \neq 0$ in Z for each $f \in \mathcal{F}(k)$, then $\mathcal{P}_k$ will need to be finite.

Suppose there is a k such that $S(g_k, f) = 0$ in Z for some $f \in \mathcal{F}(k)$. So, if p does not divide k, then p divides $S(g_k, f)$ for this $f \in \mathcal{F}(k)$, and then there are no circular $(F_{p^n}, \mathcal{B}_k^*, \in)$. But this is contrary to (5.25). We are forced to conclude that $S(g_k, f) \neq 0$ in Z for each $f \in \mathcal{F}(k)$, and, consequently, $\mathcal{P}_k$ must be finite.

Let $\mathcal{P}_k^* = \mathcal{P}_k \cup \mathcal{P}_k'$, a finite set by (5.29) and (5.31). So

(5.32) Theorem. $p \notin \mathcal{P}_k^*$ *if and only if there is an* n *such that the BIBD* $(F_{p^n}, \mathcal{B}_k^*, \in)$ *is circular.*

When the preceding table of circular $(Z_p, \mathcal{B}_k^*, \in)$ for $p < 1000$ was made, it was observed that there were circles of size k for each k, $4 \leq k \leq 18$. So, it was conjectured that for each $k \geq 4$, there would be a prime p so that k divides $p - 1$ and the resulting $(Z_p, \mathcal{B}_k^*, \in)$ would be circular. We can now prove this.

(5.33) Theorem. *For $k \geq 4$, there is a prime p so that $(Z_p, \mathcal{B}_k^*, \in)$ is circular.*

Proof. As in the proof of (5.25), there is a prime in the sequence $k+1$, $2k+1, \ldots$. In fact, there are infinitely many primes in this arithmetic sequence. Let p be a prime in this sequence which does not belong to the finite set $\mathcal{P}_k^*$. So $p = ak + 1$, and so k divides $p - 1$ and $(Z_p, \mathcal{B}_k^*, \in)$ is circular, since $p \notin \mathcal{P}_k^*$.

Also, in the table of the circular $(Z_p, \mathcal{B}_k^*, \in)$, it seemed that, for a given k, once p became 'large enough' for $(Z_p, \mathcal{B}_k^*, \in)$ to be circular sometimes, then p would soon be 'large enough' for $(Z_p, \mathcal{B}_k^*, \in)$ to be circular whenever k divided $p - 1$. We can now prove this, also.

(5.34) Theorem. *For a given $k \geq 4$, there is an N_k so that whenever $p > N_k$ and k divides $p - 1$, then $(Z_p, \mathcal{P}_k^*, \in)$ is circular.*

Proof. Let N_k be the maximum prime in $\mathcal{P}_k^*$. So, if $p > N_k$ and k divides $p - 1$, then $p \notin \mathcal{P}_k^*$, so $(Z_p, \mathcal{B}_k^*, \in)$ must be circular.

For relatively small values of k, we can certainly compute the $\mathcal{P}_k'$ part of $\mathcal{P}_k^*$. One can design a computer program to calculate the $\mathcal{P}_k$, and if k is not too large, the program can do its work in a reasonable amount of time. We shall list the elements for various $\mathcal{P}_k^*$, but first, let us notice that we can sometimes compute a subset of some of the $\mathcal{P}_k^*$ relatively easily, in addition to that provided by (5.30). For a fixed $k \geq 4$, define

$$T_k = \{p \mid p \text{ is a prime, there is an } e \text{ with } p^e \equiv 1 \pmod{k},$$
$$\text{and } p^e < (k-1)(k-2) + 2\}.$$

(5.35) Proposition. $T_k \subseteq \mathcal{P}_k \subset \mathcal{P}_k^*$.

Proof. In (5.24), let $l = p^e$ and $k = r$.

Sometimes it might be useful to know quickly if $p \notin \mathcal{P}_k^*$.

(5.36) Proposition. *If p does not divide k and if k divides $p^n + 1$ for some n, then $p \notin \mathcal{P}_k^*$. So, if k divides any $p^m - 1$, then the $(F_{p^{2m}}, \mathcal{B}_k^*, \in)$ is circular.*

Proof. $(F_{p^{2n}}, \mathcal{B}_{p^n+1}^*, \in)$ is circular by (5.9), and so $(F_{p^{2n}}, \mathcal{B}_k^*, \in)$ is circular by (5.23). Hence, $p \notin \mathcal{P}_k^*$.

Now we list the elements of $\mathcal{P}_k^*$ for some small values of k.

$\mathcal{P}_3^* = \{3\}$,
$\mathcal{P}_4^* = \{2, 5\}$,
$\mathcal{P}_5^* = \{5, 11\}$,
$\mathcal{P}_6^* = \{2, 3, 7, 13, 19\}$,
$\mathcal{P}_7^* = \{2, 7, 29, 43\}$,
$\mathcal{P}_8^* = \{2, 3, 5, 17, 41\}$,
$\mathcal{P}_9^* = \{3, 19, 37, 73, 109, 127, 271\}$,
$\mathcal{P}_{10}^* = \{2, 5, 11, 31, 41, 61, 71, 101\}$.

5.4. Ring generated designs

So, we have seen that field generated planar nearrings are rich in applications to circular planar nearrings. Ring generated planar nearrings are also useful. In considering the ring generated planar nearrings, the field generated ones become even more useful. By analogy, a *ring generated design* is any BIBD generated by a finite ring generated planar nearring.

$A_{0,1}$ is important for field generated planar nearrings, and it is also important for the ring generated ones.

(5.37) Lemma. *For a ring generated design* $(R, \mathcal{B}_\Phi^*, \in)$, *we have*

$$A_{0,1} = \{(\psi - 1)(\phi - 1)^{-1} \mid \psi, \phi \in \Phi^*, \ \phi \neq \psi\} \subseteq \mathcal{U}(R),$$

and $1 - x \in A_{0,1}$ *whenever* $x \in A_{0,1}$.

Proof. Let us quickly recall that we have a ring $(R, +, \cdot, 1)$ with identity 1, and its group of units $\mathcal{U}(R)$ defines a regular group of automorphisms Φ so that (R^+, Φ) is a Ferrero pair. Any planar nearring $(R, +, \circ)$ generated from (R^+, Φ) defines $\mathcal{B}_\Phi^* = \{R^* \circ a + b \mid a, b \in R, \ a \neq 0\}$ where $R^* = R \setminus A$ (or $R^* = R \setminus \{0\}$ if the nearring is integral).

In order for (R^+, Φ) to be a Ferrero pair, we need $\phi - 1 \in \mathcal{U}(R)$ for each $\phi \in \Phi^*$. Now, which blocks contain 0 and 1? The proof of (5.5) tells us that the blocks containing 0 are exactly the $R^* \circ a - a$. If $1 \in R^* \circ a - a$, then there is a $\phi \in \Phi^*$ such that $1 = \phi a - a = (\phi - 1)a$. Hence, $a = (\phi - 1)^{-1}$. This means that $0, 1$ belong exactly to the $R^* \circ (\phi - 1)^{-1} - (\phi - 1)^{-1}$, and the other elements are $\psi(\phi - 1)^{-1} - (\phi - 1)^{-1} = (\psi - 1)(\phi - 1)^{-1}$. Thus

$$A_{0,1} = \{(\psi - 1)(\phi - 1)^{-1} \mid \psi, \phi \in \Phi^*, \ \psi \neq \phi\} \subseteq \mathcal{U}(R).$$

Now, if $\mu \in \mathcal{N}(\Phi)$, the normalizer in $\mathrm{Aut}(R, +)$ of Φ, and $a \in R$, then $\theta(x) = \mu(x) + a$ defines an automorphism of the design $(R, \mathcal{B}_\phi^*, \in)$. This follows from θ being a bijection and $\theta(\phi c + d) = \mu(\phi c + d) + a = \mu(\phi c) + \mu d + a = \phi' \mu c + (\mu d + a)$ for some $\phi' \in \Phi$ if $\phi \in \Phi$. Hence, $\theta(R^* \circ c + d) = R^* \circ \mu c + (\mu d + a)$.

In our case here, $\mu(x) = -x$ defines $\mu \in \mathcal{N}(\Phi)$ and so $\theta(x) = \mu(x) + 1 = -x + 1$ defines $\theta \in \mathrm{Aut}(R, \mathcal{B}_\Phi^*, \in)$. By (5.11), $\theta(A_{0,1}) = A_{\theta(0), \theta(1)} = A_{1,0} = A_{0,1}$. Hence, $1 - x \in A_{0,1}$ if $x \in A_{0,1}$.

The proof of (5.37) motivates

(5.38) Theorem. *For a Ferrero pair* (N, Φ), $\text{Aut}(N, \mathcal{B}_\Phi^*, \in)$ *has a subgroup isomorphic to* $N \times_\theta \mathcal{N}(\Phi)$, *where* $\mathcal{N}(\Phi)$ *is the normalizer of* Φ *in* $\text{Aut}(N, +)$.

Proof. For $(a, \mu) \in N \times \mathcal{N}(\Phi)$, define $[a, \mu] : N \to N$ by $[a, \mu](x) = \mu(x) + a$. Certainly, $[a, \mu]$ is a bijection on N. For $\phi c + d$, $\phi \in \Phi$ and $c \neq 0$, $[a, \mu](\phi c + d) = \mu(\phi c + d) + a = \mu \phi c + (\mu d + a) = \phi' \mu c + (\mu d + a)$. Hence, $[a, \mu](N^* c + d) = N^*(\mu c) + (\mu d + a)$. Thus, $[a, \mu] \in \text{Aut}(N, \mathcal{B}_\Phi^*, \in)$.

Now $[a', \mu'] \circ [a, \mu](x) = [a', \mu'](\mu x + a) = \mu'(\mu x + a) + a' = \mu' \mu(x) + \mu' a + a'$, so $[a', \mu'] \circ [a, \mu] = [\mu' a + a', \ \mu' \mu]$. It is easy to see that $(a, \mu) \mapsto [a, \mu]$ is an isomorphism.

(5.39) Theorem. *Let* $(R, \mathcal{B}_\Phi^*, \in)$ *be the BIBD of (5.5) from a ring generated planar nearring* $(R, +, \circ)$ *from the Ferrero pair* (R, Φ), *where* R *is finite, of course. Let* $k = |\Phi|$. *Then* $(R, \mathcal{B}_\Phi^*, \in)$ *is circular if and only if* $|A_{0,1}| = (k - 1)(k - 2)$ *and* $a - b \in \mathcal{U}(R)$ *whenever* $a, b \in A_{0,1}$ *and* $a \neq b$.

Proof. We will first show that $(R, \mathcal{B}_\Phi^*, \in)$ is circular by showing that each $|A_{x,y}| = (k - 1)(k - 2)$, and then apply (5.12). To do this, we will prove that $|A_{0,1}| = |(A_{0,1})m + n|$ for all $m, n \in R$, $m \neq 0$. Then we will prove that $A_{n,m+n} = (A_{0,1})m + n$. Hence, $A_{0,1}(y - x) + x = A_{x,y}$, and we will have $|A_{x,y}| = |A_{0,1}(y - x) + x| = |A_{0,1}| = (k - 1)(k - 2)$.

Certainly, $|A_{0,1}| \geq |(A_{0,1})m + n|$. If the inequality is strict, then there are distinct $a, b \in A_{0,1}$ so that $am + n = bm + n$ and consequently $0 = am - bm = (a - b)m$. But $a - b \in \mathcal{U}(R)$, so $m = 0$, a contradiction. So we are assured that $|A_{0,1}| = |(A_{0,1})m + n|$.

Since $\lambda = k - 1$, let $\Phi a_1 - a_1, \Phi a_2 - a_2, \ldots, \Phi a_\lambda - a_\lambda$ be the λ distinct blocks containing 0 and 1. Then each $\Phi a_i m - a_i m$ contains 0 and m, and each $\Phi a_i m - a_i m + n$ contains n and $m + n$. These $\Phi a_i m - a_i m + n$ are subsets of $[(A_{0,1})m + n] \cup \{n, m + n\}$ and also of $A_{n,m+n} \cup \{n, m + n\}$. If these blocks containing n and $m + n$ were not distinct, then we would have $|(A_{0,1})m + n| < |A_{0,1}|$, which cannot be, so $A_{n,m+n} = (A_{0,1})m + n$. This makes $(R, \mathcal{B}_\Phi^*, \in)$ circular.

Assume that $(R, \mathcal{B}_\Phi^*, \in)$ is circular. By (5.12), we then have $|A_{0,1}| = (k-1)(k-2)$, so we only need to show that $a - b \in \mathcal{U}(R)$ for any $a, b \in A_{0,1}$, $a \neq b$. If some $a, b \in A_{0,1}$, $a \neq b$, and $a - b \notin \mathcal{U}(R)$, then there is an $m \in R$, $m \neq 0$, such that $(a - b)m = am - bm = 0$. This is because R is finite. When we show that $|A_{0,m}| < (k - 1)(k - 2)$, we will have a contradiction, since (5.12) assures us that $|A_{x,y}| = (k - 1)(k - 2)$ for all $x, y \in R$, $x \neq y$.

Let $\Phi = \{\phi_1, \phi_2, \ldots, \phi_{k-1}, \phi_k = 1\}$. So the $\lambda = k - 1$ blocks containing 0 and 1 are exactly the $\Phi(\phi_i - 1)^{-1} - (\phi_i - 1)^{-1}$, $1 \leq i \leq k - 1$. This means that each $\Phi(\phi_i - 1)^{-1}m - (\phi_i - 1)^{-1}m$ contains 0 and m. We now show that these are distinct, so they will be the $\lambda = k - 1$ blocks which contain 0 and m. By (5.37), $(\phi_i - 1)(\phi_j - 1)^{-1} \in \mathcal{U}(R)$, and if $i \neq j$, then $1 - (\phi_i - 1)(\phi_j - 1)^{-1} \in A_{0,1} \subseteq \mathcal{U}(R)$. This gives us the equation $[1 - (\phi_i - 1)(\phi_j - 1)^{-1}]m = m - (\phi_i - 1)(\phi_j - 1)^{-1}m \neq 0$, and, subsequently, $(\phi_i - 1)^{-1}m \neq (\phi_j - 1)^{-1}m$. By (5.4), we are assured that the above blocks containing 0 and m are distinct, and since there are $\lambda = k - 1$ of them, they are all of those containing 0 and m. We can now conclude that $A_{0,m} = (A_{0,1})m$. But $am = bm$, with $a \neq b$, so $|A_{0,m}| = |(A_{0,1})m| < |A_{0,1}| = (k-1)(k-2)$.

(5.40) Corollary. *If a ring generated design* $(R, \mathcal{B}_\Phi^*, \in)$ *is circular, then for* $m, n \in R$ *with* $m \neq 0$, *we have* $(A_{0,1})m + n = A_{n,m+n}$.

(5.41) Examples. Consider $(Z_{65}, \mathcal{B}_{(8)}^*, \in)$, where $\Phi = \langle 8 \rangle = \{8, 64, 57, 1\}$. Now $k = 4$ and $A_{0,1} = \{8, 9, 19, 37, 57, 58\}$, so we have $|A_{0,1}| = (4-1)(4-2)$. But $57 - 37 \notin \mathcal{U}(Z_{65})$, so this design is not circular. In fact, $\Phi 39 - 39 = \{0, 13, 39, 52\}$ and $\Phi 26 - 26 = \{0, 13, 26, 52\}$.

For a contrast, consider $(Z_{221}, \mathcal{B}_{(21)}^*, \in)$, where $\Phi = \langle 21 \rangle = \{21, 220, 200, 1\}$. Then $A_{0,1} = \{11, 21, 22, 200, 201, 211\}$, so $|A_{0,1}| = (k-1)(k-2)$, and $\{a - b \mid a, b \in A_{0,1}, a \neq b\} = \{\pm1, \pm10, \pm11, \pm178, \pm179, \pm180, \pm189, \pm190, \pm200\} \subseteq \mathcal{U}(Z_{221})$, so the design is circular.

5.5. Other constructions

Direct products are useful to construct more algebraic objects from given ones. It works here also. Let $(N_1, \Phi_1), (N_2, \Phi_2), \ldots, (N_s, \Phi_s)$ be Ferrero pairs, where the Φ_i are all isomorphic. Let $f_i : \Phi_i \to \Phi_{i+1}$, $1 \leq i \leq s - 1$, be isomorphisms. Let $N = N_1 \oplus N_2 \oplus \cdots \oplus N_s$, the direct product, and for $\phi_1 \in \Phi_1$, define $\phi : N \to N$ by

$$\phi(n_1, \ldots, n_s) = (\phi_1 n_1, \, f_1(\phi_1)n_2, \ldots, f_{s-1}(\phi_{s-1})n_s),$$

where $\phi_i = f_{i-1}(\phi_{i-1})$ for $1 < i < s$, and define $\Phi = \{\phi \mid \phi_1 \in \Phi_1\}$.

(5.42) Theorem. *For Ferrero pairs* (N_i, Φ_i), $1 \leq i \leq s$, *with isomorphic* Φ_i, *and for* N *and* Φ *defined as above,* (N, Φ) *is a Ferrero pair.*

Proof. Define $\Xi : \Phi_1 \to \Phi$ by $\Xi(\phi_1) = \phi$. We proceed now to prove that Ξ is an isomorphism and that Φ is a subgroup of $\mathrm{Aut}(N, +)$. Let $\Xi(\phi_1) = \phi$, and $\Xi(\mu_1) = \mu$. Then

$$\begin{aligned}
\mu \circ \phi(n_1, \ldots, n_s) &= \mu(\phi_1 n_1, \, f_1(\phi_1)n_2, \ldots, f_{s-1}(\phi_{s-1})n_s) \\
&= (\mu_1 \phi_1 n_1, \, f_1(\mu_1)f_1(\phi_1)n_2, \ldots, f_{s-1}(\mu_{s-1})f_{s-1}(\phi_{s-1})n_s) \\
&= (\mu_1 \phi_1 n_1, \, f_1(\mu_1 \phi_1)n_2, \ldots, f_{s-1}(\mu_{s-1}\phi_{s-1})n_s).
\end{aligned}$$

Hence, $\Xi(\mu_1 \phi_1) = \Xi(\mu_1)\Xi(\phi_1)$. Suppose that $\Xi(\Phi_1) \subseteq \mathrm{Aut}(N, +)$. Then $(n_1, n_2, \ldots, n_s) = (\phi_1 n_1, \, f_1(\phi_1)n_2, \ldots, f_{s-1}(\phi_{s-1})n_s) = \phi(n_1, n_2, \ldots, n_s)$, for all $n = (n_1, n_2, \ldots, n_s) \in N$, implies $\phi_1 n_1 = n_1$ for all $n_1 \in N_1$. This makes $\phi_1 = 1$ and, consequently, $\phi = 1$. So, $\ker \Xi = \{1\}$. Certainly, Ξ is surjective. So, we will now see that $\phi = \Xi(\phi_1) \in \mathrm{Aut}(N, +)$ for $\phi_1 \in \Phi_1$.

$$
\begin{aligned}
\phi[(n_1, \ldots, n_s) + (m_1, \ldots, m_s)] &= \phi(n_1 + m_1, \ldots, n_s + m_s) \\
&= (\phi_1(n_1 + m_1), f_1(\phi_1)(n_2 + m_2), \ldots, f_{s-1}(\phi_{s-1})(n_s + m_s)) \\
&= (\phi_1 n_1 + \phi_1 m_1, f_1(\phi_1)n_2 + f_1(\phi_1)m_2, \ldots, \\
f_{s-1}(\phi_{s-1})n_s &+ f_{s-1}(\phi_{s-1})m_s) \\
&= (\phi_1 n_1, f_1(\phi_1)n_2, \ldots, f_{s-1}(\phi_{s-1})n_s) \\
&+ (\phi_1 m_1, f_1(\phi_1)m_2, \ldots, f_{s-1}(\phi_{s-1})m_s) \\
&= \phi(n_1, \ldots, n_s) + \phi(m_1, \ldots, m_s).
\end{aligned}
$$

So ϕ is an endomorphism. If

$$
(0, \ldots, 0) = \phi(n_1, \ldots, n_s) = (\phi_1 n_1, f_1(\phi_1)n_2, \ldots, f_{s-1}(\phi_{s-1})n_s),
$$

then $\phi_1 n_1 = 0$ and each $f_i(\phi_i)n_{i+1} = 0$, $0 < i < s$. Since ϕ_1 and the $f_i(\phi_i)$ are automorphisms, we can rest assured that each $n_i = 0$. This makes ϕ a monomorphism. Certainly, $\phi(\phi_1^{-1}m_1, f_1(\phi_1)^{-1}m_2, \ldots, f_{s-1}(\phi_{s-1})^{-1}m_s) = (m_1, m_2, \ldots, m_s)$, so ϕ is an epimorphism. Now we have that Φ is a subgroup of $\mathrm{Aut}(N, +)$.

So, at this point we have a pair of groups (N, Φ) where Φ is a subgroup of $\mathrm{Aut}(N, +)$ which is isomorphic to Φ_1. It remains to show that Φ is regular and that $-1 + \phi$ is surjective for each $\phi \in \Phi^*$.

Suppose $\phi \in \Phi^*$ and that

$$
(n_1, \ldots, n_s) = \phi(n_1, \ldots, n_s) = (\phi_1 n_1, f_1(\phi_1)n_2, \ldots, f_{s-1}(\phi_{s-1})n_s).
$$

Then $\phi_1 \in \Phi_1^*$, so each $f_i(\phi_i) \in \Phi_{i+1}^*$. Hence, each $n_i = 0$. This makes Φ regular.

Let $(m_1, m_2, \ldots, m_s)$ be given. For $\phi \neq 1$, then $\phi_1 \neq 1$, so each $f_i(\phi_i) \neq 1$. Since each (N_i, Φ_i) is a Ferrero pair, $-1 + \phi_1$ and each $-1 + f_i(\phi_i)$ is surjective. Choose n_1 so that $m_1 = (-1 + \phi_1)n_1$, and n_{i+1} so that $m_{i+1} = (-1 + f_i(\phi_i))n_{i+1}$, $1 \leq i \leq s - 1$. Then

$$
\begin{aligned}
(-1 + \phi)&(n_1, \ldots, n_s) \\
&= -(n_1, \ldots, n_s) + (\phi_1 n_1, f_1(\phi_1)n_2, \ldots, f_{s-1}(\phi_{s-1})n_s) \\
&= ((-1 + \phi_1)n_1, (-1 + f_1(\phi_1))n_2, \ldots, (-1 + f_{s-1}(\phi_{s-1}))n_s) \\
&= (m_1, m_2, \ldots, m_s).
\end{aligned}
$$

Hence, $-1 + \phi$ is surjective, and now we know that (N, Φ) really is a Ferrero pair.

(5.43) Theorem. *Let each* (N_i, Φ_i), $1 \le i \le s$, *and* (N, Φ) *be as for* (5.42). *Then* $(N, \mathcal{B}_\Phi^*, \in)$ *is circular if and only if each* $(N_i, \mathcal{B}_{\Phi_i}^*, \in)$ *is circular*, $1 \le i \le s$.

Proof. Suppose first that $(N, \mathcal{B}_\Phi^*, \in)$ is circular. In testing if a planar nearring is circular, note that $x, y, z \in (N^*a + b) \cap (N^*c + d)$ if and only if $x - b$, $y - b$, $z - b \in N^*a \cap (N^*c + d - b)$. So suppose $x_i, y_i, z_i \in N_i^* a_i \cap (N_i^* b_i + c_i)$, with $c_i \ne 0$. For $u_i \in N_i$, define $u \in N$ by $u = (\delta_{i1}(u_i), \delta_{i2}(u_i), \ldots, \delta_{is}(u_i))$, where

$$\delta_{ij}(u_i) = \begin{cases} 0 \in N_i, & \text{if } i \ne j; \\ u_i \in N_i, & \text{if } i = j. \end{cases}$$

Then x, y, z are three distinct points in $N^*a \cap (N^*b + c)$, for if $\phi \in \Phi$, then

$$\begin{aligned} \phi u &= (\phi_1 \delta_{i1}(u_i), \; f_1(\phi_1)\delta_{i2}(u_i), \ldots, \; f_{s-1}(\phi_{s-1})\delta_{is}(u_i)) \\ &= (0, \ldots, 0, \; f_{i-1}(\phi_{i-1})u_i, \; 0, \ldots, 0). \end{aligned}$$

Since $(N, \mathcal{B}_\Phi^*, \in)$ is circular, we must have $c = 0$, hence $c_i = 0$, and so $N_i^* a_i = N_i^* b_i + c_i$.

For the converse, we have that each $(N_i, \mathcal{B}_{\Phi_i}^*, \in)$ is circular, $1 \le i \le s$. Suppose we have three distinct points x, y, $z \in N^*a \cap (N^*b + c)$, with $c \ne 0$.

First, note the following. If $1 \le i \le s$ and $x_i = y_i$, $x_i = z_i$, or $y_i = z_i$, then $x_i = y_i = z_i = 0$, as well as $a_i = c_i = 0$. For if we had $x_i = y_i \ne 0$, then if $x = \phi a$ and $y = \mu a$, then either $\phi_1 a_1 = \mu_1 a_1$ if $i = 1$, or $f_{i-1}(\phi_{i-1})a_i = f_{i-1}(\mu_{i-1})a_i$, if $i \ne 1$. In all cases, $\phi_1 = \mu_1$ and so $\phi = \mu$. Consequently, $x_i = y_i = 0$. This means $a_i = 0$, also, so since $z_i = f_{i-1}(\lambda_{i-1})a_i$, or $z_1 = \lambda_1 a_1$, where $\lambda a = z$, we get $z_i = 0$, also. What is more, each such $c_i = 0$ also. For if $x = \phi' b + c$, $y = \mu' b + c$, and $z = \lambda' b + c$, then either $\phi_1' b_1 + c_1 = \mu_1' b_1 + c_1 = \lambda_1' b_1 + c_1 = 0$, or $f_{i-1}(\phi'i - 1)b_i + c_i = f_{i-1}(\mu_{i-1}')b_i + c_i = f_{i-1}(\lambda_{i-1}')b_i + c_i = 0$. In all cases, either $\phi = \mu = \lambda$, so $x = y = z$, a contradiction, or $c_i = 0$. So, for the i, $1 \le i \le s$, for which $|\{x_i, y_i, z_i\}| < 3$, we have $x_i = y_i = z_i = 0$ and $a_i = c_i = 0$.

For any other i, we have $|\{x_i, y_i, z_i\}| = 3$, and $x_i, y_i, z_i \in N_i^* \cap (N_i^* b_i + c_i)$. Since each $(N_i, \mathcal{B}_{\Phi_i}^*, \in)$ is circular, we must have $c_i = 0$ for each i, also. Hence, $c = 0$, a contradiction. So, we must have that $(N, \mathcal{B}_\Phi^*, \in)$ is circular.

(5.44) Example. From (5.41), we know $(Z_{221}, \mathcal{B}_{(21)}^*, \in)$ is circular. We also know that $(Z_{13}, \mathcal{B}_{(5)}^*, \in)$ is circular. So, the results of (5.43) tell us that the ring $N = Z_{221} \oplus Z_{13}$ generates a circular ring generated design $(N, \mathcal{B}_\Phi^*, \in)$ where $\Phi \cong \langle 21 \rangle \cong \langle 5 \rangle$.

One might suspect that circularity might be a function of the parameters v, b, r, k, λ of the BIBD. We will show, by using Modisett's example, that this is not the case. The following lemma will be useful.

(5.45) Lemma. *Let F be a finite field with $|F| \equiv 1 \pmod{2^{n-1}}$ with $n \geq 3$. Then there is a subgroup Q_n of $\mathrm{Aut}(F^+ \oplus F^+)$ which is regular and is isomorphic to the generalized quaternion group of order 2^n. Thus $(F^+ \oplus F^+, Q_n)$ is a Ferrero pair.*

Proof. Choose $a \in F$ of multiplicative order 2^{n-1}. Let Q_n be the subgroup of $\mathrm{Aut}(F^+ \oplus F^+)$ generated by $A = \begin{bmatrix} a & 0 \\ 0 & a^{-1} \end{bmatrix}$ and $B = \begin{bmatrix} 0 & 1 \\ -1 & 0 \end{bmatrix}$. Since $A^i = \begin{bmatrix} a^i & 0 \\ 0 & a^{-i} \end{bmatrix}$, we have

$$A^{2^{n-2}} = \begin{bmatrix} -1 & 0 \\ 0 & -1 \end{bmatrix} = B^2,$$

and we also have $ABA = B$. Thus, Q_n is a generalized quaternion group of order 2^n and

$$Q_n = \{\, A^i B^j \mid 0 \leq i \leq 2^{n-1} - 1,\ 0 \leq j \leq 1 \,\}.$$

To see that Q_n is regular is equivalent to showing that 1 is not an eigenvalue for any $C \in Q_n$, $C \neq I$, the identity.

The eigenvalues of A are a and a^{-1}, and so the eigenvalues of A^i are a^i and a^{-i}, which are not 1 if $A^i \neq I$. If $C = A^i B$, then $C = \begin{bmatrix} 0 & a^i \\ -a^{-1} & 0 \end{bmatrix}$, whose characteristic polynomial is $x^2 + 1$. Since $|F| \equiv 1 \pmod{2^{n-1}}$, $n \geq 3$, we can be assured that 1 is not an eigenvalue for C. Thus Q_n is regular. Since $F^+ \oplus F^+$ is finite, we have that $(F^+ \oplus F^+,\ Q_n)$ is a Ferrero pair.

(5.46) Theorem. *Circularity of a design $(N, \mathcal{B}, \in)$ does not depend on the parameters v, b, r, k, λ alone.*

Proof. We shall construct two BIBDs with the same parameters, one circular and the other not. In fact, we will construct two Ferrero pairs, (N, Φ_1) and (N, Φ_2) for the same group $(N, +)$ and where $|\Phi_1| = |\Phi_2|$. One of the Φ_i, say Φ_2, will be cyclic and the resulting $(N, \mathcal{B}^*_{\Phi_2}, \in)$ will be circular, whereas the other Φ_1 will be the quaternion group of order 8, and the resulting $(N, \mathcal{B}^*_{\Phi_1}, \in)$ will not be circular. By (5.5), they will have the same parameters.

Let $F = \mathbb{Z}_{29}$, so $|F| \equiv 1 \bmod 2^{3-1}$. Let $a = 12 \in \mathbb{Z}_{29}$, so $\langle a \rangle = \{12, 28, 17, 1\}$. This yields

$$Q = Q_8 = \left\{ \begin{bmatrix} 1 & 0 \\ 0 & 1 \end{bmatrix}, \begin{bmatrix} 12 & 0 \\ 0 & 17 \end{bmatrix}, \begin{bmatrix} -1 & 0 \\ 0 & -1 \end{bmatrix}, \begin{bmatrix} 17 & 0 \\ 0 & 12 \end{bmatrix}, \right.$$

$$\left. \begin{bmatrix} 0 & 1 \\ -1 & 0 \end{bmatrix}, \begin{bmatrix} 0 & 12 \\ -17 & 0 \end{bmatrix}, \begin{bmatrix} 0 & -1 \\ 1 & 0 \end{bmatrix}, \begin{bmatrix} 0 & 17 \\ -12 & 0 \end{bmatrix} \right\}.$$

Then

$$Q(1,1) = \{(1,1), (12,17), (-1,-1), (17,12), (1,-1), (12,-17),$$
$$(-1,1), (17,-12)\}$$

and

$$Q(1,12) = \{(1,12), (12,1), (-1,-12), (17,-1), (12,-1), (-1,-17),$$
$$(-12,1), (1,-12)\}.$$

So

$$Q(1,1) - (1,1) = \{(0,0), (11,16), (27,27), (16,11), (0,27), (11,11),$$
$$(27,0), (16,16)\}$$

and

$$Q(1,12) - (1,12) = \{(0,0), (11,18), (27,5), (16,16), (11,16), (27,0),$$
$$(16,18), (0,5)\}.$$

This means

$$[Q(1,1) - (1,1)] \cap [Q(1,12) - (1,12)] = \{(0,0), (11,16), (27,0), (16,16)\},$$

and so $(Z_{29}^+ \oplus Z_{29}^+, \mathcal{B}_Q^*, \in)$ is not circular.

However, the field generated design $(F_{29^2}, \mathcal{B}_\Phi^*, \in)$ is circular for $|\Phi| = 8$, since $29 \notin \mathcal{P}_8^*$.

(5.47) Problem. Up to now, in constructing circular planar nearrings from a Ferrero pair (N, Φ), the group Φ has always been cyclic. Can there be a Ferrero pair (N, Φ) whose $(N, \mathcal{B}_\Phi^*, \in)$ is circular but yet Φ is not cyclic? In particular, can Φ be nonabelian? Perhaps the pairs $(F^+ \oplus F^+, Q_n)$ of (5.45) is a source for the answer. In particular, do all the $(F^+ \oplus F^+, Q_n)$ of (5.45) yield noncircular $(F^+ \oplus F^+, \mathcal{B}_{Q_n}^*, \in)$?

(5.48) Problem. Up to now, in constructing circular planar nearrings from a Ferrero pair (N, Φ), the group N has always been abelian if $|\Phi| > 3$. Can there be a nonabelian group $(N, +)$ with a regular Φ of order greater than 3 so that (N, Φ) is a Ferrero pair and so that $(N, \mathcal{B}_\Phi^*, \in)$ is circular?

In deciding if a BIBD $(S, \mathcal{B}, \in)$ is circular or not, (5.15) and (5.16) show that knowledge about $\Gamma = \mathrm{Aut}(S, \mathcal{B}, \in)$ is useful. For a Ferrero pair (N, Φ), (5.38) shows that $\mathrm{Aut}(N, \mathcal{B}_\Phi^*, \in)$ has a subgroup isomorphic to $N \times_\theta \mathcal{N}(\Phi)$. Thus, the normalizer of Φ in $\mathrm{Aut}(N, +)$ may be helpful in studying circularity of $(N, \mathcal{B}_\Phi^*)$. So, to provide an answer to problem 3 posed in (5.47), knowledge of $\mathcal{N}(Q_n)$ may be useful. At the very least, Modisett's [MM] computations show that these $\mathcal{N}(Q_n)$ are interesting.

As for (5.45), let F be a finite field with an element $a \in F$ of multiplicative order 2^{n-1}, for $n \geq 3$. Then

$$Q_n = \left\{ \begin{bmatrix} a^i & 0 \\ 0 & a^{-i} \end{bmatrix}, \begin{bmatrix} 0 & a^i \\ -a^{-i} & 0 \end{bmatrix} \mid 0 \leq i < 2^{n-1} \right\},$$

a set of 2^n nonsingular matrices.

Each $\begin{bmatrix} 0 & a^i \\ -a^{-i} & 0 \end{bmatrix}$ has multiplicative order 4.

For $A = \begin{bmatrix} a & 0 \\ 0 & a^{-1} \end{bmatrix}$ and $B = \begin{bmatrix} 0 & 1 \\ -1 & 0 \end{bmatrix}$, then $A^{2^{n-1}} = I = \begin{bmatrix} 1 & 0 \\ 0 & 1 \end{bmatrix}$, $A^{2^{n-2}} = B^2 = -I$, and $ABA = B$. So Q_n is a generalized quaternion group of order 2^n, and $Q_n < GL(2, F)$, where $GL(2, F)$ is the group of 2×2 nonsingular matrices over F, the general linear group. We will now study $\mathcal{N}(Q_n)$, the normalizer of Q_n in $GL(2, F)$.

(5.49) Proposition. $\mathcal{N}(Q_n)$ *contains all matrices*

$$\begin{bmatrix} u & 0 \\ 0 & v \end{bmatrix}, \begin{bmatrix} 0 & u \\ v & 0 \end{bmatrix}$$

where $uv^{-1} \in \langle a \rangle$, the group generated by a.

Proof. The elements of Q_n are of the form $A^i B^j$ where $0 \leq i < 2^{n-1}$ and $0 \leq j < 2$. Since $X A^i B^j X^{-1} = (XAX^{-1})^i (XBX^{-1})^j$, we need only show $XAX^{-1}, XBX^{-1} \in Q_n$ in order to conclude that $X \in \mathcal{N}(Q_n)$. If $X = \begin{bmatrix} u & 0 \\ 0 & v \end{bmatrix}$, then $XAX^{-1} = A$ and $XBX^{-1} = \begin{bmatrix} 0 & uv^{-1} \\ -u^{-1}v & 0 \end{bmatrix}$. If $X = \begin{bmatrix} 0 & u \\ v & 0 \end{bmatrix}$, then $XAX^{-1} = A^{-1}$, and $XBX^{-1} = \begin{bmatrix} 0 & -uv^{-1} \\ vu^{-1} & 0 \end{bmatrix}$.

(5.50) Theorem. *For $n > 3$, $\mathcal{N}(Q_n)$ consists of exactly the $2^n(|F| - 1)$ matrices of (5.49).*

Proof. A has order $2^{n-1} \geq 8$, so XAX^{-1} is *not* among the $\begin{bmatrix} 0 & a^i \\ -a^{-i} & 0 \end{bmatrix}$, since their order is always 4.

Suppose $X = \begin{bmatrix} u & v \\ w & x \end{bmatrix} \in \mathcal{N}(Q_n)$ and $0 \notin \{u, v\}$. Then $XA = A^j X$ for some j, that is, $\begin{bmatrix} ua & va^{-1} \\ wa & xa^{-1} \end{bmatrix} = \begin{bmatrix} ua^j & va^j \\ wa^{-j} & xa^{-j} \end{bmatrix}$. From $ua = ua^j$ and $va^{-1} = va^j$, we conclude $1 = j = -1$. Hence, $0 \in \{u, v\}$. Suppose $0 \notin \{w, x\}$. Then $wa = wa^{-j}$ and $xa^{-1} = xa^{-j}$, so $1 = j = -1$. Thus $0 \in$

$\{w, x\}$. This makes $X = \begin{bmatrix} u & 0 \\ 0 & x \end{bmatrix}$ or $X = \begin{bmatrix} 0 & v \\ w & 0 \end{bmatrix}$. Note that $\begin{bmatrix} u & 0 \\ 0 & x \end{bmatrix}^{-1} =$

$\begin{bmatrix} u^{-1} & 0 \\ 0 & x^{-1} \end{bmatrix}$ and $\begin{bmatrix} 0 & v \\ w & 0 \end{bmatrix}^{-1} = \begin{bmatrix} 0 & w^{-1} \\ v^{-1} & 0 \end{bmatrix}$. For $X = \begin{bmatrix} u & 0 \\ 0 & x \end{bmatrix} \in \mathcal{N}(Q_n)$,

then $XBX^{-1} = \begin{bmatrix} 0 & ux^{-1} \\ -xu^{-1} & 0 \end{bmatrix} \in Q_n$, so $ux^{-1} \in \langle a \rangle$. If $X = \begin{bmatrix} 0 & v \\ w & 0 \end{bmatrix} \in$

$\mathcal{N}(Q_n)$, then $XBX^{-1} = \begin{bmatrix} 0 & -vw^{-1} \\ wv^{-1} & 0 \end{bmatrix} \in Q_n$, so $wv^{-1} \in \langle a \rangle$. .

By (5.49), these matrices are in $\mathcal{N}(Q_n)$. One can choose u in $|F| - 1$ ways, and having chosen u, one must choose v so that $uv^{-1} \in \langle a \rangle$, so there are $2^{n-1}(|F|-1)$ choices for (u, v). One then has two choices for X, namely $\begin{bmatrix} u & 0 \\ 0 & v \end{bmatrix}$ or $\begin{bmatrix} 0 & u \\ v & 0 \end{bmatrix}$. Thus $|\mathcal{N}(Q_n)| = 2^n(|F| - 1)$.

(5.51) Theorem. *For $n = 3$, $\mathcal{N}(Q_3)$ has order $24(|F|-1)$, and consists of the $8(|F| - 1)$ elements from (5.49) together with the $16(|F| - 1)$ elements obtained by taking any scalar $v \neq 0$ times any one of the following 16 matrices:*

$$\begin{bmatrix} 1 & 1 \\ 1 & -1 \end{bmatrix}, \quad \begin{bmatrix} 1 & 1 \\ -1 & 1 \end{bmatrix}, \quad \begin{bmatrix} 1 & 1 \\ a & -a \end{bmatrix}, \quad \begin{bmatrix} 1 & 1 \\ -a & a \end{bmatrix},$$

$$\begin{bmatrix} 1 & -1 \\ 1 & 1 \end{bmatrix}, \quad \begin{bmatrix} -1 & 1 \\ 1 & 1 \end{bmatrix}, \quad \begin{bmatrix} 1 & -1 \\ a & a \end{bmatrix}, \quad \begin{bmatrix} -1 & 1 \\ a & a \end{bmatrix},$$

$$\begin{bmatrix} a & 1 \\ a & -1 \end{bmatrix}, \quad \begin{bmatrix} a & 1 \\ -a & 1 \end{bmatrix}, \quad \begin{bmatrix} a & 1 \\ -1 & -a \end{bmatrix}, \quad \begin{bmatrix} a & 1 \\ 1 & a \end{bmatrix},$$

$$\begin{bmatrix} a & -1 \\ a & 1 \end{bmatrix}, \quad \begin{bmatrix} -a & 1 \\ a & 1 \end{bmatrix}, \quad \begin{bmatrix} a & -1 \\ -1 & a \end{bmatrix}, \quad \begin{bmatrix} -a & 1 \\ -1 & a \end{bmatrix}.$$

Proof. Recall that A has order 2^{n-1}. If $n > 3$, then XAX^{-1} has order $2^{n-1} \geq 8$. But, if $n = 3$, then XAX^{-1} has order $2^{3-1} = 4$. This leads to the difference in $\mathcal{N}(Q_n)$ for $n = 3$ and for $n > 3$. For $n = 3$, $XAX^{-1} = \begin{bmatrix} 0 & a^i \\ -a^{-i} & 0 \end{bmatrix}$ is possible, whereas it was not possible when $n > 3$, since each $\begin{bmatrix} 0 & a^i \\ -a^{-1} & 0 \end{bmatrix}$ has order 4.

For $n = 3$ we have

$$Q_3 = \left\{ \pm \begin{bmatrix} 1 & 0 \\ 0 & 1 \end{bmatrix}, \quad \pm \begin{bmatrix} a & 0 \\ 0 & -a \end{bmatrix}, \quad \pm \begin{bmatrix} 0 & a \\ a & 0 \end{bmatrix}, \quad \pm \begin{bmatrix} 0 & 1 \\ -1 & 0 \end{bmatrix} \right\},$$

and all elements except $\pm \begin{bmatrix} 1 & 0 \\ 0 & 1 \end{bmatrix}$ are of order 4. The proof of (5.50), together with (5.49), give the consequences of $XAX^{-1} = A^j$. It remains to consider

$$X A X^{-1} \in \left\{ \pm \begin{bmatrix} 0 & a \\ a & 0 \end{bmatrix}, \quad \pm \begin{bmatrix} 0 & 1 \\ -1 & 0 \end{bmatrix} \right\}.$$

Each of these four possibilities in turn leads to $X = \begin{bmatrix} u & v \\ w & x \end{bmatrix}$ with

$$X \in \left\{ \begin{bmatrix} u & v \\ u & -v \end{bmatrix}, \quad \begin{bmatrix} u & v \\ -u & v \end{bmatrix}, \quad \begin{bmatrix} u & v \\ ua & -va \end{bmatrix}, \quad \begin{bmatrix} u & v \\ -ua & va \end{bmatrix} \right\}.$$

Each of these four possibilities for X, when considering $X B X^{-1}$, leads to either $u^2 = v^2$ or $u^2 = -v^2$, or, equivalently, $u = \pm v$ or $u = \pm av$. Here, $v \neq 0$ is arbitrary. These lead to $4 \cdot 4 = 16$ possibilities for X, namely, the scalar $v \neq 0$ times any one of the 16 matrices listed in the statement of the theorem. So, these are the only possible values for $X \in \mathcal{N}(Q_3)$.

One needs only check for each of these 16 matrices X listed in (5.51) that the $X A X^{-1}, X B X^{-1} \in Q_3$ to complete the proof of (5.51).

(5.52) Proposition. *The map* $f : \mathcal{N}(Q_n) \to \mathrm{Aut} Q_n$ *defined by* $f(X) = \gamma_X$, *where* $\gamma_X(C) = X C X^{-1}$, *is a homomorphism. Let* $K = \ker f$. *Then* $X \in K$ *if and only if* $X = \begin{bmatrix} u & 0 \\ 0 & u \end{bmatrix}$, $u \in F^*$, *that is,* X *is a scalar matrix.*

Proof. $\gamma_{XY}(C) = (XY)C(XY)^{-1} = XYCY^{-1}X^{-1} = \gamma_X(\gamma_Y(C)) = \gamma_X \circ \gamma_Y(C)$. Hence, $f(XY) = f(X) \circ f(Y)$, so f is a homomorphism. Now $X \in K$ if and only if $X A X^{-1} = A$ and $X B X^{-1} = B$, which is true if and only if X is a scalar matrix $\begin{bmatrix} u & 0 \\ 0 & u \end{bmatrix}$, where $u \neq 0$.

(5.53) Theorem. *If* $K = \left\{ \begin{bmatrix} u & 0 \\ 0 & u \end{bmatrix} \mid u \neq 0 \right\}$, *then:*
(a) *For* $n = 3$, $\mathcal{N}(Q_3)/K \cong \mathrm{Aut} Q_3$.
(b) *For* $n > 3$, $\mathcal{N}(Q_n)/K \cong D_{2^{n-1}}$, *the dihedral group of order* 2^n.

Proof. Since $|K| = |F| - 1$, we have $|\mathcal{N}(Q_3)/K| = 24 = |\mathrm{Aut} Q_3|$. Hence, $\mathcal{N}(Q_3)/K \cong \mathrm{Aut} Q_3$.

For $n > 3$, we have $|\mathcal{N}(Q_n)/K| = 2^n$. Let $s = K \begin{bmatrix} a & 0 \\ 0 & 1 \end{bmatrix}$ and $t = K \begin{bmatrix} 0 & a \\ 1 & 0 \end{bmatrix}$, cosets of K in $\mathcal{N}(Q_n)$. Then $s^{2^{n-1}} = K$ and s has order 2^{n-1}. Also, $t^2 = K$, so t has order 2, and $tst^{-1} = s^{-1}$. So $\mathcal{N}(Q_n)/K \cong D_{2^{n-1}}$.

6. Geometry of circular planar nearrings

We now have lots of examples of circular planar nearrings $(N, +, \cdot)$. They yield incidence structures $(N, \mathcal{B}^*, \in)$ where the points of N and the circles of $\mathcal{B}^*$ share at least two properties with the points and circles of our usual euclidean plane, that is, the plane of the complex numbers. For a circular planar nearring $(N, +, \cdot)$, we can be assured that three distinct points of N belong to *at most* one circle $B \in \mathcal{B}^*$, and we can be assured that any two distinct points of N belong to *more than one* circle in $\mathcal{B}^*$.

We must always remember that there exists *at least one* circular planar nearring having all the properties that points and circles should have, namely, $(\mathbf{C}, +, *_3)$ from example 3 of §4. This is, of course, our main example and from this example we obtain ideas of what to look for in other circular planar nearrings. Of course, we should expect other examples to have properties similar to that of $(\mathbf{C}, +, *_3)$, but we should also expect other examples to have contrasting properties. These similarities and differences provide the foundation for the richness of the theory. We must also remember that if $(N, +, \cdot)$ is a circular planar nearring with circles $N^*a + b \in \mathcal{B}^*$, then the circle $N^*a + b$ has radius a and centre b. These radii and centres are the keys to dealing with the similarities and differences in comparing with $(\mathbf{C}, +, *_3)$.

6.1. Equivalence classes

One should also remember that *every* finite planar nearring $(N, +, \cdot)$, circular or not, yields a BIBD $(N, \mathcal{B}^*, \in)$. See (5.5). This, of course, is a very significant geometric property, even though we will not deal with it here. The point is, that every finite planar nearring provides at least one nice geometric structure. Later, in §7, we shall see that there will always be another BIBD obtained from such planar nearrings. But now, we want to concentrate only on the circular $(N, +, \cdot)$ and corresponding $(N, \mathcal{B}^*, \in)$.

Consider the circles in the complex plane. Of course, these circles have the equivalence relation $\equiv$ where $C_1 \equiv C_2$ if and only if the radius of C_1 is the same as the radius of C_2. This easily holds for every circular planar nearring, also.

(6.1) Proposition. *Let $(N, +, \cdot)$ be a circular planar nearring. Define $\equiv$ on $\mathcal{B}^*$ by $(N^*a + b) \equiv (N^*c + d)$ if and only if $N^*a = N^*c$. Then $\equiv$ is an equivalence relation.*

It will be rewarding to look at equivalence classes of $\mathcal{B}^* / \equiv$. For a fixed $r \in N$, $r \neq 0$, define $\sim_r$ on the equivalence class of N^*r, denoted by $[N^*r]_\equiv$, by $(N^*r + a) \sim_r (N^*r + b)$ if and only if there is a $c \in N$ such that $a, b \in N^*c$.

(6.2) Theorem. *Let $(N, +, \cdot)$ be a circular planar nearring. For each $0 \neq r \in N$, $\sim_r$ is an equivalence relation on $[N^*r]_\equiv$.*

Proof. Certainly, $(N^*r + a) \sim_r (N^*r + a)$ since $a \in N^*a$, and if $(N^*r + a) \sim_r$

$(N^*r + b)$, then $(N^*r + b) \sim_r (N^*r + a)$. Finally, if $(N^*r + x) \sim_r (N^*r + y)$ and $(N^*r + y) \sim_r (N^*r + z)$, then there are $c, d \in N$ where $x, y \in N^*c$ and $y, z \in N^*d$. So $y \in N^*c \cap N^*d$, making $N^*c = N^*d$, so $x, z \in N^*d$ and so $(N^*r + x) \sim_r (N^*r + z)$.

If $N^*c = \{c_1, c_2, \ldots, c_k\}$, define $E_c^r = \{N^*r + c_1, N^*r + c_2, \ldots, N^*r + c_k\}$. More generally, $E_c^r = \{N^*r + d \mid d \in N^*c\}$. Note that $E_0^r = \{N^*r\}$, and if $c \neq 0$, then $|E_c^r| = |N^*c|$.

Recall that Φ is a subgroup of $(N, \cdot)$ consisting of a representative of each $[a]_{=m}$, $a \neq_m 0$, that is, $\Phi = B_a$ for some $a \neq_m 0$.

Of fundamental importance is the following elementary lemma.

(6.3) Lemma. *If $N^*r + b$, $N^*r + d \in E_c^r$, then there is a $\phi \in \Phi$ such that $\phi(N^*r + b) = N^*r + d$, and $\phi b = d$.*

Proof. Since $b, d \in N^*c$, there is a $\phi \in \Phi$ such that $\phi b = d$. So $\phi(N^*r + b) = N^*r + d$.

(6.4) Corollary. *If $\phi \in \Phi$ and $\phi b = d$, then $\phi(N^*r + b) = N^*r + d$.*

The following lemma is the basis for a very interesting concept which occurs from time to time. This concept deserves further study.

(6.5) Lemma. *If $B \in E_c^r$ and $B \cap D = \varnothing$ for each $D \in E_c^r \setminus \{B\}$, then $E \cap F = \varnothing$ for each $E, F \in E_c^r$ if $E \neq F$.*

Proof. Take $E = N^*r + e$, $F = N^*r + f \in E_c^r$ with $E \neq F$, so $e \neq f$. Let $x \in E \cap F$. For $B = N^*r + b \in E_c^r$, we have $b, e, f \in N^*c$, so there is a $\phi \in \Phi$ such that $\phi e = b$. Since $E \neq F$, we can be sure that $\phi E \neq \phi F$. But $\phi E = B$ by (6.4), and also $\phi F \in E_c^r$. Since $x \in E \cap F$, we have $\phi x \in \phi E \cap \phi F = B \cap \phi F$. This is our desired contradiction.

(6.6) Definition. *If $A, B \in E_c^r$, $c \neq 0$, implies that $A = B$ or $A \cap B = \varnothing$, then*

$$T(r; c) = T = \cup_{D \in E_c^r} D$$

is the torus with major radius c and minor radius r.

Further motivation for calling such a T a torus is provided by the following theorem.

(6.7) Theorem. *Let $T(r; c) = T$ be a torus with major radius c and minor radius r. Let $x = \mu r + d \in D = N^*r + d \in E_c^r$ be an arbitrary point of T. Then*

$$T = \cup_{\gamma r + d \in D} N^*(\gamma r + d).$$

In fact, each of these circles $N^*(\gamma r + d)$, $\gamma \in \Phi$, with centre 0, is tangent to each of the circles $E \in E_c^r$.

Proof. For $x = \mu r + d \in D \in E_c^r$, $\mu \in N^*$, we can take $\mu \in \Phi$, and $D = N^* r + d = \{\gamma r + d \mid \gamma \in \Phi\}$. Also, $E_c^r = \{N^* r + \phi d \mid \phi \in \Phi\}$. So, if $\phi \in \Phi$ and $y = \phi(\gamma r + d) = (\phi\gamma)r + \phi d \in N^* r + \phi d \in E_c^r$, we have $y \in T$. Thus each $N^*(\gamma r + d) \subseteq T$. If $t \in T$, then $t \in N^* r + \phi d$ for some $\phi \in \Phi$, so $t = \tau r + \phi d = \phi\gamma r + \phi d = \phi(\gamma r + d)$ for some $\tau, \gamma \in \Phi$. Thus $t \in N^*(\gamma r + d)$, and $T \subseteq \bigcup_{\gamma r + d \in D} N^*(\gamma r + d)$. So we have

$$T(r; c) = T = \cup_{\gamma r + d \in D} N^*(\gamma r + d) = \cup_{\gamma \in \Phi} N^*(\gamma r + d).$$

Consider now an $N^*(\gamma r + d) = \{\phi\gamma r + \phi d \mid \phi \in \Phi\}$. This circle has centre 0 and is tangent to $N^* r + \phi d$ at the point $\phi\gamma r + \phi d$.

(6.8) Corollary. *If $T = T(r; c)$ is a torus in a finite circular planar nearring, and if an arbitrary circle has k distinct points, then T has exactly k^2 points.*

Proof. From (6.6) or (6.7), T is the disjoint union of k circles each having k distinct points. So T has k^2 distinct points.

Let us now consider some examples of finite circular planar nearrings with a torus. This will also illustrate the use of the table in §5.

(6.9) Examples. A $T(1; 2)$ is found in each of the circular $(Z_{17}, \mathcal{B}_4^*, \in)$, $(Z_{41}, \mathcal{B}_5^*, \in)$, $(Z_{113}, \mathcal{B}_7^*, \in)$ and $(Z_{113}, \mathcal{B}_8^*, \in)$. A $T(6; 3)$ can be found in $(Z_{17}, \mathcal{B}_4^*, \in)$.
  Let us look at $T(1; 2)$ and $T(6; 3)$ in $(Z_{17}, \mathcal{B}_4^*, \in)$ in detail. First,

$$T(1; 2) = \cup_{D \in E_2^1} D.$$

Now, if $N = Z_{17}$, then $N^*1 = \{4, 16, 13, 1\} = \Phi$ and $N^*2 = \{8, 15, 9, 2\}$, so

$$E_2^1 = \{N^*1 + 8, \ N^*1 + 15, \ N^*1 + 9, \ N^*1 + 2\}$$

where

$$N^*1 + 8 = \{12, 7, 4, 9\},$$

$$N^*1 + 15 = \{2, 14, 11, 16\},$$

$$N^*1 + 9 = \{13, 8, 5, 10\},$$

$$N^*1 + 2 = \{6, 1, 15, 3\}.$$

Notice that $E, F \in E_2^1$ and $E \neq F$ imply $E \cap F = \varnothing$.
  To illustrate (6.7), take $x = 16 \cdot 1 + 9$. Then $x \in D = N^*1 + 9 = \{4 \cdot 1 + 9, \ 16 \cdot 1 + 9, \ 13 \cdot 1 + 9, \ 1 \cdot 1 + 9\} = \{13, 8, 5, 10\}$. So

$$N^*13 = \{1, 4, 16, 13\}, \quad N^*8 = \{15, 9, 2, 8\},$$

$$N^*5 = \{3, 12, 14, 5\}, \quad N^*10 = \{6, 7, 11, 10\}.$$

Sure enough, $T = N^*13 \cup N^*8 \cup N^*5 \cup N^*10$. As promised by (6.8), $|T| = 4^2 = 16$.

Next,

$$T(6; 3) = \cup_{D \in E_3^6} D.$$

For $N = Z_{17}$, we have $N * 6 = \{7, 11, 10, 6\}$ and $N^*3 = \{12, 14, 5, 3\}$, so

$$E_3^6 = \{N^*6 + 12, \ N^*6 + 14, \ N^*6 + 5, \ N^*6 + 3\}$$

where

$$N^*6 + 12 = \{2, 6, 5, 1\},$$

$$N^*6 + 14 = \{4, 8, 7, 3\},$$

$$N^*6 + 5 = \{12, 16, 15, 11\},$$

$$N^*6 + 3 = \{10, 14, 13, 9\}.$$

Notice that $E, F \in E_3^6$ and $E \neq F$ imply $E \cap F = \varnothing$.

To illustrate (6.7), take $x = 4 \cdot 6 + 12$. Then $x \in D = N^*6 + 12 = \{4 \cdot 6 + 12, \ 16 \cdot 6 + 12, \ 13 \cdot 6 + 12, \ 1 \cdot 6 + 12\} = \{2, 6, 5, 1\}$. So

$$N^*2 = \{8, 15, 9, 2\}, \quad N^*6 = \{7, 11, 10, 6\},$$

$$N^*5 = \{3, 12, 14, 5\}, \quad N^*1 = \{4, 16, 13, 1\}.$$

Sure enough, $T = N^*2 \cup N^*6 \cup N^*5 \cup N^*1$. Again, $|T| = 4 \cdot 4 = 16$.

To further illustrate (6.7) for $T(1; 2)$ take $N^*8 = \{15, 9, 2, 8\}$, and notice it is tangent to $N^*1 + 8$ at 9; it is tangent to $N^*1 + 15$ at 2; it is tangent to $N^*1 + 9$ at 8; and it is tangent to $N^*1 + 2$ at 15. The reader is encouraged to explore the table in §5 and find more tori. To further encourage the reader, and to show that the above examples were not accidents, we shall look at $T(1; 2)$ in $(Z_{41}, \mathcal{B}_5^*, \in)$. With $N = Z_{41}$ and with

$$T(1; 2) = \cup_{D \in E_2^1} D,$$

we have $N^*1 = \{10, 18, 16, 37, 1\}$ and $N^*2 = \{20, 36, 32, 33, 2\}$, so

$$E_2^1 = \{N^*1 + 20, \ N^*1 + 36, \ N^*1 + 32, \ N^*1 + 33, \ N^*1 + 2\}$$

where

$$N^*1 + 20 = \{30, 38, 36, 16, 21\},$$

$$N^*1 + 36 = \{5, 13, 11, 32, 37\},$$

$$N^*1 + 32 = \{1, 9, 7, 28, 33\},$$

$$N^*1 + 33 = \{2, 10, 8, 29, 34\},$$

$$N^*1 + 2 = \{12, 20, 18, 39, 3\}.$$

Again, $E, F \in E_2^1$ and $E \neq F$ imply $E \cap F = \varnothing$.

To illustrate (6.7), take $x = 16 \cdot 1 + 33$. Then $x \in D = N^*1 + 33 = \{10 \cdot 1 + 33, 18 \cdot 1 + 33, 16 \cdot 1 + 33, 37 \cdot 1 + 33, 1 \cdot 1 + 33\} = \{2, 10, 8, 29, 34\}$. So

$$N^*2 = \{20, 36, 32, 33, 2\}, \quad N^*10 = \{18, 16, 37, 1, 10\},$$

$$N^*8 = \{39, 21, 5, 9, 8\}, \quad N^*29 = \{3, 30, 13, 7, 29\},$$

$$N^*34 = \{12, 38, 11, 28, 34\},$$

and $T = N^*2 \cup N^*10 \cup N^*8 \cup N^*29 \cup N^*34$. Furthermore, N^*10 is: tangent to $N^*1 + 20$ at 16; tangent to $N^*1 + 36$ at 37; tangent to $N^*1 + 32$ at 1; tangent to $N^*1 + 33$ at 10; and tangent to $N^*1 + 2$ at 18.

This past discussion has been about any E_c^r where no two distinct circles in E_c^r intersect. A slight departure from this is when a circle in E_c^r intersects exactly one other. This situation is described in the next theorem.

(6.10) Theorem. *For some $c \neq 0$, suppose there are $A, B \in E_c^r$, $A \neq B$, where $A \cap B \neq \varnothing$, but if $D \in E_c^r \setminus \{A, B\}$, then $D \cap (A \cup B) = \varnothing$. Then:*
(1) $|A \cap B| = 2$.
(2) $B = \phi A$ for some $\phi \in \Phi$, where $\phi \neq 1$ and $\phi^2 = 1$.
(3) $D \in E_c^r$ implies the existence of a unique $F \in E_c^r$ such that $D \cap F \neq \varnothing$, $D \neq F$, and $G \cap (D \cup F) = \varnothing$ for each $G \in E_c^r \setminus \{D, F\}$.
(4) There is a bijection between cosets $\lambda \langle \phi \rangle$ of the subgroup $\langle \phi \rangle$ generated by $\phi \in \Phi$ and such pairs $\{D, F\}$ of (3).

Proof. Let $A = N^*r + a$, $B = N^*r + b$, $D = N^*r + d$, $F = N^*r + f$, and $G = N^*r + g$, where $a, b, d, f, g \in N^*c$. For $x \in A \cap B$, $x = \mu r + a = \gamma r + b$ for some $\mu, \gamma \in \Phi$. Since $a \neq b$, we apply (6.3) to get $\phi a = b$ for some $\phi \in \Phi$, and to obtain $\phi A = B$. But also, $\phi x = \phi \mu r + \phi a = \phi \gamma r + \phi b \in B \cap (N^*r + \phi b)$. But $\phi b \neq b$ since $\phi \neq 1$, so $N^*r + \phi b \neq N^*r + b$ by (5.4). This forces $N^*r + \phi b = N^*r + a = A$, for if $N^*r + \phi b \notin \{A, B\}$, then $(N^*r + \phi b) \cap (A \cup B) = \varnothing$. Hence, $a = \phi b = \phi^2 a$. We must then conclude that $\phi^2 = 1$ and $\phi \neq 1$. So $\phi x \neq x$. But $x, \phi x \in A \cap B$, and since N is circular, $|A \cap B| = 2$. This gives us (1) and (2).

For D as in (3), there is a unique $\lambda \in \Phi$ such that $\lambda a = d$, so $\lambda A = D$. Let $F = \lambda B$. This puts $\lambda x, \lambda \phi x \in D \cap F$. Since $A \neq B$, $\lambda A \neq \lambda B$, so $D \neq F$, and $|D \cap F| \leq 2$. If $\lambda = 1$, then $A = D$ and $B = F$. If $\lambda = \phi$, then $\lambda A = B$ and $\lambda B = A$. If $\lambda \notin \langle \phi \rangle$, then $\{A, B\} \cap \{D, F\} = \varnothing$, and $\lambda x \neq \lambda \phi x$. Since $\lambda x, \lambda \phi x \in D \cap F$, we obtain $|D \cap F| = 2$. If $G \in E_c^r \setminus \{D, F\}$ and $t \in G \cap (D \cup F)$, then $\lambda^{-1} t \in \lambda^{-1} G \cap (A \cup B)$ and $\lambda^{-1} G \notin \{A, B\}$. So, by hypothesis, $\lambda^{-1}G \cap (A \cup B) = \varnothing$ since $\lambda^{-1}G \in E_c^r$. As this cannot be, we cannot have $t \in G \cap (D \cup F)$ and we have (3).

Since $\phi A = B$, $\lambda A = D$, and $\lambda B = F$, then $\lambda A = D$ and $\lambda \phi A = F$, and the pair $\{D, F\}$ corresponds to the coset $\lambda \langle \phi \rangle$. Let $\lambda_1 \langle \phi \rangle, \ldots, \lambda_t \langle \phi \rangle$ be the cosets of $\langle \phi \rangle$ in Φ, where $t = [\Phi : \langle \phi \rangle]$ is the index of $\langle \phi \rangle$ in Φ.

We have seen from (3) that E_c^r is partitioned into disjoint pairs $\{A, B\}$, $\{F, G\}$, etc., and a coset $\lambda \langle \phi \rangle$ is associated with each of these pairs. With the coset $\lambda_i \langle \phi \rangle$, we associate the pair $\{\lambda_i A, \lambda_i B\}$ and this pair satisfies the hypothesis of the theorem, that is, $\{\lambda_i A, \lambda_i B\}$ is such a pair as $\{D, F\}$ of (3).

Suppose $\lambda_i \langle \phi \rangle \mapsto \{\lambda_i A, \lambda_i B\}$, $\lambda_j \langle \phi \rangle \mapsto \{\lambda_j A, \lambda_j B\}$, and $\{\lambda_i A, \lambda_i B\} = \{\lambda_j A, \lambda_j B\}$. If $\lambda_i A = \lambda_j A$, then $N^* r + \lambda_i a = N^* r + \lambda_j a$, so $\lambda_i a = \lambda_j a$ and $\lambda_i = \lambda_j$, making $\lambda_i \langle \phi \rangle = \lambda_j \langle \phi \rangle$. If, on the other hand, $\lambda_i A = \lambda_j B$, then $\lambda_i B = \lambda_j A$. So $A = \lambda_i^{-1} \lambda_j B$ and so $N^* r + a = N^* r + \lambda_i^{-1} \lambda_j \phi a$. With this, $\lambda_i^{-1} \lambda_j \phi a = a$, then $\lambda_i^{-1} \lambda_j \phi = 1$ and, consequently, $\lambda_i^{-1} \lambda_j = \phi^{-1} = \phi \in \langle \phi \rangle$. This also makes $\lambda_i \langle \phi \rangle = \lambda_j \langle \phi \rangle$, and so we have (4).

(6.11) Definition. For a pair $\{A, B\}$ satisfying the hypothesis and consequences of (6.10), we write

$$A \cdot B \cdot A \quad \text{or} \quad A^x B^{\phi x} \cdot A$$

and call it a *simple closed 2–link chain*.

A simple closed 2–link chain will be a special case of a simple closed s–link chain which we shall soon introduce.

(6.12) Remark. There is another point of view for any E_c^r satisfying (6.6) and (6.7). In euclidean 3–dimensional space, consider the circle with centre $(0,0,0)$ in the xy–plane with radius c. Now consider the circle with centre $(0, c, 0)$ in the yz–plane with radius r, where $0 < r < c$. Rotate this last circle about the z–axis, and one has a torus which motivates the $T(r; c)$. As we rotate this circle about the z–axis, we get the various circles of E_c^r, loosely speaking. What would we obtain if (i) $r = c$; or (ii) $r > c$? For $r > c$ we get each circle in E_c^r intersecting all the other circles of E_c^r in exactly the same two points. This is a strong contrast to an E_c^r satisfying (6.11), but we shall in fact see that it never happens. Later, we shall see that the case $r = c$ has meaning for various E_c^r. That is where each $A \in E_c^r$ intersects each and every other $B \in E_c^r$ at the same point, that is, if $A, B \in E_c^r$ and $A \neq B$ then $|A \cap B| = 1$, and $|\cap E_c^r| = 1$ as well.

(6.13) Examples. To be assured that simple closed 2–link chains exist, look at E_2^1 in $(Z_{13}, \mathcal{B}_4^*, \in)$ and E_2^6 in $(Z_{17}, \mathcal{B}_4^*, \in)$.

For an E_c^r, $c \neq 0$, we have considered two possibilities. First, there is a

$B \in E_c^r$ which intersects no other $D \in E_c^r$. Second, there is a $B \in E_c^r$ which intersects exactly one other $D \in E_c^r$. In each case, we have seen that the same property is valid for every circle in E_c^r. These are important special cases of the more general phenomenon described in the next theorem.

(6.14) Theorem. *Suppose there is an $A \in E_c^r$, $c \neq 0$, which is tangent to exactly m circles of $E_c^r \setminus \{A\}$, and intersects exactly n circles of $E_c^r \setminus \{A\}$ each at two distinct points. Then every $D \in E_c^r$ has this property. Moreover, if $A, D \in E_c^r$ and $r' \neq 0$, then $|A \cap N^* r'| = |D \cap N^* r'|$.*

Proof. For $B, C \in \mathcal{B}^*$ and $\phi \in \Phi$, it is easy to see that $|B \cap C| = |\phi B \cap \phi C| = |\phi^{-1} B \cap \phi^{-1} C|$. So, if A is tangent to exactly the circles $T_1, T_2, \ldots, T_m \in E_c^r \setminus \{A\}$ and $\phi A = D$, then D is tangent exactly to the circles $\phi T_1, \phi T_2, \ldots, \phi T_m \in E_c^r \setminus \{D\}$, and if A intersects only the $S_1, S_2, \ldots, S_n \in E_c^r \setminus \{A\}$ at two distinct points each, then D intersects only the $\phi S_1, \phi S_2, \ldots, \phi S_n \in E_c^r \setminus \{D\}$ at two distinct points each. Let $B = A$ and $C = N^* r'$, and suppose $\tau A = D$. Then $\tau C = C$, and we have $|A \cap N^* r'| = |B \cap C| = |\tau B \cap \tau C| = |\tau A \cap C| = |D \cap N^* r'|$. This completes the proof of (6.14).

If there is a $B \in E_c^r$, $c \neq 0$, which intersects *at most* two other circles, then exactly one of the following is true: (a) B intersects no other circles in $E_c^r \setminus \{B\}$; (b) B intersects exactly one other circle in $E_c^r \setminus \{B\}$; and (c) B intersects exactly two other circles in $E_c^r \setminus \{B\}$.

For (a), we have the torus of (6.6). For (b), we have exactly the case studied in (6.10). For (c), there are three subcases: (c1) B is tangent to two distinct circles A and D in $E_c^r \setminus \{B\}$; (c2) B intersects two distinct circles A and D of $E_c^r \setminus \{B\}$ at two points each; and (c3) B is tangent to an $A \in E_c^r \setminus \{B\}$ and intersects a $D \in E_c^r \setminus \{A, B\}$ at two distinct points. At this point, we shall investigate (c1) further.

Assume a $B \in E_c^r$, $c \neq 0$ is tangent to exactly two other circles in $E_c^r \setminus \{B\}$, but we really do *not* need to assume that B intersects no other circle of $E_c^r \setminus \{B\}$ at two points. By (6.14), each circle $C \in E_c^r$ is also tangent to exactly two other circles in $E_c^r \setminus \{D\}$. Take $B_1 \in E_c^r$ and suppose B_1 is tangent to B_0 and B_2 from $E_c^r \setminus \{B_1\}$. We denote this by $B_0 \cdot B_1 \cdot B_2$. Now, there is exactly one $B_3 \in E_c^r \setminus \{B_1, B_2\}$ so that B_3 is tangent to B_2, and we write $B_0 \cdot B_1 \cdot B_2 \cdot B_3$. Since our nearring N is finite, there is a least s so that if B_{s-2} and B' are the two circles in $E_c^r \setminus \{B_{s-1}\}$ tangent to B_{s-1}, then $B' \in \{B_0, B_1, \ldots, B_{s-3}\}$, for otherwise we get an infinite chain, $B_0 \cdot B_1 \cdot B_2 \cdot B_3 \cdot \ldots \cdot B_s \cdot \ldots$, where each B_i has B_{i-1} and B_{i+1} as the two circles of $E_c^r \setminus \{B_i\}$ tangent to it, $1 \leq i$. This means that eventually some B_{s-1} has, other than B_{s-2}, as its other tangent circle some B' among $B_0, B_1, \ldots, B_{s-3}$. If $B' = B_i$, $1 \leq i \leq s - 3$, then B_i would have B_{i-1}, B_{i+1}, and B_{s-1} tangent to it. Thus $B' = B_0$. In summary, we have

(6.15) Theorem. *If $B \in E_c^r$, $c \neq 0$, is tangent to exactly two other circles in $E_c^r \setminus \{B\}$, then there is a smallest integer $s \geq 3$ with circles B_0, B_1, $\cdots$, $B_{s-1} \in E_c^r$ with the property that B_{i-1} and B_{i+1} are the two circles in E_c^r tangent to B_i, for $1 \leq i \leq s-2$, B_{s-2} and B_0 are the two circles of E_c^r tangent to B_{s-1}, and B_{s-1} and B_1 are the two circles of E_c^r tangent to B_0.*

(6.16) Definition. For the circles B_0, B_1, ..., B_{s-1} as in (6.15), we write

$$B_0 \cdot B_1 \cdot \ldots \cdot B_{s-1} \cdot B_0$$

and call it a *simple closed s–link chain*. Compare this with (6.11).

(6.17) Remarks. By choosing any B_i from a simple closed s–link chain $B_0 \cdot B_1 \cdot \ldots \cdot B_{s-1} \cdot B_0$ to begin with, one would obtain the same set of circles $\{B_0, B_1, \ldots, B_{s-1}\}$. Likewise, starting with B_1, say, with its two tangent circles B_0 and B_2, one could interchange the role of B_0 and B_2, that is, we could have used B_2 for B_0 and B_0 for B_1. Thus, we identify the following simple closed s–link chains as one and the same:

$$B_{i-1} \cdot B_i \cdot B_{i+1} \cdot \ldots \cdot B_{i-2} \cdot B_{i-1};$$

$$B_2 \cdot B_1 \cdot B_0 \cdot B_{s-1} \cdot \ldots \cdot B_3 \cdot B_2;$$

$$B_{i+1} \cdot B_i \cdot B_{i-1} \cdot \ldots \cdot B_{i+2} \cdot B_{i+1}.$$

Note that the indices are taken from the cyclic group $Z_s = \{0, 1, \ldots, s-1\}$ and that $i+1$, $i-2$, $s-1$, etc., all take their meaning in this group.

(6.18) Theorem. *Suppose $B \in E_c^r$, $c \neq 0$, is tangent to exactly two other circles in $E_c^r \setminus \{B\}$. Let $B_0 \cdot B_1 \cdot \ldots \cdot B_{s-1} \cdot B_0$ be the resulting simple closed s–link chain, with $B_i \cap B_{i+1} = \{x_i\}$, $B_i = N^*r + b_i$, where $N^* \supseteq \{b_0, b_1, \ldots, b_{s-1}\}$. Then:*
(1) There is a cyclic subgroup $\langle \phi \rangle$ of Φ of order s such that $\phi x_i = x_{i+1}$, $\phi b_i = b_{i+1}$, and $\phi B_i = B_{i+1}$.
(2) There is a bijection between the cosets $\lambda \langle \phi \rangle$ of $\langle \phi \rangle$ in Φ and distinct simple closed s–link chains in E_c^r.
(3) There are exactly $[\Phi : \langle \phi \rangle]$ simple closed s–link chains in E_c^r.
(4) If $D \in E_c^r \setminus \{B_0, B_1, \ldots, B_{s-1}\}$, then D is not tangent to any B_i, $0 \leq i \leq s-1$, and D yields a simple closed s–link chain $D_0 \cdot D_1 \cdot D_2 \cdot \ldots \cdot D_{s-1} \cdot D_0$ with $D = D_0$ and each $D_i \in E_c^r$.

Proof. Choose $\phi \in \Phi \setminus \{1\}$ so that $\phi b_0 = b_1$. Then ϕB_1 is tangent to B_1 since $\phi(B_0 \cap B_1) = \phi(B_0) \cap \phi(B_1) = B_1 \cap \phi(B_1)$. If $\phi B_1 = B_0$, then $\phi^2 = 1$ and $\phi b_1 = b_0$. For $\{x_0\} = B_0 \cap B_1$, we have $\phi x_0 \in B_1 \cap B_0 = \{x_0\}$, so $\phi x_0 = x_0$ and $\phi = 1$. But then $b_0 = b_1$ and $B_0 = B_1$. This all means that $\phi B_1 = B_2$ and $\phi x_0 = x_1$. Assume that $\phi B_{i-1} = B_i$, $\phi b_{i-1} = b_i$, and $\phi x_{i-1} = x_i$. We want to show that $\phi B_i = B_{i+1}$, $\phi b_i = b_{i+1}$, and $\phi x_i = x_{i+1}$, for $i \in Z_s = \{0, 1, \ldots, s-1\}$. We have $\phi b_0 = b_1$, $\phi b_1 = b_2$, $\ldots$, $\phi b_{i-1} = b_i$. Now ϕB_i is tangent to B_i, so $\phi B_i \in \{B_{i-1}, B_{i+1}\}$. If $\phi B_i = B_{i-1}$, then $\phi b_i = b_{i-1} = \phi^2 b_{i-1}$, so $\phi^2 = 1$, contrary to what we proved at the beginning of this proof. So $\phi B_i = B_{i+1}$ and therefore $\phi b_i = b_{i+1}$ and $\phi x_i = x_{i+1}$. Thus, $\phi B_{s-2} = B_{s-1}$, $\phi b_{s-2} = b_{s-1}$, and $\phi x_{s-2} = x_{s-1}$, and $\phi B_{s-1} = B_0$, $\phi b_{s-1} = b_0$, and $\phi x_{s-1} = x_0$. From $x_0 = \phi x_{s-1} = \phi^2 x_{s-2} = \cdots = \phi^s x_0$, we conclude that $\phi^s = 1$, and so s is the smallest such positive integer, so $\langle \phi \rangle$ has order s. This gives us (1).

As with (6.11), we can, and do, write

$$B_0 \overset{x_0}{\underset{\cdot}{B_1}} \overset{x_1}{\underset{\cdot}{\cdots}} \overset{x_{s-2}}{\underset{\cdot}{B_{s-1}}} \overset{x_{s-1}}{\underset{\cdot}{B_0}}$$

for a simple closed s–link chain if we know that $B_i \cap B_{i+1} = \{x_i\}$. If

$$D_0 \overset{y_0}{\underset{\cdot}{D_1}} \overset{y_1}{\underset{\cdot}{\cdots}} \overset{y_{s-2}}{\underset{\cdot}{D_{s-1}}} \overset{y_{s-1}}{\underset{\cdot}{D_0}}$$

is another simple closed s–link chain in E_c^r, then there is a unique $\lambda \in \Phi$ such that $\lambda b_0 = d_0$ where $D_i = N^* r + d_i$. Thus $\lambda B_0 = D_0$. This means that $\lambda B_i = D_i$ for each $i \in Z_s = \{0, 1, \ldots, s-1\}$ and $\lambda x_i = y_i$. This corresponds to the coset $\lambda \langle \phi \rangle$. (We may possibly have to rewrite this simple closed s–link chain as

$$D_0 \overset{y_{s-1}}{\underset{\cdot}{D_{s-1}}} \overset{y_{s-2}}{\underset{\cdot}{\cdots}} \overset{y_2}{\underset{\cdot}{D_1}} \overset{y_1}{\underset{\cdot}{D_0}}$$

to make the above argument, since in reality, $\lambda B_1 \in \{D_1, D_{s-1}\}$, etc. But this is no loss, since this is regarded as the same chain. See (6.17).)

So, to each chain, there is a coset $\lambda \langle \phi \rangle$ of $\langle \phi \rangle$ in Φ, and obviously,

$$\lambda B_0 \overset{\lambda x_0}{\underset{\cdot}{\lambda B_1}} \overset{\lambda x_1}{\underset{\cdot}{\cdots}} \overset{\lambda x_{s-2}}{\underset{\cdot}{\lambda B_{s-1}}} \overset{\lambda x_{s-1}}{\underset{\cdot}{\lambda B_0}}$$

is such a chain corresponding to $\lambda \langle \phi \rangle$.

But suppose

$$\lambda B_0 \cdot \lambda B_1 \cdot \ldots \cdot \lambda B_{s-1} \cdot \lambda B_0$$

is the same as

$$\mu B_0 \cdot \mu B_1 \cdot \ldots \cdot \mu B_{s-1} \cdot \mu B_0.$$

Then $\lambda B_0 = \mu B_i$ for some i, so $\lambda b_0 = \mu b_i$, and $\mu^{-1} \lambda b_0 = b_i = \phi^i b_0$. This makes $\mu^{-1} \lambda = \phi^i$, so $\lambda \langle \phi \rangle = \mu \langle \phi \rangle$, and we have (2) and (3).

If $D \in E_c^r \setminus \{B_0, \ldots, B_{s-1}\}$ and D is tangent to B_i, then we must have $D \in \{B_{i-1}, B_{i+1}\}$. But $D = \lambda B_0$ for some $\lambda \in \Phi$, and

$$\lambda B_0 \overset{\lambda x_0}{\cdot} \lambda B_1 \overset{\lambda x_1}{\cdot} \cdots \overset{\lambda x_{s-1}}{\cdot} \lambda B_{s-1} \overset{\lambda x_{s-1}}{\cdot} \lambda B_0$$

is a simple closed s–link chain in E_c^r, and each $\lambda B_i \in E_c^r$.

(6.19) Corollary. *Suppose $B \in E_c^r$, $c \neq 0$, is tangent to exactly two other circles in $E_c^r \setminus \{B\}$. Also suppose that*

$$B_0 \cdot B_1 \cdot \ldots \cdot B_{s-1} \cdot B_0$$

is a simple closed s–link chain with the $B_i \in E_c^r$. Then if

$$D_0 \cdot D_1 \cdot \ldots \cdot D_{t-1} \cdot D_0$$

is a simple closed t–link chain with each $D_i \in E_c^r$, then $s = t$.

Proof. If $B_0 = N^*r + b_0$, $D_0 = N^*r + d_0$, and $\mu \in \Phi$ satisfies $\mu b_0 = d_0$, then $\{\mu B_{s-1}, \mu B_1\} = \{D_{t-1}, D_1\}$, $\{\mu B_{s-2}, \mu B_2\} = \{D_{t-2}, D_2\}$, etc.

(6.20) Example. In $(Z_{181}, \mathcal{B}_9^*, \in)$, E_4^1 has three simple closed 3–link chains.

(6.21) Theorem. *Suppose there is an $A \in E_c^r$, $c \neq 0$, which is tangent to exactly m circles in $E_c^r \setminus \{A\}$, all at the same point, say u. That is, there is a point $u \in A$ and there are m distinct circles $B_1, B_2, \ldots, B_m \in E_c^r \setminus \{A\}$ such that $B_i \cap A = \{u\}$ for each i, $1 \leq i \leq m$. Then each $D \in E_c^r$ has this property.*

Proof. Let $D = N^*r + d \in E_c^r$ and $A = N^*r + a$. Then if $B_i \cap A = \{u\}$ for each B_i, $1 \leq i \leq m$, and if $\phi \in \Phi$ and $\phi a = d$, then $\phi A = D$ and $\phi B_i \cap D = \{\phi u\}$ for each i, $1 \leq i \leq m$. So D has at least m circles $\phi B_1, \phi B_2, \ldots, \phi B_m \in E_c^r$ which are tangent to D, each at ϕu. If there should be another, say F, where $F \in E_c^r$ and $F \cap D = \{\phi u\}$, then $A \cap \phi^{-1}F = \{u\}$, and $\phi^{-1}F \in E_c^r$, but $\phi^{-1}F \notin \{B_1, B_2, \ldots, B_m\}$.

(6.22) Theorem. *If $|\cap_{B \in E_c^r} B| = 1$, then $N^*c = N^*(-r)$, and conversely. In this case $\cap_{B \in E_c^r} B = \{0\}$.*

Proof. Certainly, $0 \in N^*r - r$, so $0 \in N^*r - \phi r$ for each $\phi \in \Phi$, so if $c = -r$, then $\cap_{B \in E_c^r} B = \cap_{t \in N^* c}(N^*r + t) \supseteq \{0\}$.

If $\{c^0\} = \cap_{B \in E_c^r} B = \cap_{c' \in N^* c}(N^*r + c')$, then there exist $\phi_i, \mu_i \in \Phi$, $1 \leq i \leq k$, so that $c^0 = \phi_1 r + \mu_1 c = \phi_2 r + \mu_2 c = \cdots = \phi_k r + \mu_k c$, where $\Phi = \{\phi_1, \phi_2, \ldots, \phi_k\} = \{\mu_1, \mu_2, \ldots, \mu_k\}$ and is of order k. Therefore, $\lambda c^0 = \lambda \phi_i r + \lambda \mu_i c \in N^*r + c'$ for each $c' \in N^*c$, and so $\lambda c^0 \in \cap_{c' \in N^* c}(N^*r +$

$c') = \{c^0\}$, so $\lambda c^0 = c^0$ for each $\lambda \in \Phi$. This means $c^0 = 0$. Also, $c^0 = \phi_i r + \mu_i c = 0$ implies $\mu_i c = -\phi_i r = \phi_i(-r)$, and then $c = \mu_i^{-1}\phi_i(-r) \in N^*(-r)$. So we obtain $N^*c = N^*(-r)$.

But, if $0 \neq u \in \cap_{\phi \in \Phi}(N^*r - \phi r)$, then there are $\phi_i, \phi_j \in \Phi$ such that $u = \phi_1 r - \mu_1 r = \cdots = \phi_k r - \mu_k r$, where, as above, $\Phi = \{\phi_1, \ldots, \phi_k\} = \{\mu_1, \ldots, \mu_k\}$. Thus, for $\lambda \in \Phi$, $\lambda u = \lambda \phi_i r - \lambda \mu_i r \in \cap_{\phi \in \Phi}(N^*r - \phi r)$. Therefore, $\{0, \phi_1 u, \phi_2 u, \ldots, \phi_k u\} \subseteq \cap \phi \in \Phi(N^*r - \phi r)$. So this intersection contains $k + 1$ elements, but each $|N^*r - \phi r| = k$. So we are forced to conclude that there does not exist an element $u \neq 0$ such that $u \in \cap_{\phi \in \Phi}(N^*r - \phi r)$. Therefore, $|\cap_{B \in E_c^r} B| = 1$ if $N^*c = N^*(-r)$.

Immediately after the proof of (6.14), we described case (b) as when a circle B intersects exactly one other circle in $E_c^r \setminus \{B\}$. By (6.10), we know that B intersects this other circle in exactly two places. But what happens when a circle B is tangent to exactly one other circle in $E_c^r \setminus \{B\}$. By (6.10), we know that B must then intersect some circle in $E_c^r \setminus \{B\}$ in two places. But we can say more.

(6.23) Theorem. *If $B \in E_c^r$ is tangent to exactly one $A \in E_c^r \setminus \{B\}$, then $A \cap B = \{0\}$ and so $N^*c = N^*(-r)$.*

Proof. Let $A = N^*r + a$, $B = N^*r + b$, with $a, b \in N^*c$. For $\{x\} = A \cap B$, we have $\phi r + a = x = \mu r + b$ for some $\phi, \mu \in N^*$. So, there is a $\lambda \in N^*$ such that $\lambda a = b$ and $\lambda A = B$. Hence, $N^*r + \lambda a = N^*r + b$. This makes $\lambda x = \lambda \phi r + \lambda a = \lambda \mu r + \lambda b$. But then $\{\lambda x\} = B \cap \lambda B$. Since B is tangent only to $A \in E_c^r \setminus \{B\}$, and since $\lambda B \in E_c^r \setminus \{B\}$, we must have $\lambda B = A = N^*r + \lambda b = N^*r + a$. From $\lambda b = a = \lambda^2 a$, we conclude that $\lambda^2 = 1$. But $x, \lambda x \in A \cap B = \{x\}$, so $\lambda x = x$. This means $\lambda = 1$ or $x = 0$. Since $\lambda a = b$, we have $\lambda \neq 1$, since $a \neq b$. Thus $x = 0$ and so $A \cap B = \{0\}$. From $0 = \phi r + a$, we have $a = -(\phi r) = \phi(-r)$, giving us $N^*c = N^*a = N^*(-r)$.

Soon after the proof of (6.14), we described three subcases (c1), (c2), and (c3). We have discussed (c1) in detail. We will now see that (c3) cannot happen.

(6.24) Theorem. *Suppose $B \in E_c^r$ is tangent to $A \in E_c^r \setminus \{B\}$ and intersects $D \in E_c^r \setminus \{A, B\}$ twice, but yet intersects no other circle of $E_c^r \setminus \{A, B, D\}$. Then an impossibility has happened!*

Proof. With these hypotheses, we apply (6.23) to conclude $A \cap B = \{0\}$. Suppose $B \cap D = \{u, v\}$ with $D \in E_c^r \setminus \{A, B\}$. But D is tangent to exactly one $F \in E_c^r$ by (6.14). Again, by (6.23), we have $D \cap F = \{0\}$. Hence, $0, u, v \in B \cap D$, so $B = D$, which is a contradiction, or else $0 \in \{u, v\}$. So take $A \cap B = \{0\}$ and $B \cap D = \{0, u\}$. Hence, $0 \in A \cap D$. But B is tangent to A, so D cannot be tangent to A. Consequently, $|A \cap D| = 2$ and so $A \cap D = \{0, w\}$ for some $w \neq 0$. From $|A \cap D| = 2 = |B \cap D|$, we have two distinct circles intersecting D twice. Since D has this property, so does B, by (6.14).

When we have a simple closed s-link chain in an E_c^r, $s \geq 3$, then each circle $B \in E_c^r$ is tangent to exactly two other circles $A, D \in E_c^r \setminus \{B\}$. These points of tangency, all together, form a circle, just as they would in the ordinary euclidean plane. This is made more precise in the following theorem.

(6.25) Theorem. *Suppose $c \neq 0$ and that there is a $B \in E_c^r$ which is tangent to exactly two other circles in $E_c^r \setminus \{B\}$, say to A and D. Let $U = \{x \mid S \cap T = \{x\} \text{ for } S, T \in E_c^r, S \neq T\}$. Then $U = N^*x = C(x; 0)$, the circle of radius x and centre 0.*

Proof. If $A \cap B = \{x\}$ and $B \cap D = \{y\}$, we have seen that there is a unique $\phi \in \Phi$ such that $\phi x = y$. If $S, T \in E_c^r$, $S \neq T$, and $S \cap T = \{u\}$, then we have seen that $u = \lambda x$ for some unique $\lambda \in \Phi$. Hence, $U = N^*x$, and U is a circle with centre 0 and radius any $x \in U$.

Certainly, we can identify the elements $\phi \in \Phi$ with automorphisms $\phi^\bullet$ of $(N, +)$, where $\phi^\bullet(x) = \phi x$. So Φ is a subgroup of $A = \operatorname{Aut}(N, +)$, the group of automorphisms of the group $(N, +)$. Let $N_A(\Phi)$ be the normalizer of Φ in A, and let $\alpha \in N_A(\Phi)$. Since $\alpha(N^*r + c') = N^*\alpha r + \alpha c'$, we have that α is an automorphism of the BIBD $(N, \mathcal{B}^*, \in)$, hence

(6.26) Theorem. *If $\alpha \in N_A(\Phi)$, then E_c^r and $E_{\alpha c}^{\alpha r}$ have the same structure, since α is an automorphism of $(N, \mathcal{B}^*, \in)$.*

The following concept, suggested by Yeong-Nan Yeh, has proved useful, as we shall see.

(6.27) Definition. For an N^*r, $r \neq 0$, we define the *difference set* of N^*r as

$$\Delta(N^*r) = \{-\mu r + \phi r \mid \mu, \phi \in \Phi, \mu \neq \phi\}.$$

We can learn a lot about the circles of radius r by computing $\Delta(N^*r)$ in detail. In particular, we can learn about how an E_c^r relates to N^*r.

Suppose $(N, \mathcal{B}^*, \in)$ is circular, and that $N^*r \cap (N^*r + c) \neq \emptyset$. Then either $N^*r \cap (N^*r + c) = \{x\}$, or $N^*r \cap (N^*r + c) = \{x, y\}$, with $x \neq y$. If $N^*r \cap (N^*r + c) = \{x\}$, then $N^*r + c$ is tangent to N^*r, as is each $N^*r + c'$, $c' \in N^*c$. In fact, if $\phi c = c'$, then $N^*r \cap (N^*r + c') = \{\phi x\}$. Similarly, if $\phi c = c'$ and $N^*r \cap (N^*r + c) = \{x, y\}$, then $N^*r \cap (N^*r + c') = \{\phi x, \phi y\}$. So the circles of E_c^r intersect with N^*r exactly as $N^*r + c$ does, which is either not at all, or at exactly one point, or at exactly two points.

If $x \in N^*r \cap (N^*r + c)$, then $x = \phi r = \mu r + c$ for some $\phi, \mu \in \Phi$, $\phi \neq \mu$. Hence, $-\mu r + \phi r = c$, so $c \in \Delta(N^*r)$. Conversely, if $c = -\mu r + \phi r \in \Delta(N^*r)$, then $\mu r + c = \phi r$, so $x = \phi r = \mu r + c \in N^*r \cap (N^*r + c)$. Note also, that if $\lambda \in \Phi$, then $\lambda c = \lambda(-\mu r + \phi r) = -(\lambda \mu)r + (\lambda \phi)r \in \Delta(N^*r)$, so $\Delta(N^*r)$ is a union of some of the N^*c, $c \neq 0$. Since $|\Delta(N^*r)| \leq k(k-1)$, and $|N^*c| = k$, there are at most $k - 1$ of the N^*c, $c \neq 0$, whose union makes up $\Delta(N^*r)$.

But what if $N^*r \cap (N^*r + c) = \{x, y\}$, and $x \neq y$? Then not only does $x = \phi r = \mu r + c$, but $y = \phi' r = \mu' r + c$ for some $\phi', \mu' \in \Phi$, $\phi' \neq \mu'$, and $(\phi, \mu) \neq (\phi', \mu')$. But then $c = -\mu' r + \phi' r$. As above, if $\lambda \in \Phi$, then $\lambda c = -(\lambda \mu')r + (\lambda \phi')r$ also. So the points of N^*c will occur twice as one computes all the $-\mu r + \phi r$ possible in computing $\Delta(N^*r)$.

This tells us that we should let $\Phi = \{\phi_1, \phi_2, \ldots, \phi_k\}$, and in computing $\Delta(N^*r)$, we should compute the array

$$\mathcal{A}(r) = \begin{pmatrix} -\phi_1 r + \phi_2 r & -\phi_1 r + \phi_3 r & \cdots & -\phi_1 r + \phi_k r \\ -\phi_2 r + \phi_1 r & -\phi_2 r + \phi_3 r & \cdots & -\phi_2 r + \phi_k r \\ \cdots & \cdots & \cdots & \cdots \\ -\phi_k r + \phi_1 r & -\phi_k r + \phi_2 r & \cdots & -\phi_k r + \phi_{k-1} r \end{pmatrix}.$$

If the points of N^*c do not occur in this array, then $|N^*r \cap (N^*r + c')| = 0$ for each $c' \in N^*c$. If the points of N^*c occur in this array exactly once, then $|N^*r \cap (N^*r + c')| = 1$ for each $c' \in N^*c$. And if the points of N^*c occur in this array exactly twice, then $|N^*r \cap (N^*r + c')| = 2$ for each $c' \in N^*c$. The converse of each of these three statements is also true. In summary we have

(6.28) Theorem. *Let $(N, +, \cdot)$ be a finite circular planar nearring. The circles of radius $r \neq 0$, other than N^*r, are partitioned into the three classes*

$$S = \{N^*r + c \mid |N^*r \cap (N^*r + c)| = 0\};$$

$$T = \{N^*r + c \mid |N^*r \cap (N^*r + c)| = 1\};$$

$$U = \{N^*r + c \mid |N^*r \cap (N^*r + c)| = 2\}.$$

Let $s = |S|/k$, $t = |T|/k$, and $u = |U|/k$. Then $t + 2u = k - 1$ and $s + t + u = (|N| - 1)/k$, where $k = |N^*r|$. In fact, tk is exactly the number of N^*c whose points occur exactly once in the array $\mathcal{A}(r)$, and uk is exactly the number of N^*c whose points occur exactly twice in the array $\mathcal{A}(r)$. So sk is exactly the number of N^*c whose points do not occur at all in the array $\mathcal{A}(r)$.

(6.29) Corollary.

$$\Delta(N^*r) = \bigcup_{N^*r + c \in T \cup U} (N^*r + c).$$

(6.30) Corollary. *There are at most* $k - 1$ *of the* E_c^r *whose circles are tangent to* N^*r.

While studying the E_c^r in reference to $C(r; 0)$, we have seen that it is significant to know if a $B \in E_c^r$ intersects exactly one other $D \in E_c^r$, and it is significant to know if a $B \in E_c^r$ is tangent to exactly two other circles A and D in E_c^r. Now (6.14), (6.21), and the proof of (6.28) show that what happens for one $B \in E_c^r$ happens for each $D \in E_c^r$. The array $\mathcal{A}(r)$ has proved useful to tell us how a given E_c^r is related to $C(r; 0)$. We shall now see that $\mathcal{A}(r)$ and $\Delta(N^*r)$ are also useful for telling us if a $B \in E_c^r$ intersects exactly one other $D \in E_c^r$, and if a $B \in E_c^r$ is tangent to exactly two other circles $A, D \in E_c^r$.

(6.31) Theorem. *Consider* $\Delta(N^*r) \cap (N^*c - c)$. *Any* $x \in \Delta(N^*r) \cap (N^*c - c)$ *corresponds to a circle* $N^*r + c' \in E_c^r$ *which intersects* $N^*r + c$. *If* x *occurs exactly once in* $\mathcal{A}(r)$, *then the circle* $N^*r + c'$ *is tangent to* $N^*r + c$. *If* x *occurs exactly twice in* $\mathcal{A}(r)$, *then the circle* $N^*r + c'$ *intersects* $N^*r + c$ *twice. In particular, if* $\Delta(N^*r) \cap (N^*c - c) = \{x\}$, *and* x *occurs exactly twice in* $\mathcal{A}(r)$, *then* $N^*r + c$ *and* $N^*r + c'$ *form a simple closed 2-link chain. If* $\Delta(N^*r) \cap (N^*c - c) = \{x, y\}$, *and* x *and* y *each occurs exactly once in* $\mathcal{A}(r)$, *then* $N^*r + c$ *is a circle in a simple closed s-link chain, where* s *is the order of* $\phi \in \Phi$ *and* $\phi c = c'$ *and* x *corresponds to* $N^*r + c'$.

Proof. Suppose $u \in (N^*r + c) \cap (N^*r + c')$, where $c' \in N^*c$. Then $u = \phi r + c = \mu r + c'$ for some $\phi, \mu \in \Phi$. Hence, $0 \neq x = -\mu r + \phi r = c' - c \in \Delta(N^*r) \cap (N^*c - c)$. If $v \in (N^*r + c) \cap (N^*r + c')$ also, and $u \neq v$, then $v = \phi'r + c = \mu'r + c'$ for some other $\phi', \mu' \in \Phi$. Hence, $-\mu'r + \phi'r = c' - c = x$, also, and so $-\mu r + \phi r = -\mu'r + \phi'r = x$ occurs twice in the array $\mathcal{A}(r)$. So a circle $N^*r + c' \in E_c^r$ which intersects $N^*r + c$ twice produces an element of $\mathcal{A}(r)$ which occurs twice in $\mathcal{A}(r)$. Conversely, if $x \in \Delta(N^*r) \cap (N^*c - c)$, then there are $\mu, \phi \in \Phi$ and $c' \in N^*c$ such that $0 \neq x = -\mu r + \phi r = c' - c$. This makes $u = \phi r + c = \mu r + c' \in (N^*r + c) \cap (N^*r + c')$, where $N^*r + c \neq N^*r + c'$. If there were different $\mu', \phi' \in \Phi$ such that $x = -\mu'r + \phi'r$, then $v = \phi'r + c = \mu'r + c' \in (N^*r + c) \cap (N^*r + c')$ also, with $u \neq v$. So we have the correspondence between points in $\Delta(N^*r) \cap (N^*c - c)$ and circles $N^*r + c' \in E_c^r$ which intersect $N^*r + c$, and the number of times the point of $\Delta(N^*r) \cap (N^*c - c)$ occurs in $\mathcal{A}(r)$ tells us the number of times that $N^*r + c'$ intersects $N^*r + c$. To complete the proof, use (6.10), (6.15), and (6.18). ∎

(6.32) Example. The following example not only illustrates the power of (6.28), (6.30), and (6.31) in a practical way, but illustrates the richness and variety of what can happen in a circular planar nearring. This demonstrates a considerable contrast to the $(Z_q, \mathcal{B}_p^*, \in)$, where $q = 2^p - 1$ is a

Mersenne prime. The above-mentioned theorems will be used later in a very theoretical way.

The example we consider here is $(N, \mathcal{B}^*, \in) = (Z_{73}, \mathcal{B}_8^*, \in)$. For this circular planar nearring, there are $72/8 = 9$ circles with centre 0, namely,

$$N^*1 = \{10, 27, 51, 72, 63, 46, 22, 1\}, \quad N^*2 = \{20, 54, 29, 71, 53, 19, 44, 2\},$$
$$N^*3 = \{30, 8, \ 7, \ 70, 43, 65, 66, 3\}, \quad N^*4 = \{40, 35, 58, 69, 33, 38, 15, 4\},$$
$$N^*5 = \{50, 62, 36, 68, 23, 11, 37, 5\}, \quad N^*6 = \{60, 16, 14, 67, 13, 57, 59, 6\},$$
$$N^*9 = \{17, 24, 21, 64, 56, 49, 52, 9\}, \quad N^*12 = \{47, 32, 28, 61, 26, 41, 45, 12\},$$
$$N^*18 = \{34, 48, 42, 55, 39, 25, 31, 18\}.$$

The key to applying the above-mentioned theorems is to compute $\mathcal{A}(1)$. Doing so, one obtains

$$
\mathcal{A}(1) = \begin{pmatrix}
56 & 32 & 11 & 20 & 37 & 61 & 9 \\
17 & 49 & 28 & 37 & 54 & 5 & 26 \\
41 & 24 & 52 & 61 & 5 & 29 & 50 \\
62 & 45 & 21 & 9 & 26 & 50 & 71 \\
53 & 36 & 12 & 64 & 17 & 41 & 62 \\
36 & 19 & 68 & 47 & 56 & 24 & 45 \\
12 & 68 & 44 & 23 & 32 & 49 & 21 \\
64 & 47 & 23 & 2 & 11 & 28 & 52
\end{pmatrix}.
$$

Applying (6.28), one obtains $S = E_1^1 \cup E_3^1 \cup E_4^1 \cup E_6^1 \cup E_{18}^1$, $T = E_2^1$, and $U = E_5^1 \cup E_9^1 \cup E_{12}^1$. Note that T consists of only E_2^1, as it would in the usual euclidean plane. This will be in strong contrast to the T of $(Z_q, \mathcal{B}_p^*, \in)$ later. There, T will consist of the maximum $k - 1$ of the E_c^r guaranteed by (6.30).

We next want to apply (6.31). We have $\Delta(N^*1) \cap (N^*c - c) = A_1 \cup A_2$ where $x \in A_i$ means that x occurs exactly i times in $\mathcal{A}(1)$. The following list identifies the c and the elements of A_1 and the elements of A_2.

$c = 1$:	$A_1 = \{71\}$;	$A_2 = \{9, 26, 50, 62, 45, 21\}$.
$c = 2$:	$A_1 = \varnothing$;	$A_2 = \{52, 17\}$.
$c = 3$:	$A_1 = \varnothing$;	$A_2 = \{5, 62\}$.
$c = 4$:	$A_1 = \{54, 29\}$;	$A_2 = \{36, 11\}$.
$c = 5$:	$A_1 = \varnothing$;	$A_2 = \{45, 32\}$.
$c = 6$:	$A_1 = \{54, 53\}$;	$A_2 = \{61\}$.
$c = 9$:	$A_1 = \varnothing$;	$A_2 = \{12, 47\}$.
$c = 12$:	$A_1 = \{20, 29\}$;	$A_2 = \{49\}$.
$c = 18$:	$A_1 = \varnothing$;	$A_2 = \{24, 37, 21\}$.

Each of the E_c^1 yields a signed graph. The *vertices* of E_c^1 are the centres of the circles in E_c^1. An *edge* exists between two vertices x and y, $x \neq y$, if $(N^*1 + x) \cap (N^*1 + y) \neq \varnothing$. The edge is *even* if $|(N^*1 + x) \cap (N^*1 + y)| = 2$, and the edge is *odd* if $|(N^*1 + x) \cap (N^*1 + y)| = 1$. In a diagram of such a graph, we indicate even edges by a solid line segment, for example, $\overset{15}{\bullet}\rule[0.5ex]{2em}{0.4pt}\overset{4}{\bullet}$, and we indicate an odd edge by a broken line segment, for example, $\overset{60}{\bullet}\text{-----}\overset{6}{\bullet}$.

We now illustrate and discuss the graphs of the various E_c^1. The graphs for E_2^1 and E_3^1 are isomorphic.

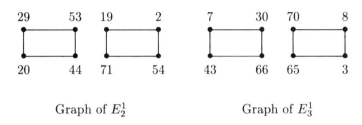

Graph of E_2^1 Graph of E_3^1

The graphs of E_5^1 and E_9^1 are also isomorphic.

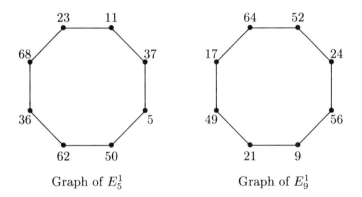

Graph of E_5^1 Graph of E_9^1

The graph of E_{12}^1 has some properties analogous to those of E_2^1 and E_3^1, but notice the simple 4-link chains as well as the simple 2-link chains.

Graph of E_{12}^1

The graphs of E_6^1, E_{18}^1, E_4^1, and E_1^1 also share some analogous properties, but notice the simple closed 8-link chain and the simple closed 2-link chains in E_6^1.

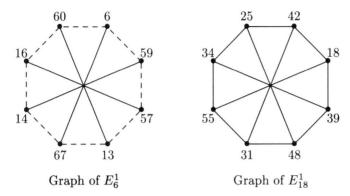

Graph of E_6^1 Graph of E_{18}^1

The graph of E_{18}^1 illustrates the unusual property that each node is connected to exactly three other nodes, and in this case, the edges are all even.

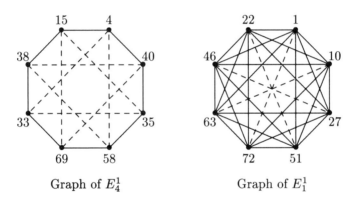

Graph of E_4^1 Graph of E_1^1

Even though it is not visually obvious from this displayed graph of E_4^1, it is also easy to find a simple closed 8-link chain in E_4^1. The graph of E_1^1 seems to be the most complex, but it is in reality the most natural. E_1^1 represents the circles of radius 1 whose centres are on the circle of radius 1 and centre 0. In the usual plane, each intersects all of the others at 0, and, except for the one directly opposite, it intersects the others at an additional point. This is what the graph of E_1^1 illustrates for this example also.

6.2. The $(Z_q, \mathcal{B}_p^*, \in)$

Since 1644, considerable interest has been directed toward the Mersenne numbers $M_p = 2^p - 1$, where p is a prime. In particular, if M_p is itself a prime, then $2^{p-1} M_p$ is a perfect number, and all even perfect numbers are of this form [B & P]. Perfect numbers were of particular interest to the ancient Greeks. So every time a new Mersenne prime is discovered, we also get a new perfect number. Not only do we not know if there are infinitely many Mersenne primes or not, but we only know of about 30 Mersenne primes [B *et al.*].

While all the above about Mersenne primes is very well known, it is also very curious—and not well known at all—that the Mersenne primes may be as intimately connected with some highly unusual geometric properties as they are intimately connected with the above rare properties of classical number theory. The intricate elementary number theory work here was provided by Yeong-Nan Yeh.

Throughout this section, $q = M_p$ will denote a Mersenne prime, and to have nontrivial meaning, p should be 5 or larger throughout. For $(Z_q, \mathcal{B}_p^*, \in)$ to make sense, we need this easy

(6.33) Lemma. *If $q = M_p = 2^p - 1$ is a Mersenne prime, then the multiplicative subgroup of the prime field $(Z_q, +, \cdot)$ has a subgroup $\Phi = \{1, 2, 2^2, \ldots, 2^{p-1}\}$ of order p.*

(6.34) Corollary. *If $q = M_p$ is a Mersenne prime, then p divides $2^p - 2$.*

Our next goal is to prove that each $(Z_q, \mathcal{B}_p^*, \in)$ is circular. The proof will be based on (5.19). Our proof that $|A_{0,1}| = (p-1)(p-2)$ is rather technical and involves the consideration of numerous cases. So to aid in the proof that $|A_{0,1}| = (p-1)(p-2)$, we will develop several elementary number theoretic lemmas.

Let $a, b, c, x, y, z, k \in Z$, the integers, and suppose $k > a \geq b \geq c \geq 1$ and $k > x \geq y \geq z \geq 1$ with $a \geq x$. These assumptions will be valid through (6.39).

(6.35) Lemma. $2^a + 2^b = 2^x + 2^y$ *if and only if $a = x$ and $b = y$.*

Proof. If $a > x$, then $2^a + 2^b > 2^a = 2^{a-1} + 2^{a-1} \geq 2^x + 2^y$. So $a = x$, and $b = y$.

(6.36) Lemma. $2^a = 2^x + 2^y$ *if and only if $a = x + 1 = y + 1$.*

Proof. $2^a = 2^x + 2^y > 2^x$, so $a > x$. Then $2^a = 2^{a-1} + 2^{a-1} \geq 2^x + 2^y = 2^a$ implies $a = x + 1 = y + 1$ by (6.35).

(6.37) Lemma. $2^a + 2^b + 2^c = 2^x + 2^y + 2^z$ *if and only if either* (i) $(a, b, c) = (x, y, z)$, *or* (ii) $a = x + 1 = y + 1$ *and* $b + 1 = c + 1 = z$.

Proof. If $a = x$, then by (6.35), $b = y$ and $c = z$. If $a = x + 1$, then either $x > y$ or $x = y$. If $x > y$, then $2^a + 2^b + 2^c > 2^a \geq 2^{a-1} + 2^{a-2} + 2^{a-2} \geq 2^x + 2^y + 2^z$. So now take $x = y$. Then $a = x + 1 = y + 1$. From $2^a + 2^b + 2^c = 2^x + 2^y + 2^z$, we have $2^b + 2^c = 2^z$, so $b + 1 = c + 1 = z$ by

(6.36).

Finally, if $a > x + 1$, then $a - 2 \geq x$, and $2^a + 2^b + 2^c > 2^a = 2^{a-2} + 2^{a-2} + 2^{a-2} + 2^{a-2} > 2^x + 2^y + 2^z$.

Remark. For (6.37), we have assumed $c \geq 1$ and $z \geq 1$. It is easy to see that (6.37) will remain valid if we have $c \geq 0$ and $z \geq 0$ instead.

(6.38) Lemma. *If $k \geq 2$, and $x, y, z \in \{0, 1, \ldots, k-1\}$, then $2(2^k - 1) \geq 2^x + 2^y + 2^z$.*

Proof. $2(2^k - 1) = 2^{k-1} + 2^{k-1} + 2^{k-1} + 2^{k-1} - 2 \geq 2^x + 2^y + 2^z$.

Lemma (6.40) will be awkward to prove. We shall frequently use the above four lemmas, and, in particular, (6.37). To facilitate using (6.37), note that we have assumed $a \geq b \geq c$ and $x \geq y \geq z$ and $a \geq x$. Suppose we have $(a, b, c) \neq (x, y, z)$, $a \geq b \geq c$, and $x \geq y \geq z$, only. Then, by (6.37), (1) if $a > x$, then $a = x + 1 = y + 1$ and $z = b + 1 = c + 1$; and (2) if $a < x$, then $x = a + 1 = b + 1$ and $c = y + 1 = z + 1$. It will be convenient to formalize this.

(6.39) Corollary. *Suppose $(a, b, c) \neq (x, y, z)$, $a \geq b \geq c$, $x \geq y \geq z$, and $2^a + 2^b + 2^c = 2^x + 2^y + 2^z$. Then:*
(1) If $a > x$, then $a = x + 1 = y + 1$, $z = b + 1 = c + 1$, $a > b = c$, and $x = y \geq z$.
(2) If $a < x$, then $x = a + 1 = b + 1$, $c = y + 1 = z + 1$, $a = b \geq c$, and $x > y = z$.

We must now attack

(6.40) Lemma. *Consider positive integers a, b, c, d, k where $a + b \geq c + d$, $k > a \geq b \geq 1$, and $k > c \geq d \geq 1$. Then $(2^a - 1)(2^b - 1) \equiv (2^c - 1)(2^d - 1)$ (mod $2^k - 1$) if and only if $(a, b) = (c, d)$.*

Proof. As with some of the above lemmas, one direction of the proof is trivial. Now

$$(2^a - 1)(2^b - 1) \equiv (2^c - 1)(2^d - 1) \pmod{2^k - 1}$$
$$\iff 2^{a+b} - 2^a - 2^b + 1 \equiv 2^{c+d} - 2^c - 2^d + 1 \pmod{2^k - 1}$$
$$\iff 2^{a+b} + 2^c + 2^d \equiv 2^{c+d} + 2^a + 2^b \pmod{2^k - 1}.$$

To prove that this last statement is equivalent to $(a, b) = (c, d)$, we consider three cases:
Case (1) $k > a + b \geq c + d$.
Case (2) $a + b \geq k > c + d$.
Case (3) $a + b \geq c + d \geq k$.
Each of these three cases will have numerous subcases.

Beginning with case (1), by (6.38), $2(2^k - 1) \geq 2^{a+b} + 2^c + 2^d > 0$ and $2(2^k - 1) \geq 2^{c+d} + 2^a + 2^b > 0$. If $2^{a+b} + 2^c + 2^d = 2^{c+d} + 2^a + 2^b$, then we want to apply (6.37) and (6.39). We have $a + b \geq c \geq d$, so the left hand side is in the correct format. To apply (6.37), we need to know the ordering of $c + d$, a, and b, if we do not have $\{a + b, c, d\} = \{c + d, a, b\}$. Then, either $a = c + d \geq b$, or $a = b > c + d$. If $a = c + d \geq b$, then $a + b = c + d + 1 = a + 1$ and $c + 1 = d + 1 = b$, so $b = 1$ and $c = d = 0$, a contradiction. If $a = b > c + d$, then $a + b = a + 1 = b + 1$ and $c + 1 = d + 1 = c + d$. But then $a = b = c = d = 1$, contrary to $b > c + d$. Thus we have $\{a + b, c, d\} = \{c + d, a, b\}$, $a + b = c + d$, and therefore $(a, b) = (c, d)$.

Next, suppose $2^{a+b} + 2^c + 2^d = 2^{c+d} + 2^a + 2^b + 2^k - 1$. But $2^k + 2^{c+d} + 2^a + 2^b - 1 > 2^k + 2^{c+d-1} + 2^{c+d-1} \geq 2^{a+b} + 2^c + 2^d$, so this cannot be. Similarly, if $2^k - 1 + 2^{a+b} + 2^c + 2^d = 2^{c+d} + 2^a + 2^b$, we have $2^{a+b} + 2^k + 2^c + 2^d - 1 > 2^{c+d} + 2^a + 2^b$, which also cannot be. So all possibilities for case (1) lead us to conclude $(a, b) = (c, d)$.

Turning now to case (2), if $a + b \geq k > c + d$, then $a \geq b > a + b - k$. Also, $2^{a+b} + 2^c + 2^d \equiv 2^{c+d} + 2^a + 2^b \pmod{2^k - 1} \iff 2^{a+b-k} + 2^c + 2^d \equiv 2^{c+d} + 2^a + 2^b \pmod{2^k - 1}$. By (6.38), $2(2^k - 1) \geq 2^{a+b-k} + 2^c + 2^d > 0$ and $2(2^k - 1) \geq 2^{c+d} + 2^a + 2^b > 0$. This means that exactly one of the following three is valid:
(i) $2^{a+b-k} + 2^c + 2^d = 2^{c+d} + 2^a + 2^b$.
(ii) $2^{a+b-k} + 2^c + 2^d = 2^{c+d} + 2^a + 2^b + 2^k - 1$.
(iii) $2^k - 1 + 2^{a+b-k} + 2^c + 2^d = 2^{c+d} + 2^a + 2^b$.

We proceed now to show that each of these three leads only to a contradiction.

Should (i) be true, we would have $c \geq d$ and $a \geq b > a + b - k$. Assume first that $c + d \geq a$. Then $c + d > c \geq d$, and $c + d \geq a > a + b - k$. So, either $c + d = a + b - k + 1 = c + 1$ and $a + 1 = b + 1 = d$, or $c + d = c + 1 = d + 1$ and $a + b - k = a + 1 = b + 1$. The former means that $d = 1$ and $a = b = 0$, which cannot be. The latter means that $a = b = k + 1$, which cannot be.

The alternative, if (i) is to be true, is that $a > c + d$. Then either $a + b - k \geq c$ or $c > a + b - k$. The former means, since $a > a + b - k$, that $b + 1 = c + d + 1 = d$, and so $c = -1$. The latter leads to $c = -1$ or $a = k + 1$, since $a > c + d > c \geq d$. So (i) cannot be true.

Also, (ii) cannot be. Since $2^{c+d} \geq 2^c + 2^d$ and $2^a > 2^{a+b-k}$, we have $2^{c+d} + 2^a > 2^{a+b-k} + 2^c + 2^d$, and therefore $2^{c+d} + 2^a + 2^b + 2^k - 1 > 2^{a+b-k} + 2^c + 2^d$.

To finish case (2), we now assume (iii) to be true. Since $2|(2^{c+d} + 2^a + 2^b)$ and $2|(2^k + 2^c + 2^d)$, we have $2|(2^{a+b-k} - 1)$, which means that $a + b = k$, and, consequently, $2^k + 2^c + 2^d = 2^{c+d} + 2^a + 2^b$. If either $c + d \geq a \geq b$ or $a \geq c + d \geq b$, then $b = 1$ and $c = d = 0$, which cannot be. If $a \geq b \geq c + d$, then $a + b = k = a + 1 = b + 1$ and $c + 1 = d + 1 = c + d$. Hence, $a = b = c = d = 1$, which also cannot be.

Since none of (i), (ii), or (iii) can be true, we must conclude that case (2) cannot happen, if we insist upon $2^{a+b} + 2^c + 2^d \equiv 2^{c+d} + 2^a + 2^b$ (mod $2^k - 1$) and $(a, b) \neq (c, d)$.

We finish the proof of (6.40) with case (3), where $a + b \geq c + d \geq k$. Therefore, $a \geq b > a + b - k \geq c + d - k$ and $c \geq d > c + d - k$. Also, $2^{a+b} + 2^c + 2^d \equiv 2^{c+d} + 2^a + 2^b$ (mod $2^k - 1$) $\iff 2^{a+b-k} + 2^c + 2^d \equiv 2^{c+d-k} + 2^a + 2^b$ (mod $2^k - 1$). By (6.38), $2(2^k - 1) \geq 2^{a+b-k} + 2^c + 2^d > 0$ and $2(2^k - 1) \geq 2^{c+d-k} + 2^a + 2^b > 0$. So exactly one of the following three is valid:

(i) $2^{a+b-k} + 2^c + 2^d = 2^{c+d-k} + 2^a + 2^b$.
(ii) $2^{a+b-k} + 2^c + 2^d = 2^{c+d-k} + 2^a + 2^b + 2^k - 1$.
(iii) $2^k - 1 + 2^{a+b-k} + 2^c + 2^d = 2^{c+d-k} + 2^a + 2^b$.

As with (ii) of case (2), we will see that (ii) and (iii) cannot be valid. Since $2^k \geq 2^c + 2^d$, and $2^a > 2^{a+b-k}$, and $2^b - 1 > 0$, we have $2^a + 2^b + 2^k - 1 > 2^{a+b-k} + 2^c + 2^d$, so $2^{c+d-k} + 2^a + 2^b + 2^k - 1 > 2^{a+b-k} + 2^c + 2^d$, so (ii) cannot be valid. Since $2^k \geq 2^a + 2^b$, $2^c > 2^{c+d-k}$, and $2^d - 1 > 0$, we have $2^k + 2^c + 2^d - 1 > 2^{c+d-k} + 2^a + 2^b$, and therefore $2^k - 1 + 2^{a+b-k} + 2^c + 2^d > 2^{c+d-k} + 2^a + 2^b$, so (iii) cannot be valid.

Finally, if (i) is valid, then

$$2^{a+b-k} + 2^c + 2^d = 2^a + 2^b + 2^{c+d-k}.$$

If $a+b-k \geq c$, then since $a > a+b-k$, we have $a = a+b-k+1 = c+1$ and $d = b+1 = c+d-k+1$. Therefore, $b = k-1$, $a+b = k+c$, and $c = k-1$. Since $k > a \geq b$, then $a = k-1$, also, so $(k-1)+(k-1) = k+(k-1)$, which we cannot accept. If $c \geq a+b-k \geq d$ and $(a, b) \neq (c, d)$, then either $d = k$ or $c = k+1$, which we cannot allow. If $c \geq d \geq a+b-k$ and $(a, b) \neq (c, d)$, then either $b = k+1$ or $a+b < c+d$, which is contrary to hypothesis. So, if (i) is valid, we must accept $(a, b) = (c, d)$. This completes the proof of (6.40).

(6.41) Corollary. *In the field* $(Z_q, +, \cdot)$*, where* $q = M_p = 2^p - 1$ *is a Mersenne prime, and for* $d \neq 0$*,* $b \neq 0$*, and* $a \neq d$*,*

$$\frac{2^a - 1}{2^d - 1} = \frac{2^c - 1}{2^b - 1},$$

if and only if $a = c$ *and* $b = d$.

(6.42) Theorem. *If* $q = M_p$*, a Mersenne prime, and* $\Phi = \{1, 2, \ldots, 2^{p-1}\}$*, then* $|A_{0,1}| = (p-1)(p-2)$*. Hence,* $(Z_q, \mathcal{B}_p^*, \in)$ *is circular.*

Proof. This is a consequence of (6.33), (6.41) and the remarks immediately following (6.34).

We want to apply (6.26). Here $N_A(\Phi) = (Z_q^*, \cdot)$, the multiplicative group of the field $(Z_q, +, \cdot)$. So if $\alpha \in Z_q^*$, then $E_c^r = E_{\alpha c}^{\alpha r}$. This means we need only consider the E_c^1.

(6.43) Theorem. *Let* $(N, \mathcal{B}^*, \in) = (Z_q, \mathcal{B}_p^*, \in)$. *Then, for each* s, $1 \leq s \leq p-1$, *the circles of* $E_{2^s-1}^1$ *are tangent to* N^*1. *In particular,* $N^*1 + 2^i(2^s - 1)$ *is tangent to* N^*1 *at* 2^{s+i}.

Proof. First, we shall see that $N^*1 + (2^s - 1)$ is tangent to N^*1 at $x = 2^s$. Certainly, $1 + (2^s - 1) = 2^s \in N^*1 \cap [N^*1 + (2^s - 1)]$. If $x \in N^*1 \cap [N^*1 + (2^s - 1)]$, then $x = 2^v = 2^u + (2^s - 1)$. If $2^u = 1$, then $s = v$, and we have $x = 2^s$ as previously. If $2^u \neq 1$, then $v = 0$ and $2^u + 2^s = 2$. Thus $u = s = 0$, a contradiction. So the only point of $N^*1 \cap [N^*1 + (2^s - 1)]$ is $x = 2^s$. Certainly, 2^{s+i} is then the only point of $N^*1 \cap [N^*1 + 2^i(2^s - 1)]$. $\blacksquare$

From (6.30), we see that in $(Z_q, \mathcal{B}_p^*, \in)$, there is the maximum number $p(p-1)$ of circles of radius $r = 1$ which are tangent to the circle $C(1; 0)$ of radius $r = 1$ and centre $c = 0$. By (6.26), and the comments before (6.43), this is true in $(Z_q, \mathcal{B}_p^*, \in)$ for any radius $r \neq 0$. From (6.28), this means that no circle of radius $r = 1$ intersects $N^*1 = C(1; 0)$ in two places. This property is not peculiar to the $(Z_q, \mathcal{B}_p^*, \in)$, however, since it is also the case for $(Z_{41}, \mathcal{B}_5^*, \in)$ and $(Z_{71}, \mathcal{B}_7^*, \in)$. In contrast, $(Z_{37}, \mathcal{B}_6^*, \in)$ has only the circles in E_2^1 tangent to $C(1; 0)$, as does the usual euclidean plane. $(Z_{37}, \mathcal{B}_6^*, \in)$ also has the circles of $E_1^1 \cup E_9^1$ intersecting $C(1; 0)$ twice.

One property that sets the $(Z_q, \mathcal{B}_p^*, \in)$ off from the others is

(6.44) Theorem. *The circles of* $E_{2^{p-1}-1}^1 = E_{-1}^1$, *in addition to being tangent to* $C(1; 0)$, *are mutually tangent at the same point* 0.

Proof. By (6.22), if the circles of some E_c^1 are mutually tangent at some point, then $E_c^1 = E_{-1}^1$, and here $E_{-1}^1 = E_{2^{p-1}-1}^1$. The proof of (6.22) shows that the point in common to the circles of $E_{2^{p-1}-1}^1$ would need to be 0. A circle in $E_{2^{p-1}-1}^1$ is of the form $N^*1 + 2^i(2^{p-1}-1)$, and $2^{i-1} + 2^i(2^{p-1}-1) = 0$ is on this circle. But suppose two of the circles in $E_{2^{p-1}-1}^1$ have another point in common. Let $0 \neq x \in [N^*1 + 2^i(2^{p-1} - 1)] \cap [N^*1 + 2^j(2^{p-1} - 1)]$, where $i \neq j$. Then $x = 2^s + 2^i(2^{p-1} - 1) = 2^t + 2^j(2^{p-1} - 1)$, so $x = 2^s + 2^{i-1} - 2^i = 2^t + 2^{j-1} - 2^j$, and $2^s + 2^{j-1} = 2^t + 2^{i-1}$. Since $i \neq j$, then by (6.35), $s = i - 1$ and $t = j - 1$. This forces $x = 0$. So the circles of $E_{2^{p-1}-1}^1$ are mutually tangent at 0. $\blacksquare$

This next theorem also represents an unusual property of the $(Z_q, \mathcal{B}_p^*, \in)$.

(6.45) Theorem. *The circles of* E_1^1, *in addition to being tangent to* $C(1; 0)$, *are mutually tangent, but at distinct points. That is, if* $N^*1 + 2^i =$

$\{1+2^i, 2+2^i, \ldots, 2^{p-1}+2^i\} \neq \{1+2^j, 2+2^j, \ldots, 2^{p-1}+2^j\} = N^*1+2^j$,
then N^*1+2^i and N^*1+2^j are tangent at 2^i+2^j.

Proof. This follows immediately from (6.35).

From (6.43), the circles of each $E^1_{2^s-1}$ are tangent to $C(1;0)$, $1 \leq s \leq p-1$. At the two extremes, for $s = 1$ and $s = p-1$, the circles of $E^1_{2^s-1}$ are also mutually tangent; for $s = p-1$ they are mutually tangent at 0, but for $s = 1$, they are mutually tangent but at distinct points. Between these two extreme values for s, nothing like this happens, but we do have simple closed p-link chains in each of these remaining cases. The proof of this is our next goal. Toward that end, we first prove the following proposition, making liberal use of the early lemmas in this section.

(6.46) Proposition. *For $1 < s < p-1$, $N^*1+(2^s-1)$ intersects exactly two of the $N^*1+2^t(2^s-1)$, $0 \leq t \leq p-1$, namely, $N^*1+2(2^s-1)$ and $N^*1+2^{p-1}(2^s-1)$.*

Proof. Using theorem (6.31), we can prove this proposition by proving that $|\Delta(N^*1) \cap (N^*(2^s-1) - (2^s-1))| = 2$. Now $x \in \Delta(N^*1)$ implies $x = -2^i+2^j$, with $i \neq j$, and $x \in N^*(2^s-1) - (2^s-1)$ implies $x = 2^t(2^s-1) - (2^s-1)$, where $t, i, j \in \{0, 1, \ldots, p-1\}$. So, if x is in the intersection, then $x \neq 0$ and so $t \neq 0$. Also, $-2^i+2^j \equiv 2^t(2^s-1) - (2^s-1)$ (mod 2^p-1) $\Longleftrightarrow 2^s+2^t+2^j \equiv 2^{s+t}+2^i+1$ (mod 2^p-1). Either $s+t < p$ or $s+t \geq p$. Suppose, first, that $s+t < p$. Then, since $1 < s < p-1$, $t \leq p-3$. Thus $2^s + 2^t + 2^j < 2^p$. Also, $2^{s+t} + 2^i + 1 = 2^p + 1$ or $2^{s+t} + 2^i + 1 < 2^p$. So either $2^s + 2^t + 2^j \equiv 2^p + 1$ (mod $2^p - 1$) or $2^s + 2^t + 2^j = 2^{s+t} + 2^i + 1$ as integers. The former implies $2^s + 2^t + 2^j = 2$, which cannot be, so $2^s + 2^t + 2^j = 2^{s+t} + 2^i + 1$, as integers.

From $2^s + 2^t + 2^j = 2^{s+t} + 2^i + 1$ with $s + t < p$, we have $0 \in \{i, j\}$. If $j = 0$, then $\{s, t\} = \{s+t, i\}$. But $s + t \notin \{s, t\}$. So $j \neq 0$ and $i = 0$. For $i = 0$, we have $2^s + 2^t + 2^j = 2^{s+t} + 2$, and $2(2^{s-1} + 2^{t-1} + 2^{j-1}) = 2(2^{s+t-1} + 1)$. Since $2^{s+t-1} + 1$ is odd, so is $2^{s-1} + 2^{t-1} + 2^{j-1}$. But 2^{s-1} is even, so $2^{t-1} + 2^{j-1}$ is odd, which would imply that either $t = 1$ or $j = 1$, but not both. With $j = 1$, we have $2^{s-1} + 2^{t-1} + 1 = 2^{s+t-1} + 1$, so $2^{s-1} + 2^{t-1} = 2^{s+t-1}$, which implies $s - 1 = s + t - 2 = t - 1$, or, equivalently, $s = t = s+t-1$. But then $t = 1$ and so both $t = 1$ and $j = 1$, which cannot be. With $t = 1$, we have $2^{s-1} + 1 + 2^{j-1} = 2^{s+t-1} + 1$, so $2^{s-1} + 2^{j-1} = 2^{s+t-1}$ and so $s + 1 = j + 1 = s + t$ and $s = j$. So $i = 0$ yields $x = 2^s - 1 \in \Delta(N^*1) \cap (N^*(2^s-1) - (2^s-1))$, and then $s + t < p$ yields exactly one value in the intersection, namely,

$$-2^0 + 2^s = 2(2^s-1) - (2^s-1). \tag{6:1}$$

This means that $2^s + (2^s - 1) = 2^0 + 2(2^s - 1)$, so

$$[N^*1 + (2^s-1)] \cap [N^*1 + 2(2^s-1)] \supseteq \{2^0 + 2(2^s-1)\}. \tag{6:2}$$

Next, suppose that $s+t \geq p$. Then $2^s + 2^t + 2^j \equiv 2^{s+t} + 2^i + 1 \pmod{2^p - 1} \iff 2^s + 2^t + 2^j \equiv 2^{s+t-p} + 2^i + 1 \pmod{2^p - 1}$. Now $s + t - p < p$. From $2^s + 2^t + 2^j \equiv 2^{s+t-p} + 2^i + 1 \pmod{2^p - 1}$, we have exactly one of the following:

Case (i) $2^s + 2^t + 2^j = 2^{s+t-p} + 2^i + 1$.

Case (ii) $2^s + 2^t + 2^j + 2^p - 1 = 2^{s+t-p} + 2^i + 1$.

Case (iii) $2^s + 2^t + 2^j = 2^{s+t-p} + 2^i + 1 + 2^p - 1 = 2^{s+t-p} + 2^i + 2^p$.

We will eliminate case (ii) very quickly. First, $2^{s+t-p} + 2^i + 1 \leq 2^p + 1$, and next, $2^s + 2^t + 2^j + 2^p - 1 \geq 5 + 2^p$. So case (ii) is impossible.

We now show that case (i) is also impossible. With $2^s + 2^t + 2^j = 2^{s+t-p} + 2^i + 1$, we have either $\{s, t, j\} = \{s+t-p, i, 0\}$ or $\{s, t, j\} \neq \{s+t-p, i, 0\}$. The former implies $j = 0$ since $s + t - p \notin \{s, t\}$ and $0 \notin \{s, t\}$. Hence, $s + t = p$, $(s, t, j) = (s, t, 0)$, and $(s + t - p, i, 0) = (0, i, 0)$. This makes $s = t = i$ and $p = s + t = 2i$. But $1 < s < p$, so p is odd. We are left with $\{s, t, j\} \neq \{s+t-p, i, 1\}$, and so $\{s+1, t+1, j+1\} \neq \{s+t-p+1, i+1, 2\}$. We wish to apply (6.35) to $2^{s+1} + 2^{t+1} + 2^{j+1} = 2^{s+t-p+1} + 2^{i+1} + 2$. Since $1 \in \{s + t - p + 1, i + 1\}$ and $0 \in \{s + t - p, i\}$, either $i = 0$ or $s + t = p$. With $i = 0$, we have $2^{s+1} + 2^{t+1} + 2^{j+1} = 2^{s+t-p+1} + 2 + 2$. Now $3 \leq s + 1$, so $2 + 2 \neq 2^{s+1}$. This means that $s + t - p + 1 = s + 2$ and $t = p + 1$. But $1 \leq t \leq p-1$. With $s + t = p$, then $2^{s+1} + 2^{t+1} + 2^{j+1} = 2^{i+1} + 2 + 2$. Again, $2 + 2 \neq 2^{s+1}$, so $s + 2 = i + 1$ and $s + 1 = i$. From $2^i + 2^{t+1} + 2^{j+1} = 2^{i+1} + 2 + 2$, we have either $t + 1 = 2$ or $j + 1 = 2$, so either $t = 1$ or $j = 1$. If $t = 1$, then $s = p - 1$ and $s + 1 = i = p$, which cannot be. Alternatively, if $j = 1$, then we get $2^i + 2^{t+1} + 2^2 = 2^{i+1} + 2 + 2$ and so $t + 2 = i + 1$, and $t + 1 = i$. But $s + 1 = i$, so $p = s + t = (i - 1) - (i - 1) = 2i - 2$. But $1 < s < p$, so p is odd. Thus case (i) is also impossible.

Finally, we show that case (iii) yields exactly one element in $\Delta(N^*1) \cap (N^*(2^s - 1) - (2^s - 1))$. So we start with $2^s + 2^t + 2^j = 2^{s+t-p} + 2^i + 2^p$, with p greater than each of s, t, and j. Since $s \leq p - 2$, we have $t = j = p - 1$ and $s + t - p + 1 = i + 1 = s$. Thus

$$x = -2^{s-1} + 2^{p-1} = 2^{p-1}(2^s - 1) - (2^s - 1) \tag{6:3}$$

is in the intersection $\Delta(N^*1) \cap (N^*(2^s - 1) - (2^s - 1))$. From (6:3) we compute $2^{p-1} + (2^s - 1) = 2^{s-1} + 2^{p-1}(2^s - 1)$, so

$$[N^*1 + (2^s - 1)] \cap [N^*1 + 2^{p-1}(2^s - 1)] \supseteq \{2^{p-1} + (2^s - 1)\}. \tag{6:4}$$

(6.47) Theorem. $N^*1 + (2^s - 1)$ *is tangent only to* $N^*1 + 2^{p-1}(2^s - 1)$ *at* $u = 2^{p-1} + (2^s - 1)$, *and to* $N^*1 + 2(2^s - 1)$ *at* $v = 2u = 2^0 + 2(2^s - 1)$.

Proof. Using (6.31) and the proof of (6.46), we need only show that $-2^0 + 2^s$ and $-2^{s-1} + 2^{p-1}$ each occurs exactly once in $\mathcal{A}(1)$. If $-2^m + 2^n = -2^i + 2^j \neq 0$, then $2^i + 2^n = 2^m + 2^j$ and $\{i, n\} = \{m, j\}$. If $i = j$, then

$-2^i + 2^j = 0$. So $i = m$. Similarly, $n = j$. Thus an arbitrary $-2^i + 2^j$ in $\mathcal{A}(1)$ occurs exactly once.

(6.48) Theorem. *For $1 < s < p - 1$, the $E^1_{2^s-1}$ form a simple closed p-link chain.*

Proof. This is immediate from (6.18) and (6.47).

From looking at $(Z_{31}, \mathcal{B}^*_5, \in)$, one might guess that the remaining E^1_c will also form a simple closed p-link chain for these $(Z_q, \mathcal{B}^*_p, \in)$. However, in $(Z_{127}, \mathcal{B}^*_7, \in)$, the E^1_{13} form a torus $T(1; 13)$.

6.3. Compound closed chains

After the proof of (6.14), we raised the question as to what could happen if a $B \in E^r_c$, $c \neq 0$, intersected at most two other circles. There were three possibilities and two have been studied. The third had three possibilities itself. One has been studied (see (6.18) and (6.19)), and another has been eliminated due to its impossibility (see (6.24)), and now we consider the third, namely, (c2), the possibility where a $B \in E^r_c$, $c \neq 0$, intersects exactly two other circles $A, D \in E^r_c \setminus \{B\}$, each at two points. The graphs of E^1_2 and E^1_3 of (6.32) show that this is a real possibility.

Take $B \in E^r_c$, $c \neq 0$, with distinct circles $A, D \in E^r_c \setminus \{B\}$, where $|B \cap A| = |B \cap D| = 2$. Let $A \cap B = \{a_1, a_2\}$ and $B \cap D = \{b_1, b_2\}$, and suppose $A = N^*r + a$, $B = N^*r + b$, and $D = N^*r + d$. This case (c2) also has three possibilities:

(c2.1) $\{a_1, a_2\} = \{b_1, b_2\}$.
(c2.2) $|\{a_1, a_2\} \cap \{b_1, b_2\}| = 1$.
(c2.3) $\{a_1, a_2\} \cap \{b_1, b_2\} = \varnothing$.

We proceed now to show that (c2.1) cannot happen, that (c2.2) can happen, but only under rather special restrictive conditions, and that (c2.3) can happen and has a rather nice connection with a group and a subgroup as in (c1).

Using (6.3), there is a $\phi \in \Phi$ such that $\phi a = b$ and $\phi A = B$.

(6.49) Lemma. *Suppose $B \in E^r_c$, $c \neq 0$, intersects exactly two other circles $A, D \in E^r_c \setminus \{B\}$, each at two distinct points. Suppose $A = N^*r + a$, $B = N^*r + b$, and $D = N^*r + d$. Also suppose $A \cap B = B \cap D$. Then the $\phi \in \Phi$ of (6.3) where $\phi a = b$ and $\phi A = B$ has the property that $\phi^2 \neq 1$.*

Proof. Suppose $\phi^2 = 1$. Then $\phi B = A$. By (6.3), there is also a $\lambda \in \Phi$ such that $\lambda d = b$ and $\lambda D = B$. Since $\lambda D = B$ and since $B \cap D = \{b_1, b_2\}$, we have $\lambda b_1, \lambda b_2 \in \lambda B \cap \lambda D = \lambda B \cap B$. Hence, $\lambda B \in \{A, D\}$. If $\lambda B = A$, then $\lambda b = a = \phi b$, and so $\lambda = \phi$ and $\lambda^2 = 1$. If $\lambda B = D$, then $\lambda^2 = 1$, also. We obtain $\lambda^2 = 1$ in either case.

Now with $A \cap B = \{a_1, a_2\}$, we have $\phi a_1, \phi a_2 \in \phi A \cap \phi B = B \cap A =$

$\{a_1, a_2\}$. If $\phi a_1 = a_1$, then $\phi a_2 = a_2$ and so $\phi = 1$, which cannot be. Hence, $\phi a_1 = a_2$ and $\phi a_2 = a_1$. Similarly, $\lambda b_1, \lambda b_2 \in \lambda B \cap \lambda D = D \cap B = \{b_1, b_2\}$. If $\lambda b_1 = b_1$, then $\lambda b_2 = b_2$ and $\lambda = 1$. So we must have $\lambda b_1 = b_2$ and $\lambda b_2 = b_1$.

From $A \cap B = B \cap D$ we have $\{a_1, a_2\} = \{b_1, b_2\}$. Hence, $\lambda a_1 = \phi a_1 = a_2$ and $\lambda a_2 = \phi a_2 = a_1$. This makes $\lambda = \phi$. But $\lambda B = D$ and $\phi B = A$, so $A = D$, a contradiction.

(6.50) Proposition. *Case* c2.1 *cannot happen.*

Proof. If case (c2.1) were valid, then we would have $A \cap B \neq \varnothing$, $A \cap D \neq \varnothing$, and $B \cap D \neq \varnothing$. Hence, any of $\{A, B, D\}$ has the other two as its two circles in E_c^r which intersect it twice. Since the $\phi \in \Phi$ where $\phi A = B$ has $\phi^2 \neq 1$, we have $\phi A = B$, $\phi B = D$, and $\phi D = A$. Also, $\phi a = b$, $\phi b = d$, and $\phi d = a$. Hence, $\phi^3 = 1$. But $\phi a_1, \phi a_2 \in \phi A \cap \phi B = B \cap D = \{b_1, b_2\} = \{a_1, a_2\}$. If $\phi a_1 = a_1$, then $\phi a_2 = a_2$ and $\phi = 1$. Otherwise, $\phi a_1 = a_2$ and $\phi a_2 = a_1$. From this, we get $\phi^2 a_1 = \phi a_2 = a_1$ and $\phi^2 a_2 = \phi a_1 = a_2$. Hence, $\phi^2 = 1$, a contradiction. So (c2.1) cannot happen.

(6.51) Lemma. *Suppose case* (c2.2) *is valid. Then the* $\phi \in \Phi$ *where* $\phi a = b$ *has the property that* $\phi b = d$, $\phi d = a$, *and* $\phi^3 = 1$. *Also,* $\{a_1, a_2\} \cap \{b_1, b_2\} \cap \{d_1, d_2\} = \{0\}$, *and* $|E_c^r| = 3$ *with* $E_c^r = E_{-r}^r$.

Proof. First notice that any one of $\{A, B, D\}$ intersects the other two, so each has the other two as the circles in E_c^r which intersect it twice. Let $D \cap A = \{d_1, d_2\}$. Now $|A \cap B \cap D| = 1$ so let $a_2 = b_2 = d_2$. Now $\phi a_1, \phi a_2 \in \phi A \cap \phi B = B \cap \phi B$. So $\phi B \in \{A, D\}$. If $\phi B = A$, then $\{\phi a_1, \phi a_2\} = \{a_1, a_2\}$. Again, $\phi a_1 = a_1$ implies $\phi a_2 = a_2$ and $\phi = 1$. Hence, $\phi a_1 = a_2$ and $\phi a_2 = a_1$. With $\phi a_1 = a_2 = b_2$, we have $a_1 = \phi a_2 = \phi b_2 \in \phi B \cap \phi D = A \cap \phi D$, so $\phi D \in \{B, D\}$ and $\phi D \neq D$. Hence, $\phi D = B$ and $\phi d = b = \phi a$. Hence, we cannot have $\phi B = A$. The only other possibility is $\phi B = D$. With $\phi B = D$ and $\phi a_1, \phi a_2 \in \phi A \cap \phi B = B \cap D = \{b_1, b_2\}$, we have $\{\phi a_1, \phi a_2\} = \{b_1, b_2\}$. But $a_2 = b_2$, and if $\phi a_2 = b_2$, then $\phi a_1 = b_1$. Also, $\phi a_2 = b_2 = a_2$ and $\phi \neq 1$ implies $a_2 = b_2 = 0$. Hence, $0 = \phi b_2 \in \{\phi b_2, \phi b_1\} \subseteq \phi B \cap \phi D = D \cap \phi D$. Now $\phi D \in \{A, B\}$. If $\phi D = B$, then $\phi^2 = 1$, $\phi A = B$, and $\phi B = A$. So we are left with $\phi D = A$, and so $\phi^3 = 1$, $\phi d = a$, and $\phi b = d$.

But what happens when $\phi a_2 \neq b_2$? We have $\phi B = D$ and $\{\phi a_1, \phi a_2\} = \{b_1, b_2\}$ with $\phi a_2 \neq b_2$. Hence, $\phi a_2 = b_1$ and $\phi a_1 = b_2 = a_2$. But then $\phi^2 a_1 = \phi a_2 = b_1$. Since $\phi B = D$, we have the same argument as above. Let $D \cap A = \{d_1, d_2\}$. Now $b_2 = a_2 \in A \cap B \cap D$, so we have from $\phi b_1, \phi b_2 \in \phi B \cap \phi D = D \cap \phi D$ that $\phi D \in \{A, B\}$. As above, $\phi D = B$ cannot be, so $\phi D = A$. With $\phi b_1, \phi b_2 \in D \cap A = \{d_1, d_2\}$, we obtain $\{\phi b_1, \phi b_2\} = \{d_1, d_2\}$. We have $a_2 = b_2 = d_2$. If $\phi b_2 = d_2$, then $a_2 = b_2 = d_2 = 0$, as above. So $\phi b_2 = d_1$ and $\phi b_1 = d_2$.

From $\phi a_1 = a_2 = b_2 = d_2$ and $\phi b_2 = d_1$, we have $\phi a_2 = b_1$ and $\phi b_1 = d_2 = b_2 = a_2$. Hence, $\phi a_1 = a_2 = b_2$ and $\phi^2 a_1 = \phi a_2 = \phi b_2 = d_1$. We conclude that $a_1 = \phi^3 a_1 = \phi d_1 = a_2$.

With $D \cap A = \{d_1, d_2\}$, we obtain $\phi d_1, \phi d_2 \in \phi D \cap \phi A = A \cap B = \{a_1, a_2\}$. If $\phi d_2 = a_2$, then $a_2 = b_2 = d_2 = 0$, as above. So $\phi d_2 = a_1$ and $\phi d_1 = a_2$, as promised. We have $a_1 = \phi^3 a_1 = a_2$, a contradiction.

So, the only possibility is with $\{a_2, b_2, d_2\} = \{0\}$, $\phi a_1 = b_1$, $\phi b_1 = d_1$, and $\phi d_1 = a_1$. This means that $0 \in X$ if $X \in E_c^r$, so $c \in N^*(-r)$, $E_c^r = E_{-r}^r$, and $|E_c^r| = 3$.

By (6.26), we need only consider E_1^{-1} in the field generated case.

(6.52) Theorem. *Suppose case (c2.2) is valid with a field generated circular $(F, +, \cdot)$ and $(F, \mathcal{B}_3^*, \in)$. Then $|F| = 2^{2u}$ for some integer u if $|F| < \infty$. Also, for such F, case (c2.2) occurs.*

Proof. Let $(F, +, \cdot)$ be field generated where E_c^r, $c \neq 0$, satisfies (c2.2). Then $|E_c^r| = 3$ and each circle has exactly three points, and $N^*c = N^*(-r)$. Hence, we need only consider E_1^{-1}. Now $F^*1 = \{a, a^2, 1\}$, so $F^*(-1) = \{-a, -a^2, -1\}$. With $F^*(-1) + 1 = \{1 - a, 1 - a^2, 0\} = A$, $F^*(-1) + a = \{0, a(1-a), a-1\} = B$, and $F^*(-1) + a^2 = \{a(a-1), 0, a^2 - 1\} = D$, we have $E_1^{-1} = \{A, B, D\}$.

If $A \cap B = \{0, x\}$, then exactly one of the following is true: (i) $1 - a = a(1-a)$; (ii) $1 - a = a - 1$; (iii) $1 - a^2 = a(1-a)$; and (iv) $1 - a^2 = a - 1$. From (i) we have $1 - a = a(1-a)$, so $a = 1$, which cannot be. From (ii) we have that F is a field of characteristic 2. From (iii), $(1-a)(1+a) = 1 - a^2 = a(1-a)$, so $1 + a = a$ and $1 = 0$. From (iv), $(1-a)(1+a) = 1 - a^2 = a - 1 = -(1-a)$. Hence, $1 + a = -1$ so $a = -2$.

Now we must have $A \cap B = \{0, x\}$ for some $x \neq 0$. This implies that F is a field of characteristic 2, or $F^*1 = \{a, a^2, 1\}$ where $a = -2$. If $a = -2$, then $a^2 = 4$ and $1 = a^3 = -8$. But $1 \equiv -8 \pmod{p}$, p a prime, if and only if $9 \equiv 0 \pmod{p}$ if and only if $p = 3$. Also, we must have that 3 divides $p^u - 1$ where $|F| = p^u$. But 3 cannot divide $3^u - 1$ for $u \geq 1$. So we only have the possibility for F to be a field of characteristic 2.

Now 3 does not divide $1 + 2 + 4 = 7$. Assume that 3 does not divide $1 + 2 + 2^2 + \cdots + 2^{2u}$. Note that 3 *does* divide $2^{2u+1} + 2^{2u+2} = 2^{2u+1}(1+2)$, so 3 does not divide $1 + 2 + 2^2 + \cdots + 2^{2u} + 2^{2u+1} + 2^{2u+2} = (2-1)(2^{2u+2} + 2^{2u+1} + 2^{2u} + \cdots + 2^2 + 2 + 1) = 2^{2u+3} - 1$. Hence, 3 does not divide $2^{2u+1} - 1 = |F| - 1 = |F^*|$ if $|F| = 2^{2u+1}$. So (c2.2) will be impossible for $|F| = 2^{2u+1}$.

Suppose $|F| = 2^{2u}$. Then $|F^*| = 2^{2u} - 1 = (2^2)^u - 1 = (2^2 - 1)[(2^2)^{u-1} + (2^2)^{u-2} + \cdots + 2^2 + 1]$, so 3 divides $2^{2u} - 1$.

Let $|F| = 2^{2u}$ and $a \in F$ with $a \neq 1$ and $a^3 = 1$. Then $A = F^*(-1) + 1 = F^*1 + 1 = \{1 + a, 1 + a^2, 0\}$, $B = F^*(-1) + a = F^*1 + a = \{0, a(1+a), a\}$, and $D = F^*(-1) + a^2 = F^*1 + a^2 = \{a(1 + a), 0, a^2\}$. But $0 = a^3 - 1 = (a - 1)(a^2 + a + 1)$. Hence, $a^2 + a + 1 = 0$ and $a^2 = a + 1$. So $D = \{a(1 + a), 0, a + 1\}$. Also, $1 + a^2 = a$. Hence, $A = \{a^2, a, 0\}$, $B = \{a(1 + a), a, 0\}$, and $D = \{a(1 + a), a^2, 0\}$. Now we see that (c2.2) does indeed happen for $|F| = 2^{2u}$.

Suppose $(N, +, \cdot)$ is a planar nearring with $|\Phi| = 3$. What would it take to satisfy (c2.2) for some E_c^r, $c \neq 0$? From (6.51), we know that we must have $c = -r$. Let $N^* r = \{r, \phi r, \phi^2 r\}$, so $N^*(-r) = \{-r, -\phi r, -\phi^2 r\}$. Let $A = N^* r - r = \{0, \phi r - r, \phi^2 r - r\}$, $B = N^* r - \phi r = \{r - \phi r, 0, \phi^2 r - \phi r\}$, and $D = N^* r - \phi^2 r = \{r - \phi^2 r, \phi r - \phi^2 r, 0\}$. If $A \cap B = \{0, x\}$, then one of the following four cases must be valid: (i) $\phi r - r = r - \phi r$; (ii) $\phi r - r = \phi^2 r - \phi r$; (iii) $\phi^2 r - r = r - \phi r$; (iv) $\phi^2 r - r = \phi^2 r - \phi r$. If (i) is true, then $\phi r - r = -(\phi r - r)$, so $\phi r - r$ is an element of order 2. If (ii) is true, then $\phi r - r = \phi^2 r - \phi r = \phi[\phi r - r]$. But $\phi \neq 1$, so $\phi r - r = 0$. But then $\phi r = r$, which cannot be. This eliminates (ii). If (iii) is true, then $\phi^2 r - r = r - \phi r = \phi^3 r - \phi r = \phi(\phi^2 r - r)$. Again, $\phi \neq 1$, so $\phi^2 r = r$, which cannot be. Now we have eliminated (iii). Finally, if (iv) is true, then $-r = -\phi r$, or $\phi r = r$, which we know cannot happen. So (iv) is also eliminated. In summary, there is only one possibility, and that is when $\phi r - r$ is an element of order 2.

If $\phi r - r$ is an element of order 2, then $\phi r - r = r - \phi r$. Hence, $A \cap B = \{0, \phi r - r\}$. Also $\phi^2 r - \phi r = \phi r - \phi^2 r$, so $B \cap D = \{0, \phi^2 r - \phi r\}$. Finally, $\phi^3 r - \phi^2 r = \phi^2 r - \phi^3 r$, or $r - \phi^2 r = \phi^2 r - r$. Hence, $A \cap D = \{0, \phi^2 r - r\}$. This gives us

(6.53) Theorem. *Suppose $(N, +, \cdot)$ is a planar nearring with $|\Phi| = 3$. Consider E_c^r, with $c \neq 0$. Then (c2.2) is valid for E_c^r if and only if there is a $\phi \in \Phi$ such that $\phi r - r$ has additive order 2. In this case, $E_c^r = E_{-r}^r$.*

(6.54) Problem. What are the Ferrero pairs (N, Φ) where $|\Phi| = 3$ and a resulting planar nearring has an E_{-r}^r where (c2.2) is valid?

Towards solving (6.54), we note that N must have an element of order 2 and that 3 divides $|N| - 1$ if $|N| < \infty$. It is then quite easy to see that $|N| = 6u + 4$ for some nonnegative integer u if N is finite.

So now we turn our attention to (c2.3). By (6.14), if one $B_1 \in E_c^r$, $c \neq 0$, intersects exactly two other circles $B_0, B_2 \in E_c^r \setminus \{B_1\}$ each in exactly two places, then each circle $X \in E_c^r$ has this property. So B_2 has exactly two circles $B_1, B_3 \in E_c^r \setminus \{B_2\}$ each of which intersects B_2 in exactly two places. Repeat this for B_3, etc. If $|N| < \infty$, then there is a least s so that if B_{s-2}, B_s are the two circles in $E_c^r \setminus \{B_{s-1}\}$ which intersect B_{s-1} in exactly two places, then $B_s \in \{B_0, B_1, \cdots, B_{s-3}\}$. If $B_s = B_t$, $1 \leq t \leq s - 3$, then B_t will be intersecting B_{t-1}, B_{t+1}, and B_{s-1}, contrary to the minimality of s. Hence, $B_s = B_0$. We denote this by

$$B_0 : B_1 : B_2 : \cdots : B_{s-1} : B_0.$$

If $B_i \cap B_{i+1} = \{a_i, c_i\}$, we also denote this by

$$
\begin{array}{ccccccc}
a_0 & a_1 & a_2 & & a_{s-2} & & a_{s-1} \\
B_0 : & B_1 : & B_2 : & \cdots & : & B_{s-1} & : & B_0 \\
c_0 & c_1 & c_2 & & c_{s-2} & & c_{s-1}
\end{array}
$$

and take our subscripts to have meaning in the cyclic group $(Z_s, +)$. Summarizing we have

(6.55) Theorem. *For a finite circular planar nearring $(N, +, \cdot)$, suppose a $B_1 \in E_c^r$, $c \neq 0$, intersects exactly two other circles $B_0, B_2 \in E_c^r \setminus \{B_1\}$, each in exactly two places, and suppose $(B_0 \cap B_1) \cap (B_1 \cap B_2) = \varnothing$. That is, condition (c2.3) is satisfied. Then there is a smallest integer $s \geq 3$ with circles $B_0, B_1, \ldots, B_{s-1} \in E_c^r$ with the property that for $i \in Z_s$, B_i intersects exactly the B_{i-1} and the B_{i+1}.*

(6.56) Definition. For the circles $B_0, B_1, \ldots, B_{s-1}$ as in (6.55), we write

$$B_0 : B_1 : B_2 : \cdots : B_{s-1} : B_0,$$

or

$$
\begin{array}{ccccccc}
a_0 & a_1 & a_2 & & a_{s-2} & & a_{s-1} \\
B_0 : & B_1 : & B_2 : & \cdots & : & B_{s-1} & : & B_0 \\
c_0 & c_1 & c_2 & & c_{s-2} & & c_{s-1}
\end{array}
$$

if $B_i \cap B_{i+1} = \{a_i, c_i\}$, and call it a *compound closed s-link chain*. Compare this with (6.16) and (6.11).

(6.57) Remark. As we did in (6.17), we identify each of the following:

$$B_{i-1} : B_i : B_{i+1} : \cdots : B_{i-2} : B_{i-1};$$

$$B_2 : B_1 : B_0 : B_{s-1} : \cdots : B_3 : B_2;$$

$$B_{i+1} : B_i : B_{i-1} : \cdots : B_{i+2} : B_{i+1}.$$

Again, the indices take their meaning within the group $(Z_s, +)$.

(6.58) Theorem. *For a finite circular planar nearring $(N, +, \cdot)$, suppose a $B_1 \in E_c^r$, $c \neq 0$, intersects exactly two other circles $B_0, B_2 \in E_c^r \setminus \{B_1\}$ each in exactly two places, and suppose $(B_0 \cap B_1) \cap (B_1 \cap B_2) = \varnothing$. That is, condition (c2.3) is satisfied. Let*

$$
\begin{array}{cccccc}
a_0 & a_1 & & a_{s-2} & & a_{s-1} \\
B_0 : & B_1 : & \cdots & : & B_{s-1} & : & B_0 \\
c_0 & c_1 & & c_{s-2} & & c_{s-1}
\end{array}
$$

be the resulting compound closed s-link chain guaranteed by (6.55). *Let* $B_i = N^*r + b_i$, *where* $\{b_0, b_1, \ldots, b_{s-1}\} \subseteq N^*c$ *and suppose* $\phi b_0 = b_1$ *with* $\phi \in \Phi$. *If* $\phi^2 \neq 1$, *then the following hold:*

(1) *There is a cyclic group* $\langle \phi \rangle$ *of* Φ *of order* s *such that* $\phi a_i = a_{i+1}$, $\phi b_i = b_{i+1}$, $\phi c_i = c_{i+1}$, *and* $\phi B_i = B_{i+1}$.

(2) *There is a bijection between the cosets* $\lambda \langle \phi \rangle$ *of* $\langle \phi \rangle$ *in* Φ *and distinct compound closed s-link chains in* E_c^r.

(3) *There are exactly* $[\Phi : \langle \phi \rangle]$ *compound closed s-link chains in* E_c^r.

(4) *If* $D \in E_c^r \setminus \{B_0, B_1, \ldots, B_{s-1}\}$, *then* D *does not intersect any* B_i, $0 \leq i \leq s-1$, *and* $D = D_0$ *yields a compound closed s-link chain* $D_0 : D_1 : \cdots : D_{s-1} : D_0$ *with each* $D_i \in E_c^r$.

Proof. From $\phi b_0 = b_1$, we have $\phi B_0 = B_1$. Hence, $\phi a_0, \phi c_0 \in \phi B_0 \cap \phi B_1 = B_1 \cap \phi B_1$. This puts $\phi B_1 \in \{B_0, B_2\}$. Since $\phi^2 \neq 1$, we have $\phi B_1 = B_2$ and $\phi b_1 = b_2$. Also, $\{\phi a_0, \phi c_0\} = \{a_1, c_1\}$, so, without loss of generality, we may take $\phi a_0 = a_1$ and $\phi c_0 = c_1$, renaming the a_1 and c_1 if necessary.

Having $\phi B_i = B_{i+1}$, $\phi b_i = b_{i+1}$, $\phi a_{i-1} = a_i$, and $\phi c_{i-1} = c_i$, we can use the same argument to conclude that $\phi B_{i+1} = B_{i+2}$, $\phi b_{i+1} = b_{i+2}$, $\phi a_i = a_{i+1}$, and $\phi c_i = c_{i+1}$. This makes $\phi^i \neq 1$ for $1 \leq i \leq s-1$ and $\phi^s = 1$, and gives us (1).

If $\langle \phi \rangle = \Phi$, we are finished. If $\langle \phi \rangle \neq \Phi$, then $\langle \phi \rangle$ has another coset $\lambda \langle \phi \rangle$ in Φ. With such a $\lambda \in \Phi$, we define $D_i = \lambda B_i$. Since $\lambda a_i, \lambda c_i \in \lambda B_i \cap \lambda B_{i+1}$, we have, with $x_i = \lambda a_i$ and $y_i = \lambda c_i$, a compound closed s-link chain

$$
\begin{array}{ccccccc}
x_0 & x_1 & & x_{s-2} & & x_{s-1} & \\
D_0 : & D_1 : & \cdots & : & D_{s-1} & : & D_0 \\
y_0 & y_1 & & y_{s-2} & & y_{s-1} &
\end{array}
$$

in E_c^r.

If some $D_i \in \{B_0, \ldots, B_{s-1}\}$, then $D_0 \in \{B_0, \ldots, B_{s-1}\}$ and so $\lambda = \phi^i$ for some i. Hence, $\{D_0, \ldots, D_{s-1}\} \cap \{B_0, \ldots, B_{s-1}\} = \varnothing$.

If $D \notin E_c^r \setminus \{B_0, \ldots, B_{s-1}\}$, then there is a $\lambda \in \Phi$ such that $\lambda B_0 = D$. Let $D_0 = D = \lambda B_0$ and conclude that with $\lambda B_i = D_i$, we have $D_0 : D_1 : \cdots : D_{s-1} : D_0$ as a compound closed s-link chain in E_c^r associated with the coset $\lambda \langle \phi \rangle$ in Φ. Also D cannot intersect any B_i, $0 \leq i \leq s-1$, for if it did, then $D \in \{B_{i-1}, B_{i+1}\}$. This gives us (4).

Suppose that

$$\lambda B_0 : \lambda B_1 : \cdots : \lambda B_{s-1} : \lambda B_0$$

and

$$\mu B_0 : \mu B_2 : \cdots : \mu B_{s-1} : \mu B_0$$

are the same compound closed s-link chain in E_c^r, with $\lambda, \mu \in \Phi$. Then $\lambda B_0 = \mu B_i$ for some i. Hence, $\lambda b_0 = \mu b_i$ and $\mu^{-1} \lambda b_0 = b_i = \phi^i b_0$. Hence,

$\mu\langle\phi\rangle = \lambda\langle\phi\rangle$. We now have (2) and (3), and we have completed the proof of (6.58).

(6.59) Corollary. *With the same hypotheses as* (6.58) *we have that if*

$$B_0 : B_1 : \cdots : B_{s-1} : B_0$$

and

$$D_0 : D_1 : \cdots : D_{t-1} : D_0$$

are compound closed s- and t-link chains, respectively, in E_c^r, $c \neq 0$, *then* $s = t$.

(6.60) Example. We have already shown in (6.32) that E_2^1 and E_3^1 each have two compound closed 4-link chains.

There remains to consider what happens when all the hypotheses of (6.58) are satisfied except $\phi^2 \neq 1$. Can we really have such a $\phi^2 = 1$? Since $B_0 \in E_c^r$ is arbitrary, we would really need $\phi_i^2 = 1$ for each i, $0 \leq i \leq s-1$, where $\phi_i B_i = B_{i+1}$. For if some $\phi_i^2 \neq 1$, then one can replace B_0 and B_1 with B_i and B_{i+1} respectively. Now ϕ_i assumes the role of ϕ in (6.58), and with $\phi_i^2 \neq 1$ we have that (6.58) is valid. Hence, each $\phi_j = \phi_i$, so $\phi^2 = 1$. With $\phi^2 = 1$, we have each $\phi_i^2 = 1$ where $\phi_i B_i = B_{i+1}$, and, consequently, $\phi_i b_i = b_{i+1}$. Now $\phi = \phi_0$. If $\phi_0 = \phi_1$, then from $\phi_0^2 = \phi_1^2 = 1$, we have $\phi_0 b_0 = b_1 = \phi_1 b_2$ and so $b_0 = \phi_0 b_1 = \phi_1 b_1 = b_2$, which implies $B_0 = B_2$. But $B_0 \neq B_2$, so we cannot have $\phi_0 = \phi_1$.

On the other hand, for a Ferrero pair (N, Φ), we cannot have two distinct elements of order 2 in Φ. For if $\phi \in \Phi \setminus \{1\}$ and $\phi^2 = 1$, then since $-1 + \phi$ is a bijection on N, each $y \in N$ has an $x \in N$ where $y = -x + \phi x$. So $y + \phi y = (-x + \phi x) + \phi(-x + \phi x) = 0$, which means $\phi y = -y$, and this defines such a $\phi \in \Phi \setminus \{1\}$ where $\phi^2 = 1$, so there cannot be two such elements. In summary, we have

(6.61) Theorem. *Suppose the hypotheses of* (6.55) *are satisfied. Let* $\phi B_0 = B_1$ *with* $\phi \in \Phi$. *Then* $\phi^2 \neq 1$. *Hence,* (6.58) *applies.*

7. Other geometric structures from planar nearrings

As of this writing, there seem to be considerably more geometric applications of the $(N, \mathcal{B}^*, \in)$ than either the $(N, \mathcal{B}, \in)$ or the $(N, \mathcal{B}^-, \in)$. But these latter two structures also have their geometric applications, as we shall develop in this section. In addition to BIBDs from each, we shall obtain substructures which are affine planes.

7.1. Tactical configurations

Traditionally, a tactical configuration has been a finite structure, but we have reason to extend the concept to the infinite.

(7.1) Definition. Let N be a nonempty set, and let $\mathcal{T}$ be a nonempty family of subsets of N. Suppose there are cardinal numbers k and r satisfying the following:

(a) If $T \in \mathcal{T}$, then the cardinality of T is $|T| = k$.

(b) If $x \in N$, then the number of $T \in \mathcal{T}$ for which $x \in T$ is $[x] = r$.

Then $(N, \mathcal{T}, \in)$ is a *tactical configuration*.

From (a), we say that each 'block' or 'line' of $\mathcal{T}$ has the same number of 'points', and from (b), we say that each 'point' of N is incident to the same number of 'blocks' or 'lines' of $\mathcal{T}$.

Of fundamental importance is to know when two blocks $Na + b$ and $Na' + b'$ of $\mathcal{B}$ are the same.

(7.2) Theorem. *Let $(N, +, \cdot)$ be a planar nearring with blocks $\mathcal{B} = \{Na + b \mid a, b \in N, a \neq 0\}$. Let $Na + b$, $Na' + b' \in \mathcal{B}$. Then $Na + b = Na' + b'$ if and only if either*

(a) $Na = Na'$, *Na is a subgroup of N^+, and $b' - b \in Na$;*

or

(b) $Na = Na'$, *Na is not a subgroup of N^+, and $b = b'$.*

Proof. If (a) is true, then, certainly, $Na + b = Na' + b'$. Likewise if (b) is true. So we can turn to the converse.

Suppose $Na + b = Na' + b'$, so $Na = Na' + c$, where $c = b' - b$. Either Na is a subgroup of N^+ or it is not. If Na is a subgroup of N^+, then $Na - c = Na'$ is a coset of Na containing 0, since $0 \in Na'$, so $Na - c = Na$, and, consequently, $Na = Na'$ and $c = b' - b \in Na$. Hence, (a).

Now suppose Na is not a subgroup of N^+ and $c \neq 0$. We will show that Na *must* be a subgroup of N^+, so this contradiction will force $c = 0$, and $b = b'$, if we insist that Na is *not* a subgroup of N^+. Once we obtain $b = b'$, we immediately have $Na = Na'$. Thus (b). So we proceed with the assumption that Na is *not* a subgroup of N^+, and $b \neq b'$, so $Na = Na' + c$ with $c = b' - b \neq 0$.

From $Na = Na' + c$, we obtain $c \in Na$ and so $Na = Nc$. Similarly, from $Na - c = Na'$, we obtain $N(-c) = Na'$, and so we have $Nc = N(-c) + c$. Take $m, n \in N$ such that $mc - nc \neq 0$ and $0 \notin \{mc, nc\}$. Then by multiplying $Nc = N(-c) + c$ on the left, we have $Nc = N(-c) + mc$ and $Nc = N(-c) + nc$. From the first of these we have $nc = s(-c) + mc$ for some $s \in N$, and so $nc - mc = s(-c)$ and, consequently, $mc - nc = sc \in Nc$. From the second, we have $mc = t(-c) + nc$ for some $t \in N$, and therefore $mc - nc = t(-c) \in N(-c)$. With $0 \neq mc - nc \in Nc \cap N(-c)$, we are assured that $Nc = N(-c)$.

The equation $Nc = N(-c) + c$ now becomes $Nc = Nc + c$, so for all $u, v \in N$, $Nc = Nc + uc = Nc + vc$. We can then write $Nc - vc = Nc + (uc - vc) = Nc$ and conclude that $vc - uc \in Nc = Na$. Of course, this means that Na is a subgroup of N^+, and, as we indicated earlier in the proof, our proof is now complete.

A particular 'basic block' $Na \in \mathcal{B}$ is either a subgroup of N^+ or it is not. Those which are will be denoted by Ng_i, where $i \in S$, and S is an index set with $|S| = s$. Similarly, those which are not will be denoted by Ne_j, where $j \in T$, and T is an index set with $|T| = t$.

(7.3) Theorem. *The incidence structure $(N, \mathcal{B}, \in)$ is a tactical configuration with parameters $k = |N/{=}_m|$ and $r = s + tk$.*

Proof. Since each $|Na| = |N/{=}_m|$, $a \neq 0$, we have each $|Na + b| = |N/{=}_m|$. Thus, $k = |N/{=}_m|$. An $x \in N$ belongs to s distinct $Ng_i + x$, $i \in S$. For each $j \in T$, x belongs to exactly $|Ne_j| = k$ blocks of the form $Ne_j + y$, since for each $u_j \in Ne_j$, there is a unique $y_j \in N$ such that $x = u_j + y_j \in Ne_j + y_j$. Hence, x belongs to exactly tk blocks of the form $Ne_j + y_j$. Altogether, x belongs to $s + tk$ blocks of $\mathcal{B}$, so $[x] = s + tk$. By (7.2), all these blocks are distinct.

(7.4) Corollary. *If $v = |N| < \infty$, then $|\mathcal{B}| = (vs)/k + tv$.*

Proof. Apply (7.2). Each of the s basic blocks Ng_i has index v/k, thus yielding $(vs)/k$ distinct blocks from the Ng_i. Each Ne_j yields v distinct blocks, and there are t of these Ne_j.

(7.5) Corollary. *If $\mathcal{B}_g = \{Ng_i + x \mid i \in S, x \in N\}$, then $(N, \mathcal{B}_g, \in)$ is a tactical configuration with parameters $r = |S|$ and $k = |Ng_i|$. If $\mathcal{B}_e = \{Ne_j + x \mid j \in T, x \in N\}$, then $(N, \mathcal{B}_e, \in)$ is a tactical configuration with parameters $r = k|T|$ and $k = |Ne_j|$.*

Proof. From (7.3) and its proof, we have $\mathcal{B} = \mathcal{B}_g \cup \mathcal{B}_e$ with $\mathcal{B}_g \cap \mathcal{B}_e = \varnothing$. So the r for $(N, \mathcal{B}, \in)$ is the sum of the r for $(N, \mathcal{B}_g, \in)$ and the r for $(N, \mathcal{B}_e, \in)$, which is $s + kt = |S| + k|T|$. Also, $k = |N/{=}_m| = |Ng_i| = |Ne_j|$.

(7.6) Lemma. *Suppose $z \neq 0$ and $0, z \in Ne_j$. Then $0, z$ belong to exactly $k = |Ne_j|$ blocks of $\mathcal{B}$, and none are of the form $Ng_i + b \in \mathcal{B}_g$.*

Proof. Certainly, $0, z \in Nz \cap (N(-z) + z)$ and $Nz \neq N(-z) + z$. So $0, z$ belong to at least these two distinct blocks.

We will obtain a bijection between $\Phi \setminus \{1\}$ and the remaining blocks containing 0 and z. For $\phi \in \Phi \setminus \{1\}$, $-\phi + 1$ defines a bijection $(-\phi + 1)x = -\phi x + x = \phi(-x) + x$, so there is a unique $a_\phi \in N$ such that $\phi(-a_\phi) + a_\phi = z$. Certainly, $0, z \in N(-a_\phi) + a_\phi$.

If $N(-a_\phi) + a_\phi = N(-a_\lambda) + a_\lambda$, then $a_\phi = a_\lambda$ since $N(-a_\phi) + a_\phi = Ng_i$ would imply $0 \neq z \in Ng_i \cap Ne_j$, which cannot be. From $\phi(-a_\phi) + a_\phi = z = \lambda(-a_\lambda) + a_\lambda$ we obtain $\phi(a_\phi) = \lambda(a_\lambda) = \lambda(a_\phi)$. This gives us $\phi = \lambda$.

So the map $\phi \mapsto N(-a_\phi) + a_\phi$ is injective.

Suppose now that $0, z \in Na + b$, where $b \notin \{0, z\}$. Then $-b = \phi(a)$ for some $\phi \in \Phi$, and so $Na = N(-b)$. This means $z = \mu(-b) + b = (-\mu + 1)b$ for some $\mu \in \Phi \setminus \{1\}$, and that $a_\mu = b$. As a consequence, $Na + b = N(-b) + b = N(-a_\mu) + a_\mu$. Thus, $\phi \mapsto N(-a_\phi) + a_\phi$ is our desired bijection. These $|\Phi| - 1$ blocks, together with the two produced at the beginning of the proof, give us $k = |Ne_j|$ blocks of $\mathcal{B}$ which contain 0 and z. In the preceding paragraph, we saw that none of the $N(-a_\phi) + a_\phi$ could be an Ng_i, so none are in $\mathcal{B}_g$. (If $0, z \in Ng_i + b$, then $Ng_i + b = Ng_i$, since Ng_i is a subgroup and $0 \in Ng_i$.)

(7.7) Corollary. *If $|N| < \infty$ and $(N, \mathcal{B}_e, \in)$ is a BIBD, then $S = \varnothing$, that is, $s = 0$.*

Proof. By the lemma, if $0, z \in Ne_j$, then 0 and z belong to exactly $k = |Ne_j|$ blocks of $\mathcal{B}_e$. So $\lambda = k = |Ne_j|$. From the equation $r(k-1) = \lambda(v-1)$, where $v = |N|$, $|\mathcal{B}_e| = vt$, $r = kt$, and $v - 1 = (s + t)(k - 1)$, we obtain $k = \lambda = r(k - 1)/(v - 1) = kt/(s + t)$, from which $s = 0$ follows.

(7.8) Lemma. *For $x, y \in N$, $x \neq y$, the number of blocks of $\mathcal{B}$ containing both x and y is the same as the number of blocks of $\mathcal{B}$ containing both 0 and $y - x$.*

Proof. If $x, y \in Na + b$, then $0, y - x \in Na + (b - x)$. If $0, y - x \in Na + b$, then $x, y \in Na + (b + x)$. So, the map $\Psi(Na + b) = Na + (b - x)$ and the map $\Lambda(Na + b) = Na + (b + x)$ have the properties that $\Psi^{-1} = \Lambda$ and $\Lambda^{-1} = \Psi$.

(7.9) Theorem. *If $|N| < \infty$ and $s = 0$, then the tactical configuration $(N, \mathcal{B}, \in)$ is a BIBD with parameter $\lambda = k$.*

Proof. For $x, y \in N$, $x \neq y$, there are, by (7.8), exactly $[0, y - x]$ blocks containing both x and y, and then by (7.6) there are exactly $\lambda = k = |Ne_i|$ blocks containing both 0 and $y - x$.

(7.10) Corollary. *If $|N| < \infty$, then $(N, \mathcal{B}_e, \in)$ is a BIBD if and only if $s = 0$.*

Proof. If $(N, \mathcal{B}_e, \in)$ is a BIBD and $s \neq 0$, then any $0, z \in Ng_i$ could only belong to Ng_i.

Now, let us concentrate on the $(N, \mathcal{B}_g, \in)$.

(7.11) Theorem. *If $|N| < \infty$, then $(N, \mathcal{B}_g, \in)$ is a BIBD if and only if $t = |T| = 0$. In this case, the parameter $\lambda = 1$, and the group $(N, +)$ is elementary abelian of order p^m. If there is an $Nx \neq N$, with $x \neq 0$, then m is composite.*

Proof. Suppose $t = 0$. Then each Nx is a subgroup of $(N, +)$ for $0 \neq x \in N$. For $x, y \in N$, $x \neq y$, there is exactly one g_i so that $x - y \in Ng_i$, hence, $x, y \in Ng_i + y$, and we have that each pair of distinct points $x, y \in N$ belongs to at least one block of $\mathcal{B}_g = \mathcal{B}$.

Now suppose $x, y \in Ng_j + c$, with $x \neq y$. So there are $\phi, \mu \in N$ such that $x = \phi g_j + c$ and $y = \mu g_j + c$. Hence, $x - y = \phi g_j - \mu g_j = \tau g_j$ for some $\tau \in N$, since Ng_i is a subgroup. Hence, $0 \neq x - y \in Ng_i \cap Ng_j$, so $Ng_i = Ng_j$. Furthermore, $Ng_i + y = Ng_j + c$. We now have that two distinct elements $x, y \in N$ belong to exactly $\lambda = 1$ blocks of $\mathcal{B}_g = \mathcal{B}$, and so we have a BIBD.

We continue with $(N, \mathcal{B}_g, \in) = (N, \mathcal{B}, \in)$, a BIBD with $\lambda = 1$ and $(Nx, +)$ a subgroup for each $0 \neq x \in N$. Since each $(Ng_i, +)$ is a subgroup, we have that each $(Ng_i, +, \cdot)$ could be a finite nearfield. By (3.66), there is a prime p and a positive integer u so that each $|Ng_i| = p^u$, and each element $x \in N \setminus \{0\}$ has additive order p.

If $p = 2$, then $(N, +)$ is abelian. If p is an odd prime, then $|\Phi|$ is even, where (N, Φ) is the Ferrero pair giving rise to $(N, +, \cdot)$. So, let $\phi \in \Phi$ be of order 2. Then $\phi(x) = -x$ for each $x \in N$. The reason for this is as follows. If $0 \neq x \in N$, then $(Nx, +, \cdot)$ could be a nearfield, and $\phi(x) \in Nx$. So $x + \phi(x) \in Nx$. But then $\phi[x + \phi(x)] = \phi(x) + \phi^2(x) = \phi(x) + x = x + \phi(x)$, since $(Nx, +)$ is abelian. Since $x + \phi(x)$ is a fixed point for ϕ, we have $0 = x + \phi(x)$, so $\phi(x) = -x$. Now take $a, b \in N$. Then $a + b = \phi(-a) + \phi(-b) = \phi[(-a) + (-b)] = -[(-a) + (-b)] = [-(-b)] + [-(-a)] = b + a$. Hence, $(N, +)$ is abelian.

Since Φ is regular, $|\Phi| \mid (|N| - 1)$, so if $|Nx| = p^n$ and $|N| = p^m$, then $(p^n - 1) \mid (p^m - 1)$, and so m is composite if $n \neq m$.

Conversely, suppose that $(N, \mathcal{B}_g, \in)$ is a BIBD. If $x \neq y$ and both belong to $Na + c \in \mathcal{B}_g$, then $x = \phi a + c$ and $y = \mu a + c$ for some $\phi, \mu \in N$. Hence, $x - y = \phi a - \mu a$. Since $(Na, +)$ is a subgroup of $(N, +)$, then $x - y = \tau a$ for some $\tau \in N$, and so $Na = N(x - y)$. If $t \neq 0$, then there would be $x, y \in N$, $x \neq y$, such that $x - y = e_1$. Then $Ne_1 = N(x - y)$. But then x and y could not belong to any $Na + c \in \mathcal{B}_g$, contrary to $(N, \mathcal{B}_g, \in)$ being a BIBD.

(7.12) Definitions. If $(N, \mathcal{B}, \in)$ is a BIBD with $\lambda = 1$, then it is also an *incidence space*, that is, if $x, y \in N$ and $x \neq y$, then there is exactly one 'line' $B \in \mathcal{B}$ with $x, y \in B$, and each $B \in \mathcal{B}$ has $|B| \geq 2$. If $N' \subseteq N$ and $\mathcal{B}' \subseteq \mathcal{B}$, then $(N', \mathcal{B}', \in)$ is a *subspace* of $(N, \mathcal{B}, \in)$ if for each $x, y \in N'$, $x \neq y$, the $B \in \mathcal{B}$ with $x, y \in B$ is also in $\mathcal{B}'$, and each $B \in \mathcal{B}'$ is a subset of N'. Three distinct points $x, y, z \in N$ are *collinear* if there is a $B \in \mathcal{B}$ such that $x, y, z \in B$. If $S \subseteq N$, then the *hull* of S is the intersection of all subspaces $(N', \mathcal{B}', \in)$ of $(N, \mathcal{B}, \in)$ where $S \subseteq N'$. For a family of subspaces $(N_i, \mathcal{B}_i, \in)$, the intersection $\cap(N_i, \mathcal{B}_i, \in) = (\cap N_i, \cap \mathcal{B}_i, \in)$. So, if $x, y \in \cap N_i$, then the

$B \in \mathcal{B}$ with $x, y \in B$ is in each $\mathcal{B}_i$, hence it is also in $\cap \mathcal{B}_i$. This makes $(\cap N_i, \cap \mathcal{B}_i, \in)$ a subspace of $(N, \mathcal{B}, \in)$ also. A *plane* $(E, \mathcal{E}, \in)$ in $(N, \mathcal{B}, \in)$ is the hull of some set of three noncollinear points of N. Two 'lines' $G, H \in \mathcal{B}$ are *parallel*, and we write $G \parallel H$, if $G = H$ or $G \cap H = \varnothing$ and there is a plane $(E, \mathcal{E}, \in)$ of $(N, \mathcal{B}, \in)$ with $G, H \in \mathcal{E}$. Our incidence space is called an *affine space* provided every plane $(E, \mathcal{E}, \in)$ in $(N, \mathcal{B}, \in)$ is an affine plane. And just what is an affine plane?

An incidence space $(E, \mathcal{E}, \in)$ is an *affine plane* provided there are at least three noncollinear points in E, and if $x \in E$ and $B \in \mathcal{E}$ with $x \notin B$, then there is a unique $B' \in \mathcal{B}$ such that $x \in B'$ and $B \cap B' = \varnothing$, that is, $B' \parallel B$.

Two incidence spaces $(N, \mathcal{B}, \in)$ and $(N', \mathcal{B}', \in)$ are *isomorphic* if there is a bijection $f : N \to N'$ so that if $B \in \mathcal{B}$, then $f(B) \in \mathcal{B}'$. So, $x \in B$ if and only if $f(x) \in f(B)$. Two subspaces $(N_1, \mathcal{B}_1, \in)$ and $(N_2, \mathcal{B}_2, \in)$ are *isomorphic* if there is a bijection $f : N_1 \to N_2$ so that if $B_1 \in \mathcal{B}_1$, then $f(B_1) \in \mathcal{B}_2$.

(7.13) Theorem. *If* $|N| < \infty$ *and* $t = 0$, *then* $(N, \mathcal{B}, \in)$ *is an affine space.*

Proof. Take $a, b \in N \setminus \{0\}$ with $Na \neq Nb$. The subgroups Na and Nb have $Na \cap Nb = \{0\}$, and since $(N, +)$ is abelian, then $P(a, b) = Na + Nb$ is a subgroup, a direct sum of Na and Nb. Let $\mathcal{B}(a, b) = \{Nc + x \in \mathcal{B} \mid Nc + x \subseteq P(a, b), c \neq 0\}$. Our proof will be given in six steps.

Step 1. If $(N', \mathcal{B}', \in)$ is a subspace, define $N' + a = \{n' + a \mid n' \in N'\}$ and $\mathcal{B}' + a = \{B' + a \mid B' \in \mathcal{B}'\}$. Then $(N' + a, \mathcal{B}' + a, \in)$ is a subspace, and the map $f_a : N' \to N' + a$, defined by $f_a(n') = n' + a$, is an isomorphism of subspaces.

Proof. Take $x + a, y + a \in N' + a$, and let B be the unique line in $\mathcal{B}$ containing $x + a$ and $y + a$. Then $x, y \in N'$ and $B - a$ is the unique line in $\mathcal{B}$ containing x and y. So $B - a \subseteq N'$ and $B - a \in \mathcal{B}'$. Hence, $B \subseteq N' + a$ and $B \in \mathcal{B}' + a$. Certainly, f_a is a bijection, since $(N, +)$ is a group, and if $B \in \mathcal{B}'$, then $f_a(B) = B + a \in \mathcal{B}' + a$.

Step 2. $(P(a, b), \mathcal{B}(a, b), \in)$ is a subspace.

Proof. For $x, y \in P(a, b)$, $x \neq y$, $x, y \in Nc + d \in \mathcal{B}$, then there are $\mu_x, \phi_x, \mu_y, \phi_y \in N$ so that $x = \mu_x a + \phi_x b$ and $y = \mu_y a + \phi_y b$. Also, there are $\gamma_x, \gamma_y \in N$ so that $x - d = \gamma_x c$ and $y - d = \gamma_y c$. From this, we obtain $0 \neq x - y = \mu_x a - \mu_y a + \phi_x b - \phi_y b = \gamma_x c - \gamma_y c$. Since Na, Nb, and Nc are subgroups, there are $\mu, \phi, \gamma \in N$ so that $0 \neq x - y = \mu a + \phi b = \gamma c \in P(a, b)$. Hence, $Nc \subseteq P(a, b)$. From $x = \gamma_x c + d$, we have $x - \gamma_x c = d \in P(a, b)$, and so $Nc + d \subseteq P(a, b)$. Hence, $(P(a, b), \mathcal{B}(a, b), \in)$ is a subspace.

Step 3. $(P(a, b), \mathcal{B}(a, b), \in)$ is the hull of the noncollinear points $0, a$, and b, and hence is a plane.

Proof. We have that $(P(a,b), \mathcal{B}(a,b), \in)$ is a subspace, that $0, a, b \in P(a,b)$, and that $0, a$, and b are noncollinear. Let $(N', \mathcal{B}', \in)$ be a subspace of $(N, \mathcal{B}, \in)$ with $0, a, b \in N'$. Then $0, a \in Na \in \mathcal{B}'$ and $0, b \in Nb \in \mathcal{B}'$. So $Na \cup Nb \subseteq N'$. We need to show that $P(a,b) \subseteq N'$ and $\mathcal{B}(a,b) \subseteq \mathcal{B}'$. If $\mu a + \phi b \in P(a,b)$ and $0 \in \{\mu a, \phi b\}$, then $\mu a + \phi b \in N'$. So, we will assume $0 \notin \{\mu a, \phi b\}$.

We will identify representatives of $N^*/{=_m}$ with their automorphism in Φ, namely, for $\alpha \in N^*$, we have $\alpha(x) = \alpha x$. For $\alpha, \beta \in \Phi$, $\alpha - \beta : N \to N$ is defined by $(\alpha - \beta)x = \alpha x - \beta x$. Since $t = 0$, $(\alpha - \beta) : Nx \to Nx$ for each $x \neq 0$. For $\mu a \in Na \setminus \{0\}$, $(1 - \mu) : Nb \to Nb$ is a bijection, so $(1 - \mu)^{-1}$ exists. Certainly, $\phi b \in Nb$, so $(1 - \mu)^{-1}\phi b \in Nb$. Now $B = N(a - (1 - \mu)^{-1}\phi(b) + (1 - \mu)^{-1}\phi b$ is the unique line in $\mathcal{B}$ containing $(1 - \mu)^{-1}\phi b$ and a. Since $(1 - \mu)^{-1}\phi b, a \in Na \cup Nb \subseteq N'$, then $B \subseteq N'$. The nice thing about this is that now $\mu a - \mu (1 - \mu)^{-1}\phi b + (1 - \mu)^{-1}\phi b = \mu a + (1 - \mu)(1 - \mu)^{-1}\phi b = \mu a + \phi b \in N'$. Hence, $P(a,b) \subseteq N'$, and so $\mathcal{B}(a,b) \subseteq \mathcal{B}'$.

Step 4. $(P(a,b), \mathcal{B}(a,b), \in)$ is an affine plane.

Proof. We start by noting that $0, a$ and b are three noncollinear points in $P(a,b)$. Take $x = \mu a + \phi b \in P(a,b)$, and a line $B = Nc + d$ with $x \notin B$. Then $x \in B' = Nc + x$ and $B' \cap B = \varnothing$, that is, $B' \parallel B$, since Nc is a subgroup of $(N, +)$.

Suppose $x \in F = Nf + x$ and $F \neq B'$. Then $Nc \neq Nf$. We want to show that $F \cap B \neq \varnothing$. If $F \cap B = \varnothing$, then $(F - d) \cap (B - d) = (Nf + x - d) \cap Nc = \varnothing$. If $Nc = \{\gamma_1 c, \gamma_2 c, \ldots, \gamma_k c\}$, where $k = |\Phi| + 1 = |Nc|$, then $(F - d + \gamma_i(c) \cap (Nc + \gamma_i(c)) = (F - d + \gamma_i(c) \cap Nc = \varnothing$ for each $\gamma_i c \in Nc$. But the index of Nf in $P(a,b)$ is $[P(a,b) : Nf] = k$, so then since $[\cup_{\gamma_i c \in Nc}(F - d + \gamma_i c)] \cap Nc = \varnothing$, we must have $\gamma_i c, \gamma_j c \in Nc$, with $\gamma_i c \neq \gamma_j c$, such that $Nf + x - d + \gamma_i c = F - d + \gamma_i c = F - d + \gamma_j c = Nf + x - d + \gamma_j c$. Hence, $0 \neq \gamma_i c - \gamma_j c \in Nf \cap Nc$. But, since $Nc \neq Nf$, then $Nf \cap Nc = \{0\}$. So there is only the one line $B' \in \mathcal{B}(a,b)$ with $x \in B'$ and $B' \cap B = \varnothing$. This makes $(P(a,b), \mathcal{B}(a,b), \in)$ an affine plane.

Step 5. For $c \in N$, $(P(a,b) + c, \mathcal{B}(a,b) + c, \in)$ is an affine plane.

Proof. By Step 1, $f_a : P(a,b) \to P(a,b) + c$ is an isomorphism from the subspace $(P(a,b), \mathcal{B}(a,b), \in)$ onto $(P(a,b) + c, \mathcal{B}(a,b) + c, \in)$.

Step 6. If $(E, \mathcal{E}, \in)$ is a plane in $(N, \mathcal{B}, \in)$, then $(E, \mathcal{E}, \in) = (P(a,b) + c, \mathcal{B}(a,b) + c, \in)$ for some $a, b, c \in N$. Hence, $(E, \mathcal{E}, \in)$ is an affine plane.

Being a plane in $(N, \mathcal{B}, \in)$, $(E, \mathcal{E}, \in)$ is the hull of three noncollinear points $x, y, z \in E$. So $0, y - x$, and $z - x$ are noncollinear in the plane $(E - x, \mathcal{E} - x, \in) = (P(y - x, z - x), \mathcal{B}(y - x, z - x), \in)$. Let $a = y - x, b = z - x$, and $c = x$. Then $(E, \mathcal{E}, \in) = (P(a,b) + c, \mathcal{B}(a,b) + c, \in)$.

With Step 6, we have finished the proof of (7.13).

(7.14) Theorem. *For $|N| < \infty$, $(N, \mathcal{B}, \in)$ is a BIBD if and only if $0 \in \{s, t\}$.*

Proof. From (7.10) and (7.11), we have that $0 \in \{s, t\}$ implies that $(N, \mathcal{B}, \in)$ is a BIBD. Suppose that $0 \notin \{s, t\}$, but yet $(N, \mathcal{B}, \in)$ is a BIBD. For $0, z \in Ne_1$, $z \neq 0$, $[0, z] = \lambda = k = |Ne_1|$ by (7.6). From the equations $v - 1 = (s+t)(k-1)$ and $r(k-1) = \lambda(v-1)$, we obtain, since $r = s + kt$ by (7.3), that $(s+kt)(k-1) = k(s+t)(k-1)$, and, consequently, $s + kt = ks + kt$, which implies $s = 0$ or $k = 1$. Either case gives a contradiction, since $k = |Ne_1| \geq 3$.

The easiest way to obtain BIBDs $(N, \mathcal{B}, \in)$ where $\lambda = k$ is to choose Φ so that $|\Phi| + 1$ does not divide $|N|$, so then the Nx cannot be subgroups of $(N, +)$.

(7.15) Examples. (a) Let $(A, +)$ be a finite abelian group with $(|A|, 6) = 1$. If $\Phi = \{1, \phi\}$, where $\phi(x) = -x$, then each $|Ax| = 3$, for $0 \neq x \in A$, and $(A, +, \cdot)$ is a planar nearring produced by the Ferrero pair and 'The Factory'. So, $s = 0$, and $(A, \mathcal{B}, \in)$ is a BIBD with parameters $(v, b, r, k, \lambda) = (|A|, v(v-1)/6, 3(v-1)/2, 3, 3)$.

(b) Let $(F, +, \cdot)$ be a finite field with $|F| = p^m$, p a prime. Choose Φ as a nontrivial subgroup of $(F^*, \cdot)$ of order d, and construct a planar nearring from the Ferrero pair (F, Φ) referred to in (4.16). As long as d is not of the form $p^n - 1$, then no $(Fa, +)$, $0 \neq a \in F$, can be a subgroup of $(F, +)$. Hence, $s = 0$ and $(F, \mathcal{B}, \in)$ is a BIBD with parameters $v = p^m$, $b = p^m(p^m - 1)/d$, $r = (d+1)(p^m - 1)/d$, and $k = \lambda = d+1$. If $d = p^n - 1$ for some n, then each $(Fa, +)$ is a subgroup of $(F, +)$, $t = 0$, $(F, \mathcal{B}, \in)$ is a BIBD with parameters $v = p^m$, $b = p^m(p^m - 1)/p^n(p^n - 1)$, $r = (p^m - 1)/(p^n - 1)$, $k = p^n$, and $\lambda = 1$.

The *efficiency* E of a BIBD is measured by $E = \lambda v/rk$. If $\lambda = k$, then $E = v/r$. Let p be an odd prime and let $d = (p^m - 1)/2$. If $p^m \neq 3$, we obtain $r = p^m + 1$, so $E = p^m/(p^m + 1)$. Thus E can be arbitrarily close to 1. Note that $0 < E < 1$ for any BIBD. Some have said that it is *good* if $0.9 < E < 1.0$, and others have said that it is *good* if $0.75 < E < 1.00$. Using either of these, we can make some *good* BIBDs.

(c) Helmut Karzel made a very important observation about the following example, as we shall explain later. Start with the cyclic group of order 9, $(Z_9, +)$. Then $\Phi = \{1, \phi\}$, where $\phi(x) = -x$, defines a Ferrero pair (Z_9, Φ). Let $N = Z_9$. The basic blocks are $N3 = Ng_1 = \{0, 3, 6\}$, $N1 = Ne_1 = \{0, 1, 8\}$, $N2 = Ne_2 = \{0, 2, 7\}$, and $N4 = Ne_3 = \{0, 4, 5\}$. Hence, $s = 1$ and $t = 3$

From what we have seen, $(N, \mathcal{B}, \in)$ will be a BIBD if and only if $0 \in \{s, t\}$. So what geometry can come out of a $(N, \mathcal{B}, \in)$ with $0 \notin \{s, t\}$? Well, it so happens that there seems to be plenty, but that will come later. Let us now see that our example with Z_9 can be readily generalized to any $(Z_{p^2}, +)$, where p is an odd prime.

Consider the ring of integers $(Z_{p^{r+1}}, +, \cdot)$ modulo p^{r+1}, where p is an odd prime. We want a Φ of order $p-1$ so that $(Z_{p^{r+1}}, \Phi)$ will be a Ferrero pair. Of course, such a Φ must come from $(\mathcal{U}(Z_{p^{r+1}}), \cdot)$, the group of units of the ring $(Z_{p^{r+1}}, +, \cdot)$. Now $\mathcal{U}(Z_{p^{r+1}}) = Z_{p^{r+1}} \setminus \{0, p, 2p, \ldots, (p^r - 1)p\}$ and has order $p^{r+1} - p^r = p^r(p-1)$, and $u \in \mathcal{U}(Z_{p^{r+1}})$ defines a fixed point free automorphism $u^\bullet$, where $u^\bullet(x) = ux$, if and only if $u - 1 \in \mathcal{U}(Z_{p^{r+1}})$. Certainly, $C(p) = \{1, p+1, 2p+1, \ldots, (p^r - 1)p + 1\}$ is a subgroup of $\mathcal{U}(Z_{p^{r+1}})$, and since $\mathcal{U}(Z_{p^{r+1}}) = C(p)\Phi$, a direct product, where Φ is a subgroup of $\mathcal{U}(Z_{p^{r+1}})$ of order $p-1$, and of course, $C(p) \cap \Phi = \{1\}$, then Φ is essentially a group of fixed point free automorphisms of $(Z_{p^{r+1}}, +)$. Hence, $(Z_{p^{r+1}}, \Phi)$ is a Ferrero pair. With $N = Z_{p^{r+1}}$, we have $Np^r = Ng_1 = \{0, p^r, 2p^r, \ldots, (p-1)p^r\}$, and all other basic blocks $Nx = Ne_j$. So $s = 1$ and $t = p^r + p^{r-1} + \cdots + p$. Karzel's observation about $(Z_9, \mathcal{B}, \in)$, referred to above, extends to these $(Z_{p^2}, \mathcal{B}, \in)$, p being an odd prime.

But before we tell you what Karzel's observation was, we shall see that we can even obtain BIBDs from $(N, \mathcal{B}, \in)$ when $0 \notin \{s, t\}$, but we will have to modify our definition of a BIBD to include the broader statistical point of view.

Actually, the idea of a BIBD developed from statistical considerations for the design of agricultural experiments. The value of BIBDs for the designs of experiments is certainly a strong motivating force for so much active research concerning BIBDs, but the balance and symmetry of these structures give rise to so many interesting results and problems that their study is justified independently of statistical considerations.

Given a solution to the equations $vr = bk$ and $\lambda(v-1) = r(k-1)$, there is the desire to find a BIBD $(S, \mathcal{S}, \in)$ with these parameters. Statisticians are even willing to take certain blocks and repeat them, but consider them distinct in order to have a BIBD with such parameters. Our definition has not allowed such repetition, which seems the thing to do for geometric considerations. A line, circle, or segment should be determined by the points on it. But, for a finite planar nearring $(N, +, \cdot)$, if $0 \notin \{s, t\}$ for $(N, \mathcal{B}, \in)$, we can still construct a BIBD if we adopt the statistical point of view. We will do so presently for any $(N, \mathcal{B}, \in)$.

(7.16) Definitions. Both the geometric and the statistical points of view begin with a set N. The geometrical view regards an $x \in N$ as a 'point', and the statistical point of view regards an $x \in N$ as a 'variety'. A *statistical* BIBD is a finite set N of v 'varieties', and an indexed set $\{B_i \mid i \in I\}$ of subsets $B_i \subset N$. The B_i are called 'blocks', and there are positive integers k, r, and λ satisfying the following:

(i) For each $i \in I$, $|B_i| = k$.
(ii) For each $x \in N$, we have $|\{i \in I \mid x \in B_i\}| = r$.
(iii) For each pair $x, y \in N$, $x \neq y$, we have $|\{i \in I \mid x, y \in B_i\}| = \lambda$.
A *geometric* BIBD is a statistical BIBD satisfying in addition:
(iv) If $i, j \in I$ and $i \neq j$, then $B_i \neq B_j$.

For both points of view, there is the additional parameter $b = |I|$, and these parameters are, in each case, related by the equations $vr = bk$ and $(v - 1)\lambda = r(k - 1)$. So the geometric BIBD is more restrictive than the statistical BIBD. Condition (iv) for the geometric BIBD shows that a block B_i is determined by its points. In contrast, for a statistical BIBD, it may be that $i \neq j$, but yet $B_i = B_j$. So, a block B_i is not necessarily determined by its varieties (points).

In the future, as it was in the past, a BIBD is a *geometrical* BIBD. If we are referring to a *statistical* BIBD, then we will say so.

We start with a finite planar nearring $(N, +, \cdot)$ and consider the $(N, \mathcal{B}, \in)$, a tactical configuration in all cases. The basic blocks are the subgroups of $(N, +)$, namely, $Ng_1, Ng_2, \ldots, Ng_s$, together with those which are not subgroups of $(N, +)$, namely, $Ne_1, Ne_2, \ldots, Ne_t$. With this notation, we also have that $\{g_1, g_2, \ldots, g_s, e_1, e_2, \ldots, e_t\}$ is a complete set of representatives of the set of basic blocks of N, and that each g_i and each e_j is nonzero.

We want to construct a statistical BIBD with $\lambda = k$. Our index set will be

$$I = \{(g_i, b) \mid 1 \leq i \leq s, \, b \in N\} \cup \{(e_j, b) \mid 1 \leq j \leq t, \, b \in N\}.$$

Our blocks will be $B_{(x,b)} = Nx + b$, with $(x, b) \in I$. So $\{B_{(x,b)} \mid (x, b) \in I\}$ is an indexed set of subsets of N. Certainly, $v = |N|$ and $k = |B_{(x,b)}| = |N/{=_m}|$, and $b = (s + t)v$, since our repeated blocks come only from repeating each coset $Ng_i + b$ exactly k times. Realizing this, one also gets $r = (s + t)k$. It remains to show that $\lambda = k$, and we will now show this.

Take any $x, y \in N$ with $x \neq y$. The blocks containing x and y depend upon where $y - x$ is. Since $y - x \neq 0$, there is exactly one basic block containing $y - x$. Suppose that $y - x \in Ng_i$. Then $x, y \in Ng_i + x$. If it should be that the same x and y belong to $Ne_j + b$, then $0, y - x \in Ne_j + (b - x)$. This makes $Ne_j = N(x - b)$, and $y - x = \tau(x - b) + (b - x)$ for some $\tau \in N$. (If $x = b$, then $Ne_j = \{0\}$, and so $y = x$. Hence, $x \neq b$.) Thus $\tau(b - x) = (b - x) + (x - y)$. Since all the planar nearrings constructed from a Ferrero pair yield the same blocks $\mathcal{B}$, we may, without loss of generality, assume that each $(Ng_i, +, \cdot)$ is a planar nearfield. Now $(Ng_i, +, \cdot)$ is a planar nearfield, and since Ng_i contains exactly one representative of each equivalence class in $N/{=_m}$, the equation $\tau(b - x) = (b - x) + (x - y)$ may be replaced by $\mu(b - x) = 1_i(b - x) + (x - y)$, where $\mu, 1_i \in Ng_i$, and 1_i is the multiplicative identity in the nearfield $(Ng_i, +, \cdot)$. Consider the equation $\mu w = 1_i w + (x - y)$. It has a unique solution in Ng_i as well as in N, and since $b - x$ is a solution, we have $b - x$ and $x - b$ in Ng_i. So $Ng_i = N(x - b) = Ne_j$, an unacceptable circumstance. So, if $y - x \in Ng_i$, we cannot tolerate $x, y \in Ne_j + b$. So, if $x, y \in Ne_j + b$ for some j, then 0 and $y - x$ cannot both belong to some Ng_i.

For a pair x, y, $x \neq y$, with $y - x \in Ng_i$ for some i, then $x, y \in Ng_i + x$, and $x, y \in Ng_i + (a + x)$ for each $a \in Ng_i$. So this gives us exactly k indices $(g_i, b) \in I$ with $x, y \in B(g_i, b) = Ng_i + b$, where $k = |Ng_i|$.

For a pair x, y, $x \neq y$, with $y - x \in Ne_j$, then $0, y - x \in Ne_j$. By (7.6), $0, y - x$ belong to exactly k blocks, and none are of the form $Ng_i + b$, so they are all of the form $Ne_j + b$. By (7.8), x, y also belong to exactly k blocks, and they are all of the form $Ne_j + c$ for some $(e_j, c) \in I$. Thus (iii) of (7.16) is satisfied, as are (i) and (ii), and this makes N, together with $\{B_{(x,b)} \mid (x, b) \in I\}$, a statistical BIBD.

What are the parameters? Certainly, $v = |N|$ and $k = |N/{=_m}| = |Ng_i| = |Ne_j|$. Also, $b = |I| = v(s + t) = v(v - 1)/(k - 1)$, since $s + t = (v - 1)/(k - 1)$. Finally, $r = k(s + t) = k(v - 1)/(k - 1)$, and $\lambda = k$. We have

(7.17) Theorem. *Let $(N, +, \cdot)$ be a finite planar nearring whose $s + t$ basic blocks are $Ng_1, Ng_2, \ldots, Ng_s$ and $Ne_1, Ne_2, \ldots, Ne_t$, where each $(Ng_i, +)$ is a subgroup of $(N, +)$ and each Ne_i is not a subgroup of $(N, +)$. Let $I = \{(x, b) \mid b \in N, x \in \{g_1, g_2, \ldots, g_s, e_1, e_2, \ldots, e_t\}\}$ be an index set, and define $B_{(x,b)} = Nx + b$. Then N, together with the indexed set $\{B_{(x,b)} \mid (x, b) \in I\}$, is a statistical BIBD with parameters $v = |N|$, $k = |N/{=_m}|$, $b = v(v - 1)/(k - 1)$, $r = k(v - 1)/(k - 1)$, and $\lambda = k$. This statistical BIBD is a geometric BIBD if and only if $s = 0$.*

(7.18) Examples. The larger s, the more repetition of blocks we have in a statistical BIBD described in (7.17). Below are examples of BIBD with the same parameters (v, b, r, k, λ), but where $s = 1$, $s = p + 1$, and $s = 2$, respectively.

(a) For an odd prime p, consider a planar nearring $(Z_{p^2}, +, *)$ arising from (c) of (7.15). Here, $s = 1$.

(b) For the same odd prime p, let F be a field of order p^2. Let $|\Phi| = p - 1$, and consider a planar nearring $(F, +, *)$ arising from (b) of (7.15). Here, $s = p + 1$.

(c) For the same odd prime p, let $(N, +) = (Z_p \oplus Z_p, +)$. If $\Phi_1 = \{(a, b) \mid a^2 - b^2 = 1\}$, then for $(a, b) \in \Phi_1$, define $(a, b)^*(c, d) = (ac + bd, ad + bc)$. If $\Phi = \{(a, b)^* \mid (a, b) \in \Phi_1\}$, then it is easy to see that $(Z_p \oplus Z_p, \Phi)$ is a Ferrero pair. How big is Φ? From $a^2 - b^2 = 1 = (a - b)(a + b)$, we get $a - b = c$ and $a + b = c^{-1}$ for some $c \in Z_p^*$. Thus $\Phi_1 = \{((c^2 + 1)/2c, (c^2 - 1)/2c) \mid c \in Z_p^*\}$. It is also easy to show that $|\Phi_1| = |\Phi| = p - 1$, hence $k = p$. If $N = Z_p \oplus Z_p$, then $(N(c, d), +)$ is a subgroup of $(N, +)$ if and only if $c = d$ or $c = -d$.

Hence, $s = 2$.

(7.19) Problem. The statistical BIBDs from the three nonisomorphic examples of (7.18) will have the same parameter (v, b, r, k, λ). From a statistical point of view, does any one of these BIBDs have any advantage over any of the others?

7.2. Karzel's observation

In 1979, Helmut Karzel made the following observation about (Z_9, Φ) of (c) of (7.15). In $(Z_9, \mathcal{B}, \in)$, there are three substructures $(Z_9, \mathcal{L}_i, \in)$, $1 \leq i \leq 3$, such that each is an affine plane and $\cap_{i=1}^3 \mathcal{L}_i = Z_9/\{0, 3, 6\}$, the cosets of the subgroup $\{0, 3, 6\}$ in Z_9. From this observation, Karzel developed the following theory.

What we need for this development is a finite planar nearring $(N, +, \cdot)$ with $0 \notin \{s, t\}$. That is, there is a $g \in N \setminus \{0\}$ such that Ng is a subgroup of $(N, +)$, and there is an $e \in N \setminus \{0\}$ such that Ne is not a subgroup of $(N, +)$. From N, we construct $\mathcal{F} = \{F \mid F$ is a subgroup of $(N, +)$ and $NF \subseteq F\}$, and call members of $\mathcal{F}$ the N-*comodules of* $(N, +)$. (See (13.2).) Note that $\{0\}$, Ng, $N \in \mathcal{F}$, and for $\phi \in \Phi \setminus \{1\}$, $(-1 + \phi)(F) = (\phi - 1)(F) = F$ for each $F \in \mathcal{F}$. We select from $\mathcal{F}$ a chain of subgroups such that:
(a) $\{0\} = F_0 < F_1 < \cdots < F_r < F_{r+1} = N$ and $r \geq 1$.
(b) For $x \in F_1$, we have $Nx \in \mathcal{F}$, and there is a $y \in F_2 \setminus F_1$ such that $Ny \notin \mathcal{F}$, so Nx is a subgroup of $(N, +)$ and Ny is not a subgroup of $(N, +)$.
(c) If $F \in \mathcal{F}$ and $F_i \leq F \leq F_{i+1}$ for some i, $1 \leq i \leq r$, then either $F_i = F$ or $F = F_{i+1}$.

For $1 \leq i \leq r$, let $R_i \subseteq F_{i+1}$ be a set of right coset representatives of F_i in F_{i+1}, with $0 \in R_i$. We select the following blocks from $\mathcal{B}$:
$\mathbf{L_1} = \{Nx + t \mid x \in F_1 \setminus \{0\}, \ t \in N\}$;
for $2 \leq i \leq r$,
$\mathbf{L_i} = \{Nx + t_{i-1} + t_i + \cdots + t_r \mid x \in F_i \setminus F_{i-1}, t_{i-1} \in F_{i-1}, t_j \in R_j, i \leq j \leq r\}$;
$\mathbf{L_{r+1}} = \{Nx + t_r \mid x \in N \setminus F_r, \ t_r \in F_r\}$.
Now let $\mathbf{L} = \cup_{i=1}^{r+1} \mathbf{L_i}$, and call the blocks of $\mathbf{L}$ *lines*.

(7.20) Proposition. *If* $F \in \mathcal{F}$ *and* $x \in N \setminus F$, *then* $Nx \cap F = \{0\}$. *If* $f \in F \setminus \{0\}$, *then* $Nx \cap (Nx + f) = Nx \cap (f + Nx) = \varnothing$.

Proof. If $y \in Nx \cap F$ and $y \neq 0$, then $Ny = Nx$ and so $Ny \subseteq F$ since $F \in \mathcal{F}$. Hence, $x \in F$. Suppose $y \in Nx \cap (Nx + f)$. Then $y = \phi x + f$ for some $\phi \in N$. Now $y, \phi x \in Nx$. If $y = \phi x$, then $f = 0$, so $y \neq \phi x$. Since $x \in N \setminus F$, then $Nx \cap F = \{0\}$. If $\phi x = 0$, then $y = f$, and $y \in Nx \cap F = \{0\}$, so $y = 0$ and again $y = \phi x$. So $\phi x \neq 0$. If $y = 0$, then $\phi x = -f \in F$ and $\phi x \in Nx \cap F = \{0\}$, so $\phi x = 0$. Thus $0 \notin \{\phi x, y\}$ and $\phi \neq_m 0$. With $y \in Nx$, there is a $\theta \in N$, $\theta \neq_m 0$, $\theta \neq_m \phi$, such that $\theta x = y$. Now $f = -\phi x + y = (-1 + \theta \phi^{-1})(\phi x)$, so $\phi x \in Nx \cap F = \{0\}$. This makes

$Nx \cap (Nx + f) = \varnothing$. Nearly the same proof will give us $Nx \cap (f + Nx) = \varnothing$.

(7.21) Proposition. *If $F \in \mathcal{F}$, $f \in F$, and $c, d \in N \setminus F$ have $|Nc \cap (Nd + f)| \geq 2$, then $f = 0$ and $Nc = Nd$.*

Proof. Let $c_1, c_2 \in Nc \cap (Nd + f)$ with $c_1 \neq c_2$ and $c_1 \neq 0$. Then there is a $\phi \in N$ such that $d_2 + f = c_2 = \phi c_1 = \phi(d_1 + f) = \phi d_1 + \phi f$ for appropriate $d_1, d_2 \in Nd$, where $d_i + f = c_i$. This makes $\phi d_1 = d_2 + f - \phi f \in Nd \cap (Nd + (f - \phi f))$. Since $F \in \mathcal{F}$ and $d \in N \setminus F$, we have $\phi f = f$. If $\phi =_m 0$, then $f = 0$, and if $\phi \neq_m 0$, then $f = 0$ or $\phi =_m 1_u$ for some u. If $\phi =_m 1_u$, then $c_2 = \phi c_1 = c_1$. Hence, $\phi \neq_m 1_u$, and so $f = 0$. This makes $|Nc \cap Nd| \geq 2$, and so $Nc = Nd$.

(7.22) Proposition. *For $F \in \mathcal{F}$, $f \in F$, $c \in N \setminus F$, and $t \in N$, we have $|(Nc + f) \cap (F + t)| \leq 1$.*

Proof. Suppose $c_1, c_2 \in Nc$, $f_1, f_2 \in F$, and $c_i + f = f_i + t \in (Nc + f) \cap (F + t)$, for $i = 1, 2$. Then $t - f = -f_i + c_i$, so $-f_1 + c_1 = -f_2 + c_2$ and we have $c_1 = f_1 - f_2 + c_2 \in Nc \cap ((f_1 - f_2) + Nc)$. Since $f_1 - f_2 \in F$ and $c \in N \setminus F$, if $f_1 - f_2 \neq 0$, then $Nc \cap ((f_1 - f_2) + Nc) = \varnothing$ by (7.20). Thus $f_1 = f_2$ and $c_1 = c_2$. This puts at most one point in the intersection.

(7.23) Theorem. *Any two distinct lines $X, Y \in \mathbf{L}$ have at most one point in common.*

Proof. There are three types of lines: (1) those in $\mathbf{L}_1$; (2) those in some $\mathbf{L}_i$, $2 \leq i \leq r$; and (3) those in $\mathbf{L}_{r+1}$. So there are six cases: Case 1: X and Y are of type (1); Case 2: X is of type (3) and Y is of type (1); Case 3: X and Y are each of type (3); Case 4: X is of type (1) and Y is of type (2); Case 5: X is of type (3) and Y is of type (2); and Case 6: X and Y are each of type (2). We shall consider each case separately.

Case 1: Here, $X = Nx + t$ and $Y = Ny + u$, where Nx and Ny are subgroups of $(N, +)$. So, if $a, b \in X \cap Y$, then $Nx + t = Nx + a = Nx + b$ and $Ny + u = Ny + a = Ny + b$. This puts $a - b \in Nx \cap Ny = \{0\}$, since $X \neq Y$. Hence, $a = b$.

Case 2: Here, $X = Nx + t_r$ and $Y = Ny + u$, where $x \in N \setminus F_r$, $t_r \in F_r$, $y \in F_1$, and $u \in N$. Hence, $Ny \subseteq F_r$, so $|X \cap Y| = |(Nx + t_r) \cap (Ny + u)| \leq |(Nx + t_r) \cap (F_r + u)| \leq 1$ by (7.22).

Case 3: Here, $X = Nx + t_r$ and $Y = Ny + u_r$ with $x, y \in N \setminus F_r$, $F_r \in \mathcal{F}$, and $u_r, t_r, u_r - t_r \in F_r$. If $|X \cap Y| \geq 2$, then $|Nx \cap (Ny + u_r - t_r)| \geq 2$. By (7.21), we have $u_r = t_r$ and $Nx = Ny$, hence $X = Y$.

Case 4: Here, $X = Nx + t$, with $x \in F_1$ and $t \in N$, and $Y = Ny + u_{j-1} + u_j + \cdots + u_r \in \mathbf{L}_j$ for some j, $2 \leq j \leq r$. Hence, $y \in F_j \setminus F_{j-1}$, $u_{j-1} \in F_{j-1}$, and each $u_k \in R_k$, $j \leq k \leq r$. Let $w = u_{j-1} + u_j + \cdots + u_r$. Now $x \in F_1 \subseteq F_{j-1}$, so $x \in F_{j-1}$ and $y \in F_j \setminus F_{j-1}$. By (7.22), $|(Ny + 0) \cap$

$(F_{j-1}+t-w)| = |Ny \cap (F_{j-1}+t-w)| \leq 1$. Hence, $|Ny \cap (Nx+t-w)| \leq 1$, and $|(Ny+w) \cap (Nx+t)| = |Y \cap X| \leq 1$.

Case 5: Here, $X = Nx + t_r$ with $t_r \in F_r$ and $x \in N \setminus F_r$, and $Y = Ny + u_{j-1} + u_j + \cdots + u_r$, with $y \in F_j \setminus F_{j-1}$. So $Ny \subseteq F_j \subseteq F_r$. This makes $|X \cap Y| \leq |X \cap (F_r + u_{j-1} + u_j + \cdots + u_r)| = |(Nx + t_r) \cap (F_r + u_{j-1} + u_j + \cdots + u_r)| \leq 1$, by (7.22).

Case 6: This case is considerably more involved. Without loss of generality, we will take $X \in \mathbf{L}_i$, $Y \in \mathbf{L}_j$, with $2 \leq i \leq j \leq r$. So we have $X = Nx + t_{i-1} + t_i + \cdots + t_r$ and $Y = Ny + u_{j-1} + u_j + \cdots + u_r$. If there is an m such that $j \leq m \leq r$ and $t_m \neq u_m$, we may take m to be the largest such index.

Now $Nx + t_{i-1} + t_i + \cdots + t_m \subseteq F_m + t_m$ since $x \in F_i \subseteq F_m$, $t_{i-1} \in F_{i-1} \subseteq F_m$, and each $t_k \in R_k \subseteq F_{k+1} \subseteq F_m$, for $i \leq k \leq m-1$. Similarly, $Ny + u_{j-1} + u_j + \cdots + u_m \subseteq F_m + u_m$. Since $u_m \neq t_m$ and the $F_m + t_m$ and $F_m + u_m$ are cosets of F_m, and the $u_m, t_m \in R_m$ are distinct coset representatives of F_m in F_{m+1}, we have $(F_m + t_m) \cap (F_m + u_m) = \varnothing$. Since $w = t_{m+1} + \cdots + t_r = u_{m+1} + \cdots + u_r$, we have $(X - w) \cap (Y - w) = \varnothing$, and then $X \cap Y = \varnothing$.

So now we are faced with $t_j = u_j$, $t_{j+1} = u_{j+1}$, $\cdots$, $t_r = u_r$. Let $w = u_j + \cdots + u_r = t_j + \cdots + t_r$. Hence, $X - w = Nx + t_{i-1} + t_i + \cdots + t_{j-1}$ and $Y - w = Ny + u_{j-1}$. First, let $i < j$. Now $u_{j-1} \in F_{j-1}$ and $y \in F_j \setminus F_{j-1}$, so $y \in N \setminus F_{j-1}$ and $F_{j-1} \in \mathcal{F}$. By (7.22), we have $1 \geq |(Ny + u_{j-1}) \cap (F_{j-1} + t_{i-1} + t_i + \cdots + t_{j-1})| \geq |(Ny + u_{j-1}) \cap (Nx + t_{i-1} + t_i + \cdots + t_{j-1})| = |(Y - w) \cap (X - w)|$, since $x \in F_i \subseteq F_{j-1}$. This forces $|Y \cap X| \leq 1$.

Finally, let $i = j$. Then $(X - w) \cap (Y - w) = (Nx + t_{i-1}) \cap (Ny + u_{i-1})$, with $x, y \in F_i \setminus F_{i-1}$ and $t_{i-1}, u_{i-1} \in F_{i-1}$. So $x, y \in N \setminus F_{i-1}$ and $u_{i-1} - t_{i-1} \in F_{i-1}$. If $|Nx \cap (Ny + u_{i-1} - t_{i-1})| \geq 2$, then, by (7.21), we have $u_{i-1} = t_{i-1}$ and $Nx = Ny$. Hence, $X = Y$, and this completes the proof of (7.23). $\blacksquare$

We now define a relation $\|$ of $\mathbf{L}$ by $X \| Y$ if and only if there is an i, $1 \leq i \leq r+1$, such that $X, Y \in \mathbf{L}_i$, and if $X = Nx + a$ and $Y = Ny + b$, then $Nx = Ny$.

(7.24) Proposition. *If $X, Y \in \mathbf{L}$, $X \| Y$, and $X \neq Y$, then $X \cap Y = \varnothing$.*

Proof. If $X, Y \in \mathbf{L}_1$, then they are cosets of the same subgroup Nx, so $X \neq Y$ implies $X \cap Y = \varnothing$. Now let $X, Y \in \mathbf{L}_i$, where $2 \leq i \leq r$. Then $X = Nx + t_{i-1} + t_i + \cdots + t_r$ and $Y = Nx + u_{i-1} + u_i + \cdots + u_r$. If there is an m, $i \leq m \leq r$, such that $t_m \neq u_m$, then $X \cap Y = \varnothing$ as shown early in Case 6 of the proof of (7.23). If $t_k = u_k$ for each k, $i \leq k \leq r$, then $X \neq Y$ implies $t_{i-1} \neq u_{i-1}$, since $Nx + t_{i-1} \neq Nx + u_{i-1}$ forces $t_{i-1} \neq u_{i-1}$. By (7.20), $Nx \cap (Nx + u_{i-1} - t_{i-1}) = \varnothing$ since $x \in F_i \setminus F_{i-1}$, and $u_{i-1}, t_{i-1}, u_{i-1} - t_{i-1} \in F_{i-1}$. Hence, $(Nx + t_{i-1}) \cap (Nx + u_{i-1}) = \varnothing$, and $X \cap Y = \varnothing$. Finally, let $X, Y \in \mathbf{L}_{r+1}$. So $X = Nx + t_r$ and $Y = Nx + u_r$, where $x \in N \setminus F_r$, and $t_r, u_r, u_r - t_r \in F_r$. Again, by (7.20), $Nx \cap (Nx + u_r - t_r) = \varnothing$, so $X \cap Y = \varnothing$.

(7.25) Proposition. *Let $X \in \mathbf{L}$ and let $X_{\parallel} = \{Y \in \mathbf{L} \mid Y \parallel X\}$. So $X_{\parallel}$ is an equivalence class of the equivalence relation $\parallel$ of $\mathbf{L}$. Then: if $X \in \mathbf{L}_1$, we obtain $\cup X_{\parallel} = N$ and $|X_{\parallel}| = v/k$; if $X \in \mathbf{L}_i$, with $2 \le i \le r$, then $|X_{\parallel}| = |F_{i-1}||R_i| \cdots |R_r|$; and if $X \in \mathbf{L}_{r+1}$, then $|X_{\parallel}| = |F_r|$.*

Proof. The proof of (7.25) will be quite easy once we prove two statements, each of which might answer a question the reader has had since the definitions of the sets $\mathbf{L}_i$, $1 \le i \le r + 1$. These two statements are: (1) if $i \ne j$, then $\mathbf{L}_i \cap \mathbf{L}_j = \varnothing$; and (2) for $2 \le i \le r + 1$, the representation of a line $X = Nx + t_{i-1} + t_i + \cdots + t_r \in \mathbf{L}_i$, and $X = Nx + t_r \in \mathbf{L}_{r+1}$, as in the definitions of each $\mathbf{L}_i$, is unique. We will be proving these statements somewhat concurrently.

Suppose $Nx + t$, $Ny + u \in \mathbf{L}$ and $Nx + t \in \mathbf{L}_1$, with $x \in F_1 \setminus \{0\}$, and suppose $Nx + t = Ny + u$. Then $Nx = Ny + (u - t)$, a subgroup of $(N, +)$. Hence, $-(u - t) \in Ny$ and $u - t \in Nx$. So $-(u - t) \in Nx \cap Ny$. We either have $Nx = Ny$, or $u = t$. If $u = t$, we also get $Nx = Ny$. Hence, $Ny + u \in \mathbf{L}_1$, also.

Next, take $X = Nx + t_r \in \mathbf{L}_{r+1}$ with $x \in N \setminus F_r$ and $t_r \in F_r$. If $X \in \mathbf{L}_i$, for some $i \le r + 1$, then, as we have seen above, $2 \le i \le r + 1$. Suppose $Nx + t_r = Ny + u_r$, where $y \in N \setminus F_r$ and $u_r \in F_r$. Then $u_r - t_r \in F_r$ and $Nx = Ny + (u_r - t_r)$. Hence, $3 \le |Nx \cap [Ny + (u_r - t_r)]|$ with $x, y \in N \setminus F_r$ and $u_r - t_r \in F_r$. By (7.21), we may conclude that $u_r = t_r$. Hence, $Nx = Ny$.

We move on to the possibility that $Nx + t_r = Ny + u_{i-1} + u_i + \cdots + u_r \in \mathbf{L}_i$, and $2 \le i \le r$. Then $Nx = Ny + [u_{i-1} + u_i + \cdots + u_r - t_r]$ with $x \in N \setminus F_r$ and $y \in F_i \setminus F_{i-1}$. Hence, $3 \le |(Nx + t_r) \cap (Ny + u_{i-1} + u_i + \cdots + u_r)|$. With $y \in F_i$, we have $Ny \subseteq F_i \subseteq F_r$, and so, by (7.22), $|(Nx + t_r) \cap (F_r + u_{i-1} + u_i + \cdots + u_r)| \le 1$. Hence, $\mathbf{L}_{r+1} \cap \mathbf{L}_i = \varnothing$ if $i < r + 1$. (We also have $\mathbf{L}_1 \cap \mathbf{L}_i = \varnothing$ for $2 \le i$.)

Next is the case where there are $i, j \in \{2, \ldots, 4\}$, $i \le j$, such that $X = Nx + t_{i-1} + t_i + \cdots + t_r \in \mathbf{L}_i$ and $Y = Ny + u_{j-1} + u_j + \cdots + u_r \in \mathbf{L}_j$, and $X = Y$. Let $a = t_{i-1} + \cdots + t_r$, $b = u_{j-1} + \cdots + u_r$, and $c = a - b$. Then $Ny = Nx + c$, so $Ny = Ny \cap (Nx + c)$. But $Nx \subseteq F_i \subseteq F_{j-1}$ if $i < j$, and $Nx + c \subseteq F_{j-1} + c$. We also obtain $Ny = (Ny + 0) \cap (F_{j-1} + c)$, and we have the uncomfortable consequence that $3 \le |Ny| = |(Ny + 0) \cap (F_{j-1} + c)| \le 1$ from (7.22). So we cannot have $i < j$, and thus we can conclude that $i \ne j$ implies $\mathbf{L}_i \cap \mathbf{L}_j = \varnothing$.

Now take $i = j$. We may, without loss of generality, assume, for some m, $t_m \ne u_m$, but that $t_{m+1} = u_{m+1}, \ldots, t_r = u_r$. Then we may write $U = Nx + t_{i-1} + \cdots + t_m$, and $V = Ny + u_{i-1} + \cdots + u_m$, with $i - 1 \le m$, and assume $U = V$.

If $i \leq m$, then with $x, y \in F_i \setminus F_{i-1}$, $t_{i-1}, u_{i-1} \in F_{i-1}$, and $t_k, u_k \in R_k \subseteq F_{k+1}$, we have $(Nx + t_{i-1} + \cdots + t_{m-1}) \cup (Ny + u_{i-1} + \cdots + u_{m-1}) \subseteq F_m$. With $U \subseteq F_m + t_m$ and $V \subseteq F_m + u_m$, and $U = V$, we have no choice but to conclude that $t_m = u_m$. So, we may assume that $Nx + t_{i-1} = Ny + u_{i-1}$ and $u_{i-1} \neq t_{i-1}$. But then $u_{i-1} - t_{i-1} \in F_{i-1}$, and $Nx = Ny + (u_{i-1} - t_{i-1})$ with $x, y \in F_{i-1}$ and $u_{i-1} - t_{i-1} \neq 0$. So $3 \leq |Nx \cap (Ny + (u_{i-1} - t_{i-1}))|$, and from (7.21) we must conclude that $u_{i-1} = v_{i-1}$ and that $Nx = Ny$.

With the validity of (1) and (2), one can easily show that $\parallel$ is an equivalence relation on **L**. So we shall proceed with the remainder of the proof. First, if $X \in \mathbf{L}_1$, then $X = Nx + t$, a coset of the subgroup Nx, and $|Nx| = k$, so $|X_\parallel| = |N/Nx| = [N : Nx] = v/k$. Certainly, $X \parallel Y$ if and only if $Y = Nx + u$, so $\cup X_\parallel = N$.

Next, if $2 \leq i \leq r$ and $X \in \mathbf{L}_i$, then $X = Nx + t_{i-1} + t_i + \cdots + t_r$. We can choose $t_{i-1} \in F_{i-1}$ in $|F_{i-1}|$ ways, $t_i \in R_i$ in $|R_i|$ ways, $\ldots$, $t_r \in R_r$ in $|R_r|$ ways. Hence, $|X_\parallel| = |F_{i-1}||R_i| \cdots |R_r|$. Similarly, if $X \in \mathbf{L}_{r+1}$, then $X = Nx + t_r$, and we can choose $t_r \in F_r$ in $|F_r|$ ways. Hence, $|X_\parallel| = |F_r|$.

(7.26) Definitions. An incidence structure $(P, \mathcal{S}, \in)$ is a *partial plane* if $A, B \in \mathcal{S}$ and $A \neq B$ implies $|A \cap B| \leq 1$. From (7.23), our $(N, \mathbf{L}, \in)$ is a partial plane. A partial plane $(P, \mathcal{S}, \in)$ has *parallelism* if there is an equivalence relation $\parallel$ on $\mathcal{S}$ having the property that if $A \parallel B$ and $A \neq B$, then $A \cap B = \varnothing$. By (7.24) and (7.25), $(N, \mathbf{L}, \in, \parallel)$ is a partial plane with parallelism. Referring back to (7.12), a partial plane $(P, \mathcal{S}, \in)$ is an *incidence space* if, for any $a, b \in P$, $a \neq b$, there is a unique 'line' $B \in \mathcal{S}$ with $a, b \in B$. Suppose $(P, \mathcal{S}, \in, \parallel)$ is a partial plane with parallelism. Then it fulfils *Euclid's axiom of parallelism* if for any $x \in P$ and any $A \in \mathcal{S}$, there is exactly one 'line' $B \in \mathcal{S}$ such that $x \in B$ and $B \parallel A$.

In our development here, take any F_j in our chain and construct the following substructures: $(F_j, \mathbf{L}(F_j), \in, \parallel_j)$ of $(N, \mathbf{L}, \in, \parallel)$. Let $\mathbf{L}(F_j) = \{X \in \mathbf{L} \mid X \subseteq F_j\}$, and let $\parallel_j \subseteq \parallel$ be defined by $A \parallel_j B$ if and only if $A, B \in \mathbf{L}(F_j)$ and $A \parallel B$.

(7.27) Lemma. *Each $(F_j, \mathbf{L}(F_j), \parallel_j)$ is a partial plane with parallelism.*

Proof. It is easy, as well as direct, to show that each $\parallel_j$ is an equivalence relation on $\mathbf{L}(F_j)$. If $A, B \in \mathbf{L}(F_j)$, then $A \cup B \subseteq F_j$ and $A, B \in \mathbf{L}$. So if $A \neq B$, then $|A \cap B| \leq 1$ by (7.23). Hence, each $(F_j, \mathbf{L}(F_j), \in)$ is a partial plane. Finally, take $A, B \in \mathbf{L}(F_j)$ with $A \parallel_j B$ and $A \neq B$. Then $A, B \in \mathbf{L}$ and $A \neq B$. By (7.24), $A \cap B = \varnothing$, and this completes the proof of (7.27).

(7.28) Corollary. *$(F_1, \mathbf{L}(F_1), \in, \parallel_1)$ is an affine space, $(F_1, +)$ is an elementary abelian p-group, and $|F_1| = k^n$ for some n, where $k = |F_1 x|$ for any nonzero $x \in F_1$.*

Proof. Recall that F_1 was chosen so that Nx is a subgroup of F_1 for each $x \in F_1$, and $F_1 \neq \{0\}$. Hence, $Nx = F_1x$ if we assume $a =_m 0$ if and only if $a = 0$, and there is no loss in generality in doing so. So (7.13) is applicable and we get that $(F_1, \mathbf{L}(F_1), \in)$ is an affine space, since $\mathbf{L}(F_1) = \{F_1x + t \mid 0 \neq x \in F_1, t \in F_1\}$. By (7.11), $(F_1, +)$ is an elementary abelian p-group, and so $(F_1, +)$ is the direct sum of some of the subgroups F_1x. This makes $|F_1| = k^n$ where $k = |F_1x|$ for any nonzero $x \in F_1$.

(7.29) Theorem. *The following are equivalent.*
(a) $(N, \mathbf{L}, \in, \|)$ *fulfils Euclid's axiom of parallelism.*
(b) $X \in \mathbf{L}$ *implies* $\cup X_\| = N$.
(c) *For each i, $2 \leq i \leq r$, $v = |F_{i-1}||R_i| \cdots |R_r|k = |F_r|k$, where $v = |N|$ and $k = |Nx|$ for any nonzero $x \in N$.*
(d) $k = |R_1| = |R_2| = \cdots = |R_r|$ *and* $v = |F_1| \cdot k^r$.
(e) $(N, \mathbf{L}, \in)$ *is an incidence space.*
    (*As a consequence of (7.28), $v = |N| = k^{n+r}$ and $(N, +)$ is a p-group if any of the above are satisfied.*)

Proof. We will first see that (a) and (b) are equivalent. If (a) is true, and $X \in \mathbf{L}$, and $y \in N$, then there is exactly one $Y \in \mathbf{L}$ such that $y \in Y$ and $Y \| X$. Hence, $\cup X_\| = N$. So (a) implies (b). Now suppose (b) is true. Let $y \in N$ and $X \in \mathbf{L}$. Since $\cup X_\| = N$, there is a $Y \in \mathbf{L}$ such that $Y \| X$ and $y \in Y$. If $Z \in \mathbf{L}$, $Z \| X$, and $y \in Z$, then $Z \| Y$ and $y \in Y \cap Z$. By (7.24), it follows that $Y = Z$. So, (b) implies (a).

Next, we will see that (b) and (c) are equivalent. If (b), then by (7.25), $v = |N| = |F_{i-1}||R_i| \cdots |R_r|k$ for $2 \leq i \leq r$, and $v = |F_r|k$, since $|X| = k$. Hence, (b) implies (c). Conversely, if $X \in \mathbf{L}_1$, then $\cup X_\| = N$ by (7.25). If $X \in \mathbf{L}_i$, $2 \leq i \leq r$, then $|\cup X_\|| = \sum_{Y \in X_\|} |Y| = tk$, where $t = |X_\||$. From (7.25), $t = |F_{i-1}||R_i| \cdots |R_r|$, so $|\cup X_\|| = v$, by (c). Hence, $\cup X_\| = N$. If $X \in \mathbf{L}_{r+1}$, then $t = |F_r|$, so $|\cup X_\|| = v$, by (c). Hence, $\cup X_\| = N$. This makes (a), (b), and (c) equivalent.

To see that (c) and (d) are equivalent, we start with (c). Since $v = |F_r||R_r|$ and $v = |F_r|k$, we have $|R_r| = k$. Now $|F_i| = |F_{i-1}||R_{i-1}|$, so $v = |F_{i-1}||R_{i-1}||R_i| \cdots |R_r|$ and $v = |F_{i-1}||R_i| \cdots |R_r|k$. Hence, $|R_{i-1}| = k$ for $2 \leq i \leq r$ and $v = |F_1|k^r$. Conversely, if $k = |R_i|$ for each i, $1 \leq i \leq r$, and $v = |F_1|k^r$, then since $|F_i| = |F_{i-1}||R_{i-1}|$, we have $v = |F_1||R_2| \cdots |R_r|k = (|F_1||R_2|)|R_3| \cdots |R_r|k = (|F_1||R_1|)|R_3| \cdots |R_r|k = |F_2||R_3| \cdots |R_r|k = \cdots = |F_{i-1}||R_i| \cdots |R_r|k = \cdots = |F_r|k$. So (c) and (d) are equivalent.

Finally, to show that (a) and (e) are equivalent, we will show that each is equivalent to (7:1), where (7:1) is now explained. For $a \in N$, define $\mathbf{L}(a) = \{X \in \mathbf{L} \mid a \in X\}$ and $P(a) = \cup \mathbf{L}(a)$. Let $\mathbf{B}_0 = \{Nx \mid x \in N \setminus \{0\}\}$. Then we have a condition

$$|\mathbf{L}(a)| = |\mathbf{B}_0| \text{ for each } a \in N. \tag{7:1}$$

Each $|Nx \setminus \{0\}| = k - 1$, so $v = (k - 1)|\mathbf{B}_0| + 1$. Similarly, $|P(a)| = (k-1)|\mathbf{L}(a)| + 1$ by authority of (7.23). Suppose (e) is true. If $p, q \in N$ and $p \neq q$, then there is a line $X \in \mathbf{L}$ with $p, q \in X$. Hence, for each $a \in N$, $P(a) = N$ and so $v = (k - 1)|\mathbf{L}(a)| + 1 = (k - 1)|\mathbf{B}_0| + 1$, from which we conclude that $|\mathbf{L}(a)| = |\mathbf{B}_0|$ for each $a \in N$. Thus (e) implies (7:1). If (7:1) is true, then $v = |P(a)|$ for each $a \in N$, so if $p, q \in N$ and $p \neq q$, then there is a line $X \in \mathbf{L}$ with $p, q \in X$. Hence, (7:1) implies (e).

Now we will see that (a) and (7:1) are equivalent. If $X \in \mathbf{L}$, then $X = Na + u$ and $Na \in \mathbf{L}$, so if $X \in \mathbf{L}_i$, then $Na \in \mathbf{L}_i$ and so $X \parallel Na$. But $Na \in \mathbf{B}_0$, so any line $X \in \mathbf{L}$ is parallel to a line in $\mathbf{B}_0$.

Assume (a) to be true. If $X, Y \in \mathbf{L}(a)$ and $X \neq Y$, then $X \parallel Nx$ and $Y \parallel Ny$ imply $Nx \neq Ny$, and $Nx, Ny \in \mathbf{B}_0$. Hence, $|\mathbf{L}(a)| \leq |\mathbf{B}_0|$ for each $a \in N$. But $\mathbf{B}_0 \subseteq \mathbf{L}(0)$, so $|\mathbf{B}_0| \leq |\mathbf{L}(0)|$. Therefore, $|\mathbf{B}_0| = |\mathbf{L}(0)|$. For each $X \in \mathbf{L}(a)$, there is a unique $X_0 \in \mathbf{L}(0)$ such that $X \parallel X_0$. Hence, $|\mathbf{L}(a)| \leq |\mathbf{L}(0)|$. For each $X \in \mathbf{L}(0)$, there is a unique $X_a \in \mathbf{L}(a)$ such that $X \parallel X_a$. Hence, $|\mathbf{L}(0)| \leq |\mathbf{L}(a)|$, and so $|\mathbf{L}(a)| = |\mathbf{L}(0)| = |\mathbf{B}_0|$ for each $a \in N$. This says that (a) implies (7:1).

If (7:1) is true, then $(N, \mathbf{L}, \in)$ is an incidence space, since (7:1) is equivalent to (e). By (7.27), $(F_{r+1}, \mathbf{L}_{r+1}, \parallel_{r+1}) = (N, \mathbf{L}, \parallel)$ is a partial plane with parallelism. Take $y \in N$ and any $X \in \mathbf{L}$. Then $X = Nx + u$, so $X \parallel Nx$, $Nx \in \mathbf{B}_0$, and $Nx \in \mathbf{L}$. Since $|\mathbf{L}(y)| = |\mathbf{B}_0|$, and since each line in $\mathbf{L}(y)$ is parallel to a line in $\mathbf{B}_0$, there is a line $Y \in \mathbf{L}(y)$ parallel to $Nx \in \mathbf{B}_0$. Hence, $y \in Y$ and $Y \parallel X$. If $Z \in \mathbf{L}(y)$ with $Z \parallel X$, then $y \in Y \cap Z$ forces $Y = Z$, by virtue of (7.24). Hence, (a) is valid, so (7:1) implies (a). Now we have (a) and (e) each equivalent to (7:1), and this completes the proof of (7.29).

(7.30) Theorem. *Suppose that $|Nx| = k = |R_i| = [F_{i+1} : F_i]$ for each i, $1 \leq i \leq r$, and for each $x \in N \setminus \{0\}$. Then the substructure $(N, \mathbf{L}, \in, \parallel)$ of $(N, \mathcal{B}, \in)$ is an incidence space with parallelism that fulfils Euclid's axiom of parallelism, N is a p-group of order k^{n+r} for some n, and if $n = r = 1$, then $(N, \mathbf{L}, \in, \parallel)$ is an affine plane. If $r \geq n = 1$, then $(F_2, \mathbf{L}(F_2), \parallel_2)$ is an affine plane.*

Proof. With each $|R_i| = k$, the conditions of (7.29) are valid, and so $(N, \mathbf{L}, \in)$ is an incidence space with parallelism that fulfils Euclid's axiom of parallelism. With (7.28) we know that N is a p-group of order k^{n+r} for some n. If $r \geq n = 1$, then $(F_2, \mathbf{L}(F_2), \parallel_2)$ has three noncollinear points $0, x \in F_1$ and $y \in F_2 \setminus F_1$, and by Euclid's axiom of parallelism and (7.27), we have that $(F_2, \mathbf{L}(F_2), \parallel_2)$ is an affine plane. So, if $r = n = 1$, then $(N, \mathbf{L}, \in, \parallel)$ is an affine plane.

A permutation δ of N is a *dilatation* of $(N, \mathbf{L}, \in, \parallel)$ if, for any $X \in \mathbf{L}$, we have $\delta(X) \in \mathbf{L}$ and $\delta(X) \parallel X$.

(7.31) Proposition. *The set Δ of all dilatations of $(N, \mathbf{L}, \in, \|)$ is a group. For a normal subgroup $(F_1, +)$ of $(N, +)$, let $F_1^+ = \{f^+ : N \to N \mid f^+(x) = x + f,\ f \in F_1\}$. Then F_1^+ is a subgroup of Δ.*

Proof. Certainly, $1_N \in \Delta$. If $\delta_1, \delta_2 \in \Delta$, then for $X \in \mathbf{L}$, $\delta_2(X) \in \mathbf{L}$ and $\delta_2(X) \parallel X$. Hence, $\delta_1(\delta_2(X)) \in \mathbf{L}$ and $\delta_1(\delta_2(X)) \parallel \delta_2(X)$. So $\delta_1 \circ \delta_2 \in \Delta$. Since the group of permutations of N is finite, we have that Δ is a group. For $f \in F_1$ and $f^+(x) = x + f$, we have that f^+ is a permutation of N and for $X \in \mathbf{L}_i$, $2 \leq i \leq r$, $X = Nx + t_{i-1} + t_i + \cdots + t_r$, then $f^+(X) = Nx + t_{i-1} + t_i + \cdots + t_r + f$. Since $f \in F_1$ and F_1 is normal in N, we have $f^+(X) = Nx + t_{i-1} + t_i + \cdots + t_{r-1} + f_r + t_r = \cdots = Nx + t_{i-1} + f_i + t_i + \cdots + t_r$. Since $t_{i-1} + f_i \in F_{i-1}$, we have $f^+(X) \in \mathbf{L}_i$ also, and $f^+(X) \parallel X$. If $X \in \mathbf{L}_1$, then $X = Nx + t$, so $f^+(X) = Nx + t + f \in \mathbf{L}_1$ and $f^+(X) \parallel X$. If $X \in \mathbf{L}_{r+1}$, then $X = Nx + t_r$ and $f^+(X) = Nx + t_r + f \in \mathbf{L}_{r+1}$. Again, $f^+(X) \parallel X$. So each $f^+ \in \Delta$. If $f_1^+, f_2^+ \in F_1^+$, then $f_1^+ \circ f_2^+(X) = f_1^+(X + f_2) = X + f_2 + f_1$, and so $f_1^+ \circ f_2^+ = (f_2 + f_1)^+ \in F_1^+$. This makes F_1^+ a subgroup of Δ.

Let us return to $(N, \mathcal{B}, \in)$, but with F_1 as a normal subgroup of N. For $t \in N$, define $t^+(x) = x + t$. So if $X \in \mathbf{L}_1$, then $t^+(X) \in \mathbf{L}_1$ and $t^+(X) \parallel X$. However, if $X \in \mathbf{L} \backslash \mathbf{L}_1$, then we have no assurance that $t^+(X) \in \mathbf{L}$ if $t \notin F_1$. Define $(N, \mathbf{L}, \|)_t = (N, t^+(\mathbf{L}), \|_t)$, where $t^+(\mathbf{L}) = \{t^+(X) \mid X \in \mathbf{L}\}$, and where $t^+(X) \parallel_t t^+(Y)$ if and only if $X \parallel Y$. One easily obtains

(7.32) Theorem. *Each $(N, \mathbf{L}, \|)_t$ is a partial plane with parallelism and is isomorphic to $(N, \mathbf{L}, \|)$.*

Proof. This is because $(N, \mathbf{L}, \|)$ is a partial plane with parallelism and t^+ is the required isomorphism.

(7.33) Corollary. *Each $(N, \mathbf{L}, \|)_t$ shares the lines in $\mathbf{L}_1$.*

Proof. $X \in \mathbf{L}_1$ if and only if $t^+(X) \in \mathbf{L}_1$.

(7.34) Corollary. *For F_1 normal in N and $t \in F_1$, then $t^+(\mathbf{L}_i) = \mathbf{L}_i$ for $1 \leq i \leq r + 1$, and $(N, \mathbf{L}, \|)_t = (N, \mathbf{L}, \|)$.*

Proof. Since $t \in F_1$, $t^+(\mathbf{L}_i) = \mathbf{L}_i$ follows as in the proof of (7.31).

(7.35) Proposition. *If F_1 is normal in N and $u - t \in F_1$, then $t^+(\mathbf{L}) = u^+(\mathbf{L})$, and so $(N, \mathcal{B}, \in)$ contains at most $v/|F_1|$ distinct partial planes $(N, \mathbf{L}, \|)_t$.*

Proof. Since $u - t \in F_1$, then $(u - t)^+(\mathbf{L}) = \mathbf{L}$. But $(u - t)^+ = (u + (-t))^+ = (-t)^+ \circ u^+ = (t^+)^{-1} \circ u^+$.

If there is an $X \in \mathbf{L}$ such that $t^+(X) \notin u^+(\mathbf{L})$, then $t^+(X) \neq u^+(Y)$ for each $Y \in \mathbf{L}$. Hence, $X \neq (t^+)^{-1} \circ u^+(Y)$ for each $Y \in \mathbf{L}$. Hence, $X \notin \mathbf{L}$ since $(t^+)^{-1} \circ u^+ = (u - t)^+$, and $(u - t)^+(\mathbf{L}) = \mathbf{L}$. So $t^+(\mathbf{L}) = u^+(\mathbf{L})$.

Hence, if $(N, \mathbf{L}, \|)_t \neq (N, \mathbf{L}, \|)_u$, then $t^+(\mathbf{L}) \neq u^+(\mathbf{L})$, which forces $u - t \notin F_1$. If $u - t \in F_1$, then $t^+(\mathbf{L}) = u^+(\mathbf{L})$ and so $(N, \mathbf{L}, \|)_t = (N, \mathbf{L}, \|)_u$. So each coset $F_1 + t$ in N is associated with an $(N, \mathbf{L}, \|)_t$, and there are $[N : F_1] = v/|F_1|$ such cosets, hence there are at most $v/|F_1|$ such $(N, \mathbf{L}, \|)_t$.

(7.36) Theorem. *If F_1 is normal in N and $R = R_1 + \cdots + R_r = \{t_1 + \cdots + t_r \mid$ each $t_i \in R_i\}$, then*

$$\mathcal{B} = \mathbf{L}_1 \cup \left[\bigcup_{t \in R} t^+(\mathbf{L} \setminus \mathbf{L}_1) \right]. \tag{7:2}$$

Proof. The right hand side of (7:2) is certainly a subset of $\mathcal{B}$. For $Nx + u \in \mathcal{B}$, we want to show that $Nx + u \in \mathbf{L}_1$ or $Nx + u \in t^+(\mathbf{L} \setminus \mathbf{L}_1)$ for some $t \in R$. If $x \in F_1$, we have our wish. So let $x \in F_i \setminus F_{i-1}$, $u \in F_j$, $2 \leq i \leq r$. If $j \leq i - 1$, then $u \in F_{i-1}$, and so $Nx + u \in \mathbf{L}_i$. Since $0 \in R$, $0^+(Nx + u) = Nx + u \in 0^+(\mathbf{L} \setminus \mathbf{L}_1)$. If $i - 1 < j$, then $u \in F_{j-1} + t_j$ for some $t_j \in R_j$, so $u = f_{j-1} + t_j$ for some $f_{j-1} \in F_{j-1}$. Similarly, $u = f_{j-2} + t_{j-1} + t_j$ for some $f_{j-2} \in F_{j-2}$ and $t_{j-1} \in R_{j-1}$. Continuing, we obtain $u = f_{i-1} + t_i + \cdots + t_j$, $f_{i-1} \in F_{i-1}$, $t_i \in R_i$. So $Nx + u = Nx + f_{i-1} + t_i + \cdots + t_j = (t_i + \cdots + t_j)^+(Nx + f_{j-1}) \in (t_i + \cdots + t_j)^+(\mathbf{L} \setminus \mathbf{L}_1)$, where $t_i + \cdots + t_j \in R$. Now suppose $x \in F_{r+1} \setminus F_r = N \setminus F_r$ and $u \in F_r$. Then $Nx + u \in \mathbf{L}_{r+1}$ and $0^+(Nx + u) = Nx + u \in 0^+(\mathbf{L} \setminus \mathbf{L}_1)$. Finally, suppose $x \in N \setminus F_r = F_{r+1} \setminus F_r$ and $u \in F_{r+1} \setminus F_r$. So $u = f_r + t_r$ with $f_r \in F_r$ and $t_r \in R_r$. Thus $Nx + u = t_r^+(Nx + f_r) \in t_r^+(\mathbf{L} \setminus \mathbf{L}_1)$. Hence, the left hand side of (7:2) is a subset of the right hand side, and this gives us equality in (7:2).

(7.37) Lemma. *Fix $t \in N \setminus F_1$. If $\phi \neq_m \lambda$, then $F_1 + \phi t \neq F_1 + \lambda t$.*

Proof. From $F_1 + \phi t = F_1 + \lambda t$, we conclude that $\lambda t - \phi t \in F_1$, so $(1 - \phi \lambda^{-1})(\lambda t) \in F_1$, and so λt, $t \in F_1$. This works if $\phi \neq_m 0$ and $\lambda \neq_m 0$. If $\phi =_m 0$, then $F_1 + \phi t = F_1$ and $F_1 + \lambda t \neq F_1$, since $\lambda t \in F_1$ implies $t \in F_1$.

(7.38) Theorem. *Suppose $k = |F_1|$, $r = 1$, F_1 is normal in N, and $(N, \mathbf{L}, \in, \|)$ fulfils Euclid's axiom of parallelism. Then $v = k^2$ and the tactical configuration $(N, \mathcal{B}, \in)$ contains exactly k distinct affine planes $(N, \mathbf{L}, \|)_{\phi t}$ where ϕ varies through representatives of $N/=_m$ and $t \in N \setminus F_1$ is fixed. These k affine planes share the pencil $\mathbf{L}_1 = \{F_1 + \phi t \mid \phi \in N\}$. Here $\mathbf{L} = \mathbf{L}_1 \cup \mathbf{L}_2$, where $\mathbf{L}_2 = \{Nx + f \mid x \in N \setminus F_1, f \in F_1\}$ and $(\phi t)^+(\mathbf{L}_2) = \{Nx + f + \phi t \mid x \in N \setminus F_1, f \in F_1\}$.*

Proof. Since $r = 1$, apply (7.29) and conclude $|N/F_1| = k$ and $|N| = v = k^2$. From (7.35) we have that $(N, \mathcal{B}, \in)$ contains at most k distinct partial planes $(N, \mathbf{L}, \in)_t$. From (7.30) and (7.32), we learn that the $(N, \mathbf{L}, \in)_t$ are affine planes.

For a fixed $t \in N \setminus F_1$ and $\phi, \lambda \in N$ with $\phi \neq_m \lambda$, we want to show that $(N, \mathbf{L}, \|)_{\phi t} \neq (N, \mathbf{L}, \|)_{\lambda t}$. Now $\mathbf{L}$ consists of $\mathbf{L}_1$, which is the set of cosets of F_1 in N, and of $\mathbf{L}_2 = \{Nx + t_1 \mid x \in N \setminus F_1, \, t_1 \in F_1\}$. So $(N, \mathbf{L}, \|)_{\phi t}$ and $(N, \mathbf{L}, \|)_{\lambda t}$ have $\mathbf{L}_1 \subseteq (\phi t)^+(\mathbf{L}) \cap (\lambda t)^+(\mathbf{L})$. Hence, the affine planes share the pencil $\mathbf{L}_1$.

For $Nx + t_1 + \phi t \in (\phi t)^+(\mathbf{L})$, suppose there is a $y \in N \setminus F_1$, $t'_1 \in F_1$, so that $Nx + t_1 + \phi t = Ny + t'_1 + \lambda t$. By (7.21), $Nx = Ny$ and $t'_1 + \lambda t = t_1 + \phi t$, so $F_1 + \lambda t = F_1 + \phi t$, and so $\phi = \lambda$ by (7.37). Thus, $(N, \mathbf{L}, \|)_{\phi t} \neq (N, \mathbf{L}, \|)_{\lambda t}$ if $\phi, \lambda \in N$ and $\phi \neq_m \lambda$. This gives us at least $k = |N/=_m|$ distinct affine planes. So we have exactly k affine planes and they all share $\mathbf{L}_1$.

(7.39) Applications. (a) For an odd prime p and positive integer r, we have the $(Z_{p^{r+1}}, \Phi)$ discussed in (c) of (7.15). Karzel's theory developed here applies to any planar nearring constructed from a $(Z_{p^{r+1}}, \Phi)$. For $1 \leq i \leq r$, define $F_i = \{0, \, p^{r+1-i}, \, 2p^{r+1-i}, \, \cdots, \, (p^i - 1)p^{r+1-i}\}$. Then $\{0\} < F_1 < F_2 < \cdots < F_r < F_{r+1} = N = Z_{p^{r+1}}$ is a chain with $|F_1| = p$ which meets conditions (a), (b), and (c) described just before (7.20), and certainly F_1 is normal in the abelian group $(Z_{p^{r+1}}, +)$. So if $r = 1$, then (7.38) applies. This means that $(Z_{p^2}, \mathcal{B}, \in)$ has p substructures $(Z_{p^2}, \mathbf{L}, \|)_{\phi t}$ which are affine planes and each of these substructures has the pencil $\mathbf{L}_1 = Z_{p^2}/F_1$ as an equivalence class of parallel lines.

(b) For $p = 2$, consider the examples as described in (a) of (4.20). We have $GF(2^n)$, a finite field of order 2^n, with n odd, so $N = GF(2^n) \times GF(2^n)$ has exactly 2^{2n} elements. The binary operation on N is defined by $(a, x) + (b, y) = (a + b + xy^2, \, x + y)$. Let d belong to A, the multiplicative group of $GF(2^n)$, and define $\sigma_d : N \to N$ by $\sigma_d(a, x) = (d^3 a, dx)$. By (4.19), $\Phi_A = \{\sigma_d \mid d \in A\}$ is a regular group of automorphisms of the nonabelian group $(N, +)$. Let $(N, +, \cdot)$ be a planar nearring constructed from the Ferrero pair (N, Φ_A). Certainly, $F_1 = N(1, 0) = \{(a, 0) \mid a \in GF(2^n)\}$ is a subgroup of $(N, +)$ of order 2^n. But for $x \neq 0$, $N(a, x) = \{(0, 0), \, (a, x), \, (d^3 a, dx), \, \cdots, \, (d^{3(2^n - 2)} a, d^{2^n - 2} x)\}$ where $d \in A$ has order $2^n - 1$. If such an $N(a, x)$ were a subgroup of $(N, +)$, then $(a, x) + (a, x) = (a + a + xx^2, x + x) = (x^3, 0)$ would be in $N(a, x)$. Since $x^3 \neq 0$ and none of the $d^i x = 0$, we have a contradiction. It is easy to see that F_1 is normal in N, and with $|F_1| = 2^n$, it follows that $r = 1$. So our $(N, \mathbf{L}, \in, \|)$ satisfy the conditions of (7.38), so each $(N, \mathcal{B}, \in)$ has 2^n distinct substructures $(N, \mathbf{L}, \|)_{\phi t}$ which are affine planes, and each of these has the pencil $\mathbf{L}_1 = N/F_1$ as an equivalence class of parallel lines.

(7.40) Problem. Are there any other nearrings satisfying Karzel's theory? In particular, are there any other $(N, \mathcal{B}, \in)$ with $s = 1$ and $r = 1$? Can the theory be modified to create an infinite example with $s = 1$ and $r = 1$? Could the affine planes ever be isomorphic to the euclidean plane?

(7.41) Remark. For the examples in (7.39) with $s = 1$ and $r = 1$, the tactical configurations $(N, \mathcal{B}, \in)$ have k substructures $(N, \mathcal{L}_i, \in)$, each of which is an affine plane, and for $i \neq j$, $\mathcal{L}_i \cap \mathcal{L}_j = X_\parallel$ for some X. In fact, X could be Ng_1 and $X_\parallel = N/Ng_1$, the cosets of the only basic block Ng_1 which is a subgroup of N. We suggest that we say that these planar nearrings $(N, +, \cdot)$ have an *affine configuration with one pencil.*

7.3. The van der Walt connection

In 1989, Andries P. J. van der Walt developed a family of planar nearrings, finite and infinite, which are close to having affine configurations with one pencil. More descriptively, they would have affine configurations with *two pencils.* The theory developed by Karzel does *not* apply, however. The nearrings of van der Walt, when finite, are of order q^2, they have $2q$ substructures which are affine planes, and if $(N, \mathcal{L}_1, \in)$ and $(N, \mathcal{L}_2, \in)$ denote two of these substructures, with $\mathcal{L}_1 \neq \mathcal{L}_2$, then $\mathcal{L}_1 \cap \mathcal{L}_2 \supseteq N/Ng_1 \cup N/Ng_2$, where Ng_1 and Ng_2 are the two basic blocks which are subgroups of N. So here, $s = 2$. Also, if $X = Ng_1$ and $Y = Ng_2$, then $X_\parallel = N/Ng_1$ and $Y_\parallel = N/Ng_2$.

As a rather amusing application of van der Walt's examples, we construct some rather surprising affine planes. Fix any horizontal line $y = k$ in $\mathbf{R} \times \mathbf{R}$, where $\mathbf{R}$ denotes the real numbers. Consider all cubic polynomials $y - k = b(x - h)^3$, having no relative maximum or minimum, but having exactly one horizontal tangent line, and having an inflection point at some point (h, k) on the line $y = k$. Throw in all vertical lines $x = c$ and all horizontal lines $y = d$. It would not be difficult for the reader to show that the graphs of these can be considered as lines of an affine plane. But could the reader have discovered this easily?

With van der Walt's examples, there will be numerous planar nearrings $(N, +, \cdot)$, both finite and infinite, and each will have exactly two basic blocks $Na = Ng_1$ and $Nb = Ng_2$ as subgroups of $(N, +)$, and they will both be normal. *One* subgroup Na will be used to sift out from $(N, \mathcal{B}, \in)$ a substructure $(N, \mathcal{L}_0, \in)$, which will be an affine plane. Using this $(N, \mathcal{L}_0, \in)$ the *other* subgroup, Nb, will be used to sift out further substructures $(N, \mathcal{L}_t, \in)$ of $(N, \mathcal{B}, \in)$, one for each $t \in Nb$, and these $(N, \mathcal{L}_t, \in)$ will also be affine planes. Then the roles of Na and Nb will be reversed. We will use Nb to sift out from $(N, \mathcal{B}, \in)$ still another affine plane $(N, \mathcal{L}_0', \in)$, and then use the other subgroup Na to sift out further affine planes $(N, \mathcal{L}_u', \in)$, one for each $u \in Na$.

As always, the planar nearrings here have associated with them a Ferrero pair (N, Φ). One gets a natural action of Φ on $\mathcal{B}$, and N can be reconstructed from the nontrivial orbits of Φ on $\mathcal{B}$. In fact, the nontrivial orbits of Φ can be thought of as points of a Möbius plane.

We start with a field F and let $N = F \oplus F$ as a group. So $(N, +)$ is the external direct sum of $(F, +)$ and $(F, +)$. Let α denote an automorphism of $(F^*, \cdot)$, the multiplicative group of F. Extend α to F by letting $\alpha(0) = 0$. For $a \in F^*$, we define $a' : N \to N$ by $a'(x, y) = (ax, \alpha(a)y)$ for all $(x, y) \in N$. Let $\Phi_\alpha = \{a' \mid a \in F^*\}$. One easily obtains

(7.42) Proposition. (N, Φ_α) *is a Ferrero pair.*

(7.43) Proposition. *Let* $(N, +, \cdot)$ *be a planar nearring constructed from the Ferrero pair* (N, Φ_α). *Then the basic blocks of* $(N, +, \cdot)$ *are* $N(1, 0)$, $N(0, 1)$, *and the* $N(1, b)$ *for* $b \neq 0$.

Proof. If $a \neq 0$, then $N(a, 0) = \{(xa, 0) \mid x \in F\} = \{(x, 0) \mid x \in F\} = N(1, 0)$. Similarly, if $b \neq 0$, then $N(0, b) = N(0, 1)$. Suppose $0 \notin \{c, d\}$ and consider $N(c, d) = \{(ac, \alpha(a)d) \mid a \in F\}$. Let $b = \alpha(c)^{-1}d$, so $N(1, b) = \{(a, \alpha(a)b) \mid a \in F\}$. For $a = c$, we obtain $(c, \alpha(c)b) = (c, \alpha(c)\alpha(c)^{-1}d) = (c, d) \in N(1, b)$. Hence, $N(c, d) = N(1, b)$.

(7.44) Proposition. *For* $b \in F$, $b \neq 0$, *the basic block* $N(1, b)$ *is a subgroup of* N *if and only if* α *is an automorphism of the field* F.

Proof. If α is a field automorphism and $(a, \alpha(a)b), (c, \alpha(c)b) \in N(1, b)$, then $(a, \alpha(a)b) - (c, \alpha(c)b) = (a - c, \alpha(a)b - \alpha(c)b) = (a - c, \alpha(a - c)b) \in N(1, b)$, so $N(1, b)$ is a subgroup. Conversely, if $N(1, b)$ is a subgroup, then $(a, \alpha(a)b) + (c, \alpha(c)b) = (a + c, [\alpha(a) + \alpha(c)]b) \in N(1, b)$. But $(a + c, \alpha(a + c)b) \in N(1, b)$, also, so we must have $[\alpha(a) + \alpha(c)]b = \alpha(a + c)b$, and, since $b \neq 0$, we obtain $\alpha(a + c) = \alpha(a) + \alpha(b)$, which makes α a field automorphism.

We shall be interested now only in the case where the $N(1, b)$, $b \neq 0$, are *not* subgroups of N, so we shall assume from now on that

$$\alpha \text{ is not a field automorphism of } F.$$

(7.45) Corollary. *Suppose* α *is not an automorphism of the field* F. *Then the basic blocks* $N(1, 0)$ *and* $N(0, 1)$ *are subgroups of* N, *and for* $b \neq 0$, $N(1, b)$ *is not a subgroup of* N.

From the tactical configuration $(N, \mathcal{B}, \in)$, construct the index set $I = \{(\delta, u) \mid u \in N, \delta \in \{(1, 0), (0, 1)\} \cup \{(1, b) \mid b \in F, b \neq 0\}\}$ and the indexed family $\mathcal{B}' = \{B_{(\delta, u)} \mid (\delta, u) \in I\}$, where $B_{(\delta, u)} = N\delta + u$.

(7.46) Theorem. *For a finite field* F *with* $|F| = q$, *then* $(N, \mathcal{B}', \in)$ *is a statistical BIBD with parameters* $v = q^2$, $k = q - 1$, $b = v(v - 1)/(k - 1)$, $r = k(v - 1)/(k - 1)$, *and* $\lambda = k$.

Proof. Apply (7.17)

Following the plan outlined earlier, we let $Ng_1 = Na = N(1, 0)$, a basic block which is also a subgroup of N. Let $\mathcal{L}_0 = \mathcal{L}(1, 0) = N/N(1, 0) \cup$

$N/N(0,1) \cup \{N(1,b) + (x,0) \mid b \in F^*, (x,0) \in N(1,0)\}$. So we have used $N(1,0)$ to sift out the substructure $(N, \mathcal{L}(1,0), \in)$ of $(N, \mathcal{B}, \in)$. We want to show that $N/N(1,0)$, $N/N(0,1)$, and for $b \neq 0$, $\mathcal{P}_b = \{N(1,b) + (x,0) \mid (x,0) \in N(1,0)\}$ are distinct families of parallel lines in an affine plane $(N, \mathcal{L}(1,0), \in)$.

(7.47) Theorem. $(N, \mathcal{L}(1,0), \in)$ *is an affine plane.*

Proof. Certainly, $(0,0)$, $(0,1)$, and $(1,0)$ are noncollinear. The proof will be completed after a few lemmas and corollaries.

(7.48) Lemma. *Each of $N/N(1,0)$, $N/N(0,1)$, and the $\mathcal{P}_b$, $b \neq 0$, defines a partition of N.*

Proof. The $N(1,0)$ and $N(0,1)$ are subgroups, and their cosets in N give us $N/N(1,0)$ and $N/N(0,1)$, respectively. Now take any $b \neq 0$ and consider $\mathcal{P}_b$. If $(x,y) \in N$, then we want to show that (x,y) belongs to exactly one $N(1,b) + (c,0) \in \mathcal{P}_b$. To be in such an $N(1,b) + (c,0)$, we must have an a where $(x,y) = (a + c, \alpha(a)b)$. Then $a = \alpha^{-1}(yb^{-1})$ and $c = x - \alpha^{-1}(yb^{-1})$. So (x,y) belongs to at most one $N(1,b) + (c,0) \in \mathcal{P}_b$. In fact, $(\alpha^{-1}(yb^{-1}), \alpha(\alpha^{-1}(yb^{-1}))b) + (x - \alpha^{-1}(yb^{-1}), 0) = (x,y) \in N(1,b) + (x - \alpha^{-1}(yb^{-1}), 0)$.

(7.49) Corollary. (a) *Each of $N/N(1,0)$, $N/N(0,1)$, and the $\mathcal{P}_b$, $b \neq 0$, is a 'pencil' of parallel lines.*
  (b) *Given a line L in one of the 'pencils' and an $(x,y) \in N$, but $(x,y) \notin L$, then there is at least one M in this same 'pencil' so that $(x,y) \in M$ but $L \cap M = \varnothing$, that is, $L \parallel M$.*
  (c) *Given $(x,y) \in N$, there is a unique M for each 'pencil' with $(x,y) \in M$.*

(7.50) Lemma. *Any line L in any of the 'pencils' $N/N(1,0)$, $N/N(0,1)$, and the $\mathcal{P}_b$, $b \neq 0$, intersects each M in any other 'pencil' from the same $N/N(1,0)$, $N/N(0,1)$, and the $\mathcal{P}_b$, $b \neq 0$, at exactly one point.*

Proof. An $L \in N/N(1,0)$ has the form $L = N(1,0) + (0,d)$. An $M \in N/N(0,1)$ has the form $M = N(0,1) + (c,0)$. Certainly, $L \cap M = \{(c,d)\}$.
  If $X \in \mathcal{P}_b$, $b \neq 0$, then $X = N(1,b) + (e,0)$, so $(a, \alpha(a)b) + (e,0) = (u,0) + (0,c) \in X \cap L$ if and only if $(a + e, \alpha(a)b) = (u,d)$. Hence, $X \cap L = \{(\alpha^{-1}(db^{-1}) + e, d)\}$. Also, $(a, \alpha(a)b) + (e,0) = (0,v) + (c,0) \in X \cap M$ if and only if $(a + e, \alpha(a)b) = (c,v)$. Hence, $X \cap M = \{(c, \alpha(c - e)b)\}$.
  Now take $Y = N(1,b') + (e',0)$, where $b \neq b'$. So $(x,y) \in X \cap Y$ if and only if $(x,y) = (a, \alpha(a)b) + (e,0) = (a', \alpha(a')b') + (e',0)$ for some $a, a' \in N$, if and only if $(x,y) = (a + e, \alpha(a)b) = (a' + c', \alpha(a')b')$. Hence, $X \cap Y = \{((c' - c)[\alpha^{-1}(b'b^{-1}) - 1]^{-1} + c', \alpha[(c' - c)[\alpha^{-1}(b'b^{-1}) - 1]^{-1}]b')\}$.

(7.51) Corollary. *If $x \in N$ and $L \in \mathcal{L}(1,0)$, then there is a unique $M \in \mathcal{L}(1,0)$ such that $x \in M$ and $L \parallel M$.*

Proof. This follows from (7.48) and (7.50).

To complete the proof of (7.47), we must show that two distinct points of N lie in exactly one line $L \in \mathcal{L}(1,0)$. Take two distinct points $(x,y), (u,v) \in N$. If $x = u$, then $(x,y) - (u,v) = (0, y-v) \in N(0,1)$, and so $(x,y), (u,v) \in N(0,1) + (x,y)$, and (x,y) and (u,v) cannot both be in any other coset of $N(0,1)$. By (7.50), (x,y) and (u,v) cannot both be in any other line from $\mathcal{L}(1,0)$. A similar argument is valid if $y = v$.

Suppose $x \neq u$ and $y \neq v$. We will see that $(x,y), (u,v) \in N(1,b) + (e,0)$ for some $b, e \in F$, $b \neq 0$. So, we must show the existence of $b \neq 0$, d, m, and n so that $(x,y) = (m, \alpha(m)b) + (e,0)$ and $(u,v) = (n, \alpha(n)b) + (e,0)$. Let $m = (x - u)(1 - \alpha^{-1}(vy^{-1}))^{-1}$, if $y \neq 0$, which we can readily assume since $y \neq v$. Then let $n = \alpha^{-1}(vy^{-1})m$, $e = x - m$, and $b = \alpha(m)^{-1}y$. If $m = 0$, then either $x = u$ or $y = v$. Hence, $m \neq 0$, and so $b \neq 0$. One can compute directly that $(x,y), (u,v) \in N(1,b) + (c,0)$. Again, (7.50) assures us that (x,y) and (u,v) can be in no other line of $\mathcal{L}(1,0)$. This completes the proof of (7.47).

We have used one subgroup $N(1,0)$ to construct $(N, \mathcal{L}(1,0), \in)$, an affine plane. We will now use $(N, \mathcal{L}(1,0), \in)$ and the other subgroup $N(0,1)$ to construct other affine planes $(N, \mathcal{L}(1,0) + (0,d), \in)$, one for each $(0,d) \in N(0,1)$.

For $(c,d) \in N$, define $\mathcal{L}(1,0) + (c,d) = \{L + (c,d) \mid L \in \mathcal{L}(1,0)\}$. When is $\mathcal{L}(1,0) + (c,d) = \mathcal{L}(1,0) + (e,f)$? If $L \in N/N(1,0)$, then $L + (c,d) \in N/N(1,0)$, and $L \in N/N(0,1)$ implies $L + (c,d) \in N/N(0,1)$. So, these two families of cosets are in both $\mathcal{L}(1,0) + (c,d)$ and $\mathcal{L}(1,0) + (e,f)$. Since $N(1,b) + (x,0) + (c,d) = N(1,b) + (x+c,0) + (0,d)$, we obtain $\mathcal{L}(1,0) + (c,d) \subseteq \mathcal{L}(1,0) + (0,d)$. Also, $N(1,0) + (x+c,0) + (0,d) = N(1,0) + (x,0) + (c,d)$, so $\mathcal{L}(1,0) + (c,d) = \mathcal{L}(1,0) + (0,d)$. Therefore, $\mathcal{L}(1,0) + (e,f) = \mathcal{L}(1,0) + (0,f)$. Now the question is, 'when is $\mathcal{L}(1,0) + (0,d) = \mathcal{L}(1,0) + (0,f)$'? If $N(1,b) + (0,d) \in \mathcal{L}(1,0) + (0,f)$, then there is a $N(1,b') + (c',0) + (0,f) \in \mathcal{L}(1,0) + (0,f)$ such that $N(1,b) + (0,d) = N(1,b') + (c',0) + (0,f)$. By (7.2), $(0,d) = (c',f)$, hence $d = f$. So we have

(7.52) Proposition. $\mathcal{L}(1,0) + (c,d) = \mathcal{L}(1,0) + (e,f)$ *if and only if $d = f$. We therefore have $\mathcal{L}(1,0) + (c,d) = \mathcal{L}(1,0) + (0,d)$.*

So, for each $(0,d) \in N(0,1)$, we have the map $T_d : N \to N$ defined by $T_d(x,y) = (x,y) + (0,d) = (x, y+d)$. Certainly, T_d is a bijection of N, which maps the lines of $\mathcal{L}(1,0)$ bijectively onto the lines of $\mathcal{L}(1,0) + (0,d)$, and so each T_d is an isomorphism from $(N, \mathcal{L}(1,0), \in)$ onto $(N, \mathcal{L}(1,0) + (0,d), \in)$. This, along with (7.52), gives us

(7.53) Theorem. *For each distinct* $(0, d) \in N(0, 1)$, *there is the distinct affine plane* $(N, \mathcal{L}(1, 0) + (0, d), \in)$ *isomorphic to* $(N, \mathcal{L}(1, 0), \in)$

(7.54) Corollary. $\mathcal{L}(1, 0) + (0, 0) = \mathcal{L}(1, 0)$.

(7.55) Remark. From $N(1, 0)$, we have constructed $|N(0, 1)| = |F| = q$ affine planes $(N, \mathcal{L}(1, 0) + (0, d), \in)$, and the bijection is simply $(0, d) \mapsto (N, \mathcal{L}(1, 0) + (0, d), \in)$. So we have q such substructures of $(N, \mathcal{B}, \in)$, and any two of them have exactly $N/N(1, 0) \cup N/N(0, 1)$ as their intersection.

We next interchange the roles of $N(1, 0)$ and $N(0, 1)$ and follow an analogous procedure. So let $Ng_2 = Nb = N(0, 1)$, the other basic block which is also a subgroup of N. Let $\mathcal{L}'_0 = \mathcal{L}(0, 1) = N/N(1, 0) \cup N/N(0, 1) \cup \{N(1, b) + (0, y) \mid b \in F^*, (0, y) \in N(0, 1)\}$. We have now used $N(0, 1)$ to sift out the substructure $(N, \mathcal{L}(0, 1), \in)$ of $(N, \mathcal{B}, \in)$. One needs to show that $N/N(1, 0)$, $N/N(0, 1)$, and for each $b \neq 0$, $\mathcal{P}'_b = \{N(1, b) + (0, y) \mid (0, y) \in N(0, 1)\}$ are distinct families of parallel lines in an affine plane $(N, \mathcal{L}(0, 1), \in)$.

(7.56) Theorem. $(N, \mathcal{L}(0, 1), \in)$ *is an affine plane.*

Proof. The reader should be able to construct an easier but analogous proof to that of (7.47).

As above, we define for $(c, d) \in N$, the family $\mathcal{L}(0, 1) + (c, d) = \{L + (c, d) \mid L \in \mathcal{L}(0, 1)\}$, and we obtain in an analogous manner,

(7.57) Proposition. $\mathcal{L}(0, 1) + (c, d) = \mathcal{L}(0, 1) + (e, f)$ *if and only if* $c = e$. *So we have* $\mathcal{L}(0, 1) + (c, d) = \mathcal{L}(0, 1) + (c, 0)$.

So for each $(c, 0) \in N(1, 0)$, we have the map $T'_c : N \to N$ defined by $T'_c(x, y) = (x, y) + (c, 0) = (x + c, y)$. Each T'_c defines an isomorphism from $(N, \mathcal{L}(0, 1), \in)$ onto $(N, \mathcal{L}(0, 1) + (c, 0), \in)$. From this and (7.57) we obtain

(7.58) Theorem. *For each distinct* $(c, 0) \in N(1, 0)$, *there is the distinct affine plane* $(N, \mathcal{L}(0, a) + (c, 0), \in)$ *isomorphic to* $(N, \mathcal{L}(0, 1), \in)$.

(7.59) Corollary. $\mathcal{L}(0, 1) + (0, 0) = \mathcal{L}(0, 1)$.

(7.60) Remark. From $N(0, 1)$, we have $|N(1, 0)| = |F| = q$ affine planes $(N, \mathcal{L}(0, 1) + (c, 0), \in)$, and the bijection is simply $(c, 0) \mapsto (N, \mathcal{L}(0, 1) + (c, 0), \in)$. So we have q such substructures of $(N, \mathcal{B}, \in)$, and any two of them have exactly $N/N(1, 0) \cup N/N(0, 1)$ as their intersection.

We have constructed $q + q = 2q$ different substructures of $(N, \mathcal{B}, \in)$, each of which is an affine plane, and each of which contains the two pencils $N/N(1, 0)$ and $N/N(0, 1)$ of distinct families of parallel lines. But $\mathcal{B}$ has more structure in terms of the $\mathcal{L}(1,0) + (0,d)$ and the $\mathcal{L}(0,1) + (c,0)$. This is the content of our next theorem.

(7.61) Theorem. (a) $\mathcal{B} = \cup_{(0,d)\in N(0,1)}(\mathcal{L}(1,0) + (0,d))$, *and* $d \neq d'$ *implies* $(\mathcal{L}(1,0) + (0,d)) \cap (\mathcal{L}(1,0) + (0,d')) = N/N(1,0) \cup N/N(0,1)$.

(b) $\mathcal{B} = \cup_{(c,0)\in N(1,0)}(\mathcal{L}(0,1)+(c,0))$, *and* $c \neq c'$ *implies* $(\mathcal{L}(0,1)+(c,0))\cap$ $(\mathcal{L}(0,1) + (c',0)) = N/N(1,0) \cup N/N(0,1)$.

(c) $(\mathcal{L}(0,1)+(c,0))\cap(\mathcal{L}(1,0)+(0,d)) = N/N(1,0)\cup N/N(0,1)\cup\{N(1,b)+ (c,d) \mid b \neq 0\}$.

Proof. These follow from (7.2) and the equation $N(1,b) + (c,d) = N(1,b) + (c,0) + (0,d) = N(1,b) + (0,d) + (c,0)$.

The group Φ_α acts on $\mathcal{B}$ as well as N. For $a' \in \Phi_\alpha$, $a'(N(x,y)+(c,d)) = N(x,y) + (ac, \alpha(a)d)$. Certainly, $(a' \circ b')(L) = a'(b'(L))$, and $1'(L) = L$. The action of Φ_α on $\mathcal{B}$ has each basic block $N(0,1)$, $N(1,0)$, and $N(1,b)$, $b \neq 0$ as a fixed point, or trivial orbit. The remaining orbits each have $q-1$ blocks, and these orbits are:

$$\Phi_\alpha[N(1,b) + (1,b')] \equiv (b',b); \quad \Phi_\alpha[N(1,b) + (1,0)] \equiv (0,b);$$

$$\Phi_\alpha[N(1,b) + (0,1)] \equiv (b,0); \quad \Phi_\alpha[N(1,0) + (0,1)] \equiv (0,0);$$

$$\Phi_\alpha[N(0,1) + (1,0)] \equiv \infty.$$

Thus, there is a natural bijection between the families of nontrivial orbits of Φ_α's action on $\mathcal{B}$ and the elements of $N \cup \{\infty\}$. In particular, N can be recovered from $\mathcal{B}$ as a result of Φ_α's action on $\mathcal{B}$. Thus, we have

(7.62) Theorem. (a) *There is a bijection between* Φ_α *'s trivial orbits in* $\mathcal{B}$ *and the set* $F \cup \{\infty\}$, *where* $N(1,b) \mapsto b \in F$, *and* $N(0,1) \mapsto \infty$.

(b) *There is a bijection between* Φ_α *'s nontrivial orbits in* $\mathcal{B}$ *and the set* $N\cup\{\infty\}$ *described in the paragraph just before the statement of this theorem.*

(c) N *can be reconstructed from* $\mathcal{B}$ *by considering the nontrivial orbits of* Φ_α *'s action on* $\mathcal{B}$, *except for the orbit* $\Phi_\alpha[N(0,1) + (1,0)]$, *which is associated with* ∞.

(7.63) Corollary. *In* $N = F \oplus F$, *the graph of each* $y = b\alpha(x)$ *is left fixed by the action of* Φ_α *on* N.

Proof. The graph of $y = b\alpha(x)$ is exactly $\{(x, b\alpha(x)) \mid x \in F\} = N(1,b)$.

The bijections in (7.62) between the trivial orbits of Φ_α in $\mathcal{B}$ and $F\cup\{\infty\}$, and between the nontrivial orbits of Φ_α in $\mathcal{B}$ and $N \cup \{\infty\}$, could make one wonder about further geometric applications of planar nearrings with $0 \notin \{s,t\}$, that is, where some but not all of the basic blocks are subgroups. This is because the $N \cup \{\infty\}$ are nice settings for *Möbius*, or *inversive planes*, which are, in the finite case, some very nice BIBDs. This also gives us a nice opportunity to introduce a topic that has been traditionally associated with nearfields, and we will have more to say about that in §9.

The setting for this diversion is with a field K, a quadratic extension field $L = K(\omega)$ of K, and an additional symbol ∞. The reader may recall experiences with the extended complex number plane $\mathbf{C} \cup \{\infty\}$. In fact, we let $K' = K \cup \{\infty\}$ and $L' = L \cup \{\infty\}$.

Let $GL(2, L)$ be the set of all 2 by 2 nonsingular matrices over L. For a $\begin{pmatrix} a & b \\ c & d \end{pmatrix} \in GL(2, L)$, we define a transformation $\Gamma \begin{pmatrix} a & b \\ c & d \end{pmatrix} = f$, where

$$f(x) = \frac{ax + b}{cx + d}$$

for all $x \in L'$. The reader should be satisfied that $f : L' \to L'$ is well defined. If $\Gamma \begin{pmatrix} a' & b' \\ c' & d' \end{pmatrix} = g$ where $g(x) = (a'x + b')/(c'x + d')$, then it is easy to see that Γ is a semigroup homomorphism. But since $\begin{pmatrix} a & b \\ c & d \end{pmatrix}, \begin{pmatrix} a' & b' \\ c' & d' \end{pmatrix} \in GL(2, L)$, then one can conclude that f and g are permutations of L', and that the image of Γ is a group of permutations of L', which we denote by $lf(L)$, and call the group of *linear fractional transformations* or the group of *Möbius transformations* of L'. If $GL(2, K)$ denotes the subgroup of $GL(2, L)$ consisting of those $\begin{pmatrix} a & b \\ c & d \end{pmatrix} \in GL(2, L)$ with $a, b, c, d \in K$, then the image of $GL(2, K)$ by Γ is a subgroup of $lf(L)$, which we shall denote by $lf(K)$, and call the group of *linear fractional transformations* or the group of *Möbius transformations* of L' over K.

One of our main reasons for this diversion is to show that if $\mathcal{C} = \{f(K') \mid f \in lf(L)\}$, then $(L', \mathcal{C}, \in)$ is a *Möbius plane* or *inversive plane*. An incidence structure $(P, \mathcal{S}, \in)$ is a *Möbius (inversive) plane* if the following five conditions are satisfied. (1) $|P| \geq 4$. (2) $|A| > 0$ if $A \in \mathcal{S}$. (3) There is an $A \in \mathcal{S}$ and an $x \in P$ such that $x \notin A$. (4) If $x, y, z \in P$ are three distinct points, then there is a unique $A \in \mathcal{S}$ such that $x, y, z \in A$. Hence, $A \in \mathcal{S}$ is called the *circle* determined by $x, y, z \in P$. (5) If $x, y \in P$, $A \in \mathcal{S}$, $x \in A$, and $y \notin A$, then there exists a unique circle $B \in \mathcal{S}$ such that $x, y \in B$ and $A \cap B = \{x\}$.

Notice that the group homomorphism $\Gamma : GL(2, L) \to lf(L)$ has nontrivial kernel $\ker \Gamma = \left\{ \begin{pmatrix} \alpha & 0 \\ 0 & \alpha \end{pmatrix} \mid \alpha \in L^* \right\}$. Hence, an $f \in lf(L)$ may have a formula for each $\alpha \in L^*$, namely, $f(x) = (\alpha a x + \alpha b)/(\alpha c x + \alpha d) = (ax+b)/(cx+d)$. The group $lf(L)$ acts on $\mathcal{C}$ as well as on L'. Of importance to us will be the subgroup of $lf(L)$ which fixes $K' \in \mathcal{C}$.

(7.64) Proposition. *We have $K' \in \mathcal{C}$, and the subgroup of $lf(L)$ which fixes K' is $lf(K)$. If $A \in \mathcal{C}$ and $g(K') = A$ for $g \in lf(L)$, then $\{h \in lf(L) \mid h(K') = A\} = g(lf(K))$, a left coset of the subgroup $lf(K)$ in $lf(L)$.*

Proof. Since $i(x) = x$ defines $i \in lf(L)$, we have $K' \in \mathcal{C}$. Suppose $f(K') = K'$ and $f(x) = (ax + b)/(cx + d)$. Either $a \neq 0$ or $a = 0$.

If $a \neq 0$, then $f(x) = (x + (b/a))/((c/a)x + (d/a))$, so, without loss of generality, we take $f(x) = (x + b)/(cx + d)$. If $d = 0$, then $c \neq 0$ and $\infty = f(0)$. But $1/c = f(\infty)$ makes $c \in K$, and $f(1) = (1 + b)/c$ makes $b \in K$. If $d \neq 0$, then $f(0) = b/d = \lambda \in K$. If $c = 0$, then $f(x) = (1/d)x + (b/d)$, and $f(1) = (1/d) + \lambda$, so $d \in K$, and then $b \in K$. If $c \neq 0$, then $f(\infty) = 1/c$ forces $c \in K$. Also, $f(1) = (1+b)/(c+d) = \mu \in K$ means that $1 + b = \mu(c + d) = \mu c + \mu d$. So $\mu c - 1 = b - \mu d = \lambda d - \mu d = (\lambda - \mu)d$. This puts $d \in K$ as well as $b \in K$. So if $a \neq 0$, then we may take $f \in lf(K)$.

If $a = 0$, then $f(x) = b/(cx + d) = 1/((c/b)x + (d/b))$, and, without loss of generality, we may take $f(x) = 1/(cx + d)$. If $d = 0$, then $f(1) = 1/c$ and so $c \in K$. If $d \neq 0$, then $f(0)$ forces $d \in K$. If $c = -d$, we are through. If $c \neq -d$, then $f(1) = 1/(c + d) = \mu \in K$ means $c + d \in K$, and since $d \in K$, we also have $c \in K$. So if $f(K') = K'$, we may take $f \in lf(K)$, that is, $f(x) = (ax + b)/(cx + d)$, with $a, b, c, d \in K$.

If $h_1(K') = A = h_2(K')$, then $h_2^{-1} \circ h_1(K') = K'$, and so $h_2^{-1} \circ h_1 \in lf(K)$ and so $h_1(lf(K)) = h_2(lf(K))$, a left coset of $lf(K)$ in $lf(L)$.

Let G be a group of permutations on a set $X \neq \varnothing$. It is said that G is *k-transitive* on X if, for every pair of k-tuples $(x_1, \dots, x_k)$ and $(y_1, \dots, y_k)$, each having distinct entries from X, there is a $g \in G$ with $gx_i = y_i$ for each i, $1 \leq i \leq k$. If each such pair $(x_1, \dots, x_k)$ and $(y_1, \dots, y_k)$ has a unique such $g \in G$, then G is *sharply k-transitive* on X. If G is k-transitive, this is equivalent to saying that only the identity in G fixes k distinct elements of X.

(7.65) Lemma. *If $f \in lf(L)$ and $f \neq i$, the identity, then f fixes at most two points in L.*

Proof. If $f(x) = (ax + b)/(cx + d) = x$, then $cx^2 + (d - a)x - b = 0$. Since L is a field, there are at most two solutions to this last polynomial equation.

(7.66) Lemma. *If $a, b, c \in L'$ are three distinct points, then there is a unique $f \in lf(L)$ such that $f(a) = 1$, $f(b) = 0$, and $f(c) = \infty$. Hence, there is a unique $g \in lf(L)$ such that $g(1) = a$, $g(0) = b$, and $g(\infty) = c$.*

Proof. If $a, b, c \in L$, then

$$f(x) = \left(\frac{x - b}{x - c}\right) \bigg/ \left(\frac{a - b}{a - c}\right)$$

does the job. If $a = \infty$, then take $f(x) = (x-b)/(x-c)$. If $b = \infty$, then let $f(x) = (a-c)/(x-c)$. Finally, if $c = \infty$, then $f(x) = (x-b)/(a-b)$. This gives existence. If $h(a) = 1$, $h(b) = 0$, and $h(c) = \infty$, then $h^{-1} \circ f(a) = a$, $h^{-1} \circ f(b) = b$, and $h^{-1} \circ f(c) = c$. By (7.65), $h = f$. This gives uniqueness. Let $g = f^{-1}$. Then $g(1) = a$, $g(0) = b$, and $g(\infty) = c$.

(7.67) Theorem. *The group $lf(L)$ is sharply 3-transitive on L'.*

Proof. Let (x_1, x_2, x_3) and (y_1, y_2, y_3) be a pair of 3-tuples, each having distinct entries in L'. Then there is a unique $f \in lf(L)$ such that $f(x_1) = 1$, $f(x_2) = 0$, and $f(x_3) = \infty$. Also, there is a unique $g \in lf(L)$ such that $g(1) = y_1$, $g(0) = y_2$, and $g(\infty) = y_3$. Hence, $lf(L)$ is 3-transitive on L'. If $h \in lf(L)$ and $h(x_1) = y_1$, $h(x_2) = y_2$, and $h(x_3) = y_3$, then $h^{-1} \circ g \circ f = i$, and so $g \circ f = h$. Hence, $lf(L)$ is sharply 3-transitive on L'.

(7.68) Proposition. *The only $A \in \mathcal{C}$ with $0, 1, \infty \in A$ is $A = K'$.*

Proof. Certainly, $0, 1, \infty \in K'$. If $0, 1, \infty \in A = f(K')$ for $f \in lf(L)$, then there exist $x_1, x_2, x_3 \in K'$ so that $f(x_1) = 0$, $f(x_2) = 1$, and $f(x_3) = \infty$. By (7.66), f is unique, and from the proof of (7.66), we see that $f \in lf(K)$, hence, $f(K') = K' = A$.

(7.69) Proposition. *If $a, b, c \in L'$ are three distinct points, then there is a unique $A \in \mathcal{C}$ with $a, b, c \in A$.*

Proof. If $a, b, c \in A = f(K')$, and if $a, b, c \in B = g(K')$, with $f, g \in lf(L)$, then there is a unique $h \in lf(L)$ such that $h(a) = 1$, $h(b) = 0$, and $h(c) = \infty$. Hence, $0, 1, \infty \in h \circ f(K') = h(A)$ and $0, 1, \infty \in h \circ g(K') = h(B)$. From (7.68), we know that $h(A) = h(B)$, and since $h \in lf(L)$, we know that $A = B$. This gives us uniqueness. Existence follows from (7.66).

(7.70) Proposition. *If $x, y \in L'$, $A \in \mathcal{C}$, $x \in A$, and $y \notin A$, then there exists a $B \in \mathcal{C}$ such that $x, y \in B$ and $A \cap B = \{x\}$.*

Proof. First, let $A = K'$, $x = 0$, and $a = y$. Consider $f(x) = ax/(x + a)$. Then $f(0) = 0$ and $f(\infty) = a$. Let $B = f(K')$. So $x, y \in B$. Now $\infty \notin f(K') = B$. If $u, t \in K$ and $t = f(u) = au/(u+a)$, then $t(u+a) = au$. Since $y = a \in L \setminus K$, we have $a = a_1 + a_2\omega$ where $a_1, a_2 \in K$ and $a_2 \neq 0$. From $t(u + a_1 + a_2\omega) = u(a_1 + a_2\omega)$ we can conclude that $u = t$ and that $tu + ta_1 = ua_1$. Hence, $tu = 0$, so $t = u = 0$. Hence, $A \cap B = \{0\} = \{x\}$.

Now we remove all the extra restrictions on A and x. By (7.66) and (7.69), there is an $f \in lf(L)$ such that $f(x) = 0$ and $f(A) = K'$. Since $y \notin A$, then $f(y) \notin K'$. By the first paragraph of this proof, we have the existence of a $C \in \mathcal{C}$ such that $f(y) \in C$ and $K' \cap C = \{0\}$. So $f^{-1}(C) \in \mathcal{C}$, $y \in f^{-1}(C)$, and $\{x\} = \{f^{-1}(0)\} = f^{-1}(K') \cap f^{-1}(C) = A \cap f^{-1}(C)$. So let $B = f^{-1}(C)$.

(7.71) Lemma. *Let $a \in L \setminus K'$, $f(x) = ax/(x + a)$, and $B = f(K')$. Suppose $g \in lf(L)$, $g(0) = 0$, $g(\infty) = a$, and $g(K') = B$. Then $g(x) = ax/(x + \lambda a)$ for some $\lambda \in K$, and conversely.*

Proof. From $g(0) = 0$ and $g(\infty) = a$, we can write $g(x) = ax/(x+d)$. Since $a/(1 + a) \in B$, there is a $\lambda \in K$ such that $a/(1 + a) = a\lambda/(\lambda + d)$. Hence, $a(\lambda + d) = a\lambda(1 + a)$. This makes $\lambda a^2 = ad$, and since $a \neq 0$, $d = \lambda a$. If $g(x) = ax/(x+\lambda a)$, then $g(\lambda) = a/(1+a)$. Hence, $0, a, a/(1+a) \in g(K') \cap B$. From (7.69), we obtain $g(K') = B$.

(7.72) Lemma. *Let $a \in L \setminus K'$ and $g(x) = ax/(x + \lambda)$ for some $\lambda \in K$. Then $0, \infty \in K' \cap g(K')$.*

Proof. We have $g(0) = 0$ and $g(-\lambda) = \infty$.

(7.73) Lemma. *Let $a \in L \setminus K'$, $g(x) = ax/(x + d)$, where $d \in L \setminus K'$ but $d \notin Ka$. Then $|K' \cap g(K')| = 2$.*

Proof. Certainly, $0 \in K' \cap g(K')$. Let $a = a_1 + a_2\omega$ and $d = d_1 + d_2\omega$, with the $a_i, d_i \in K$, $i = 1, 2$. Certainly, $0 \notin \{a_2, d_2\}$. Now $a_1 d_2 - a_2 d_1 \neq 0$, for if $a_1 d_2 - a_2 d_1 = 0$, then $a_1/a_2 = d_1/d_2 = \lambda \in K$, and so $a_1 = \lambda a_2$, $d_1 = \lambda d_2$, $a = a_2(\lambda + \omega)$, and $d = d_2(\lambda + \omega)$. This in turn makes $d \in Ka$. So $t = (a_1 d_2 - a_2 d_1)/d_2 \neq 0$ and $u = (a_1 d_2 - a_2 d_1)/a_2 \neq 0$. Since $g(u) = t$, we have $t \neq 0$ and $t \in K' \cap g(K')$. But $g(\infty) = a \notin K'$, so by (7.69), we have $|K' \cap g(K')| = 2$.

(7.74) Proposition. *The B of (7.70) is unique.*

Proof. First, let $A = K'$, $x = 0$, and $a = y$, as in the first paragraph of the proof of (7.70). Then $0, a, a/(1 + a) \in B$. If $0, a \in g(K') = C$, then there is an $h \in lf(K)$ such that $h(0) = x_1$ and $h(\infty) = x_2$, where $x_1, x_2 \in K'$ and $g(x_1) = 0$ and $g(x_2) = a$. Hence, $g \circ h(K') = g(K') = C$ and $g \circ h(0) = 0$ and $g \circ h(\infty) = a$. Hence, $g \circ h(x) = ax/(x + d)$. Using (7.71), (7.72), and (7.73), we have $B = C$, $K' \cap C = \{0, \infty\}$, and $|K' \cap C| = 2$ respectively. This gives the uniqueness of B for this specialized case of $A = K'$ and $x = 0$.

As in the proof of (7.70), if, when we consider the general case, we had $B, D \in \mathcal{C}$ with $x, y \in B \cap D$ and $A \cap B = A \cap D = \{x\}$, then an $f \in lf(L)$ such that $f(x) = 0$ and $f(A) = K'$ would yield $f(B), f(D) \in \mathcal{C}$ with $f(y) \in f(B) \cap f(D)$, and $K' \cap f(B) = K' \cap f(D) = \{0\}$. Our first paragraph of this proof assures us that $f(B) = f(D)$, so we must have $B = D$ since f is a permutation of L'.

We can now easily provide a proof to the promised

(7.75) Theorem. *$(L', \mathcal{C}, \in)$ is a Möbius plane.*

Proof. With $|L'| \geq 5$, we have (1). With $|A| = |f(K')| = |K'| \geq 3$, we have (2). With $L = K(\omega)$, we have $\omega \in L'$, and $\omega \notin K'$, so (3). From (7.69) we have (4), and from (7.70) and (7.74), we obtain (5). Hence, $(L', \mathcal{C}, \in)$ is a Möbius plane.

(7.76) Theorem. *If L is finite and $|K| = q$, then $(L', \mathcal{C}, \in)$ is a BIBD with parameters $v = q^2 + 1$, $b = q(q^2 + 1)$, $r = q(q + 1)$, and $k = \lambda = q + 1$.*

Proof. Certainly, there are $|L'| = q^2 + 1$ points, and since $|K'| = q + 1$, each $|f(K')| = |K'| = q + 1$, so $k = q + 1$. If 0 belongs to r distinct $B_1 = f_1(K'), \ldots, B_r = f_r(K')$, and $g(0) = x$ for some $g \in lf(L)$, then x belongs to $g(B_1) = g \circ f_1(K'), \ldots, g(B_r) = g \circ f_r(K')$, and conversely. Hence, each point belongs to exactly r blocks. For $a, b \in L'$, $a \neq b$, and any $x \in L' \setminus \{a, b\}$, then a, b, x belong to exactly one $B = \{a, b, x_{11}, x_{12}, \ldots, x_{1,q-1}\}$. Let λ be the number of blocks containing a and b. Then $q^2 - 1 = \lambda(q - 1)$, and so $\lambda = q + 1 = k$. So an arbitrary pair of distinct points $a, b \in L'$ belongs to exactly $q + 1 = \lambda = k$ distinct blocks. Hence, $(N', \mathcal{C}, \in)$ is a BIBD. From $vr = bk$, and $\lambda(v - 1) = r(k - 1)$, we compute $r = q(q + 1)$ and then $b = q(q^2 + 1)$.

The natural bijection of (7.62) of Φ_α's nontrivial orbits and the set $N \cup \{\infty\}$ induces the structure of a Möbius plane, and when N is finite, a BIBD. This is because the additive group of $N = F \oplus F$ is isomorphic to the additive group of a quadratic extension field of the field F. So we have

(7.77) Theorem. *The nontrivial orbits of Φ_α in $\mathcal{B}$ are the points of a Möbius plane. If $|N| < \infty$, then these nontrivial orbits are the points of a BIBD.*

(7.78) Note. From the prime field $(Z_5, +, \cdot)$, there is the Ferrero pair (Z_5, Φ) where $|\Phi| = 2$. The resulting BIBD $(Z_5, \mathcal{B}, \in)$ is also a Möbius plane.

(7.79) Exploratory problem. Consider a planar nearring $(N, +, \cdot)$ and its Ferrero pair (N, Φ). Suppose $0 \notin \{s, t\}$, that is, suppose some, but not all basic blocks of N are subgroups. Then Φ acts on $\mathcal{B}$ and has trivial and nontrivial orbits. Other than what has been observed here with the van der Walt connection, does one get further geometric structures from these orbits of Φ acting on $\mathcal{B}$?

(7.80) Problem Other than the example mentioned in (7.78), does one ever get a Möbius plane from an $(N, \mathcal{B}, \in)$?

Early in our discussion of the van der Walt connection, we discussed a little about cubic polynomials $y - k = b(x - h)^3$. It is interesting to use these to illustrate our theory developed here. What we need is any field F

in which $a \mapsto a^3$ describes an automorphism of F^* but not of the field F. We can certainly let $F = \mathbf{R}$, the real number field, or for a finite example, we could take $F = Z_5$, the prime field of order five.

For $N = F \oplus F$, then $N(1,0)$ is just the x-axis and $N(0,1)$ is the y-axis, but $N(1,b)$ consists of the graph of $y = bx^3$. $\mathcal{L}(1,0)$ consists of the vertical lines, given by $N/N(0,1)$, the horizontal lines, given by $N/N(1,0)$, and for each $b \neq 0$, the family of 'parallel lines' $\mathcal{P}_b = \{y = b(x - c)^3 \mid c \in F\}$, which, of course, is a *horizontal translation* of the graph of $y = bx^3$. When we construct $\mathcal{L}(1,0) + (0,d)$, we just translate everything to the line $y = d$. So in $\mathcal{L}(1,0) + (0,d)$, we still have the horizontal lines and the vertical lines, but the $\mathcal{P}_b$ become $\mathcal{P}_b + (0,d) = \{y - d = b(x - c)^3 \mid c \in F\}$, which consists of all cubic equations having no relative maximum or relative minimum, but having exactly one horizontal tangent line, and each having its inflection point on the line $y = d$ somewhere. By (7.63), the action of Φ_α on N leaves the graph of each $y = bx^3$ fixed.

In an analogous way, $\mathcal{L}(0,1)$ consists of the same horizontal and vertical lines, but for each $b \neq 0$, we get the family of 'parallel lines' $\mathcal{P}'_b = \{y - d = bx^3 \mid c \in F\}$, which is, of course, a *vertical translation* of the graph of $y = bx^3$. When we construct $\mathcal{L}(0,1) + (c,0)$, we just translate everything over to the line $x = c$. So, in $\mathcal{L}(0,1) + (c,0)$, we still have the usual horizontal and vertical lines, but $\mathcal{P}'_b$ becomes $\mathcal{P}'_b + (c,0) = \{y - d = b(x - c)^3 \mid c \in F\}$, which consists of all cubics with no relative maximum or minimum, but having exactly one horizontal tangent line, and each having its inflection point on the line $x = c$ somewhere.

For a fixed (c,d), we have $(\mathcal{L}(0,1) + (c,0)) \cap (\mathcal{L}(1,0) + (0,d)) = N/N(1,0) \cup N/N(0,1) \cup \{N(1,b) + (c,d) \mid b \neq 0\}$. Now $\{N(1,b) + (c,d) \mid b \neq 0\}$ consists of all $y - d = b(x - c)^3$, $b \neq 0$, all having an inflection point at (c,d).

Notice that van der Walt's examples result from modifying a Ferrero pair (N, Φ) to the Ferrero pair (N, Φ_α). Here, N is still $F \oplus F$ and Φ becomes Φ_α with α an automorphism of $(F^*, \cdot)$. It is easy to see that Φ and Φ_α are isomorphic. So we basically have Ferrero pairs (N, Φ) and (N, Φ_α) where Φ and Φ_α are isomorphic. Need the resulting nearrings be isomorphic, or are the resulting tactical configurations isomorphic? The examples developed here with the van der Walt connection provide negative answers to these questions.

We have $\Phi = \{a^* \mid a \in F^*\}$ where $a^*(x,y) = (ax, ay)$. Let $\mathcal{B}$ denote the blocks arising from (N, Φ) and $\mathcal{B}_\alpha$ denote the blocks arising from (N, Φ_α). As long as α is not a field automorphism, $\mathcal{B}_\alpha$ has only the two basic blocks $N(1,0)$ and $N(0,1)$ as subgroups. However, every basic block in $\mathcal{B}$ is a subgroup. So the resulting nearrings are not isomorphic.

If N is finite, $\mathcal{B}$ has far fewer blocks than does $\mathcal{B}_\alpha$, so the resulting tactical configurations cannot be isomorphic. In summary,

(7.81) Theorem. *For Ferrero pairs* (N, Φ) *and* (N, Φ_α) *with* Φ *isomorphic to* Φ_α, *it may be that* $(N, \mathcal{B}, \in)$ *and* $(N, \mathcal{B}_\alpha, \in)$ *are not isomorphic, and so a planar nearring from* (N, Φ) *need not be isomorphic to any planar nearring arising from* (N, Φ_α).

7.4. Karzel's hyperbolae

During the summer of 1988, Helmut Karzel responded to the author's request for an infinite circular planar nearring. His first attempt did not quite work out, but his examples do provide a very interesting family of planar nearrings. They have exactly two basic blocks which are subgroups. These two basic blocks play an analogous role to the two basic blocks which are also subgroups as described in the van der Walt connection. Hence, Karzel's examples will also be affine configurations with two pencils.

As we develop Karzel's examples, we will want to refer continually to the usual euclidean plane $\mathbf{R} \times \mathbf{R}$. As a certain class of cubic polynomials, together with the horizontal and vertical lines, were used to construct an affine plane with $\mathbf{R} \times \mathbf{R}$, in the van der Walt connection, Karzel's examples use certain hyperbolae together with lines with slope ± 1. Let one family of parallel lines be the lines $y = x + b$, $b \in \mathbf{R}$. Another will be the lines $y = -x + b$, $b \in \mathbf{R}$. For $a > 0$ and $h \in \mathbf{R}$, let $X(a, h) = \{(x, y) \mid (x - h)^2/a^2 - (y - h)^2/a^2 = 1\} \cup \{(h, h)\}$, and $Y(a, h) = \{(x, y) \mid (y - h)^2/a^2 - (x - h)^2/a^2 = 1\} \cup \{(h, h)\}$. Then $X(a, 0)_\| = \{X(a, h) \mid h \in \mathbf{R}\}$ and $Y(a, 0)_\| = \{Y(a, h) \mid h \in \mathbf{R}\}$ are families of parallel lines, two for each $a > 0$.

It is elementary but tedious to show that $\mathbf{R} \times \mathbf{R}$, together with all these families as lines, satisfy the requirements for an affine plane. As with the cubics described in the van der Walt connection, the author believes that these 'pointed hyperbolae' together with the lines of slope ± 1, are most easily discovered in the context of a planar nearring.

Our setting for Karzel's examples is $N = F \oplus F$, where F is a field with characteristic $\neq 2$ and $\neq 3$. We also need $S = \{x^2 \mid x \in F^*\}$ to be a subgroup of F^* of index 2, where for each $x \in F^*$, either $x \in S$ or $-x \in S$. Define $\Gamma : F^* \to GL(2, K)$ by

$$\Gamma(c) = \begin{pmatrix} \frac{c^2+1}{2c} & \frac{c^2-1}{2c} \\ \frac{c^2-1}{2c} & \frac{c^2+1}{2c} \end{pmatrix}.$$

Then Γ is a group monomorphism with image

$$\Phi = \left\{ \begin{pmatrix} a & b \\ b & a \end{pmatrix} \,\middle|\, a^2 - b^2 = 1 \right\}.$$

(7.82) Proposition. (N, Φ) *is a Ferrero pair.*

Proof. It is easy to see that Γ is a group monomorphism and that Φ is a subgroup of $GL(2, F)$. If $a^2 - b^2 = 1 = (a - b)(a + b)$, we let $a + b = c$, $a - b = c^{-1}$, and solve for a and b obtaining $a = (c^2 + 1)/2c$ and $b = (c^2 - 1)/2c$, so Γ defines an isomorphism between F^* and Φ. Since

$$\begin{pmatrix} a & b \\ b & a \end{pmatrix} \begin{pmatrix} x \\ y \end{pmatrix} = \begin{pmatrix} ax + by \\ bx + ay \end{pmatrix},$$

we identify $\begin{pmatrix} a & b \\ b & a \end{pmatrix} = (a, b)^*$, and define $(a, b)^* : N \to N$ by $(a, b)^*(x, y) = (ax + by, bx + ay)$. Since Φ is a subgroup of $GL(2, F)$, we have that Φ consists of automorphisms of N. If $(x, y) \neq (0, 0)$ and $(a, b)^*(x, y) = (x, y)$, then from $(ax + by, bx + ay) = (x, y)$, we have a nontrivial solution to

$$\begin{cases} (a - 1)x + by = 0 \\ bx + (a - 1)y = 0. \end{cases}$$

Hence, $(a - 1)^2 - b^2 = a^2 - 2a + 1 - b^2 = 2 - 2a = 2(1 - a) = 0$. This makes $a = 1$ and $b = 0$, so $(a, b)^* = (1, 0)^*$, the identity of Φ. So Φ is regular.

To finish the proof that (N, Φ) is a Ferrero pair, we need only show that $-(1, 0)^* + (a, b)^*$ is surjective for any $(a, b)^* \neq (1, 0)^*$. This follows since

$$-\begin{pmatrix} 1 & 0 \\ 0 & 1 \end{pmatrix} + \begin{pmatrix} a & b \\ b & a \end{pmatrix} = \begin{pmatrix} -1 + a & b \\ b & -1 + a \end{pmatrix}$$

is nonsingular. Hence, (N, Φ) is a Ferrero pair.

(7.83) Proposition. *Let $(N, +, \cdot)$ be a planar nearring constructed from the Ferrero pair (N, Φ) of (7.82). Then the basic blocks of $(N, +, \cdot)$ are $N(1, 1)$, $N(1, -1)$, the $N(c, 0)$, $c \neq 0$, and the $N(0, c)$, $c \neq 0$. Only $N(1, 1)$ and $N(1, -1)$ are subgroups of N. We have:*
(1) $N(1, 1) = \{(x, x) \mid x \in F\}$, the 'straight line' with slope $m = 1$ passing through $(0, 0)$.
(2) $N(1, -1) = \{(x, -x) \mid x \in F\}$, the 'straight line' with slope $m = -1$ passing through $(0, 0)$.
(3) If $c \neq 0$, then $N(c, 0) = \{(0, 0)\} \cup \{(x, y) \mid x^2/c^2 - y^2/c^2 = 1\}$.
(4) If $c \neq 0$, then $N(0, c) = \{(0, 0)\} \cup \{(x, y) \mid y^2/c^2 - x^2/c^2 = 1\}$.

Proof. For $u \neq 0$, $N(u, u) = \{(0, 0)\} \cup \{(au + bu, bu + au) \mid a^2 - b^2 = 1\}$. Let $t \neq 0$, and consider the equations $\begin{cases} a + b = tu^{-1} \\ a - b = t^{-1}u \end{cases}$. They have the solution $(a, b) = ((tu^{-1} + t^{-1}u)/2, (tu^{-1} - t^{-1}u)/2)$, and $a^2 - b^2 = 1$. Hence, $(t, t) \in N(u, u)$, and so $N(1, 1) = N(u, u) = \{(x, x) \mid x \in F\}$, a subgroup of N. Similarly, $N(1, -1) = N(u, -u) = \{(x, -x) \mid x \in F\}$, which is also a subgroup of N.

Suppose $u \neq v$, so $u^2 - v^2 \neq 0$, and either $u^2 - v^2 = c^2 \in S$ or $v^2 - u^2 = c^2 \in S$. In either case, $N(u,v) = \{(0,0)\} \cup \{(au + bv, bu + av) \mid a^2 - b^2 = 1\}$. If $u^2 - v^2 = c^2$, then $(au + bv)^2/c^2 - (bu + av)^2/c^2 = 1$, and so $N(u,v) \subseteq \{(0,0)\} \cup \{(x,y) \mid x^2/c^2 - y^2/c^2 = 1\}$. Similarly, if $v^2 - u^2 = c^2$, then $N(u,v) \subseteq \{(0,0)\} \cup \{(x,y) \mid y^2/c^2 - x^2/c^2 = 1\}$. If $u^2 - v^2 = c^2$ and $x^2/c^2 - y^2/c^2 = 1$, then we will see that there is a unique $(a,b)^* \in \Phi$ such that $(a,b)^*(u,v) = (au + bv, bu + av) = (x,y)$. This follows since $\begin{cases} ua + vb = x \\ va + ub = y \end{cases}$ has a unique solution for (a,b). Hence, $N(u,v) = \{(0,0)\} \cup \{(x,y) \mid x^2/c^2 - y^2/c^2 = 1\}$. Similarly, if $v^2 - u^2 = c^2$, then $N(u,v) = \{(0,0)\} \cup \{(x,y) \mid y^2/c^2 - x^2/c^2 = 1\}$. Notice that $u^2 - v^2 = c^2$ implies $(c,0), (-c,0) \in N(u,v)$, so $N(c,0) = N(u,v)$. Also, $v^2 - u^2 = c^2$ implies $(0,c), (0,-c) \in N(u,v)$, and then $N(0,c) = N(u,v)$.

If some $N(c,0)$, $c \neq 0$, could be a subgroup, then we would need $(c,0) + (c,0) = (2c,0) \in N(c,0)$. If there were a $(a,b)^* \in \Phi$ such that $(ac, bc) = (2c,0)$, then it would have to be $(2,0)^*$. If $((c^2+1)/2c, (c^2-1)/2c) = (2,0)$, then $c = -1$ and $c^{-1} = 2$. This would require the characteristic of F to be 3. Since this is not allowed, we can be assured that no $N(c,0)$ could be a subgroup of N. Similarly, if $c \neq 0$, then $N(0,c)$ cannot be a subgroup of N.

(7.84) Exercise. Here, if we construct $(N, \mathcal{B}', \in)$ similarly to the way we did in (7.46), we would obtain a statistical BIBD with the same parameters as we did for the corresponding statistical BIBD of (7.46). Are these two BIBDs isomorphic?

Of course, we want to find substructures of $(N, \mathcal{B}, \in)$ which are affine planes. Following the guide of the van der Walt connection, let

$$\mathcal{L}(1,1) = N/N(1,1) \cup N/N(1,-1) \cup$$
$$\{N(c,0) + (x,x), \; N(0,c) + (x,x) \mid c \neq 0, \; (x,x) \in N(1,1)\}.$$

We have used the subgroup $N(1,1)$ to sift out a substructure $(N, \mathcal{L}(1,1), \in)$ of $(N, \mathcal{B}, \in)$. We want to show that $N/N(1,1)$, $N/N(1,-1)$, and for $c \neq 0$, $\mathcal{P}_{c,1} = \{N(c,0) + (x,x) \mid (x,x) \in N(1,1)\}$ and $\mathcal{P}_{c,2} = \{N(0,c) + (x,x) \mid (x,x) \in N(1,1)\}$ are each distinct families of parallel lines in an affine plane $(N, \mathcal{L}(1,1), \in)$. Our proof, as with the van der Walt connection, will be given as a result of several lemmas and corollaries.

(7.85) Lemma. *Each of* $N/N(1,1)$, $N/N(1,-1)$, *the* $\mathcal{P}_{c,1}$, $c \neq 0$, *and the* $\mathcal{P}_{c,2}$, $c \neq 0$, *defines a partition of* N.

Proof. Since $N(1,1)$ and $N(1,-1)$ are subgroups, the $N/N(1,1)$ and $N/N(1,-1)$ are the corresponding cosets. Take a $c \neq 0$ and consider first $\mathcal{P}_{c,1}$. Now $N(c,0)+(h,h) = \{(h,h)\} \cup \{(x,y) \mid (x-h)^2/c^2 - (y-h)^2/c^2 = 1\}$, so any (h, h) belongs to exactly the one $N(c, 0) + (h, h) \in \mathcal{P}_{c,1}$. Take $(a, b) \in N$ with $a \neq b$. Let $h = (c^2 - a^2 + b^2)/2(b - a)$. Then $(a - h)^2/c^2 - (b - h)^2/c^2 = 1$, and so $(a, b) \in N(c, 0) + (h, h)$.

If $(u, v) \in N$, then either $u = v$ or $u \neq v$. If $u = v$, then $(u, v) = (u, u) \in N(c, 0) + (u, u)$ only. If $u \neq v$ and $(u, v) \in [N(c, 0) + (h, h)] \cap [N(c, 0) + (k, k)]$, then

$$\frac{(u - h)^2}{c^2} - \frac{(v - h)^2}{c^2} = 1 = \frac{(u - k)^2}{c^2} - \frac{(v - k)^2}{c^2}.$$

Hence, $h = k$. So each $\mathcal{P}_{c,1}$ defines a partition of N. The proof that each $\mathcal{P}_{c,2}$ defines a partition of N is similar.

A contribution of (7.85) is that the partitions $N/N(1, 1)$ and $N/N(1, -1)$, the $\mathcal{P}_{c,1}$, $c \neq 0$, and the $\mathcal{P}_{c,2}$, $c \neq 0$, will each be an equivalence class of parallel lines, that is, a 'pencil' of parallel lines. In fact, (7.85) tells us that for each line $L \in \mathcal{L}(1, 1)$ and each point $x \in N$, with $x \notin L$, there is a line $M \in \mathcal{L}(1, 1)$ such that $x \in M$ and $L \cap M = \varnothing$, that is, $L \parallel M$. This next lemma will give the uniqueness of M.

(7.86) Lemma. *Consider the four types of lines in $\mathcal{L}(1, 1)$: (I) those in $N/N(1, 1)$; (II) those in $N/N(1, -1)$; (III) those in a $\mathcal{P}_{c,1}$, $c \neq 0$; and (IV) those in a $\mathcal{P}_{c,2}$, $c \neq 0$. Let $\mathcal{D} = \{N/N(1, 1), N/N(1, -1)\} \cup \{\mathcal{P}_{c,1} \mid c \in F^*\} \cup \{\mathcal{P}_{c,2} \mid c \in F^*\}$. If $\mathcal{S}, \mathcal{T} \in \mathcal{D}$, $L \in \mathcal{S}$, and $\mathcal{S} \neq \mathcal{T}$, then $|L \cap M| = 1$ for each $M \in \mathcal{T}$. Hence, $\mathcal{D}$ will be the family of all equivalence classes of parallel lines in $\mathcal{L}(1, 1)$.*

Proof. We proceed case by case through the many possibilities. It will prove useful to note the following about the different types of lines.
(I) If $L \in N/N(1, 1)$, then there is a $b \in F$ so that $L = \{(x, x + b) \mid x \in F\}$.
(II) If $L \in N/N(1, -1)$, then there is a $b \in F$ so that $L = \{(x, -x + b) \mid x \in F\}$.
(III) If $L \in \mathcal{P}_{c,1}$, $c \neq 0$, then there is an $h \in F$ so that $L = \{(h, h)\} \cup \{(x, y) \mid (x - h)^2/c^2 - (y - h)^2/c^2 = 1\}$.
(IV) If $L \in \mathcal{P}_{c,2}$, $c \neq 0$, then there is a $k \in F$ so that $L = \{(k, k)\} \cup \{(x, y) \mid (y - k)^2/c^2 - (x - k)^2/c^2 = 1\}$

First, let $\mathcal{S} = N/N(1, 1)$ and let $\mathcal{T}$ vary. If $\mathcal{T} = N/N(1, -1)$, then $L = \{(x, x + b) \mid x \in F\}$ and $M = \{(x, -x + b') \mid x \in F\}$ for some $b, b' \in F$. Certainly, $L \cap M = \{((b' - b)/2, (b' + b)/2)\}$.

If $\mathcal{T} = \mathcal{P}_{c,1}$, $c \neq 0$, then $M = \{(h, h)\} \cup \{(x, y) \mid (x - h)^2/c^2 - (y - h)^2/c^2 = 1\}$ for some $h \in F$. If $b = 0$, then $L \cap M = \{(h, h)\}$. If $b \neq 0$, then $\begin{cases} y = x + b \\ (x - h)^2/c^2 - (y - h)^2/c^2 = 1 \end{cases}$ has a unique solution $(x, y) = ((2bh - c^2 - b^2)/2b, (2bh - c^2 + b^2)/2b)$, so $L \cap M = \{((2bh - c^2 - b^2)/2b, (2bh - c^2 + b^2)/2b)\}$.

If $\mathcal{T} = \mathcal{P}_{c,2}$, $c \neq 0$, then $M = \{(k, k)\} \cup \{(x, y) \mid (y - k)^2/c^2 - (x - $

$k)^2/c^2 = 1\}$ for some $k \in F$. If $b = 0$, then $L \cap M = \{(k, k)\}$. If $b \neq 0$,
then $\begin{cases} y = x + b \\ (y - k)^2/c^2 - (x - k)^2/c^2 = 1 \end{cases}$ has a unique solution and $L \cap M = \{((2bk + c^2 - b^2)/2b, (2bk + c^2 + b^2))\}$.

Next, let $\mathcal{S} = N/N(1, -1)$ and let $\mathcal{T}$ vary. We need not consider $\mathcal{T} = N/N(1, 1)$, since that case was considered above. So let $M = \{(h, h)\} \cup \{(x, y) \mid (x - h)^2/c^2 - (y - h)^2/c^2 = 1\}$ for some $h \in F$. If $b = 2h$, then $L \cap M = \{(h, h)\}$. If $b \neq 2h$, then $\begin{cases} y = -x + b \\ (x - h)^2/c^2 - (y - h)^2/c^2 = 1 \end{cases}$ has a unique solution and $L \cap M = \{((b^2 + c^2 - 2bh)/(2b - 4h), (b^2 - c^2 - 2bh)/(2b - 4h))\}$.

Suppose $\mathcal{T} = \mathcal{P}_{c,2}$, $c \neq 0$. Then $M = \{(k, k)\} \cup \{(x, y) \mid (y - k)^2/c^2 - (x - k)^2/c^2 = 1\}$ for some $k \in F$. If $b = 2k$, then $L \cap M = \{(k, k)\}$. If $b \neq 2k$, then $\begin{cases} y = -x + b \\ (y - k)^2/c^2 - (x - k)^2/c^2 = 1 \end{cases}$ has a unique solution and $L \cap M = \{((b^2 - c^2 - 2bk)/(2b - 4k), (b^2 + c^2 - 2bk)/(2b - 4k))\}$.

Next let $\mathcal{S} = \mathcal{P}_{c,1}$, $c \neq 0$. We need not let $\mathcal{T} \in \{N/N(1, 1), N/N(1, -1)\}$, but we must consider $\mathcal{T} = \mathcal{P}_{d,1}$, $d \neq c$ and $d \neq 0$. Now $L = \{(h, h)\} \cup \{(x, y) \mid (x - h)^2/c^2 - (y - h)^2/c^2 = 1\}$ and $M = \{(k, k)\} \cup \{(x, y) \mid (x - k)^2/d^2 - (y - k)^2/d^2 = 1\}$. If $h = k$, then $L \cap M = \{(h, h)\}$. If $h \neq k$, then $\begin{cases} (x - h)^2/c^2 - (y - h)^2/c^2 = 1 \\ (x - k)^2/d^2 - (y - k)^2/d^2 = 1 \end{cases}$ has a unique solution and $L \cap M = \{((2hA - A^2 - c^2)/2A, (2hA + A^2 - c^2)/2A)\}$ where $A = (c^2 - d^2)/2(h - k)$.

If $\mathcal{T} = \mathcal{P}_{d,2}$, $d \neq 0$, then $M = \{(k, k)\} \cup \{(x, y) \mid (y - k)^2/d^2 - (x - k)^2/k^2 = 1\}$ for some $k \in F$. If $h = k$, then $L \cap M = \{(h, h)\}$. If $h \neq k$, then $\begin{cases} (x - h)^2/c^2 - (y - h)^2/c^2 = 1 \\ (y - k)^2/d^2 - (x - k)^2/d^2 = 1 \end{cases}$ has a unique solution, and $L \cap M = \{((2hB - B^2 - c^2)/2B, (2hB + B^2 - c^2)/2B)\}$ where $B = (c^2 + d^2)/2(h - k)$.

Finally, let $\mathcal{S} = \mathcal{P}_{c,2}$, $c \neq 0$. We only need to consider $\mathcal{T} = \mathcal{P}_{d,2}$ with $d \neq 0$ and $d \neq c$. If $L = \{(h, h)\} \cup \{(x, y) \mid (y - h)^2/c^2 - (x - h)^2/c^2 = 1\}$ and $M = \{(k, k)\} \cup \{(x, y) \mid (y - k)^2/d^2 - (x - k)^2/d^2 = 1\}$, then $L \cap M = \{(h, h)\}$ if $h = k$. If $h \neq k$, then $\begin{cases} (y - h)^2/c^2 - (x - h)^2/c^2 = 1 \\ (y - k)^2/d^2 - (x - k)^2/d^2 = 1 \end{cases}$ has a unique solution, and $L \cap M = \{((2hE - E^2 + c^2)/2E, (2hE + E^2 + c^2)/2E)\}$ where $E = (c^2 - d^2)/2(k - h)$.

This completes the proof of (7.86), but the reader is requested to note the essential role that the extra points (h, h) or (k, k) played for lines in a $\mathcal{P}_{c,1}$ or $\mathcal{P}_{c,2}$. These points (h, h) or (k, k) seem rather awkward or unnatural when attached to these hyperbolae. They dramatically disrupt the 'connectedness' of these 'lines'. However, these points are a natural part of any $N(c, 0) + (h, h)$ or $N(0, c) + (k, k)$, as seen from a planar nearring point of view.

With (7.85) and (7.86), we have that the lines in $\mathcal{L}(1,1)$ satisfy the traditional parallel properties, and so for an $L \in \mathcal{L}(1,1)$ and $x \in N$, $x \notin L$, there is a unique $M \in \mathcal{L}(1,1)$ such that $x \in M$ and $M \parallel L$. It really only remains to show that two distinct points of N lie on exactly one line $L \in \mathcal{L}(1,1)$. We will obtain this as a consequence of the next two lemmas.

(7.87) Lemma. *Let (a,b) and (c,d) be two distinct points in N. Then there is a line $L \in \mathcal{L}(1,1)$, with $(a,b),(c,d) \in L$.*

Proof. Again, there are several cases to consider. If $c \neq a$, compute $m = (d-b)/(c-a)$. If $m = 1$, then $(a,b),(c,d) \in \{(x, x+(d-c)) \mid x \in F\}$, and if $m = -1$, then $(a,b),(c,d) \in \{(x, -x+(d+c)) \mid x \in F\}$. If $(c,d) = (c,c)$, compute $n = (a-c)^2 - (b-c)^2$. Either $n = e^2 \in S$ or $-n = e^2 \in S$. If $n = e^2$, then $(a,b),(c,c) \in \{(c,c)\} \cup \{(x,y) \mid (x-c)^2/e^2 - (y-c)^2/e^2 = 1\}$. If $-n = e^2$, then $(a,b),(c,c) \in \{(c,c)\} \cup \{(x,y) \mid (y-c)^2/e^2 - (x-c)/e^2 = 1\}$.

Now suppose $a \neq b$ and $c \neq d$, and if $a \neq c$, we also assume that $m \neq \pm 1$. Compute $h = [(a^2 - b^2) - (c^2 - d^2)]/2[(a-b)-(c-d)]$. If it should be that $(a-b)-(c-d) = 0$, then $m = 1$, so h is well defined. With this h, one can show that $(a-h)^2 - (b-h)^2 = (c-h)^2 - (d-h)^2$. Let this common value be n and note that $n \neq 0$. Either $n = e^2 \in S$ or $-n = e^2 \in S$. If $n = e^2$, then $(a,b),(c,d) \in \{(h,h)\} \cup \{(x,y) \mid (x-h)^2/e^2 - (y-h)^2/e^2 = 1\}$. If $-n = e^2$, then $(a,b),(c,d) \in \{(h,h)\} \cup \{(x,y) \mid (y-h)^2/e^2 - (x-h)^2/e^2 = 1\}$, and this completes the proof of (7.87).

(7.88) Lemma. *Let (a,b) and (c,d) be two distinct points in N. Then there is a unique line $L \in \mathcal{L}(1,1)$ with $(a,b),(c,d) \in L$.*

Proof. Existence comes from (7.87). Here we show uniqueness. For an arbitrary planar nearring, if $x, y \in Na+b$, then $0, y-x \in Na+(b-x)$, and conversely. So we need only show that $(0,0),(a,b) \in N$ belong to a unique line $L \in \mathcal{L}(1,1)$, for $(a,b) \neq (0,0)$. If $a = b$, then $(0,0),(a,b) \in N(1,1)$ only. If $a = -b$, then $(0,0),(a,b) \in N(1,-1)$ only. If $a \neq \pm b$, then $n = a^2 - b^2 \neq 0$. If $n = e^2 \in S$, then $(0,0),(a,b) \in N(e,0)$ only. If $-n = e^2 \in S$, then $(0,0),(a,b) \in N(0,e)$ only.

(7.89) Theorem. *$(N, \mathcal{L}(1,1), \in)$ is an affine plane.*

Proof. By (7.88), $(N, \mathcal{L}(1,1), \in)$ is an incidence space. Now $N(1,1) \in \mathcal{L}(1,1)$, and $(1,0) \in N \setminus N(1,1)$. Since $(0,0),(1,1) \in N(1,1)$, we have three noncollinear points in N. From (7.85) and (7.86), if $(x,y) \in N$ and $L \in \mathcal{L}(1,1)$ with $(x,y) \notin L$, then there is a unique $M \in \mathcal{L}(1,1)$ with $(x,y) \in M$ and $M \parallel L$.

As in the van der Walt connection, we have used one basic block subgroup $N(1,1)$ to construct $(N, \mathcal{L}(1,1), \in)$, an affine plane. We will now use this

$(N, \mathcal{L}(1,1), \in)$ and the other basic block subgroup $N(1,-1)$ to construct other affine planes $(N, \mathcal{L}(1,1)+(d,-d), \in)$, one for each $(d,-d) \in N(1,-1)$.

For $(u,v) \in N$, define $\mathcal{L}(1,1) + (u,v) = \{L + (u,v) \mid L \in \mathcal{L}(1,1)\}$. When is $\mathcal{L}(1,1) + (u,v) = \mathcal{L}(1,1) + (x,y)$?

(7.90) Lemma. $\mathcal{L}(1,1) + (u,v) = \mathcal{L}(1,1) + (x,y)$ *if and only if* $u - v = x - y$. *We have* $\mathcal{L}(1,1) + (u,v) = \mathcal{L}(1,1) + ((u-v)/2, (v-u)/2)$ *with* $((u-v)/2, (v-u)/2) \in N(1,-1)$.

Proof. If $L \in N/N(1,1) \cup N/N(1,-1)$, then so is $L + (u,v)$ and $L + (x,y)$. Hence, $N/N(1,1) \cup N/N(1,-1) \subseteq \mathcal{L}(1,1) + (u,v)$ for each $(u,v) \in N$. Now $N(c,0) + (h,h) + (u,v) = N(c,0) + (h,h) + ((u+v)/2, (u+v)/2) + ((u-v)/2, (v-u)/2) = N(c,0) + (h+(u+v)/2, h+(u+v)/2) + ((u-v)/2, (v-u)/2)$. Similarly, $N(0,c) + (h,h) + (u,v) = N(0,c) + (h + (u+v)/2, h + (u+v)/2) + ((u-v)/2, (v-u)/2)$. So if $u - v = x - y$, then $\mathcal{L}(1,1) + (u,v) = \mathcal{L}(1,1) + (x,y)$.

Conversely, if $\mathcal{L}(1,1) + (u,v) = \mathcal{L}(1,1) + (x,y)$, then any $N(c,0) + (h,h) + (u,v) = N(c,0) + (h + (u+v)/2, h + (u+v)/2) + ((u-v)/2, (v-u)/2)$ must equal some $N(c,0) + (k,k) + (x,y) = N(c,0) + (k+(x+y)/2, k+(x+y)/2) + ((x-y)/2, (y-x)/2)$. By (7.2), we have $(h + u, h + v) = (k + x, k + y)$. Hence, $(h + u) - (h + v) = (k + x) - (k + y)$, so $u - v = x - y$.

(7.91) Theorem. *For each distinct* $(d,-d) \in N(1,-1)$, *there is a distinct affine plane* $(N, \mathcal{L}(1,1) + (d,-d), \in)$ *isomorphic to* $(N, \mathcal{L}(1,1), \in)$.

Proof. The isomorphism is $T_d : N \to N$ defined by $T_d(x,y) = (x,y) + (d,-d)$. As for (7.52), our T_d here is also a bijection on N which maps the lines of $\mathcal{L}(1,1)$ onto the lines of $\mathcal{L}(1,1) + (d,-d)$.

As we did in the van der Walt connection, we interchange the roles of $N(1,1)$ and $N(1,-1)$. Let

$$\mathcal{L}(1,-1) = N/N(1,1) \cup N/N(1,-1)$$
$$\cup \{N(c,0) + (x,-x), N(0,c) + (x,-x) \mid c \neq 0, (x,-x) \in N(1,-1).\}$$

The cosets in $N/N(1,1) \cup N/N(1,-1)$ remain the same, of course. But $N(c,0) + (h,-h) = \{(h,-h)\} \cup \{(x,y) \mid (x-h)^2/c^2 - (y+h)^2/c^2 = 1\}$ and $N(0,c) + (k,-k) = \{(k,-k)\} \cup \{(x,y) \mid (y-k)^2/c^2 - (x+k)^2/c^2 = 1\}$. We obtain

(7.92) Theorem. *The incidence structure* $(N, \mathcal{L}(1,-1), \in)$ *is an affine plane isomorphic to* $(N, \mathcal{L}(1,1), \in)$.

Proof. Since $\begin{pmatrix} 0 & -1 \\ 1 & 0 \end{pmatrix} \begin{pmatrix} x \\ y \end{pmatrix} = \begin{pmatrix} -y \\ x \end{pmatrix}$ we define $A : N \to N$ by $A(x,y) = (-y,x)$. Certainly, A is a bijection. Noting that $A(N(1,1)) = N(1,-1)$,

$A(N(1,-1)) = N(1,1)$, $A(N(c,0)) = N(0,c)$, $A(N(0,c)) = N(c,0)$, and $N(x,y) = (-y,x)$, we have $A(\mathcal{L}(1,1)) = \mathcal{L}(1,-1)$ and A is an isomorphism from $(N, \mathcal{L}(1,1), \in)$ onto $(N, \mathcal{L}(1,-1), \in)$.

Having used the basic block subgroup $N(1,-1)$ to construct the affine plane $(N, \mathcal{L}(1,-1), \in)$, we now use $(N, \mathcal{L}(1,-1), \in)$ and the other basic block subgroup $N(1,1)$ to construct other affine planes $(N, \mathcal{L}(1,-1) + (d,d), \in)$, one for each $(d,d) \in N(1,1)$.

For $(u,v) \in N$, define $\mathcal{L}(1,-1) + (u,v) = \{L + (u,v) \mid L \in \mathcal{L}(1,-1)\}$. When is $\mathcal{L}(1,-1) + (u,v) = \mathcal{L}(1,-1) + (x,y)$?

(7.93) Lemma. $\mathcal{L}(1,-1) + (u,v) = \mathcal{L}(1,-1) + (x,y)$ *if and only if* $u+v = x+y$. *We have* $\mathcal{L}(1,-1) + (u,v) = \mathcal{L}(1,-1) + ((u+v)/2, (u+v)/2)$.

Proof. One proof is analogous to that of (7.90). Alternatively, one can take the A in the proof of (7.92). Since $A(\mathcal{L}(1,1)) = \mathcal{L}(1,-1)$ and $A(u,v) = (-v,u)$, we have $A(\mathcal{L}(1,1) + (u,v)) = A(\mathcal{L}(1,1) + ((u-v)/2, (v-u)/2)) = \mathcal{L}(1,-1) + ((u-v)/2, (u-v)/2) = \mathcal{L}(1,-1) + (-v,u)$.

(7.94) Lemma. *For each distinct* $(d,d) \in N(1,1)$, *there is a distinct affine plane* $(N, \mathcal{L}(1,-1) + (d,d), \in)$ *isomorphic to* $(N, \mathcal{L}(1,-1), \in)$.

Proof. The isomorphism is $T_d : N \to N$ defined by $T_d(x,y) = (x,y) + (d,d)$.

(7.95) Theorem. (a) $\mathcal{B} = \cup_{(d,-d) \in N(1,-1)}(\mathcal{L}(1,1) + (d,-d))$, *and* $d \neq d'$ *implies* $(\mathcal{L}(1,1) + (d,-d)) \cap (\mathcal{L}(1,1) + (d',-d')) = N/N(1,1) \cup N/N(1,-1)$.
(b) $\mathcal{B} = \cup_{(d,d) \in N(1,1)}(\mathcal{L}(1,-1) + (d,d))$, *and* $d \neq d'$ *implies* $(\mathcal{L}(1,-1) + (d,d)) \cap (\mathcal{L}(1,-1) + (d',d')) = N/N(1,1) \cup N/N(1,-1)$.
(c) $(\mathcal{L}(1,1) + (h,-h)) \cap (\mathcal{L}(1,-1) + (k,k)) = N/N(1,1) \cup N/N(1,-1) \cup \{N(c,0) + (k+h, k-h), N(0,c) + (k+h, k-h) \mid c \in F^*\}$.

Proof. These follow from (7.2) and the equation $N(a,b) + (h,-h) + (k,k) = N(a,b) + (k,k) + (h,-h) = N(a,b) + (k+h, k-h)$.

7.5. The $(N, \mathcal{B}^-, \in)$

So far in §7 we have been studying $(N, \mathcal{B}, \in)$ exclusively. In §5 and §6, we studied the $(N, \mathcal{B}^*, \in)$. From time to time we have promised that the $(N, \mathcal{B}^-, \in)$ can also have applications. Recall Example 2 introduced in (4.5). We have seen that for this $(\mathbf{C}, +, *_2)$, each $\mathbf{C} *_2 a + b \in \mathcal{B}$ is a ray from b through $b+a$. This suggests the construction $\mathcal{B}^- = \{\mathbf{C} *_2 \{a, -a\} + b \mid a, b \in \mathbf{C}, a \neq 0\}$, realizing that each $\mathbf{C} *_2 \{a, -a\} + b \in \mathcal{B}^-$ is the line passing through $b + a$ and $b - a$.

For an arbitrary planar nearring $(N, +, \cdot)$, we let $\mathcal{B}^- = \{N\{a, -a\} + b \mid a, b \in N, a \neq 0\}$ and ask the question, 'when does $(N, \mathcal{B}^-, \in)$ have a nice geometric interpretation'? Very little has been done for the $(N, \mathcal{B}^-, \in)$, but we shall see presently that some of our previous ideas can be extended to the $(N, \mathcal{B}^-, \in)$. Our first goal is to make a BIBD from $(N, \mathcal{B}^-, \in)$. For this and future use, we will need the following proposition whose use will parallel that of (5.4) and (7.2).

(7.96) Proposition. *Let* $(N, +, \cdot)$ *be a planar nearring with blocks* $\mathcal{B}^- = \{N\{a, -a\} + b \mid a, b \in N, a \neq 0\}$. *For* $a \in N \setminus \{0\}$, *suppose* $Na \cap N(-a) = \{0\}$. *Let* $N\{a, -a\} + b$, $N\{a', -a'\} + b' \in \mathcal{B}^-$. *Then* $N\{a, -a\} + b = N\{a', -a'\} + b'$ *if and only if either*

(a) $N\{a, -a\} = N\{a', -a'\}$ *is a subgroup of* N^+, *and* $b' - b \in N\{a, -a\}$;

or

(b) $N\{a, -a\} = N\{a', -a'\}$, $N\{a, -a\}$ *is not a subgroup of* N^+, *and* $b = b'$.

Proof. If (a) is true, then certainly $N\{a, -a\} + b = N\{a', -a'\} + b'$. Also, if (b) is true, then $N\{a, -a\} + b = N\{a', -a'\} + b'$. So we can turn to the converse. Our proof will parallel that of (7.2), but there is enough variation to warrant inclusion.

Suppose $N\{a, -a\} + b = N\{a', -a'\} + b'$, so $N\{a, -a\} = N\{a', -a'\} + c$, where $c = b' - b$. Either $N\{a, -a\}$ is a subgroup of N^+ or it is not. If $N\{a, -a\}$ is a subgroup of N^+, then $N\{a, -a\} - c = N\{a', -a'\}$ is a coset of $N\{a, -a\}$ containing 0, since $0 \in N\{a', -a'\}$, so $N\{a, -a\} - c = N\{a, -a\}$ and, consequently, $N\{a, -a\} = N\{a', -a'\}$ and $c = b' - b \in N\{a, -a\}$. Hence, (a).

Now suppose $N\{a, -a\}$ is *not* a subgroup of N^+ and $c \neq 0$. We will show that $N\{a, -a\}$ *must* be a subgroup of N^+, so this contradiction will force $c = 0$, and $b = b'$, if we insist that $N\{a, -a\}$ is *not* a subgroup of N^+. Once we obtain $b = b'$, we immediately have $N\{a, -a\} = N\{a', -a'\}$. Thus (b). So we proceed with the assumption that $N\{a, -a\}$ is *not* a subgroup of N^+, and $b \neq b'$, so $N\{a, -a\} = N\{a', -a'\} + c$ with $c = b' - b \neq 0$.

From $N\{a, -a\} = N\{a', -a'\} + c$, we obtain $c \in N\{a, -a\}$, and so $c \in Na$ or $c \in N(-a)$. If $c = ua$, then $-c = -(ua) = u(-a)$, and so $-c \in N(-a)$. If $c = u(-a)$, then $-c = -[u(-a)] = ua$, so $-c \in Na$. Hence, $N\{a, -a\} = N\{c, -c\}$. Similarly, from $N\{a, -a\} - c = N\{a', -a'\}$, we obtain $N\{a', -a'\} = N\{c, -c\}$. Consequently, $N\{a, -a\} = N\{c, -c\} = N\{a', -a'\}$ and we have the equation $N\{c, -c\} = N\{c, -c\} + c$. So for all $u, v \in N$, $N\{c, -c\} = N\{c, -c\} + uc = N\{c, -c\} + vc$. We can then write $N\{c, -c\} - vc = N\{c, -c\} + (uc - vc) = N\{c, -c\}$ and conclude that $uc - vc \in N\{c, -c\}$. We also have $N\{c, -c\} = N\{c, -c\} - c$, and in a similar way we can conclude that $u(-c) - v(-c) \in N\{c, -c\}$. Similarly, $uc - v(-c)$, $v(-c) - uc \in N\{c, -c\}$. This makes $N\{c, -c\} = N\{a, -a\}$ a subgroup of N^+. As indicated earlier in our proof, our proof is now complete.

(7.97) Proposition. *Let* $(N, +, \cdot)$ *be a finite planar nearring with* $v = |N|$ *and* $k = 2|\Phi| + 1$, *where* $\Phi = B_c$ *for some* $c \neq_m 0$. *Suppose, for* $0 \neq a \in N$,

$Na \cap N(-a) = \{0\}$ *and no* $N\{a, -a\}$ *is a subgroup of* N^+. *Then the number of blocks in* $\mathcal{B}^-$ *containing an arbitrary point* $x \in N$ *is* $r = k(v-1)/(k-1)$.

Proof. If $a \in N \setminus \{0\}$, then $|N\{a, -a\} \setminus \{0\}| = k - 1$, and so there are $(v - 1)/(k - 1)$ 'basic blocks' $N\{a, -a\}$, each of which contains exactly $k = 2|\Phi| + 1$ points. For each of the basic blocks $N\{a, -a\}$, as b varies through N, there will be k distinct blocks $N\{a, -a\} + b_i$, $1 \le i \le k$, with $x \in N\{a, -a\} + b_i$. Hence, $r = k(v-1)/(k-1)$.

(7.98) Proposition. *With the same conditions of (7.97), if* $x, y \in N$ *and* $x \ne y$, *then the number of blocks* $B \in \mathcal{B}^-$ *with* $x, y \in B$ *is the same as the number of blocks* $D \in \mathcal{B}^-$ *with* $0, y - x \in D$.

Proof. If $x, y \in N\{a, -a\} + b$, then $y = ua + b$ or $y = u(-a) + b$ for some $u \in N$. So $y - x = ua + (b - x)$ or $y - x = u(-a) + (b - x)$, which means $0, y - x \in N\{a, -a\} + (b - x)$. Conversely, if $0, y - x \in N\{a, -a\} + b$, then $x, y \in N\{a, -a\} + (b + x)$. This gives our desired bijection.

(7.99) Theorem. *Let* $(N, +, \cdot)$ *be a finite planar nearring with* $v = |N|$ *and* $k = 2|\Phi| + 1$, *where* $\Phi = B_c$ *for some* $c \ne_m 0$. *Suppose, for* $0 \ne a \in N$, *that* $Na \cap N(-a) = \{0\}$, *no* $N\{a, -a\}$ *is a subgroup of* N^+, *and the map* $t_a : N \to N$ *defined by* $t_a(x) = x + ax$ *is a bijection. Then* $(N, \mathcal{B}^-, \in)$ *is a BIBD with parameters* $b = v(v-1)/(k-1)$, $r = k(v-1)/(k-1)$, *and* $\lambda = k$.

Proof. We easily have $k = 2|\Phi| + 1 = |B|$ for each $B \in \mathcal{B}^-$. Each of the $(v - 1)/(k - 1)$ basic blocks $N\{a, -a\}$ yields v distinct blocks $N\{a, -a\} + b \in \mathcal{B}^-$ by (7.96), hence $b = v(v - 1)/(k - 1)$. From (7.97), we have $r = k(v-1)/(k-1)$. So, it remains to show that $\lambda = k$, and by the force of (7.98), we only need to show that 0 and z belong to exactly k distinct blocks $B \in \mathcal{B}^-$ for arbitrary $z \in N \setminus \{0\}$.

Certainly, 0 and z belong to $N\{z, -z\}$ and to $N\{z, -z\} + z$, which are two distinct blocks of $\mathcal{B}^-$ by (7.96). For $1_c \in B_c$, $t_{1_c}(x) = x + x$ is a bijection, so there is a unique $a_z \in N$ such that $z = a_z + a_z$. So $0, z \in N\{a_z, -a_z\} + a_z$. This gives a third block in $\mathcal{B}^-$ containing 0 and z.

Let $\Phi^* = \Phi \setminus \{1\}$. For each $u \in \Phi$, we will find two blocks in $\mathcal{B}^-$ containing 0 and z. First, there is a unique $a_u \ne 0$ such that $z = ua_u - a_u$ and there is a unique $b_u \ne 0$ such that $-z = b_u + ub_u$. Hence, $z = u(-b_u) - b_u$. Now $0, z \in (N\{a_u, -a_u\} - a_u) \cap (N\{b_u, -b_u\} - b_u)$.

If $N\{a_u, -a_u\} - a_u = N\{a_v, -a_v\} - a_v$, then $a_u = a_v$ and $ua_u = va_v = va_u$. Hence, $a_u = u^{-1}va_u$, and since $a_u \ne 0$, we have $u = v$. If $N\{b_u, -b_u\} - b_u = N\{b_v, -b_v\} - b_v$, then $b_u = b_v$ and so $ub_u = vb_v = vb_u$. Hence, $b_u = u^{-1}vb_u$, and since $b_u \ne 0$, we have $u = v$. If $N\{a_u, -a_u\} - a_u = N\{b_v, -b_v\} - b_v$, then $a_u = b_v$. From $z = ua_u - a_u = v(-b_v) - b_v = v(-a_u) - a_u$, we obtain $ua_u = v(-a_u)$, so $a_u = u^{-1}v(-a_u) \in Na_u \cap N(-a_u) = \{0\}$, contrary to $a_u \neq 0$.

In summary, there are $2|\Phi^*| = 2|\Phi| - 2$ distinct blocks $N\{a_u, -a_u\} - a_u$, $N\{b_u, -b_u\} - b_u$, each containing 0 and z. Now we need to show that $N\{z, -z\}$, $N\{z, -z\} + z$, and $N\{a_z, -a_z\} + a_z$ are not among these. From (7.96), we certainly know that $N\{z, -z\}$ is not among these. If $N\{z, -z\} + z = N\{a_u, -a_u\} - a_u$, then $z = -a_u$, and $z = ua_u - a_u$, so $ua_u = 0$, which cannot be. If $N\{z, -z\} + z = N\{b_u, -b_u\} - b_u$, then $-b_u = z = u(-z) + z$, so $u(-z) = 0$, which cannot be. From $N\{a_z, -a_z\} + a_z = N\{a_u, -a_u\} - a_u$, we have $a_z = -a_u$ and $z = ua_u - a_u = a_z + a_z$. Hence, $ua_u = u(-a_z) = a_z \in Na_z \cap N(-a_z) = \{0\}$, which cannot be. Finally, from $N\{a_z, -a_z\} + a_z = N\{b_u, -b_u\} - b_u$, we have $a_z = -b_u$ and $z = a_z + a_z = u(-b_u) - b_u$. Hence, $a_z = u(-b_u) = ua_z$. Since $a_z \neq 0$, then $u = 1_c$. But b_u came from $u \in \Phi^*$, so $u \neq 1_c$.

We now have $2|\Phi| + 1 = k$ distinct blocks of $\mathcal{B}^-$ containing 0 and z. Now we need to see that there are no other blocks containing 0 and z. So suppose $0, z \in N\{a, -a\}$. Then $N\{a, -a\} = N\{z, -z\}$, and we already have this one. Suppose $0, z \in N\{a, -a\} + b$ with $b \neq 0$. Either $0 = ua + b$ or $0 = u(-a) + b$. Hence, $u(-a) = b$ or $ua = b$, and so $N\{a, -a\} + b = N\{b, -b\} + b$. We have $N\{z, -z\} + z$ and $N\{a_z, -a_z\} + a_z$ already, so we must assume $b \notin \{0, z, a_z\}$. If $z = u(-b) + b = u(-b) - (-b)$, then $-b = a_u$ and $N\{b, -b\} + b = N\{a_u, -a_u\} - a_u$, which we already have. If $z = ub + b = u(-(-b)) - (-b)$, then $-z = (-b) + u(-b)$, and so $-b = b_u$, and $N\{b, -b\} + b = N\{b_u, -b_u\} - b_u$, which we also already have. Certainly, $z \neq 1_c(-b) + b$ and $z \neq (-1_c')(b) + b$. If $z = 1_c b + b$ or $z = (-1_c')(-b) + b = b + b$, then $z = a_z$. Having exhausted all ways to express $z \in N\{b, -b\} + b$, we conclude that we have accounted for all blocks containing 0 and z. So if $0, z \in N\{a, -a\} + b$, then $N\{a, -a\} + b$ is among $N\{z, -z\}$, $N\{z, -z\} + z$, $N\{a_z, -a_z\} + a_z$, the $N\{a_u, -a_u\} - a_u$, and the $N\{b_u, -b_u\} - b_u$, and so 0 and z belong to exactly $k = 2|\Phi| + 1$ blocks in $\mathcal{B}^-$. Since $z \in N \setminus \{0\}$ was arbitrarily chosen, we have $\lambda = k$, and our proof of (7.99) is complete.

(7.100) Examples. Numerous examples of Frobenius groups due to N. Itô [Hu, p.499] provide planar nearrings satisfying the hypothesis of (7.99). Let $K = GF(p^f)$, a Galois field of order p^f, where p is a prime. Consider the group

$$N = \left\{ \begin{pmatrix} 1 & b & c \\ 0 & 1 & a \\ 0 & 0 & 1 \end{pmatrix} \middle| a, b, c \in K \right\},$$

a multiplicative nonabelian group of order p^{3f}. Let q be a prime divisor of $p^f - 1$ with $q \geq 3$. Then K has three distinct elements d_1, d_2, and d_3 such that $d_i^q = 1$, $1 \leq i \leq 3$. Let $\alpha = \begin{pmatrix} d_1 & 0 & 0 \\ 0 & d_2 & 0 \\ 0 & 0 & d_3 \end{pmatrix}$ and let $\Phi' = \langle \alpha \rangle$, the subgroup of $GL(3, K)$ generated by α. If $1 \neq \beta \in \Phi'$, then $\bar{\beta} : N \to N$ defined by $\bar{\beta}(n) = \beta n \beta^{-1}$ defines a fixed point free automorphism $\bar{\beta}$ of N. Let $\Phi = \{\bar{\beta} \mid \beta \in \Phi'\}$, so (N, Φ) is a Ferrero pair.

Since $\Phi' \cong \Phi$ and $|\Phi'| = q$, we have that $2|\Phi| + 1 = 2q + 1$. For a planar nearring constructed from (N, Φ), we want to ensure that no $N\{a, -a\}$, $a \neq 0$, is a subgroup of N. We need only ensure that $(2q + 1)$ does not divide p^{3f}, or make sure that $2q + 1$ is not a power of p. Examples are $(p, f, q) \in \{(5, 2, 3), (7, 2, 19)\}$. These two examples for (p, f, q) also assure us that $Na \cap N(-a) = \{0\}$ for each $a \in N \setminus \{0\}$. (We are using rather loose language, since N^+ is really a multiplicative group, and so $0 \in N$ denotes $\begin{pmatrix} 1 & 0 & 0 \\ 0 & 1 & 0 \\ 0 & 0 & 1 \end{pmatrix}$.) It remains to show that each $x \mapsto x + ax$ is a bijection, $a \in N$. Again, since N is multiplicative and $ax = \phi_a(x)$ for some $\phi_a \in \Phi$, this amounts to showing that each $X \mapsto X\beta X \beta^{-1}$ is bijective, for $\beta \in \Phi'$. Since N is finite, we need only show that $X \mapsto X\beta X\beta^{-1}$ is injective.

Suppose $A = \begin{pmatrix} 1 & b & c \\ 0 & 1 & a \\ 0 & 0 & 1 \end{pmatrix}$, $X = \begin{pmatrix} 1 & y & z \\ 0 & 1 & x \\ 0 & 0 & 1 \end{pmatrix}$, $\beta = \begin{pmatrix} d_1 & 0 & 0 \\ 0 & d_2 & 0 \\ 0 & 0 & d_3 \end{pmatrix}$ and $A\beta A\beta^{-1} = X\beta X\beta^{-1}$. Then

$$\begin{pmatrix} 1 & (d_1 d_2^{-1} + 1)b & (d_1 d_3^{-1} + 1)c + d_2 d_3^{-1} ab \\ 0 & 1 & (d_2 d_3^{-1} + 1)a \\ 0 & 0 & 1 \end{pmatrix} =$$

$$\begin{pmatrix} 1 & (d_1 d_2^{-1} + 1)y & (d_1 d_3^{-1} + 1)z + d_2 d_3^{-1} xy \\ 0 & 1 & (d_2 d_3^{-1} + 1)x \\ 0 & 0 & 1 \end{pmatrix}.$$

If $b \neq y$, then $d_1 d_2^{-1} + 1 = 0$, and so $d_1 d_2^{-1}$ has multiplicative order 2, which is false. Hence, $b = y$. Similarly, $a = x$. We then conclude that $(d_1 d_3^{-1} + 1)c = (d_1 d_3^{-1} + 1)z$, and so $c = z$. Hence, $A\beta A\beta^{-1} = X\beta X\beta^{-1}$ implies $A = X$, and so $X \mapsto X\beta X\beta^{-1}$ is injective. So the hypotheses of (7.99) are satisfied.

For A as above, define $\Gamma(A) = [c, (a, b)]$. So $\Gamma(X) = [z, (x, y)]$. Since $AX = \begin{pmatrix} 1 & b+y & c+z+xb \\ 0 & 1 & a+x \\ 0 & 0 & 1 \end{pmatrix}$, we have $\Gamma(AX) = [c+z+xb, (a+x, b+y)]$. Now take a fresh look at (4.21) with $A = B = C = K^+$ and $\pi(x, b) = xb$. We see that Γ is an isomorphism, and so the constructions of (4.21) generalize these examples of N. Itô.

(7.101) Examples. In truth, the examples provided by (4.19) were included with (7.99) in mind. Consider (a) of (4.20). Let $F = Z_{13}$ and

$|A| = 3$. Since any basic block has $|N\{a, -a\}| = 2 \cdot 3 + 1 = 7$, we know that no such $N\{a, -a\}$ is a subgroup of N^+. Since $|A| = 3$, we know that each $x \mapsto x + ax$ is a bijection and that each $Na \cap N(-a) = \{0\}$ for $a \neq 0$. Now (7.99) is applicable.

Next, take $F = Z_{31}$ and $|A| = 5$. Since any basic block has $|N\{a, -a\}| = 2 \cdot 5 + 1 = 11$, we know that no such $N\{a, -a\}$ is a subgroup of N^+. Since $|A| = 5$, we know that each $x \mapsto x + ax$ is a bijection as well as each $Na \cap N(-a) = \{0\}$ for $a \neq 0$.

(7.102) Examples. For a contrast, let F be a field for which $S = \{x^2 \mid x \in F^*\}$ is a multiplicative subgroup of F^* of index 2. In (4.19), let $R = F$, $A = S$, and $\sigma = 1$. If $\beta \in A^*$, then $\beta^\sigma = \beta = \beta^{-1}$ would require that β^{-1} be a square, and that $\beta^2 = 1$, so F^* would have solutions to the equation $x^4 = 1$ other than ± 1. So we want to make sure that our F has no elements of multiplicative order 4. For example, F could be any subfield of the real field **R** or F could be any finite field where $|K^*| \neq 0 \pmod 4$. Since S has index 2 in K^*, then the characteristic of F is $\neq 2$. Hence, (vi) and (vii) of (4.19) are applicable. So we are nearly set up for (7.99). If $(a, b) \in N \setminus \{(0,0)\}$, then $-(a, b) = (b^2 - a, -b)$, and so $N(a, b) \cap N(-(a, b)) = \{(0,0)\}$. But $N\{(1,0), (-1,0)\}$ and $N\{(b^2/2, b), (b^2/2, -b)\}$ are subgroups of N^+, and these are the only basic blocks of N^+ which are subgroups. Hence, (7.99) is not applicable. Notice that $N\{(b^2/2, b), (b^2/2, -b)\}$ is a parabola when $F = \mathbf{R}$. It is interesting to compute some of the other $N\{(a, b), -(a, b)\}$.

(7.103) Exercise. Follow the examples in (7.101) and find more situations where (4.19) provides examples where (7.99) is applicable.

(7.104) Exercise. Follow the examples in (7.100) and find more situations where (4.21) provides examples where (7.99) is applicable.

(7.105) Problem The examples in (7.102) should make one recall the van der Walt connection and Karzel's hyperbolae. Can one use the basic block subgroups $N\{(1,0), -(1,0)\}$ and $N\{(b^2/2, b), -(b^2/2, b)\}$ to sift out substructures of $(N, \mathcal{B}^-, \in)$ which are affine planes?

(7.106) Problem The finite examples in (7.102) should also make one wonder if statistical BIBD can be constructed with $(N, \mathcal{B}^-, \in)$.

7.6. Partially balanced incomplete block designs (PBIBD)

One should have a fresh supply of courage and a benevolent attitude as this topic is approached. Partially balanced incomplete block designs are considerably more complex than BIBDs, and the author is unable to provide adequate motivation. There is, however, considerable interest directed towards these structures. For such complex structures to command so much interest, there must be some important applications to statistics, or designs of experiments. Perhaps the day will come when someone who knows will be able to communicate effectively to the rest of us something more than the definition.

Part of a PBIBD is an *association scheme*. In (7.107) we state what we mean by an association scheme. In Chapter 21 of [M & S], there is another definition of an association scheme. A casual inspection will show that these two definitions are related, if not equivalent. In [M & S], one gets the idea that association schemes have applications to coding theory. So perhaps the association schemes we get here from planar nearrings will have some nice applications to coding theory. We support this in (8.25).

Marshall Hall, Jr, developed a powerful method to construct PBIBDs, and G. Ferrero has shown that his factory is particularly suitable for applying Hall's method to finite planar nearrings, or Ferrero pairs.

(7.107) Definitions. Start with a finite tactical configuration $(N, \mathcal{T}, \in)$ and let $\mathcal{P} = \{A \mid A \subseteq N, |A| = 2\}$. Suppose $\mathcal{A} = \{A_1, A_2, \ldots, A_m\}$ is a partition of $\mathcal{P}$. Then $\mathcal{A}$ is an *association scheme* on N if, given $\{x, y\} \in A_h$, the number of $z \in N$ such that $\{x, z\} \in A_i$ and $\{y, z\} \in A_j$ depends only upon h, i, and j, and *not* upon x and/or y. That is, there is a number p_{ij}^h such that for $\{x, y\} \in A_h$, there are exactly p_{ij}^h distinct elements $z \in N$ such that $\{x, z\} \in A_i$ and $\{y, z\} \in A_j$. Association schemes with $m = 1$ or $m = v(v-1)/2$, where $v = |N|$, are declared 'uninteresting'.

Suppose $(N, \mathcal{T}, \in, \mathcal{A})$ is a finite tactical configuration with association scheme $\mathcal{A}$. This structure is a *partially balanced incomplete block design* (PBIBD) if:
(a) to each $A_i \in \mathcal{A}$, there is a number n_i such that for each $x \in N$, there are exactly n_i distinct elements $y \in N$ such that $\{x, y\} \in A_i$;
(b) to each $A_i \in \mathcal{A}$, there is a number λ_i such that $\{x, y\} \in A_i$ implies x and y belong to exactly λ_i blocks of $\mathcal{T}$.

So a PBIBD $(N, \mathcal{T}, \in, \mathcal{A})$ has parameters p_{ij}^h, λ_i, and n_i in addition to those of the tactical configuration $(N, \mathcal{T}, \in)$.

The reader is now invited to sit down and come up with numerous examples of PBIBDs, but he should not feel a bit discouraged if he fails to do so. Perhaps he will have a little more luck after we present Hall's method, but if there is still no success, try again after we present Ferrero's adaptation.

(7.108) Hall's eight-step Method for Constructing PBIBDs.
1. Start with a finite nonempty set N with a transitive permutation group G acting on N, with an intransitive subgroup S. (A group of permutations G on a set N is *transitive* provided G has exactly one orbit on N, so if $(x, y) \in N \times N$, then there is a $g \in G$ such that $g(x) = y$. Of course, such a group is *intransitive* if it is not transitive.)

2. Since S is intransitive on N, it has more than one orbit on N. Let B_1 be *any* union of orbits of S.

3. Let S_1 be the stabilizer of B_1, that is, $S_1 = \{g \in G \mid g(B_1) = B_1\}$. Then S_1 is a subgroup of G.

4. As a subgroup of G, S_1 has finitely many cosets $x_1 S_1, x_2 S_1, \ldots, x_b S_1$, with $x_1 = 1$, and $[G : S_1] = b$. Choose such representatives $x_1, x_2, \ldots, x_b$.

5. Compute $\mathcal{B} = \{B_j = x_j(B_1) \mid 1 \le j \le b\}$. (Then $(N, \mathcal{B}, \in)$ is an *orbital design* of Higman [Hi].)

6. Choose any $a_1 \in N$ and compute G_1, the stabilizer of a_1, that is, $G_1 = \{g \in G \mid g(a_1) = a_1\}$, a subgroup of G.

7. Compute and designate the orbits of G_1 on N by $\{a_1\}; \Delta_1, \Delta_2, \ldots, \Delta_u$; $\Delta_{u+1}, \Delta'_{u+1}, \ldots, \Delta_{u+w}, \Delta'_{u+w}$. We must explain this notation. First of all, if Δ is any orbit of G_1, then $\Delta' = \{x(a_1) \mid x \in G \text{ and } a_1 \in x(\Delta)\}$. Now $(\Delta')' = \Delta$ and Δ' is also an orbit of G_1 on N. Now we know what a Δ' is. In the above listing of orbits of G_1 on N, $\Delta_1, \Delta_2, \ldots, \Delta_u$ are exactly those orbits Δ where $\Delta' = \Delta$. Such an orbit is 'self paired'. The remaining orbits, not $\{a_1\}$, are 'paired' as $\Delta_{u+i}, \Delta'_{u+i}$. See [Ha, 1971] and [Wi, p. 44].

8. For our final step, we must make our association scheme $\mathcal{A}$. Construct each A_k as follows. First put in all pairs $\{a_1, a_i\}$ if $k \le u$ and $a_i \in \Delta_k$, or if $k = u + j$ and $a_i \in \Delta_k \cup \Delta'_k$. Next put into A_k all pairs $\{x(a_1), x(a_i)\}$ where $x \in G$. Then $\mathcal{A} = \{A_1, A_2, \ldots, A_{u+w}\}$ will be an association scheme, and, believe it or not, $(N, \mathcal{B}, \in, \mathcal{A})$ will be a PBIBD!

The raw materials needed to apply Hall's method are a finite nonempty set N, a transitive permutation group G on N, and an intransitive subgroup S.

(7.109) Ferrero's adaptation for planar nearrings. Start with a finite planar nearring $(N, +, \cdot)$ and obtain its Ferrero pair (N, Φ). Of course, one could start with a Ferrero pair (N, Φ). Let $G = N \times_\theta \Phi$, the natural semidirect product where $(a, \phi) \cdot (b, \lambda) = (a + \phi(b), \phi\lambda)$. Then G acts transitively on N by $(a, \phi)n = a + \phi(n)$. Then $S = \{(0, \phi) \mid \phi \in \Phi\}$ is an intransitive subgroup of G.

After we have given an easy application of Ferrero's adaptation, anyone should be able to make PBIBDs. Why not take a little time and try your hand at constructing a nontrivial partially balanced incomplete block design?

(7.110) Application. Perhaps the easiest meaningful example comes from the Ferrero pair (Z_9, Φ) where $\Phi = \{1, \phi\}$ and $\phi(x) = -x$. So $N = Z_9$, $G = Z_9 \times_\theta \Phi$ and $S \cong \Phi$. Let $B_1 = \{0\} \cup \Phi(1) \cup \Phi(2) = \{0, 1, 8, 2, 7\}$. The stabilizer of B_1 is $S_1 = \Phi$, and the cosets of S_1 in G are $\bar{0}\Phi, \bar{1}\Phi, \ldots, \bar{8}\Phi$

where $\bar{a} = (a, 0)$. Hence, $\mathcal{B} = \{B_1, B_2, \ldots, B_9\}$ where

$$B_2 = \{1, 2, 0, 3, 8\} \qquad B_3 = \{2, 3, 1, 4, 0\}$$
$$B_4 = \{3, 4, 2, 5, 1\} \qquad B_5 = \{4, 5, 3, 6, 2\}$$
$$B_6 = \{5, 6, 4, 7, 3\} \qquad B_7 = \{6, 7, 5, 1, 6\}$$
$$B_8 = \{7, 8, 6, 0, 5\} \qquad B_9 = \{8, 0, 7, 1, 6\}$$

The parameters of the tactical configuration $(N, \mathcal{B}, \in)$ are $v = 9$, $b = 9$, $r = 5$, and $k = 5$.

Let $a_1 = 0$, so the stabilizer of a_1 is $G_1 = \Phi$ and the orbits of Φ are $\{0\}$, $\{1, 8\}$, $\{2, 7\}$, $\{3, 6\}$, and $\{4, 5\}$. It is elementary to see that the relevant orbits are all self paired, so $u = 4$. Let $\Delta_1 = \{1, 8\}$, $\Delta_2 = \{2, 7\}$, $\Delta_3 = \{3, 6\}$, and $\Delta_4 = \{4, 5\}$. Consequently,

$$A_1 = \{\{0, 1\}, \{0, 8\}, \{1, 2\}, \{2, 3\}, \{3, 4\}, \{4, 5\}, \{5, 6\}, \{6, 7\}, \{7, 8\}\},$$
$$A_2 = \{\{0, 2\}, \{0, 7\}, \{1, 3\}, \{2, 4\}, \{3, 5\}, \{4, 6\}, \{5, 7\}, \{6, 8\}, \{8, 1\}\},$$
$$A_3 = \{\{0, 3\}, \{0, 6\}, \{1, 4\}, \{2, 5\}, \{3, 6\}, \{4, 7\}, \{5, 8\}, \{7, 1\}, \{8, 2\}\},$$
$$A_4 = \{\{0, 4\}, \{0, 5\}, \{1, 5\}, \{2, 6\}, \{3, 7\}, \{4, 8\}, \{6, 1\}, \{7, 2\}, \{8, 3\}\}.$$

As a result of tedious checking, one obtains the following parameters: $(\lambda_1, \lambda_2, \lambda_3, \lambda_4) = (4, 3, 2, 1)$; $n_i = 2$ for $1 \le i \le 4$; and $p_{ij}^h = 1$ if (h, i, j) is any permutation of $(1, 1, 2)$, $(1, 2, 3)$, $(1, 3, 4)$, or $(1, 4, 4)$, and $p_{ij}^h = 0$ otherwise.

(7.111) Exercise and/or exploratory problem. Take the ordered pair (Z_9, Φ) of (7.110), and choose alternative B_1 and/or a_1. Now investigate the resulting PBIBD. When this game grows old, move on to other suitable Ferrero pairs (N, Φ), or finite planar nearrings $(N, +, \cdot)$. Keep a journal of all your discoveries. Perhaps you will discover and prove some theorems worth sharing with others.

7.7. Double planar nearrings

We return to $(\mathbf{C}, +, *_2)$ and $(\mathbf{C}, +, *_3)$ from (4.5) and (4.6). With $(N, +, \cdot)$ $= (\mathbf{C}, +, *_2)$, an $Na + b \in \mathcal{B}_{*_2}$ is the ray from b through $a + b$, and with $(N, +, \cdot) = (\mathbf{C}, +, *_3)$, an $N^* a + b \in \mathcal{B}_{*_3}^*$ is the circle with centre b and radius a, or $|a|$, traditionally. If we want to consider these structures simultaneously, we have $(N, +, *_2, *_3) = (\mathbf{C}, +, *_2, *_3)$ and $(N, \mathcal{B}_{*_2}, \mathcal{B}_{*_3}^*)$.

First, notice that $a *_2 (b *_3 c) = |a| \cdot ((b/|b|) \cdot c) = (|a|bc)/|b|$, and $(a *_2 b) *_3 (a *_2 c) = (|a|b) *_3 (|a|c) = (|a|b/||a|b|) \cdot (|a|c) = |a|^2 bc/|a||b| = (|a|bc)/|b|$. Hence, $*_2$ is left distributive over $*_3$. Next, note that $a *_3 (b *_2 c) = (a/|a|) \cdot (|b|c) = (|b|ac)/|a|$, and $(a *_3 b) *_2 (a *_3 c) = [(a/|a|)b] *_2 [(a/|a|)c] = |(a/|a|)b|[(a/|a|)c] = (|a||b|/|a|)((a/|a|)c) = (|b|ac)/|a|$. So we also get that $*_3$ is left distributive over $*_2$. This motivates

(7.112) Definition. An algebraic structure $(N, +, \star, \circ)$ is a *(left) double planar nearring* if each of $(N, +, \star)$ and $(N, +, \circ)$ is a (left) planar nearring, and each of $\star$ and $\circ$ is left distributive over the other.

The existence of the double planar nearring $(\mathbf{C}, +, *_2, *_3)$ is curious enough, but it is really the relationships between circles and rays in the usual euclidean plane that actually motivates the further investigation. Let us take $(\mathbf{C}, +, *_2, *_3)$ back to the Factory and see how it works. First, the nearring $(\mathbf{C}, +, *_2)$ comes from the Ferrero pair $(\mathbf{C}, \Phi)$ where Φ is essentially the group of positive real numbers. Next, $(\mathbf{C}, +, *_3)$ comes from the Ferrero pair $(\mathbf{C}, \Gamma)$ where Γ is essentially the unit circle. Note that $\mathbf{C}^* = \Phi\Gamma$, a direct product and that $(\mathbf{C}, \mathbf{C}^*)$ is also a Ferrero pair.

Suppose (N, Λ) is a Ferrero pair and that $\Lambda = \Phi\Gamma$, a direct product of nontrivial subgroups Φ and Γ. So (N, Φ) and (N, Γ) are also Ferrero pairs. Choose a representative for each nontrivial orbit of Λ, and let $\Lambda(e)$ be any such orbit with representative e. Let the elements of $\{\phi(e) \mid \phi \in \Phi\}$ be representatives of orbits of Γ, and the elements of $\{\gamma(e) \mid \gamma \in \Gamma\}$ be representatives of orbits of Φ which are contained in $\Lambda(e)$. For $\phi \in \Phi$ and $\gamma \in \Gamma$, we have $\phi \circ \gamma = \gamma \circ \phi$. Use (N, Φ) and (N, Γ) to construct planar nearrings $(N, +, *_\Phi)$ and $(N, +, *_\Gamma)$, respectively, with these representatives of orbits $\phi(e)$ and $\gamma(e)$.

(7.113) Proposition. *If $a \in \Gamma(e) \neq \{0\}$, then $a =_\Phi \gamma(a)$ for every $\gamma \in \Gamma$. That is, a and $\gamma(a)$ are equivalent multipliers for $(N, +, *_\Phi)$. Consequently, $a =_\Gamma \phi(a)$ for each $\phi \in \Phi$.*

Proof. Let $a = \phi_1 \circ \gamma_1(e)$, where $\phi_1 \in \Phi$ and $\gamma_1 \in \Gamma$. Then for $\gamma \in \Gamma$, $\gamma(a) *_\Phi x = [\gamma \circ \phi_1 \circ \gamma_1(e)] *_\Phi x = [\phi_1 \circ (\gamma \circ \gamma_1)](e) *_\Phi x = \phi_1(x) = a *_\Phi x$ since $(\gamma \circ \gamma_1)(e)$ is the representative for the orbit of Φ containing a. Of course, the proof that $a =_\Gamma \phi(a)$ is virtually the same.

(7.114) Theorem. $(N, +, *_\Phi, *_\Gamma)$ *is a double planar nearring.*

Proof. Take $a, b, c \in N$. Then $a *_\Phi (b *_\Gamma c) = \phi_a(\gamma_b(c)) = \phi_a \circ \gamma_b(c)$. Also, $(a *_\Phi b) *_\Gamma (a *_\Phi c) = \phi_a(b) *_\Gamma \phi_a(c) = \gamma_{\phi_a(b)}(\phi_a(c)) = \gamma_{\phi_a(b)} \circ \phi_a(c)$. Now $\gamma_{\phi_a(b)} \circ \phi_a = \phi_a \circ \gamma_{\phi_a(b)}$ and $\gamma_{\phi_a(b)} = \gamma_b$ since $\phi_a(b) =_\Gamma b$ by (7.113). Hence, $*_\Phi$ is left distributive over $*_\Gamma$. And of course, the proof that $*_\Gamma$ is left distributive over $*_\Phi$ is virtually the same.

For examples, we need only take any field $(F, +, \cdot)$ with Φ and Γ subgroups of the multiplicative group $(F^*, \cdot)$ and with $\Phi \cap \Gamma = \{1\}$. Let $\Lambda = \Phi\Gamma$, the direct product. Of course $(\mathbf{C}, +, *_2, *_3)$ is our main example and $(\mathbf{C}, \mathcal{B}_{*_2}, \mathcal{B}_{*_3}^*)$ will be our source for motivation and our standard for comparison.

(7.115) Definitions (*tentative*). Let $(N, +, \star, \circ)$ be a double planar nearring and suppose $(N, +, \circ)$ is a circular planar nearring with circles $N^* \circ a + b \in \mathcal{B}_\circ^*$. Then $c \in N$ is an *interior* point of the circle $N^* \circ a + b$ if (1) $c \notin N^* \circ a + b$, and (2) every ray $N \star a + b \in \mathcal{B}_\star$ from c intersects the circle $N^* \circ a + b$. So an *exterior* point $e \in N$ of $N^* \circ a + b$ is a point which is not in $N^* \circ a + b$ and is not an interior point. (At the time of writing, the author is not convinced that these are the best definitions for interior and exterior points. An alternative for each would be as follows. A point $c \in N$ is an *exterior point* of the circle $N^* \circ a + b$ if (1) there is a ray from c which intersects $N^* \circ a + b$ exactly twice, (2) there is a ray from c which intersects $N^* \circ a + b$ exactly once, and (3) there is a ray from c which does not intersect $N^* \circ a + b$. A point $c \in N$ is an *interior point* of the circle $N^* \circ a + b$ if every ray from c intersects $N^* \circ a + b$ in at most one point.)

Take the prime field $(Z_{13}, +, \circ)$ of order 13. Let $N = Z_{13}$, $N^* = \Phi\Gamma$ where $|\Phi| = 4$ and $|\Gamma| = 3$. Then $(N, +, *_\Phi)$ is circular, so $(N, +, *_\Gamma, *_\Phi)$ is a reasonable place to look for an interesting example. The circles with centre 0 are $N_{*_\Phi}^* 1 = \{5, 12, 8, 1\}$, $N_{*_\Phi}^* 2 = \{10, 11, 3, 2\}$, and $N_{*_\Phi}^* 4 = \{7, 9, 6, 4\}$. The rays from 0 are $N_{*_\Gamma} 1 = \{0, 3, 9, 1\}$, $N_{*_\Gamma} 2 = \{0, 6, 5, 2\}$, $N_{*_\Gamma} 4 = \{0, 12, 10, 4\}$, and $N_{*_\Gamma} 7 = \{0, 8, 11, 7\}$. The structure $(N, \mathcal{B}_{*_\Gamma}, \mathcal{B}_{*_\Phi}^*)$ has the following pleasant property.

(P) Given a circle $A \in \mathcal{B}_{*_\Phi}^*$ and a point $x \in N$ exterior to A, then there are exactly two rays from x tangent to the circle A.

In our finite example here, the only interior point of a circle $N_{*_\Phi}^* a + b$ is its centre b. For example, the point 1 is an exterior point for the circle $N_{*_\Phi}^* 2 = \{10, 11, 3, 2\}$. The rays from 1 are $N_{*_\Gamma} 1 + 1 = \{1, 4, 10, 2\}$, $N_{*_\Gamma} 2 + 1 = \{1, 7, 6, 3\}$, $N_{*_\Gamma} 4 + 1 = \{1, 0, 11, 5\}$, and $N_{*_\Gamma} 7 + 1 = \{1, 9, 12, 8\}$.

(7.116) Exploratory problem. How extensive is property (P) in double planar nearrings $(N, +, \star, \circ)$ where $(N, +, \circ)$ is circular? Are there such finite double planar nearrings where circles have more than their centre as interior points? Are there any such double planar nearrings where the topology defined by interiors to circles leads to any interesting mathematics?

(7.117) Exploratory problem. What other relationships between rays and circles in $(\mathbf{C}, \mathcal{B}_{*_2}, \mathcal{B}_{*_3}^*)$ are found among other double planar nearrings? What interesting relationships are there between rays and circles in some double planar nearrings which are in contrast to those found in $(\mathbf{C}, \mathcal{B}_{*_2}, \mathcal{B}_{*_3}^*)$?

8. Coding, cryptography, and combinatorics

As the 1990s begin, coding theory and cryptography are two very fashionable fields of study. There is a vast body of literature on both subjects, and we will not even make a pretence that we are including an adequate introduction to either subject. All we want to do is to show that ideas from planar nearrings have some interesting applications to each subject.

Coding theory and cryptography have in common the idea of sending information from point A to point B. Whereas coding theory is primarily concerned with efficiency and accuracy, with no regard to secrecy, cryptography is primarily concerned with secrecy.

With coding theory, as information is transmitted from A to B, there really is no concern if the information also arrives at a point C. If the information being sent from A to B is distorted in some way, then coding theory is concerned with (1) knowing that is has been distorted, and (2) correcting the distortion. At the same time, coding theory wants to be able to do its job quickly, efficiently, and accurately.

With cryptography, as information is transmitted from A to B, there is the concern that information will arrive at some point C. If the information does arrive at C, when it was intended only to arrive at B, then it is desirable that the information will be of no use at C. Also, if the information arrives at its intended destination B, it is desirable that the information will be of use at B.

In this section we examine each situation in more detail and see how planar nearrings can be used in both situations. Applications of planar nearrings to binary codes were first explored by Modisett [MM, 1988, 1989] and by Fuchs, Hofer, and Pilz [FHP]. Applications of planar nearrings to cryptography were also first noticed and explored by Modisett [MM, 1988, 1989]. Many of the ideas here are highly influenced by, or are directly from these two references.

Recently, I-Hsing Chen and Tayuan Huang [I-H] took a more traditional approach to finding applications of planar nearrings to geometry. Their efforts led to constructing nets from arbitrary finite planar nearrings. This automatically leads to the construction of transversal designs, families of mutually orthogonal Latin squares, and to strongly regular graphs. Chen and Huang also constructed a family of association schemes and PBIBDs, which we have extended here.

8.1. Planar nearrings and coding theory

Today, a great deal of information is transmitted from point A to point B in the form of 0s and 1s. A sequence $a_1 a_2 \ldots a_n$ of 0s and 1s represents a datum. If this sequence $a_1 a_2 \ldots a_n$ leaves A, one hopes that the same sequence $a_1 a_2 \ldots a_n$ will arrive at B. But, it may be that $a_1 a_2 \ldots a_n$ is transmitted from A and $b_1 b_2 \ldots b_n$ is received at B, and $a_i \neq b_i$ for some i. In transmitting a_i, there is the possibility that b_i is received, and $a_i \neq b_i$.

As an example, suppose 0010101 represents the letter x. If 0010101 is sent from A and 1010101 is received at B, then there is an error in transmission. If 1010101 represents the letter y, then the recipient at B must assume that the letter y was sent, even though, in truth, x was sent. In this example, only one small error was made, but yet false information was received, and it was not detected.

How could the receiver at B (1) know that false information had been received, and (2) correct the false information? An elementary example will quickly illustrate a possibility.

Suppose a communication system is designed to transmit exactly one of two values at a given time, a y for 'yes' and an n for 'no'. Suppose 11110000 represents y and 00001111 represents n. If one wants to transmit a y, then one transmits 11110000. But what if 11110001 is received? Obviously an error has occurred, and if only one error has occurred then interpreting 11110001 at B as 11110000 will correct the error. So error detection and error correction can take place. If the communication system is highly reliable, then one is reasonably assured that a received 11110001 was meant to be a 11110000. But suppose 11001010 was received at B. The receiver at B cannot be confident of what was sent from A. Errors can be detected, but not necessarily corrected.

Certainly, there are 2^n distinct sequences $a_1 a_2 \ldots a_n$ of 0s and 1s. Perhaps this is significantly more than we need. For example, perhaps all we need are the 26 letters of the Roman alphabet, the 10 digits, 13 punctuation symbols, and one symbol for a blank. So, with a total of 50 symbols required, we could take $n = 6$, and with $2^6 = 64$, we have more than enough sequences $a_1 a_2 \ldots a_n$ to represent the 50 symbols needed for effective communication. However, with $n = 10$, we have 1024 such sequences $a_1 a_2 \ldots a_{10}$. In order to detect and correct errors, we will want to isolate 50 of these 1024 sequences as much as possible.

Exactly what do we mean when we say 'we want to isolate' a sequence? Let $\mathbf{a} = a_1 a_2 \ldots a_n$ and $\mathbf{b} = b_1 b_2 \ldots b_n$ be two sequences of 0s and 1s of length n. Define

$$d(\mathbf{a}, \mathbf{b}) = |\{i \mid 1 \leq i \leq n, \ a_i \neq b_i\}|. \qquad (8:1)$$

So $d(\mathbf{a}, \mathbf{b})$ counts the number of places where $\mathbf{a}$ differs from $\mathbf{b}$, and so it is a measure of how much $\mathbf{a}$ differs from $\mathbf{b}$. If $\mathbf{a}$ is transmitted from A and $\mathbf{b}$ is received at B, then $d(\mathbf{a}, \mathbf{b})$ errors have occurred. Let Z_2^n denote all sequences $a_1 a_2 \ldots a_n$ of 0s and 1s. Then d is a metric on Z_2^n. That is, for all $\mathbf{a}, \mathbf{b}, \mathbf{c} \in Z_2^n$, (i) $d(\mathbf{a}, \mathbf{b}) = d(\mathbf{b}, \mathbf{a})$, (ii) $d(\mathbf{a}, \mathbf{c}) \leq d(\mathbf{a}, \mathbf{b}) + d(\mathbf{b}, \mathbf{c})$, (iii) $d(\mathbf{a}, \mathbf{a}) = 0$, and (iv) if $d(\mathbf{a}, \mathbf{b}) = 0$, then $\mathbf{a} = \mathbf{b}$. To see that d is a metric on Z_2^n is immediate except for (ii). Suppose $\mathbf{a}$ and $\mathbf{b}$ differ at $i_1, i_2, \ldots, i_k$, so $d(\mathbf{a}, \mathbf{b}) = k$. Suppose $\mathbf{b}$ and $\mathbf{c}$ differ at $j_1, j_2, \ldots, j_l$, so $d(\mathbf{b}, \mathbf{c}) = l$. Also, suppose $\mathbf{a}$ and $\mathbf{c}$ differ at $s_1, s_2, \ldots, s_m$, so $d(\mathbf{a}, \mathbf{c}) = m$. If $a_s \neq c_s$, then we cannot have both $a_s = b_s$ and $b_s = c_s$. So $s \in \{i_1, i_2, \ldots, i_k\}$ or

$s \in \{j_1, j_2, \ldots, j_l\}$, or $s \in \{i_1, i_2, \ldots, i_k\} \cap \{j_1, j_2, \ldots, j_l\}$. Hence, $m \leq k+l$, or $d(\mathbf{a}, \mathbf{c}) \leq d(\mathbf{a}, \mathbf{b}) + d(\mathbf{b}, \mathbf{c})$. If $\mathbf{a}_1, \mathbf{a}_2, \ldots, \mathbf{a}_M$ are M of the 2^n distinct sequences $a_1 a_2 \ldots a_n$ of 0s and 1s, then we want a positive r so that if $1 \leq d(\mathbf{b}, \mathbf{a}_i) \leq r$, then $\mathbf{b} \notin \{\mathbf{a}_1, \ldots, \mathbf{a}_M\} \setminus \{\mathbf{a}_i\}$ for each i, $1 \leq i \leq M$. Then the $\mathbf{a}_1, \mathbf{a}_2, \ldots, \mathbf{a}_M$ are 'isolated'. We will return to this idea shortly.

Let $\mathcal{C}(n)$ denote a nonempty set of sequences $\mathbf{a} = a_1 a_2 \ldots a_n$ of 0s and 1s, where n is a positive integer measuring the *length* of the sequence. So $1 \leq |\mathcal{C}(n)| \leq 2^n$. If $\mathbf{a}, \mathbf{b} \in \mathcal{C}(n)$ and $\mathbf{a} \neq \mathbf{b}$, then $1 \leq d(\mathbf{a}, \mathbf{b}) \leq n$, and so

$$D = \min\{d(\mathbf{a}, \mathbf{b}) \mid \mathbf{a}, \mathbf{b} \in \mathcal{C}(n), \mathbf{a} \neq \mathbf{b}\}$$

exists, and we are assured that $1 \leq D \leq n$. If $M = |\mathcal{C}(n)|$, then we refer to $\mathcal{C}(n)$ as an (n, M, D)-*(binary) code*. (For applications of nearrings, planar and nonplanar, to nonbinary coding theory, see the interesting paper by H. Karzel and A. Oswald [K & O].) Our codes will be binary, in that each element of $\mathcal{C}(n)$ is a sequence of 0s and 1s. There is a function $w : \mathcal{C}(n) \to \{0, 1, \ldots, n\}$ defined by $w(a_1 a_2 \ldots a_n) = |\{i \mid 1 \leq i \leq n, a_i = 1\}|$. So $w(a_1 a_2 \ldots a_n)$ is the *weight* of the *codeword* $\mathbf{a} = a_1 a_2 \ldots a_n \in \mathcal{C}(n)$.

Of all the possible codes $\mathcal{C}(n)$, some have advantages over others. When one has a finite tactical configuration $(N, \mathcal{B}, \in)$, one can easily construct two codes, a *row code* $\mathcal{C}^A(v)$ and a *column code* $\mathcal{C}_A(b)$. Following the conventions in MacWilliams and Sloane [M & S], we define for a finite tactical configuration $(N, \mathcal{B}, \in)$ an *incidence matrix* A, a $b \times v$ matrix of 0s and 1s. Let $\mathcal{B} = \{B_1, B_2, \ldots, B_b\}$ and $N = \{x_1, x_2, \ldots, x_v\}$. Define $A = (a_{ij})$ where $a_{ij} = \begin{cases} 1, & \text{if } x_j \in B_i; \\ 0, & \text{otherwise.} \end{cases}$ Let $\mathcal{C}^A(v)$ consist of the codewords $\mathbf{a}_1, \mathbf{a}_2, \ldots, \mathbf{a}_b$, where $\mathbf{a}_i = a_{i1} a_{i2} \ldots a_{iv}$. Let $\mathcal{C}_A(b)$ consist of the distinct codewords $\mathbf{b}_1, \mathbf{b}_2, \ldots \mathbf{b}_v$, where $\mathbf{b}_j = a_{1j} a_{2j} \ldots a_{bj}$. That is, $\mathcal{C}^A(v)$ consists of the b rows of A and $\mathcal{C}_A(b)$ consists of the distinct columns from the v columns of A. Then $\mathcal{C}^A(v)$ is a (v, b, D_v)-code for some D_v, and $\mathcal{C}_A(b)$ is a (b, v', D_b)-code for some D_b and for some v', $1 \leq v' \leq v$. Actually, any (n, M, D)-binary code $\mathcal{C}(n)$ has several incidence matrices A. Let $\mathcal{C}(n) = \{\mathbf{a}_1, \mathbf{a}_2, \ldots, \mathbf{a}_M\}$ with $\mathbf{a}_i = a_{i1} a_{i2} \ldots a_{in}$, and then let $A = (a_{ij})$. Now $\mathcal{C}^A(n) = \mathcal{C}(n)$ and $\mathcal{C}_A(M)$ are as before. If one takes any $s \times t$ matrix $A = (a_{ij})$, where each $a_{ij} \in \{0, 1\}$, then let $\mathcal{C}^A(t)$ consist of the distinct rows of A and let $\mathcal{C}_A(s)$ consist of the distinct columns of A. So $\mathcal{C}^A(t)$ and $\mathcal{C}_A(s)$ are binary codes.

The codes $\mathcal{C}^A(v)$ and $\mathcal{C}_A(b)$ are nice in that they are constant weight codes. A code $\mathcal{C}(n)$ is a *constant weight code* if there is a number W so that $w(\mathbf{a}) = W$ for each $\mathbf{a} \in \mathcal{C}(n)$. Hence, $W = k$ for $\mathcal{C}^A(v)$ and $W = r$ for $\mathcal{C}_A(b)$. This is because each block $B \in \mathcal{B}$ has exactly k elements, and each $x \in N$ belongs to exactly r blocks.

We have occasion to use the following proposition.

(8.1) Proposition. *Let $\mathcal{C}(n)$ be a constant weight code. Then $D = 2\delta$, an even positive integer.*

Proof. For $\mathbf{a} = a_1 a_2 \ldots a_n$, $\mathbf{b} = b_1 b_2 \ldots b_n \in \mathcal{C}(n)$, $\mathbf{a} \neq \mathbf{b}$, if $1 = a_i \neq b_i$ for some i, $1 \leq i \leq n$, then there is a j, $1 \leq j \leq n$, such that $a_j \neq b_j = 1$, since $w(\mathbf{a}) = w(\mathbf{b})$, and conversely. Hence, $\mathbf{a}$ differs from $\mathbf{b}$ in an even number of places, that is, $d(\mathbf{a}, \mathbf{b})$ is even. This forces D to be even also.

Let us return now to the idea of 'isolation'. For a code $\mathcal{C}(n)$, suppose $D \geq 2$. Let $\mathcal{C}(n) = \{\mathbf{a}_1, \mathbf{a}_2, \ldots, \mathbf{a}_M\}$. If $\mathbf{a} \in \mathcal{C}(n)$ is transmitted from A and $\mathbf{b}$ is received at B, then one tries to find an $\mathbf{a}_i$ so that $\mathbf{b} = \mathbf{a}_i$. If this search is successful, one assumes that $\mathbf{a}_i = \mathbf{a}$, that is, $\mathbf{a}_i$ was transmitted from A. Of course, this could be incorrect, but to be incorrect, at least D errors would have occurred in transmission. The next best thing is to find the $\mathbf{a}_i$ closest to $\mathbf{b}$. It is enough to have the existence of an $\mathbf{a}_i$ such that $d(\mathbf{a}_i, \mathbf{b}) < D/2$. For if $i \neq j$ and $d(\mathbf{a}_i, \mathbf{b}) = d(\mathbf{a}_j, \mathbf{b})$, then $D \leq d(\mathbf{a}_i, \mathbf{a}_j) \leq d(\mathbf{a}_i, \mathbf{b}) + d(\mathbf{b}, \mathbf{a}_j) = 2d(\mathbf{a}_i, \mathbf{b})$, and so $D/2 \leq d(\mathbf{a}_i, \mathbf{b})$. Of course, if we find an $\mathbf{a}_i$ such that $d(\mathbf{a}_i, \mathbf{b}) < D/2$, then we assume that $\mathbf{a}_i = \mathbf{a}$. This means that we can correct fewer that $D/2$ errors. If $\mathbf{b} \notin \mathcal{C}(n)$, then at least one error occurred during transmission. If fewer than D errors were made, then $\mathbf{b} \notin \mathcal{C}(n)$. Hence, we can detect $D - 1$ or fewer errors. In summary,

(8.2) Theorem. *Let $\mathcal{C}(n)$ be a binary code with $D \geq 2$. Then fewer than $D/2$ errors can be corrected, and fewer than D errors can be detected. One traditionally writes that up to $[(D-1)/2]$ errors can be corrected, and up to $D - 1$ errors can be detected, where $[x]$ is the integer satisfying $[x] \leq x < [x] + 1$.*

In our study of finite planar nearrings, we frequently obtained a BIBD. In particular, every time we obtained an affine plane, we also got a BIBD with $\lambda = 1$. Hence, the following theorem and its corollaries are applicable to the BIBDs obtained from finite planar nearrings. In other words, whenever a finite planar nearring generates a BIBD, then the row code $\mathcal{C}^A(v)$ and the column code $\mathcal{C}_A(b)$ will have properties supplied by the following theorem and its corollaries.

(8.3) Theorem. *Let $(N, \mathcal{B}, \in)$ be a BIBD with incidence matrix A and parameters (v, b, r, k, λ). Let $\mathcal{C}^A(v)$ and $\mathcal{C}_A(b)$ denote the row code of A and the column code of A respectively. Let $\mu = \max\{|B_i \cap B_j| \mid B_i, B_j \in \mathcal{B}, B_i \neq B_j\}$. Then D_v for $\mathcal{C}^A(v)$ is $2(k - \mu)$, and D_b for $\mathcal{C}_A(b)$ is $2(r - \lambda)$. In fact, $d(\mathbf{a}, \mathbf{b}) = 2(r - \lambda)$ for any two distinct $\mathbf{a}, \mathbf{b} \in \mathcal{C}_A(b)$.*

Proof. Now B_i is used to make $\mathbf{a}_i$ in the incidence matrix and in $\mathcal{C}^A(v)$. Suppose $|B_i \cap B_j| = \mu_{ij}$. Then $d(\mathbf{a}_i, \mathbf{a}_j) = 2k - 2\mu_{ij}$. Hence, $D_v = 2k - 2\mu = 2(k - \mu)$.

Now $\mathbf{b}_j \in \mathcal{C}_A(b)$ corresponds to the jth column of the incidence matrix. This tells by 1s of exactly which $B_i \in \mathcal{B}$ contain $x_j \in N$, and there are exactly r such 1s in $\mathbf{b}_j$. Since $x_s, x_t \in N$, $x_s \neq x_t$, have exactly λ blocks $B_i \in \mathcal{B}$ with $x_s, x_t \in B_i$, there will be exactly λ identical positions in $\mathbf{b}_s$ and in $\mathbf{b}_t$ which are 1s. Hence, $d(\mathbf{b}_s, \mathbf{b}_t) = 2r - 2\lambda = 2(r - \lambda) = D_b$.

An affine plane has $v = n^2$ points and $b = n^2 + n$ lines, so $r = n + 1$, $k = n$, and $\mu = \lambda = 1$. Hence, we can easily prove

(8.4) Corollary. *A finite affine plane* $(N, \mathbf{L}, \in)$ *with* $v = n^2$ *points and incidence matrix* A *has* D *for* $\mathcal{C}^A(n^2)$ *equal to* $2(k-1)$, *and* D *for* $\mathcal{C}_A(n^2+n)$ *equal to* $2(r - 1)$. *In particular, for* $\mathcal{C}^A(n^2)$, *up to* $n - 2$ *errors can be corrected and up to* $2n - 3$ *errors can be detected. For* $\mathcal{C}_A(n^2 + n)$, *up to* $n - 1$ *errors can be corrected and up to* $2n - 1$ *errors can be detected.*

Proof. Apply (8.2), (8.3), and the remarks made just before the statement of (8.4).

(8.5) Corollary. *For a BIBD* $(N, \mathcal{B}, \in)$ *with parameters* (v, b, r, k, λ) *and* μ *of (8.3), let* A *be the incidence matrix. Then* $\mathcal{C}^A(v)$ *is a* $(v, b, 2(k - \mu))$-*code, and* $\mathcal{C}_A(b)$ *is a* $(b, v, 2(r - \lambda))$-*code.*

Proof. Apply (8.3) and remarks made before (8.1). But we need to say more about $\mathcal{C}_A(b)$. How can we be assured that $v = |\mathcal{C}_A(b)|$? That is, why are the columns of A distinct? If two columns are identical, say $\mathbf{b}_i = \mathbf{b}_j$, with $i \neq j$, then $x_i, x_j \in N$ are two distinct points which belong to exactly the same blocks. Hence, $\lambda = r$. From $\lambda(v - 1) = r(k - 1)$, we conclude that $v = k$. Hence, $b = 1$ and so $\mathcal{B} = \{N\}$. This makes $A = (1 \ \ 1 \ \ \cdots \ \ 1)$, which we really do not want.

(8.6) Corollary. *Let* $(N, +, \cdot)$ *be a finite circular planar nearring. Let* A *be the incidence matrix for the BIBD* $(N, \mathcal{B}^*, \in)$. *Then* $\mathcal{C}^A(v)$ *is a* $(v, v(v - 1)/k, 2(k - 2))$-*code, and* $\mathcal{C}_A(b)$ *is a* $(v(v - 1)/k, v, 2(v - k))$-*code.*

Proof. Apply (8.5), (5.5), and the fact that circularity implies that the μ of (8.3) is 2.

(8.7) Corollary. *Let* $(N, +, \cdot)$ *be a finite planar nearring with BIBD* $(N, \mathcal{B}, \in)$ *with no basic blocks as subgroups. Let* A *be the incidence matrix for* $(N, \mathcal{B}, \in)$. *Then* $\mathcal{C}^A(v)$ *is a* $(v, v(v - 1)/(k - 1), 2(k - \mu))$-*code, and* $\mathcal{C}_A(b)$ *is a* $(v(v - 1)/(k - 1), v, 2k(v - k)/(k - 1))$-*code.*

Proof. Apply (8.5) and (7.17).

Let n be a positive integer and consider the set Z_2^n. From (8.2), we learn that D controls the error correcting and error detecting capabilities of a binary code $\mathcal{C}(n)$. So, let us assume that a D, $1 \leq D \leq n$, is specified. Next choose a W, $1 \leq W \leq n$. For $1 \leq M \leq 2^n$, does there exist an (n, M, D)-code $\mathcal{C}(n)$ with constant weight W? Let $\mathcal{M}$ be the set of all M, $1 \leq M \leq 2^n$, such that there is an (n, M, D)-code $\mathcal{C}(n)$ with constant weight W. If $\mathcal{M} = \varnothing$, define $A(n, D, W) = 0$. If $\mathcal{M} \neq \varnothing$, define $A(n, D, W) = \max \mathcal{M}$. So, if $\mathcal{M} \neq \varnothing$, then there is a code $\mathcal{C}(n)$ with constant weight W and minimum distance D with exactly $A(n, D, W)$ codewords and there can be no larger code with constant weight W and minimum distance D in Z_2^n. An $(n, A(n, D, W), D)$-code $\mathcal{C}(n)$ is the best one can obtain with n, D, and W fixed, if one is looking to maximize the number of codewords. Just how large can $A(n, D, W)$ be? There is a bound, called *Johnson's bound* [M & S], which is attained by some of the codes constructed from planar nearrings. So, in some sense, these codes are among the best. For a constant weight code, (8.1) assures us that $D = 2\delta$, an even integer, so $A(n, 2t + 1, W)$ is not positive for odd integers $2t + 1$, a point overlooked in Theorem 1, page 525, of [M & S].

(8.8) Theorem (Johnson's bound). *If $W^2 - Wn + \delta n > 0$, then*

$$A(n, 2\delta, W) \leq \frac{\delta n}{W^2 - Wn + \delta n}.$$

Proof. Let $\mathcal{C}(n)$ be an $(n, M, 2\delta)$-binary code of constant weight W, where $M = A(n, 2\delta, W)$. Let $\mathcal{C}(n) = \{\mathbf{a}_1, \mathbf{a}_2, \ldots, \mathbf{a}_M\}$, with code words $\mathbf{a}_i = a_{i1}a_{i2}\ldots a_{in}$, have incidence matrix $A = (a_{ij})$. For $\mathbf{a}_i, \mathbf{a}_j \in \mathcal{C}(n)$, the inner product $\mathbf{a}_i \cdot \mathbf{a}_j = \sum_{k=1}^{n} a_{ik}a_{jk}$. Notice that $\mathbf{a}_i \cdot \mathbf{a}_j = W$ if and only if $d(\mathbf{a}_i, \mathbf{a}_j) = 0$; $\mathbf{a}_i \cdot \mathbf{a}_j = W - 1$ if and only if $d(\mathbf{a}_i, \mathbf{a}_j) = 2$; $\ldots$; $\mathbf{a}_i \cdot \mathbf{a}_j = W - t$ if and only if $d(\mathbf{a}_i, \mathbf{a}_j) = 2t$. If $i \neq j$, then $d(\mathbf{a}_i, \mathbf{a}_j) = 2t$ cannot happen for $2t < 2\delta = D$. Hence, since $d(\mathbf{a}_i, \mathbf{a}_j) \geq D = 2\delta$ for $i \neq j$, we have for $i \neq j$, $\mathbf{a}_i \cdot \mathbf{a}_j \leq W - \delta$.

For each i, $1 \leq i \leq M$, we want to take all $\mathbf{a}_i \cdot \mathbf{a}_j$, $i \neq j$, and add the results. The sum is

$$S = \sum_{\substack{i=1}}^{M} \sum_{\substack{j=1 \\ j \neq i}}^{M} \sum_{k=1}^{n} a_{ik}a_{jk}. \tag{8:2}$$

Since each $\mathbf{a}_i \cdot \mathbf{a}_j = \sum_{k=1}^{n} a_{ik}a_{jk} \leq W - \delta$, we have

$$S \leq M(M-1)(W - \delta). \tag{8:3}$$

We can rewrite S as

$$S = \sum_{k=1}^{n} \sum_{i=1}^{M} \sum_{\substack{j=1 \\ j \neq i}}^{M} a_{ik}a_{jk}. \tag{8:4}$$

Let s_k be the number of 1s in the kth column of A. Then the kth column contributes $s_k(s_k - 1)$ to the sum in (8:4). Hence

$$S = \sum_{k=1}^{n} s_k(s_k - 1) = \sum_{k=1}^{n} s_k^2 - \sum_{k=1}^{n} s_k. \qquad (8:5)$$

But $\sum_{k=1}^{n} s_k = WM$. Using the method of Lagrange multipliers from calculus, one minimizes $\sum_{k=1}^{n} s_k^2$ by setting each $s_k = WM/n$, and so $W^2 M^2/n \le \sum_{k=1}^{n} s_k^2$. From (8:5) and (8:3) we can write

$$\frac{W^2 M^2}{n} - WM \le S \le M(M - 1)(W - \delta), \qquad (8:6)$$

which simplifies to

$$M(W^2 - Wn + \delta n) \le \delta n, \qquad (8:7)$$

from which our theorem follows.

Our next theorem proves that every time we have a BIBD, then we get a column code $\mathcal{C}_A(b)$ which satisfies Johnson's bound! Planar nearrings give BIBDs in various ways, so planar nearrings are useful in constructing binary equal weight codes which reach Johnson's bound, hence they are among the best in this sense.

(8.9) Theorem. *Let $(N, \mathcal{B}, \in)$ be a BIBD with parameters (v, b, r, k, λ). If $\mathcal{B} = \{B_1, B_2, \dots, B_b\}$, let A be the incidence matrix for $(N, \mathcal{B}, \in)$. Then $\mathcal{C}_A(b)$ is a $(b, v, 2(r - \lambda))$-binary code with constant weight $W = r$, and*

$$v = \frac{\delta n}{W^2 - Wn + \delta n} = A(b, 2(r - \lambda), r) = \frac{(r - \lambda)b}{r^2 - \lambda b}. \qquad (8:8)$$

Proof. First apply (8.5) to recall that $\mathcal{C}_A(b)$ is a $(b, v, 2(r - \lambda))$-binary code. We noticed just before (8.1) that $W = r$. So (8:8) is all we have to prove. With $W = r$, $\delta = r - \lambda$, and $n = b$, we obtain that $A(v, 2(r - \lambda), r) \le \delta n/(W^2 - Wn + \delta n) = (r - \lambda)b/(r^2 - \lambda b)$, and certainly $|\mathcal{C}_A(b)| = v$. We will be finished upon showing $v = (r - \lambda)b/(r^2 - \lambda b)$.

Recall that a BIBDs parameters satisfy $bk = vr$ and $\lambda(v - 1) = r(k - 1)$. So $b = vr - bk + b = vr - b(k - 1)$. Hence, $rb = vr^2 - br(k - 1) = vr^2 - b\lambda(v - 1) = vr^2 + \lambda b - \lambda bv$. So $rb - \lambda b = vr^2 - \lambda bv = (r^2 - \lambda b)v$, giving us the required $v = (r - \lambda)b/(r^2 - \lambda b)$.

Notice that (8.9) not only tells us that $\mathcal{C}_A(b)$ is maximal but it gives an explicit formula for $A(v, 2(r - \lambda), r)$ in terms of the parameters of the BIBD. It is a bonus whenever an explicit formula for an $A(n, 2\delta, W)$ is known.

(8.10) Theorem. $A(n, 2\delta, W) \leq [\frac{n}{W} A(n-1, 2\delta, W-1)] \leq \frac{n}{W} A(n-1, 2\delta, W-1)$.

Proof. Let $\mathcal{C}(n)$ be an $(n, A(n, 2\delta, W), 2\delta)$-code with constant weight W and incidence matrix A. Hence, $\mathcal{C}^A(n) = \mathcal{C}(n)$. Let $M = A(n, 2\delta, W) = |\mathcal{C}(n)|$ and let $\mathcal{C}^A(n) = \{\mathbf{a}_1, \mathbf{a}_2, \ldots, \mathbf{a}_M\}$. First note that A has exactly $W A(n, 2\delta, W)$ entries which are 1. We now count these 1s another way and get an upper bound for them.

Take the kth column of A. Let M_k denote the number of 1s in the kth column. Take the $\mathbf{a}_i = a_{i1} a_{i2} \ldots a_{in}$ with $a_{ik} = 1$, eliminate a_{ik}, and let $\mathbf{b}_i$ be the resulting codeword in Z_2^{n-1} of length $n-1$ and weight $W-1$. Altogether, these codewords $\mathbf{b}_i$ form a code $\mathcal{C}_k^A(n-1)$ and $\mathcal{C}_k^A(n-1)$ is an $(n-1, M_k, 2\delta')$-code since $d(\mathbf{b}_i, \mathbf{b}_j) \geq 2\delta' \geq 2\delta$, if $i \neq j$. Hence, $M_k = |\mathcal{C}_k^A(n-1)| \leq A(n-1, 2\delta', W-1)$. Also, the number of 1s in A is $\sum_{k=1}^n M_k$. Hence, $W A(n, 2\delta, W) = \sum_{k=1}^n M_k \leq \sum_{k=1}^n A(n-1, 2\delta', W-1) = nA(n-1, 2\delta', W-1)$, from which we infer that $A(n, 2\delta, W) \leq (n/W)A(n-1, 2\delta', W-1)$.

We will be through when we have $A(n-1, 2\delta', W-1) \leq A(n-1, 2\delta, W-1)$. What we will show is the lemma that if we keep n and W constant, then $A(n, 2\delta, W)$ is a decreasing function of $D = 2\delta$, that is, if $D' = 2\delta' \geq 2\delta = D$, then $A(n, D', W) \leq A(n, D, W)$. Let $\mathcal{C}(n) = \{\mathbf{a}_1, \mathbf{a}_2, \ldots, \mathbf{a}_M\}$ be a constant weight binary code with $D = 2\delta$, and, without loss of generality, we will insist that $d(\mathbf{a}_1, \mathbf{a}_2) = 2\delta$. We will replace $\mathbf{a}_1$ with an $\mathbf{a}_1'$. Choose s and t where $a_{1s} = 0$ and $a_{2s} = 1$, and where $a_{1t} = 1$ and $a_{2t} = 0$. Set $a_{1s}' = 1$, $a_{1t}' = 0$, and $a_{1i}' = a_{1i}$ if $i \notin \{s, t\}$. Let $\mathbf{a}_1' = a_{11}' a_{12}' \ldots a_{1n}'$. Now $d(\mathbf{a}_1', \mathbf{a}_2) = 2(\delta - 1) = d(\mathbf{a}_1, \mathbf{a}_2) - 2$. What about $d(\mathbf{a}_1', \mathbf{a}_i)$ for $i \geq 2$? There are four cases for $a_{is}, a_{it} \in \{0, 1\}$. If $a_{is} = a_{it}$, then $d(\mathbf{a}_1', \mathbf{a}_i) = d(\mathbf{a}_1, \mathbf{a}_i)$. If $a_{is} \neq a_{it}$, then $d(\mathbf{a}_1', \mathbf{a}_i) \in \{d(\mathbf{a}_1, \mathbf{a}_i) \pm 2\}$. So $\mathcal{C}'(n) = \{\mathbf{a}_1', \mathbf{a}_2, \ldots, \mathbf{a}_M\}$ has minimal distance $D' = 2\delta' = 2(\delta - 1) = D - 2$. As long as $D = 2\delta \geq 4$, one can replace $\mathcal{C}(n)$ by $\mathcal{C}'(n)$, and $|\mathcal{C}'(n)| = |\mathcal{C}(n)|$, and the minimal distance of $\mathcal{C}'(n)$ is $D' = 2\delta' = 2(\delta - 1) = D - 2$, where D is the minimal distance of $\mathcal{C}(n)$. Hence, if $M = A(n, 2\delta, W)$, then $M \leq A(n, 2(\delta - 1), W)$. A direct application now assures us that we really do have $A(n-1, 2\delta', W-1) \leq A(n-1, 2\delta, W-1)$, as required, and this completes the proof of (8.10).

(8.11) Lemma. $\delta \leq W$.

Proof. If codewords $\mathbf{a}$ and $\mathbf{b}$, $\mathbf{a} \neq \mathbf{b}$, have 1s in exactly τ common positions, then $2(W - \tau) = d(\mathbf{a}, \mathbf{b}) \geq D = 2\delta$. Hence, $2W \geq 2\delta + 2\tau$ and $W \geq \delta + \tau$. Since $\tau \geq 0$, we obtain $W \geq \delta$.

(8.12) Corollary. $A(n, 2\delta, \delta) = [n/\delta]$.

Proof. If $W = \delta$ and $D = 2\delta$, then for $\mathbf{a} \neq \mathbf{b}$, the τ in the proof of (8.11) is 0. That is, no two distinct codewords $\mathbf{a}$ and $\mathbf{b}$ have a 1 in a common position. Hence, $0 \leq n - \delta A(n, 2\delta, \delta) < \delta$, and so $0 \leq (n/\delta) - A(n, 2\delta, \delta) < 1$.

(8.13) Corollary.

$$A(n, 2\delta, W) \leq \left[\frac{n}{W}\left[\frac{n-1}{W-1}\left[\cdots\left[\frac{n-W+\delta}{\delta}\right]\cdots\right]\right]\right]$$
$$\leq \frac{n(n-1)\cdots(n-W+\delta+1)}{W(W-1)\cdots(\delta+1)}\left[\frac{n-W+\delta}{\delta}\right].$$

Proof. From (8.10), we have $A(n, 2\delta, W) \leq [(n/W)A(n-1, 2\delta, W-1)] \leq (n/W)A(n-1, 2\delta, W-1)$. Applying (8.10) to $A(n-1, 2\delta, W-1)$, we obtain $A(n, 2\delta, W) \leq [(n/W)[((n-1)/(W-1))A(n-2, 2\delta, W-2)]] \leq (n(n-1)/W(W-1))A(n-2, 2\delta, W-2)$. Continuing we obtain

$$A(n, 2\delta, W) \leq \left[\frac{n}{W}\left[\frac{n-1}{W-1}\left[\cdots\left[\frac{n-W+\delta+1}{\delta+1}A(n-W+\delta, 2\delta, \delta)\right]\cdots\right]\right]\right]$$
$$\leq \frac{n(n-1)\cdots(n-W+\delta+1)}{W(W-1)\cdots(\delta+1)}A(n-W+\delta, 2\delta, \delta).$$

Now apply (8.12) to get the desired results.

Since any finite affine plane $(N, \mathbf{L}, \in)$ is a BIBD with the parameters $(v, b, r, k, \lambda) = (m^2, m^2 + m, m+1, m, 1)$ for some m, from (8.9) we know that the column code $\mathcal{C}_A(m^2 + m)$ of the incidence matrix A for $(N, \mathbf{L}, \in)$ is maximal in the sense that $A(m^2 + m, 2m, m+1) = m^2 = |\mathcal{C}_A(m^2 + m)|$. But the row code $\mathcal{C}^A(m^2)$ of A is also maximal in the sense that $A(m^2, 2(m-1), m) = m^2 + m$.

(8.14) Theorem. *Let $(N, \mathbf{L}, \in)$ be a finite affine plane, hence a BIBD with parameters $(v, b, r, k, \lambda) = (m^2, m^2 + m, m+1, m, 1)$ for some m. Let A be the incidence matrix for $(N, \mathbf{L}, \in)$. Then*

$$A(m^2 + m, 2(m-1), m+1) = |\mathcal{C}^A(m^2)| = m^2 + m.$$

Proof. We have $m^2 + m = |\mathcal{C}^A(m^2)| \leq A(m^2 + m, 2(m-1), m+1) \leq (m^2/m)A(m^2 - 1, 2(m-1), m-1) = m[(m^2 - 1)/(m-1)] = m(m+1) = m^2 + m$, from (8.13) and (8.12). Hence, $m^2 + m = |\mathcal{C}^A(m^2)| = A(m^2 + m, 2(m-1), m+1)$.

Again, we get an explicit formula for $A(m^2 + m, 2(m-1), m+1)$ where m is the number of points in a line of an affine plane.

All the affine planes which we get from finite planar nearrings give optimal $\mathcal{C}^A(m^2)$ and $\mathcal{C}_A(m^2 + m)$. Actually, (8.14) should be the corollary of the more general

(8.15) Theorem. *Let $(N, \mathcal{B}, \in)$ be a BIBD with parameters (v, b, r, k, λ) with $\lambda = 1$. Let A be the incidence matrix for $(N, \mathcal{B}, \in)$. Then*

$$b = |\mathcal{B}| = |\mathcal{C}^A(v)| = A(v, 2(k-1), k).$$

Proof. Again, with $\lambda = 1$, we have $D = 2(k-1)$ and $W = k$. Now apply (8.13) and (8.12) and recall that $bk = vr$ and $\lambda(v-1) = v - 1 = r(k-1)$, so $b = |\mathcal{B}| = |\mathcal{C}^A(v)| \leq A(v, 2(k-1), k) \leq (v/k)A(v-1, 2(k-1), k-1) = (v/k)[(v-1)/(k-1)] = vr/k = bk/k = b$. Hence, $b = |\mathcal{B}| = |\mathcal{C}^A(v)| = A(v, 2(k-1), k)$, as was to be shown.

All the affine spaces constructed in §7 from finite planar nearrings are BIBD with $\lambda = 1$, and so (8.15) applies to each of these.

Johnson's bound was not applicable for the proof of (8.14) since $W^2 - Wn + \delta n = 0$. For (8.15), $W^2 - Wn + \delta n = k^2 - v$, and so we have no immediate assurance that $W^2 - Wn + \delta n > 0$. The techniques of the proofs of (8.14) and (8.15) are also applicable for finite Möbius planes, and, in particular, the finite Möbius planes described in the van der Walt connection in §7.

Start with a BIBD $(N, \mathcal{B}, \in)$ having parameters $(v, b, r, k, \lambda) = (n^2 + 1, n(n^2 + 1), n(n + 1), n + 1, n + 1)$ with the additional property that for every three distinct points $x, y, z \in N$, there is a unique $B \in \mathcal{B}$ with $x, y, z \in B$. Hence, the μ of (8.3) is 2. Such a BIBD is also called a *Möbius plane*. Let A be the incidence matrix for $(N, \mathcal{B}, \in)$, so $\mathcal{C}^A(n^2 + 1)$ is an $(n^2 + 1, n(n^2 + 1), 2(n - 1))$-code with constant weight $W = n + 1$. Then $n(n^2 + 1) = b \leq A(n^2 + 1, 2(n - 1), n + 1) \leq ((n^2 + 1)/(n + 1))A(n^2, 2(n - 1), n) \leq ((n^2 + 1)n^2/(n + 1)n)A(n^2 - 1, 2(n - 1), n - 1) = (n^2 + 1)n^2(n^2 - 1)/(n + 1)n(n - 1) = n(n^2 + 1)$. So we have

(8.16) Theorem. *Let a BIBD $(N, \mathcal{B}, \in)$ be a Möbius plane with $v = n^2 + 1$. Let A be the incidence matrix for $(N, \mathcal{B}, \in)$. Then*

$$|\mathcal{C}^A(v)| = A(n^2 + 1, 2(n - 1), n + 1) = n(n^2 + 1).$$

Again, Johnson's bound is not applicable since the denominator of Johnson's bound is $(n+1)^2 - (n+1)(n^2+1) + (n-1)(n^2+1) = -(n+1)^2 < 0$.

We havee pointed out that the denominator for Johnson's bound is not positive for the row codes $\mathcal{C}^A(v)$ of the BIBDs of (8.14) and (8.16). The same is true for the row codes $\mathcal{C}^A(v)$ of the BIBDs $(N, \mathcal{B}^*, \in)$ for finite circular planar nearrings $(N, +, \cdot)$.

(8.17) Proposition. *Let $(N, +, \cdot)$ be a finite circular planar nearring of order v with BIBD $(N, \mathcal{B}^*, \in)$. Let A be the incidence matrix for $(N, \mathcal{B}^*, \in)$ and $\mathcal{C}^A(v)$ the corresponding row code which has constant weight $W = k$ and distance $D = 2\delta = 2(k - \mu) = 2(k - 2)$. Then $W^2 - Wv + \delta v < 0$.*

Proof. Since $k = |B|$ for each $B \in \mathcal{B}^*$ and k divides $v - 1$, there is an integer t such that $v = kt + 1$. Now $W^2 - Wv + \delta v = k^2 - 2(kt + 1)$. So the answer to when $W^2 - Wv + \delta v > 0$ is equivalent to when $k^2 - 2tk - 2 > 0$, which is, of course, when $k < t - \sqrt{t^2 + 2} < 0$ or $t + \sqrt{t^2 + 2} < k$. Since $0 < k$, we concentrate on the latter. With $t = (v - 1)/k$, we then seek to answer the question when is $(v - 1)/k + \sqrt{((v - 1)/k)^2 + 2} < k$? This will happen exactly when $\sqrt{2v} < k$.

We also have $k \leq (3 + \sqrt{4v - 7})/2$ since $(N, +, \cdot)$ is circular. So we will need $2\sqrt{2v} < 3 + \sqrt{4v - 7}$, which means $2 < v < 8$. But $v - 1 = kt$ and $1 \notin \{k, t\}$, so $v \in \{5, 7\}$. With $v = 5$, then $k = t = 2$, but $k = 2 > 2 + \sqrt{4 + 2} = t + \sqrt{t^2 + 2}$ is not valid. With $v = 7$ and $k = 2$, we obtain $t = 3$, and $k = 2 > 3 + \sqrt{9 + 2} = t + \sqrt{t^2 + 2}$ is not valid. Finally, for $v = 7$ and $k = 3$, we have $t = 2$, and $k = 3 > 2 + \sqrt{4 + 2} > 4$ is not valid. We have now exhausted all the possibilities.

(8.18) Problem. Do we have $b = v(v - 1)/k = A(v, 2(k - 2), k)$ for the row code $\mathcal{C}^A(v)$ of any BIBD $(N, \mathcal{B}^*, \in)$ from a finite circular planar nearring? If so, which ones?

Even though (8.17) is a negative result about the row code $\mathcal{C}^A(v)$ of a BIBD $(N, \mathcal{B}^*, \in)$ of a finite circular planar nearring, we know that the codes $\mathcal{C}^A(v)$ can correct up to $k - 3$ errors and k can be arbitrarily large. Also, $b = |\mathcal{C}^A(v)| = v(v - 1)/k$ can be very large. We can construct a new code from $\mathcal{C}^A(v)$, and, in doing so, even improve on its error correcting and error detecting capabilities. The price we pay is in the number of code words. But if $|\mathcal{C}^A(v)|$ is sufficiently large, perhaps we can pay the price.

Let $(N, \mathcal{B}^*, \in)$ be the BIBD of a finite circular planar nearring $(N, +, \cdot)$ of order v with incidence matrix A. Start with the row code $\mathcal{C}^A(v)$. Take an arbitrary $B_1 \in \mathcal{B}^*$ and list all the blocks $B \in \mathcal{B}^*$ which intersect B_1 in two or more places. That is, we include B_1 in this list. For $x, y \in B_1$, there are $\lambda - 1 = k - 2$ blocks $B \in \mathcal{B}^*$ with $B \neq B_1$ and $|B \cap B_1| = 2$. For $\{u, v\} \neq \{x, y\}$, $u, v \in B_1$, we again obtain $\lambda - 1 = k - 2$ new blocks $C \in \mathcal{B}^*$ with $C \neq B_1$ and $|C \cap B_1| = 2$. Let $\mathcal{B}_1^* = \{B \in \mathcal{B}^* \mid |B \cap B_1| = 2\} \cup \{B_1\}$. Then $|\mathcal{B}_1^*| = \binom{k}{2}(k - 2) + 1$. So $\mathcal{B}_1^*$ consists of a list of $\binom{k}{2}(k - 2) + 1$ blocks which intersect B_1 twice, together with B_1 itself.

Next, take $B_2 \in \mathcal{B}^* \setminus \mathcal{B}_1^*$ and repeat the procedure. That is, let $\mathcal{B}_2^* = \{B \in \mathcal{B}^* \setminus \mathcal{B}_1^* \mid |B \cap B_2| = 2\} \cup \{B_2\}$. Obviously, $|\mathcal{B}_2^*| \leq \binom{k}{2}(k - 2) + 1$. The third step is to take $B_3 \in \mathcal{B}^* \setminus (\mathcal{B}_1^* \cup \mathcal{B}_2^*)$ and repeat the procedure. Let $\mathcal{B}_3^* = \{B \in \mathcal{B}^* \setminus (\mathcal{B}_1^* \cup \mathcal{B}_2^*) \mid |B \cap B_3| = 2\} \cup \{B_3\}$. In general, let

$B_u \in \mathcal{B}^* \setminus (\mathcal{B}_1^* \cup \cdots \cup \mathcal{B}_{u-1}^*)$ and let $\mathcal{B}_u^* = \{B \in \mathcal{B}^* \setminus (\mathcal{B}_1^* \cup \cdots \cup \mathcal{B}_{u-1}^*) \mid |B \cap B_u| = 2\} \cup \{B_u\}$. Since $|\mathcal{B}^*| < \infty$, we eventually must come to an end. When we do, we let $\mathcal{C} = \{B_1, B_2, \ldots, B_u\}$ where u is the smallest positive integer such that $\mathcal{B}^* = \mathcal{B}_1^* \cup \cdots \cup \mathcal{B}_u^*$. Let $\mathcal{C}(v)$ be the row code corresponding to the incidence matrix constructed from $\mathcal{C}$.

Let us call the procedure described above to construct a code $\mathcal{C}(v)$ from the $\mathcal{C}^A(v)$ for a finite circular planar nearring the *FHP-procedure* since it came from [FHP]. We can obtain upper and lower bounds for a $|\mathcal{C}(v)|$. First note that if $\mathcal{C} = \{B_1, B_2, \ldots, B_u\}$ and $\mathcal{C}(v) = \{\mathbf{b}_1, \mathbf{b}_2, \ldots, \mathbf{b}_u\}$, then $d(\mathbf{b}_i, \mathbf{b}_j) \geq 2(k-1)$, if $i \neq j$, since $|B_i \cap B_j| \leq 1$. Hence, for $\mathcal{C}(v)$, we have $D = 2\delta \geq 2(k-1)$, if $i \neq j$. So a $\mathcal{C}(v)$ can correct at least one more error than can $\mathcal{C}^A(v)$. From (5.5), (8.10), and (8.12), we obtain $v(v-1)/k = b = |\mathcal{C}^A(v)| \leq v(v-1)(v-2)/k(k-1)(k-2)$ and $|\mathcal{C}(v)| \leq v(v-1)/k(k-1)$. Hence, $v(v-1)/k\left(\binom{k}{2}(k-2)+1\right) \leq |\mathcal{C}(v)| \leq v(v-1)/k(k-1)$. So we have

(8.19) Theorem. *For a finite circular planar nearring of order v with circles having k points each, let $\mathcal{C}(v)$ be a code produced by the FHP-procedure. Then*

$$\frac{v(v-1)}{k\left(\binom{k}{2}(k-2)+1\right)} \leq |\mathcal{C}(v)| \leq \frac{v(v-1)}{k(k-1)},$$

and $\mathcal{C}(v)$ can correct at least one more error than $\mathcal{C}^A(v)$ can.

(8.20) Problem. Is the u for the FHP-procedure an invariant of the finite circular planar nearring? That is, does u depend upon the choices of $B_1, B_2, \ldots, B_u$? If not, is there an algorithm which will maximize u?

In [FHP], the question is raised as to whether the denominator $W^2 - Wn + \delta N$ of Johnson's bound for the row code of a BIBD is always nonpositive. We can give here only one example where $W^2 - Wn + \delta n > 0$ and, in doing so, give infinitely many more examples where $W^2 - Wn + \delta n < 0$.

In (5.8) and (5.9), let $q = p^n$. Then for any circular planar nearring $(N, +, \cdot)$ constructed from a Ferrero pair (K, Φ) of (5.9), we obtain a BIBD $(N, \mathcal{B}^*, \in)$ with parameters $(v, b, r, k, \lambda) = (q^2, q^2(q-1), q^2-1, q+1, q)$, and since $(N, +, \cdot)$ is circular, the μ of (8.3) is 2. So for $W^2 - Wn + \delta n$, we have $W = q+1$, $n = q^2$, and $\delta = q-1$. Asking when $W^2 - Wn + \delta n > 0$ translates into asking when $-q^2 + 2q + 1 > 0$. Since $q > 0$, the answer is when $0 < q < 1 + \sqrt{2}$. Hence, $q = 2$ is our only possibility. For $q = 2$, then $W = 3$, $n = r$, and $\delta = 1$ make $W^2 - Wn + \delta n = 1 > 0$. Hence, $b = 4$, but $A(n, 2\delta, W) = A(4, 2, 3) \leq (1 \cdot 4)/1 = 4$, so again we have $b = |\mathcal{C}^A(4)| = A(4, 2, 3)$.

With this one example of a $\mathcal{C}^A(v)$ with $W^2 - Wn + \delta n > 0$, there is added intrigue to the following problem suggested in [FHP].

(8.21) Problem. Does there exist a relatively large family of BIBD $(N, \mathcal{B}, \in)$ whose incidence matrix A yields a row code $\mathcal{C}^A(v) = \mathcal{C}^A(n)$ with positive denominator $W^2 - Wn + \delta n$ for Johnson's bound?

We have seen several times already that $|\mathcal{C}^A(n)| = A(n, 2\delta, W)$ even though $W^2 - Wn + \delta n \leq 0$. One naturally raises the question

(8.22) Problem. For a BIBD $(N, \mathcal{B}, \in)$, with $n = v = |N|$, when is $|\mathcal{C}^A(n)| = A(n, 2\delta, W)$?

All the codes we have dealt with so far are constant weight codes. On page 530 of [M & S], it is mentioned that 'constant weight codes have a number of practical applications'. Then numerous references are given. This brings up

(8.23) Problem. For the constant weight codes arising from finite planar nearrings, which have practical applications and why?

And then in [M & S], on page 530, we also see a 'Research Problem' which we shall borrow and list here with appropriate modifications.

(8.24) Problem. 'Find good methods of encoding and decoding constant weight codes' arising from finite planar nearrings.

In Chapter 21 of [M & S], one is lead to believe that association schemes are useful for coding theory. Now we have met with association schemes in Ferrero's adaptation to finite planar nearrings of Hall's method to construct partially balanced incomplete block designs. The two definitions of association schemes are not quite the same, but they may in fact be equivalent, and certainly there is common ground. So we have

(8.25) Problem. Do any of the association schemes produced by Ferrero's adaptation to finite planar nearrings of Hall's method of constructing partially balanced incomplete block designs have any nice applications to coding theory?

Exercise. Look up the definition in [M & S] of an association scheme. Is it equivalent to the one included in these notes?

8.2. Planar nearrings and cryptography

In recent years, Beutelspacher has lobbied for applications of finite geometry to cryptography [Be 1988, Be 1989, Be & R, Be & V]. It has been rumoured that such applications are used on the new German currency. In [Be 1989], Beutelspacher suggests that only the classical finite geometric structures are useful for such applications, but we hope that the reader will see here that alternative geometric structures also promise nice applications.

How does one send a message from point A to point B in such a way that if the message should also reach point C, then the information will be of no use at point C? There are numerous ways to do this, and it is also easy for one to invent new ways. There are three desirable properties of such a method. First, it should be easy for the sender at A to construct the cryptogram; second, it should be easy for the desired receiver at B to understand the cryptogram, and finally, it should be difficult for any undesired receiver at C to understand the cryptogram. A very simple example would be if A and B can communicate in German and C cannot. Then the three desired properties are met. However, with effort, C can learn to communicate in German, also, and then that system would no longer be effective. Or, C could take the cryptogram to a knowledgeable fourth party and discover the contents of the cryptogram. This would take time, however, and so perhaps the information in the cryptogram would no longer be of value.

All the applications of planar nearrings to cryptography which we shall discuss here are variations of a single principle. Let $\mathbf{B}$ be the set of symbols with which it is easy to communicate. For example, $\mathbf{B}$ could consist of all the symbols used by your favourite typewriter, or all the symbols used by your keyboard, or all the symbols used in your chosen font. Let $\mathbf{A}$ be the set of symbols to be used in writing your cryptogram. For example, $\mathbf{A} = \{0, 1\}$, or $\mathbf{A} = \{0, 1, 2, \ldots, 8, 9\}$. Let n be a positive integer so that $|\mathbf{A}^n| \geq |\mathbf{B}|$. Any surjective function $f : \mathbf{A}^n \to \mathbf{B}$ could be an encrypting function. If $\mathbf{B}$ consists of the symbols on my keyboard, $\mathbf{A} = \{0, 1, \ldots, 9\}$, and $n = 2$, then part of such a function $f : \mathbf{A}^2 \to \mathbf{B}$ is defined by

01	02	$\cdots$	26	27	28	$\cdots$	36	37	38	$\cdots$
A	B	$\cdots$	Z	0	1	$\cdots$	9	.		$\cdots$

Then 200805 380301 203809 193802 010311 373838 is a cryptogram for the message THE CAT IS BACK. Even with this short message, an expert cryptanalyst could quickly understand the cryptogram.

If F is any finite field, then one obtains an affine plane $(F \oplus F, \mathcal{L}, \in)$ with lines $L(m, b) = \{(x, mx + b) \mid x \in F\}$ and $L(a) = \{(a, y) \mid y \in F\}$. So a line with slope m and y-intercept b is determined by any two points (x_1, y_1) and (x_2, y_2) with $x_1 \neq x_2$. That is, the parameters (m, b) can be determined by any two distinct points $(x_1, y_1), (x_2, y_2) \in L(m, b)$. Similarly, the parameter a can be determined by any two distinct points $(a, y_1), (a, y_2) \in L(a)$. There are $|F|^2$ distinct lines $L(m, b)$, and $|F|$ distinct lines $L(a)$.

Let $\mathbf{B}$ consist of the symbols on my keyboard, let $\mathbf{C}$ consist of all pairs $(m, b) \in F \times F$ together with all elements $a \in F$, where $|F|^2 + |F| \geq |\mathbf{B}|$. Let $g : \mathbf{C} \to \mathbf{B}$ be any surjective map. Now let $\mathbf{A}$ consist of all pairs of points $[(x_1, y_1), (x_2, y_2)]$ where each $(x_i, y_i) \in F \times F$ and where $(x_1, y_1) \neq (x_2, y_2)$. Let $f : \mathbf{A} \to \mathbf{C}$ be the surjective map defined by $f[(x_1, y_1), (x_2, y_2)] =$

(m, b) if $(x_1, y_1), (x_2, y_2) \in L(m, b)$, and $f[(x_1, y_1), (x_2, y_2)] = a$ whenever $(x_1, y_1), (x_2, y_2) \in L(a)$, that is, if $x_1 = x_2 = a$. Then $g \circ f : \mathbf{A} \to \mathbf{B}$ is an encrypting function that would be considerably more difficult for a cryptanalyst to analyse.

For example, let $F = Z_7$, the prime field with seven elements. Then there are $7^2 + 7 = 56$ lines in the affine plane $(F \oplus F, \mathcal{L}, \in)$ described above. Make believe $56 \geq |\mathbf{B}|$. Part of a function $g : \mathbf{C} \to \mathbf{B}$ is defined by

00	01	$\cdots$	06	10	11	$\cdots$	16	$\cdots$	30	31	$\cdots$
A	B	$\cdots$	G	H	I	$\cdots$	N	$\cdots$	V	W	$\cdots$

34	35	36	40	41	$\cdots$	46	50	51	52	$\cdots$
Z	0	1	2	3	$\cdots$	8	9	.		$\cdots$

Hence

05102244   34241025   42021020   22342533
56230445   64415121   60401222   51253255

is a cryptogram for the message THE CAT IS BACK. How did we obtain this cryptogram? From our partial definition of g, we see that $g(25) = T$, $g(10) = H$, $g(04) = E$, and $g(52)$ is a blank. That is, the line $y = 2x + 5$ represents T, the line $y = x$ represents H, the line $y = 4$ represents E, and the line $y = 5x + 2$ represents a blank. The points $(0, 5), (1, 0) \in L(2, 5)$, hence 0510 denotes T. The points $(2, 2), (4, 4) \in L(1, 0)$, hence 2244 denotes H. The points $(3, 4), (2, 4) \in L(0, 4)$, hence 3424 denotes E. The points $(1, 0), (2, 5) \in L(5, 2)$, hence 1025 denotes a blank. This means 05102244 34241025 denotes THE . Notice that we have also used 2234 for T, 6040 for A, and 2533 for a blank. We could also have used 3422 for T and 4060 for A, etc.

Exercise. Using what is given for $g : \mathbf{C} \to \mathbf{B}$ above and the $f : \mathbf{A} \to \mathbf{C}$, construct a very different cryptogram than that given above for the message THE CAT IS BACK.

Even though cryptograms constructed from these affine planes $(F \oplus F, \mathcal{L}, \in)$ are considerably more complex than the one constructed using $f : \mathbf{A}^2 \to \mathbf{B}$ above, the author feels that a clever cryptanalyst could probably understand a cryptogram of considerable length. That is because the functions f and g are relatively straight forward. The thing to do is to choose affine planes that are not so well known, and/or to make g more complicated.

In §7, planar nearrings $(Z_{p^2}, +, \cdot)$ were shown to exist whose tactical configurations $(Z_{p^2}, \mathcal{B}, \in)$ have p substructures $(Z_{p^2}, \mathbf{L}, \|)_{\phi t}$ which are affine planes. We should be able to do something with each of these affine planes similar to what we have done above. This can add confusion for the crypt-analyst. Affine planes of the form $(Z_{p^2}, \mathbf{L}, \|)_{\phi t}$ are not well known to mathematicians. Also, one has p such affine planes $(Z_{p^2}, \mathbf{L}, \|)_{\phi t}$ and so one could change from one to another while constructing a cryptogram.

Let $N = Z_{p^2}$. The affine plane $(Z_{p^2}, \mathbf{L}, \|)$ is as follows. The cosets of $Np = \{0, p, 2p, \ldots, (p-1)p\}$ form one family of parallel lines. If $m \notin Np$, then $\{Nm, Nm + p, Nm + 2p, \ldots, Nm + (p-1)p\}$ is also a family of parallel lines. Think of the cosets $Np + a$, $0 \le a \le p - 1$, as playing a role similar to the $L(a)$ above. Next, choose a set of representatives $M = \{m_0, m_1, \ldots, m_{p-1}\}$ for the basic blocks $Nm_j \neq Np$. Then each line $Nm_j + b_k p$ is determined by m_j and b_k with $j, k \in \{0, 1, \ldots, p-1\}$. So a line $Nm + bp$ is determined by the two parameters m and b. Think of the line $Nm + bp$ as playing a similar role to the $L(m, b)$ above.

If $x, y \in N$ and $x \neq y$, then there is a unique line in $\mathbf{L}$ containing x and y. If $x - y \in Np$, then $Np + x = Np + y$, so there is a unique a, $0 \le a \le p - 1$, such that $x, y \in Np + a$. If $x - y \notin Np$, then there is a unique ordered pair $(m, b) \in M \times N$ such that $x, y \in Nm + bp$. So either the pair $\{x, y\}$ determines a or the ordered pair (m, b), just as with the affine planes $(F \oplus F, \mathcal{L}, \in)$ above.

Exercise. Consider one of the affine planes $(Z_{p^2}, \mathbf{L}, \|)_t$ from $(Z_{p^2}, \mathcal{B}, \in)$. How can one associate with each pair $x, y \in Z_{p^2} = N$, $x \neq y$, either a parameter a or an ordered pair (m, b) for the line containing x and y?

We have seen that planar nearrings can be used to construct infinitely many unusual finite affine planes in numerous ways. Each finite affine plane has n^2 points and $n^2 + n$ lines. If we can associate with each of these n lines a parameter a, and with each of these n^2 lines an ordered pair (m, b), then one can design encrypting functions similar to $g \circ f : \mathbf{A} \to \mathbf{C}$ above.

Of course, there is an obvious alternative in designing encrypting functions for an affine plane. Above, we associate each line with a symbol in **B**. Alternatively, we could assign each point x of the affine plane to a symbol in **B**. Again, a pair of distinct points a and b determine a line. So two pairs of distinct points $\{a, b\}$ and $\{a', b'\}$ determine lines $l(a, b)$ and $l(a', b')$ respectively. If $l(a, b) = l(a', b')$ or $l(a, b) \| l(a', b')$, then we do not associate any symbol in **B** with $(\{a, b\}, \{a', b'\})$. But if $l(a, b)$ and $l(a', b')$ intersect at a point x, then we associate with $(\{a, b\}, \{a', b'\})$ the symbol in **B** associated with x.

There is another use we can make of affine planes. Suppose knowing a fixed line will give access to something, for example, a datum, entry to a safe, access to a file. Of course, two distinct points are enough to identify a line, but if more participants are desired, one can insist that all participants, or at least enough participants, provide a point on the desired line. For example, suppose a computer file is accessible from numerous

terminals. Access to this file could be restricted to providing at least two
distinct points identifying the correct line L. That is, if a point $x \in L$
is entered from terminal A and a different point $y \in L \setminus \{x\}$ is entered
from terminal B, then L is identified and access could be permitted. To
be more restrictive, suppose a point $x_i \in L$ must be entered from terminal
A_i, $1 \leq i \leq n$, where $n \geq 2$. If y_j is entered from terminal A_j and $y_j \notin L$,
then access could be denied. Less restrictive is the case where there are n
terminals, and at least $k \geq 2$ distinct $x_{i_1}, x_{i_2}, \ldots, x_{i_k} \in L$ are entered at
terminals $A_{i_1}, A_{i_2}, \ldots, A_{i_k}$ respectively. Then access to the file from each
terminal could be permitted.

Next consider some applications of circular planar nearrings to cryptography. One will see considerable transfer of thought from the affine plane
applications to the circular planar nearring applications.

While it is still fresh in our minds, let us consider such applications as
the last one mentioned for affine planes, where access is permitted upon
correctly identifying the fixed line. Let $(N, +, \cdot)$ be a circular planar nearring with circles in $\mathcal{B}^*$. Three distinct points $x, y, z \in N$ are in at most one
circle $A \in \mathcal{B}^*$. So if access is permitted upon knowing a designated circle
A, three distinct points of A is enough. Again, access can be restricted
unless all n entries, $n \geq 3$, belong to the correct circle A.

Recall that a finite circular planar nearring $(N, +, \cdot)$ always gives a BIBD
with parameters $b = v(v-1)/k$, $r = v - 1$, and $\lambda = k - 1$, where $v = |N|$
and $k = |\Phi| = |N^*/{=_m}|$. See (5.5). Let $\mathbf{B}$ denote the set of symbols
on my keyboard, and let $\mathbf{C}$ denote the points in N, where we must have
$|N| \geq |\mathbf{B}|$. Let $g : \mathbf{C} \to \mathbf{B}$ be any surjective mapping. So each point $x \in N$
is associated with exactly one symbol of $\mathbf{B}$. There are $r = v - 1$ distinct
circles $A_1, A_2, \ldots, A_r \in \mathcal{B}^*$ containing an arbitrary point $x \in N$. If $k \geq 4$,
then the points of each $A_i \setminus \{x\}$ will not only identify A_i, by circularity, but
will also identify x, by taking the complement in A_i of $A_i \setminus \{x\}$. So we let
$\mathbf{A}$ consist of all $A \setminus \{t\}$ where $A \in \mathcal{B}^*$ and $t \in A$. The function $f : \mathbf{A} \to \mathbf{C}$
is as follows. For $A \setminus \{t\} \in \mathbf{A}$, let $f(A \setminus \{t\}) = t$. Then $g \circ f : \mathbf{A} \to \mathbf{B}$ is
our encrypting function.

A variation of this last idea is to let $\mathbf{C}$ denote the set of pairs $\{x, y\} \subset N$,
with $x \neq y$. We must have $\binom{v}{2} \geq |\mathbf{B}|$. Let $g : \mathbf{C} \to \mathbf{B}$ be any surjective
mapping. So each pair $\{x, y\} \subset N$, $x \neq y$, is associated with exactly one
symbol of $\mathbf{B}$. There are exactly $\lambda = k - 1$ distinct circles $A_1, A_2, \ldots, A_\lambda \in$
$\mathcal{B}^*$ containing x and y, for any pair $x, y \in N$, $x \neq y$. If $k \geq 5$, then the
points of each $A_i \setminus \{x, y\}$ will not only identify A_i, by circularity, but will
also identify the pair $\{x, y\}$, by taking the complement in A_i of $A_i \setminus \{x, y\}$.
We let $\mathbf{A}$ consist of all $A \setminus \{x, y\}$ where $A \in \mathcal{B}^*$, $x, y \in A$, and $x \neq y$. The
function $f : \mathbf{A} \to \mathbf{C}$ is defined by $f(A \setminus \{x, y\}) = \{x, y\}$. So $g \circ f : \mathbf{A} \to \mathbf{B}$
is our encrypting function.

Let us consider variations on these last two ideas. First, consider the case where $\mathbf{C}$ denotes the points in N and $g : \mathbf{C} \to \mathbf{B}$ is surjective. We need $\mathbf{A}$ to consist of ways to identify a point $x \in N = \mathbf{C}$. If circles $A, B \in \mathcal{B}^*$ are tangent at x, then we could take $f\{A, B\} = x$. So let $\mathbf{A}$ consist of all pairs $\{A, B\}$, with $A, B \in \mathcal{B}^*$, and $|A \cap B| = 1$. Then $g \circ f : \mathbf{A} \to \mathbf{B}$ is our encrypting function.

Next, consider the case where $\mathbf{C}$ consists of the $\binom{v}{2}$ pairs $\{x, y\} \subset N$ with $x \neq y$. We need $\mathbf{A}$ to consist of ways to identify the pair $\{x, y\} \in \mathbf{C}$. If circles $A, B \in \mathcal{B}^*$ have $A \cap B = \{x, y\}$, then we could take $f\{A, B\} = \{x, y\}$. So let $\mathbf{A}$ consist of all pairs $\{A, B\}$, with $A, B \in \mathcal{B}^*$, and $|A \cap B| = 2$. Then $g \circ f : \mathbf{A} \to \mathbf{B}$ is our encrypting function.

A circle $N^*m + b \in \mathcal{B}^*$, where $(N, +, \cdot)$ is a circular planar nearring, can be represented by two parameters, a radius m and the centre b. Suppose we fix a set M of representatives m for the basic blocks N^*m. We now have a set $M \times N$ of representatives of the circles of $\mathcal{B}^*$. Each $(m, b) \in M \times N$ corresponds to exactly one circle $N^*m + b$, and each circle $A \in \mathcal{B}^*$ corresponds to exactly one ordered pair $(m, b) \in M \times N$ where $A = N^*m + b$.

With $|M \times N| \geq |\mathbf{B}|$, we take any surjective mapping $g : M \times N \to \mathbf{B}$. Three distinct points $x, y, z \in N$ identify at most one circle $A \in \mathcal{B}^*$ with $x, y, z \in A$. Now $A = N^*m + b$ for exactly one ordered pair $(m, b) \in M \times N$. So let $\mathbf{A}$ consist of all the triples $\{x, y, z\}$ which are subsets of a circle $A \in \mathcal{B}^*$. If $f\{x, y, z\} = (m, b)$, where $x, y, z \in N^*m + b$, then $g \circ f : \mathbf{A} \to \mathbf{B}$ is an encrypting function.

In place of $f\{x, y, z\} = (m, b)$ in the last paragraph, define $\phi\{x, y, z\} = b$, where $g : N \to \mathbf{B}$ is surjective. Then $g \circ \phi : \mathbf{A} \to \mathbf{B}$ is an encrypting function.

8.3. Planar nearrings and combinatorics

During the winter of 1989, the author ran a workshop on planar nearrings at National Cheng Kung University. (It was during this workshop that the concept of a double planar nearring was formulated. See (7.112).) Fortunately, Tayuan Huang was a most faithful participant. With a student, I-Hsing Chen, Tayuan Huang discovered more applications of planar nearrings to combinatorics, and these are included in I-Hsing Chen's thesis at National Chiao Tung University [I-H]. During the summer of 1991, Tayuan Huang invited the author to run another workshop on planar nearrings at National Chiao Tung University. It was during this time that the author became aware of I-Hsing Chen's thesis, and the ideas we present here are taken from, or motivated by, this thesis.

There are five areas of applications, with four of them being strongly interrelated. First, there is an easy way to construct PBIBDs. The strongly interrelated topics are (a) nets, (b) transversal designs, (c) mutually orthogonal Latin squares, and (d) strongly regular graphs.

In [I-H], I-Hsing Chen constructs a class of PBIBDs, and we extend his construction method here.

Let $(N, +, \cdot)$ be a finite planar nearring with basic blocks $Ng, Ne_1, \ldots,$ Ne_t, and with $|N/_{=m}| = k + 1$. So Ng is the only basic block which is a subgroup of $(N, +)$. With $\mathcal{B} = \{Na + b \mid a, b \in N, a \neq 0\}$, we wish to construct a PBIBD $(N, \mathcal{B}, \in, \mathcal{A})$, so we need an association scheme $\mathcal{A}$.

Define $A_1 = \{\{a, b\} \mid a, b \in N, a \neq b, a - b \in Ng\}$ and $A_2 = \{\{a, b\} \mid a, b \in N, a \neq b, a - b \notin Ng\}$, and set $\mathcal{A} = \{A_1, A_2\}$. To see that $\mathcal{A}$ is an association scheme, we need to show the existence of the parameters p_{ij}^h, $h, i, j \in \{1, 2\}$. See (7.107).

For any $x \in N$, we can display the elements of N by

$$x; \quad z_1, \ldots, z_k; \quad u_1, \ldots, u_{v-1-k}, \tag{8:9}$$

where each $\{x, z_i\} \in A_1$, and each $\{x, u_j\} \in A_2$. If $\{x, y\} \in A_1$, then y is one of the z_i, say $y = z_k$. With each $x - z_i \in Ng$ and $y - x \in Ng$, we obtain $y - z_i \in Ng$, so each $\{y, z_i\} \in A_1$, $1 \leq i \leq k - 1$. Hence, $p_{11}^1 = k - 1$. With each $\{x, u_j\} \in A_2$, and each $\{y, z_i\} \in A_1$, $1 \leq i \leq k - 1$, along with $\{y, x\} \in A_1$, we also have each $\{y, u_j\} \in A_2$, $1 \leq i \leq v - 1 - k$. Consequently, $p_{22}^1 = v - 1 - k$.

Now each $\{x, z_i\} \in A_1$ along with $\{y, x\} \in A_1$ and each $\{y, z_i\} \in A_1$, $1 \leq i \leq k - 1$, so we have $p_{12}^1 = 0$. Similarly, $p_{21}^1 = 0$.

Now consider $\{x, y\} \in A_2$ with $y = u_{v-1-k}$. If $\{x, z_i\} \in A_1$ and $\{y, z_i\} \in A_1$, then $\{x, y\} \in A_1$. Consequently, $p_{11}^2 = 0$. So each $\{y, z_i\} \in A_2$, and we have $p_{12}^2 = k$. A similar argument provides us with $p_{21}^2 = k$.

There is a similar display of elements of N for y, namely,

$$y; \quad z_1', \ldots, z_k'; \quad u_1', \ldots, u_{v-1-k}', \tag{8:10}$$

with each $\{y, z_i'\} \in A_1$ and each $\{y, u_j'\} \in A_2$. With $\{x, y\} \in A_2$, we have each $\{x, z_i'\} \in A_2$, so $p_{21}^2 = k$, as we promised in the last paragraph.

With $\{x, y\} \in A_2$, we have already noted that each $\{y, z_i'\} \in A_1$, so $\{z_1', \ldots, z_k'\} \cap \{z_1, \ldots, z_k\} = \emptyset$. Hence, $\{z_1', \ldots, z_k'\} \subseteq \{u_1, \ldots, u_{v-2-k}\}$. So the $u_j \in \{u_1, \ldots, u_{v-2-k}\} \setminus \{z_1', \ldots, z_k'\}$ are exactly those elements of N where $\{x, u_j\} \in A_2$ and $\{u_j, y\} \in A_2$. Hence, $p_{22}^2 = v - 2 - 2k$.

Summarizing, we have

(8.26) Theorem. *Let $(N, +, \cdot)$ be a finite planar nearring with exactly one basic block Ng as a subgroup of $(N, +)$. Then $\mathcal{A} = \{A_1, A_2\}$, as defined above, is an association scheme for $(N, \mathcal{B}, \in)$ with parameters*

$$p_{11}^1 = k - 1; \quad p_{12}^1 = p_{21}^1 = 0; \quad p_{22}^1 = v - 1 - k;$$

$$p_{11}^2 = 0; \quad p_{12}^2 = p_{21}^2 = k; \quad p_{22}^2 = v - 2 - 2k.$$

Exercise. Take a finite group $(G, +)$ with a proper subgroup $(B, +)$. What requirements must be put on G and B so that one can construct an association scheme $\mathcal{A} = \{A_1, A_2\}$ modelled on our construction for (8.26)?

(8.27) Theorem. *Let* $(N, +, \cdot)$ *and* $\mathcal{A}$ *be as in* (8.26). *Then* $(N, \mathcal{B}, \in, \mathcal{A})$ *is a PBIBD with parameters* p_{ij}^h *as in* (8.26), *and* $n_1 = k$, $n_2 = v - 1 - k$, $\lambda_1 = 1$, *and* $\lambda_2 = k + 1$.

Proof. One can see directly from (8:9) or (8:10) that $n_1 = k$ and $n_2 = v - 1 - k$.

For $x, y \in N$, $x \neq y$, either $\{x, y\} \in A_1$, or $\{x, y\} \in A_2$. The blocks containing $\{x, y\}$ are obtained from those containing $\{0, y - x\}$, as seen in the proof of (7.8). Moreover, the number of blocks containing $\{x, y\}$ is the number of blocks containing $\{0, y - x\}$, that is, $[x, y] = [0, y - x]$. If $y - x \in Ne_i$ for some i, $1 \leq i \leq t$, then $[x, y] = [0, y - x] = k + 1$, and none are of the form Ng_b. See (7.6). By the proof of (7.10), if $y - x \in Ng$, then $\{0, y - x\}$ is a subset of Ng only. In this case, $\{x, y\}$ cannot be a subset of any $Ne_i + b$. Hence, $\lambda_1 = 1$ and $\lambda_2 = k + 1$.

We have seen numerous examples of planar nearrings with only one basic block Ng as a subgroup of $(N, +)$. For example, see (7.39). But (8.27) gives a renewed importance to such finite planar nearrings. For example, we have the Ferrero pair $(N, \Phi) = (Z_{15}, \Phi)$ with $\Phi = \{\pm 1\}$ and with $N5 = \{0, 5, 10\}$. We also have $(N, \Phi) = (Z_{91}, \Phi)$ with $\Phi = \{\pm 1, \pm 9, \pm 10\}$ and with $N13 = \{0, 13, 26, 39, 52, 65, 78\}$.

Suppose we have a partial plane $(X, \mathcal{B}, \in)$. Then it is an $(s, r, 1)$-*net* if $\mathcal{B}$ can be partitioned into r classes $\{\mathcal{B}_1, \mathcal{B}_2, \ldots, \mathcal{B}_r\}$, $r \geq 3$, such that (a) each $\mathcal{B}_i$, $1 \leq i \leq r$, is a partition of X; (b) $|A_i \cap B_j| = 1 - \delta_{ij}$ if $A_i \in \mathcal{B}_i$ and $B_j \in \mathcal{B}_j$ with $A_i \neq B_j$; and (c) $|B| = s$ for each $B \in \mathcal{B}$.

Exercise. Suppose $(X, \mathcal{B}, \in, \|)$ is a partial plane with parallelism which fulfils Euclid's axiom of parallelism. Suppose that $r = |\mathcal{B}/\,\| \,| \geq 3$ and that $|X| < \infty$ with $|B| = s$ for each $B \in \mathcal{B}$. Prove that $(X, \mathcal{B}, \in)$ is an $(s, r, 1)$-net.

Finite planar nearrings $(N, +, \cdot)$ provide easy examples of such nets. Take $X = N \times N$, the cartesian product. For each $a \in N$, define $L(a) = \{(a, y) \mid y \in N\}$ and $\mathcal{B}_\infty = \{L(a) \mid a \in N\}$. Let M be a complete set of representatives of $N/=_m$, with $0 \in M$ representing $[0]=_m$, of course. For $(\alpha, a) \in M \times N$, define $L(\alpha, a) = \{(x, \alpha x + a) \mid x \in N\}$ and $\mathcal{B}_\alpha = \{L(\alpha, a) \mid a \in N\}$. Now set $\mathcal{B} = \mathcal{B}_\infty \cup [\cup_{\alpha \in M} \mathcal{B}_\alpha]$. We are now set up to prove

(8.28) Theorem. *Let* $(N, +, \cdot)$ *be a finite planar nearring with* $k + 1 = |N/=_m|$ *and with* $v = |N|$ *and* $X = N \times N$. *Then for* $\mathcal{B}$ *as defined above,* $(X, \mathcal{B}, \in)$ *is a* $(v, k + 2, 1)$-*net.*

Proof. We number the $\mathcal{B}_i$ as follows: $\mathcal{B}_1 = \mathcal{B}_0$, $\mathcal{B}_2 = \mathcal{B}_{\alpha_1}, \dots, \mathcal{B}_{k+1} = \mathcal{B}_{\alpha_k}$, $\mathcal{B}_{k+2} = \mathcal{B}_\infty$. Then $\mathcal{B} = \mathcal{B}_1 \cup \mathcal{B}_2 \cup \cdots \cup \mathcal{B}_{k+2}$. It is easy to see that $\mathcal{B}$ is partitioned into $k + 2$ classes $\{\mathcal{B}_1, \dots, \mathcal{B}_{k+2}\}$, that $k + 2 \geq 3$, and that each $\mathcal{B}_i$ is a partition of X. Certainly, $|B| = v$ for each $B \in \mathcal{B}$.

If $A_i, B_i \in \mathcal{B}_i$ and $A_i \neq B_i$, then $A_i \cap B_i = \varnothing$, and so $|A_i \cap B_i| = 1 - \delta_{ii}$. Now suppose $A_i \in \mathcal{B}_i$, $B_j \in \mathcal{B}_j$, and $i \neq j$. We first consider $A_i = L(a)$, and $B_j = L(\beta, b)$. Then $A_i \cap B_j = \{(a, \beta a + b)\}$, so $|A_i \cap B_j| = 1 = 1 - \delta_{ij}$. With $A_i = L(\alpha, a)$ and $B_j = L(\beta, b)$, then for $(u, v) \in L(\alpha, a) \cap L(\beta, b)$, we have $(u, v) = (x, \alpha x + a) = (y, \beta y + b)$, and so $u = x = y$, and $\alpha x + b = \beta x + b$. With $i \neq j$, we are assured that $\alpha \neq_m \beta$, and so there is a unique x satisfying $\alpha x = \beta x + (b - a)$. Hence, $|A_i \cap B_j| = 1 = 1 - \delta_{ij}$.

We have done everything except to see that $(X, \mathcal{B}, \in)$ is a partial plane. So take two distinct points (a, b) and (a', b'). If $a = a'$, then $(a, b), (a', b') \in L(a)$, and are not on any $L(c)$ if $c \neq a$. If $a = a'$, then it is impossible for $(a, b), (a', b') \in L(\gamma, c)$. So in this case, (a, b) and (a', b') are on exactly one line.

If $a \neq a'$, then if there is to be an $L(\gamma, c)$ with $(a, b), (a', b') \in L(\gamma, c)$, there are $x, x' \in N$ such that $(a, b) = (x, \gamma x + c)$ and $(a', b') = (x', \gamma x' + c)$. Of course, $x = a$ and $x' = a'$. So the big question is for the existence of (γ, c) so that $\begin{cases} \gamma a + c = b \\ \gamma a' + c = b'. \end{cases}$ From $c = -\gamma a + b = -\gamma a' + b'$, we have $\gamma a' - \gamma a = \gamma(a' - a) = b' - b$. If $b = b'$, let $\gamma = 0$. If $b \neq b'$, there may not be a γ such that $\gamma(a' - a) = b' - b$, but if there is one, there is only one, since $\gamma \in M$. So again, there is at most one $L(\gamma, c)$ with $(a, b), (a', b') \in L(\gamma, c)$. Hence, $(X, \mathcal{B}, \in)$ is a partial plane, and our proof of (8.28) is complete.

To explain a transversal design with r plots of size s, we start with an incidence structure $(X, \mathcal{B}, \in)$. Suppose there is a partition of X with r *plots of size* s, that is, a partition $\mathcal{G} = \{G_1, G_2, \dots, G_r\}$ of X with each $|G_i| = s$, $1 \leq i \leq r$. So $|X| = rs$.

Next, suppose each block $B \in \mathcal{B}$ intersects each G_i at exactly one point. So $B \in \mathcal{B}$ and $G_i \in \mathcal{G}$ imply $|B \cap G_i| = 1$. Finally, suppose two distinct points $x_i, y_j \in X$ belong to at most one block in such a way that $x_i \in G_i$ and $y_j \in G_j$ have $[x_i, y_j] = 1 - \delta_{ij}$ where $[x_i, y_j]$ is the number of distinct blocks of $\mathcal{B}$ containing $\{x_i, y_j\}$. If we have such a structure $(X, \mathcal{B}, \in, \mathcal{G})$, it is called a *transversal design with r plots of size s*, or simply a $TD[r; s]$.

Perhaps something more than the symbolism $1 - \delta_{ij}$ made the reader suspect that transversal designs and nets are related. To see the relationship, let us first talk about the *dual* of an incidence structure $(\mathbf{N}, \mathcal{B}, \in)$. For $n \in \mathbf{N}$ and $B \in \mathcal{B}$, we may have $n \in B$. Now we write $N \in' b$, and finally $b \in N$. So $n \in B$ is equivalent to $b \in N$. Consider the structure $(\mathbf{B}, \mathcal{N}, \in)$ where $\mathbf{B} = \{b \mid B \in \mathcal{B}\}$ and $\mathcal{N} = \{N \mid n \in \mathbf{N}\}$. Then $(\mathbf{B}, \mathcal{N}, \in)$ is the *dual* of $(\mathbf{N}, \mathcal{B}, \in)$. Of course, this is more intuitive than rigorous.

More rigorously, suppose $(\mathbf{N}, \mathcal{B}, I)$ is an incidence structure, so $I \subseteq \mathbf{N} \times \mathcal{B}$, and we write nIB if $(n, B) \in I$. See the remarks just before (5.1). The dual of $(\mathbf{N}, \mathcal{B}, I)$ is $(\mathcal{B}, \mathbf{N}, J)$ where $J = I^{-1}$, that is, $(B, n) \in J$ if and only if $(n, B) \in I$.

With a little patience, one can see that the proof of the following theorem is really quite easy.

(8.29) Theorem. *Let N be a finite planar nearring, and let $(X, \mathcal{B}, \in)$ be the $(v, k + 2, 1)$-net of (8.28). Let $(\mathbf{B}, \mathcal{B}', \in)$ be the dual of $(X, \mathcal{B}, \in)$. Then $(\mathbf{B}, \mathcal{B}', \in)$ is a transversal design with $k+2$ plots of size v, that is, $(\mathbf{B}, \mathcal{B}', \in)$ is a $TD[k + 2; v]$.*

Proof. We will first state our requirements in the language of $(\mathbf{B}, \mathcal{B}', \in)$ and then translate it to the language of $(X, \mathcal{B}, \in)$ in order to see that our requirements are met. We require a partition of $\mathbf{B}$. We require a partition of $\mathcal{B}$. Now $\{\mathcal{B}_1, \ldots, \mathcal{B}_{k+2}\}$ is a partition of $\mathcal{B}$. So let $G_1 = \mathcal{B}_1, \ldots, G_{k+2} = \mathcal{B}_{k+2}$, and then $\{G_1, \ldots, G_{k+2}\}$ is a partition of $\mathbf{B}$. We require each $|G_i|$ to be the same size. Do we have each $|\mathcal{B}_i|$ the same size? Each $|\mathcal{B}_i| = |N| = v$, so each $|G_i| = v$. Hence, $|\mathbf{B}| = v(k + 2) = |\mathcal{B}|$.

We require that $B \in \mathcal{B}'$ and $G_i \in \{G_1, \ldots, G_{k+2}\}$ have $|B \cap G_i| = 1$. We need each $x \in X$ to belong to exactly one line $L \in \mathcal{B}_i$, and this is true since each $\mathcal{B}_i$ is a partition of X.

Finally, we require that two distinct $x_i, y_j \in \mathbf{B}$ belong to at most one $B' \in \mathcal{B}'$ in such a way that $x_i \in G_i$ and $y_j \in G_j$ have $[x_i, y_j] = 1 - \delta_{ij}$. We require that two distinct lines $A_i, B_j \in \mathcal{B}$ have at most one point $x \in X$ in common in such a way that $|A_i \cap B_j| = 1 - \delta_{ij}$.

Hence, $(\mathbf{B}, \mathcal{B}', \in)$ is a $TD[k + 2; v]$.

Our next goal is to construct a family of k distinct mutually orthogonal Latin squares from any $(v, k + 2, 1)$-net of (8.28). Perhaps a few definitions would be helpful.

A $v \times v$-matrix $B = (b_{ij})$ is a *Latin square of order v* if the set of elements of each row $\{b_{s1}, \ldots, b_{sv}\} = \{1, 2, \ldots, v\}$ and the set of elements of each column $\{b_{1t}, \ldots, b_{vt}\} = \{1, 2, \ldots, v\}$. So each row and each column of B defines a permutation of $\{1, 2, \ldots, v\}$. Two Latin squares $B = (b_{st})$ and $C = (c_{st})$ of order v are *orthogonal* if $\{(b_{st}, c_{st}) \mid s, t \in \{1, 2, \ldots, v\}\} = \{(i, j) \mid i, j \in \{1, 2, \ldots, v\}\}$. If $B_1, B_2, \ldots, B_k$ are k distinct Latin squares of order v, and if each pair $B_i, B_j, i \neq j$, is orthogonal, then $B_1, B_2, \ldots, B_k$ are said to be *mutually orthogonal*.

Let $N = \{0 = n_1, n_2, \ldots, n_v\}$ be the elements of a finite planar nearring $(N, +, \cdot)$ of order v with Ferrero pair (N, Φ) having $|\Phi| = k$. Let $M = \{0 = \mu_0, \mu_1, \ldots, \mu_k\}$ be a complete set of representatives of $N/{=_m}$. For our purpose of constructing k mutually orthogonal Latin squares of order k, we introduce the following notation for the $(v, k + 2, 1)$-net $(X, \mathcal{B}, \in)$ of

(8.28).

For $\mu \in M$, let $L(\mu, i) = \{(x, \mu x + n_i) \mid x \in N\}$, $L(\infty, i) = \{(n_i, y) \mid y \in N\}$, $\mathcal{L}(\mu) = \{L(\mu, 1), \ldots, L(\mu, v)\}$, and $\mathcal{L}(\infty) = \{L(\infty, 1), \ldots, L(\infty, v)\}$. Then $\mathcal{B} = \mathcal{L}(0) \cup \mathcal{L}(\mu_1) \cup \cdots \cup \mathcal{L}(\mu_k) \cup \mathcal{L}(\infty)$. So each $\mathcal{L}(\mu)$ corresponds to the family of all 'lines' $y = \mu x + b$, $b \in N$, of 'slope' μ, and $\mathcal{L}(\infty)$ corresponds to the family of all 'vertical' lines $x = a$, $a \in N$.

The next step will be to construct a $(k + 2)$ by v^2 matrix A with entries only from $\{1, 2, \ldots, v\}$. Label the rows $\mathcal{L}(0), \mathcal{L}(\mu_1), \ldots, \mathcal{L}(\mu_k)$ and $\mathcal{L}(\infty)$, respectively. Label the columns with the v^2 elements of $X = N \times N$, namely with $x_1, x_2, \ldots, x_{v^2}$ where

$$x_1 = (n_1, n_1), \quad x_2 = (n_1, n_2), \quad \cdots, \quad x_v = (n_1, n_v),$$

$$x_{v+1} = (n_2, n_1), \quad x_{v+2} = (n_2, n_2), \quad \cdots, \quad x_{2v} = (n_2, n_v),$$

$$\cdots$$

$$x_{(v-1)v+1} = (n_v, n_1), \quad x_{(n-1)v+2}, \quad \cdots, \quad x_{v^2} = (n_v, n_v).$$

Now we must define the $(\mathcal{L}(\mu), (n_s, n_t))$–entry of our matrix A, for $\mu \in \{0, \mu_1, \ldots, \mu_k, \infty\}$. Now $\mathcal{L}(\mu) = \{L(\mu, 1), \ldots, L(\mu, v)\}$ is a partition of X, so there is a unique $L(\mu, u)$ with $(n_s, n_t) \in L(\mu, u)$. The $(\mathcal{L}(\mu), (n_s, n_t))$-entry of A is u. Hence, our $(k + 2)$ by v^2 matrix A has the form

$$\begin{pmatrix} 1 & 2 & \cdots & v & 1 & 2 & \cdots & v & \cdots & 1 & 2 & \cdots & v \\ \cdots & \cdots & \cdots & \cdots & \cdots & \cdots & \cdots & \cdots & \cdots & \cdots & \cdots & \cdots & \cdots \\ 1 & 1 & \cdots & 1 & 2 & 2 & \cdots & 2 & \cdots & v & v & \cdots & v \end{pmatrix}.$$

Perhaps the following proposition will assist in understanding our construction.

(8.30) Proposition. *Each row* **a** *of our matrix A contains each of the values $u \in \{1, 2, \ldots, v\}$ exactly v times.*

Proof. A row **a** of A is labelled with some $\mathcal{L}(\mu)$, $\mu \in \{0, \mu_1, \ldots, \mu_k, \infty\}$, and each $\mathcal{L}(\mu) = \{L(\mu, 1), \ldots, L(\mu, v)\}$ defines a partition of X. Each $L(\mu, u)$ contains exactly v distinct points, and the columns of A labelled with the v distinct points of $L(\mu, u)$ have the entry u in the row **a** corresponding to $\mathcal{L}(\mu)$. This is true for each $u \in \{1, 2, \ldots, v\}$.

The following proposition plays an important role.

(8.31) Proposition. *Let* $\mathbf{a} = (a_1, a_2, \ldots, a_{v^2})$ *and* $\mathbf{b} = (b_1, b_2, \ldots, b_{v^2})$ *be any two distinct rows of A. Then*

$$\{(a_i, b_i) \mid 1 \le i \le v^2\} = \{(i, j) \mid i, j \in \{1, 2, \ldots, v\}\}.$$

Proof. Since each $a_i, b_i \in \{1, 2, \ldots, v\}$, one only needs to show that $(a_i, b_i) = (a_j, b_j)$ for $i \neq j$ is impossible. Suppose $i \neq j$ and $(a_i, b_i) = (a_j, b_j)$. Now $a_i = a_j$ implies the points x_i and x_j of X belong to the same line $L \in \mathcal{L}(\mu)$, where **a** is the row labelled with $\mathcal{L}(\mu)$. Similarly, $b_i = b_j$ implies these points x_i and x_j of X belong to the same line $L' \in \mathcal{L}(\mu')$, where **b** is the row labelled with $\mathcal{L}(\mu') \neq \mathcal{L}(\mu)$. Hence, $|L \cap L'| \geq 2$, contrary to $(X, \mathcal{B}, \in)$ being the $(v, k+2, 1)$-net of (8.28). This completes the proof of (8.31).

We are ready to construct our k distinct v by v matrices $B_1, B_2, \ldots, B_k$. The $\mathcal{L}(\mu_i)$-row of A, $1 \leq i \leq k$ can be written as

$$(b^i_{11}, b^i_{12}, \ldots, b^i_{1v}, b^i_{21}, b^i_{22}, \ldots, b^i_{2v}, \ldots, b^i_{v1}, b^i_{v2}, \ldots, b^i_{vv}), \qquad (8{:}11)$$

which defines $B_i = (b^i_{st})$, $1 \leq i \leq k$. We are ready now to reach our goal.

(8.32) Theorem. *The matrices $B_1, B_2, \ldots, B_k$ are k distinct mutually orthogonal Latin squares of order v.*

Proof. We will first show that each B_i is a Latin square. Consider the elements $\{b^i_{s1}, \ldots, b^i_{sv}\}$ of the sth row of B_i. Since they came from A, each $b^i_{st} \in \{1, 2, \ldots, v\}$. We will see that there are no duplications. If $b^i_{sa} = b^i_{sc} = m$, then b^i_{sa} is in the (n_s, n_a)-column and b^i_{sc} is in the (n_s, n_c)-column of the row for $\mathcal{L}(\mu_i)$. Hence, $n_a = \mu_i n_s + n_m$ and $n_c = \mu_i n_s + n_m$, from which we conclude that $\{b^i_{s1}, b^i_{s2}, \ldots, b^i_{sv}\}$ has v distinct entries, namely, the elements of $\{1, 2, \ldots, v\}$.

Consider the entries $\{b^i_{1t}, b^i_{2t}, \ldots, b^i_{vt}\} \subseteq \{1, 2, \ldots, v\}$ of the tth column of B_i. If $b^i_{at} = b^i_{ct} = m$, then b^i_{at} is in the (n_a, n_t)-column, and b^i_{ct} is in the (n_c, n_t)-column of the row for $\mathcal{L}(\mu_i)$. Hence, $n_t = \mu_i n_c + n_m$, which tells us that $\mu_i n_a = \mu_i n_c$, and since $\mu_i \notin \{0, \infty\}$, we can cheerfully conclude that $a = c$, and so $\{b^i_{1t}, b^i_{2t}, \ldots, b^i_{vt}\} = \{1, 2, \ldots, v\}$. We can now also conclude that each B_i is a Latin square.

For $i \neq j$, we want to show that B_i and B_j are orthogonal. The elements of B_i correspond to the $\mathcal{L}(\mu_i)$-row of A as displayed in (8:11). The elements of B_j have a similar display using the $\mathcal{L}(\mu_j)$-row of A. From (8.31) we conclude that B_i and B_j are orthogonal. This completes the proof of (8.32).

Exercise. Choose a planar nearring $(N, +, \cdot)$ whose order v is not too large. Use the procedure here to construct k mutually orthogonal Latin squares.

In [B & P], one finds another algorithm for constructing mutually orthogonal Latin squares. The authors of [B & P] also promise the reader that families of mutually orthogonal Latin squares are of interest to applied statistics. So this is a particularly nice application of any finite planar nearring.

It only remains to discuss strongly regular graphs obtainable from finite planar nearrings. In order to do this, we will continue to use some of the ideas and notation we used to construct our mutually orthogonal Latin squares, *and*, we need to introduce more terminology.

We will look at graphs more rigorously than we did in §6. Consider an incidence structure $\Gamma = (V, \mathcal{E}, I)$. We say that Γ is a *graph* with *vertices* the elements of V, and with *edges* the elements of $\mathcal{E}$, provided (1) each edge $E \in \mathcal{E}$ is incident with exactly two distinct vertices of V, and (2) for each pair of distinct vertices from V, there is at most one edge of $\mathcal{E}$ incident with each element of this pair of vertices. If $E \in \mathcal{E}$ and $(x, E), (y, E) \in I$, with $x \neq y$, we write $v(E) = \{x, y\}$. If $x, y \in V$ and $x \neq y$, there may or may not be an edge $E \in \mathcal{E}$ with $v(E) = \{x, y\}$. If there is, we write $e\{x, y\} = E$ and also $e\{x, y\} \in \mathcal{E}$. If there is no such edge, we write $e\{x, y\} \notin \mathcal{E}$.

Given a vertex x of a graph $\Gamma = (V, \mathcal{E}, I)$, we say that a vertex $y \in V \setminus \{x\}$ is *adjacent* to x if $e\{x, y\} \in \mathcal{E}$. The graph Γ is *complete* if $e\{x, y\} \in \mathcal{E}$ for every pair $x, y \in V$, $x \neq y$. The graph is *null* if $e\{x, y\} \notin \mathcal{E}$ for every pair $x, y \in V$, $x \neq y$.

Fix $x \in V$ for a graph $\Gamma = (V, \mathcal{E}, I)$. Let $V_\Gamma(x) = \{y \in V \setminus \{x\} \mid e\{x, y\} \in \mathcal{E}\}$. The *valency of x* is $v_\Gamma(x) = |V_\Gamma(x)|$, so $v_\Gamma(x)$ is the number of vertices which are adjacent to x. If there is a constant v_Γ such that $v_\Gamma(x) = v_\Gamma$ for each $x \in V$, then v_Γ is the *valency of Γ*, and Γ is said to be *regular*.

Some regular graphs are *strongly regular*. Suppose $\Gamma = (V, \mathcal{E}, I)$ is a regular graph which is neither complete nor null. We say that Γ is *strongly regular* if there are constants α and β such that, for $x, y \in V$, $x \neq y$,

$$|V_\Gamma(x) \cap V_\Gamma(y)| = \begin{cases} \alpha, & \text{if } e\{x, y\} \in \mathcal{E}; \\ \beta, & \text{if } e\{x, y\} \notin \mathcal{E}. \end{cases} \tag{8:12}$$

That is, the number of vertices adjacent to both x and y depends only upon whether x and y are adjacent or not.

For a finite planar nearring $(N, +, \cdot)$, let $(X, \mathcal{B}, \in)$ be the $(v, k+2, 1)$-net of (8.28). We want to construct a strongly regular graph $\Gamma = (V, \mathcal{E}, I)$ from $(X, \mathcal{B}, \in)$. First, let $V = X$. For $x, y \in V = X$, $x \neq y$, we say that $e\{x, y\} \in \mathcal{E}$ if and only if there is a 'line' $L \in \mathcal{B}$ such that $x, y \in L$. So $\mathcal{E} = \{\{x, y\} \mid x, y \in X, x \neq y, \text{ and } \{x, y\} \subseteq L \text{ for some } L \in \mathcal{B}\}$. Since $\mathcal{B} \neq \varnothing$ and each line $L \in \mathcal{B}$ has $|L| = v > 2$, we are assured that $\mathcal{E} \neq \varnothing$ and so Γ is not null. If $(X, \mathcal{B}, \in)$ is an affine plane, then Γ is complete, so we do not want $(X, \mathcal{B}, \in)$ to be an affine plane. To be assured of this, we need only ensure that $k < v - 1$, so there will not be enough families of parallel lines $\mathcal{L}(\mu)$. Now, with $k < v - 1$, there are points $x, y \in X = V$, $x \neq y$, not on any line $L \in \mathcal{B}$. Hence, $e\{x, y\} \notin \mathcal{E}$, and so Γ is not complete.

There are v^2 vertices in $V = X$. For a fixed $x \in V$, x belongs to exactly one line from each of the $k + 2$ $\mathcal{L}(\mu)$, $\mu \in \{0, \mu_1, \mu_2, \dots, \mu_k, \infty\}$, and each of these lines containing x contains $v - 1$ additional points. Two distinct points belong to *at most* one line of $\mathcal{B}$, so Γ has valency $v_\Gamma = (k+2)(v-1)$. Hence, Γ is regular.

Suppose $e\{x, y\} \notin \mathcal{E}$. Let $\mathcal{B}(x) = \{L_1, \dots, L_{k+2}\}$ be the $k + 2$ lines of $\mathcal{B}$ on x, and let $\mathcal{B}(y) = \{M_1, \dots, M_{k+2}\}$ be the $k + 2$ lines of $\mathcal{B}$ on y in such a way that each $L_i \cap M_i = \varnothing$, that is, L_i and M_i are parallel. If $i \neq j$, then $L_i \cap M_j = \{p_{ij}\}$ for some unique vertex $p_{ij} \in V = X$. So $p_{ij} \in V_\Gamma(x) \cap V_\Gamma(y)$.

If it should be that $\{p_{ij}\} = \{p_{st}\}$, then we would have $x, p_{ij} \in L_i \cap L_s$ and $y, p_{ij} \in M_j \cap M_t$. Hence, $L_i = L_s$ and $M_j = M_t$, so $(i, j) = (s, t)$. So distinct pairs (i, j), $i \neq j$, yield distinct parameters p_{ij}. Hence, $\beta = (k + 2)(k + 1)$, that is, $|V_\Gamma(x) \cap V_\Gamma(y)| = (k + 2)(k + 1)$ if $e\{x, y\} \notin \mathcal{E}$.

We turn now to the case where $e\{x, y\} \in \mathcal{E}$. With $\mathcal{B}(x)$ and $\mathcal{B}(y)$ as above, except that now we take $L_1 = M_1$, and $L_i \cap M_i = \varnothing$ for $2 \leq i \leq k+2$, we have $L_i \cap M_j = \{p_{ij}\}$ for $i, j \in \{2, 3, \dots, k + 2\}$, and $i \neq j$. The same argument puts $p_{ij} \in V_\Gamma(x) \cap V_\Gamma(y)$ and gives us distinct parameters p_{ij} for distinct pairs (i, j), $i \neq j$. These parameters p_{ij} account for $(k+1)k$ points of $V_\Gamma(x) \cap V_\Gamma(y)$, and none of these p_{ij} are in $L_1 = M_1$. Of course, the $v - 2$ points of $L_1 \setminus \{x, y\}$ are also in $V_\Gamma(x) \cap V_\Gamma(y)$, and this accounts for all such points. Hence, $|V_\Gamma(x) \cap V_\Gamma(y)| = (v - 2) + (k + 1)k = \alpha$. In summary,

(8.33) Theorem. *Let $(N, +, \cdot)$ be a finite planar nearring. From the $(v, k + 2, 1)$-net of (8.28), we construct the graph $\Gamma = (V, \mathcal{E}, I)$ as described above. If $k < v - 1$, then Γ is a strongly regular graph with valency $v_\Gamma = (k+2)(v-1)$ and parameters α, β of $(8:12)$ given by $\alpha = (v-2)+(k+1)k$ and $\beta = (k + 2)(k + 1)$.*

The strongly regular graphs of (8.33) are *connected*. That is, if $x, y \in V$ and $x \neq y$, then there is a sequence of vertices $(x_0, x_1, \dots, x_n)$ where $x_0 = x$, $x_n = y$, and $e\{x_i, x_{i+1}\} \in \mathcal{E}$ for each i, $0 \leq i \leq n-1$. To see that our graphs are connected, one only needs to consider two cases. First, if $e\{x, y\} \in \mathcal{E}$, we let $n = 1$ and (x_0, x_1) is the required sequence. Second, if $e\{x, y\} \notin \mathcal{E}$, we take any line $L \in \mathcal{B}$ on x and any line $M \in \mathcal{B}$ on y which is not parallel to L. Then $L \cap M = \{x_1\}$. In this case, let $n = 2$, and (x_0, x_1, x_2) is the required sequence.

9. Sharply transitive groups and nondesarguesian geometry from nearfields

In this section, we wish to provide an introduction to two very nice topics, the use of nearfields in constructing sharply transitive groups and the use of nearfields in geometry. One could argue that the applications to geometry came first, but we shall find it more comfortable to reverse the order in our presentation here.

Sharply transitive groups

For about 120 years, the topic of sharply transitive groups has been of interest to mathematicians. Research efforts have produced a very nice theory, but the author knows of no single source that brings all the theory together. This introduction will not change that point. Let us begin with a little terminology.

Let G be a group of permutations on a set X, and let k be a positive integer. We say that G is k-*transitive* on X if for every pair of k-tuples $(x_1, x_2, \ldots, x_k)$ and $(y_1, y_2, \ldots, y_k)$ having distinct entries in X, there is a $g \in G$ with $gx_i = y_i$ for each i, $1 \leq i \leq k$. If each such pair $(x_1, x_2, \ldots, x_k)$ and $(y_1, y_2, \ldots, y_k)$ has a *unique* $g \in G$ with $gx_i = y_i$ for each i, $1 \leq i \leq k$, then one says that G is *sharply k-transitive*.

An obvious problem is to determine all the sharply k-transitive groups. It is rather easy to summarize the results for $k \geq 4$, and especially easy if $k \geq 6$.

There are no infinite sharply k-transitive groups for $k \geq 4$. See [W].

The only finite sharply k-transitive groups for $k \geq 6$, are the symmetric groups S_k and S_{k+1}, and the alternating group A_{k+2}. See [W] and [B & H].

The only finite sharply 5-transitive groups are S_5, S_6, A_7, and the Mathieu group M_{11}. The only finite sharply 4-transitive groups are S_4, S_5, A_6, and the Matheiu group M_{10}. See [W], [B & H], and [R].

On the face of it, it is rather easy to summarize the results for $k = 2$ and $k = 3$.

The sharply 2-transitive groups are exactly the permutation groups

$$T_2(F) = \{[b, m] : F \to F \mid [b, m]x = b + mx,\ m \in F^*,\ b \in F\}$$

where $(F, +, \cdot)$ is a neardomain, which shall be defined shortly.

So the problem of finding the sharply 2-transitive groups is now the problem of finding the neardomains $(F, +, \cdot)$. As we shall see in this section, the finite neardomains are exactly the nearfields. All nearfields are neardomains, but it is not known if there even is a neardomain which is not a nearfield. See [W]. So, what are the nearfields? This is not easy to answer, but Zassenhaus determined all the finite ones [Z]. This alone is a very interesting theory. The best exposition of constructing the finite nearfields is to be found in [W].

The sharply 3-transitive groups correspond to KT-fields, and a KT-field is a neardomain $(F, +, \cdot)$ with an automorphism ϵ of the group $(F^*, \cdot)$, where $F^* = F \setminus \{0\}$, and where $\epsilon(1 - \epsilon(x)) = 1 - \epsilon(1 - x)$ for all $x \in$

$F \setminus \{0, 1\}$. As we saw in §7, the linear fractional transformations, or Möbius transformations, provide examples of sharply 3-transitive groups. As shown in [W], all sharply 3-transitive groups are constructed in the same spirit.

Every group $(G, \cdot)$ is a sharply 1-transitive permutation group on itself. For $g \in G$, the map $g \mapsto \pi_g$, where $\pi_g x = gx$, is an isomorphism of G onto a subgroup G' of S_G, the group of all permutations on G. If $a, b \in G$, then the equation $ya = b$ has a unique solution, so π_y is the unique permutation from G' such that $\pi_y a = b$. This completes our summary of sharply k-transitive groups.

A neardomain is a generalization of a nearfield. One relaxes the associativity for $+$. The material presented here about neardomains and nearfields is highly influenced by material found in [K] and [W]. Karzel's influence here is profound.

(9.1) Definition. Suppose we start with $(F, +, \cdot)$, where $+$ and $\cdot$ are binary operations on a set F with $|F| \geq 2$. We shall require the following to be true.

1. $(F, +)$ is a loop, that is, there is an identity $0 \in F$ for $+$, and if $a, b \in F$ are arbitrary, then the equations $a + x = b$ and $y + a = b$ have unique solutions in F.
2. $(F^*, \cdot)$ is a group with identity 1, where $F^* = F \setminus \{0\}$.
3. $a \cdot (b + c) = (a \cdot b) + (a \cdot c)$ for all $a, b, c \in F$.
4. $0 \cdot a = 0$ for all $a \in F$.
5. For $a, b \in F$, let $d_{a,b}$ be the solution to

$$a + (b + 1) = (a + b) + x.$$

Then $a + (b + c) = (a + b) + d_{a,b}c$ for each $c \in F$, and $d_{a,b} = 1$ if $a + b = 0$. If $(F, +, \cdot)$ satisfies the above, then $(F, +, \cdot)$ is called a *neardomain*.

Certainly, every nearfield is a neardomain, and a neardomain will be a nearfield exactly when $d_{a,b} = 1$ for all $a, b \in F$.

The value of neardomains is partly explained by the following theorem.

(9.2) Theorem. *Let $(F, +, \cdot)$ be a neardomain, and let*

$$T_2(F) = \{[b, m] : F \to F \mid [b, m]x = b + mx, \ b \in F, \ m \in F^*\}.$$

Then $T_2(F)$ is a sharply 2-transitive permutation group on the set F. The $[b, m]$ are called linear mappings on F.

Proof. Now $a \cdot 0 = a \cdot (0 + 0) = (a \cdot 0) + (a \cdot 0)$, so 0 and $a \cdot 0$ are solutions to $(a \cdot 0) + x = a \cdot 0$. Hence, $a \cdot 0 = 0$. Suppose b is the solution to $a + x = 0$, so $a + b = 0$ and $d_{a,b} = 1$. Then $a + (b + a) = (a + b) + d_{a,b}a = 0 + 1 \cdot a = a$. Also, $a + 0 = a$, hence $b + a = 0$. Let $-a$ denote the common solution to the equations $a + x = 0$ and $y + a = 0$.

Since $[0,1]x = 0 + 1x = x$, we have $[0,1] = 1_F \in T_2(F)$. From $[b,n] \circ [a,m]x = b + n(a + mx) = b + (na + nmx) = (b + na) + d_{b,na}nmx = [b + na, d_{b,na}nm]x$, we obtain

$$[b,n] \circ [a,m] = [b + na, d_{b,na}nm] \tag{9:1}$$

and that $\circ$ is a binary operation on $T_2(F)$, that is, $T_2(F)$ is closed with respect to $\circ$.

Using (9:1), we have $[a,m][-m^{-1}a, m^{-1}] = [0,1] = [-m^{-1}a, m^{-1}][a,m]$. Hence,

$$[a,m]^{-1} = [-m^{-1}a, m^{-1}] \tag{9:2}$$

and the $[a,m]$ are permutations of F and $T_2(F)$ is a group.

Suppose $a, b \in F$ with $a \neq b$. Then $a + x = b$ has a unique solution $x = m$. Now $[a,m]0 = a$ and $[a,m]1 = b$. So $[a,m] \in T_2(F)$ sends 0 to a and 1 to b. If $[a',m']$ does also, then $a = [a',m']0 = a' + m'0 = a'$ and $b = [a',m']1 = [a,m']1 = a + m'$. Hence, $m' = m$, and so $[a,m]$ is the unique element of $T_2(F)$ sending 0 to a and 1 to b.

If $c, d \in F$ with $c \neq d$, then there is a unique $[c,n] \in T_2(F)$ such that $[c,n]0 = c$ and $[c,n]1 = d$. Hence, $[a,m][c,n]^{-1}$ is the unique element of $T_2(F)$ sending c to a and d to b. This makes $T_2(F)$ sharply 2-transitive.

This next theorem, a converse of (9.2), closes a chapter on the story of sharply 2-transitive groups. Our proof will be a consequence of numerous lemmas, however.

(9.3) Theorem. *Let G be a sharply 2-transitive permutation group on a set M. Then there are two binary operations, $+$ and $\cdot$, on M so that $(M, +, \cdot)$ is a neardomain, and from M we obtain $T_2(M) = G$.*

Proof. Let $J = \{\alpha \in G \mid \alpha^2 = 1, \alpha \neq 1\}$. The proof will now be given in several steps. The hypotheses of (9.3) will continue until we have completed the proof.

(9.4) Lemma. *For $p, q \in M$, $p \neq q$, there is a unique $\alpha \in J$ with $\alpha p = q$.*

Proof. Applying the sharply 2-transitivity of G to (p,q) and (q,p), we have the existence of a unique $\alpha \in G$ such that $\alpha p = q$ and $\alpha q = p$. Hence, $\alpha^2 p = p$ and $\alpha^2 q = q$. But $1_M p = p$ and $1_M q = q$, so $\alpha^2 = 1_M$. Since $p \neq q$ and $\alpha p = q$, we have $\alpha \neq 1_M$. Hence, $\alpha \in J$.

(9.5) Lemma. *If $\alpha, \beta \in J$, $p \in M$, and $\alpha p = \beta p$, then $\alpha = \beta$.*

Proof. First, consider the case where $\alpha p \neq p$. By (9.4), there is a unique $\beta \in J$ such that $\beta p = \alpha p$. But $\alpha \in J$ and $\alpha p = \alpha p$, so $\alpha = \beta$.

Next, suppose $\alpha p = \beta p = p$. For $q \in M \setminus \{p\}$, we have $\alpha q \neq q$, for if $\alpha p = p$ and $\alpha q = q$, then $\alpha = 1_M$. So $\alpha q \neq q$ for each $q \in M \setminus \{p\}$. With

$q \neq p$, there is a unique $\gamma \in G$ such that $\gamma q = q$ and $\gamma(\alpha q) = \beta q$, since $\beta q \neq q$ for the same reasons that $\alpha q \neq q$. We apply sharp 2-transitivity to $(q, \alpha q)$ and $(q, \beta q)$ to obtain this γ. Note that $\gamma \alpha \gamma^{-1} \in J$, and $\gamma \alpha \gamma^{-1} q = \gamma \alpha q = \beta q$, so by (9.4), $\gamma \alpha \gamma^{-1} = \beta$, which means that $\gamma \alpha = \beta \gamma$. This in turn means that $\gamma \alpha p = \gamma p = \beta \gamma p$, and so γp is a fixed point for β. But p is β's only fixed point, so $\gamma p = p$. Since $p \neq q$, $\gamma p = p$, and $\gamma q = q$, we have $\gamma = 1_M$. From $\gamma \alpha \gamma^{-1} = \beta$, we now obtain $\alpha = \beta$.

(9.6) Lemma. *For $\alpha, \beta \in J$ with $\alpha(\beta p) = p$, we obtain $\alpha \beta = 1_M$ and $\alpha = \beta$.*

Proof. If $\beta p = p$, then $p = \alpha(\beta p) = \alpha p$, so $\alpha = \beta$ and $\alpha \beta = 1_M$. If $\beta p = q \neq p$, then $\beta q = p = \alpha(\beta p) = \alpha q$, so by (9.5), $\alpha = \beta$.

(9.7) Lemma. *All elements in J are conjugate, that is, if $\alpha, \beta \in J$, then there is a $\gamma \in G$ such that $\beta = \gamma \alpha \gamma^{-1}$.*

Proof. Suppose $p \neq \beta p$ and $p \neq \alpha p$. Then there is a unique $\gamma \in G$ such that $\gamma p = p$ and $\gamma \alpha p = \beta p$. With $\gamma \alpha \gamma^{-1} \in J$ and $\gamma \alpha \gamma^{-1} p = \gamma \alpha p = \beta p$, we obtain $\gamma \alpha \gamma^{-1} = \beta$ by (9.5).

Next, suppose $\alpha p = p$ and $\beta p \neq p$. If $q \in M \setminus \{p\}$ implies $\beta q \neq q$, then since $\alpha q \neq q$, we can apply the argument of the preceding paragraph to obtain $\gamma \alpha \gamma^{-1} = \beta$. So we may assume the existence of a $q \in M \setminus \{p\}$ such that $\beta q = q$. Now there is a $\gamma \in J$ such that $\gamma p = q$, so $\gamma \alpha \gamma^{-1} q = \gamma \alpha p = \gamma p = q = \beta q$, and since $\gamma \alpha \gamma^{-1} \in J$, we have $\gamma \alpha \gamma^{-1} = \beta$.

(9.8) Lemma. *If $\alpha \in J$ has a fixed point, then every $\beta \in J$ has a fixed point, and to each $q \in M$, there is a unique $\lambda \in J$ with $\lambda q = q$.*

Proof. Suppose $\alpha p = p$, $\beta \in J$, and $\gamma \in G$ with $\beta = \gamma \alpha \gamma^{-1}$. Then $\beta \gamma = \gamma \alpha$ and $\beta \gamma p = \gamma \alpha p = \gamma p$, so γp is a fixed point for β. Let $q \in M$. Then there is a unique $\gamma \in J$ with $\gamma p = q$ by (9.4). Let $\lambda = \gamma \alpha \gamma^{-1}$, so $\lambda q = \gamma \alpha \gamma^{-1} q = \gamma \alpha p = \gamma p = q$, and Certainly, $\lambda \in J$. Hence, $q \in M$ is a fixed point for $\lambda \in J$.

(9.9) Lemma. *For $p \in M$, define $H_p = \{\lambda \in G \mid \lambda p = p\}$. Then H_p is a subgroup of G and H_p is regular on $M \setminus \{p\}$, that is, if $q_1, q_2 \in M \setminus \{p\}$, then there is a unique $\lambda \in H_p$ such that $\lambda q_1 = q_2$.*

Proof. For $q_1, q_2 \in M \setminus \{p\}$, there is a unique $\gamma \in G$ with $\gamma p = p$ and $\gamma q_1 = q_2$, and so $\gamma \in H_p$. Also, $1_M \in H_p$. If $\lambda \in H_p$, then $\lambda p = p$ and so $\lambda^{-1} p = p$ putting $\lambda^{-1} \in H_p$. If $\mu \in H_p$, then $\mu \lambda p = \mu p = p$, so $\mu \lambda \in H_p$. Now we have (9.9).

(9.10) Definition. Choose $0, 1 \in M$, $0 \neq 1$. If each element of J has a fixed point (see (9.8)), let $\omega \in J$ have $\omega(0) = 0$, and then define $A = J\omega$.

If no element of J has a fixed point (see (9.8) again), define $A = J \cup \{1_M\}$. So

$$A = \begin{cases} J\omega, & \text{if there is a } \omega \in J \text{ such that } \omega 0 = 0; \\ J \cup \{1_M\}, & \text{otherwise.} \end{cases}$$

(9.11) Lemma. *For each $a \in M$, there is a unique $a^+ \in A$ such that $a^+(0) = a$.*

Proof. First assume $A = J \cup \{1_M\}$. If $a = 0$, then $a^+ = 0^+ = 1_M$, since $0^+(0) = 0$ implies 0 is a fixed point for 0^+. If $a \neq 0$, then there is a unique $\alpha \in J$ such that $\alpha 0 = a$ by (9.4). Let $a^+ = \alpha \in J \subset A$. Now assume $A = J\omega$, so for $a = 0$, let $a^+ = \omega\omega = 1_M$. If $a \neq 0$, then, by (9.4), there is a unique $\alpha \in J$ such that $\alpha 0 = a$. Let $a^+ = \alpha\omega$, so $a^+(0) = \alpha\omega 0 = \alpha 0 = a$.

(9.12) Lemma. *If $a^+ p = p$ for some $p \in M$, then $a^+ = 1_M$.*

Proof. For $A = J \cup \{1_M\}$, then $a^+ p = p$ implies $a^+ \in J$ or $a^+ = 1_M$. If $a^+ \in J$, then a^+ has a fixed point, contrary to $A = J \cup \{1_M\}$. For $A = J\omega$, then $a^+ = \alpha\omega$ for some $\alpha \in J$. From $p = a^+ p = \alpha\omega p = \alpha(\omega p)$, we apply (9.6) to conclude $1_M = \alpha\omega = \alpha^+$.

(9.13) Lemma. (a) *For each $a \in M \setminus \{0\} = M^*$, there is a unique $a^\bullet \in H_0$ such that $a^\bullet(1) = a$.*
(b) *For each $\gamma \in H_0$, there is a unique $c \in M^*$ such that $\gamma = c^\bullet$.*

Proof. From (9.9), the regularity of H_0 on M^* gives a unique $\lambda \in H_0$ such that $\lambda 1 = a$. Let $a^\bullet = \lambda$ for $a \neq 0$. This gives us (a). For $g \in H_0$, let $c = \gamma 1$. Then $0 \neq c \in M^*$, and by (a), $c^\bullet(1) = c$ and $c^\bullet \in H_0$. Again by regularity from (9.9), we conclude $\gamma = c^\bullet$.

(9.14) Definition. Define $+$ and $\cdot$ on M by $a + b = a^+(b)$, and

$$a \cdot b = \begin{cases} 0, & \text{if } a = 0; \\ a^\bullet(b), & \text{if } a \neq 0. \end{cases}$$

(9.15) Lemma. $(M^*, \cdot)$ *is a group.*

Proof. Define $\Psi : M^* \to H_0$ by $\Psi a = a^\bullet$, a well defined function by (9.13). For $a, b \in M$, we have $\Psi(a \cdot b) = \Psi(a^\bullet(b)) = [a^\bullet(b)]^\bullet$ and $\Psi(a) \circ \Psi(b) = a^\bullet \circ b^\bullet$. Now $a^\bullet, b^\bullet, [a^\bullet(b)]^\bullet \in H_0$, since $a, b, a^\bullet(b) \in M^*$. Hence, $a^\bullet \circ b^\bullet \in H_0$. Note that $[a^\bullet(b)]^\bullet 1 = a^\bullet(b)$, and $a^\bullet \circ b^\bullet 1 = a^\bullet(b^\bullet 1) = a^\bullet(b)$, so by (9.9), we have $\Psi(a \cdot b) = [a^\bullet(b)]^\bullet = a^\bullet \circ b^\bullet = \Psi(a) \circ \Psi(b)$ and Ψ is a homomorphism. From $\Psi(a) = a^\bullet = b^\bullet = \Psi(b)$, we have $a^\bullet(1) = a$ and $b^\bullet(1) = b$, so $a = b$, and, therefore, Ψ is injective. From (b) of (9.13), we know that Ψ is surjective. Hence, Ψ is an isomorphism making $(M^*, \cdot)$ a group.

(9.16) Lemma. $(M, +)$ *is a loop.*

Proof. From $0 + a = o^+(a) = 1_M a = a$ and $a + 0 = a^+(0) = a$, we know that 0 is an identity for $+$. Since a^+ is a permutation on M, there is a unique c such that $a^+(c) = b$, where b is an arbitrary but fixed element of M. That is to say, the equation $a + x = b$ has a unique solution for x, no matter which $a, b \in M$ are chosen.

For $a, b \in M$, we now consider the equation $y + a = b$. We will treat this with the two possibilities for A of (9.10). First, let $A = J\omega$, so each element in J has a fixed point, and so there is a unique $\alpha \in J$ such that $\alpha(\omega(a)) = b$ unless $\omega(a) = b$, and even if $\omega(a) = b$, then by (9.8), there is a unique $\alpha \in J$ such that $\alpha\omega(a) = b$. Let $y = \alpha\omega(0)$. So $y + a = y^+(a)$ and $y^+ = \alpha\omega$, so $y^+(a) = \alpha(\omega(a)) = b$. Hence, $y + a = b$. Since α is unique, $y + a = b$ has a unique solution for y, namely, $y = \alpha(\omega(0))$.

Now let $A = J \cup \{1_M\}$. We have already seen that $y + a = a$ has $y = 0$ as a solution. If z is also a solution, then $z^+(a) = a$ and so $z^+ = 1_M$ since each element of J does not have a fixed point. But $0^+ = 1_M$, so $z = 0$. This means that the solution to $y + a = a$ is unique. If $a \neq b$, then by (9.4), there is a unique $\alpha \in J$ such that $\alpha a = b$. Let $y = \alpha(0)$. Then $y^+ = \alpha \in A$ and $b = \alpha a = y^+(a) = y + a$. Since α is unique, then y is unique, and so $y + a = b$ has a unique solution for y.

Hence, $(M, +)$ is a loop.

(9.17) Lemma. *For arbitrary* $a, b, c \in M$, $a \cdot (b + c) = (a \cdot b) + (a \cdot c)$.

Proof. Since $0 \cdot (b + c) = 0 = 0 + 0 = (0 \cdot b) + (0 \cdot c)$, we may assume that $a \neq 0$. Now $a(b+c) = a^\bullet(b^+(c)) = a^\bullet b^+(c)$ and $(ab)+(ac) = a^\bullet(b)+a^\bullet(c) = [a^\bullet(b)]^+(a^\bullet(c)) = [a^\bullet(b)]^+ \circ a^\bullet(c)$. So we see that we will be finished once we have proven that $a^\bullet b^+ = [a^\bullet(b)]^+ \circ a^\bullet$.

First note that $[a^\bullet(b)]^+ \in A$ and that $[a^\bullet(b)]^+(0) = a^\bullet(b) = ab$, and $a^\bullet b^+ a^{\bullet -1}(0) = a^\bullet b^+(0) = a^\bullet(b) = ab$. If we show that $a^\bullet b^+ a^{\bullet -1} \in A$, then we will have $a^\bullet b^+ a^{\bullet -1} = [a^\bullet(b)]^+$, or, equivalently, $a^\bullet b^+ = [a^\bullet(b)]^+ a^\bullet$, and our proof of (9.17) will be complete.

To see that $a^\bullet b^+ a^{\bullet -1} \in A$, first consider the case where $A = J \cup \{1_M\}$. If $b^+ = 0^+$, that is, if $b = 0$, then $b^+ = 1_M$ and $a^\bullet b^+ a^{\bullet -1} = 1_M \in A$. If $b^+ \neq 0^+$, then $b^+ \in J$, and so $a^\bullet b^+ a^{\bullet -1} \in J$ since $a^\bullet \in G$. Hence, $a^\bullet b^+ a^{\bullet -1} \in A$ if $A = J \cup \{1_M\}$. Next, consider the case where $A = J\omega$. Since $b^+ \in A$, then $b^+ = \beta\omega$ for some $\beta \in J$. Also, $a^\bullet \in H_0$, so $a^\bullet \omega a^{\bullet -1} \in J \cap H_0$, hence $a^\bullet \omega a^{\bullet -1} = \omega$ by (9.5). Also $a^\bullet \beta a^{\bullet -1} \in J$ since $\beta \in J$. Hence, $a^\bullet b^+ a^{\bullet -1} = a^\bullet \beta\omega a^{\bullet -1} = a^\bullet \beta a^{\bullet -1} a^\bullet \omega a^{\bullet -1} = a^\bullet \beta a^{\bullet -1}\omega \in A$. In all cases, $a^\bullet b^+ a^{\bullet -1} \in A$, and, as noted in the preceding paragraph, we can now accept the lemma as being true.

(9.18) Lemma. *From* $a + (b + 1) = (a + b) + d_{a,b}$, *we obtain* $a + (b + c) = (a + b) + d_{a,b}c$ *for all* $c \in M$.

Proof. Now $a^+ b^+(0) = a^+(b) = a + b = (a + b)^+(0) = (a^+(b))^+(0)$. Hence,

$[a^+(b)]^{+-1}a^+b^+(0) = 0$, so $[a^+(b)]^{+-1}a^+b^+ \in H_0$. Let $\delta = [a^+(b)]^{+-1}a^+b^+$ and $d = \delta(1)$. So $d^\bullet = \delta$. Now $[a^+(b)]^+\delta = a^+b^+$ and $a^+b^+(1) = a^+(b+1) = a+(b+1)$, and $[a^+(b)]^+\delta(1) = [a^+(b)]^+(d) = a^+(b)+d = (a+b)+d$, so $d = d_{a,b}$ is the solution to $(a+b)+x = a+(b+1)$. But $[a^+(b)]^+\delta = a^+b^+$, so for any $c \in M$, $a^+b^+(c) = a^+(b+c) = a+(b+c)$, and $[a^+(b)]^+\delta(c) = [a^+(b)]^+d_{a,b}^\bullet(c) = [a^+(b)]^+(d_{a,b}c) = a^+(b)+d_{a,b}c = (a+b)+d_{a,b}c$. Hence, (9.18) is true.

(9.19) Lemma. *If $a + b = 0$, then $d_{a,b} = 1$.*

Proof. If $a+b = 0$, then $1_M = 0^+ = (a+b)^+ = (a^+(b))^+$, so $(a^+(b))^+ = 1_M$.

If $A = J \cup \{1_M\}$, then from $a + b = 0$, we get $a^+b^+(0) = a^+(b+0) = a^+(b) = a + b = 0$. Now $a^+, b^+ \in A$ imply $a^+, b^+ \in J$, or perhaps $1_M \in \{a^+, b^+\}$. If $1_M \in \{a^+, b^+\}$, then $0 \in \{a, b\}$, which implies $a = b = 0$, and so $a^+ = b^+ = 0^+ = 1_M$. Then $a^+b^+ = 1_M$. If $1_M \notin \{a^+, b^+\}$, then $a^+, b^+ \in J$, so by (9.6), $a^+b^+ = 1_M$. From the proof of (9.18), we have $[a^+(b)]^{+-1}\delta = a^+b^+ = 1_M$ and $\delta = [a^+(b)]^{+-1}a^+b^+ = [a^+(b)]^{+-1}$. Since $a^+b^+ = 1_M$ and $a^+, b^+ \in J$, we obtain $a^+ = b^+$, and, consequently, $a = b$.

We have $a^+b^+ = 1_M$ and $(a^+(b))^+ = 1_M$, and $[a^+(b)]^+\delta = a^+b^+$, so $\delta = 1_M = 0^+$. Hence, $d_{a,b} = \delta(1) = 1_M(1) = 1$, and so $d_{a,b} = 1$.

If $A = J\omega$, then $a^+ = \alpha\omega$ and $b^+ = \beta\omega$ for some $\alpha, \beta \in J$. Now $a^+b^+(0) = a^+(b+0) = a + b = 0$, so $\alpha\omega\beta\omega(0) = 0$ and $\omega\beta\omega \in J$ since $\omega^2 = 1_M$ and $\beta \in J$. Also, $\alpha(\omega\beta\omega(0)) = 0$, so by (9.6), $\alpha\omega\beta\omega = 1_M$. Therefore, $a^+b^+ = 1_M$. We have $[a^+(b)]^+ = 1_M$, so $[a^+(b)]^+ = 1_M = a^+b^+ = [a^+(b)]^+\delta = 1_M\delta = \delta$. With $\delta = 1_M$, we obtain $d_{a,b} = \delta(1) = 1_M(1) = 1$.

In all cases, $d_{a,b} = 1$, as intended.

We now have that $(M, +, \cdot)$ is a neardomain. It remains to show that $G = T_2(M)$.

(9.20) Lemma. $G = T_2(M)$.

Proof. For $a \in M$, $m \in M^*$, the linear mapping $[a, m]x = a + mx = a^+(m^\bullet x) = a^+ \circ m^\bullet(x)$, so $[a, m] = a^+ \circ m^\bullet \in G$, and so $T_2(M) \subseteq G$.

Let $\gamma \in G$ and let $a = \gamma(0)$. Then $a + x = \gamma(1)$ has exactly one solution, since $(M, +)$ is a loop and $a, \gamma(1) \in M$. Let m be the unique solution. If $m = 0$, then $\gamma(1) = a + m = a + 0 = a = \gamma(0)$, and so γ is not a permutaion on M. Hence, $m \neq 0$. Now $a^+, m^\bullet \in G$ and $a^+m^\bullet(0) = a^+(m \cdot 0) = a^+(0) = a = \gamma(0)$. Also, $a^+m^\bullet(1) = a + m \cdot 1 = a + m = \gamma(1)$. Since $\gamma, [a, m]$, and $a^+m^\bullet$ are all in G, and agree at 0 and 1, they must be the same. In particular, $\gamma = [a, m] \in T_2(M)$ and we conclude that $G \subseteq T_2(M)$, and then $T_2(M) = G$. This completes the proof of (9.20) and of (9.3).

The following proposition will make subsequent calculations easier.

(9.21) Proposition. *Let $(F, +, \cdot)$ be a neardomain. Then the following calculation rules are valid for elements of F.*
(a) $-(-a) = a$, *where $-a$ is the solution to the equation $y + a = 0$.*
(b) $a + (-a) = 0$ *for each $a \in F$.*
(c) $a \cdot 0 = 0$ *for each $a \in F$.*
(d) *If $a, b \in F$, then $a(-b) = -(ab)$.*
(e) $-1 \in Z(F^*)$, *the centre of the group $(F^*, \cdot)$, and $(-1)(-1) = 1$.*
(f) *For $a, b \in F$, $(-a)b = -(ab)$.*

Proof. For $a \in F$, $-(-a) = -(-a) + 0 = -(-a) + (-a + a) = [-(-a) + (-a)] + d_{-(-a),-a}a = 0 + 1 \cdot a = a$. Hence, (a). Since $-(-a) + (-a) = 0$ and $-(-a) = a$, we have $a + (-a) = 0$. Hence, (b). From $a \cdot 0 = a \cdot (0 + 0) = (a \cdot 0) + (a \cdot 0)$, we obtain $(a \cdot 0) + 0 = a \cdot 0$ and $(a \cdot 0) + (a \cdot 0) = a \cdot 0$. Hence, $a \cdot 0 = 0$, and we have (c). Continuing, $0 = a \cdot 0 = a \cdot (-b + b) = a \cdot (-b) + a \cdot b$, so we have $0 = -(a \cdot b) + (a \cdot b) = a \cdot (-b) + (a \cdot b)$, which implies $a(-b) = -(ab)$, and so we have (d).

Next, $(-1)(-1) = -[(-1)1] = -(-1) = 1$. It is not quite so easy to obtain $-1 \in Z(F^*)$. Consider $[0, -1]$. Then $[0, -1]^2 = [0, 1]$. Also, $[0, -1]0 = 0$, so 0 is a fixed point for $[0, -1]$. Now take $c \in F^*$ and consider the linear map $[0, c(-1)c^{-1}]$. We have $[0, c(-1)c^{-1}]^2 = [0, 1]$ and $[0, c(-1)c^{-1}]0 = 0$, so 0 is also a fixed point for $[0, c(-1)c^{-1}]$. Thus, either $[0, c(-1)c^{-1}] = 1_F$ or $[0, c(-1)c^{-1}] \in J$. If $[0, c(-1)c^{-1}] = 1_F$, then $c = [0, c(-1)c^{-1}]c = c(-1) = -c$, and so $c + c = 0 = c \cdot 1 + c \cdot 1 = c \cdot (1 + 1)$. Hence, $1 + 1 = 0$ and $1 = -1$. If $1 = -1$, then $-1 \in Z(F^*)$. Otherwise $[0, c(-1)c^{-1}] \in J$ as is $[0, -1]$. They both fix 0. By (9.5), $[0, c(-1)c^{-1}] = [0, -1]$, and therefore $c(-1)c^{-1} = -1$. This yields $c(-1) = (-1)c$ and puts $-1 \in Z(F^*)$, and we have (e).

For (f), $(-a)b = [a(-1)]b = a[(-1)b] = a[b(-1)] = a(-b) = -(ab)$. This completes the proof of (9.21).

(9.22) Proposition. *For a neardomain $(F, +, \cdot)$, let $J = \{[a, m] \in T_2(F) \mid [a, m]^2 = [0, 1], [a, m] \neq [0, 1]\}$. Then $[a, m] \in J$ if and only if $m = -1$.*

Proof. For any $a \in F$, $[a, -1]^2 = [0, 1]$ by (9:1). Hence, all $[a, -1] \in J$. Suppose $[0, 1] = [a, m]^2$. Then $0 = [a, m]^2 0 = [a, m](a + m0) = a + ma$. Hence, $ma = -a = a(-1) = (-1)a$. If $a \neq 0$, then $m = -1$, and so $[a, m] = [a, -1]$. If $a = 0$, we have by (9:1) that $[0, 1] = [0, m^2]$. So what are the solutions to the equation $m^2 = 1$? This question is answered by the next lemma, and the answer will complete the proof of (9.22).

(9.23) Lemma. *For a neardomain $(F, +, \cdot)$, the solution to the equations $x^2 = 1$ are $x = \pm 1$.*

Proof. Certainly, $1^2 = 1$, and by (9.21), $(-1)^2 = 1$.
Suppose $1 + 1 \neq 0$, that $e^2 = 1$, and that $e \in F \setminus \{1, -1\}$. Then there is

a unique $\alpha \in T_2(F)$ such that $\alpha 1 = 1$ and $\alpha e = -1$. Let $\beta = \alpha[0, e]\alpha^{-1}$. Then $\beta^2 = 1_F$, and $\beta 1 = -1$, and $\beta(-1) = 1$. But also $[0, -1]1 = -1$, and $[0, -1](-1) = 1$. Hence, $\beta = [0, -1]$. From $0 = [0, -1]0 = \beta 0 = \alpha[0, e]\alpha^{-1}0$, we obtain $\alpha^{-1}0 = [0, e]\alpha^{-1}0$, so $\alpha^{-1}0$ is a fixed point for $[0, e]$. But $[0, e] \neq 1_F$ and $[0, e]0 = 0$, so $\alpha^{-1}0 = 0$. Since 0 and 1 are fixed points for α, we have $\alpha = 1_F$, and therefore $e = \alpha e = -1$, a contradiction.

Suppose now that $1 + 1 = 0$, that $e^2 = 1$, and that $e \in F \setminus \{1, -1\} = F \setminus \{1\}$, since $1 = -1$. Let a be the solution to $y + 1 = e$, so $a + 1 = e$. Certainly, $a \neq 0$, so $a \in F^*$. We have $[0, e]^2 = 1_F$ and $[0, e] \neq 1_F$. Also, $[a, 1]^2 = [a + a, d_{a,a}] = [0, 1]$ since $0 = a \cdot 0 = a \cdot (1 + 1) = a + a$. Since $[a, 1]1 = a + 1 = e = [0, e]1$ and $[a, 1]e = a + e = 1 = [0, e]e$, we conclude that $[a, 1] = [0, e]$, and therefore $a = 0$ and $e = 1$, a contradiction. This completes the proof of (9.23) as well as (9.22).

(9.24) Theorem. *For a neardomain $(F, +, \cdot)$, let J be as in (9.22). Then $(F, +, \cdot)$ is a nearfield if and only if $J^2 = \{\alpha\beta \mid \alpha, \beta \in J\}$ is a subgroup of $T_2(F)$.*

Proof. Suppose first that $(F, +, \cdot)$ is a nearfield. Take $\alpha, \beta \in J$. From (9.22), there are $a, b \in F$ such that $\alpha = [a, -1]$ and $\beta = [b, -1]$. Now $\beta \alpha x = [b, -1](a + (-1)x) = b + (-1)(a + (-1)x) = (b - a) + x$. So, if $\gamma \in J^2$, then there is a $c \in F$ such that $\gamma = [c, 1]$. Suppose $\delta \in J^2$ and $\delta^2 = [d, 1]$. Then $[c, 1][d, 1] = [c + d, 1]$. From $\alpha\beta = [a, -1][b, -1] = [b - a, 1]$, we have all $[c, 1] \in J^2$, where $c \in F$. Hence, $J^2 = \{[c, 1] \mid c \in F\}$ and J^2 is closed with respect to $\circ$. If $[c, 1] \in J^2$, then $[-c, 1] \in J^2$ and $[-c, 1] = [c, 1]^{-1} \in J^2$. This makes J^2 a subgroup of $T_2(F)$.

We now assume that J^2 is a subgroup of $T_2(F)$. By (9.8), either each element of J has a fixed point, or no element of J has a fixed point. It will be important for our proof to show that in either case, $J^2 = A$.

Suppose first that each element of J has a fixed point. From (9.10), we have $A = J\omega$. We proceed to show that $J^2 = A$. Certainly, $A = J\omega \subseteq J^2$, since $\omega \in J$. Now we need only show that $J^2 \subseteq J\omega = A$. Take $\alpha, \beta \in J$. Then $\alpha\beta 0 \in F$, so by (9.4), there is a unique $\gamma \in J$ such that $\gamma 0 = \alpha\beta 0$, if $\alpha\beta 0 \neq 0$. If $\alpha\beta 0 = 0$, then $\omega \in J$ and $\omega 0 = 0$. So, in all cases, there is a unique $\gamma \in J$ such that $\alpha\beta 0 = \gamma 0$. Hence, $0 = \gamma\alpha\beta 0 = \omega 0 = \omega\gamma\alpha\beta 0$. With $\omega, \gamma, \alpha, \beta \in J$, and J^2 a subgroup, we have $(\omega\gamma)(\alpha\beta) \in J^2$. So there are $\sigma, \tau \in J$ such that $(\omega\gamma)(\alpha\beta) = \sigma\tau$, and $0 = \sigma\tau 0 = \sigma(\tau(0))$. From (9.6) we learn that $\sigma\tau = 1_F = \omega\gamma\alpha\beta$. Thus $\gamma\alpha\beta = \omega$ and $\alpha\beta = \gamma\omega \in J\omega$. This puts $J^2 \subseteq J\omega = A$, as required, and so $J^2 = J\omega = A$.

Next, we assume that no element of J has a fixed point, so from (9.10), we have $A = J \cup \{1_F\}$. We again proceed to show that $J^2 = A$.

First, we will see that $J^2 \subseteq A$. Take $\alpha, \beta \in J$ with $\alpha \neq \beta$. (If $\alpha = \beta$, then $\alpha\beta = 1_F \in A$.) If $\alpha\beta 0 \neq 0$, then from $A = J \cup \{1_F\}$ and (9.11), there is a unique $\gamma \in J$ such that $\gamma 0 = \alpha\beta 0$, so $0 = \gamma^2 0 = \gamma\alpha\beta 0$. If $\gamma 0 = \alpha\beta 0 = 0$, then since $\beta 0 \neq 0$, there is a unique $\tau \in J$ such that $\tau \beta 0 = 0$. But $\alpha \beta 0 = 0$, $\beta \beta 0 = 0$, and $\tau \beta 0 = 0$ imply $\tau = \alpha = \beta$. But $\alpha \neq \beta$, so $\alpha \beta 0 \neq 0$. In any case, since $\alpha \beta 0 \neq 0$, there is a unique $\gamma \in J$ such that $\gamma 0 = \alpha \beta 0$.

We proceed next to show that $\gamma \alpha \beta = 1_F$. If $\gamma \alpha \beta \neq 1_F$, then there is a p such that $\gamma \alpha \beta p \neq p$. From (9.4) we obtain the existence of a unique $\delta \in J$ with $\delta p = \gamma \alpha \beta p$. Hence, $p = \delta \gamma \alpha \beta p = \sigma \tau p = \sigma(\tau(p))$ for some $\sigma, \tau \in J$ such that $(\delta \gamma)(\alpha \beta) = \sigma \tau$. From (9.6), we obtain $\sigma \tau = 1_F$, so $\delta \gamma \alpha \beta = 1_F$, and, consequently, $\gamma \alpha \beta = \delta^{-1} = \delta$. But $0 = \gamma^2 0 = \gamma \alpha \beta 0 = \delta 0$, so δ has a fixed point. We have assumed that no element of J has a fixed point, so our only option is to accept $\gamma \alpha \beta = 1_F$. Hence, $\alpha \beta = \gamma$, and so $J^2 \subseteq A$.

It will be easier to show that $A \subseteq J^2$. Certainly, $1_F \in J^2$, so we need only show $J \subseteq J^2$. If $\alpha, \beta \in J$, then $\gamma = \alpha \beta \in T_2(F)$. We have $\alpha \beta = \gamma \in J^2 \subseteq A = J \cup \{1_F\}$. If $\alpha \beta = 1_F$, then $\alpha = \beta$. If $\alpha \beta \neq 1_F$, then $\alpha \beta = \gamma \in J$, so $\alpha = \gamma \beta \in J^2$, hence $\alpha \in J^2$. We have $1_F \in J^2$ and $\alpha \in J^2$ if $\alpha \in J$. So $A = J \cup \{1_F\} \subseteq J^2$.

With $A \subseteq J^2$ and $J^2 \subseteq A$, we can conclude that $A = J^2$, and remember that we are assuming that J^2 is a subgroup of $T_2(F)$. But our goal is to show that $(F, +, \cdot)$ is a nearfield. What we need is for $(F, +)$ to be a group. As a result of having an isomorphism $\Psi : (F, +) \to A = J^2$, we will reach our goal.

Define $\Psi(a) = a^+$, and so (9.11) assures us that Ψ is well defined. For $a^+, b^+ \in A = J^2$, we have $a^+ \circ b^+ \in A = J^2$, since $A = J^2$ is a group. Now $(a+b)^+ 0 = a+b$ and $a^+ \circ b^+ 0 = a^+(b+0) = a+b$. So $[(a+b)^+]^{-1} a^+ b^+ 0 = 0$. But $a^+, b^+, (a+b)^+, [(a+b)^+]^{-1}, a^+ b^+, [(a+b)^+]^{-1} a^+ b^+$ are all in A since $A = J^2$ and J^2 is a group. From (9.11), we obtain $(a+b)^+ = a^+ b^+$, or $\Psi(a+b) = (a+b)^+ = a^+ \circ b^+ = \Psi(a) \circ \Psi(b)$. Now we have that Ψ is a homomorphism.

If $\Psi(a) = a^+ = b^+ = \Psi(b)$, then $a = a^+ 0 = b^+ 0 = b$, so Ψ is injective. If $\gamma \in A$ and $\gamma 0 = c \in F$, then $\gamma = c^+$ and $\Psi(c) = c^+ = \gamma$, so Ψ is surjective. Now we can conclude that Ψ is an isomorphism, and so $(F, +)$ is a group, and $(F, +, \cdot)$ is a nearfield.

Our next goal is to show that every finite neardomain is a nearfield, but we will need a little more preparation.

(9.25) Definition. A neardomain $(F, +, \cdot)$ is *planar* when for any $a, b \in F$, $b \neq 1$, the equation $a + bx = x$ has a solution.

(9.26) Proposition. *For a neardomain $(F, +, \cdot)$, the following are equivalent.*

(a) $(F, +, \cdot)$ *is planar.*

(b) $(F, +, \cdot)$ *is a planar nearfield.*

(c) *The elements of $T_2(F)$ having no fixed point, together with 1_F, form a subgroup A' of $T_2(F)$.*

Proof. Assume (a) to be true. To obtain (b), it is enough to show that each $d_{a,b} = 1$. Now $a + (b + x) = (a + b) + d_{a,b}x = x$ has a solution, since $(F, +, \cdot)$ is planar. Now $a + (-a + x) = (a + (-a)) + d_{a,-a}x = 0 + x = x$, so $b + x = -a + x$ and, consequently, $b = -a$. Then $d_{a,b} = d_{a,-a} = 1$, and so $(F, +, \cdot)$ is a nearfield.

To see that $(F, +, \cdot)$ is a planar nearfield, consider $a + bx = cx$ with $b \neq c$. If $c = 0$, then $a + bx = 0$ has the unique solution $x = b^{-1}(-a)$. If $b = 0$, then $x = c^{-1}a$ is the unique solution to $a = cx$. If $0 \notin \{b, c\}$, then $cx = cc^{-1}a + cc^{-1}bx = c(c^{-1}a + c^{-1}bx)$, and so $c^{-1}a + (c^{-1}b)x = x$, which has a solution by (a). If $c^{-1}a + (c^{-1}b)y = y$, then $c^{-1}a = x - (c^{-1}b)x = y - (c^{-1}b)y$, so $-y + x = (c^{-1}b)(-y + x)$, and since $c^{-1}b \notin \{0, 1\}$, and since $(F, +, \cdot)$ is a nearfield, we must conclude that $x = y$. So the solution to $a + bx = cx$ is unique, and so the nearfield $(F, +, \cdot)$ is planar. Hence, (a) implies (b).

Next, assume (b) to be true. If $x = [a, m]x = a + mx$, then there is a unique solution for x if $m \neq 1$. So, for $[a, m]$ to have no fixed point, we must take $m = 1$. Hence, the elements of A' have the form $[a, 1]$ for some $a \in F$. Conversely, if $x = [a, 1]x = a + x$, then $a = 0$ and so $[a, 1] = 1_F$. Hence, $A' = \{[a, 1] \mid a \in F\}$. Now $[a, 1][b, 1] = [a + b, 1]$ by (9:1), and $[a, 1]^{-1} = [-a, 1]$ by (9:2). Hence, A' is a subgroup of $T_2(F)$, and so (b) implies (c).

Of course, we will be finished with the proof of (9.26) when we show that (c) implies (a). Let A' be the elements of $T_2(F)$ with no fixed point, together with $1_F = [0, 1]$. Take $[a, b] \in A'$, $[a, b] \neq 1_F$. Then $1 \neq [a, b]1 = a + b$. Let c be the solution to $y + (a + b) = 1$, so $c + (a + b) = 1$. If $c = 0$, then $1 = c + (a + b) = 0 + (a + b) = a + b$, and so $1 = [a, b]1$, contrary to $[a, b] \in A'$ and $[a, b] \neq 1_F$. Hence, $c \neq 0$. Now $[c, 1]x = c + x = x$ implies $c = 0$, so $[c, 1] \in A'$. But $[c, 1][a, b]1 = c + (a + b) = (c + a) + d_{c,a}b = 1$, and $[c, 1][a, b] \in A'$, so $[c, 1][a, b] = [0, 1] = 1_F$. Thus $0 = [c, 1][a, b]0 = c + a$ and $1 = [c, 1][a, b]1$. From $c + a = 0$, we have $d_{c,a} = 1$, and so $1 = (c + a) + b = 0 + b = b$. Hence, $[a, b] = [-c, 1]$. So, if $[a, b]$ has no fixed point, then $b = 1$. Again, if $b \neq 1$, $[a, b]$ has a fixed point. So there is an x such that $x = [a, b]x = a + bx$. This is exactly what we need to conclude that $(F, +, \cdot)$ is planar. Hence, (c) implies (a), and the proof of (9.26) is complete.

The following theorem is very significant.

(9.27) Theorem. *Every finite neardomain is a nearfield.*

Proof. Let $(F, +, \cdot)$ be a finite neardomain. Suppose $[a, b] \in T_2(F)$ has no fixed point. Take any $x \in F$. Then $x \neq [a, b]x = a + bx$. So there is a unique $c_x \in F$ such that $c_x + (a + bx) = x$. If $c_x = 0$, then $x = a + bx = [a, b]x$, and so $[a, b]$ has a fixed point. This means that the map $x \mapsto c_x$ is not injective. Suppose $x \neq y$ and $c_x = c_y = c$. Then $[c, 1][a, b]x = [c, 1](a + bx) =$

$c + (a + bx) = c_x + (a + bx) = x$ and $[c, 1][a, b]y = c_y + (a + by) = y$. With $[c, 1][a, b] \in T_2(F)$ having two fixed points, x and y, we must conclude that $[c, 1][a, b] = [0, 1] = 1_F$. But $[c, 1]^{-1} = [-c, 1]$, so $[-c, 1] = [a, b]$, and so $b = 1$. In summary, if $[a, b]$ has no fixed point, then $b = 1$. Thus, if $b \neq 1$, then $[a, b]$ has a fixed point, so there is an x such that $x = [a, b]x = a + bx$. Hence, $(F, +, \cdot)$ is a planar neardomain. From (9.26) we conclude that $(F, +, \cdot)$ is a planar nearfield.

(9.28) Problem. Does there exist an infinite neardomain that is not a nearfield? This problem has remained unsolved for nearly 30 years. See [K].

Theorems (9.2) and (9.3) show that the problem of finding all sharply 2-transitive groups is equivalent to finding all neardomains. From (9.27), we know that, for the finite case, one can focus on finding all finite nearfields. In the infinite case, finding a nearfield is also a discovery of a neardomain, hence a sharply 2-transitive group. Because of (9.28), one will never be sure that all sharply 2-transitive groups have been found until one knows if there is a neardomain that is not a nearfield.

So what are the finite nearfields? Dickson [Di, 1905a] developed a factory that would eventually produce all but seven of them. Zassenhaus [Z] took over the factory and actually produced all but these seven with it. Foulser [Fou] and Lüneburg [Lü] sorted out the production by isomorphism types, and Wähling [W] presented the production in polished form.

But what about those seven strays? Actually, Dickson [Di, 1905a] knew of five of these, and Zassenhaus [Z] found two more, and assured us that these seven are exactly those finite nearfields not capable of being produced by the Dickson factory.

The theory of finite nearfields is very beautiful, with a rather intricate beauty. The author would like to note that the applications of nearfields to planar nearrings has not even begun, and that he expects many rewarding discoveries. See (4.17).

To provide a beginning to the study of nearfields, we shall outline Dickson's production process and indicate how to construct the notorious seven strays.

The Dickson process

For our purpose, the raw materials for the Dickson process will always include a known nearring $(N, +, \cdot)$. One can obtain something with as little as a mapping $k : N \to \text{End}(N, +)$. For any such mapping, define $*$ on N by

$$a * b = a \cdot k(a)(b). \tag{9:3}$$

Then $(N, +, *)$ is a *nonassociative nearring*, that is, $*$ is left distributive over $+$. For if $a, b, c \in N$, then $a * (b+c) = a \cdot k(a)(b+c) = a \cdot [k(a)(b) + k(a)(c)] =$

$[a \cdot k(a)(b)] + [a \cdot k(a)(c)] = (a * b) + (a * c)$. But if all one wants is a nonassociative nearring, it really only requires a group $(G, +)$ and a function $\phi : G \to \mathrm{End}(G, +)$. One then defines $a *_\phi b = \phi(a)(b)$. As shown above, one obtains that $*_\phi$ is left distributive over $+$, and so $(G, +, *_\phi)$ is a nonassociative nearring. For a given group $(G, +)$ of nearly any size, there are an enormous number of such nonassociative nearrings.

The mathematician who can appreciate beauty in an algebraic structure that is not loaded with obvious symmetry, for example, commutativity and both distributive laws, can find beauty in structures generated by the Dickson process. Whereas nonassociative nearrings may be too chaotic for some, one way to find more order is by requiring associativity. Then one obtains a good old fashioned nearring.

(9.29) Theorem. *For a nearring* $(N, +, \cdot)$, *let* $*$ *be the left distributive multiplication defined by* (9:3) *using a mapping* $k : N \to \mathrm{End}(N, +, \cdot)$. *Then if*

$$k(a * b) = k(a) \circ k(b) \qquad (9\!:\!4)$$

for all $a, b \in N$, *we can conclude that* $(N, +, *)$ *is a nearring.*

Proof. Note that each $k(a)$ is a nearring endomorphism, whereas in (9:3), we only required $k(a)$ to be a group homomorphism. So, we only need to show that $*$ is associative. For $a, b, c \in N$, $a * (b * c) = a \cdot k(a)[b \cdot k(b)(c)] = a \cdot [k(a)(b) \cdot k(a)(k(b)(c))] = [a \cdot k(a)(b)] \cdot k(a)(k(b)(c)) = (a * b) \cdot [k(a) \circ k(b)](c) = (a * b) \cdot k(a * b)(c) = (a * b) * c$. That is all there is to it!

Such nearrings as we get in (9.29) are called *Dickson nearrings* and have been studied by Maxson [Ma, 1970, 1971].

The Dickson process is used to construct all finite nearfields except for the notorious seven. So what refinements must be made? Take a good look at (9:3) and (9:4). Since a nearfield $(F, +, *)$ must have $0 * a = a * 0 = 0$ for all $a \in F$, we will have to put that limitation on $k : F \to \mathrm{End}(F, +, \cdot)$. Since $(F^*, *)$ must be a group, we cannot just let $k(a)$ be any old endomorphism of $(F, +, \cdot)$. In fact, if $a \neq 0$, we will require that $k(a)$ be an automorphism of $(F, +, \cdot)$. Again, since $(F^*, *)$ must be a group, we will require each $a \neq 0$ to be a unit with respect to $\cdot$.

So let us start with a nearfield $(F, +, \cdot)$, which, of course, could be a field, or even a skewfield. For the group of nearfield automorphisms $\mathrm{Aut}(F, +, \cdot)$, let $k : F^* \to \mathrm{Aut}(F, +, \cdot)$ be any mapping satisfying (9:4). Define $*$ on F by

$$a * b = \begin{cases} 0, & \text{if } a = 0; \\ a \cdot k(a)(b), & \text{if } a \neq 0. \end{cases} \qquad (9\!:\!5)$$

Nothing more is required from the Dickson process to obtain a nearfield.

(9.30) Theorem. *Let* $k : F^* \to \mathrm{Aut}(F, +, \cdot)$ *be a mapping from the multiplicative group* $(F^*, \cdot)$ *of a nearfield* $(F, +, \cdot)$, *and let* $*$ *be the left distributive multiplication defined by* (9:5). *Then* $(F, +, *)$ *is a nearfield, if*

$$k(a * b) = k(a) \circ k(b) \tag{9:6}$$

for all $a, b \in F^*$.

Proof. If one thinks of $k' : F \to \mathrm{End}(F, +, \cdot)$ as extending $k : F^* \to \mathrm{Aut}(F, +, \cdot)$ by defining $k(0) = \zeta$, where $\zeta(x) = 0$ for all $x \in F$, then $*$ of (9:5) is exactly the $*$ of (9:3), so by (9.29), $(F, +, *)$ is a nearring.

Since $(F, +, \cdot)$ is a nearfield, and $k(a) \in \mathrm{Aut}(F, +, \cdot)$ for each $a \in F^*$, the definition of $*$ in (9:5) assures us that $*$ defines a binary operation on F^*, that is, if $a, b \in F^*$, then $a * b \in F^*$. We have associativity for $*$, so we only need an identity e for $*$ on F^* and an inverse for each $a \in F^*$ with respect to e and $*$.

Our identity e for $*$ will be the identity 1 for $\cdot$. For if $a \in F$, then $a * 1 = a \cdot k(a)(1) = a \cdot 1 = a$, if $a \neq 0$, and $0 * 1 = 0$. So 1 is Certainly, a right identity for $*$. Now all we need for $a \in F^*$ is a right inverse with respect to 1 and $*$. But $k(a) \in \mathrm{Aut}(F, +, \cdot)$, so there is an element $a' \in F^*$ so that $k(a)(a') = a^{-1}$, where a^{-1} is the inverse of a with respect to 1 and $\cdot$. Hence, $a * a' = a \cdot k(a)(a') = a \cdot a^{-1} = 1$. We can now conclude that $(F, +, *)$ is also a nearfield.

In the nearfields of (2.18), we started with a field $(F, +, \cdot)$ with a subgroup U of F^* of index 2. We had an automorphism σ of $(F, +, \cdot)$ of order 2. Define $k : F^* \to \mathrm{Aut}(F, +, \cdot)$ by $k(a) = \begin{cases} 1_F, & \text{if } a \in U; \\ \sigma, & \text{if } a \notin U. \end{cases}$ Then k satisfies the condition of (9.30), and that is another reason why $(F, +, *)$ is a nearfield. (Our $*$ here is the $\circ$ of (2.18).)

Such nearfields as we get from (9.30) are called *Dickson nearfields*. There is extensive literature on Dickson nearfields. The reader is encouraged to study Wähling's book [W], and, in particular, to refer to the bibliography of that book.

(9.31) Exploratory problem. Find interesting subgroups of Dickson nearfields to construct interesting planar nearfields, especially those with interesting geometric interpretations. Each such subgroup gives various planar nearrings. See (4.17). Actually, the Dickson process applied to planar nearrings might also produce something interesting. Again, see Wähling's book [W] for information about the multiplicative group structure of nearfields as well as the automorphism group of various nearfields.

In Wähling [W], one can eventually sort out all the finite nearfields.

The notorious seven

I have referred to the seven finite nearfields not produced by the Dickson process as 'the seven strays', or 'the notorious seven'. Wähling calls them 'Ausnahmefastkörper', or 'exceptional nearfields'. Actually, I have used the terms 'notorious' and 'strays' to disguise a considerable amount of affection for these nearfields. They really are quite a mathematical curiosity, or in other words, they each represent a mathematical marvel with attractive individualism. I strongly suspect that we do not fully understand their beauty and secrets.

Each is of order p^2, where $p \in \{5, 7, 11, 23, 29, 59\}$. No, I did not leave one out. There are two of order 11^2. Each has, as its additive group, $Z_p \oplus Z_p$. Since each is a planar nearring, I need only provide the appropriate Φ so that $(Z_p \oplus Z_p, \Phi)$ is a Ferrero pair. The appropriate Φ will be defined by generators in the form of nonsingular matrices for the corresponding vector spaces $Z_p \oplus Z_p$ over the prime field $(Z_p, +, \cdot)$. We use the numbering attributed to Zassenhaus. In each case, $\Phi = \langle A, B, C \rangle$ has three generators A, B, and C, with $A = \begin{pmatrix} 0 & -1 \\ 1 & 0 \end{pmatrix}$.

I.	$p = 5$,	$B = \begin{pmatrix} 1 & -2 \\ -1 & -2 \end{pmatrix}$,	$C = \begin{pmatrix} 1 & 0 \\ 0 & 1 \end{pmatrix}$.	
II.	$p = 11$,	$B = \begin{pmatrix} 1 & 5 \\ -5 & -2 \end{pmatrix}$,	$C = \begin{pmatrix} 4 & 0 \\ 0 & 4 \end{pmatrix}$.	
III.	$p = 7$,	$B = \begin{pmatrix} 1 & 3 \\ -1 & -2 \end{pmatrix}$,	$C = \begin{pmatrix} 1 & 0 \\ 0 & 1 \end{pmatrix}$.	
IV.	$p = 23$,	$B = \begin{pmatrix} 1 & -6 \\ 12 & -2 \end{pmatrix}$,	$C = \begin{pmatrix} 2 & 0 \\ 0 & 2 \end{pmatrix}$.	
V.	$p = 11$,	$B = \begin{pmatrix} 2 & 4 \\ 1 & -3 \end{pmatrix}$,	$C = \begin{pmatrix} 1 & 0 \\ 0 & 1 \end{pmatrix}$.	
VI.	$p = 29$,	$B = \begin{pmatrix} 1 & -7 \\ -12 & -2 \end{pmatrix}$,	$C = \begin{pmatrix} -4 & 0 \\ 0 & -4 \end{pmatrix}$.	
VII.	$p = 59$,	$B = \begin{pmatrix} 9 & 15 \\ -10 & -10 \end{pmatrix}$,	$C = \begin{pmatrix} 4 & 0 \\ 0 & 4 \end{pmatrix}$.	

The unnatural order to this listing is better explained in Wähling [W], but in I through IV, the groups Φ are solvable, and in V through VII, the groups Φ are not solvable. Also, the group Φ for I is isomorphic to a direct factor of the group Φ for II, and the group Φ for III is isomorphic to a direct factor of the group Φ for IV. The group Φ for V is isomorphic to a direct factor of the groups Φ for VI and VII. The group for I is called the *binary tetrahedral group*; the group for III is called the *binary octahedral group*; and the group for V is called the *binary icosahedral group*.

(9.32) **Exploratory problem.** Use the 'notorious seven' and their subgroups to construct planar nearrings. Discover interesting and unusual geometric properties of these planar nearrings. Relate these to other known geometric properties of the various Φ's.

A nondesarguesian plane

Desargues' property for an affine plane gives the affine plane a lot of nice qualities. This property can be rather confusing to describe, especially when pictures are provided. The reader can easily get confused with the nine lines and seven points involved. We shall attempt to provide the reader with an understanding of Desargues' property, but to do so, we ask the reader to take pen in hand and go through the procedure outlined below, perhaps several times with variations, until he is confident of an understanding. It may be useful to refer to the diagram of The Veblen and Wedderburn example at the end of this chapter, but remember that these lines really need not have any shape at all!

(9.33) **Definition.** An affine plane $(N, \mathcal{L}, \in)$ is *desarguesian* provided the following steps can always be carried out, theoretically at least.

Step 1. Take any point $p \in N$ and any three distinct lines $L_1, L_2, L_3 \in \mathcal{L}$ passing through p.

Step 2. Choose points q_1, q_2, q_3 on L_1, L_2, L_3, respectively, ensuring that $p \notin \{q_1, q_2, q_3\}$.

Step 3. Let L_4 be the line through q_1 and q_2, let L_5 be the line through q_2 and q_3, and let L_6 be the line through q_1 and q_3.

Step 4. Choose one of the lines L_1, L_2, L_3, say L_2, and then choose any point r_2 on L_2 such that $r_2 \notin \{q_2, p\}$.

Step 5. Let L_7 be the line through r_2 such that L_4 and L_7 are parallel. Similarly, let L_8 be the line through r_2 such that L_5 and L_8 are parallel.

Step 6. Let r_1 be the point where L_1 and L_7 meet, and let r_3 be the point where L_3 and L_8 meet.

Step 7. Let L_9 be the line through r_1 and r_3.

Step 8. So if our affine plane $(N, \mathcal{L}, \in)$ is to be desarguesian, we can be assured that L_6 and L_9 are parallel.

For any finite nearfield $(F, +, \cdot)$, we let $N = F \times F$, and let the lines in $\mathcal{L}$ be all those with 'slope', namely, all $L(m, b) = \{(x, mx + b) \mid x \in F\}$, where $(m, b) \in F \times F$, and all the 'vertical lines' $L(\infty, a) = \{(a, y) \mid y \in F\}$, where $a \in F$. The reader may want to assure herself that $(N, \mathcal{L}, \in)$ is indeed an affine plane, and that if F is a field, then $(N, \mathcal{L}, \in)$ is indeed desarguesian.

Finite nearfields which are not fields can be used to construct nondesarguesian affine planes [VW]. We shall illustrate this using a nearfield $(F, +, \circ)$ of order 9. This nice easy example was kindly provided for us by F. W. Stevenson [St] in exchange for the author's rights to purchase a popular season ticket to the University of Arizona's home basketball games.

Start with the quotient ring $Z_3[x]/(1+x^2)$ using the ideal $(1+x^2)$ generated by the irreducible polynomial $1 + x^2 \in Z_3[x]$. If we identify the coset $c_1 + c_2 x + (1 + x^2) \equiv c_1 + c_2 a$, then $Z_3[x]/(1+x^2) = \{0, 1, 2, a, 1 + a, 2 + a, 2a, 1 + 2a, 2 + 2a\}$. Let $(F, +, \cdot)$ be this field $Z_3[x]/(1+x^2)$. The squares of elements in F^* constitute $U = \{1, 2, a, 2a\}$. Define $\circ$ on F, as we did in (2:2), by $s \circ t = \begin{cases} s \cdot t, & \text{if } s \in U; \\ s \cdot t^3, & \text{if } s \notin U. \end{cases}$ (At this point, it may be helpful to construct a multiplication table for $\circ$, or perhaps one could get by with just noting that $1 + a^2 = 0$, so $a^2 = 2$, hence $a^3 = 2a$.) Also, it may help to be influenced by the diagram of The Veblen and Wedderburn example on the next page.

Consider the three lines L_1, defined by $y = 2ax + 1$, L_2, defined by $y = ax$, and L_3, defined by $y = (2 + a)x + (1 + 2a)$. Note that $p = (a, 2)$ is on all three lines. Choose $q_1 = (0, 1)$ on L_1, $q_2 = (0, 0)$ on L_2, and $q_3 = (1, 0)$ on L_3. Let L_4 be the line through q_1 and q_2, so L_4 is the y-axis defined by $x = 0$. Let L_5 be the x-axis, defined by $y = 0$, and note that q_2 and q_3 are on L_5. Now if L_6 is the line through q_1 and q_3, then L_6 has equation $y = 2x + 1$.

Corresponding to Step 4, we take $r_2 = (2 + 2a, 1 + 2a)$ on L_2. If L_7 is the line through r_2 and parallel to L_4, then L_7 has equation $x = 2 + 2a$. Similarly, If L_8 is the line through r_2 and parallel to L_5, then L_8 has equation $y = 1 + 2a$.

Corresponding to Step 6, compute that L_1 and L_7 intersect at $r_1 = (2 + 2a, a)$, and that L_3 and L_8 intersect at $r_3 = (0, 1 + 2a)$. The line L_9 passing through r_1 and r_3 has equation $y = x + (1 + 2a)$. Noting its slope is 1, which is different than the slope 2 for L_6, we anticipate that L_6 and L_9 are not parallel. Indeed, $(2a, 1 + a)$ is exactly the point where the lines L_6 and L_9 intersect.

Had the affine plane for F been desarguesian, then we would have that the line L_6 and the line L_9 are parallel. Since they are not parallel, we must conclude that this affine plane is really nondesarguesian. So, with all due thanks to Fred Stevenson, we have one of the simpler, if not the simplest, presentations in the literature that nearfields actually provide nondesarguesian affine planes. But perhaps even simpler presentations are possible. Readers are invited in the following exercise to try their hand at constructing nondesarguesian affine planes from finite nearfields. In so doing, something new or interesting might be discovered.

Exercise. Construct another finite nearfield F which is not a field. Find lines and points showing that the affine plane for F is not desarguesian. Then see if you can develop an algorithm to do this for any honest nearfield.

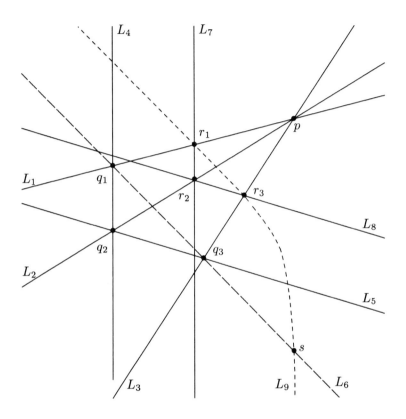

The Veblen and Wedderburn example

Don't mess with my circles!

Archimedes

CHAPTER 3

The great unifier

One of the main points of this chapter is to demonstrate a very special place that traditional ring theory and module theory play within algebra, and even within mathematics. This is due to an abundance of symmetry in rings and their associated modules. More specifically, a ring enjoys the consequences of *both* distributive laws: the left, where $a \cdot (b + c) = (a \cdot b) + (a \cdot c)$, and the right, where $(a + b) \cdot c = (a \cdot c) + (b \cdot c)$. We shall see that each of these is a consequence of different properties.

For a fixed ring R, the category of left R-modules $_R\mathcal{M}$ is very special when compared to categories in general. In $_R\mathcal{M}$, each object is both a group object and a cogroup object, and the resulting binary operations $\oplus_1$ and $\oplus_2$ defined on $\mathbf{hom}(M, M) = \mathrm{Hom}_R(M, M)$ coincide. One, say $\oplus_1$, gives the left distributive law, and the other, $\oplus_2$, gives the right distributive law.

Group and cogroup objects of a category $\mathcal{C}$ were first studied by Eckmann and Hilton [EH, 1962, 1963, 1963a]. A group object in a category $\mathcal{C}$ is the result of trying to find 'groups' entirely within the objects and morphisms of $\mathcal{C}$. A cogroup object is simply the dual idea to a group object. Sometimes group objects actually are closely identified with groups, but sometimes they are more general. It is in this latter setting that the theory becomes more interesting. Likewise, cogroup objects often provide a source for interesting mathematics.

If G is either a group or a cogroup object of a category $\mathcal{C}$, then there is a natural way to define $\oplus$ on $\mathbf{hom}(G, G)$ so that $(\mathbf{hom}(G, G), \oplus, \circ)$ is a nearring. If G is a group object, then $(\mathbf{hom}(G, G), \oplus, \circ)$ is a right nearring, but if G is a cogroup object, then $(\mathbf{hom}(G, G), \oplus, \circ)$ is a left nearring. This is why $(\mathrm{Hom}_R(M, M), \oplus, \circ)$ is a ring, since each M is a group object and a cogroup object, and since the corresponding additions coincide.

In most concrete categories, for $f, g \in \mathbf{hom}(G, G)$, addition, defined by $(f + g)(x) = f(x) + g(x)$, does not yield $f + g \in \mathbf{hom}(G, G)$. But there are numerous nice mathematical structures which are group objects or cogroup objects of appropriate categories. In this setting $f \oplus g \in \mathbf{hom}(G, G)$. For example, if R is a commutative ring with identity, then the R-algebra of polynomials $(R[x], +, \cdot)$ cannot have its R-algebra endomorphisms added in the usual pointwise way. Since $R[x]$ is a cogroup object in the category of R-algebras, there is a natural $\oplus$ for $\text{Hom}_R(R[x], R[x])$ and one obtains that the nearring of endomorphisms of the R-algebra $R[x]$ is isomorphic to the nearring of polynomials $(R[x], +, \circ)$. That is, $(\text{Hom}_R[(R[x], +, \cdot), (R[x], +, \cdot)], \oplus, \circ) \cong (R[x], +, \circ)$.

Group and cogroup objects play a role in the study of Hopf algebras as well as in topology. However, the mathematical community has not exploited the full natural structure of the $\textbf{hom}(G, G)$. Essentially, it has only used the group structure $(\textbf{hom}(G, G), \oplus)$, and seems to be unaware of the richer nearring structure $(\textbf{hom}(G, G), \oplus, \circ)$ and its action upon groups $(\textbf{hom}(X, G), \oplus)$ or $(\textbf{hom}(G, X), \oplus)$. Hanna Neumann [NH] and A. Fröhlich [F, 1958] noticed the tip of the iceberg in the 1950s. Later, Fröhlich and M. Holcombe [Ho, 1977] realized that the $(\textbf{hom}(G, G), \oplus, \circ)$ are nearrings for each group or cogroup object G. Fröhlich presented these ideas to the author in the early 1970s.

In addition to being a source of some nice interesting mathematics with important applications, the author firmly believes that these nearrings $(\textbf{hom}(G, G), \oplus, \circ)$ arising from group and cogroup objects promise to be a rich area for future research.

10. A little category theory

We will not need to know much about category theory, but we shall be well rewarded for learning a little. It would not be unusual for the reader to be familiar already with much of what we present here.

The concepts and terminology of category theory has had a mixed reception by the mathematical community during the past 30 years or so. As could be predicted, there are those who enjoy category theory as an end in itself, and there are those who consider it 'abstract nonsense'. Somewhere in the middle are those who accept a limited amount of category theory as a tool kit for various purposes. In this work, we take the position that there are numerous ideas from category theory that help one to focus upon unifying concepts. This has lead to the discovery of several very interesting nearrings and related structures. First, we need to know exactly what a category is.

(10.1) Definition. Suppose one starts with a class $\textbf{Ob}\,\mathcal{C}$ whose elements are called *objects* of $\mathcal{C}$, and for each ordered pair (A, B), with $A, B \in \textbf{Ob}\,\mathcal{C}$, there is a set $\textbf{hom}(A, B)$ of *morphisms* from A to B. Furthermore, for each ordered triple (A, B, C), with $A, B, C \in \textbf{Ob}\,\mathcal{C}$, suppose there is a function $\circ = \circ(A, B, C)$, depending upon $A, B,$ and C, where $\circ : \textbf{hom}(B, C) \times \textbf{hom}(A, B) \rightarrow \textbf{hom}(A, C)$, and where one denotes $\circ(f, g)$ by $f \circ g$. These functions $\circ$ must satisfy $(f \circ g) \circ h = f \circ (g \circ h)$ for all $h \in \textbf{hom}(A, B)$, for all $g \in \textbf{hom}(B, C)$, and for all $f \in \textbf{hom}(C, D)$, where $A, B, C,$ and D are arbitrarily chosen from $\textbf{Ob}\,\mathcal{C}$. In addition, each $A \in \textbf{Ob}\,\mathcal{C}$ is to have a $1_A \in$ **hom**(A, A) satisfying $f \circ 1_A = f$ and $1_A \circ g = g$ for each $f \in$ **hom**(A, B) and for each $g \in$ **hom**(B, A), where $B \in$ **Ob**$\mathcal{C}$ is arbitrary. When all this happens, the **Ob** $\mathcal{C}$, together with all the **hom**(A, B) and with all the $\circ = \circ(A, B, C)$, constitute a *category* $\mathcal{C}$.

To obtain the full benefit of this chapter, it is important that the reader be familiar with the following examples. It should be rather easy for the reader to verify rigorously that each of these is *really* an example of a category. We shall have occasion to refer to each of these examples except example 6.

(10.2) Examples: The 'Dirty Dozen'
   Example 1. *The category of sets* $\mathcal{S}$. Let the objects in **Ob**$\mathcal{S}$ be sets X, and if $X, Y \in$ **Ob** $\mathcal{S}$, let **hom**(X, Y) consist of all mappings with domain X and codomain Y. The $\circ = \circ(X, Y, Z)$ denote the usual composition of mappings. It is particularly important to understand that this is a category, since most of the remaining examples are conceptually dependent upon this example.

   Example 2. *The category of pointed sets* $\mathcal{S}_0$. Let the objects in **Ob** $\mathcal{S}_0$ be ordered pairs (X, x), where $X \in$ **Ob** $\mathcal{S}$ and $x \in X$. The elements of **hom**$[(X, x), (Y, y)]$ are those mappings $f : X \rightarrow Y$ which satisfy $f(x) = y$, and $\circ$ is the usual composition of mappings.

   Example 3. *The category of topological spaces* $\mathcal{T}$. Let the objects of **Ob** $\mathcal{T}$ be the ordered pairs (X, T), where $X \in$ **Ob** $\mathcal{S}$ and T is a topology for X. The elements of **hom**$[(X, T), (X', T')]$ are exactly the continuous mappings $f : X \rightarrow X'$, and $\circ$ is always a composition of mappings.

   Example 4. *The category of groups* $\mathcal{G}$. The objects in **Ob** $\mathcal{G}$ are the groups $(G, +)$, and the elements of **hom**$[(G, +), (H, +)]$ are exactly the group homomorphisms $f : G \rightarrow H$. As usual, $\circ$ is composition of mappings.

   Example 5. *The category of rings* $\mathcal{R}$. You are correct if you guessed that the objects in **Ob** $\mathcal{R}$ are simply the rings $(R, +, \cdot)$, and the morphisms of **hom**$[(R, +, \cdot), (R', +, \cdot)]$ are the ring homomorphisms, with $\circ$ denoting composition.

   Example 6. For something different, let **Ob** $\mathcal{C} = N = \{1, 2, 3, \ldots\}$, the positive integers. For $m, n \in$ **Ob** $\mathcal{C}$, let **hom**$(m, n) = N$ also, and let $m \circ n = mn$, the usual multiplication of positive integers.

   Note that the following categories are made by first fixing an object of one of the above categories.

   Example 7. *The category of left R-modules* $_R\mathcal{M}$. For a fixed $R \in$

Ob $\mathcal{R}$, let **Ob** $_R\mathcal{M}$ consist of all left R-modules, let **hom**(M, N) be the set $\mathrm{Hom}_R(M, N)$, all R-module homomorphisms $f : M \to N$, with $\circ$ denoting the expected composition.

Example 8. *The category of abstract R-affine spaces* $\mathrm{Aff}(R)$. As in example 7, we start with a fixed $R \in$ **Ob** $\mathcal{R}$, and let **Ob** $\mathrm{Aff}(R) =$ **Ob** $_R\mathcal{M}$, the class of all left R-modules. But this time our morphisms will be different. For $M, N \in$ **Ob** $\mathrm{Aff}(R)$, we let **hom**$(M, N) = \{A \mid A(x) = L(x) + \alpha, \ L \in \mathrm{Hom}_R(M, N), \ \alpha \in N\}$, where $\mathrm{Hom}_R(M, N)$ consists of all R-homomorphisms of the R-module M into the R-module N. That is, $A \in$ **hom**(M, N) is an *abstract affine transformation*.

Example 9. *The category of S-actions* $\mathcal{A}_S$. Fix a semigroup $(S, \cdot)$. As usual, we say that S *acts* on a set $X \neq \varnothing$, if there is a function, called an *action*, $* : X \times S \to X$ satisfying $x * (ab) = (x * a) * b$ for all $x \in X$ and for all $a, b \in S$. If $(S, \cdot, 1)$ is a monoid, then we will usually insist that $x * 1 = x$ for all $x \in X$. The objects $(X, *) \in$ **Ob** $\mathcal{A}_S$ are ordered pairs where $X \in$ **Ob** $\mathcal{S}$ and $*$ is an action of S on X. For objects $(X, *)$ and $(Y, \bullet)$, we let $f : X \to Y$ be an element of **hom**$[(X, *), (Y, \bullet)]$ if and only if $f(x * a) = f(x) \bullet a$ for all $x \in X$ and for all $a \in S$.

Example 10. *The category of commutative R-algebras with identity* $\mathcal{A}(R)$. For a fixed commutative ring with identity $(R, +, \cdot, 1)$, let **Ob** $\mathcal{A}(R)$ consist of all commutative R-algebras with identity. If $A, B \in$ **Ob** $\mathcal{A}(R)$, then **hom**(A, B) consists of all maps $f : A \to B$ such that f is a ring homomorphism, f is an R-module homomorphism, and $f(1_A) = 1_B$. So **hom**$(A, B) = \mathrm{Hom}_R(A, B)$.

Examples 11. Let $\mathcal{C} \in \{\mathcal{S}, \mathcal{S}_0, \mathcal{T}, \mathcal{G}, \mathcal{R}, {}_R\mathcal{M}, \mathrm{Aff}(R), \mathcal{A}_S\}$ and fix an $A \in$ **Ob**$\mathcal{C}$. We will construct a new category from $\mathcal{C}$ and A, and call it $\mathcal{C}(A)$. The objects $(X, \phi) \in$ **Ob** $\mathcal{C}(A)$ will be ordered pairs where $X \in$ **Ob** $\mathcal{C}$ and $\phi \in$ **hom**(X, A) and ϕ is a surjective mapping. For $(X, \phi), (Y, \lambda) \in$ **Ob** $\mathcal{C}(A)$, we will admit $f \in$ **hom**$[(X, \phi), (Y, \lambda)]$ if and only if $f \in$ **hom**(X, Y) for $\mathcal{C}$ and $\lambda \circ f = \phi$.

Examples 12. These are dual to those in examples 11. For the same $\mathcal{C}$ and $A \in$ **Ob** $\mathcal{C}$, we construct an alternative category from $\mathcal{C}$ and A, and call it $(A)\mathcal{C}$. The objects $(\phi, X) \in$ **Ob** $(A)\mathcal{C}$ will be ordered pairs where ϕ is the insertion map $\phi : A \to X$ with $\phi(a) = a$, $\phi \in$ **hom**(A, X), and $X \in$ **Ob** $\mathcal{C}$ contains A as a subset. Here, $f \in$ **hom**$[(\phi, X), (\lambda, Y)]$ if and only if $f \circ \phi = \lambda$ and $f \in$ **hom**(X, Y) for $\mathcal{C}$.

Exercise. Do all values for $\mathcal{C}$ in examples 11 and 12 really produce categories as outlined?

Examples of type 11 and 12 work because these $\mathcal{C}$ are examples of a *concrete category*. Specifically, a category $\mathcal{C}$ is *concrete* if to each $A \in \mathbf{Ob}\mathcal{C}$, there is a set $U(A)$, called the *underlying* set of A, such that $f \in \mathbf{hom}(A, B)$ implies that $f : U(A) \to U(B)$ is a function, $1_A \in \mathbf{hom}(A, A)$ equals $1_{U(A)}$, the identity function on the set $U(A)$, and $\circ$ of $\mathcal{C}$ really is a composition of functions. Example 6 is not a concrete category.

A category $\mathcal{C}$ may have an $I \in \mathbf{Ob}\mathcal{C}$ with the property that $\mathbf{hom}(I, X) = \{i_X\}$, a singleton, for each $X \in \mathbf{Ob}\,\mathcal{C}$. Examples 1, 2, 3, 4, 5, 7, and 8 have such objects. Example 6 obviously does not. What about the others? Such an object I is an *initial object*.

Similarly, a category $\mathcal{C}$ may have a $T \in \mathbf{Ob}\,\mathcal{C}$ with the property that $\mathbf{hom}(X, T) = \{t_X\}$, a singleton, for each $X \in \mathbf{Ob}\,\mathcal{C}$. Examples 1, 2, 3, 4, 5, 7, and 8 each have such objects. Example 6 does not. You may wish to decide if examples 9, 10, 11, and 12 have such objects. We shall call such a T a *terminal object*.

You may feel that the initial objects for any one of these categories are essentially the same. Also, you may feel that the terminal objects for any one category are essentially the same. We say the objects $A, B \in \mathbf{Ob}\,\mathcal{C}$ are *isomorphic* or *equivalent* if there is an $f \in \mathbf{hom}(A, B)$ and a $g \in \mathbf{hom}(B, A)$ such that $f \circ g = 1_B \in \mathbf{hom}(B, B)$ and $g \circ f = 1_A \in \mathbf{hom}(A, A)$. The morphisms f and g are *isomorphisms* and we write $f = g^{-1}$. This agrees with our usual ideas of isomorphic and homeomorphic, and one can easily confirm that initial objects of a category $\mathcal{C}$ are isomorphic, as are the terminal objects of a category.

The categories of interest to us in this work will have initial objects or terminal objects. They will also have *products* or *coproducts (sums)*. Actually, all we will need is a product or coproduct of two, and of three objects. For objects $A, B \in \mathbf{Ob}\mathcal{C}$, a *product* $(P(A, B), \pi_A, \pi_B)$ is an ordered triple, where $P(A, B) \in \mathbf{Ob}\mathcal{C}$, $\pi_A \in \mathbf{hom}(P(A, B), A)$, $\pi_B \in \mathbf{hom}(P(A, B), B)$, and the following universal property is valid. If $X \in \mathbf{Ob}\,\mathcal{C}$, $f_A \in \mathbf{hom}(X, A)$, $f_B \in \mathbf{hom}(X, B)$, then there is a unique $[f_A, f_B] \in \mathbf{hom}(X, P(A, B))$ satisfying $\pi_A \circ [f_A, f_B] = f_A$ and $\pi_B \circ [f_A, f_B] = f_B$. A *coproduct* or *sum* is a dual idea. It is also an ordered triple $(S(A, B), e_A, e_B)$, where $S(A, B) \in \mathbf{Ob}\,\mathcal{C}$, $e_A \in \mathbf{hom}(A, S(A, B))$, $e_B \in \mathbf{hom}(B, S(A, B))$, and the following universal property is valid. If $X \in \mathbf{Ob}\,\mathcal{C}$, $g_A \in \mathbf{hom}(A, X)$, $g_B \in \mathbf{hom}(B, X)$, then there is a unique $(g_A, g_B) \in \mathbf{hom}(S(A, B), X)$ satisfying $(g_A, g_B) \circ e_A = g_A$ and $(g_A, g_B) \circ e_B = g_B$.

The definitions for three objects are analogous. We will have $(P(A, B, C),$ $\pi_A, \pi_B, \pi_C)$ and $[f_A, f_B, f_C]$ for the product, and $(S(A, B, C), e_A, e_B, e_C)$ and (g_A, g_B, g_C) for the sum or coproduct.

We have spoken of 'a product' and 'a coproduct'. Indeed, objects $A, B \in \mathbf{Ob}\,\mathcal{C}$ may have various products or coproducts, but they are all equivalent, or isomorphic. Suppose $(P'(A, B), \pi'_A, \pi'_B)$ is also a product. First, letting $(P(A,B), \pi_A, \pi_B)$ be the product, and $(P'(A,B), \pi'_A, \pi'_B)$ be the (X, f_A, f_B), we obtain the $[f_A, f_B] = [\pi'_A, \pi'_B]$, and so $\pi_A \circ [\pi'_A, \pi'_B] = \pi'_A$ and $\pi_B \circ [\pi'_A, \pi'_B] = \pi'_B$. Reversing the roles, we obtain $\pi'_A \circ [\pi_A, \pi_B] = \pi_A$ and $\pi'_B \circ [\pi_A, \pi_B] = \pi_B$. Obviously, $\pi'_A \circ 1_{P'(A,B)} = \pi'_A$ and $\pi'_B \circ 1_{P'(A,B)} = \pi'_B$, and also $\pi_A \circ 1_{P(A,B)} = \pi_A$ and $\pi_B \circ 1_{P(A,B)} = \pi_B$. But $\pi_A \circ [\pi'_A, \pi'_B] \circ [\pi_A, \pi_B] = \pi'_A \circ [\pi_A, \pi_B] = \pi_A$, and $\pi_B \circ [\pi'_A, \pi'_B] \circ [\pi_A, \pi_B] = \pi'_B \circ [\pi_A, \pi_B] = \pi_B$. So $[\pi'_A, \pi'_B] \circ [\pi_A, \pi_B] = 1_{P(A,B)}$. Similarly, $[\pi_A, \pi_B] \circ [\pi'_A, \pi'_B] = 1_{P'(A,B)}$. This makes $P(A,B)$ and $P'(A,B)$ equivalent. I believe the reader can now prove that products of three objects are equivalent, and the same for coproducts.

Perhaps our notation suggests that a product $(P(A,B), \pi_A, \pi_B)$ might be different from a product $(P^*(B,A), \pi^*_B, \pi^*_A)$, and that the $[f_A, f_B]$ might be different from $[f_B, f_A]$. But study the roles of $P(A,B)$ and $P^*(B,A)$, of π_A and π^*_A, of π_B and π^*_B, and of $[f_A, f_B]$ and $[f_B, f_A]$. Each can be replaced by the other, so the order of the A and B is not important in $P(A,B)$. Similarly, for $A, B, C \in \mathbf{Ob}\,\mathcal{C}$, any permutation of A, B and C makes no difference for $(P(A,B,C), \pi_A, \pi_B, \pi_C)$.

There is a little problem when $A = B$. We cannot conclude that $\pi_A = \pi_B$, for if so, and if $f_A \neq f_B$, we would have $f_A = \pi_A \circ [f_A, f_B] = \pi_B \circ [f_A, f_B] = f_B$. When $A = B$, it will be convenient to use such notation as $(P(A,A), \pi_1, \pi_2)$ for the product, and similarly, if $A = B = C$, we use $(P(A,A,A), \pi_1, \pi_2, \pi_3)$. Of course, similar statements apply when considering coproducts.

What about $(P(A,B,C), \pi_A, \pi_B, \pi_C)$, $(P(P(A,B),C), \pi_{P(A,B)}, \pi_C)$, and $(P(A, P(B,C)), \pi_A, \pi_{P(B,C)})$? It should be comforting to know that the objects $P(A,B,C)$, $P(P(A,B),C)$, and $P(A, P(B,C))$ are all equivalent. One can see this by making a product of A, B, and C from each of the latter two. For example, starting with $f_A \in \mathbf{hom}(X,A)$, $f_B \in \mathbf{hom}(X,B)$, and $f_C \in \mathbf{hom}(X,C)$, and with $(P(P(A,B),C), \pi_{P(A,B)}, \pi_C)$, we define $\pi'_C = \pi_C$, $\pi'_A = \pi_A \circ \pi_{P(A,B)}$, and $\pi'_B = \pi_B \circ \pi_{P(A,B)}$. We have a unique $[f_A, f_B] \in \mathbf{hom}(X, P(A,B))$ satisfying $\pi_A \circ [f_A, f_B] = f_A$ and $\pi_B \circ [f_A, f_B] = f_B$. And then we have a unique $[[f_A, f_B], f_C] \in \mathbf{hom}(X, P(P(A,B),C))$ satisfying $\pi'_C \circ [[f_A, f_B], f_C] = \pi_C \circ [[f_A, f_B], f_C] = f_C$, and $\pi_{P(A,B)} \circ [[f_A, f_B], f_C] = [f_A, f_B]$. Consequently, $\pi'_A \circ [[f_A, f_B], f_C] = \pi_A \circ \pi_{P(A,B)} \circ [[f_A, f_B], f_C] = f_A$, and $\pi'_B \circ [[f_A, f_B], f_C] = \pi_B \circ \pi_{P(A,B)} \circ [[f_A, f_B], f_C] = f_B$. So, the quadruple $(P(P(A,B),C), \pi'_A, \pi'_B, \pi'_C)$ does exactly what a product of A, B, and C should do. Alternatively, one could show that $[[\pi_A, \pi_B], \pi_C] \in \mathbf{hom}(P(A,B,C), P(P(A,B),C))$, as well as the morphism $[\pi_A, [\pi_B, \pi_C]] \in \mathbf{hom}(P(A,B,C), P(A, P(B,C)))$, are isomorphisms.

It will be convenient to have some special notation. For $f \in \mathbf{hom}(A, A')$ and $g \in \mathbf{hom}(B, B')$, let $f \times g = [f \circ \pi_A, g \circ \pi_B] \in \mathbf{hom}(P(A,B), P(A', B'))$ and $f \vee g = (e_{A'} \circ f, e_{B'} \circ g) \in \mathbf{hom}(S(A,B), S(A', B'))$. Also, let $\Delta = [1_A, 1_A]$ and $\nabla = (1_A, 1_A)$.

For our purpose, we shall be interested in products for $\mathcal{C}$ only when $\mathcal{C}$ has a terminal object, and, dually, we shall be interested in coproducts for $\mathcal{C}$ only when $\mathcal{C}$ has an initial object. We shall have occasion to refer to the products and coproducts of the examples of (10.2).

Products and coproducts from the 'Dirty Dozen'

Example 1. $\mathcal{S}$ has $\varnothing$ as an initial object, and any single element set $X = \{x\}$ is a terminal object. A product of two objects $A, B \in \mathbf{Ob}\,\mathcal{S}$ is simply the cartesian product $A \times B$, together with the two corresponding projections π_A and π_B.

Example 2. Objects $(\{x\}, x)$ are both initial and terminal objects for $\mathcal{S}_0$. The product of (A, a) and (B, b) is $((A \times B, (a, b)), \pi_A, \pi_B)$. The sum is more complicated. Let A_0 be $A \times \{0\}$ with $(a, 0)$ replaced by $[(a, 0), (b, 1)]$, and let B_1 be $B \times \{1\}$ with $(b, 1)$ replaced by $[(a, 0), (b, 1)]$. Given $f \in \mathbf{hom}((A, a), (X, x))$ and $g \in \mathbf{hom}((B, b), (X, x))$, define $(f, g) : A_0 \cup B_1 \to X$ by $(f, g)(s, 0) = f(s)$, $(f, g)(t, 1) = g(t)$, and $(f, g)[(a, 0), (b, 1)] = x$. If $e_A(s) = (s, 0)$, $e_A(a) = [(a, 0), (b, 1)]$, and $e_B(t) = (t, 1)$, $e_B(b) = [(a, 0), (b, 1)]$, then $((A_0 \cup B_1, [(a, 0), (b, 1)]), e_A, e_B)$ is a coproduct of (A, a) and (B, b).

Example 3. $\mathcal{T}$ has initial object $(\varnothing, \{\varnothing\})$, and single point spaces $(\{x\}, \{\{x\}, \varnothing\})$ are terminal objects. A product of (A, T_A) and (B, T_B) is $(A \times B, T)$, where T is the product topology, together with the natural projections π_A and π_B.

Example 4. Groups of order 1 are both initial objects and terminal objects in $\mathcal{G}$. A product of A and B is $(A \times B, \pi_A, \pi_B)$, where $A \times B$ is the external direct product. Coproducts are not quite so easy and are somewhat less familiar. A coproduct of A and B is the free product $A * B$, together with e_A, where $e_A(a) = a$, and with e_B, where $e_B(b) = b$. For further details, see [R, th.12.30].

Example 5. Rings of order 1 are both initial objects and terminal objects. The direct sum $A \oplus B$, together with the projections π_A and π_B, give us a product, but not a coproduct.

Example 6. $\mathcal{C}$ has neither an initial object nor a terminal object.

Example 7. A trivial module $\{0\}$ is both an initial object and a terminal object. For modules $M, N \in \mathbf{Ob}\,_R\mathcal{M}$, the direct sum $M \oplus N$ is the object of both the product and the coproduct. For the product, we also need the projections π_M and π_N. For the coproduct, we need the insertions e_M and e_N, where $e_M(m) = (m, 0)$, and $e_N(n) = (0, n)$.

Example 8. Again, a trivial module is a terminal object, but it is no longer an initial object. Products are exactly as for example 7.

Example 9. Any singleton $\{x\}$ gives us a terminal object. For $(X, \cdot)$, $(Y, \cdot) \in \mathbf{Ob}\,\mathcal{A}_S$, the object of the product will be $(X \times Y, \cdot)$, where $(x, y)a = (xa, ya)$. With the usual projections, π_X and π_Y, one obtains a product.

Example 10. R itself is an initial object. Let $A, B \in \mathbf{Ob}\,\mathcal{A}(R)$. The tensor product $A \otimes_R B$ will be used for their coproduct. The morphisms e_A and e_B are defined by $e_A(a) = a \otimes 1$ and $e_B(b) = 1 \otimes b$.

Examples 11. Let $\mathcal{C} = \mathcal{S}$. The object $(A, 1_A)$ is a terminal object. Let $(S, \sigma), (T, \tau) \in \mathbf{Ob}\mathcal{C}(A)$. We will now construct a product, called the *fibered product*, of (S, σ) and (T, τ). Let $S \times_A T = \{(s, t) \in S \times T \mid \sigma(s) = \tau(t)\}$, and let $\phi : S \times_A T \to A$ be defined by $\phi(s, t) = \sigma(s) \; (= \tau(t))$. The morphisms π_S and π_T will be the natural projections $\pi_S(s, t) = s$ and $\pi_T(s, t) = t$. It is elementary to see that $((S \times_A T, \phi), \pi_S, \pi_T)$ is a desired product.

Exercise. For which of the other possible values for $\mathcal{C}$ in examples 11 is $((S \times_A T, \phi), \pi_S, \pi_T)$ a product?

Examples 12. Let $\mathcal{C} = \mathcal{S}$. The object $(1_A, A) \in \mathbf{Ob}\,(A)\mathcal{S}$ is an initial object. We will now construct a coproduct of objects $(\lambda, L), (\mu, M) \in \mathbf{Ob}\,(A)\mathcal{S}$. Suppose $(l, 0) \notin A$ if $l \in L \setminus A$, and $(m, 1) \notin A$ if $m \in M \setminus A$. If this is not the case, then replace 0 and/or 1 by appropriate symbols. Let $L_0 = \{(l, 0) \mid l \in L \setminus A\}$, $M_1 = \{(m, 1) \mid m \in M \setminus A\}$ and $S(L, M) = A \cup L_0 \cup M_1$, a disjoint union. Define e_L by $e_L(l) = (l, 0)$ and $e_L(a) = a$ for $a \in A$ and $l \in L \setminus A$. Similarly, $e_M(m) = (m, 1)$ and $e_M(a) = a$ for $a \in A$ and $m \in M \setminus A$. Then $(S(L, M), e_L, e_M)$ is a coproduct.

We shall need to know that initial and terminal objects are used to construct further equivalent objects.

(10.3) Theorem. *For a terminal object* $T \in \mathbf{Ob}\,\mathcal{C}$ *and any* $A \in \mathbf{Ob}\,\mathcal{C}$, *if the product* $(P(A, T), \pi_A, \pi_T)$ *exists for* $\mathcal{C}$, *then the objects* $P(A, T)$ *and* A *are equivalent.*

Proof. Let $1_A \in \mathbf{hom}(A, A)$ and $t_A \in \mathbf{hom}(A, T)$ be given. Then there is a unique $[1_A, t_A] \in \mathbf{hom}(A, P(A, T))$ such that $\pi \circ [1_A, t_A] = 1_A$. It remains to show that $[1_A, t_A] \circ \pi_A = 1_{P(A, T)}$.

Note that $\pi_T = t_{P(A, T)}$. Starting with the product $(P(A, T), \pi_A, t_{P(A, T)})$, and given $1_A \circ \pi_A = \pi_A \in \mathbf{hom}(P(A, T), A)$ and $t_A \circ \pi_A = t_{P(A, T)} \in \mathbf{hom}(P(A, T), T)$, there is a unique $[1_A \circ \pi_A, t_A \circ \pi_A] = [\pi_A, t_{P(A, T)}] \in \mathbf{hom}(P(A, T), P(A, T))$ satisfying $\pi_A \circ [\pi_A, t_{P(A, T)}] = \pi_A = 1_A \circ \pi_A$ and $\pi_T \circ [\pi_A, t_{P(A, T)}] = t_{P(A, T)} \circ [1_A \circ \pi_A, t_A \circ \pi_A] = t_{P(A, T)} = \pi_T$. But

$\pi_A \circ 1_{P(A,T)} = \pi_A$ and $\pi_T \circ 1_{P(A,T)} = \pi_T$, so $1_{P(A,T)} = [\pi_A, t_{P(A,T)}] = [1_A \circ \pi_A, t_A \circ \pi_A]$.

On the other hand, $\pi_A \circ [1_A, t_A] \circ \pi_A = 1_A \circ \pi_A = \pi_A$ and $\pi_T \circ [1_A, t_A] \circ \pi_A = t_A \circ \pi_A = t_{P(A,T)}$, so $[1_A, t_A] \circ \pi_A = [1_A \circ \pi_A, t_A \circ \pi_A] = 1_{P(A,T)}$, as was intended.

The reader is now invited to provide a proof to

(10.4) Theorem. *For an initial object $I \in$ **Ob** C and any $A \in$ **Ob** C, if the coproduct $(S(A, I), e_A, e_T)$ exists for C, then the objects $S(A, I)$ and $A are equivalent.*

We complete this section with two related ideas. Viewing a category itself as a mathematical structure, it is natural to have a 'morphism' between two categories. Let C and D be two categories. A *(covariant) functor F* from C to D, often denoted by $F : C \to D$, is really a pair of functions, an *object function*, and a *morphism function*. Each of these functions will also be denoted by F. If $A \in$ **Ob** C, we must have $F(A) \in$ **Ob** D. If $f \in$ **hom**(A, B) in C, then $F(f) \in$ **hom**$(F(A), F(B))$ in D. In addition, $F(1_A) = 1_{F(A)}$ for each $A \in$ **Ob** C, and $F(g \circ f) = F(g) \circ F(f)$ for any $f \in$ **hom**(A, B) and $g \in$ **hom**(B, C) in C. A *contravariant functor F* from C to D, also denoted by $F : C \to D$, has a similar definition. It is also an object function and a morphism function. However, if $f \in$ **hom**(A, B) in C, a contravariant functor F has $F(f) \in$ **hom**$(F(B), F(A))$ in D. In addition to $F(1_A) = 1_{F(A)}$ for each $A \in$ **Ob** C, we have $F(g \circ f) = F(f) \circ F(g)$ for any $f \in$ **hom**(A, B) and $g \in$ **hom**(B, C) in C.

(10.5) Examples. Several examples of a covariant functor can be obtained by modifying the following covariant functor $F : S \to G$ appropriately. For $X \in$ **Ob** S, let $F(X) = \mathcal{F}(X)$ denote the free group on the set X. Let $\iota_X : X \to F(X)$ denote the corresponding insertion map, where $\iota_X(x) = x$ for all $x \in X$. For $\varnothing \neq Y \in$ **Ob** S, and $f \in$ **hom**(X, Y) in S, let $F(f)$ denote the extension of $\iota_Y \circ f$ to the unique group homomorphism $F(f) \in$ **hom**$(F(X), F(Y))$. Then it is easy to see that F is a covariant functor.

Take any category C and fix any $X \in$ **Ob** C. Define $H : C \to S$ by $H(A) =$ **hom**$(A, X) \in$ **Ob** S for any $A \in$ **Ob** C. If $\phi \in$ **hom**(A, B) in C, define $H(\phi) \in$ **hom**$(H(B), H(A))$ by $H(\phi)(f) = f \circ \phi$. Then H is a contravariant functor.

Actually, H has a covariant counterpart. Take any category C and fix any $X \in$ **Ob** C. Define $G : C \to S$ by $G(A) =$ **hom**$(X, A) \in$ **Ob** S for any $A \in$ **Ob** C. If $\phi \in$ **hom**(A, B) in C, define $G(\phi) \in$ **hom**$(G(A), G(B))$ by $G(\phi)(f) = \phi \circ f$. Then G is a covariant functor.

We shall use variations of G and H later in this chapter.

11. Group and cogroup objects

Well, exactly what are these group and cogroup objects? Group objects are the results of trying to describe something like a group in a categorical setting, using only objects and morphisms. Cogroup objects are the results of taking the dual idea of a group object.

For a group $(G, +)$, we need (1) a set $G \neq \varnothing$, and (2) a binary operation $+$. Since $+ : G \times G \to G$, we start out with $G \in \mathbf{Ob}\,\mathcal{C}$ and $\pi \in \mathbf{hom}(P(G,G),G)$. Now $+$ is associative. What does this mean for π? It means that the following diagram is commutative, where f and g are appropriate isomorphisms. In particular, $f = [[\pi_1, \pi_2], \pi_3]$ and $g = [\pi_1, [\pi_2, \pi_3]]$ for the product $(P(G,G,G), \pi_1, \pi_2, \pi_3)$.

$$
\begin{array}{ccccc}
P(G,G,G) & \xrightarrow{\ f\ } & P(P(G,G),G) & \xrightarrow{\pi \times 1_G} & P(G,G) \\
\big\downarrow{\scriptstyle g} & & & & \\
P(G,P(G,G)) & & & & \big\downarrow{\scriptstyle \pi} \qquad (11:1) \\
\big\downarrow{\scriptstyle 1_G \times \pi} & & & & \\
P(G,G) & & \xrightarrow{\ \pi\ } & & G
\end{array}
$$

The associative diagram.

That is, $\pi \circ (\pi \times 1_G) \circ f = \pi \circ (1_G \times \pi) \circ g$. If π satisfies (11:1), then (G, π) is something like a semigroup.

Next, we need an identity. For a group $(G, +)$, any $\mu \in \mathbf{hom}(\{0\}, G)$ has $\mu(0) = 0 \in G$. In a categorical setting, we require a terminal object T and a $\mu \in \mathbf{hom}(T, G)$ satisfying the following diagram.

$$
\begin{array}{ccccc}
P(T,G) & \xrightarrow{\mu \times 1_G} & P(G,G) & \xleftarrow{1_G \times \mu} & P(G,T) \\
\big\downarrow{\scriptstyle [t_G,1_G]^{-1}} & & \big\downarrow{\scriptstyle \pi} & & \big\downarrow{\scriptstyle [1_G,t_G]^{-1}} \quad (11:2) \\
G & \xrightarrow{1_G} & G & \xleftarrow{1_G} & G
\end{array}
$$

The identity diagram.

That is, $[t_G, 1_G]^{-1} = \pi \circ (\mu \times 1_G)$ and $[1_G, t_G]^{-1} = \pi \circ (1_G \times \mu)$. If π satisfies (11:1) and μ satisfies (11:2), then (G, π, μ) is like a monoid.

Finally, we need an inverse. Now $g \mapsto -g$ is a map. Hence, we need $\alpha \in \mathbf{hom}(G, G)$ satisfying the following diagram.

$$
\begin{array}{ccccc}
P(G,G) & \xrightarrow{\alpha \times 1_G} & P(G,G) & \xleftarrow{1_G \times \alpha} & P(G,G) \\
\big\uparrow{\scriptstyle \Delta} & & \big\downarrow{\scriptstyle \pi} & & \big\uparrow{\scriptstyle \Delta} \\
G & & & & G \qquad (11:3) \\
\big\downarrow{\scriptstyle t_G} & & & & \big\downarrow{\scriptstyle t_G} \\
T & \xrightarrow{\ \mu\ } & G & \xleftarrow{\ \mu\ } & T
\end{array}
$$

The inverse diagram.

That is, $\pi \circ (\alpha \times 1_G) \circ \Delta = \mu \circ t_G = \pi \circ (1_G \times \alpha) \circ \Delta$. We can now state

(11.1) Definition. For a category $\mathcal{C}$ with products $P(A, B)$ and a terminal object T, a *group object* is a quadruple (G, π, μ, α) where $G \in \mathbf{Ob}\,\mathcal{C}$, $\pi \in \mathbf{hom}(P(G, G), G)$, $\mu \in \mathbf{hom}(T, G)$, $\alpha \in \mathbf{hom}(G, G)$, and the diagrams (11:1), (11:2), and (11:3) are all commutative.

Now what about the dual of a group object? Our category $\mathcal{C}$ must have sums and an initial object I. Diagram (11:1) gives us the following commutative diagram where $f = ((e_1, e_2), e_3)$ and $g = (e_1, (e_2, e_3))$ for the coproduct $(S(G, G, G), e_1, e_2, e_3)$.

$$
\begin{array}{ccccc}
S(G, G, G) & \xleftarrow{\;f\;} & S(S(G, G), G) & \xleftarrow{\pi \vee 1_G} & S(G, G) \\
\big\uparrow{\scriptstyle g} & & & & \\
S(G, S(G, G)) & & & & \big\uparrow{\scriptstyle \pi} \qquad (11:4) \\
\big\uparrow{\scriptstyle 1_G \vee \pi} & & & & \\
S(G, G) & & \xleftarrow{\;\pi\;} & & G
\end{array}
$$

The coassociative diagram.

That is, $f \circ (\pi \vee 1_G) \circ \pi = g \circ (1_G \vee \pi) \circ \pi$, where $\pi \in \mathbf{hom}(G, S(G, G))$. The identity diagram also gives us a commutative diagram.

$$
\begin{array}{ccccc}
S(I, G) & \xleftarrow{\mu \vee 1_G} & S(G, G) & \xrightarrow{1_G \vee \mu} & S(G, I) \\
\big\uparrow{\scriptstyle (i_G, 1_G)^{-1}} & & \big\uparrow{\scriptstyle \pi} & & \big\uparrow{\scriptstyle (1_G, i_G)^{-1}} \quad (11:5) \\
G & \xleftarrow{1_G} & G & \xrightarrow{1_G} & G
\end{array}
$$

The coidentity diagram.

That is, $(\mu \vee 1_G) \circ \pi = (i_G, 1_G)^{-1}$ and $(1_G \vee \mu) \circ \pi = (1_G, i_G)^{-1}$ where $\mu \in \mathbf{hom}(G, I)$. Now (11:3) gives us the following commutative diagram.

$$
\begin{array}{ccccc}
S(G, G) & \xleftarrow{\alpha \vee 1_G} & S(G, G) & \xrightarrow{1_G \vee \alpha} & S(G, G) \\
\big\downarrow{\scriptstyle \nabla} & & & & \big\downarrow{\scriptstyle \nabla} \\
G & & \big\uparrow{\scriptstyle \pi} & & G \qquad (11:6) \\
\big\uparrow{\scriptstyle i_G} & & & & \big\uparrow{\scriptstyle i_G} \\
I & \xleftarrow{\mu} & G & \xrightarrow{\mu} & I
\end{array}
$$

The coinverse diagram.

That is, $\nabla \circ (\alpha \vee 1_G) \circ \pi = i_G \circ \mu = \nabla \circ (1_G \vee \alpha) \circ \pi$, where $\alpha \in \mathbf{hom}(G, G)$. So, we state

(11.2) Definition. For a category $\mathcal{C}$ with coproducts $S(A, B)$ and an initial object I, a *cogroup object* is a quadruple (G, π, μ, α) where $G \in \mathbf{Ob}\mathcal{C}$, $\pi \in \mathbf{hom}(G, S(G, G))$, $\mu \in \mathbf{hom}(G, I)$, $\alpha \in \mathbf{hom}(G, G)$, and the diagrams (11:4), (11:5), and (11:6) are all commutative.

What is a $\oplus$ for $\mathbf{hom}(G, G)$ for a group object or a cogroup object? Actually, we are not restricted to $\mathbf{hom}(G, G)$.

Let (G, π, μ, α) be a group object for a category $\mathcal{C}$. Let $X \in \mathbf{Ob}\,\mathcal{C}$, and $f, g \in \mathbf{hom}(X, G)$. Define $\oplus$ by

$$f \oplus g = \pi \circ [f, g]. \tag{11:7}$$

Dually, let (G, π, μ, α) be a cogroup object for a category $\mathcal{C}$. For $X \in \mathbf{Ob}\mathcal{C}$, and $f, g \in \mathbf{hom}(G, X)$, define $\oplus$ on $\mathbf{hom}(G, X)$ by

$$f \oplus g = (f, g) \circ \pi. \tag{11:8}$$

We then obtain the following very important theorem.

(11.3) Theorem. *For a group object* (G, π, μ, α) *of a category* $\mathcal{C}$, *we obtain that* $(\mathbf{hom}(X, G), \oplus)$ *is a group with identity* $\mu \circ t_X$ *and inverse* $-f = \alpha \circ f$.
For a cogroup object (G, π, μ, α) *of a category* $\mathcal{C}$, $(\mathbf{hom}(G, X), \oplus)$ *is a group with identity* $i_X \circ \mu$ *and inverse* $-f = f \circ \alpha$.

Proof. Let (G, π, μ, α) be a group object of $\mathcal{C}$. Certainly, $f \oplus g$ of (11:7) defines a binary operation on $\mathbf{hom}(X, G)$. We need to show that $\oplus$ is associative, that $\mu \circ t_X$ is the identity for $\oplus$, and that $-f = \alpha \circ f$ for each $f \in \mathbf{hom}(X, G)$. This will be done partially in a series of steps. Recall that T is a terminal object and that $\mathbf{hom}(X, T) = \{t_X\}$.

1. *For* $f \in \mathbf{hom}(X, G)$ *and* $g, h \in \mathbf{hom}(G, G)$, *we have* $[g \circ f, h \circ f] = [g, h] \circ f$.

Proof. For the product $(P(G, G), \pi_1, \pi_2)$, we have $\pi_1 \circ [g, h] \circ f = g \circ f = \pi_1 \circ [g \circ f, h \circ f]$ and $\pi_2 \circ [g, h] \circ f = h \circ f = \pi_2 \circ [g \circ f, h \circ f]$. By uniqueness, $[g, h] \circ f = [g \circ f, h \circ f]$.

2. *For* $f \in \mathbf{hom}(X, G)$, $t_X = t_G \circ f$.

Proof. Certainly, $t_G \circ f \in \mathbf{hom}(X, T) = \{t_X\}$.

3. $(1_G \times \mu) \circ [1_G, t_G] = [1_G, \mu \circ t_G]$.

Proof. Similar to the proof of **1**, $\pi_1 \circ (1_G \times \mu) \circ [1_G, t_G] = \pi_1 \circ [1_G \circ \pi_G, \mu \circ \pi_T] \circ [1_G, t_G] = (1_G \circ \pi_G) \circ [1_G, t_G] = \pi_G \circ [1_G, t_G] = 1_G$, and $\pi_2 \circ (1_G \times \mu) \circ [1_G, t_G] = \pi_2 \circ [1_G \circ \pi_G, \mu \circ \pi_T] \circ [1_G, t_G] = (\mu \circ \pi_T) \circ [1_G, t_G] = \mu \circ t_G$. So $(1_G \times \mu) \circ [1_G, t_G]$ does exactly what $[1_G, \mu \circ t_G]$ should do, so they must be the same.

4. $1_G \oplus (\mu \circ t_G) = 1_G$.

Proof. $1_G \oplus (\mu \circ t_G) = \pi \circ [1_G, \mu \circ t_G] = \pi \circ (1_G \times \mu) \circ [1_G, t_G]$. Now $\pi \circ (1_G \times \mu) = [1_G, t_G]^{-1}$ by (11:2).

5. *For $f \in \mathbf{hom}(X, G)$, we obtain $f \oplus (\mu \circ t_X) = f$.*

Proof. $f \oplus (\mu \circ t_X) = \pi \circ [f, \mu \circ t_X] = \pi \circ [1_G, \mu \circ t_G] \circ f$ by **1** and **2**. But $\pi \circ [1_G, \mu \circ t_G] = 1_G \oplus (\mu \circ t_G) = 1_G$ by **4**.

At this point, we know that $\mu \circ t_X$ is a right identity for $\oplus$.

6. *For $f \in \mathbf{hom}(X, G)$, $\alpha \circ f = -f$, a right inverse of f with respect to $\oplus$ and $\mu \circ t_X$.*

Proof. We must show that $f \oplus (\alpha \circ f) = \mu \circ t_X$. Now $f \oplus (\alpha \circ f) = \pi \circ [f, \alpha \circ f] = \pi \circ [1_G, \alpha] \circ f$, $\mu \circ t_X = \mu \circ t_G \circ f$ by **2**, and $\mu \circ t_G = \pi \circ (1_G \times \alpha) \circ \Delta$ by (11:3). So, when we have $(1_G \times \alpha) \circ \Delta = [1_G, \alpha]$, we will be finished.

But $(1_G \times \alpha) \circ \Delta = [1_G \circ \pi_1, \alpha \circ \pi_2] \circ [1_G, 1_G]$. Now $\pi_1 \circ [1_G \circ \pi_1, \alpha \circ \pi_2] \circ [1_G, 1_G] = (1_G \circ \pi_1) \circ [1_G, 1_G] = 1_G$ and $\pi_1 \circ [1_G, \alpha] = 1_G$. Also, $\pi_2 \circ [1_G \circ \pi_1, \alpha \circ \pi_2] \circ [1_G, 1_G] = (\alpha \circ \pi_2) \circ [1_G, 1_G] = \alpha$ and $\pi_2 \circ [1_G, \alpha] = \alpha$. By uniqueness, we must have $(1_G \times \alpha) \circ \Delta = [1_G, \alpha]$.

Finally, to show that $\oplus$ is associative, one must be prepared to draw several diagrams and be very patient while going through lots of detail. It will be convenient to let $G_1 = G_2 = G_3 = G$, and to consider the following five products: (1) $(P(G_1, G_2, G_3), \pi_1, \pi_2, \pi_3)$; (2) $(P(G_2, G_3), \pi_2', \pi_3')$; (3) $(P(G_1, G_2), \pi_1^*, \pi_2^*)$; (4) $(P(G_1, P(G_2, G_3)), \pi_1', \pi_{23}')$; and finally the product (5) $(P(P(G_1, G_2), G_3), \pi_{12}^*, \pi_3^*)$. Start with $f_1, f_2, f_3 \in \mathbf{hom}(X, G)$. We must show that $\pi \circ [f_1, \pi \circ [f_2, f_3]] = f_1 \oplus (f_2 \oplus f_3) = (f_1 \oplus f_2) \oplus f_3 = \pi \circ [\pi \circ [f_1, f_2], f_3]$.

First, we will show that $[f_1, \pi \circ [f_2, f_3]] = 1_{G_1} \times \pi \circ [\pi_1, [\pi_2, \pi_3]] \circ [f_1, f_2, f_3]$, and then we will show that $[\pi \circ [f_1, f_2], f_3] = \pi \times 1_{G_3} \circ [[\pi_1, \pi_2], \pi_3] \circ [f_1, f_2, f_3]$. So, we will conclude that $\pi \circ 1_{G_1} \times \pi \circ [\pi_1, [\pi_2, \pi_3]] \circ [f_1, f_2, f_3] = \pi \circ [f_1, \pi \circ [f_2, f_3]] = f_1 \oplus (f_2 \oplus f_3)$ and $\pi \circ \pi \times 1_{G_3} \circ [[\pi_1, \pi_2], \pi_3] \circ [f_1, f_2, f_3] = (f_1 \oplus f_2) \oplus f_3$. From (11:1), we know that $\pi \circ 1_{G_1} \times \pi \circ [\pi_1, [\pi_2, \pi_3]] = \pi \circ \pi \times 1_{G_3} \circ [[\pi_1, \pi_2], \pi_3]$, so we will be able to conclude that that $\oplus$ is associative.

Define $\pi_{12} = [\pi_1, \pi_2]$, and $\pi_{23} = [\pi_2, \pi_3]$. Note that $\pi_i \circ [f_1, f_2, f_3] = f_i$, $\pi_2' \circ [f_2, f_3] = f_2$, $\pi_3' \circ [f_2, f_3] = f_3$, $\pi_2' \circ \pi_{23} = \pi_2$, $\pi_3' \circ \pi_{23} = \pi_3$, $\pi_1^* \circ \pi_{12} = \pi_1$, and $\pi_2^* \circ \pi_{12} = \pi_2$.

Now $\pi_1^* \circ (\pi_{12} \circ [f_1, f_2, f_3]) = \pi_1 \circ [f_1, f_2, f_3] = f_1$ and $\pi_2^* \circ (\pi_{12} \circ [f_1, f_2, f_3]) = \pi_2 \circ [f_1, f_2, f_3] = f_2$. Hence,

$$[f_1, f_2] = \pi_{12} \circ [f_1, f_2, f_3]. \tag{11:9}$$

Similarly, $\pi_2' \circ (\pi_{23} \circ [f_1, f_2, f_3]) = \pi_2 \circ [f_1, f_2, f_3] = f_2$ and $\pi_3' \circ (\pi_{23} \circ$

$[f_1, f_2, f_3]) = \pi_3 \circ [f_1, f_2, f_3] = f_3$. Hence,

$$[f_2, f_3] = \pi_{23} \circ [f_1, f_2, f_3]. \qquad (11:10)$$

Next, we compute $\pi_1^* \circ 1_{G_1} \times \pi \circ [\pi_1, [\pi_2, \pi_3]] \circ [f_1, f_2, f_3] = \pi_1^* \circ [1_{G_1} \circ \pi_1', \pi \circ \pi_{23}'] \circ [\pi_1, [\pi_2, \pi_3]] \circ [f_1, f_2, f_3] = 1_{G_1} \circ \pi_1' \circ [\pi_1, [\pi_2, \pi_3]] \circ [f_1, f_2, f_3] = \pi_1 \circ [f_1, f_2, f_3] = f_1$. And $\pi_2^* \circ 1_{G_1} \times \pi \circ [\pi_1, [\pi_2, \pi_3]] \circ [f_1, f_2, f_3] = \pi_2^* \circ [1_{G_1} \circ \pi_1', \pi \circ \pi_{23}'] \circ [\pi_1, [\pi_2, \pi_3]] \circ [f_1, f_2, f_3] = \pi \circ \pi_{23}' \circ [\pi_1, [\pi_2, \pi_3]] \circ [f_1, f_2, f_3] = \pi \circ [\pi_2, \pi_3] \circ [f_1, f_2, f_3] = \pi \circ \pi_{23} \circ [f_1, f_2, f_3] = \pi \circ [f_2, f_3]$ by (11:10).
But we also have $\pi_1^* \circ [f_1, \pi \circ [f_2, f_3]] = f_1$, and $\pi_2^* \circ [f_1, \pi \circ [f_2, f_3]] = \pi \circ [f_2, f_3]$. So, by uniqueness, we must conclude that

$$[f_1, \pi \circ [f_2, f_3]] = 1_{G_1} \times \pi \circ [\pi_1, [\pi_2, \pi_3]] \circ [f_1, f_2, f_3]. \qquad (11:11)$$

With a similar argument, $\pi_2' \circ \pi \times 1_{G_3} \circ [[\pi_1, \pi_2], \pi_3] \circ [f_1, f_2, f_3] = \pi_2' \circ [\pi \circ \pi_{12}^*, 1_{G_3} \circ \pi_3^*] \circ [[\pi_1, \pi_2], \pi_3] \circ [f_1, f_2, f_3] = \pi \circ \pi_{12}^* \circ [[\pi_1, \pi_2], \pi_3] \circ [f_1, f_2, f_3] = \pi \circ [\pi_1, \pi_2] \circ [f_1, f_2, f_3] = \pi \circ [f_1, f_2]$ by (11:9). And $\pi_3' \circ \pi \times 1_{G_3} \circ [[\pi_1, \pi_2], \pi_3] \circ [f_1, f_2, f_3] = \pi_3' \circ [\pi \circ \pi_{12}', 1_{G_3} \circ \pi_3'] \circ [[\pi_1, \pi_2], \pi_3] \circ [f_1, f_2, f_3] = 1_{G_3} \circ \pi_3' \circ [[\pi_1, \pi_2], \pi_3] \circ [f_1, f_2, f_3] = \pi_3 \circ [f_1, f_2, f_3] = f_3$.
But we also have $\pi_2' \circ [\pi \circ [f_1, f_2], f_3] = \pi \circ [f_1, f_2]$ and $\pi_3' \circ [\pi \circ [f_1, f_2], f_3] = f_3$. By uniqueness, we have

$$[\pi \circ [f_1, f_2], f_3] = \pi \times 1_{G_3} \circ [[\pi_1, \pi_2], \pi_3] \circ [f_1, f_2, f_3]. \qquad (11:12)$$

This completes what we must do to show that $\oplus$ is associative. Hence, $(\mathbf{hom}(X, G), \oplus)$ really is a group.

We have followed tradition up to this point by leaving the proofs of dual statements to the diligent reader. However, in this case, we want to emphasize the role of cogroup objects, so we will proceed with the proof that $(\mathbf{hom}(G, X), \oplus)$ is also a group. This should also help the reader to provide proofs to dual statements when we continue to follow tradition.

When (G, π, μ, α) is a cogroup object, our proof *will* parallel that which has been done. Recall that I is an initial object and that $\mathbf{hom}(I, X) = \{i_X\}$.

1d. *For $f \in \mathbf{hom}(G, X)$ and $g, h \in \mathbf{hom}(G, G)$, we have $f \circ (g, h) = (f \circ g, f \circ h)$.*

Proof. For the coproduct $(S(G, G), e_1, e_2)$, we have $f \circ (g, h) \circ e_1 = f \circ g$ and $f \circ (g, h) \circ e_2 = f \circ h$. By uniqueness, $f \circ (g, h) = (f \circ g, f \circ h)$.

2d. *For $f \in \mathbf{hom}(G, X)$, $i_X = f \circ i_G$.*

Proof. Certainly, $f \circ i_G \in \mathbf{hom}(I, X) = \{i_X\}$.

3d. $(1_G, i_G) \circ (1_G \vee \mu) = (1_G, i_G \circ \mu)$.

Proof. $(1_G, i_G) \circ (1_G \vee \mu) \circ e_1 = (1_G, i_G) \circ (e_G \circ 1_G, e_I \circ \mu) \circ e_1 = (1_G, i_G) \circ e_G = 1_G$ and $(1_G, i_G) \circ (1_G \vee \mu) \circ e_2 = (1_G, i_G) \circ (e_G \circ 1_G, e_I \circ \mu) \circ e_2 = (1_G, i_G) \circ e_I \circ \mu = i_G \circ \mu$.

4d. $1_G \oplus (i_G \circ \mu) = 1_G$.

Proof. $1_G \oplus (i_G \circ \mu) = (1_G, i_G \circ \mu) \circ \pi = (1_G, i_G) \circ (1_G \vee \mu) \circ \pi = (1_G, i_G) \circ (1_G, i_G)^{-1} = 1_G$ by (11:5).

5d. *For $f \in \mathbf{hom}(G, X)$, we obtain $f \oplus (i_X \circ \mu) = f$.*

Proof. $f \oplus (i_X \circ \mu) = (f, i_X \circ \mu) \circ \pi = f \circ (1_G, i_G \circ \mu) \circ \pi = f \circ [1_G \oplus (i_G \circ \mu)] = f \circ 1_G = f$.

Now, we know that $i_X \circ \mu$ is a right identity for $\oplus$.

6d. *For $f \in \mathbf{hom}(G, X)$, $f \circ \alpha = -f$, a right inverse of f with respect to $\oplus$ and $i_X \circ \mu$.*

Proof. We must show that $f \oplus (f \circ \alpha) = i_X \circ \mu$. Now $f \oplus (f \circ \alpha) = (f, f \circ \alpha) \circ \pi = f \circ (1_G, \alpha) \circ \pi$. Also, $i_X \circ \mu = f \circ i_G \circ \mu$ and $i_G \circ \mu = \nabla \circ (1_G \vee \alpha) \circ \pi$ by (11:6). So when we obtain $\nabla \circ (1_G \vee \alpha) = (1_G, \alpha)$, we will be finished.

But $\nabla \circ (1_G \vee \alpha) = (1_G, 1_G) \circ (e_G \circ 1_G, e_I \circ \alpha)$. Now $(1_G, 1_G) \circ (e_G \circ 1_G, e_I \circ \alpha) \circ e_1 = (1_G, 1_G) \circ e_G \circ 1_G = 1_G$ and $(1_G, 1_G) \circ (e_G \circ 1_G, e_I \circ \alpha) \circ e_2 = (1_G, 1_G) \circ e_I \circ \alpha = 1_G \circ \alpha = \alpha$. By uniqueness, $(1_G, \alpha) = (1_G, 1_G) \circ (e_G \circ 1_G, e_I \circ \alpha) = \nabla \circ (1_G \vee \alpha)$.

So it remains to brace ourselves and attack the associativity of $\oplus$. Again, we shall let $G_1 = G_2 = G_3 = G$ and consider the five coproducts (1) $(S(G_1, G_2, G_3), e_1, e_2, e_3)$; (2) $(S(G_2, G_3), e'_2, e'_3)$; (3) $(S(G_1, G_2), e^*_1, e^*_2)$; (4) $(S(G_1, S(G_2, G_3)), e'_1, e'_{23})$; and (5) $(S(S(G_1, G_2), G_3), e^*_{12}, e^*_3)$. Take arbitrary $f_1, f_2, f_3 \in \mathbf{hom}(G, X)$. We must show that $(f_1, (f_2, f_3) \circ \pi) \circ \pi = f_1 \oplus (f_2 \oplus f_3) = (f_1 \oplus f_2) \oplus f_3 = ((f_1, f_2) \circ \pi, f_3) \circ \pi$.

First, we will show that $(f_1, (f_2, f_3) \circ \pi) = (f_1, f_2, f_3) \circ (e_1, (e_2, e_3)) \circ (1_{G_1} \vee \pi)$ and then we will show that $((f_1, f_2) \circ \pi, f_3) = (f_1, f_2, f_3) \circ ((e_1, e_2), e_3) \circ (\pi \vee 1_{G_3})$. So we will conclude that $(f_1, f_2, f_3) \circ (e_1, (e_2, e_3)) \circ (1_{G_1} \vee \pi) \circ \pi = (f_1, (f_2, f_3) \circ \pi) \circ \pi$ and $(f_1, f_2, f_3) \circ ((e_1, e_2), e_3) \circ (\pi \vee 1_{G_3}) \circ \pi = ((f_1, f_2) \circ \pi, f_3) \circ \pi$. From (11:4), we know that $((e_1, e_2), e_3) \circ (\pi \vee 1_{G_3}) \circ \pi = (e_1, (e_2, e_3)) \circ (1_{G_1} \vee \pi) \circ \pi$, so we will be able to conclude that $\oplus$ is associative.

Define $e_{12} = (e_1, e_2)$, and $e_{23} = (e_2, e_3)$. Note that $(f_1, f_2, f_3) \circ e_i = f_i$, $(f_2, f_3) \circ e'_2 = f_2$, $(f_2, f_3) \circ e'_3 = f_3$, $e_{23} \circ e'_2 = e_2$, $e_{23} \circ e'_3 = e_3$, $e_{12} \circ e^*_1 = e_1$, and $e_{12} \circ e^*_2 = e_2$.

Now $((f_1, f_2, f_3) \circ e_{12}) \circ e^*_1 = (f_1, f_2, f_3) \circ e_1 = f_1$ and $((f_1, f_2, f_3) \circ e_{12}) \circ e^*_2 = (f_1, f_2, f_3) \circ e_2 = f_2$. Hence,

$$(f_1, f_2) = (f_1, f_2, f_3) \circ e_{12}. \tag{11:13}$$

Similarly, $((f_1, f_2, f_3) \circ e_{23}) \circ e_2' = f_2$ and $((f_1, f_2, f_3) \circ e_{23}) \circ e_3' = f_3$. Hence,

$$(f_2, f_3) = (f_1, f_2, f_3) \circ e_{23}. \qquad (11:14)$$

Next, we compute $(f_1, f_2, f_3) \circ (e_1, (e_2, e_3)) \circ (1_{G_1} \vee \pi) \circ e_1^* = (f_1, f_2, f_3) \circ (e_1, (e_2, e_3)) \circ (e_1' \circ 1_{G_1}, e_{23}' \circ \pi) \circ e_1^* = (f_1, f_2, f_3) \circ (e_1, (e_2, e_3)) \circ e_1' \circ 1_{G_1} = (f_1, f_2, f_3) \circ e_1 = f_1$. And $(f_1, f_2, f_3) \circ (e_1, (e_2, e_3)) \circ (1_{G_1} \vee \pi) \circ e_2^* = (f_1, f_2, f_3) \circ (e_1, (e_2, e_3)) \circ e_{23}' \circ \pi = (f_1, f_2, f_3) \circ (e_2, e_3) \circ \pi = (f_1, f_2, f_3) \circ e_{23} \circ \pi = (f_2, f_3) \circ \pi$ by (11:14).

But we also have $(f_1, (f_2, f_3) \circ \pi) \circ e_1^* = f_1$, and $(f_1, (f_2, f_3) \circ \pi) \circ e_2^* = (f_2, f_3) \circ \pi$. By uniqueness, we conclude that

$$(f_1, (f_2, f_3) \circ \pi) = (f_1, f_2, f_3) \circ (e_1, (e_2, e_3)) \circ (1_{G_1} \vee \pi). \qquad (11:15)$$

By a similar argument, $(f_1, f_2, f_3) \circ ((e_1, e_2), e_3) \circ (\pi \vee 1_{G_3}) \circ e_2' = (f_1, f_2, f_3) \circ ((e_1, e_2), e_3) \circ (e_{12}^* \circ \pi, e_3^* \circ 1_{G_3}) \circ e_2' = (f_1, f_2, f_3) \circ ((e_1, e_2), e_3) \circ e_{12}^* \circ \pi = (f_1, f_2, f_3) \circ (e_1, e_2) \circ \pi = (f_1, f_2, f_3) \circ e_{12} \circ \pi = (f_1, f_2) \circ \pi$ by (11:13). And $(f_1, f_2, f_3) \circ ((e_1, e_2), e_3) \circ (\pi \vee 1_{G_3}) \circ e_3' = (f_1, f_2, f_3) \circ ((e_1, e_2), e_3) \circ e_3^* \circ 1_{G_3} = (f_1, f_2, f_3) \circ e_3 = f_3$.

But we also have $((f_1, f_2) \circ \pi, f_3) \circ e_2' = (f_1, f_2) \circ \pi$ and $((f_1, f_2) \circ \pi, f_3) \circ e_3' = f_3$. By uniqueness, we have

$$((f_1, f_2) \circ \pi, f_3) = (f_1, f_2, f_3) \circ ((e_1, e_2), e_3) \circ (\pi \vee 1_{G_3}). \qquad (11:16)$$

This completes the proof that $\oplus$ is associative. Hence, $(\mathbf{hom}(G, X), \oplus)$ is, in fact, a group. This also completes the proof of (11.3).

The proof of (11.3) should remind one that associativity can be difficult to prove.

If (G, π, μ, α) is either a group or a cogroup object, then $(\mathbf{hom}(G, G), \oplus)$ is a group. It is significant that $\mathbf{hom}(G, G)$ interacts with $\mathbf{hom}(X, G)$ if (G, π, μ, α) is a group object, and with $\mathbf{hom}(G, X)$ if (G, π, μ, α) is a cogroup object. This interaction is expressed in the next theorem.

(11.4) Theorem. *Let (G, π, μ, α) be a group object of a category $\mathcal{C}$, and let $X \in \mathbf{Ob}\,\mathcal{C}$. Take $f \in \mathbf{hom}(X, G)$ and $g_1, g_2 \in \mathbf{hom}(G, G)$. Then*

$$(g_1 \oplus g_2) \circ f = (g_1 \circ f) \oplus (g_2 \circ f)$$

and $(g_1 \circ g_2) \circ f = g_1 \circ (g_2 \circ f)$ with $1_G \circ f = f$.

Let (G, π, μ, α) be a cogroup object of a category $\mathcal{C}$, and let $X \in \mathbf{Ob}\,\mathcal{C}$. Take $f \in \mathbf{hom}(G, X)$ and $g_1, g_2 \in \mathbf{hom}(G, G)$. Then

$$f \circ (g_1 \oplus g_2) = (f \circ g_1) \oplus (f \circ g_2)$$

and $f \circ (g_1 \circ g_2) = (f \circ g_1) \circ g_2$ *with* $f \circ 1_G = f$.

Proof. For the group object, $(g_1 \oplus g_2) \circ f = \pi \circ [g_1, g_2] \circ f = \pi \circ [g_1 \circ f, g_2 \circ f] = (g_1 \circ f) \oplus (g_2 \circ f)$. We have used **1** from the proof of (11.3). For the cogroup object, $f \circ (g_1 \oplus g_2) = f \circ (g_1, g_2) \circ \pi = (f \circ g_1, f \circ g_2) \circ \pi = (f \circ g_1) \oplus (f \circ g_2)$. We have used **1d** from the proof of (11.3).

The climax of this section is the following theorem.

(11.5) Theorem. *For a group object* (G, π, μ, α) *of* $\mathcal{C}$, $(\mathbf{hom}(G, G), \oplus, \circ)$ *is a right nearring.*

For a cogroup object (G, π, μ, α) *of* $\mathcal{C}$, $(\mathbf{hom}(G, G), \oplus, \circ)$ *is a left nearring.*

Proof. Apply (11.3) and (11.4). Associativity of $\circ$ comes from the definition of a category.

Note 1. Some will argue that a right nearring is *the* natural consequence of placing the variable to the right of a function, as in $f(x)$, whereas a left nearring is *the* natural consequence of placing the variable to the left of a function, as in xf. This is not an adequate argument, as (11.5) will demonstrate as we obtain examples of both group and cogroup objects.

Note 2. For a group object (G, π, μ, α) of $\mathcal{C}$ and an $X \in \mathbf{Ob}\,\mathcal{C}$, we have from (11.4) that the elements of the right nearring $(\mathbf{hom}(G, G), \oplus, \circ)$ act on the left of elements of the group $(\mathbf{hom}(X, G), \oplus)$. We have $(g_1 \oplus g_2) \circ f = (g_1 \circ f) \oplus (g_2 \circ f)$, that is, the group elements distribute over the nearring elements. In addition, $(g_1 \circ g_2) \circ f = g_1 \circ (g_2 \circ f)$ as well as $1_G \circ f = f$. So $(\mathbf{hom}(X, G), \oplus)$ is something like a *left* module for the *right* nearring $(\mathbf{hom}(G, G), \oplus, \circ)$.

But, in contrast, the elements of the nearring $\mathbf{hom}(G, G)$ might also act on the left of the elements of G, or $U(G)$, if $\mathcal{C}$ is a concrete category. For example, if $\mathcal{C} = \mathcal{G}$, the category of groups, then $g_1(x + y) = (g_1 x) + (g_1 y)$, that is, the nearring elements distribute over the group elements. In addition, $(g_1 \circ g_2)x = g_1(g_2 x)$ as well as $1_G x = x$. So G is also something like a left module for the right nearring $(\mathbf{hom}(G, G), \oplus, \circ)$.

Of course, there are dual statements for a cogroup object.

This illustrates a difficulty in developing a concept of a nearring module.

Up to now, we have only promised the existence of group and cogroup objects. At the beginning of this chapter, we suggested that the nice symmetry in $_R\mathcal{M}$ might be the easiest place to find examples, which is true indeed.

Take any $M \in \mathbf{Ob}\,_R\mathcal{M}$. First, we will make a group object from M, and then a cogroup object. After this, we will see that the addition $\oplus$ defined by (11:7) and (11:8) each coincide with pointwise addition in $\mathbf{hom}(M, M) =$

$\mathrm{Hom}_R(M, M)$.

Let $\pi \in \mathbf{hom}(M \oplus M, M)$ be defined by $\pi(a, b) = a + b$. Let $\mu \in \mathbf{hom}(\{0\}, M) = \{i_M\}$ be the only candidate, and $\alpha \in \mathbf{hom}(M, M)$ be defined by $\alpha(a) = -a$. Then (M, π, μ, α) is a group object. There could not be a more natural choice for π, μ, and α.

Let us go through the details. Take $(a, b, c) \in M \oplus M \oplus M$. Then $\pi \circ (\pi \times 1_M) \circ [[\pi_1, \pi_2], \pi_3](a, b, c) = \pi \circ (\pi \times 1_M)((a, b), c) = \pi(\pi(a, b), c) = (a + b) + c$, and $\pi \circ (1_M \times \pi) \circ [\pi_1, [\pi_2, \pi_3]](a, b, c) = \pi \circ (1_G \times \pi)(a, (b, c)) = \pi(a, \pi(b, c)) = a + (b + c)$. So π satisfies (11:1).

Next, $\pi \circ (\mu \times 1_M)(0, a) = \pi(0, a) = 0 + a = a$ and $[t_M, 1_M](a) = (0, a)$, so $\pi \circ (\mu \times 1_M) = [t_M, 1_M]^{-1}$. Also, $\pi \circ (1_M \times \mu)(a, 0) = \pi(a, 0) = a + 0 = a$ and $[1_M, t_M](a) = (a, 0)$, so $\pi \circ (1_M \times \mu) = [1_M, t_M]^{-1}$. Hence, π and μ satisfy (11:2).

Finally, for the inverse diagram (11:3), $\pi \circ (\alpha \times 1_M) \circ \Delta(a) = \pi \circ (\alpha \times 1_M)(a, a) = \pi(-a, a) = -a + a = 0$, $\mu \circ t_M(a) = \mu(0) = 0$, and $\pi \circ (1_M \times \alpha) \circ \Delta(a) = \pi \circ (1_M \times \alpha)(a, a) = \pi(a, -a) = a + (-a) = 0$. Hence, π, μ, and α satisfy (11:3). This will make (M, π, μ, α) a group object of $_R\mathcal{M}$ as soon as we assure ourselves that π, μ, and α are morphisms of $_R\mathcal{M}$. But that is so easy that we will not bother here.

An important point here is that $(f \oplus g)(a) = \pi \circ [f, g](a) = \pi(f(a), g(a)) = f(a) + g(a)$. So $\oplus$ coincides with pointwise addition.

Now for a cogroup object, let $\pi(a) = (a, a)$, let $\mu \in \mathbf{hom}(M, \{0\}) = \{t_M\}$ be the only candidate, and $\alpha(a) = -a$. Again, we will check the details.

Take $(a, b, c) \in M \oplus M \oplus M$. Then $((e_1, e_2), e_3) \circ (\pi \vee 1_M) \circ \pi(a) = ((e_1, e_2), e_3) \circ (\pi \vee 1_M)(a, a) = ((e_1, e_2), e_3)((a, a), a) = (a, a, a)$, and similarly, $(e_1, (e_2, e_3)) \circ (1_M \vee \pi) \circ \pi(a) = (e_1, (e_2, e_3))(a, (a, a)) = (a, a, a)$. So π satisfies (11:4).

Next, $(\mu \vee 1_M) \circ \pi(a) = (\mu \vee 1_M)(a, a) = (0, a)$, $(1_M \vee \mu) \circ \pi(a) = (a, 0)$, $(i_M, 1_M)(0, a) = a$, and $(1_M, i_M)(a, 0) = a$. So π and μ satisfy (11:5).

Finally, for (11:6), $\nabla \circ (\alpha \vee 1_M) \circ \pi(a) = (1_M, 1_M)(-a, a) = -a + a = 0$, $i_M \circ \mu(a) = i_M(0) = 0$, and $\nabla \circ (1_M \vee \alpha) \circ \pi(a) = (1_M, 1_M)(a, -a) = a + (-a) = 0$. Hence, π, μ, and α satisfy (11:6), and this will make (M, π, μ, α) a cogroup object of $_R\mathcal{M}$.

Again, an important point here is that $(f \oplus g)(a) = (f, g) \circ \pi(a) = (f, g)(a, a) = f(a) + g(a)$. So $\oplus$ coincides with pointwise addition, as well as with the $\oplus$ for the group object. This is why we obtain

(11.6) Theorem. *For any $M \in \mathbf{Ob}\ _R\mathcal{M}$, $(\mathbf{hom}_R(M, M), \oplus, \circ)$ is a ring.*

Proof. This is because M is the object of a group object and a cogroup object in $_R\mathcal{M}$, where the $\oplus$ of (11:7) and of (11:8) each coincide with pointwise addition of functions. Since $\mathbf{hom}(M, M) = \mathrm{Hom}_R(M, M)$, (11.5) tells us that $(\mathrm{Hom}_R(M, M), +, \circ)$ is both a left nearring and a right nearring.

In the future, in general, an arbitrary $G \in \mathbf{Ob}\,\mathcal{C}$ will not be the object of both a group object and a cogroup object. So we will not have the nearrings $(\mathbf{hom}(G, G), \oplus, \circ)$ enjoying the rich symmetry of a ring. But we must always remember that the $\oplus$ of (11:7) and of (11:8) are each a generalization of the pointwise addition $(f + g)(x) = f(x) + g(x)$ of two functions into a group.

We have looked at example 7 from §10 in detail. We now turn our attention to some of the other examples from §10, some with group objects or cogroup objects which can be described quite easily. We postpone until §12 some of the more intricate or involved examples.

Example 1. $\mathcal{S}$ has lots of group objects (G, π, μ, α). Take any $\varnothing \neq G \in$ **Ob** $\mathcal{S}$. Consider $\pi \in \mathbf{hom}(G \times G, G)$. If π satisfies (11:1), then π is simply an associative binary operation, and conversely. A $\mu \in \mathbf{hom}(\{x\}, G)$ can only select an element $\mu(x) \in G$. If π and μ satisfy (11:2), then $\mu(x)$ is an identity for π, and conversely. Similarly, $\alpha \in \mathbf{hom}(G, G)$ together with π and μ satisfy (11:3) if and only if $\alpha(a)$ is an inverse for a with respect to π and $\mu(x)$. Hence, (G, π, μ, α) is a group. Conversely, if $(G, +, 0, -)$ is a group, then it is also a group object for $\mathcal{S}$. Since there are groups of every positive cardinality, every nonempty $G \in$ **Ob** $\mathcal{S}$ is an object of a group object (G, π, μ, α).

For a group object (G, π, μ, α) in the category $\mathcal{S}$, what are the nearrings $(\mathbf{hom}(G, G), \oplus, \circ)$ and what are the groups $(\mathbf{hom}(X, G), \oplus)$? Now $\pi(a, b) = a + b$, so $\pi \circ [f, g](a) = \pi(f(a), g(a)) = f(a) + g(a)$. Hence, $\mathbf{hom}(G, G) = M(G)$, and $\mathbf{hom}(X, G) = G^X$, the group of all mappings of X into G.

Once again, we see where the $\oplus$ of (11:7) coincides with pointwise addition of functions. Also, we have two extremely contrasting and important nearrings arising as $(\mathbf{hom}(G, G), \oplus, \circ)$ in suitable categories, namely, $M(G)$ for $\mathcal{S}$, and $\mathrm{Hom}_R(M, M)$ for $_R\mathcal{M}$.

Exercise. Why can't $\mathcal{S}$ have cogroup objects?

In Chapter 1, we saw that the $M(G)$ are really special cases of $\mathcal{C}(G, G)$, the continuous functions $f : G \to G$, where G is a topological group, so it may not be surprising to learn what we get from

Example 3. For $\mathcal{T}$, one proceeds as for $\mathcal{S}$ to obtain that the group objects are $((G, T), \pi, \mu, \circ) = ((G, T), +, 0, -)$ where $((G, +), T)$ is a topological group. Hence, $(\mathbf{hom}(G, G), \oplus.\circ) = (\mathcal{C}(G, G), \oplus, \circ)$, and we obtain another important class of nearrings arising as a $(\mathbf{hom}(G, G), \oplus, \circ)$.

Exercise. How about $\mathcal{T}$, can it have a cogroup object?

There are those who feel more comfortable with zero-symmetric nearrings, that is, nearrings where $0 \cdot x = x \cdot 0 = 0$ for all x. We will obtain some of these from

Example 2. For $\mathcal{S}_0$, since morphisms $f \in \mathbf{hom}[(G, g), (H, h)]$ must satisfy $f(g) = h$, the μ will be more restricted superficially. But remember, for a set $G \neq \varnothing$, if $u \in G$, then $(G, u) \in \mathbf{Ob}\mathcal{S}_0$. One proceeds exactly as in $\mathcal{S}$ to obtain that each (G, g) is the object of a group object $((G, g), \pi, \mu, \alpha)$, and each (G, π) is a group whose identity is g. But in this case, by our opening remark for this example, $f \in \mathbf{hom}[(G, g), (H, h)]$ must send the identity of (G, π) to the identity of (H, π). This means $\mathbf{hom}[(G, g), (G, g)] = M_0(G)$, and $\mathbf{hom}[(G, g), (H, h)] = M_0(G, H) = \{f : G \to H \mid f(g) = h\}$, the set of all zero or identity preserving maps of the group (G, π) into the group (H, π). So again, we get another important class of nearrings as a $(\mathbf{hom}(G, G), \oplus, \circ)$.

Exercise. Can $\mathcal{S}_0$ have a cogroup object?

The category of groups $\mathcal{G}$ offers a surprise!

Example 4. What are the group objects for the category of groups $\mathcal{G}$? In the category of sets $\mathcal{S}$, and the category of pointed sets $\mathcal{S}_0$, we get very comforting and nice results, namely, for $\mathcal{S}$, the group objects are essentially all groups $(G, +, 0, -)$, and $\mathbf{hom}(G, G) = M(G)$, and for $\mathcal{S}_0$, the group objects are again essentially all groups $((G, 0), +, 0, -)$, and $\mathbf{hom}[(G, 0), (G, 0)] = M_0(G)$.

Let $(G, +) \in \mathbf{Ob}\,\mathcal{G}$. Suppose $((G, +), \pi, \mu, \alpha)$ is a group object for $\mathcal{G}$. As with $\mathcal{S}$ and $\mathcal{S}_0$, we see that (G, π, μ, α) must be a group with binary operation π, with identity $\mu(0)$, and with $-a = \alpha(a)$. But, $\pi \in \mathbf{hom}(P(G, G), G)$, so π is a group homomorphism from $G \oplus G$ into G. Let $+_\pi$ be defined by $a +_\pi b = \pi(a, b)$ for all $(a, b) \in G \oplus G$. So $(G, +)$ and $(G, +_\pi)$ are groups. How are these two groups related? Since $0 = \pi(0, 0) = 0 +_\pi 0$, we know that 0 is the identity for $(G, +_\pi)$ as well as for $(G, +)$. Since π is a homomorphism, $\pi(a + x, b + y) = \pi[(a, b) + (x, y)] = \pi(a, b) + \pi(x, y)$, or, equivalently, $(a + x) +_\pi (b + y) = (a +_\pi b) + (x +_\pi y)$. This being true for all $(a, b), (x, y) \in G \oplus G$, let $a = y = 0$. Then $\pi(x, b) = x +_\pi b = b + x$. This defines π! From $\pi(a + x, b + y) = \pi[(a, b) + (x, y)] = \pi(a, b) + \pi(x, y)$ we obtain $(b + y) + (a + x) = b + a + y + x$ for all $a, b, x, y \in G$. Hence, $y + a = a + y$. The group $(G, +)$ is abelian! Since $\pi(a, b) = a +_\pi b = b + a = a + b$, we obtain that $+ = +_\pi$, and so (G, π) is an abelian group. We have already mentioned that $\mu(0) = 0$, and we also have $\alpha(a) = -a$, of course, since $+ = +_\pi$. So all group objects (G, π, μ, α) are abelian groups $(G, +, 0, -)$. It is easy to see that any abelian group $(G, +, 0, -)$ is also a group object. Hence, in the category of groups $\mathcal{G}$, the group objects are the *abelian* groups $(G, +, 0, -)$.

In $_R\mathcal{M}$, everybody is a group object as well as a cogroup object, so each $\mathbf{hom}(M, M) = \mathrm{Hom}_R(M, M)$ is a ring. If $(G, +, 0, -)$ is a group object for $\mathcal{G}$, then $\mathbf{hom}(G, G) = \mathrm{Hom}_Z(G, G)$ is also a ring. But what about cogroup objects (G, π, μ, α) for $\mathcal{G}$? We shall see that each free group $\mathcal{F}(X)$ is the object of a cogroup object for $\mathcal{G}$, so, as long as $|X| > 1$, then $\mathcal{F}(X)$ will be nonabelian, and $\mathbf{hom}(\mathcal{F}(X), \mathcal{F}(X))$ will not be a ring.

Let $X \in \mathbf{Ob}\mathcal{S}$, and let $\mathcal{F}(X)$ be a free group on X. Define the morphism $\pi \in \mathbf{hom}(\mathcal{F}(X), \mathcal{F}(X) * \mathcal{F}(X))$ by $\pi(x) = e_1(x) + e_2(x) \in \mathcal{F}(X) * \mathcal{F}(X)$, where $(\mathcal{F}(X) * \mathcal{F}(X), e_1, e_2)$ is a coproduct. We have no other choice for $\mu \in \mathbf{hom}(\mathcal{F}(X), \{0\})$. Take $\alpha(x) = -x$ for all $x \in X$ to define $\alpha \in \mathbf{hom}(\mathcal{F}(X), \mathcal{F}(X))$. We want to see that $(\mathcal{F}(X), \pi, \mu, \alpha)$ is a cogroup object. First, look at (11:4). For $x \in X$, $f \circ (\pi \vee 1_{\mathcal{F}(X)}) \circ \pi(x) = f \circ (\pi \vee 1_{\mathcal{F}(X)})[e_2(x) + e_3(x)] = ((e_1, e_2), e_3)[(e_1(x) + e_2(x)) + e_3(x)] = e_1(x) + e_2(x) + e_3(x) \in \mathcal{F}(X) * \mathcal{F}(X) * \mathcal{F}(X)$ and $g \circ (1_{\mathcal{F}(X)} \vee \pi) \circ \pi(x) = g \circ (1_{\mathcal{F}(X)} \vee \pi)[e_1(x) + e_2(x)] = (e_1, (e_2, e_3))[e_1(x) + (e_2(x) + e_3(x))] = e_1(x) + e_2(x) + e_3(x)$. So (11:4) is satisfied by π.

For (11:5), take $x \in X$. Then $(\mu \vee 1_{\mathcal{F}(X)}) \circ \pi(x) = (\mu \vee 1_{\mathcal{F}(X)})(e_1(x) + e_2(x)) = e_2(x)$, and $(i_{\mathcal{F}(X)}, 1_{\mathcal{F}(X)})e_2(x) = x$. Similarly, $(1_{\mathcal{F}(X)} \vee \mu) \circ \pi(x) = (1_{\mathcal{F}(X)} \vee \mu)(e_1(x) + e_2(x)) = e_1(x)$, and $(1_{\mathcal{F}(X)}, i_{\mathcal{F}(X)})e_1(x) = x$. So (11:5) is satisfied by π and μ.

Finally, let us see what happens for (11:6). Take any $x \in X$. Then $\nabla \circ (\alpha \vee 1_{\mathcal{F}(X)}) \circ \pi(x) = \nabla \circ (\alpha \vee 1_{\mathcal{F}(X)})[e_1(x) + e_2(x)] = \nabla(e_1(-x) + e_2(x)) = -x + x = 0$, and $\nabla \circ (1_{\mathcal{F}(X)} \vee \alpha) \circ \pi(x) = \nabla \circ (1_{\mathcal{F}(X)} \vee \alpha)(e_1(x) + e_2(x)) = \nabla(e_1(x) + e_2(-x)) = x + (-x) = 0$. Also, $i_{\mathcal{F}(X)} \circ \mu(x) = i_{\mathcal{F}(X)}(0) = 0$. So (11:6) is satisfied. This makes $(\mathcal{F}(X), \pi, \mu, \alpha)$ a cogroup object.

What are the $\mathbf{hom}(\mathcal{F}(X), \mathcal{F}(X))$? An $f \in \mathbf{hom}(\mathcal{F}(X), \mathcal{F}(X))$ is identified with a map $f : X \to \mathcal{F}(X)$. For $f, g \in \mathbf{hom}(\mathcal{F}(X), \mathcal{F}(X))$, $f \oplus g = (f, g) \circ \pi$. So for $x \in X$, $(f \oplus g)(x) = (f, g) \circ \pi(x) = (f, g)(e_1(x) + e_2(x)) = f(x) + g(x)$. Hence, $(\mathbf{hom}(\mathcal{F}(X), \mathcal{F}(X)), \oplus) \cong \mathcal{F}(X)^X$, the complete direct product of $\mathcal{F}(X)$ with itself $|X|$ times. Hence, if $|X| > 1$, then $\mathcal{F}(X)$ is not abelian, thus $\mathcal{F}(X)^X$ and $(\mathbf{hom}(\mathcal{F}(X), \mathcal{F}(X)), \oplus)$ are nonabelian. So $(\mathbf{hom}(\mathcal{F}(X), \mathcal{F}(X)), \oplus, \circ)$ cannot be a ring.

Are there any other cogroup objects in $\mathcal{G}$. In [EH, 1962], one learns that these are all the cogroup objects of $\mathcal{G}$. So there are only the two groups, $\{0\}$ and the integers Z, which are both group and cogroup objects in $\mathcal{G}$. This is a big contrast to any $_R\mathcal{M}$.

Note. For a nonabelian group $(G, +)$, one cannot reasonably expect $f + g \in \mathbf{hom}(G, G)$ if $f, g \in \mathbf{hom}(G, G)$ and $+$ is pointwise addition. But, for the free groups $\mathcal{F}(X)$, one can define a natural $\oplus$ so that $f \oplus g \in \mathbf{hom}(\mathcal{F}(X), \mathcal{F}(X))$. Of course, we will see that this is valid for objects in other categories as well.

Example 5. The category of rings $\mathcal{R}$ provides some amusing group objects. Let (R, π, μ, α) be a group object in $\mathcal{R}$, where $(R, +, \cdot)$ is a ring of **Ob** $\mathcal{R}$. Then $\pi \in \mathbf{hom}(R \oplus R, R)$ in $\mathcal{R}$, so $\pi \in \mathbf{hom}(R \oplus R, R)$ in $\mathcal{G}$. Similarly, μ and α are also morphisms in $\mathcal{G}$. Hence, (R, π, μ, α) is a group object in $\mathcal{G}$. From what we have seen about group objects in $\mathcal{G}$, we know that $(R, \pi, \mu, \alpha) = (R, +, 0, -)$. Since $\pi = +$ and $\pi \in \mathbf{hom}(R \oplus R, R)$ in $\mathcal{R}$, then $\pi(ax, by) = \pi[(a, b)(x, y)] = \pi(a, b)\pi(x, y)$, for all $a, b, x, y \in R$. This means that $ax + by = ax + ay + bx + by$, so $ay + bx = 0$ for all $a, b, x, y \in R$. Hence, $ay = 0$ for all $a, y \in R$. Hence, $\cdot$ is the trivial multiplication defined on the abelian group $(R, +)$. So the group objects are essentially among the trivial rings $(R, +, \cdot)$, where $a \cdot b = 0$ for all $a, b \in R$. If one starts with such a trivial ring $(R, +, \cdot)$, it is easy to see that $(R, +, 0, -)$ is not only a group object for $\mathcal{G}$ but also for $\mathcal{R}$. Hence, the group objects in $\mathcal{R}$ are essentially the trivial rings $(R, +, \cdot)$ and $\mathbf{hom}(R, R) = \mathrm{Hom}_Z[(R, +), (R, +)]$, a ring.

Exercise. Before $\mathcal{R}$ can hope to have a cogroup object, it must have a coproduct. Can you find a coproduct in $\mathcal{R}$?

Example 8. We have already looked at $_R\mathcal{M}$ in detail. For $\mathrm{Aff}(R)$, we have the same objects of $_R\mathcal{M}$, and all the morphisms of $_R\mathcal{M}$ are morphisms of $\mathrm{Aff}(R)$. So a group object of $_R\mathcal{M}$ is also a group object of $\mathrm{Aff}(R)$. But since $\mathrm{Aff}(R)$ has more morphisms than $_R\mathcal{M}$, the $(\mathbf{hom}(M, M), \oplus, \circ)$ will be different. For $A_1, A_2 \in \mathbf{hom}(M, M)$ in $\mathrm{Aff}(R)$, there are $L_1, L_2 \in \mathrm{Hom}_R(M, M)$ in $_R\mathcal{M}$ and constants $\alpha_1, \alpha_2 \in M$ so that $A_i(x) = L_i(x) + \alpha_i$ for each $x \in M$. Hence, $(A_1 \oplus A_2)(x) = \pi \circ [A_1, A_2](x) = \pi(A_1(x), A_2(x)) = A_1(x) + A_2(x) = (L_1 + L_2)(x) + (\alpha_1 + \alpha_2)$. This means that $\mathbf{hom}(M, M) \cong \mathrm{Hom}_R(M, M) \oplus_A M$, the abstract affine nearring constructed from the $\mathrm{Hom}_R(M, M)$-module M. It also means that still another important class of nearrings arises from the $(\mathbf{hom}(G, G), \oplus, \circ)$.

Example 9. What one learns from the group objects in $\mathcal{A}_S$ should enlarge the way in which researchers in nearring theory have looked at an important class of nearrings. Objects and morphisms of $\mathcal{A}_S$ are closely related to objects and morphisms in $\mathcal{S}$. The same is true for products in $\mathcal{A}_S$. If we have a group object $((\Gamma, \cdot), \pi, \mu, \alpha)$ of $\mathcal{A}_S$, then $\Gamma \in \mathbf{Ob}\mathcal{S}$ and π, μ, and α are all morphisms of $\mathcal{S}$. Hence, $(\Gamma, \pi, \mu, \alpha) = (\Gamma, +, 0, -)$, a group. What does this mean for $f \in \mathbf{hom}[(\Gamma, \cdot), (\Gamma, \cdot)]$? Since $f(\gamma \cdot g) = f(\gamma) \cdot g$ for all $\gamma \in \Gamma$ and for all $g \in S$, we have $f \in M_S(\Gamma) = \{f : \Gamma \to \Gamma \mid f(\gamma \cdot g) = f(\gamma) \cdot g$ for all $\gamma \in \Gamma$ and for all $g \in S\}$. Conversely, $f \in M_S(\Gamma)$ implies $f \in \mathbf{hom}[(\Gamma, \cdot), (\Gamma, \cdot)]$, so $\mathbf{hom}[(\Gamma, \cdot), (\Gamma, \cdot)] = M_S(\Gamma)$. As with $\mathcal{S}$, $\oplus$ is just pointwise addition, so $(\mathbf{hom}[(\Gamma, \cdot), (\Gamma, \cdot)], \oplus, \circ) = (M_S(\Gamma), +, \circ)$. The $M_S(\Gamma)$ are very important nearrings, and they arise as $(\mathbf{hom}(G, G), \oplus, \circ)$.

So how should this enlarge the view of researchers in nearring theory? It has been traditional for researchers to look at $S = G \cup \{0\}$, where G is a group of automorphisms of a group Γ, and 0 is the trivial endomorphism of Γ mapping every element $\gamma \in \Gamma$ to the identity $0_\Gamma \in \Gamma$. In particular, it is common for G to be a regular group of automorphisms of Γ, and for the action of S upon Γ to be the natural one, that is, $\gamma \cdot g = (\gamma)g$. From what we see in $\mathcal{A}_S$, (1) one need not take G as a group of automorphisms of Γ, and (2) the action of $S = G \cup \{0\}$ need not be the natural one even if G is a group of automorphisms of Γ. We will look at this point in further detail later on.

If S has a zero, denoted by 0, it is traditional to assume that $\gamma \cdot 0 = 0_\Gamma$ for all $\gamma \in \Gamma$. From $f(\gamma \cdot 0) = f(\gamma) * 0$, we obtain $f(0) = 0$ for all $f \in M_S(\Gamma)$, and so $M_S(\Gamma) \subseteq M_0(\Gamma)$.

Let us summarize a few important points of this section. If (G, π, μ, α) is a group object of a category $\mathcal{C}$, then $(\mathbf{hom}(G, G), \oplus, \circ)$ is a right nearring. If (G, π, μ, α) is a cogroup object of $\mathcal{C}$, then $(\mathbf{hom}(G, G), \oplus, \circ)$ is a left nearring. This is why $(\mathrm{Hom}_R(M, M), +, \circ)$ is a ring for each $M \in \mathbf{Ob}_R\mathcal{M}$, since each $M \in \mathbf{Ob}_R\mathcal{M}$ is the object of both a group object and a cogroup object in $_R\mathcal{M}$, and since the right nearring $(\mathbf{hom}(M, M), \oplus_r, \circ)$ and the left nearring $(\mathbf{hom}(M, M), \oplus_l, \circ)$ each coincide with $(\mathrm{Hom}_R(M, M), +, \circ)$. Thus, pointwise addition $+$ for the R-endomorphisms of an R-module agrees with that defined by either (11:7) or (11:8).

For the category of sets $\mathcal{S}$, the group objects are all groups $(G, +)$ and $(\mathbf{hom}(G, G), \oplus, \circ) = (M(G), +, \circ)$, where, again, $\oplus$ of (11:7) coincides with pointwise addition of functions $+$. For pointed sets in $\mathcal{S}_0$, the group objects are once again all groups $(G, +)$, but this time, $(\mathbf{hom}(G, G), \oplus, \cdot) = (M_0(G), +, \circ)$, where $\oplus$ of (11:7) coincides with $+$. For $\mathcal{T}$, the category of topological spaces, we have $(\mathbf{hom}(G, G), \oplus, \circ) = (\mathcal{C}(G, G), +, \circ)$.

In $\mathrm{Aff}(R)$, we have for group objects (M, π, μ, α) that the right nearring $(\mathbf{hom}(M, M), \oplus, \circ) \cong (\mathrm{Hom}_R(M, M) \oplus_A M, +, \circ)$, an abstract affine nearring, and in $\mathcal{A}_S$, if $(\Gamma, \cdot)$ is the object of a group object, then the nearring $(\mathbf{hom}(\Gamma, \Gamma), \oplus, \circ) = (M_S(\Gamma), +, \circ)$.

In the category of groups $\mathcal{G}$, there is a surprising contrast between group and cogroup objects, with only the two groups $\{0\}$ and Z in their intersection. The cogroups, being identified with the free groups $\mathcal{F}(X)$, give us examples where the endomorphisms of a nonabelian group can be added in a natural way. Even though $(\mathbf{hom}(\mathcal{F}(X), \mathcal{F}(X)), \oplus) \cong (\mathcal{F}(X)^X, +)$, very little is known about the nearrings $(\mathbf{hom}(\mathcal{F}(X), \mathcal{F}(X)), \oplus, \circ)$.

Examples 10, 11, and 12 require considerably more attention, and this will be given in §12. Of course, $(\mathrm{Hom}_R(M, M), +, \circ)$ is extremely well studied. The right nearrings $(M(G), +, \circ)$, $(M_0(G), +, \circ)$, $(\mathcal{C}(G, G), +, \circ)$, $(\mathrm{Hom}_R(M, M) \oplus_A M, +, \circ)$, and $(M_S(\Gamma), +, \circ)$ have also been well studied, and we shall return to each of these later.

Before we study the examples of 10, 11, and 12 in more detail, let us make another observation about morphisms involving a group or cogroup object.

(11.7) Theorem. *Let (G, π, μ, α) and $(G', \pi', \mu', \alpha')$ be group objects of a category $\mathcal{C}$. Then for $h, h_1, h_2 \in \mathbf{hom}(G, G)$, and $f, f_1, f_2 \in \mathbf{hom}(G, G')$ we have*

$$(f_1 \oplus f_2) \circ h = (f_1 \circ h) \oplus (f_2 \circ h)$$

and $f \circ (h_1 \circ h_2) = (f \circ h_1) \circ h_2$, with $f \circ 1_G = f$.

Let (G, π, μ, α) and $(G', \pi', \mu', \alpha')$ be cogroup objects of a category $\mathcal{C}$. Then for $f, f_1, f_2 \in \mathbf{hom}(G, G')$ and $h, h_1, h_2 \in \mathbf{hom}(G', G')$, we have

$$h \circ (f_1 \oplus f_2) = (h \circ f_1) \oplus (h \circ f_2)$$

and $(h_1 \circ h_2) \circ f = h_1 \circ (h_2 \circ f)$ with $1_{G'} \circ f = f$.

Proof. The proof here will be for the cogroup objects, and we will leave the analogous proof for the group objects to the reader.

From (11:8) and **1d**, we have $h \circ (f_1 \oplus f_2) = h \circ (f_1, f_2) \circ \pi = (h \circ f_1, h \circ f_2) \circ \pi = (h \circ f_1) \oplus (h \circ f_2)$. The definition of a category assures us that $(h_1 \circ h_2) \circ f = h_1 \circ (h_2 \circ f)$ and $1_{G'} \circ f = f$.

Déjà vu would be an understandable reaction, but there is an important nuance between (11.7) and (11.4). In (11.4), the group elements distribute over any sum of nearring elements, and in (11.7), the nearring elements distribute over any sum of group elements. This should cause ambivalence when developing the idea of a nearring module. Also, see (12.26) and (12.29) below.

12. Examples

In §10, we introduced the 'dirty dozen'. In §11, we discussed some group and cogroup objects of the 'dozen' and introduced their nearrings of morphisms. In this section, we look at some group and cogroup objects coming from examples 10, 11, and 12, and their nearrings of morphisms. The reason for segregating examples 10, 11, and 12 are (1) they are considerably more involved, and (2) each represents new and promising directions for research.

12.1. Example 10

For a commutative R-algebra A with identity 1_A, one cannot expect that $f + g \in \mathbf{hom}(A, A)$ for arbitrary $f, g \in \mathbf{hom}(A, A)$. But for cogroup objects (A, π, μ, α) in $\mathcal{A}(R)$, we shall have $f \oplus g \in \mathbf{hom}(A, A)$ if $\oplus$ is defined by (11:8). In particular, if $A = R[x]$, the R-algebra of polynomials over R, we shall see that $(\mathbf{hom}(R[x], R[x]), \oplus, \circ) \cong (R[x], +, \circ)$, that is, the nearring of endomorphisms of the R-algebra of polynomials $R[x]$ is isomorphic to the nearring of polynomials $(R[x], +, \circ)$.

Let X be a nonempty set of indeterminates, and let $R[X]$ be the R-algebra of polynomials in indeterminates from X. To avoid having anti-isomorphisms, it will be desirable to write the variable to the left of the function while discussing example 10.

(12.1) Theorem. *In $\mathcal{A}(R)$, $R[X]$ is a free R-algebra on the set X.*

Proof. Let $i : X \to R[X]$ be the insertion map $x \mapsto xi = x$. Let $A \in$ **Ob** $\mathcal{A}(R)$ and let $f' : X \to A$ be any function. Define $f : R[X] \to A$ by

$$\left(\sum a_{n_1, n_2, \ldots, n_k} x_1^{n_1} x_2^{n_2} \cdots x_k^{n_k} \right) f$$
$$= \sum a_{n_1, n_2, \ldots, n_k} (x_1 f')^{n_1} (x_2 f')^{n_2} \cdots (x_k f')^{n_k}.$$

Then f is the unique morphism in **hom**$(R[X], A)$ such that $i \circ f = f'$. (Remember, the variable goes to the left throughout example 10.)

(12.2) Corollary. *There is a natural bijection between the morphisms of* **hom**$(R[X], A)$ *and the functions in A^X.*

(12.3) Theorem. *Each $(R[X], \pi, \mu, \alpha)$ is a cogroup object in $\mathcal{A}(R)$, where π, μ, and α are defined by $x\pi = x \otimes 1 + 1 \otimes x$, $x\mu = 0$, and $x\alpha = -x$, for each $x \in X$.*

Proof. Looking at (11:4), take $x \in X$. Then $x\pi \circ (\pi \vee 1_{R[X]}) \circ f = (x \otimes 1 + 1 \otimes x)(\pi \vee 1_{R[X]}) \circ f = [(x \otimes 1) \otimes 1 + (1 \otimes x) \otimes 1 + (1 \otimes 1) \otimes x]((e_1, e_2), e_3) = x \otimes 1 \otimes 1 + 1 \otimes x \otimes 1 + 1 \otimes 1 \otimes x$, and $x\pi \circ (1_{R[X]} \vee \pi) \circ g = (x \otimes 1 + 1 \otimes x)(1_{R[X]} \vee \pi) \circ g = [x \otimes (1 \otimes 1) + 1 \otimes (x \otimes 1) + 1 \otimes (1 \otimes x)](e_1, (e_2, e_3)) = x \otimes 1 \otimes 1 + 1 \otimes x \otimes 1 + 1 \otimes 1 \otimes x$. So π satisfies (11:4).

Next, consider (11:5). We have for $x \in X$, $x\pi \circ (\mu \vee 1_{R[X]}) = (x \otimes 1 + 1 \otimes x)(\mu \vee 1_{R[X]}) = 1 \otimes x$, and $1 \otimes x(i_{R[X]}, 1_{R[X]}) = x$. Similarly, $x\pi \circ (1_{R[X]} \vee \mu) = (x \otimes 1 + 1 \otimes x)(1_{R[X]} \vee \mu) = x \otimes 1$, and $x \otimes 1(1_{R[X]}, i_{R[X]}) = x$. So π and μ satisfy (11:5).

Finally, for (11:6), take $x \in X$. Then $x\pi \circ (\alpha \vee 1_{R[X]}) \circ \nabla = (x \otimes 1 + 1 \otimes x)(\alpha \vee 1_{R[X]}) \circ \nabla = [(-x) \otimes 1 + 1 \otimes x]\nabla = -x + x = 0$, $x\pi \circ (1_{R[X]} \vee \alpha) \circ \nabla = (x \otimes 1 + 1 \otimes x)(1_{R[X]} \vee \alpha) \circ \nabla = [x \otimes 1 + 1 \otimes (-x)]\nabla = x + (-x) = 0$, and $x\mu \circ i_{R[X]} = 0$. So π, μ, and α satisfy (11:6), and this completes the proof of (12.3).

Taking guidance from (12.3), let T be a nonempty set of indeterminates where $T \cap T^{-1} = \emptyset$ and $T^{-1} = \{t^{-1} \mid t \in T\}$. Let $R[T, T^{-1}]$ be all polynomials with indeterminates in $T \cup T^{-1}$. So $p \in R[T, T^{-1}]$ if and only if p is a finite sum of monomials $a t_1^{n_1} t_2^{n_2} \cdots t_k^{n_k}$, where $a \in R$, each $t_i \in T$, each $n_i \in \mathbb{Z}$, and k is a nonnegative integer.

(12.4) Theorem. *There is a natural bijection between morphisms in* $\mathbf{hom}(R[T, T^{-1}], A)$ *in* $\mathcal{A}(R)$ *and mappings in* $\mathcal{U}(A)^T$, *where* $\mathcal{U}(A)$ *denotes the group of units of the commutative R-algebra A.*

Proof. For $f \in \mathbf{hom}(R[T, T^{-1}], A)$, we must have $1_{R[T,T^{-1}]}f = 1_A$, and so $1_A = (tt^{-1})f = (tf)(t^{-1}f)$. Hence, $tf \in \mathcal{U}(A)$ for each $t \in T$. Conversely, if $t \mapsto tf$ is a map in $\mathcal{U}(A)^T$, it can be extended to an $f \in \mathbf{hom}(R[T, T^{-1}], A)$.

(12.5) Theorem. *Each* $(R[T, T^{-1}], \pi, \mu, \alpha)$ *is a cogroup object in* $\mathcal{A}(R)$, *where* π, μ, *and* α *are defined by* $t\pi = t \otimes t$, $t\mu = 1$, *and* $t\alpha = t^{-1}$ *for each* $t \in T$.

Proof. The proof is direct, but we will sketch it. Let $B = R[T, T^{-1}]$ for these calculations. For (11:4), and $t \in T$, $t\pi \circ (\pi \vee 1_B) \circ f = t \otimes t(\pi \vee 1_B) \circ f = (t \otimes t) \otimes t((e_1, e_2), e_3) = t \otimes t \otimes t$, and $t\pi \circ (1_B \vee \pi) \circ g = t \otimes (t \otimes t)(e_1, (e_2, e_3)) = t \otimes t \otimes t$. Hence, (11:4). For (11:5), and $t \in T$, $t\pi \circ (\mu \vee 1_B) = t \otimes t(\mu \vee 1_B) = 1 \otimes t$, and $1 \otimes t(i_B, 1_B) = t$. Also, $t\pi \circ (1_B \vee \mu) = t \otimes 1$, and $t \otimes 1(1_B, i_B) = t$. Hence, (11:5). Finally, for (11:6) and $t \in T$, $t\pi \circ (\alpha \vee 1_B) \circ \nabla = t \otimes t(\alpha \vee 1_B) \circ \nabla = t^{-1} \otimes t\nabla = 1$, $t\pi \circ (1_B \vee \alpha) \circ \nabla = t \otimes t^{-1}\nabla = 1$, and $t\mu \circ i_B = 1$. Hence, (11:6), and we have (12.5).

Studying the proof of (12.5) leads us to a stronger result, of which (12.5) is an important corollary. First, we prove (12.6) which generalizes (12.4).

(12.6) Theorem. *Let* $(G, \cdot)$ *be a multiplicative abelian group, and let* RG *denote the group algebra constructed from R and G. If* $f' \in \mathbf{hom}(G, \mathcal{U}(A))$ *in* $\mathcal{G}$, *the category of groups, then* f' *can be extended uniquely to an* $f \in \mathbf{hom}(RG, A)$ *in* $\mathcal{A}(R)$. *Hence, there is a natural bijection between the group homomorphisms in* $\mathbf{hom}(G, \mathcal{U}(A))$ *and the R-algebra homomorphisms in* $\mathbf{hom}(RG, A)$.

Proof. For $f \in \mathbf{hom}(RG, A)$, $1_A = 1_{RG}f = (gg^{-1})f = (gf)(g^{-1}f)$, hence $f|G : G \to \mathcal{U}(A)$ is a group homomorphism. Conversely, if $f' : G \to \mathcal{U}(A)$ is a group homomorphism, and $\sum_i r_i g_i \in RG$, then $(\sum_i r_i g_i)f = \sum_i r_i(g_i f')$ defines $f : RG \to A$. It is easy to see that $f \in \mathbf{hom}(RG, A)$ in $\mathcal{A}(R)$. Certainly, $f' = f|G$ and the correspondence $f' \leftrightarrow f$ is a bijection.

Note. That (12.4) is a special case of (12.6) comes from realizing that all monomials $t_1^{n_1} t_2^{n_2} \cdots t_k^{n_k}$ describe the multiplicative free abelian group $F_a(T)$ on the set T, so mappings in $\mathcal{U}(A)^T$ correspond bijectively to group homomorphisms from $F_a(T)$ into A, and bijectively to R-algebra homomorphisms in $\mathbf{hom}(R[T, T^{-1}], A)$.

(12.7) Theorem. *Let* RG *denote the group algebra in* $\mathcal{A}(R)$ *constructed from R and a multiplicative abelian group* $(G, \cdot)$. *Define* π, μ, *and* α *by*

$g\pi = g \otimes g$, $g\mu = 1$, *and* $g\alpha = g^{-1}$ *for each* $g \in G$. *Then* (RG, π, μ, α) *is a cogroup object of* $\mathcal{A}(R)$.

Proof. One can easily check that π, μ, and α are each morphisms in $\mathcal{G}$, which can be extended to morphisms in $\mathcal{A}(R)$ by (12.6). Then the proof parallels that of (12.5).

From (12.3), each $(R[X], \pi, \mu, \alpha)$ is a cogroups in $\mathcal{A}(R)$. From (12.7), each (RG, π, μ, α) is a cogroups in $\mathcal{A}(R)$ of which each $(R[T, T^{-1}], \pi, \mu, \alpha)$ of (12.5) is a special case. Of course, this means that each $(RG[X], \pi, \mu, \alpha)$ is a cogroup in $\mathcal{A}(RG)$ and that each $(R[X]G, \pi, \mu, \alpha)$ is a cogroup in $\mathcal{A}(R[X])$. But one can say more. Really, since the elements of R, X, and G all commute with one another, with respect to multiplication, we have $RG[X] = R[X]G$.

(12.8) Theorem. *For a multiplicative abelian group G and a nonempty set of indeterminates X, consider the commutative R-algebra $RG[X] = R[X]G$. If $A \in \mathbf{Ob}\,\mathcal{A}(R)$ also, then the morphisms $f \in \mathbf{hom}(RG[X], A)$ are in one-to-one correspondence with ordered pairs (f_1, f_2) where $f_1 \in \mathbf{hom}(G, \mathcal{U}(A))$ in $\mathcal{G}$, and $f_2 \in \mathbf{hom}(X, A)$ in $\mathcal{S}$.*

Proof. Certainly, $f \in \mathbf{hom}(RG[X], A)$ defines $f_1 = f|G$ and $f_2 = f|X$. For appropriate f_1 and f_2, define $f : RG[X] \to A$ by $(agx_1^{n_1} x_2^{n_2} \cdots x_k^{n_k})f = a(gf_1)(x_1 f_2)^{n_1}(x_2 f_2)^{n_2} \cdots (x_k f_2)^{n_k}$ for $a \in R$, $g \in G$, and each $x_i \in X$. Here we have defined f for each monomial in $RG[X]$, and to finish the definition, just extend f 'linearly' since f needs to be a group homomorphism and an R-module homomorphism. Also $f|G = f_1$ and $f|X = f_2$. It is easy to see that f is an R-module homomorphism, and certainly $1_{RG[X]}f = 1_A$. It remains to see that $(uv)f = (uf)(vf)$ for all $u, v \in RG[X]$.

A little notation will help. Let $\mathbf{x} = (x_1, x_2, \ldots, x_k)$, where each $x_i \in X$, and $i \neq j$ implies $x_i \neq x_j$, and $\mathbf{x}f_2 = (x_1 f_2, x_2 f_2, \ldots, x_k f_2)$. Let $\mathbf{n} = (n_1, n_2, \ldots, n_k)$, where each n_i is a nonnegative integer. Let $\mathbf{x^n} = x_1^{n_1} x_2^{n_2} \cdots x_k^{n_k}$. Then for $u, v \in RG[X]$, each is a finite sum of terms $r_{\mathbf{n}} g_{\mathbf{n}} \mathbf{x^n}$ where $r_{\mathbf{n}} \in R$, $g_{\mathbf{n}} \in G$. So $u = \sum_{\mathbf{n}} a_{\mathbf{n}} g_{\mathbf{n}} \mathbf{x^n}$ and $v = \sum_{\mathbf{m}} b_{\mathbf{m}} h_{\mathbf{m}} \mathbf{x^m}$ where the $a_{\mathbf{n}}, b_{\mathbf{m}} \in R$, and $g_{\mathbf{n}}, h_{\mathbf{m}} \in G$. Then

$$(uv)f = \left[\left(\sum_{\mathbf{n}} a_{\mathbf{n}} g_{\mathbf{n}} \mathbf{x^n}\right)\left(\sum_{\mathbf{m}} b_{\mathbf{m}} h_{\mathbf{m}} \mathbf{x^m}\right)\right]f = \left[\sum_{\mathbf{n}} \sum_{\mathbf{m}} a_{\mathbf{n}} b_{\mathbf{m}} g_{\mathbf{n}} h_{\mathbf{m}} \mathbf{x^{n+m}}\right]f$$

$$= \sum_{\mathbf{n}} \sum_{\mathbf{m}} a_{\mathbf{n}} b_{\mathbf{m}} (g_{\mathbf{n}} f_1)(h_{\mathbf{m}} f_1)(\mathbf{x}f_2)^{\mathbf{n+m}}$$

$$= \left[\sum_{\mathbf{n}} a_{\mathbf{n}} (g_{\mathbf{n}} f_1)(\mathbf{x}f_2)^{\mathbf{n}}\right]\left[\sum_{\mathbf{m}} b_{\mathbf{m}} (h_{\mathbf{m}} f_1)(\mathbf{x}f_2)^{\mathbf{m}}\right]$$

$$= \left(\sum_{\mathbf{n}} a_{\mathbf{n}} g_{\mathbf{n}} \mathbf{x^n}\right)f\left(\sum_{\mathbf{m}} b_{\mathbf{m}} h_{\mathbf{m}} \mathbf{x^m}\right)f = (uf)(vf).$$

(By doing this 'proof', the author indulges in a hypocrisy common to numerous mathematicians. *When rigor demands extraordinary patience, with long tedious arguments, and involving extremely cumbersome notation, then hide all these problems by waving one's hands, or by intimidating the reader and/or student.*)

Denote the homomorphism defined by (f_1, f_2) by $(f_1; f_2)$.

(12.9) Theorem. *For $RG[X]$ as in (12.8), define π, μ, and α by $g\pi_1 = g \otimes g$ and $x\pi_2 = x \otimes 1 + 1 \otimes x$, by $g\mu_1 = 1$ and $x\mu_2 = 0$, and by $g\alpha_1 = g^{-1}$ and $x\alpha_2 = -x$. Then $(RG[X], \pi, \mu, \alpha)$ is a cogroup object in $\mathcal{A}(R)$.*

Proof. We have already seen that π_1, μ_1, and α_1 define morphisms in $\mathcal{G}$, and that π_2, μ_2, and α_2 define morphisms in $\mathcal{S}$. So by (12.8), $\pi = (\pi_1; \pi_2)$, $\mu = (\mu_1; \mu_2)$, and $\alpha = (\alpha_1; \alpha_2)$ define morphisms in $\mathcal{A}(R)$. That π, μ, and α satisfy (11:4), (11:5), and (11:6) follow from combining arguments such as those of (12.3) and (12.7) or (12.5).

Exercise. Are there products in the category $\mathcal{A}(R)$? If so, are there group objects?

(12.10) Problem. Have all the cogroup objects in $\mathcal{A}(R)$ been described here?

Let us now turn our attention to describing some $(\mathbf{hom}(G, A), \oplus)$ and some $(\mathbf{hom}(G, G), \oplus, \circ)$ for some of our cogroup objects (G, π, μ, α) of $\mathcal{A}(R)$.

First consider the $(R[X], \pi, \mu, \alpha)$ of (12.3). From (12.2) we identify $f, g \in \mathbf{hom}(R[X], A)$ with $f, g \in A^X$. So for $x \in X$, $x(f \oplus g) = x\pi \circ (f, g) = [x \otimes 1 + 1 \otimes x](f, g) = xf + xg$. Hence,

(12.11) Theorem. $(\mathbf{hom}(R[X], A), \oplus) \cong (A^X, \oplus) \cong (\sum^*_{|X|} A, +)$ *where $\sum^*_{|X|} A$ denotes the complete direct sum of $|X|$ copies of A, and where $x(f \oplus g) = xf + xg$ for all $x \in X$ and for all $f, g \in A^X$.*

When $A = R[X]$, we have the right nearring $(\mathbf{hom}(R[X], R[X]), \oplus, \circ)$. In (12.11), we described the group, but what does $\circ$ induce on $R[X]^X$ or $\sum^*_{|X|} R[X]$? It seems rather messy to describe this in general, but the idea is really quite simple. One should be able to understand what is going on by considering the case where $X = \{x, y\}$. For $f, g \in \mathbf{hom}(R[x, y], R[x, y])$, one has $xf = f_x(x, y) = f_x$, $xg = g_x(x, y) = g_x$, $yf = f_y(x, y) = f_y$, and $yg = g_y(x, y) = g_y \in R[x, y]$, and one identifies $f = (f_x, f_y)$ and $g = (g_x, g_y)$ via (12.2).

In describing $f \circ g$, note that for $x \in X$, $x(f \circ g) = (xf)g = (f_x(x, y))g =$

$f_x(g_x, g_y)$, and $y(f \circ g) = (yf)g = (f_y(x,y))g = f_y(g_x, g_y)$. Hence,

$$f \circ g = (f_x(x,y), f_y(x,y)) \circ (g_x(x,y), g_y(x,y))$$
$$= (f_x(g_x(x,y), g_y(x,y)), f_y(g_x(x,y), g_y(x,y))).$$

Now look at what happens when $X = \{x\}$. Let $xf = p(x)$ and $xg = q(x)$. Then $x(f \circ g) = (xf)g = p(x)g = p(xg) = p(q(x))$. Hence, we have the following pleasant theorem.

(12.12) Theorem. *The nearring of algebra endomorphisms of the R-algebra $R[x]$ is isomorphic to the nearring of polynomials $R[x]$.*

Note 1. The author prefers $p(q(x))$ to $(q(x))p$ for polynomials $p(x)$ and $q(x)$. This is one reason we have been writing the variable to the left of a function as in $xf = p(x)$.

Note 2. Theorem (12.12) is a result promised at the beginning of chapter 3.

Note 3. With (12.12) and the discussion before it, we obtain the near-rings $(R[x], +, \circ)$ and $(R[x,y] \times R[x,y], +, \circ)$, etc., as $(\mathbf{hom}(G, G), \oplus, \circ)$ for cogroup objects of a category $\mathcal{C}$. This represents yet another very important class of nearrings arising in this way.

The nearrings $(\mathbf{hom}(R[X], R[X]), \oplus, \circ)$ have some very interesting sub-nearrings. Let us begin by considering some 'linear' polynomials. If $X = \{x\}$, then $R[X] = R[x]$ contains the subnearring of all linear polynomials $a + bx$. Hence, $(\mathbf{hom}(R[X], R[X]), \oplus, \circ)$ contains a subnearring isomorphic to the subnearring of $R[X] = R[x]$ of all linear polynomials. Now let $X = \{x_1, x_2, \ldots, x_n\}$, a finite set. In $R[X] = R[x_1, x_2, \ldots, x_n]$, there is the set L of all polynomials of degree 1, that is, all $\alpha + a_1 x_1 + a_2 x_2 + \cdots + a_n x_n$. Then $U_0 = L^n$ is a subnearring of $(R[X]^n, +, \circ)$. An element of U_0 is an n-tuple $(\alpha_i + a_{i1} x_1 + \cdots + a_{in} x_n)$, $1 \leq i \leq n$. This subnearring U_0 is isomorphic to a subnearring of $(\mathbf{hom}(R[X], R[X]), \oplus, \circ)$. If $b = (\beta_i + b_{i1} x_1 + \cdots + b_{in} x_n)$, then $a \oplus b = ((\alpha_i + \beta_i) + (a_{i1} + b_{i1})x_1 + \cdots + (a_{in} + b_{in})x_n)$ and $a \circ b = (\alpha_i + \sum_{j=1}^n a_{ij}(\beta_j + b_{j1} x_1 + \cdots + b_{jn} x_n)) = (\alpha_i + \sum_{j=1}^n a_{ij}\beta_j + \sum_{j=1}^n (a_{i1} b_{1j} + \cdots + a_{in} b_{nj})x_j)$. We can represent the data in such an a by

$$a = \left[\begin{pmatrix} \alpha_1 \\ \alpha_2 \\ \vdots \\ \alpha_n \end{pmatrix}, \begin{pmatrix} a_{11} & a_{12} & \cdots & a_{1n} \\ a_{21} & a_{22} & \cdots & a_{2n} \\ \cdots & \cdots & \cdots & \cdots \\ a_{n1} & a_{n2} & \cdots & a_{nn} \end{pmatrix} \right] = [(\alpha_i), (a_{ij})].$$

Then $(U_0, \oplus, \circ)$ is a subnearring of the nearring $(\mathbf{hom}(R[X], R[X]), \oplus, \circ)$ with the same identity, where addition is defined by $a \oplus b = [(\alpha_i) +$

$(\beta_i), (a_{ij}) + (\beta_{ij})]$ and multiplication is defined by the familiar rule $a \circ b = [(\alpha_i) + (a_{ij})(\beta_i), (a_{ij})(b_{ij})]$. We can summarize with the following theorem.

(12.13) Theorem. *For a finite set* $X = \{x_1, x_2, \ldots, x_n\}$, *the nearring* $(\mathbf{hom}(R[X], R[X]), \oplus, \circ)$ *has the subnearring of 'linear' polynomials* $U_0 = \{a = [(\alpha_i), (a_{ij})] \mid \alpha_i, a_{ij} \in R\}$, *where* U_0 *is isomorphic to the abstract affine nearring* $(\oplus \sum_n R^+) \oplus_A \mathcal{M}(n, R)$, *where* $\mathcal{M}(n, R)$ *denotes the ring of* n *by* n *matrices over* R, *and* $\oplus \sum_n R^+$ *denotes the free* R-*module on* X, *which is also a left* $\mathcal{M}(n, R)$-*module.*

Note. In (12.13), we see that U_0 generalizes the subnearring of linear polynomials $a + bx$ in $R[x]$. This again reinforces the idea that our $\oplus$, defined by (11:8), is a correct approach.

Next, consider the $(R[T, T^{-1}], \pi, \mu, \alpha)$ of (12.4) and (12.5). We will be considering mappings $f : T \to \mathcal{U}(A)$. In fact, $(\mathbf{hom}(R[T, T^{-1}], A), \oplus) \cong (\mathcal{U}(A)^T, \oplus) \cong (\prod_{|T|}^* \mathcal{U}(A), \cdot)$. To see this, take $t \in T$ and $f, g \in \mathcal{U}(A)^T$. Then $t(f \oplus g) = t\pi \circ (f, g) = (t \otimes t)(f, g) = (tf)(tg)$. This gives us

(12.14) Theorem. *We have isomorphisms* $(\mathbf{hom}(R[T, T^{-1}], A), \oplus) \cong (\mathcal{U}(A)^T, \oplus) \cong (\prod_{|T|}^* \mathcal{U}(A), \cdot)$, *where* $\prod_{|T|}^* \mathcal{U}(A)$ *denotes the complete direct product of* $|T|$ *copies of* $\mathcal{U}(A)$, *and where* $t(f \oplus g) = (tf)(tg)$ *for all* $t \in T$ *and for all* $f, g \in \mathcal{U}(A)^T$.

If R is an integral domain, our units in $R[T, T^{-1}]$ are quite easy to determine. For comparison, take $R = Z_9$, the ring of integers modulo 9. Then in $Z_9[t, t^{-1}]$, we have $(3t^{-1} + 1 + 6t) \cdot (6t^{-1} + 1 + 3t) = 1$, so $3t^{-1} + 1 + 6t, 6t^{-1} + 1 + 3t \in \mathcal{U}(Z_9[t, t^{-1}])$. To make life easier, we let $R = K$, an integral domain. Any unit in $K[T, T^{-1}]$ is of the form $at_1^{n_1} t_2^{n_2} \cdots t_k^{n_k}$ where $a \in \mathcal{U}(K)$. Conversely, such an $at_1^{n_1} t_2^{n_2} \cdots t_k^{n_k}$, with $a \in \mathcal{U}(K)$, is a unit in $K[T, T^{-1}]$.

By (12.4), any $f \in \mathbf{hom}(K[T, T^{-1}], K[T, T^{-1}])$ can be identified with $f \in \mathcal{U}(K[T, T^{-1}])^T$. For $u \in T$, let $uf = a_u t_{u1}^{n_{u1}} t_{u2}^{n_{u2}} \cdots t_{uk_u}^{n_{uk_u}}$. So f corresponds bijectively with a pair $(\phi, (f_u)) \in \mathcal{U}(K)^T \times [Z^{(T)}]^T$. Specifically, for $u \in T$, $\phi(u) = a_u$, and $f_u(t_{uj}) = n_{uj}$. Hence, each $f_u(t) = 0$ for all but finitely many $t \in T$. This puts each $f_u \in Z^{(T)}$, and $(f_u) \in [Z^{(T)}]^T$. Thus

$$uf = \phi(u) \prod_{t \in T} t^{f_u(t)}$$

for each $u \in T$, and since each $f_u \in Z^{(T)}$, the product $\prod_{t \in T} t^{f_u(t)}$ makes sense since all but finitely many of the $f_u(t) = 0$ as t varies through all values in T.

Suppose $f, g \in \mathbf{hom}(K[T, T^{-1}], K[T, T^{-1}])$, $f = (\phi, (f_u))$, and $g = (\gamma, (g_u))$. Then $t(f \oplus g) = (tf)(tg) = \phi(i)\gamma(i) \prod_{t \in T} t^{f_u(t) + g_u(t)}$, so $f \oplus g =$

$(\phi\gamma, (f_u + g_u))$. Also,

$$u(f \circ g) = \left(\phi(u) \prod_{t \in T} t^{f_u(t)}\right)g = \phi(u) \prod_{t \in T} (tg)^{f_u(t)}$$

$$= \phi(u) \prod_{t \in T} \left[\gamma(t) \prod_{s \in T} s^{g_t(s)}\right]^{f_u(t)}$$

$$= \phi(u)\left\{\prod_{t \in T} \gamma(t)^{f_u(t)}\right\}\left[\prod_{s \in T} \prod_{t \in T} s^{f_u(t)g_t(s)}\right]$$

$$= \phi(u) \cdot \gamma^{(f_a)}(u) \cdot \prod_{s \in T}\left(s^{\sum_{t \in T} f_u(t)g_t(s)}\right)$$

$$= [\phi \cdot \gamma^{(f_a)}](u) \cdot \prod_{s \in T} s^{h_u(s)},$$

so $f \circ g = (\phi \cdot \gamma^{(f_a)}, (f_u)(g_u))$, where $\gamma^{(f_a)}(u) = \prod_{s \in T} \gamma(s)^{f_u(s)}$, and where multiplication is defined by the rule $(f_u)(g_u) = (h_u)$ with $h_u(s) = \sum_{t \in T} f_u(t)g_t(s)$.

So we have the binary operations $\oplus$ and $\cdot$ defined on $[Z^{(T)}]^T$ which make it into a ring with identity (e_u) where $e_u(s) = \delta_{us}$, the Kronecker delta symbol. It is a ring because it is essentially $(\mathrm{Hom}_Z(F_a(T), F_a(T)), +, \circ)$, where $F_a(T) = Z^{(T)}$, the free abelian group on T. We also have an 'addition' $\cdot$ on $\mathcal{U}(K)^T$ and a multiplication $(\gamma, (f_a)) \mapsto \gamma^{(f_a)}$ which make $(\mathcal{U}(K)^T, \cdot)$ a multiplicative abelian group and a unitary $[Z^{(T)}]^T$-module. In fact, the morphism nearring $(\mathbf{hom}(K[T, T^{-1}], K[T, T^{-1}]), \oplus, \circ)$ is isomorphic to the abstract affine nearring $\mathcal{U}(K)^T \oplus_A [Z^{(T)}]^T$.

It is worthwhile to consider $|T| < \infty$, say $T = \{t_1, t_2, \ldots, t_n\}$. We can write $t_i f = \phi(i) \prod_{j=1}^n t_j^{f_i(j)}$ if we will allow ourselves to abuse our notation a little in order to gain some clarity. We can identify $f = [(\phi(i)), (f_i(j))]$ where $(\phi(i))$ is an n by 1 column matrix and $(f_i(j))$ is an n by n matrix. Suppose $g = [(\gamma(i)), (g_i(j))]$. Then

$$t_i(f \circ g) = (\phi(i) \prod_{j=1}^n t_j^{f_i(j)})g$$

$$= \phi(i)\left\{\prod_{j=1}^n \gamma(j)^{f_i(j)}\right\} \prod_{k=1}^n \prod_{j=1}^n t_k^{f_i(j)g_j(k)}$$

$$= \phi(i)\left\{\prod_{j=1}^n \gamma(j)^{f_i(j)}\right\} \prod_{k=1}^n \left(t_k^{\sum_{j=1}^n f_i(j)g_j(k)}\right),$$

so

$$
f \circ g = \left[\left(\begin{array}{c} \phi(1) \\ \phi(2) \\ \vdots \\ \phi(n) \end{array} \right) \cdot \left(\begin{array}{ccc} f_1(1) & \cdots & f_1(n) \\ f_2(1) & \cdots & f_2(n) \\ \cdots & \cdots & \cdots \\ f_n(1) & \cdots & f_n(n) \end{array} \right) \left(\begin{array}{c} \gamma(1) \\ \gamma(2) \\ \vdots \\ \gamma(n) \end{array} \right), (f_i(j))(g_i(j)) \right]
$$
$$
= [(\phi(i)) \cdot (f_i(j))(\gamma(j)), \ (f_i(j))(g_i(j))].
$$

Note that the column matrix

$$
\left(\begin{array}{c} \gamma(1)^{f_1(1)} \gamma(2)^{f_1(2)} \cdots \gamma(n)^{f_1(n)} \\ \gamma(1)^{f_2(1)} \gamma(2)^{f_2(2)} \cdots \gamma(n)^{f_2(n)} \\ \cdots \\ \gamma(1)^{f_n(1)} \gamma(2)^{f_n(2)} \cdots \gamma(n)^{f_n(n)} \end{array} \right)
$$

is the multiplicative form of the column matrix

$$
\left(\begin{array}{c} f_1(1)\gamma(1) + f_1(2)\gamma(2) + \cdots + f_1(n)\gamma(n) \\ f_2(1)\gamma(1) + f_2(2)\gamma(2) + \cdots + f_2(n)\gamma(n) \\ \cdots \\ f_n(1)\gamma(1) + f_n(2)\gamma(2) + \cdots + f_n(n)\gamma(n) \end{array} \right),
$$

and this is why we have used the matrix notation

$$
\left(\begin{array}{ccc} f_1(1) & \cdots & f_1(n) \\ f_2(1) & \cdots & f_2(n) \\ \cdots & \cdots & \cdots \\ f_n(1) & \cdots & f_n(n) \end{array} \right) \left(\begin{array}{c} \gamma(1) \\ \gamma(2) \\ \vdots \\ \gamma(n) \end{array} \right) = (f_i(j))(\gamma(j))
$$

in the equation for $f \circ g$ above.

So we have that $\mathbf{hom}(K[T, T^{-1}], K[T, T^{-1}]), \oplus, \circ)$ is isomorphic to the abstract nearring $(\prod_n \mathcal{U}(K)) \oplus_A \mathcal{M}(n, Z)$, where $\mathcal{M}(n, Z)$ is the ring of n by n matrices over Z, and $\prod_n \mathcal{U}(K)$ is the direct product of n copies of $\mathcal{U}(K)$. Notice the similarities and differences with $(U_0, \oplus, \circ)$ of (12.13). We summarize and obtain

(12.15) Theorem. *Let K be an integral domain with group of units $\mathcal{U}(K)$. Then the nearring $\mathbf{hom}(K[T, T^{-1}], K[T, T^{-1}]), \oplus, \circ)$ is isomorphic to the abstract affine nearring $\mathcal{U}(K)^T \oplus_A [Z^{(T)}]^T$. The ring $[Z^{(T)}]^T$ has operations $(f_u) + (g_u) = (f_u + g_u)$ and $(f_u)(g_u) = (h_u)$ where $h_u(s) = \sum_{t \in T} f_u(t)g_t(s)$. The identity of the ring is (e_u) where $e_u(s) = \delta_{us}$, the Kronecker delta symbol. The abelian group $\mathcal{U}(K)^T$ has 'addition' defined by $(\phi \cdot \gamma)(s) = \phi(s)\gamma(s)$ and the ring $[Z^{(T)}]^T$ acts on the module $\mathcal{U}(K)^T$ by $\gamma \cdot (f_a) = \gamma^{(f_a)}$ where $\gamma^{(f_a)}(u) = \prod_{s \in T} \gamma(s)^{f_u(s)}$.*

If $|T| = n < \infty$, then $(\mathbf{hom}(K[T, T^{-1}], K[T, T^{-1}]), \oplus, \circ)$ is isomorphic to the abstract affine nearring $\prod_n \mathcal{U}(K) \oplus_A \mathcal{M}(n, Z)$.

In investigating the $(\mathbf{hom}(K[T, T^{-1}], K[T, T^{-1}]), \oplus, \circ)$ we have really been considering the $(\mathbf{hom}(KG, KG), \oplus, \circ)$ where $G = F_a(T)$, the free abelian group on T. The results in (12.15) are quite satisfying, but if we let G be a finite abelian group, we also obtain some very pleasant results with a different spirit from (12.15).

So let $G = \{1 = x_1, x_2, \ldots, x_n\}$ be a finite abelian group of order n. As an R-module, RG is just the free R-module on G, so an element of RG has a unique expression of the form $r_1 x_1 + r_2 x_2 + \cdots + r_n x_n$. With an $f \in \mathbf{hom}(G, \mathcal{U}(RG))$ in G, we have

$$x_i f = f_{i1} x_1 + f_{i2} x_2 + \cdots + f_{in} x_n$$

with $x_1 f = 1f = 1$, so $f_{11} = 1$, and $f_{1j} = 0$ for $1 < j \leq n$. So we identify f with the n by n matrix (f_{ij}). If $g = (g_{ij})$ is another group homomorphism in $\mathbf{hom}(G, \mathcal{U}(RG))$, then

$$x_i(f \circ g) = (f_{i1} x_1 + f_{i2} x_2 + \cdots + f_{in} x_n)g = \sum_{k=1}^{n} f_{ik}(x_k g)$$

$$= \sum_{k=1}^{n} f_{ik} \left(\sum_{j=1}^{n} g_{kj} x_j \right) = \sum_{k=1}^{n} \sum_{j=1}^{n} f_{ik} g_{kj} x_j$$

$$= \sum_{j=1}^{n} \sum_{k=1}^{n} f_{ik} g_{kj} x_j = \sum_{j=1}^{n} \left(\sum_{k=1}^{n} f_{ik} g_{kj} \right) x_j,$$

so $f \circ g = (\sum_{k=1}^{n} f_{ik} g_{kj}) = (f_{ij})(g_{ij})$. Thus, $\circ$ induces the usual matrix multiplication on the matrices representing the elements of $\mathbf{hom}(RG, RG)$.

The operation induced by $\oplus$ offers an interesting new way to add matrices. To describe $f \oplus g$ in terms of the (f_{ij}) and (g_{ij}), we need to look at a Cayley-like permutation of G. For each i, define a permutation σ_i by the equation

$$x_k x_{\sigma_i(k)} = x_i. \tag{12:1}$$

The equations $x_i x_j = x_{\pi_i(j)}$ define the Cayley representation of G as a subgroup of S_n, the group of all permutations on $\{1, 2, \ldots, n\}$. From (12:1) we get $x_{\sigma_i(k)} = x_i x_k^{-1}$. The map $x \mapsto x^{-1}$ also defines a $\delta \in S_n$, where $x_{\delta(i)} = x_i^{-1}$. Thus, $x_{\sigma_i(k)} = x_i x_k^{-1} = x_i x_{\delta(k)} = x_{\pi_i(\delta(k))}$, forcing $\sigma_i = \pi_i \circ \delta$. One easily sees that $\delta = \sigma_1$. If we identify G with $\{\pi_i \mid 1 \leq i \leq n\}$, then for $G' = \{\sigma_i \mid 1 \leq i \leq n\}$ we have $G' = G\sigma_1$, a coset of G in S_n.

Of course $G' = G$ if and only if $\sigma_1 \in G$. This means that $x_i x_j = x_{\pi_i(j)} = x_{\delta(j)}$ for some i, and $x_i x_j = x_{\delta(j)} = x_j^{-1}$. Exactly when does a group have an element u such that $ux = x^{-1}$ for each $x \in G$? This happens exactly when $x^2 = 1$ for each $x \in G$ and so $u = 1$. So G' is a group only for such groups.

Proceeding for $f \oplus g$, we have

$$
\begin{aligned}
x_i(f \oplus g) &= (x_i f)(x_i g) \\
&= (f_{i1}x_1 + f_{i2}x_2 + \cdots + f_{in}x_n)(g_{i1}x_1 + g_{i2}x_2 + \cdots + g_{in}x_n) \\
&= f_{i1}g_{i1}x_1 x_1 + \cdots + f_{i1}g_{in}x_1 x_n + \cdots + f_{in}g_{i1}x_n x_1 + \cdots + f_{in}g_{in}x_n x_n \\
&= (f_{i1}g_{i\sigma_1(1)}x_1 x_{\sigma_1(1)} + \cdots + f_{in}g_{i\sigma_1(n)}x_n x_{\sigma_1(n)}) \\
&\quad + (f_{i1}g_{i\sigma_2(1)}x_1 x_{\sigma_2(1)} + \cdots + f_{in}g_{i\sigma_2(n)}x_n x_{\sigma_2(n)}) + \cdots \\
&\quad + (f_{i1}g_{i\sigma_n(1)}x_1 x_{\sigma_n(1)} + \cdots + f_{in}g_{i\sigma_n(n)}x_n x_{\sigma_n(n)}) \\
&= (f_{i1}g_{i\sigma_1(1)} + f_{i2}g_{i\sigma_1(2)} + \cdots + f_{in}g_{i\sigma_1(n)})x_1 + \cdots \\
&\quad + (f_{i1}g_{i\sigma_j(1)} + f_{i2}g_{i\sigma_j(2)} + \cdots + f_{in}g_{i\sigma_j(n)})x_j + \cdots \\
&\quad + (f_{i1}g_{i\sigma_n(1)} + f_{i2}g_{i\sigma_n(2)} + \cdots + f_{in}g_{i\sigma_n(n)})x_n.
\end{aligned}
$$

So the (i, j)-entry for the matrix for $f \oplus g$ is

$$
f_{i1}g_{i\sigma_j(1)} + f_{i2}g_{i\sigma_j(2)} + \cdots + f_{in}g_{i\sigma_j(n)}
$$

which is the inner or dot product of the row vector

$$
(f_{i1}, f_{i2}, \ldots, f_{in})
$$

and the row vector

$$
(g_{i\sigma_j(1)}, g_{i\sigma_j(2)}, \ldots, g_{i\sigma_j(n)}).
$$

The first is just the ith row of (f_{ij}), call it f_i, but the second is the ith row of (g_{ij}) acted upon by the permutation σ_j, call it $g_i^{\sigma_j}$. So the (i, j)-entry of the matrix for $f \oplus g$ can be written as

$$
f_i \cdot g_i^{\sigma_j} = \sum_{k=1}^{n} f_{ik}g_{i\sigma_j(k)}.
$$

Hence,

$$
\begin{aligned}
f \oplus g &= \begin{pmatrix} f_1 \cdot g_1^{\sigma_1} & f_1 \cdot g_1^{\sigma_2} & \cdots & f_1 \cdot g_1^{\sigma_n} \\ f_2 \cdot g_2^{\sigma_1} & f_2 \cdot g_2^{\sigma_2} & \cdots & f_2 \cdot g_2^{\sigma_n} \\ \cdots & \cdots & \cdots & \cdots \\ f_n \cdot g_n^{\sigma_1} & f_n \cdot g_n^{\sigma_2} & \cdots & f_n \cdot g_n^{\sigma_n} \end{pmatrix} \\
&= (f_i \cdot g_i^{\sigma_j}) = \left(\sum_{k=1}^{n} f_{ik}g_{i\sigma_j(k)} \right).
\end{aligned}
$$

This means that $(\mathbf{hom}(RG, RG), \oplus, \circ) \cong (\mathbf{M}(n, R), \oplus_\sigma, \cdot)$ where $\mathbf{M}(n, R) = \{(f_{ij}) \mid f \in \mathbf{hom}(G, \mathcal{U}(RG))\}$, where $\cdot$ is the usual matrix multiplication, but $\oplus_\sigma$ is defined by

$$(f_{ij}) \oplus_\sigma (g_{ij}) = \left(\sum_{k=1}^{n} f_{ik} g_{i\sigma_j(k)} \right). \tag{12:2}$$

Notice that the additive identity is $z = (z_{ij})$, where each $z_{i1} = 1$ and $z_{ij} = 0$ for $j \neq 1$. Summarizing our results for these RG, we have

(12.16) Theorem. *Let R be a commutative ring with identity 1_R, and let $G = \{1 = x_1, x_2, \ldots, x_n\}$ be a finite abelian group of order n. Let $G' = \{\sigma_i \mid 1 \leq i \leq n\}$ be the permutations on $\{1, 2, \ldots, n\}$ defined by (12 : 1). Then $G' = G\sigma_1$, a coset of G as a subgroup of S_n, and G' is a subgroup of S_n if and only if $x^2 = 1$ for each $x \in G$.*

For the group algebra RG, $(\mathbf{hom}(RG, RG), \oplus, \circ) \cong (\mathbf{M}(n, R), \oplus_\sigma, \cdot)$, where $\cdot$ is the usual matrix multiplication, but $\oplus_\sigma$ is defined by (12:2). The additive identity with respect to $\oplus_\sigma$ is $z = (z_{ij})$.

It will be convenient to partition a matrix $(f_{ij}) \in \mathbf{M}(n, R)$ as $\begin{pmatrix} 1 & 0 \\ \phi & F \end{pmatrix}$

where $1 = (1)$, $\phi = \begin{pmatrix} f_{21} \\ f_{31} \\ \vdots \\ f_{n1} \end{pmatrix}$, and $F = \begin{pmatrix} f_{22} & f_{23} & \cdots & f_{2n} \\ f_{32} & f_{33} & \cdots & f_{3n} \\ \cdots & \cdots & \cdots & \cdots \\ f_{n2} & f_{n3} & \cdots & f_{nn} \end{pmatrix}$, since the

$f_{1j} = 0$ for $j \geq 2$. If $(g_{ij}) = \begin{pmatrix} 1 & 0 \\ \gamma & G \end{pmatrix}$, then $(f_{ij})(g_{ij}) = \begin{pmatrix} 1 & 0 \\ \phi + F\gamma & FG \end{pmatrix}$.

Let $N = \mathbf{hom}(RG, RG)$. Then the nearring N is decomposable into a constant part $N_c = \{m \in N \mid mz = m\}$, and a zero-symmetric part $N_0 = \{m \in N \mid mz = z\}$. See (3.11). As groups, $N^+ = N_c^+ \oplus N_0^+$, a direct sum, since N^+ is abelian. If $mz = m$, then

$$\begin{pmatrix} 1 & 0 \\ \mu & M \end{pmatrix} = \begin{pmatrix} 1 & 0 \\ \mu & M \end{pmatrix} \begin{pmatrix} 1 & 0 \\ 1 & 0 \end{pmatrix} = \begin{pmatrix} 1 & 0 \\ \mu + M\mathbf{1} & M\mathbf{0} \end{pmatrix} = \begin{pmatrix} 1 & 0 \\ \mu + M\mathbf{1} & 0 \end{pmatrix},$$

hence $M = 0$ making $m = \begin{pmatrix} 1 & 0 \\ \mu & 0 \end{pmatrix}$. If $mz = z$, then

$$\begin{pmatrix} 1 & 0 \\ 1 & 0 \end{pmatrix} = \begin{pmatrix} 1 & 0 \\ \mu & M \end{pmatrix} \begin{pmatrix} 1 & 0 \\ 1 & 0 \end{pmatrix} = \begin{pmatrix} 1 & 0 \\ \mu + M\mathbf{1} & M\mathbf{0} \end{pmatrix} = \begin{pmatrix} 1 & 0 \\ \mu + M\mathbf{1} & 0 \end{pmatrix}.$$

Hence, $\mu + M\mathbf{1} = \mathbf{1}$. This means that for $m = (m_{ij})$, we have for each i, $m_{i1} + m_{i2} + \cdots + m_{in} = 1$. So $x_i m \in \ker z$, the kernel of z, when we think of $z \in \mathbf{hom}(\mathcal{U}(RG), \mathcal{U}(RG))$ in $\mathcal{G}$, that is, $z|\mathcal{U}(RG)$.

Let us now summarize our results about $N^+ = N_c^+ \oplus N_0^+$.

(12.17) Theorem. *Let $N = \mathbf{hom}(RG, RG)$. Then*

$$N_c = \left\{ m = \begin{pmatrix} 1 & 0 \\ \mu & \mathbf{0} \end{pmatrix} \, \Big| \, \mu = \begin{pmatrix} m_{21} \\ \vdots \\ m_{n1} \end{pmatrix}, \right.$$

$$\left. \{1, m_{21}, \ldots, m_{n1}\} = m(G), \ m \in \mathbf{hom}(G, \mathcal{U}(R)) \right\},$$

and

$$N_0 = \left\{ m = \begin{pmatrix} 1 & 0 \\ \mu & M \end{pmatrix} \, \Big| \, m(G) \subseteq \ker z, \ z \in \mathbf{hom}(\mathcal{U}(RG), \mathcal{U}(RG)) \right\}.$$

Since $(N, \oplus)$ is abelian, we have $N^+ = N_c^+ \oplus N_0^+$, a direct sum.

12.2. Examples 11

Let $\mathcal{C} = \mathcal{G}$, and fix $A \in \mathbf{Ob}\,\mathcal{G}$. From examples 11, we obtain the category $\mathcal{G}(A)$. For $(S, \sigma), (T, \tau) \in \mathbf{Ob}\,\mathcal{G}(A)$, we construct $((S \times_A T, \phi), \pi_S, \pi_T)$ as in §10. Again, it is easy to see that this is a desired product.

Our first goal will be to determine the group objects of $\mathcal{G}(A)$. Suppose $((G, \gamma), \pi, \mu, \alpha)$ is such a group object, and let $K = \ker \gamma$ with insertion morphism $i : K \to G$. Since μ must satisfy $\gamma \circ \mu = 1_A$, we have that

$$0 \longrightarrow K \xrightarrow{i} G \underset{\mu}{\overset{\gamma}{\underset{\longleftarrow}{\longrightarrow}}} A \longrightarrow 0$$

is a split exact sequence. Thus G is isomorphic to a semidirect product $K \times_\theta A$ for some group homomorphism $\theta : A \to \mathrm{Aut}K$, where $(k, a) + (k', a') = (k + \theta(a)k', a + a')$ defines the binary operation on $K \times A$ for the group $K \times_\theta A$, and where we may take $\mu(a) = (0, a)$ and $\gamma(k, a) = a$. (We are essentially replacing G by $K \times_\theta A$.) A little bit has given us a lot, but it will take more to see that K must be abelian.

Consider first the identity diagram (11:2). The elements of $A \times_A G$ are (a, g) where $\gamma g = a$, so the elements of $A \times_A G$ are exactly the $(\gamma g, g)$ for $g \in G$. Now $[t_G, 1_G](g) = (\gamma g, g)$, and $\pi \circ \mu \times 1_A(\gamma g, g) = \pi(\mu \gamma g, g)$. Hence, $\pi(\mu \gamma g, g) = g$. Similarly, $\pi(g, \mu \gamma g) = g$. Recall that $(a, b) \in G \times_A G$ if and only if $\gamma a = \gamma b$. For $(a, b) \in G \times_A G$, we have $(-\mu \gamma b, -\mu \gamma a), (\mu \gamma b, b) \in G \times_A G$. Since $\gamma a = \gamma b$, we obtain $\pi(a, b) = \pi(a, \mu \gamma a) + \pi(-\mu \gamma b, -\mu \gamma a) + \pi(\mu \gamma b, b) = a + \pi(\mu \gamma \mu \gamma(-b), \mu \gamma(-b)) + b = a + \mu \gamma(-b) + b$. Take $a, b \in K = \ker \gamma$. Then $\pi(a, b) = a + b$. Also, $\pi(a, b) = \pi[(a, \mu \gamma a) + (-\mu \gamma b, -\mu \gamma a) +$

$(\mu\gamma b, b)] = \pi[(a,0)+(0,b)] = \pi[(0,b)+(a,0)] = \pi(\mu\gamma b, b)+\pi(a, \mu\gamma a) = b+a$.
So for $a, b \in \ker\gamma = K$, we obtain $a + b = \pi(a, b) = b + a$. I believe we can now agree that K is abelian, as promised.

We have that $\mu(a) = (0, a)$ and $\pi(x, y) = x + \mu\gamma(-y) + y$, but what is α? With $\mathbf{hom}[(G, \gamma), (A, 1_A)] = \{\gamma\} = \{t_{(G,\gamma)}\}$, the inverse diagram (11:3) tells us that $\mu\circ\gamma(x) = \pi\circ\alpha\times 1_G\circ[1_G, 1_G](x) = \pi(\alpha x, x) = \alpha x + \mu\gamma(-x) + x$. Hence, $\alpha(x) = \mu\gamma(x) - x + \mu\gamma(x)$. We have half of

(12.18) Theorem. *The group objects of $\mathcal{G}(A)$ are exactly the quadruples $((K \times_\theta A, \gamma), \pi, \mu, \alpha)$ where K is abelian, $\gamma(k, a) = a$, $\mu(a) = (0, a)$, $\pi(x, y) = x + \mu \circ \gamma(-y) + y$, and $\alpha(x) = \mu \circ \gamma(x) - x + \mu \circ \gamma(x)$.*

Proof. To finish the proof, we start with such a $((K \times_\theta A, \gamma), \pi, \mu, \alpha)$ as defined in the statement of the theorem. Certainly, $(K \times_\theta A, \gamma) \in \mathbf{Ob}\mathcal{G}(A)$, and $\mu \in \mathbf{hom}[(A, 1_A), (K \times_\theta A, \gamma)]$. For the π and α of (12.18), we must show that each is a morphism in $\mathcal{G}(A)$, and that the diagrams (11:1), (11:2), and (11:3) are commutative.

First, consider α. Let $x = (k, a)$, $y = (k', a') \in K \times_\theta A$ and remember that $\gamma(k, a) = a$ and $\mu(a) = (0, a)$. Then $\alpha(x + y) = \mu\gamma(x + y) - (x + y) + \mu\gamma(x + y) = (0, a + a') - (k + \theta(a)k', a + a') + (0, a + a') = -(k + \theta(a)k', 0) + (0, a + a') = (-k + \theta(a)(-k'), a + a') = (-k, a) + (-k', a') = (0, a) - (k, a) + (0, a) + (0, a') - (k', a') + (0, a') = \mu\gamma(x) - x + \mu\gamma(x) + \mu\gamma(y) - y + \mu\gamma(y) = \alpha(x) + \alpha(y)$. Also, $\gamma\circ\alpha(x) = \gamma[\mu\gamma(x) - x + \mu\gamma(x)] = \gamma(x) - \gamma(x) + \gamma(x) = \gamma(x)$. Hence, α is a morphism of $\mathcal{G}(A)$.

Now, consider π. Assume first that π is a morphism in $\mathcal{G}$. We want to show that π is a morphism of $\mathcal{G}(A)$, and to do that, we will prove that $\gamma \circ \pi = \overline{\gamma}$, where $(((K \times_\theta A) \times_A (K \times_\theta A), \overline{\gamma}), \pi_1, \pi_2)$ is the product of $(K \times_\theta A, \gamma)$ with itself in $\mathcal{G}(A)$. If $((k, a), (k', a')) \in (K \times_\theta A) \times_A (K \times_\theta A)$, then $\gamma(k, a) = \gamma(k', a')$, so $a = a'$. Now $\gamma \circ \pi((k, a), (k', a')) = \gamma[(k, a) - \mu\gamma(k', a') + (k', a')] = \gamma[(k, a) - (0, a') + (k', a')] = \gamma[(k, a) + (\theta(-a')k', 0)] = \gamma(k + \theta(a)\theta(-a')k, a) = a = \overline{\gamma}[(k, a), (k', a')]$, since $a = a'$. This makes $\gamma \circ \pi = \overline{\gamma}$.

Now we proceed to see that π is a morphism of $\mathcal{G}$. Take $((k, a), (k', a'))$, $((l, b), (l', b')) \in (K \times_\theta A) \times_A (K \times_\theta A)$, and recall that $a = a'$ and $b = b'$. Then $\pi[((k, a), (k', a')) + ((l, b), (l', b'))] = \pi((k, a) + (l, b), (k', a') + (l', b')) = \pi((k + \theta(a)l, a + b), (k' + \theta(a)l', a + b)) = (k + \theta(a)l, a + b) - (0, a + b) + (k' + \theta(a)l', a + b) = (k + \theta(a)l, 0) + (k' + \theta(a)l', a + b) = (k + \theta(a)l + k' + \theta(a)l', a + b) = (k + k' + \theta(a)(l + l'), a + b) = (k + k', a) + (l + l', b) = (k + \theta(a)\theta(-a)k', a) + (l + \theta(b)\theta(-b)l', b) = (k, a) + (\theta(-a)k', 0) + (l, b) + (\theta(-b)l', 0) = (k, a) + (0, -a') + (k', a') + (l, b) + (0, -b') + (l', b') = (k, a) + \mu\gamma[-(k', a')] + (k', a') + (l, b) + \mu\gamma[-(l', b')] + (l', b') = \pi((k, a), (k', a')) + \pi((l, b), (l', b'))$.

It remains to verify the commutativity of the various group object diagrams (11:1), (11:2), and (11:3). Looking at (11:1), we note that $\pi \circ \pi \times$

$1_{K \times_\theta A} \circ f[(k,a),(k',a'),(k'',a'')] = \pi \circ \pi \times 1_{K \times_\theta A}[[(k,a),(k',a')],(k'',a'')] =$
$\pi\big(\pi((k,a),(k',a')),(k'',a'')\big) = \pi\big((k,a) - \mu\gamma(k',a') + (k',a'),(k'',a'')\big) =$
$\pi\big((k,a) - (0,a') + (k',a'),(k'',a'')\big) = \pi\big((k,a) + (0,-a) + (k',a'),(k'',a'')\big) =$
$\pi\big((k,0) + (k',a'),(k'',a)\big) = \pi\big((k+k',a),(k'',a)\big) = (k+k'+k'',a)$, since
$a = a' = a''$. Similarly, $\pi \circ 1_{K \times_\theta A} \times \pi \circ g[(k,a),(k',a'),(k'',a'')] =$
$\pi \circ 1_{K \times_\theta A} \times \pi[(k,a),[(k',a'),(k'',a'')]] = \pi\big((k,a),\pi((k',a'),(k'',a''))\big) =$
$\pi\big((k,a),(k'+k'',a)\big) = (k+k'+k'',a)$, since $a = a' = a''$. So (11:1) is
commutative.

Next, consider (11:2). First, $\pi \circ \mu \times 1_{K \times_\theta A}(a,(k,a)) = \pi((0,a),(k,a)) = (k,a)$, and $[t_{K \times_\theta A}, 1_{K \times_\theta A}](k,a) = [\gamma, 1_{K \times_\theta A}](k,a) = (a,(k,a))$. Similarly, $\pi \circ 1_{K \times_\theta A} \times \mu((k,a),a) = \pi((k,a),(0,a)) = (k,a)$, and $[1_{K \times_\theta A}, \gamma](k,a) = ((k,a),a)$. This makes (11:2) commutative.

Turning our attention now to (11:3), first note that $\mu \circ t_{K \times_\theta A}(k,a) = \mu\gamma(k,a) = (0,a)$. Now $\pi \circ \alpha \times 1_{K \times_\theta A} \circ \Delta(k,a) = \pi \circ \alpha \times 1_{K \times_\theta A}[(k,a),(k,a)] = \pi(\alpha(k,a),(k,a)) = \pi\big(\mu\gamma(k,a) - (k,a) + \mu\gamma(k,a),(k,a)\big) = \pi\big((0,a) - (k,a) + (0,a),(k,a)\big) = \pi\big((0,a) - (0,a) - (k,0) + (0,a),(k,a)\big) = \pi((-k,a),(k,a)) = (0,a)$. Similarly, $\pi \circ 1_{K \times_\theta A} \times \alpha \circ \Delta(k,a) = (0,a)$. This makes (11:3) commutative, and completes the proof of (12.18).

For a fixed group A, an abelian group K, and a group homomorphism $\theta : A \to \operatorname{Aut} K$, we have *crossed homomorphisms* from A to K. A map $b : A \to K$ satisfying $b(a+a') = b(a) + \theta(a)b(a')$ for all $a, a' \in A$ is a *crossed homomorphism from A to K with respect to θ*. Let $Z_\theta^1(A,K)$ denote the set of all crossed homomorphisms from A to K with respect to θ. It should be easy for the reader to verify that $Z_\theta^1(A,K)$ is an abelian group with respect to pointwise addition of functions. Certainly, $(\operatorname{Hom}_Z(K,K), +, \circ)$ is a ring and $\theta(A) \subseteq \operatorname{Aut} K \subset \operatorname{Hom}_Z(K,K)$. Define $\mathcal{C}_h(K,A) = \{l \in \operatorname{Hom}_Z(K,K) \mid l \circ \theta(a) = \theta(a) \circ l$ for all $a \in A\}$. It should also be easy to prove that $\mathcal{C}_h(K,A)$ is a subring of $\operatorname{Hom}_Z(K,K)$ and that $Z_\theta^1(A,K)$ is a unitary $\mathcal{C}_h(K,A)$-module. With these, we have the abstract affine nearring $Z_\theta^1(A,K) \oplus_A \mathcal{C}_h(K,A)$.

It was very satisfying for the author to learn that this abstract affine nearring describes exactly the structure of the nearring $(\mathbf{hom}[(K \times_\theta A, \gamma),(K \times_\theta A, \gamma)], \oplus, \circ)$.

(12.19) Theorem. *For a group object $((K \times_\theta A, \gamma), \pi, \mu, \alpha)$ of $\mathcal{G}(A)$, the endomorphism nearring $(\mathbf{hom}[(K \times_\theta A, \gamma),(K \times_\theta A, \gamma)], \oplus, \circ)$ is isomorphic to the abstract affine nearring $Z_\theta^1(A,K) \oplus_A \mathcal{C}_h(K,A)$.*

Proof. Take $f \in \mathbf{hom}[(K \times_\theta A, \gamma),(K \times_\theta A, \gamma)]$. We have $\gamma \circ f = \gamma$, and, consequently, $a = \gamma(k,a) = \gamma \circ f(k,a) = \gamma(k',a') = a'$, where $f(k,a) = (k',a')$. Hence, $f(k,a) = (k',a)$. Also, $\ker f \subseteq \ker \gamma = \{(k,0) \mid k \in K\}$. Suppose $f(k,0) = (k',0) = (l(k),0)$. Then one readily obtains that $l : K \to K$ is a group homomorphism.

Suppose $f(0,a) = (b(a), a)$, where $b : A \to K$. Then $f(k,a) = f[(k,0) + (0,a)] = f(k,0) + f(0,a) = (l(k), 0) + (b(a), a) = (l(k) + b(a), a)$. Also, $(b(a+a'), a+a') = f(0, a+a') = f[(0,a) + (0,a')] = (b(a), a) + (b(a'), a') = (b(a) + \theta(a)b(a'), a + a')$. We now know that $b(a + a') = b(a) + \theta(a)b(a')$ for all $a, a' \in A$. This certainly makes b a crossed homomorphism from A to K with respect to θ, that is, $b \in Z^1_\theta(A, K)$. We also have $f(k,a) = (l(k) + b(a), a)$ for some $l \in \operatorname{Hom}_Z(K, K)$.

Are there any other conditions that l must satisfy? From $f[(k,a) + (k', a')] = f(k,a) + f(k', a') = (l(k) + b(a), a) + (l(k') + b(a'), a') = (l(k) + b(a) + \theta(a)[l(k') + b(a')], a + a')$ and $f(k + \theta(a)(k'), a + a') = (l(k + \theta(a)(k')) + b(a + a'), a + a')$, we obtain $l(k) + b(a) + \theta(a)l(k') + \theta(a)b(a') = l(k) + l(\theta(a)(k')) + b(a) + \theta(a)b(a')$, which in turn forces $\theta(a)l(k') = l \circ \theta(a)(k')$ for all $a \in A$ and for all $k' \in K$. This means that $\theta(a) \circ l = l \circ \theta(a)$ for each $a \in A$, and this puts $l \in \mathcal{C}_h(K, A)$.

Define $F : \mathbf{hom}[(K \times_\theta A, \gamma), (K \times_\theta A, \gamma)] \to Z^1_\theta(A, K) \oplus_A \mathcal{C}_h(K, A)$ by $F(f) = (b, l)$. Of course, our goal is to show that F is an isomorphism.

For $f, f' \in \mathbf{hom}[(K \times_\theta A, \gamma), (K \times_\theta A, \gamma)]$ and $F(f) = (b,l)$, $F(f') = (b', l')$, then $f \oplus f' = \pi \circ [f, f']$, so $(f \oplus f')(k,a) = \pi \circ [f, f'](k,a) = \pi[f(k,a), f'(k,a)] = f(k,a) - \mu \gamma f'(k,a) + f'(k,a) = (l(k) + b(a), a) - \mu \gamma (l'(k) + b'(a), a) + (l'(k) + b'(a), a) = (l(k) + b(a), 0) + (l'(k) + b'(a), a) = ((l + l')(k) + (b + b')(a), a)$. Hence, $F(f \oplus f') = (b + b', l + l') = F(f) + F(f')$.

Similarly, $f \circ f'(k,a) = f(l'(k) + b'(a), a) = (l \circ l'(k) + (l \circ b' + b)(a), a)$, so $F(f \circ f') = (b + l \circ b', l \circ l') = (b, l) \cdot (b', l') = F(f) \cdot F(f')$, and we have that F is a homomorphism.

If $F(f) = (b, l) = F(f')$, then $f(k,a) = (l(k) + b(a), a) = f'(k,a)$ for each $(k,a) \in K \times_\theta A$. So $f = f'$ and F is a monomorphism. For any $(b, l) \in Z^1_\theta(A, K) \oplus_A \mathcal{C}_h(K, A)$, define $f(k,a) = (l(k) + b(a), a)$. If $f \in \mathbf{hom}[(K \times_\theta A, \gamma), (K \times_\theta A, \gamma)]$, then $F(f) = (b, l)$, and F will be an isomorphism. Certainly, $\gamma \circ f = \gamma$, so it remains to see that f is a group homomorphism.

Take $(k,a), (k', a') \in K \times_\theta A$. Then $f[(k,a) + (k', a')] = f(k + \theta(a)k', a + a') = (l(k + \theta(a)k') + b(a + a'), a + a') = (l(k) + l(\theta(a)k') + b(a) + \theta(a)b(a'), a + a') = (l(k) + b(a) + l \circ \theta(a)(k') + \theta(a)b(a'), a + a') = (l(k) + b(a) + \theta(a) \circ l(k') + \theta(a)b(a'), a + a') = (l(k) + b(a) + \theta(a)[l(k') + b(a')], a + a') = (l(k) + b(a), a) + (l(k') + b(a'), a') = f(k,a) + f(k', a')$. So f is indeed a group homomorphism. Of course, this completes the proof of (12.19).

Theorem (12.19) is a very nice application of this theory of group objects, and the conclusion of (12.19) is rather tidy. If $(X, \epsilon) \in \mathbf{Ob}\, \mathcal{G}(A)$, then $(\mathbf{hom}[(X, \epsilon), (K \times_\theta A, \gamma)], \oplus)$ is a group for each group object $((K \times_\theta A, \gamma), \pi, \mu, \alpha)$. If X is the extension of an *abelian* group $\ker \epsilon = L$, then $(\mathbf{hom}[(X, \epsilon), (K \times_\theta A, \gamma)], \oplus)$ has a rather complicated but nice structure. This application of the theory nicely illustrates the beauty of some results of nearring theory, a beauty that has rather complicated structure, and is not particularly heavy with symmetry. Of course, (12.19) will be a special case when $(X, \epsilon) = (K \times_\theta A, \gamma)$, but it will be interesting to see how the ideas of (12.19) and its proof generalize to the more general case.

Using notation and terminology which is quite close to that of Rotman [R], we suppose that $(X, \epsilon) \in \mathbf{Ob}\,\mathcal{G}(A)$ is an extension of an abelian group $\ker \epsilon = L$. So there is a factor set $f : A \times A \to L$ and a group homomorphism $\lambda : A \to \mathrm{Aut}\,L$ so that X is isomorphic to $L \times_\lambda^f A$, where the set of $L \times_\lambda^f A$ is $L \times A$ and the binary operation is

$$(x, a) + (x', a') = (x + \lambda(a)x' + f(a, a'), a + a'),$$

and $\epsilon(x, a) = a$. We shall identify X with $L \times_\lambda^f A$. Let F be as in the proof of (12.19). Since $\gamma \circ F = \epsilon$, we have $F(x, a) = (\overline{x}, a)$, and $F(x, 0) = (l(x), 0)$ for some $l : L \to K$. It is easy to see that l must be a group homomorphism of these abelian groups. If we write $F(0, a) = (b(a), a)$, we obtain $F(x, a) = F[(x, 0) + (0, a)] = (l(x), 0) + (b(a), a) = (l(x) + b(a), a)$. We know that $l \in \mathrm{Hom}(L, K)$, but there will be further restrictions on l, and it should be no surprise to learn that there will be restrictions on $b : A \to K$, also.

First, consider b. We have $F[(0, a) + (0, a')] = F(f(a, a'), a + a') = (l \circ f(a, a') + b(a + a'), a + a')$ and $F(0, a) + F(0, a') = (b(a), a) + (b(a'), a') = (b(a) + \theta(a)b(a'), a + a')$. Consequently, $l \circ f(a, a') + b(a + a') = b(a) + \theta(a)b(a')$, or, equivalently,

$$b(a + a') = b(a) + \theta(a)b(a') - l \circ f(a, a'). \qquad (12:3)$$

Notice that $-l \circ f(a, a')$ stands in the way of b being a crossed homomorphism from A to K with respect to θ. But (12 : 3) is exactly what is meant by saying that $l \circ f$ is a *coboundary* of A by K with respect to θ. In particular, for a group A and an abelian group K and a homomorphism $\theta : A \to \mathrm{Aut}K$, a function $g : A \times A \to K$ satisfying

$$g(a, a') = b(a) + \theta(a)b(a') - b(a + a')$$

for some $b : A \to K$ with $b(0) = 0$ and for all $a, a' \in A$, is a *coboundary of A by K with respect to* θ. Let $B_\theta^2(A, K)$ denote all such coboundaries of A by K with respect to θ, and verify that $(B_\theta^2(A, K), +)$ is an abelian group. So (12 : 3) puts a restriction on b, but there is more to come.

Turning now to l, using $F(x, a) = (l(x) + b(a), a)$, first compute $F[(x, a) + (x', a')] = F(x + \lambda(a)(x') + f(a, a'), a + a')$ and then compute $F(x, a) + F(x', a')$. Comparing first coordinates allows us to conclude that $l \circ \lambda(a) = \theta(a) \circ l$ for each $a \in A$. If

$$\mathcal{C}_h(\lambda, \theta) = \{l \in \mathrm{Hom}(L, K) \mid \theta(a) \circ l = l \circ \lambda(a) \text{ for all } a \in A\},$$

then one easily obtains that $C_h(\lambda, \theta)$ is a subgroup of $\mathrm{Hom}(L, K)$. So we have a further restriction upon l, as well as two groups $B_\theta^2(A, K)$ and $C_h(\lambda, \theta)$. The group $B_\theta^2(A, K)$ is used to define still another group, a subgroup of $C_h(\lambda, \theta)$. Let

$$A(f) = \{l \in C_h(\lambda, \theta) \mid l \circ f \in B_\theta^2(A, K)\}.$$

It is easy to see that $A(f)$ is a subgroup of $C_h(\lambda, \theta)$.

Our immediate uses of the group $A(f)$ will be to define first what will turn out to be a coset of a group and, second, the group itself. Define

$$\mathcal{B}(l \circ f) = \{b : A \to K \mid b(a + a') = b(a) + \theta(a)b(a') - l \circ f(a, a')$$
$$\text{for all } a, a' \in A, \text{ and } b(0) = 0\}$$

and set

$$\overline{\mathcal{B}}(f) = \cup_{l \in A(f)} \mathcal{B}(l \circ f).$$

(12.20) Lemma. $\overline{\mathcal{B}}(f)$ *is an abelian group,* $\mathcal{B}(0)$ *is a subgroup, and* $\mathcal{B}(0) = Z_\theta^1(A, K)$.

Proof. Since the set of all mappings $t : A \to K$ is an abelian group, and since $\overline{\mathcal{B}}(f)$ is a nonempty subset, we need only show that $b_1 - b_2 \in \overline{\mathcal{B}}(f)$ for arbitrary $b_1, b_2 \in \overline{\mathcal{B}}(f)$. If $b_i(a + a') = b_i(a) + \theta(a)b_i(a') - l_i \circ f(a, a')$, $i = 1, 2$, for all $a, a' \in A$, then $(b_1 - b_2)(a + a') = b_1(a + a') - b_2(a + a') = [b_1(a) + \theta(a)b_1(a') - l_1 \circ f(a, a')] - [b_2(a) + \theta(a)b_2(a') - l_2 \circ f(a, a')] = (b_1 - b_2)(a) + \theta(a)(b_1 - b_2)(a') - (l_1 - l_2) \circ f(a, a')$. Since $l_1 - l_2$ is a member of the group $C_h(\lambda, \theta)$, we have $b_1 - b_2 \in \mathcal{B}((l_1 - l_2) \circ f) \subseteq \overline{\mathcal{B}}(f)$. As remarked above for $(12:3)$, if $l \circ f = 0$, then $b \in Z_\theta^1(A, K)$, and conversely. Hence, $\mathcal{B}(0) = Z_\theta^1(A, K)$, which is a group, hence a subgroup of $\overline{\mathcal{B}}(f)$.

(12.21) Lemma. *For any* $b \in \mathcal{B}(l \circ f)$, *we have* $\mathcal{B}(l \circ f) = \mathcal{B}(0) + b$, *a coset of* $\mathcal{B}(0)$ *in* $\overline{\mathcal{B}}(f)$.

Proof. For $b_1 \in \mathcal{B}(0)$, it is easy to see that $b_1 + b \in \mathcal{B}(l \circ f)$, so $\mathcal{B}(0) + b \subseteq \mathcal{B}(l \circ f)$. Likewise, if $b_2 \in \mathcal{B}(l \circ f)$, it follows that $b_2 - b \in \mathcal{B}(0)$, so $b_2 = c + b$ for some $c \in \mathcal{B}(0)$. Hence, $\mathcal{B}(l \circ f) \subseteq \mathcal{B}(0) + b$.

(12.22) Lemma. $\mathcal{B}((l_1 + l_2) \circ f) = \mathcal{B}(l_1 \circ f) + \mathcal{B}(l_2 \circ f)$.

Proof. If $b_i(a + a') = b_i(a) + \theta(a)b_i(a') - l_i \circ f(a, a')$ for all $a, a' \in A$, $i = 1, 2$, then $b_i \in \mathcal{B}(l_i \circ f)$. Hence, for $b = b_1 + b_2$ and $a, a' \in A$, we have $b(a + a') = b_1(a + a') + b_2(a + a') = (b_1 + b_2)(a) + \theta(a)(b_1 + b_2)(a') - l_1 \circ f(a, a') - l_2 \circ f(a, a') = b(a) + \theta(a)b(a') - (l_1 + l_2) \circ f(a, a')$. Hence, $\mathcal{B}(l_1 \circ f) + \mathcal{B}(l_2 \circ f) \subseteq \mathcal{B}((l_1 + l_2) \circ f)$. By (12.21), we get equality, hence (12.22) is true.

Let $n : \overline{\mathcal{B}}(f) \to \overline{\mathcal{B}}(f)/\mathcal{B}(0)$ be the natural map, and define $h : A(f) \to \overline{\mathcal{B}}(f)/\mathcal{B}(0)$ by $h(l) = \mathcal{B}(l \circ f)$. Then one easily obtains that $(\overline{\mathcal{B}}(f), n)$ and $(A(f), h)$ are objects in the category $\mathcal{G}(\mathbf{B})$ where $\mathbf{B} = \overline{\mathcal{B}}(f)/\mathcal{B}(0)$.

(12.23) Theorem. *As a group, $(\mathbf{hom}[(L \times_{\lambda}^f A, \epsilon), (K \times_{\theta} A, \gamma)], \oplus)$ is isomorphic to the product $\overline{\mathcal{B}}(f) \times_{\mathbf{B}} A(f)$ in $\mathcal{G}(\mathbf{B})$.*

Proof. Define $\Gamma : \mathbf{hom}[(L \times_{\lambda}^f A, \epsilon), (K \times_{\theta} A, \gamma)] \to \overline{\mathcal{B}}(f) \times_{\mathbf{B}} A(f)$ by $\Gamma(F) = (b, l)$, where $F(x, a) = (l(x) + b(a), a)$, as shown above. We know that $l \in A(f)$ and that $b \in \mathcal{B}(l \circ f) \subseteq \overline{\mathcal{B}}(f)$. Since $h(l) = \mathcal{B}(l \circ f)$, and $n(b) = \mathcal{B}(0) + b = \mathcal{B}(l \circ f)$ by (12.21), we have that $(b, l) \in \overline{\mathcal{B}}(f) \times_{\mathbf{B}} A(f)$, so Γ is well defined. If $\Gamma(F_1) = (b, l) = \Gamma(F_2)$, then $F_1 = F_2$ because of the formula $F(x, a) = (l(x) + b(a), a)$ above. Hence, Γ is injective.

The proof that Γ is a group homomorphism is nearly as the proof that the F of the proof of (12.19) is a group homomorphism. For any $(b, l) \in \overline{\mathcal{B}}(f) \times_{\mathbf{B}} A(f)$, define $F : L \times_{\lambda}^f A \to K \times_{\theta} A$ by $F(x, a) = (l(x) + b(a), a)$. Then $\gamma \circ F = \epsilon$, and it only remains to show that F is a group homomorphism.

Take $(x, a), (x', a') \in L \times_{\lambda}^f A$. Then we obtain $F[(x, a) + (x', a')] = F(x + \lambda(a)(x') + f(a, a'), a + a') = (l[x + \lambda(a)(x') + f(a, a')] + b(a, a'), a + a') = (l(x) + l \circ \lambda(a)(x') + l \circ f(a, a') + b(a) + \theta(a)b(a') - l \circ f(a, a'), a + a') = (l(x) + b(a) + \theta(a)[l(x')] + \theta(a)b(a'), a + a') = (l(x) + b(a), a) + (l(x') + b(a'), a') = F(x, a) + F(x', a')$. This completes the proof of (12.23). ∎

(12.24) Corollary. *If $(X, \epsilon) \in \mathcal{G}(A)$, and X is a semidirect product of an abelian group L by A, then $X \cong L \times_{\lambda} A$ for some group homomorphism $\lambda : A \to \mathrm{Aut} A$ and $(\mathbf{hom}[(L \times_{\lambda} A, \epsilon), (K \times_{\theta} A, \gamma)], \oplus)$ is isomorphic to the direct sum $Z_{\theta}^1(A, K) \oplus \mathcal{C}_h(\lambda, \theta)$.*

Proof. Here, $f = 0$, so $\overline{\mathcal{B}}(f) = Z_{\theta}^1(A, K)$, $A(f) = \mathcal{C}_h(\lambda, \theta)$, and $\mathbf{B} = \{0\}$. Hence, $\overline{\mathcal{B}}(f) \times_{\mathbf{B}} A(f) = Z_{\theta}^1(A, K) \oplus \mathcal{C}_h(\lambda, \theta)$. ∎

(12.25) Theorem. *For a group object $((K \times_{\theta} A, \gamma), \pi, \mu, \alpha)$ and an $(L \times_{\lambda}^f A, \epsilon) \in \mathbf{Ob}\, \mathcal{G}(A)$ with L abelian, the nearring $(\mathbf{hom}[(K \times_{\theta} A, \gamma), (K \times_{\theta} A, \gamma)], \oplus, \circ)$ acts on the left of the group $(\mathbf{hom}[(L \times_{\lambda}^f A, \epsilon), (K \times_{\theta} A, \gamma)], \oplus)$. If $g \in \mathbf{hom}[(K \times_{\theta} A, \gamma), (K \times_{\theta} A, \gamma)]$ and $F \in \mathbf{hom}[(L \times_{\lambda}^f A, \epsilon), (K \times_{\theta} A, \gamma)]$, then $g \circ F \in \mathbf{hom}[(L \times_{\lambda}^f A, \epsilon), (K \times_{\theta} A, \gamma)]$. If g corresponds to $(b, l) \in Z_{\theta}^1(A, K) \oplus_A \mathcal{C}_h(K, A)$ as in the proof of (12.19), and F corresponds to $(b', l') \in \overline{\mathcal{B}}(f) \times_{\mathbf{B}} A(f)$ as in the proof of (12.23), then $g \circ F$ corresponds to $(b, l) \cdot (b', l') = (b + l \circ b', l \circ l') \in \overline{\mathcal{B}}(f) \times_{\mathbf{B}} A(f)$.*

Proof. The proof is direct, but we will do the last step. For $(x, a) \in L \times_{\lambda}^f A$, $g \circ F(x, a) = g(l'(x) + b'(a), a) = (l[l'(x) + b'(a)] + b(a), a) = (l \circ l'(x) + (l \circ b' + b)(a), a)$. So $g \circ F$ corresponds to $(b + l \circ b', l \circ l')$. ∎

At this point, we take a lesson learned from Gonshor [G] when he devel-

oped the abstract affine nearring by looking carefully at the affine transformations of a vector space. Nice examples often lead one to a general
theorem. Our results of (12.25) have the elements of an abstract affine
nearring $M \oplus_A R$ acting on a group. This motivates a more general construction.

It should be easy to construct a product in $_R\mathcal{M}(B)$ for a fixed R-module
B. If desired, we may assume that B is a unitary R-module. Objects
of $_R\mathcal{M}(B)$ are pairs (S, σ), and if $(S, \sigma), (T, \tau) \in \mathbf{Ob}_R\mathcal{M}(B)$, then $\phi \in$
$\mathbf{hom}[(S, \sigma), (T, \tau)]$ is an R-homomorphism $\phi : S \to T$ satisfying $\tau \circ \phi = \sigma$.
In $_R\mathcal{M}(M)$, one obtains a product $((S \times_B T, \xi), \pi_S, \pi_T)$.

(12.26) Theorem. *Let R be a ring. Suppose B is an R-module, and let
(S, σ) and (T, τ) be objects in $_R\mathcal{M}(B)$. For an R-submodule $M \subseteq \ker \sigma$,
there is the abstract affine nearring $M \oplus_A R$. Now $M \oplus_A R$ acts on $S \times_B T$
by the rule*

$$(m, r) \cdot (s, t) = (m + rs, rt)$$

for all $(m, r) \in M \oplus_A R$ and for all $(s, t) \in S \times_B T$. One obtains that

$$[(m_1, r_1) + (m_2, r_2)] \cdot (s, t) = (m_1, r_1) \cdot (s, t) + (m_2, r_2) \cdot (s, t)$$

and

$$[(m_1, r_1) \cdot (m_2, r_2)] \cdot (s, t) = (m_1, r_1) \cdot [(m_2, r_2) \cdot (s, t)].$$

*If $1 \in R$ and all our modules are unitary, then $(0, 1)$ is the identity of
$M \oplus_A R$ and $(0, 1) \cdot (s, t) = (s, t)$.*

Proof. At this point, the reader should find the proof almost too easy.

(12.27) Note. In $\mathcal{G}(A)$, if $A = \{0\}$, then $G_1 \times_A G_2 = G_1 \times G_2$, the usual
product in $\mathcal{G}$. In $_R\mathcal{M}(B)$, if $B = \{0\}$, then $S \times_B T = S \oplus T$, the usual
product in $_R\mathcal{M}$.

In light of (11.7), (12.25) and (12.26) have analogs.

(12.28) Theorem. *For the two group objects $((K \times_\theta A, \gamma), \pi, \mu, \alpha)$ and
$((K' \times_{\theta'} A, \gamma'), \pi', \mu', \alpha')$ of $\mathcal{G}(A)$, the nearring $(\mathbf{hom}[(K \times_\theta A, \gamma), (K \times_\theta A, \gamma)], \oplus, \circ)$ acts on the right of the group $(\mathbf{hom}[(K \times_\theta A, \gamma), (K' \times_{\theta'} A, \gamma')],
\oplus)$. If $h \in \mathbf{hom}[(K \times_\theta A, \gamma), (K \times_\theta A, \gamma)]$ and $F \in \mathbf{hom}[(K \times_\theta A, \gamma), (K' \times_{\theta'}
A, \gamma')]$, then $F \circ h \in \mathbf{hom}[(K \times_\theta A, \gamma), (K' \times_{\theta'} A, \gamma')]$. If h corresponds to
$(b, l) \in Z^1_\theta(A, K) \oplus_A \mathcal{C}_h(K, A)$ as in the proof of (12.19) and F corresponds
to $(b', l') \in Z^1_{\theta'}(A, K') \oplus \mathcal{C}_h(\theta, \theta')$ as in (12.24), then $F \circ h$ corresponds to
$(b', l') \cdot (b, l) = (b' + l' \circ b, l' \circ l) \in Z^1_{\theta'}(A, K') \oplus \mathcal{C}_h(\theta, \theta').$*

Proof. See the proof of (12.25).

Generalizing (12.28) as we did (12.25), we obtain the next result.

(12.29) Theorem. *Let R be a ring with left R-module M and right R-module L. Let K be an abelian group with middle linear map $\phi : L \times M \to K$ [Hu]. Then the abstract affine nearring $M \oplus_A R$ acts on the right of the abelian group $K \oplus L$ by the rule*

$$(k, l) \cdot (m, r) = (k + \phi(l, m), lr)$$

for all $(m, r) \in M \oplus_A R$ and for all $(k, l) \in K \oplus L$. One obtains that

$$[(k, l) + (k', l')] \cdot (m, r) = (k, l) \cdot (m, r) + (k', l') \cdot (m, r)$$

and

$$(k, l) \cdot [(m, r) \cdot (m', r')] = [(k, l) \cdot (m, r)] \cdot (m', r').$$

If $1 \in R$ and all our modules are unitary, then $(0, 1)$ is the identity of $M \oplus_A R$ and $(k, l) \cdot (0, 1) = (k, l)$.

Proof. Again, the proof offers no challenge.

As mentioned before (12.26), (12.26) offers one method to construct something like a nearring module, and now (12.29) offers a contrasting method. The question really is: which elements should distribute over which sums?

From (12.29), we can make some more interesting observations. Let R, M, and L be as in (12.29). We shall construct a category $\mathcal{M}_R(L, M)$ whose objects are all pairs (ϕ, K) where $\phi : L \times M \to K$ is a middle linear map, so K must be an abelian group. For objects (ϕ, K) and (ϕ', K') of $\mathcal{M}_R(L, M)$, we have $f \in \mathbf{hom}[(\phi, K), (\phi', K')]$ if and only if $f : K \to K'$ is a group homomorphism and $f \circ \phi = \phi'$. It should be easy to see that $\mathcal{M}_R(L, M)$ is a category.

(12.30) Exploratory problem. Explore various categories of the form $\mathcal{M}_R(L, M)$. Let R, L, and M vary. In particular, let $R = \mathcal{C}_h(K, A)$ for some group A and some abelian group K, and perhaps let $M = Z_\theta^1(A, K)$ and $L = Z_\lambda^1(A, K)$, or something closely related. There are suggestions that these $\mathcal{M}_R(L, M)$ can be closely related to some $\mathcal{G}(A)$. If so, how?

What we have done here with $\mathcal{G}(A)$ again illustrates the importance of the ideas of this chapter. In particular, let the structures obtained from group and cogroup objects of an interesting category direct you to interesting constructions and research.

12.3. Examples 12

Again, let $\mathcal{C} = \mathcal{G}$ and fix $A \in \mathbf{Ob}\,\mathcal{G}$. We consider the category $(A)\mathcal{G}$ for examples 12. Objects are ordered pairs (γ, G), where $\gamma : A \to G$

is the insertion group homomorphism. So, truly speaking, our objects are the groups having A as a subgroup. (Certainly, any group G' with a subgroup A' isomorphic to A is isomorphic to a group G containing A.) For $(\sigma, S), (\tau, T) \in \mathbf{Ob}\ (A)\mathcal{G}$, $f \in \mathbf{hom}[(\sigma, S), (\tau, T)]$ if and only if $f : S \to T$ is a group homomorphism and $f \circ \sigma = \tau$. Certainly, $(1_A, A)$ is an initial object of $(A)\mathcal{G}$. But what about a coproduct in $(A)\mathcal{G}$?

The coproduct of (σ, S) and (τ, T) in $(A)\mathcal{G}$ is often called the free product with amalgamated subgroup, or simply the amalgam by Rotman [R]. We shall denote such a coproduct by $(\nu, S *_A T)$. For the reader familiar with free products, one can describe the elements of the amalgam of (σ, S) and (τ, T) as all finite sequences (formal sums)

$$s_1 + t_1 + s_2 + t_2 + \cdots + s_n + t_n,$$

$s_i \in S$, $t_i \in T$, and where we agree not to combine an $s_i + t_i$ or a $t_i + s_{i+1}$ unless each term, or summand, is in A, or unless at least one of the summands is an identity. So we think of $S \cap T = A$, even though this need not be the case. (If $S \cap T$ contains A properly, then we take two symbols, say 0 and 1, so that $\{(s, 0) \mid s \in S \backslash A\} \cap A = \{(t, 1) \mid t \in T \backslash A\} \cap A = \varnothing$. Replace S by $S_0 = \{(s, 0) \mid s \in S \setminus A\} \cup A$ and T by $T_1 = \{(t, 1) \mid t \in T \setminus A\} \cup A$, and so $S_0 \cong S$ and $T_1 \cong T$, as groups.) We then have the monomorphisms $e_S : S \to S *_A T$ defined by $e_S(s) = s + 0_T = s$ and $e_T : T \to S *_A T$ defined by $e_T(t) = 0_S + t = t$. So if $\nu(a) = a + 0_T = 0_S + a = a$, then the triple $((\nu, S *_A T), e_S, e_T)$ is a coproduct of (σ, S) and (τ, T) in the category $(A)\mathcal{G}$.

With a coproduct and an initial object, we are ready to find some cogroup objects in $(A)\mathcal{G}$. Suppose we have a cogroup object in the category $((\gamma, G), \pi, \mu, \alpha)$ in $(A)\mathcal{G}$. Now $\mu \in \mathbf{hom}[(\gamma, G), (1_A, A)]$, so $\mu \circ \gamma = 1_A$. Hence, $\mu : G \to A$ is a group epimorphism with kernel $F = \ker \mu$, and this is exactly what is needed to make the group G a split extension of F by A, or, equivalently, G is isomorphic to a semidirect product $F \times_\theta A$ for some group homomorphism $\theta : A \to \mathrm{Aut} F$. We will take $G = \{(f, a) \mid f \in F, a \in A\}$ and $(f, a) + (f', a') = (f + \theta(a)(f'), a + a')$. We identify $f = (f, 0)$ and $(0, a) = a$ so that $G = F \times_\theta A$ has A as a subgroup. Notice that $(0, a) + (f, 0) - (0, a) = (\theta(a)f, 0)$.

We have that

$$0 \longrightarrow F \overset{i}{\longrightarrow} G \underset{\underset{\gamma}{\longleftarrow}}{\overset{\mu}{\longrightarrow}} A \longrightarrow 0$$

is a split short exact sequence, and we suggest the reader compare the analogous result for Examples 11 for similarities and differences.

For $\mathcal{G}(A)$, we saw that K must be abelian, but for $(A)\mathcal{G}$, it can be seen in [C & M] that F must be free. Just as it was also sufficient for K to be abelian for $\mathcal{G}(A)$, we shall see here that it is also sufficient for F to be free for $F \times_\theta A$ to be a cogroup object of $(A)\mathcal{G}$.

Since G above is a semidirect product $F \times_\theta A$, then $\theta : A \to \mathrm{Aut}F$ is a group homomorphism, and so $\theta(a) \in \mathrm{Aut}F$. If F is a free group on the nonemptry set X, then $\theta(a)|X$ is a permutation on X. We now wish to prove the following theorem.

(12.31) Theorem. *Let* $F = \mathcal{F}(X)$ *be a free group on* $X \neq \varnothing$. *Let* $\theta : A \to \mathrm{Aut}F$ *be a group homomorphism, and set* $G = F \times_\theta A$, *the resulting semidirect product. Let* $\pi : G \to G *_A G$ *be defined by* $\pi(a) = a$ *for all* $a \in A$, *and* $\pi(x) = x(1) + x(2)$ *for all* $x \in X$. *Let* $\alpha : G \to G$ *be defined by* $\alpha(a) = a$ *for all* $a \in A$, *and* $\alpha(x) = -x$ *for all* $x \in X$. *Let* $\mu : G \to A$ *be defined by* $\mu(f, a) = a$. *Then the quadruple* $((\nu, G), \pi, \mu, \alpha)$ *is a cogroup object in the category* $(A)\mathcal{G}$.

Remark. A little explanation is in order concerning the group $G *_A G$. We have G as a factor twice in $G *_A G$, and each factor has A as a subgroup. We want A to be the only thing in common between the two representations of G in $G *_A G$ and we want the elements of the subgroup A to be represented only once in $G *_A G$. So what do we do with $g \in G \setminus A$? If we want to think of g as coming from the left or first representation of G, we denote this by $g(1)$. Likewise, if we want to think of g as being in the right or second representation of G, we shall denote this by $g(2)$. Hence, $g(1) \neq g(2)$ while $g(1), g(2) \in G *_A G$ and $g \in G \setminus A$. For this reason we have $g(1) + g(2) \neq g(2) + g(1)$.

Proof of (12.31). The first thing we want to do is to show that π, μ, and α define morphisms in the category $(A)\mathcal{G}$. We shall need some patience and understanding from the reader since we are going to indulge in a lot of ambiguous and inconsistent notation in the hope of making the ideas of the proof clearer. For an arbitrary element $a \in A$, we identify $a \equiv (0, a) \in G = F \times_\theta A$. For any element $f \in F$, we identify $f \equiv (f, 0) \in G$. We certainly have that μ is a morphism in the category $(A)\mathcal{G}$. Since $(f, a) = (f, 0) + (0, a)$, we need to define $\alpha f = \alpha(f, 0)$ because we easily have that $(0, a) = a = \alpha a = \alpha(0, a)$. If $f = \sum_{i=1}^m \epsilon_i x_i$ with $x_i \in X$ and $\epsilon_i \in \{\pm 1\}$, then $f = (f, 0) = (\sum_{i=1}^m \epsilon_i x_i, 0) = \sum_{i=1}^m (\epsilon_i x_i, 0) = \sum_{i=1}^m \epsilon_i (x_i, 0)$, and $\alpha f = \alpha(f, 0) = \alpha \sum_{i=1}^m \epsilon_i (x_i, 0) = \sum_{i=1}^m \epsilon_i (-x_i, 0) = (\sum_{i=1}^m (-\epsilon_i) x_i, 0)$. Hence, we get the equation

$$\alpha\left(\sum_{i=1}^m \epsilon_i x_i, a\right) = \left(\sum_{i=1}^m (-\epsilon_i) x_i, a\right).$$

Remember that $\theta(a)|X$ is a permutation on X, so if $x \in X$ and $a \in A$, then $\theta(a)x \in X$. Hence,

$$\alpha[(f,a) + (f',a')] = \alpha(f + \theta(a)f', a + a')$$

$$= \alpha\left(\sum_{i=1}^{m} \epsilon_i x_i + \theta(a) \sum_{i=1}^{n} \epsilon'_i x'_i, \, a + a'\right)$$

$$= \alpha\left(\sum_{i=1}^{m} \epsilon_i x_i + \sum_{i=1}^{n} \epsilon'_i \theta(a)x'_i, \, a + a'\right)$$

$$= \left(\sum_{i=1}^{m} (-\epsilon_i)x_i + \sum_{i=1}^{n} (-\epsilon'_i)\theta(a)x'_i, \, a + a'\right)$$

$$= \left(\sum_{i=1}^{m} (-\epsilon_i)x_i + \theta(a) \sum_{i=1}^{n} (-\epsilon'_i)x'_i, \, a + a'\right)$$

$$= \left(\sum_{i=1}^{m} (-\epsilon_i)x_i, a\right) + \left(\sum_{i=1}^{n} (-\epsilon'_i)x'_i, a'\right)$$

$$= \alpha(f,a) + \alpha(f',a').$$

This makes $\alpha : G \to G$ a morphism in $(A)\mathcal{G}$.

Similarly, $\pi(x) = x(1) + x(2) = (x(1), 0) + (x(2), 0) = (x(1) + x(2), 0)$, so $\pi(x, a) = (x(1) + x(2), a)$. From $\pi(f, 0) = \pi f = \pi \sum_{i=1}^{m} \epsilon_i x_i = \sum_{i=1}^{m} \epsilon_i \pi(x_i) = \sum_{i=1}^{m} \epsilon_i(x_i(1) + x_i(2), 0) = \sum_{i=1}^{m} (\epsilon_i(x_i(1) + x_i(2)), 0) = (\sum_{i=1}^{m} \epsilon_i(x_i(1) + x_i(2)), 0)$, we obtain

$$\pi(f,a) = \left(\sum_{i=1}^{m} \epsilon_i(x_i(1) + x_i(2)), \, a\right).$$

Since $\theta(a)x'_i \in X$, we have $\theta(a)x'_i = \bar{x}_i$ and $\pi\bar{x}_i = \bar{x}_i(1) + \bar{x}_i(2) = \theta(a)x'_i(1) + \theta(a)x'_i(2) = \theta(a)[x'_i(1) + x'_i(2)]$, so

$$\pi[(f,a) + (f',a')] = \pi\left(\sum_{i=1}^{m} \epsilon_i x_i + \theta(a) \sum_{i=1}^{n} \epsilon'_i x'_i, \, a + a'\right)$$

$$= \left(\sum_{i=1}^{m} \epsilon_i(x_i(1) + x_i(2)) + \sum_{i=1}^{n} \epsilon'_i \pi(\theta(a)x'_i), \, a + a'\right)$$

$$= \left(\sum_{i=1}^{m} \epsilon_i\big(x_i(1) + x_i(2)\big) + \sum_{i=1}^{n} \epsilon'_i\big(\theta(a)x'_i(1) + \theta(a)x'_i(2)\big), \, a + a'\right)$$

$$= \left(\sum_{i=1}^{m} \epsilon_i\big(x_i(1) + x_i(2)\big) + \theta(a) \sum_{i=1}^{n} \epsilon'_i\big(x'_i(1) + x'_i(2)\big), \, a + a'\right)$$

$$= \left(\sum_{i=1}^{m} \epsilon_i\big(x_i(1) + x_i(2)\big), \, a\right) + \left(\sum_{i=1}^{n} \epsilon'_i\big(x'_i(1) + x'_i(2)\big), \, a'\right)$$

$$= \pi(f,a) + \pi(f',a').$$

This makes $\pi : G \to G *_A G$ a morphism in $(A)\mathcal{G}$.

Our next task is to show that the diagrams (11:4), (11:5), and (11:6) are commutative in our present setting. For our $G = F \times_\theta A$, if $\tau \in$ $\mathbf{hom}((\nu, G), (\lambda, L))$ for some $(\lambda, L) \in (A)\mathcal{G}$, then for $(f, a) \in G$, $\tau(f, a) = \tau[(f, 0) + (0, a)] = \tau(f, 0) + \tau(0, a) = \sum_{i=1}^{m} \epsilon_i \tau(x_i, 0) + a$. So if morphisms τ and τ' agree at each $x \in X$, then $\tau = \tau'$. So, we will take an arbitrary $x \in X$ and consider each of the diagrams (11:4), (11:5), and (11:6).

First, for (11:4), $f \circ \pi \vee 1 \circ \pi(x) = f \circ \pi \vee 1(x(1) + x(2)) = f(x(1, 1) + x(1, 2) + x(2)) = x(1) + x(2) + x(3)$, and $g \circ 1 \vee \pi \circ \pi(x) = g \circ 1 \vee \pi(x(1) + x(2)) = g(x(1) + x(2, 1) + x(2, 2)) = x(1) + x(2) + x(3)$, and we are relying upon the reader for a lot of help and understanding. So we will declare (11:4) a commutative diagram.

For (11:5), first note that $(i_G, 1)x(2) = x$, so $(i_G, 1)^{-1}(x) = x(2)$. Also, $\mu \vee 1 \circ \pi(x) = \mu \vee 1(x(1) + x(2)) = \mu(x(1)) + x(2) = x(2)$. Similarly, $(1, i_G)^{-1}(x) = x(1)$ and $1 \vee \mu \circ \pi(x) = 1 \vee \mu(x(1) + x(2)) = x(1) + \mu(x(2)) = x(1)$. So we also declare (11:5) a commutative diagram.

For (11:6), $i_G \circ \mu(x) = i_G(0) = 0$, and $\nabla \circ \alpha \vee 1_G \circ \pi(x) = \nabla \circ \alpha \vee 1_G(x(1) + x(2)) = \nabla(-x(1) + x(2)) = -x + x = 0$. Similarly, $\nabla \circ 1_G \vee \alpha \circ \pi(x) = \nabla \circ 1_G \vee \alpha(x(1) + x(2)) = \nabla(x(1) + \alpha(x(2))) = \nabla(x(1) + (-x(2))) = x - x = 0$. Now we can declare (11:6) commutative, and so ends our proof of (12.21).

The next natural steps are to (1) study the endomorphism nearrings $(\mathbf{hom}[(\nu, G), (\nu, G)], \oplus, \circ)$, and (2) its modules $(\mathbf{hom}[(\nu, G), (\lambda, L)], \oplus)$, for $(\lambda, L) \in (A)\mathcal{G}$. It is a nice story with many pleasing climaxes, but it is long and requires lots of detail. It would have been nice to include it here, but I have been striving to keep the number of pages of this tome within a reasonable approximation to that set by the publisher, who in turn has tried to keep the price within a reasonable approximation of a purchaser's budget. Experience suggests that it will take a reader with unusual courage and a highly sophisticated sense of mathematical beauty to seek out the beauties of $\mathbf{hom}[(\nu, G), (\nu, G)]$ and $\mathbf{hom}[(\nu, G), (\lambda, L)]$, and there *is* an outline in [C & M] to help and encourage her. On top of this courage and sophistication, one needs a strong dose of rugged individualism to extend the work in [C & M] to other categories and/or to applications. I am very confident that such an individual, should she be willing to accept such an assignment, will experience a sense of satisfaction available only to a few, and I would truly appreciate it if such individuals would share their experiences with me within my lifetime.

Though this be madness, yet there is method in't.

Shakespeare (Hamlet)

CHAPTER 4

Some first families of nearrings and some of their ideals

In §2, we had the debut of many of the nearrings whose influence on the theory has been profound. Later, in Chapter 3, we explored an Eden for many of these influential families of nearrings. In this chapter, we wish to get better acquainted with some of these first families of nearrings. With a better understanding of these first families, one should be able to find the abstract structure theory more meaningful and alive. Also, these first families continue to provide geneses for important research.

Before we explore these first families further, we shall first find it useful to develop the concept of a nearring module. It has been rather traditional to take one particular viewpoint on this subject, even though it was known that an alternative existed. With promptings from J. D. P. Meldrum [M], Gary Grainger [GG] explored the alternative viewpoint and has found ample evidence that the alternative deserves serious attention.

13. First, what is a nearring module?

To answer the title question for this section, one should look again at ring modules. We start with *left modules*. Take an abelian group $(M, +)$ and a ring $(R, +, \cdot)$. We want a *scalar multiplication* $* : R \times M \to M$, with $*(r, m) = r * m = rm$, such that $(r_1 r_2)m = r_1(r_2 m)$ and

$$r(m_1 + m_2) = (rm_1) + (rm_2) \qquad (13:1)$$

and

$$(r_1 + r_2)m = (r_1 m) + (r_2 m) \qquad (13:2)$$

for all $r, r_1, r_2 \in R$ and for all $m, m_1, m_2 \in M$. If R has an identity 1_R, and if $1_R m = m$ for all $m \in M$, then we say that M is a *unitary* left R-module. From (13:1) and (13:2), one easily sees that $r \mapsto \phi_r$, where $\phi_r(m) = rm$, is a homomorphism from R into the ring $(\text{End}M, +, \circ)$, the ring of endomorphisms of the abelian group M. This homomorphism is a *representation* of R, and if this homomorphism is a monomorphism then M is said to be a *faithful* left R-module.

For our purposes we want to focus particularly on (1) the fact that elements of R act on the left of elements of M, (2) that the elements of the *ring* R are left distributive over the elements of the *group* $(M, +)$, as in (13:1), and (3) that the elements of the *group* $(M, +)$ are right distributive over the elements of the *ring* R, as in (13:2).

Next we do nearly the same thing all over again for a *right module*. With an abelian group $(M, +)$ and a ring $(R, +, \cdot)$, we want a *scalar multiplication* $\star : M \times R \to M$, with $(m, r)\star = m \star r = mr$, such that $m(r_1 r_2) = (mr_1)r_2$ and

$$(m_1 + m_2)r = (m_1 r) + (m_2 r) \qquad (13:3)$$

and

$$m(r_1 + r_2) = (mr_1) + (mr_2) \qquad (13:4)$$

for all $r, r_1, r_2 \in R$ and for all $m, m_1, m_2 \in M$. If R has an identity 1_R, and if $m1_R = m$ for all $m \in M$, then we say that M is a *unitary* right R-module. From (13:3) and (13:4), one readily sees that $r \mapsto \phi_r$, where $m\phi_r = mr$, is a homomorphism from R into the ring $(\mathrm{End}\, M, +, \circ)$, the ring of endomorphisms of the abelian group M. This homomorphism is a *representation* of R, and if this homomorphism is a monomorphism, then M is said to be a *faithful* right R-module.

For our *left* R-modules, we had $\phi_r(m)$, so the variable was put on the *right* of ϕ_r, but for our *right* R-modules, we had $m\phi_r$, so the variable was put on the *left* of ϕ_r. This is really the only difference.

If we take any abelian group $(A, +)$, we can think of A as a faithful unitary right $\mathrm{End}A$-module, where for $x \in A$ and $\alpha \in \mathrm{End}A$, we write $x\alpha$ for $x \star \alpha$, and at the same time, think of A as a unitary Z-module, where $(Z, +, \cdot)$ is the usual ring of integers, and where $n * x = nx$ has the usual meaning for each $n \in Z$. We also have $(n * x) \star \alpha = n * (x \star \alpha)$ or $(nx)\alpha = n(x\alpha)$ for each $n \in Z$, each $x \in A$, and each $\alpha \in \mathrm{End}A$. It is common to refer to A as a $(Z, \mathrm{End}A)$-*bimodule*. So if R and S are rings and M is an abelian group so that M is a left R-module and a right S-module, and $(rm)s = r(ms)$ for all $r \in R$, $m \in M$, and $s \in S$, then M is an (R, S)-*bimodule*.

Turning our attention now to nearrings, let us first examine the traditional point of view. When asked for an example of a nearring, most nearring researchers have a Pavlovian reaction and give $(M(G), +, \circ)$ or $(M_0(G), +, \circ)$ for an arbitrary group $(G, +)$. Of course, there can be no agreement as to where the variable should be placed with respect to the function. If one chooses $f(x)$ for $f \in M(G)$ and $x \in G$, then one obtains $(f_1 f_2)x = f_1(f_2 x)$, $1_G x = x$, and

$$(f_1 + f_2)x = (f_1 x) + (f_2 x) \qquad (13:5)$$

for all $f_1, f_2 \in M(G)$ and for all $x \in G$. Of course, $(M(G), +, \circ)$ is a *right* nearring acting on the *left* of G, and the group elements $x \in G$ are *right* distributive over the nearring elements $f_1, f_2 \in M(G)$.

If one chooses xf for $x \in G$ and $f \in M(G)$, then one obtains $x(f_1 f_2) = (xf_1)f_2$, $x1_G = x$, and

$$x(f_1 + f_2) = (xf_1) + (xf_2) \qquad (13:6)$$

for all $f_1, f_2 \in M(G)$ and for all $x \in G$. Now, $(M(G), +, \circ)$ is a *left* nearring acting on the *right* of G, and the group elements $x \in G$ are *left* distributive over the nearring elements $f_1, f_2 \in M(G)$.

Of course, these two choices, $f(x)$ or xf, are really not conceptual choices. They amount to nothing more than two different ways of 'displaying the merchandise'. The key focal point here is that the group elements $x \in G$ distribute over the nearring elements $f_1, f_2 \in M(G)$.

Next, we consider the nontraditional point of view. The pieces were there to consider way back in (1.1) and (1.2). Let $|X| \geq 2$, and consider the free group $(\mathcal{F}(X), +)$ and the left nearring $(\mathrm{Hom}(\mathcal{F}(X), \mathcal{F}(X)), \oplus, \circ)$. Here, we take $f \in \mathrm{Hom}(\mathcal{F}(X), \mathcal{F}(X)) = \mathrm{End}\mathcal{F}(X)$, and $u \in \mathcal{F}(X)$ and write $f(u) = fu$. Certainly, $(f_1 f_2)u = f_1(f_2 u)$, $1_{\mathcal{F}(X)}u = u$, and

$$f(u_1 + u_2) = (fu_1) + (fu_2) \qquad (13:7)$$

for all $f, f_1, f_2 \in \mathrm{End}\mathcal{F}(X)$ and for all $u, u_1, u_2 \in \mathcal{F}(X)$. Notice that this is different from the traditional point of view in that the nearring elements $f \in \mathrm{End}\mathcal{F}(X)$ distribute over the group elements $u_1, u_2 \in \mathcal{F}(X)$. In the same setting, we can take any group G and consider the group $(\mathrm{Hom}(\mathcal{F}(X), G), \oplus)$. If we continue with the variable on the right of the function, then for $t \in X$, $f_1, f_2 \in \mathrm{End}\mathcal{F}(X)$, and $\alpha \in \mathrm{Hom}(\mathcal{F}(X), G)$, we have $\alpha \circ (f_1 \circ f_2)(t) = (\alpha \circ f_1) \circ f_2(t)$, so $\alpha(f_1 f_2) = (\alpha f_1)f_2$, and $(\alpha \circ [f_1 \oplus f_2])(t) = \alpha(f_1(t) + f_2(t)) = \alpha(f_1(t)) + \alpha(f_2(t)) = [(\alpha \circ f_1) \oplus (\alpha \circ f_2)](t)$. This makes

$$\alpha(f_1 \oplus f_2) = (\alpha f_1) \oplus (\alpha f_2) \qquad (13:8)$$

for all $f_1, f_2 \in \mathrm{End}\mathcal{F}(X)$ and for all $\alpha \in \mathrm{Hom}(\mathcal{F}(X), G)$. So we have the traditional situation where the group elements $\alpha \in \mathrm{Hom}(\mathcal{F}(X), G)$ distribute over the nearring elements $f_1, f_2 \in \mathrm{End}\mathcal{F}(X)$.

As we saw many times in Chapter 3, the situation gets more frustrating when we look at $(\mathrm{Hom}(\mathcal{F}(X), \mathcal{F}(Y)), \oplus)$, that is, let $G = \mathcal{F}(Y)$ in the preceding paragraph. Here we are with two left nearrings $(\mathrm{End}\mathcal{F}(X), \oplus, \circ)$ and $(\mathrm{End}\mathcal{F}(Y), \oplus, \circ)$ and a group $(\mathrm{Hom}(\mathcal{F}(X), \mathcal{F}(Y)), \oplus)$. Keeping the variable on the right of the function, we have:

(L1) The nearring $\mathrm{End}\mathcal{F}(Y)$ acts on the *left* of the group $\mathrm{Hom}(\mathcal{F}(X), \mathcal{F}(Y))$.
(L2) If $g_1, g_2 \in \mathrm{End}\mathcal{F}(X)$ and $\alpha \in \mathrm{Hom}(\mathcal{F}(X), \mathcal{F}(Y))$, then $(g_1 g_2)\alpha = g_1(g_2 \alpha)$ and $1_{\mathcal{F}(Y)}\alpha = \alpha$.
(L3) If $g \in \mathrm{End}\mathcal{F}(Y)$ and $\alpha_1, \alpha_2 \in \mathrm{Hom}(\mathcal{F}(X), \mathcal{F}(Y))$, then

$$g(\alpha_1 \oplus \alpha_2) = (g\alpha_1) \oplus (g\alpha_2). \qquad (13:9)$$

(R1) The nearring of endomorphisms $\mathrm{End}\mathcal{F}(X)$ acts on the *right* of the group of homomorphisms $\mathrm{Hom}(\mathcal{F}(X), \mathcal{F}(Y))$.

(R2) If $f_1, f_2 \in \mathrm{End}\mathcal{F}(X)$ and $\alpha \in \mathrm{Hom}(\mathcal{F}(X), \mathcal{F}(Y))$, then $\alpha(f_1 f_2) = (\alpha f_1)f_2$ and $\alpha 1_{\mathcal{F}(X)} = \alpha$.

(R3) If $f_1, f_2 \in \mathrm{End}\mathcal{F}(X)$ and $\alpha \in \mathrm{Hom}(\mathcal{F}(X), \mathcal{F}(Y))$, then

$$\alpha(f_1 \oplus f_2) = (\alpha f_1) \oplus (\alpha f_2). \tag{13:10}$$

We also have for $g \in \mathrm{End}\mathcal{F}(Y)$, $\alpha \in \mathrm{Hom}(\mathcal{F}(X), \mathcal{F}(Y))$, and $f \in \mathrm{End}\mathcal{F}(X)$, that $g(\alpha f) = (g\alpha)f$. We have from (13:9) that the nearring elements of $\mathrm{End}\mathcal{F}(Y)$ distribute over the group elements of $\mathrm{Hom}(\mathcal{F}(X), \mathcal{F}(Y))$, and from (13:10) that the group elements of $\mathrm{Hom}(\mathcal{F}(X), \mathcal{F}(Y))$ distribute over the nearring elements of $\mathrm{End}\mathcal{F}(X)$. Also, the nearring $\mathrm{End}\mathcal{F}(Y)$ acts on the *left* of the group $\mathrm{Hom}(\mathcal{F}(X), \mathcal{F}(Y))$ and the nearring $\mathrm{End}\mathcal{F}(X)$ acts on the *right* of the same group $\mathrm{Hom}(\mathcal{F}(X), \mathcal{F}(Y))$.

Keeping with tradition, we have

(13.1) Definition. For a (left) nearring $(N, +, \cdot)$ and a group $(G, +)$, if there is a 'scalar multiplication' $* : G \times N \to G$, where $*(g, n) = g * n = gn$, such that $g(n_1 n_2) = (gn_1)n_2$ and

$$g(n_1 + n_2) = (gn_1) + (gn_2) \tag{13:11}$$

for all $g \in G$ and for all $n_1, n_2 \in N$, then G is a *nearring module*, or *N-module*. If $1 \in N$ and $g1 = g$ for all $g \in G$, then G is a *unitary N*-module. If there is no ambiguity, we will simply refer to G as a *module*. (Of course, there is the same concept if $(N, +, \cdot)$ is a right nearring.) In Pilz [P] and elsewhere, such G are also called *N-groups*.

The discussion earlier in this section, together with the multitude of examples given in Chapter 3, show that there could have been an alternative definition for a nearring module. What should one do when a nearring acts on a group and the nearring elements distribute over the group elements? A compromise with Grainger [GG] is

(13.2) Definition. For a (left) nearring $(N, +, \cdot)$ and a group $(G, +)$, if there is a 'scalar multiplication' $\star : N \times G \to G$, where $\star(n, g) = n \star g = ng$, such that $(n_1 n_2)g = n_1(n_2 g)$ and

$$n(g_1 + g_2) = (ng_1) + (ng_2) \tag{13:12}$$

for all $g, g_1, g_2 \in G$ and for all $n, n_1, n_2 \in N$, then G is a *nearring comodule*, or *N-comodule*. If $1 \in N$ and $1g = g$ for all $g \in G$, then G is a *unitary N*-comodule. If there is no ambiguity, such a structure G would be simply a *comodule*. (And, of course, there is the same concept if $(N, +, \cdot)$ is a right nearring.)

Up to now, very little has been done with comodules. Those to whom the term 'comodule' is offensive should do something significant with the concept and rename the structures appropriately.

We will finish this section by showing that there are other comodules. Up to now, we have seen the concept of a comodule really only in the setting of group and cogroup objects. Grainger has shown that they arise elsewhere, for example, in settings that seem to be different from these around group and cogroup objects. The remainder of this section is due to Grainger [GG] and Fröhlich [F, 1958], but it is all strongly influenced by Fröhlich [F, 1958].

With an eye on the free groups $\mathcal{F}(X)$ and what we have said about them, Fröhlich extended what was going on there and Grainger has refined it [GG].

(13.3) Definition. For groups G and H, suppose $\varnothing \neq \mathcal{H} \subseteq \mathrm{Hom}(G, H)$. A *basis* for $\mathcal{H}$ is a subset $B \subseteq G$ satisfying: (i) B is a set of generators for the group G; (ii) for $\alpha \in \mathcal{H}$, there is a unique $\alpha' \in \mathcal{H}$, so that $\alpha'(x) = -(\alpha(x))$ for each $x \in B$; (iii) for $\alpha, \beta \in \mathcal{H}$, there is a unique $\gamma \in \mathcal{H}$ so that $\gamma(x) = \alpha(x) + \beta(x)$ for each $x \in B$.

Certainly, X is a basis for $\mathcal{H} = \mathrm{Hom}(\mathcal{F}(X), H)$ for any group H. As with these $\mathcal{F}(X)$, H, $\mathcal{H} = \mathrm{Hom}(\mathcal{F}(X), H)$, and $B = X$, we define an addition on $\mathcal{H}$.

(13.4) Definition. Suppose B is a basis for $\varnothing \neq \mathcal{H} \subseteq \mathrm{Hom}(G, H)$, where G and H are groups. For $\alpha, \beta \in \mathcal{H}$, there is a unique $\gamma \in \mathcal{H}$ such that $\gamma(x) = \alpha(x) + \beta(x)$ for each $x \in B$. Let us denote γ by $\alpha \oplus \beta$, that is, $\alpha \oplus \beta = \gamma$ defines a binary operation on $\mathcal{H}$. We shall refer to $\oplus$ as the *free addition*.

(13.5) Theorem. *Suppose G and H are groups.*
(i) *If $\varnothing \neq \mathcal{H} \subseteq \mathrm{Hom}(G, H)$ has a basis B and $\oplus$ is the free addition defined from B, then $(\mathcal{H}, \oplus)$ is a group.*
(ii) *If $\mathcal{G}$ is a subsemigroup of $(\mathrm{End}G, \circ)$ with a basis B and $\oplus$ is the free addition defined from B, then $(\mathcal{G}, \oplus, \circ)$ is a left nearring and G is a $\mathcal{G}$-comodule.*

Proof. First, we will prove (i). Take $f \in \mathcal{H}$, and for $f' \in \mathcal{H}$ guaranteed by (ii) of (13.3), we will let $\zeta = f \oplus f'$. Then $(f \oplus f')(x) = \zeta(x) = 0$ for each $x \in B$. Hence, $g \oplus \zeta = \zeta \oplus g = g$ for each $g \in \mathcal{H}$, and $g' \oplus g = g \oplus g' = \zeta$ for each $g \in \mathcal{H}$. So $(\mathcal{H}, \oplus)$ has an identity ζ and each $g \in \mathcal{H}$ has an inverse $g' \in \mathcal{H}$ with respect to $\oplus$ and ζ. For $f, g, h \in \mathcal{H}$ and $x \in B$

we have $[(f \oplus g) \oplus h](x) = (f \oplus g)(x) + h(x) = [f(x) + g(x)] + h(x) = f(x) + [g(x) + h(x)] = f(x) + [g \oplus h](x) = [f \oplus (g \oplus h)](x)$. Hence, $(\mathcal{H}, \oplus)$ is indeed a group.

Now let us focus on (ii). We have that $(\mathcal{G}, \oplus)$ is a group and $(\mathcal{G}, \circ)$ is a semigroup. If $f, g, h \in \mathcal{G}$ and $x \in B$, then $[f \circ (g \oplus h)](x) = f((g \oplus h)(x)) = f(g(x) + h(x)) = f(g(x)) + f(h(x)) = f \circ g(x) + f \circ h(x) = [(f \circ g) \oplus (f \circ h)](x)$. So $f \circ (g \oplus h) = (f \circ g) \oplus (f \circ h)$ and $(\mathcal{G}, \oplus, \circ)$ is a left nearring.

For $f, g \in \mathcal{G}$ and $x, y \in G$, we have $(fg)x = f(gx)$ and $f(x+y) = fx + fy$. So G is a $\mathcal{G}$-comodule. This completes the proof of (13.5).

We have seen in Chapter 3 that $\oplus$ may coincide with pointwise addition and that it may be different. Another example where $\oplus$ will coincide with pointwise addition is the following.

(13.6) Example. Let $(V, +)$ be Klein's four group where $V = \{0, a, b, c\}$. $\mathrm{End}V$ has 16 elements and there are four of them in $\mathcal{G} = \{\zeta, \alpha, \beta, \gamma\}$, where $\zeta(x) = 0$ for all $x \in V$, where $\delta(0) = \delta(b) = 0$ for each $\delta \in \mathcal{G}$, where $\alpha\{a, c\} = \{a\}$, $\beta\{a, c\} = \{b\}$, and where $\gamma\{a, c\} = \{c\}$. Then $B = \{a, c\}$ is a basis for $\mathcal{G}$, and $(\mathcal{G}, \oplus) \cong (V, +)$. For $(\mathcal{G}, \circ)$ we have $\zeta \circ \delta = \delta \circ \zeta = \zeta$ for all $\delta \in \mathcal{G}$, $\alpha \circ \alpha = \alpha \circ \gamma = \alpha$, $\beta \circ \alpha = \beta \circ \gamma = \beta$, $\gamma \circ \alpha = \gamma \circ \gamma = \gamma$, and $\delta \circ \beta = \zeta$ for all $\delta \in \mathcal{G}$. So $(\mathcal{G}, \oplus, \circ)$ is a right nearring and V is a right $\mathcal{G}$-comodule. A weakness in this example is that $(\mathcal{G}, \oplus, \circ)$ is a ring, a subring of $(\mathrm{End}V, +, \circ)$, and V is a ring $\mathcal{G}$-module. But it is an elementary example to illustrate the ideas of this section.

Take any group $(G, +)$ and consider it an $M(G)$-module, or even an $M_0(G)$-module. If $x \in G \setminus \{0\}$, then $\{\alpha x \mid \alpha \in M(G)\} = G$ and $\{\alpha x \mid \alpha \in M_0(G)\} = G$. As an $M(G)$-module, or as an $M_0(G)$-module, we have that G has a single generator, such as a cyclic group, that is, as a Z-module.

(13.7) Definitions. For a (left) nearring $(N, +, \cdot)$ and a corresponding N-module G, suppose there is an $x \in G$ such that $\{xn \mid n \in N\} = G$. Then G is a *monogenic* N-module and x is a *generator*.

(13.8) Definitions. For a (left) nearring $(N, +, \cdot)$ and N-modules G and H, a group homomorphism $\alpha : G \rightarrow H$ is an *N-homomorphism* if $\alpha(gn) = \alpha(g)n$ for all $g \in G$ and for all $n \in N$. We shall feel free to use such terms as *N-endomorphism*, *N-isomorphism*, *N-epimorphism*, and *N-automorphism* since we are confident that the reader will know precisely what is meant.

(13.9) Definitions. For a nearring $(N, +, \cdot)$ and an N-module $(G, +)$, let $\varnothing \neq K \subseteq G$. The *(right) annihilator of K in N* is defined and denoted by $\mathrm{Ann}(K, N) = \mathrm{Ann}(K) = \{n \in N \mid kn = 0 \text{ for all } k \in K\}$. If $K = \{k\}$, we write $\mathrm{Ann}(k, N)$ or simply $\mathrm{Ann}(k)$.

For $\varnothing \neq K \subseteq G$ and $\varnothing \neq M \subseteq N$, we let $KM = \{k_1 m_1 + \cdots + k_s m_s \mid$

$k_i \in K$, $m_i \in M$, $1 \le i \le s$, $s \in Z$}. For $K = \{k\}$ and/or $M = \{m\}$, we write kM for $\{k\}M$ and Km for $K\{m\}$.

Sometimes an N-module is also a comodule for an appropriate nearring.

(13.10) Theorem. *For a nearring $(N, +, \cdot)$ with identity 1, suppose G is a monogenic unitary N-module with generator x, and suppose that $T = \{m \in N \mid m\mathrm{Ann}(x) \subseteq \mathrm{Ann}(x)\}$ is a subgroup of $(N, +)$. Then the N-endomorphisms $\mathrm{End}_N G$ of the N-module G form a left nearring $(\mathrm{End}_N G, \oplus, \circ)$ where $(f \oplus g)(x) = f(x) + g(x)$ and $(f \circ g)(x) = f(g(x))$. Also, G is an $\mathrm{End}_N G$-comodule.*

Proof. Since $\zeta \in \mathrm{End}_N G$, where $\zeta(y) = 0$ for all $y \in G$, we are assured that $\mathrm{End}_N G \ne \varnothing$. Take $f \in \mathrm{End}_N G$. Since G is monogenic, there is an $m_f \in N$ such that $f(x) = xm_f$. If $n \in \mathrm{Ann}(x)$, then $0 = f(0) = f(xn) = f(x)n = x(m_f n)$. Hence, $m_f n \in \mathrm{Ann}(x)$, and so $m_f \in T$. Conversely, choose $m \in T$ and define $f(x) = xm$. We want to see that this defines an $f \in \mathrm{End}_N G$. If $y = xn$ and $y \ne x$, then define $f(y) = x(mn)$. We had better make sure that f is well defined. If $xn = xn'$, can we be assured that $f(xn) = x(mn) = x(mn') = f(xn')$? Now $0 = xn - xn'$. As we did for (3.1), we can easily see that $y0 = 0$ for all $y \in G$, and that $-(yn) = y(-n)$ for all $y \in G$ and for all $n \in N$. Hence, $0 = xn - xn' = xn + x(-n') = x(n - n')$, and so $n - n' \in \mathrm{Ann}(x)$. Since $m \in T$, we have $m(n - n') \in \mathrm{Ann}(x)$, and so $0 = x[mn - mn'] = (xm)n - (xm)n' = f(xn) - f(xn')$. Hence, $f(xn) = f(xn')$, and f is indeed well defined. It is now easy to see that $f \in \mathrm{End}_N G$. So the elements of $f \in \mathrm{End}_N G$ correspond to $m \in T$, but perhaps $xm = xm'$ with $m \ne m'$, in which case $m - m' \in \mathrm{Ann}(x)$.

Let $f, g \in \mathrm{End}_N G$ with $f(x) = xm_f$ and $g(x) = xm_g$, where $m_f, m_g \in T$. Define $f \oplus g = h$, where $h(x) = x(m_f + m_g)$. Then $h(x) = f(x) + g(x)$. Since T is a subgroup of $(N, +)$ we have that $h \in \mathrm{End}_N G$. Also, define $f'(x) = -(xm_f) = x(-m_f)$. Since T is a subgroup of $(N, +)$, we have that $f' \in \mathrm{End}_N G$.

Certainly, $f \oplus \zeta = \zeta \oplus f = f$ and $f \oplus f' = f' \oplus f = \zeta$. For $h \in \mathrm{End}_N G$ also, with $h(x) = xm_h$, then $([f \oplus g] \oplus h)(x) = (f \oplus g)(x) + h(x) = x(m_f + m_g) + xm_h = x[(m_f + m_g) + m_h] = x[m_f + (m_g + m_h)] = xm_f + x(m_g + m_h) = f(x) + (g \oplus h)(x) = [f \oplus (g \oplus h)](x)$, and so $(\mathrm{End}_N G, \oplus)$ is a group. Since $\circ$ is always associative, it remains only to show the left distributive law.

Now $f \circ [g \oplus h](x) = f([g \oplus h](x)) = f(x(m_g + m_h)) = x[m_f(m_g + m_h)] = x[m_f m_g + m_f m_h] = (xm_f)m_g + (xm_f)m_h = f(x)m_g + f(x)m_h = f(g(x)) + f(h(x)) = (f \circ g)(x) + (f \circ h)(x) = [(f \circ g) \oplus (f \circ h)](x)$, so $f \circ [g \oplus h] = (f \circ g) \oplus (f \circ h)$, and $(\mathrm{End}_N G, \oplus, \circ)$ is a left nearring. What would it take for the right distributive law to hold? Consider $[(f \oplus g) \circ h](x) = (f \oplus g)(xm_h) = (f \oplus g)(x)m_h = [x(m_f + m_g)]m_h = x[(m_f + m_g)m_h]$ and $[(f \circ h) \oplus (g \circ h)](x) = (f \circ h)(x) + (g \circ h)(x) = f(h(x)) + g(h(x)) =$

$x(m_f m_h) + x(m_g m_h) = x[m_f m_h + m_g m_h]$. So, for the right distributive law to hold, we would need $(m_f + m_g)m_h - m_g m_h - m_f m_h \in \text{Ann}(x)$, and we cannot be assured of that.

To see that G is a comodule, just remember that each $f \in \text{End}_N G$ is a homomorphism, so $f(u + v) = fu + fv$ for all $f \in \text{End}_N G$ and for all $u, v \in G$. Also, if $f, g \in \text{End}_N G$ and $u \in G$, then $(f \circ g)u = f(gu)$.

(13.11) Example. For any group $(G, +)$, then $x = 0$ is a generator for the monogenic $M(G)$-module G, and $\text{Ann}(0) = M_0(G)$. Also, $T = \{m \in M(G) \mid m\text{Ann}(0) \subseteq \text{Ann}(0)\} = M_0(G)$, a subgroup of $M(G)$. Hence, by (13.9), $(\text{End}_{M(G)} G, \oplus, \circ)$ is a nearring. This example's merits are in showing that a nontrivial T exists, but since $\text{End}_{M(G)} G = \{\zeta\}$, the resulting nearring is of no interest.

Now $0 \in G$ is a generator of G as an $M(G)$-module, but 0 will most likely not be a generator of an interesting group G. So we could use a variation of the idea of a basis.

(13.12) Definition. Let X be a set, G be a group, and suppose that $\varnothing \neq \mathcal{M} \subseteq \text{Map}(X, G) = \{f \mid f : X \to G\}$. A *support system* for $\mathcal{M}$ is a subset $S \subseteq X$ that satisfies the following:
(i) If $f, g \in \mathcal{M}$ and $f(s) = g(s)$ for all $s \in S$, then $f = g$.
(ii) If $f \in \mathcal{M}$, there is an $f' \in \mathcal{M}$ such that $f'(s) = -(f(s))$ for all $s \in S$.
(iii) If $f, g \in \mathcal{M}$, there is an $h \in \mathcal{M}$ such that $h(s) = f(s) + g(s)$ for all $s \in S$.

With a support system we also have a free addition.

(13.13) Definition. Start with a set X, a group G, and a nonempty subset $\mathcal{M} \subseteq \text{Map}(X, G)$ with support system S. For $f, g \in \mathcal{M}$, there is a unique $h \in \mathcal{M}$ such that $f(s) + g(s) = h(s)$ for all $s \in S$. (Uniqueness is a result of (i) of (13.12).) Denote h by $f \oplus_S g$, that is, $f \oplus_S g = h$ defines a binary operation on $\mathcal{M}$. We shall call $\oplus_S$ the *free addition relative to S*.

(13.14) Theorem. *Let X be a set and G a group.*
(i) *If $\varnothing \neq \mathcal{M} \subseteq \text{Map}(X, G)$ has a support system S and $\oplus_S$ is the free addition relative to S, then $(\mathcal{M}, \oplus_S)$ is a group.*
(ii) *If $\mathcal{E}$ is a subsemigroup of $(\text{End}\,G, \circ)$ with a support system S, and $\oplus_S$ is the free addition relative to S, then $(\mathcal{E}, \oplus_S, \circ)$ is a left nearring and G is an $\mathcal{E}$-comodule.*

Proof. First we will show that $(\mathcal{M}, \oplus_S)$ is a group. By (i) of (13.12), the f' of (ii) and the h of (iii) are unique. So $\oplus = \oplus_S$ really is a well defined binary operation. Let $\zeta = f \oplus f'$ for some $f \in \mathcal{M}$. Then $\zeta(s) = f(s) + f'(s) = f(s) + [-f(s)] = 0$ for each $s \in S$, so there is a $\zeta \in \mathcal{M}$ such that $\zeta(s) = 0$ for each $s \in S$. Then note that $\zeta = g' \oplus g = g \oplus g'$ for each

$g \in \mathcal{M}$, and that $g \oplus \zeta = \zeta \oplus g = g$ for each $g \in \mathcal{M}$. Hence, ζ is the identity for $\oplus$ and g' is the inverse for g with respect to $\oplus$ and ζ. For $f, g, h \in \mathcal{M}$ and $s \in S$, $[(f \oplus g) \oplus h](s) = (f \oplus g)(s) + h(s) = [f(s) + g(s)] + h(s) = f(s) + [g(s) + h(s)] = f(s) + (g \oplus h)(s) = [f \oplus (g \oplus h)](s)$. Hence, $\oplus_S = \oplus$ is associative and $(\mathcal{M}, \oplus_S)$ is a group.

Turning now to (ii), we only have to show that the left distributive law holds to obtain a nearring. Again, take $f, g, h \in \mathcal{E}$ and $s \in S$. Then $[f \circ (g \oplus h)](s) = f((g \oplus h)(s)) = f(g(s) + h(s)) = f(g(s)) + f(h(s)) = f \circ g(s) + f \circ h(s) = [(f \circ g) \oplus (f \circ h)](s)$, and we have that $(\mathcal{E}, \oplus_S, \circ)$ is a left nearring. Since $(fg)x = f(gx)$ and $f(x + y) = fx + fy$, we have that G is an $\mathcal{E}$-comodule. This completes the proof of (13.14).

(13.15) Examples. Grainger has two elementary but curious examples of $(\text{End}, \oplus_S, \circ)$. Both come from the same subsemigroup $(\mathcal{E}, \circ)$ of $(\text{End}V, \circ)$, where $V = \{0, a, b, c\}$ is Klein's four group, but where $\mathcal{E} = \{\alpha, \beta, \gamma, \iota\}$ and the elements of $\mathcal{E}$ are defined in Table 1.

	0	a	b	c
α	0	0	c	c
β	0	c	0	c
γ	0	b	a	c
ι	0	a	b	c

Table 1.

The semigroup $(\mathcal{E}, \circ)$ is displayed in Table 2.

$\circ$	α	β	γ	ι
α	α	β	β	α
β	α	β	α	β
γ	α	β	ι	γ
ι	α	β	γ	ι

Table 2.

The key to these two examples will be that the elements of $\mathcal{E}$ are determined by what they do to the element a, and also that the elements of $\mathcal{E}$ are determined by what they do to the element b.

First we let $S = \{a\}$ and realize that $\{a\}$ is a support system for $\mathcal{E}$. So, by (13.14), $(\mathcal{E}, \oplus_{\{a\}}, \circ)$ is a left nearring. In Table 3, we display $(\mathcal{E}, \oplus_{\{a\}})$.

$\oplus_{\{a\}}$	α	β	γ	ι
α	α	β	γ	ι
β	β	α	ι	γ
γ	γ	ι	α	β
ι	ι	γ	β	α

Table 3.

Notice that $(\mathcal{E}, \oplus_{\{a\}}) \cong (V, +)$.

Next, let $S = \{b\}$ and realize that $\{b\}$ is also a support system for $\mathcal{E}$. Again, by (13.14), $(\mathcal{E}, \oplus_{\{b\}}, \circ)$ is a left nearring. Table 4 displays $(\mathcal{E}, \oplus_{\{b\}})$.

$\oplus_{\{b\}}$	β	α	γ	ι
β	β	α	γ	ι
α	α	β	ι	γ
γ	γ	ι	β	α
ι	ι	γ	α	β

Table 4.

Also notice that $(\mathcal{E}, \oplus_{\{b\}}) \cong (V, +)$. In fact, $(\mathcal{E}, \oplus_{\{a\}}, \circ) \cong (\mathcal{E}, \oplus_{\{b\}}, \circ)$ where the isomorphism is $\alpha \mapsto \beta$, $\beta \mapsto \alpha$, $\gamma \mapsto \gamma$, and $\iota \mapsto \iota$. Since neither the identity for $\oplus_{\{a\}}$ nor the identity for $\oplus_{\{b\}}$ is the zero endomorphism ζ, we know that neither binary operation is a pointwise addition of functions. Also, since the additive identity α for $\oplus_{\{a\}}$ has $\alpha \circ \beta = \beta \neq \alpha$, we know that $(\mathcal{E}, \oplus_{\{a\}}, \circ)$ is not a ring.

(13.16) Problems. The two examples of (13.15) give rise to several questions. Let $(\mathcal{E}, \circ)$ be a subsemigroup of $(\mathrm{End}\, G, \circ)$ for some group $(G, +)$.
(1) How many support systems are there for $(\mathcal{E}, \circ)$?
(2) If S and T are support systems for $(\mathcal{E}, \circ)$, when are $(\mathcal{E}, \oplus_S, \circ)$ and $(\mathcal{E}, \oplus_T, \circ)$ isomorphic?
(3) What are the necessary and sufficient conditions on a support system S so that $(\mathcal{E}, \oplus_S) \cong (G, +)$?

We have neglected at least one idea illustrated by the examples just before (13.1).

(13.17) Definition. For a group $(G, +)$, suppose $(N, +, \cdot)$ and $(M, +, \cdot)$ are nearrings such that G is an M-module and G is also an N-comodule. Also suppose $n(gm) = (ng)m$ for all $n \in N$, $g \in G$, and $m \in M$. Then G is an (M, N)-*bimodule*.

(13.18) Exploratory problem. Is there anything interesting and/or significant that can be done with (M, N)-bimodules?

(13.19) Exploratory problem. Is there anything interesting and/or significant that can be done with the T and the $(\mathrm{End}_N G, \oplus, \circ)$ of (13.10)?

(13.20) Exploratory problem. I think we have already done something interesting with comodules. But is there anything significant that can be done with comodules?

It will be useful to have the following lemma.

(13.21) Lemma. *Let $(N, +, \cdot)$ be a nearring with N-module $(G, +)$ and N-comodule $(H, +)$. Let 0_N, 0_G, and 0_H be their additive identities, respectively. Then for $n \in N$, $g \in G$, and $h \in H$, we have*
(i) $g \cdot 0_N = 0_G$ *and* $g(-n) = -(gn)$.
(ii) $n \cdot 0_H = 0_H$ *and* $n(-h) = -(nh)$.
If N is zero-symmetric, we also have
(iii) $0_G \cdot n = 0_G$.

Proof. With the exception of (iii), these proofs are as in ring and module theory. For (iii), $0_G n = (0_G 0_N)n = 0_G(0_N n) = 0_G 0_N = 0_G$.

Actually, one might be able to find some N-modules, N-comodules, and (N, N)-bimodules within a nearring $(N, +, \cdot)$. Suppose $(A, +)$ is a subgroup of $(N, +)$ with the property that $AN \subseteq A$. Then $a(n_1 n_2) = (an_1)n_2$ and $a(n_1 + n_2) = (an_1) + (an_2)$ make $(A, +)$ an N-module which is unitary if $1 \in N$. Likewise, if $(A, +)$ is a subgroup of $(N, +)$ with $NA \subseteq A$, then $(n_1 n_2)a = n_1(n_2 a)$ and $n(a_1 + a_2) = (na_1) + (na_2)$ show that $(A, +)$ is an N-comodule which is unitary if $1 \in N$. For a subgroup $(A, +)$ of $(N, +)$ having the property that $AN \subseteq A$ and $NA \subseteq A$, we have that $(A, +)$ is an (N, N)-bimodule.

These conditions, of course, describe a right ideal, a left ideal, and an ideal, respectively, for ring theory. But we need more for these concepts for nearrings. Perhaps it is better to introduce new terminology when the N-module, N-comodule, or (N, N)-bimodule $(A, +)$ is a subgroup of $(N, +)$, but to minimize the terminology, we shall use such phrases as '$(A, +)$ is an N-module *in* the nearring $(N, +, \cdot)$'.

At this point, the reader is invited to find examples of nearrings $(N, +, \cdot)$ which nicely illustrate these concepts.

We shall have occasion to use

(13.22) Theorem. *Let $(N, +, \cdot)$ be a nearring with N-module $(G, +)$. Then $\mathrm{Ann}(G, N)$ is an ideal of N.*

Proof. Take $m, n \in N$, $a, b \in \mathrm{Ann}(G, N)$, and $g \in G$. Then $g(a + b) = ga + gb = 0$, and $g(-a) = -(ga) = 0$, and $g(m + a - m) = gm + ga - gm = 0$, so

$\text{Ann}(G, N)$ is a normal subgroup of $(N, +)$. From $g(ma) = (gm)a = 0$, we obtain $N\text{Ann}(G, N) \subseteq \text{Ann}(G, N)$. From $g[(m+a)n - mn] = [g(m+a)]n - (gm)n = [gm + ga]n - (gm)n = 0$, we can now conclude that $\text{Ann}(G, N)$ is an ideal of N.

When one has an N-module $(G, +)$ for a nearring $(N, +, \cdot)$, one might be able to make another N-module if G has a suitable normal subgroup H. We want, of course, G/H to be an N-module. For $g + H \in G/H$ and $n \in N$, we will want $(g + H)n = gn + H$, from which $(g + H)(nn') = [(g + H)n]n'$ and $(g + H)(n + n') = (g + H)n + (g + H)n'$ are immediate consequences, if we can be assured that $(g + H)n = gn + H$ is well defined.

Let us suppose $(g + H)n = gn + H$ is well defined. So if $g + H = g' + H$ and $n \in N$, we have that $gn + H = g'n + H$, and also that $H + gn = H + g'n$ since $(H, +)$ is normal in $(G, +)$. The latter gives us $g'n - gn \in H$. Now there is an $h \in H$ such that $g + h = g'$, so $(g + h)n = g'n$. We then have $(g + h)n - gn = g'n - gn \in H$.

With this observation, take any $g \in G$, any $h \in H$, and any $n \in N$. Then $g + h = g' \in g + H$, and $g + H = g' + H$. Since the action of N on G/H is well defined, we have $gn + H = g'n + H$, and the preceding paragraph allows us to conclude that $(g + h)n - gn \in H$.

Conversely, suppose $(g+h)n - gn \in H$ for every $g \in G$, every $h \in H$, and every $n \in N$. Then if $g + H = g' + H$, we have $g + h = g'$ for some $h \in H$, and so, if $n \in N$, then $(g + h)n = g'n$. Since $g'n - gn = (g + h)n - gn \in H$, we have that $H + gn = H + g'n$, and as long as $(H, +)$ is normal in $(G, +)$, we obtain $gn + H = g'n + H$. So we can be assured that $(g + H)n = gn + H$ is a well defined action of N on G/H. This motivates a definition and gives us a theorem as well.

(13.23) Definition. Let $(G, +)$ be an N-module. Suppose a normal subgroup H of G has the property that $(g + h)n - gn \in H$ for each $g \in G$, for each $h \in H$, and for each $n \in N$. Then H is an *N-ideal* of the N-module $(G, +)$.

(13.24) Theorem. *For a nearring $(N, +, \cdot)$, let H be an N-ideal of an N-module $(G, +)$. Then $(g+H)n = gn+H$ is a well defined action of N on G/H, and $(G/H, +)$ is also an N-module. Conversely, if $(G, +)$ is an N-module, and $(H, +)$ is a normal subgroup of $(G, +)$, and $(g+H)n = gn+H$ is well defined, then H is an N-ideal of G.*

(13.25) Definition. Let $(G, +)$ be an N-module with $A \subseteq G$ and $B \subseteq G$. Define

$$(A : B) = \{n \in N \mid bn \in A \text{ for each } b \in B\}.$$

(13.26) Theorem. *For an N-ideal H of an N-module G, $(H : G)$ is an ideal of N.*

Proof. We have $(H : G) = \text{Ann}(G/H, N)$, so apply (13.22).

14. Centralizer and transformation nearrings

Linear transformations of a vector space $(V, +)$ over a field F into itself have been a very rich source of mathematical ideas and applications. These are functions $f : V \to V$ satisfying

$$(x + y)f = xf + yf \qquad (14 : 1)$$

for all $x, y \in V$, and

$$(x\lambda)f = (xf)\lambda \qquad (14 : 2)$$

for all $x \in V$ and all $\lambda \in F$. It is natural to wonder if functions satisfying only (14:2) could also be a source of nice mathematics.

Let us look at the functions $f : V \to V$ satisfying (14:2) in a broader sense. If $\lambda \in F \setminus \{0\}$, then λ defines an automorphism $\lambda^* : V \to V$ of the group $(V, +)$ by $x\lambda^* = x\lambda$. Also, $0 \in F$ defines $0^* : V \to V$ where $x0^* = x0 = 0$. We can rewrite (14:2) by saying that

$$\lambda^* \circ f = f \circ \lambda^* \qquad (14 : 3)$$

for each $\lambda \in F$. Let $\Phi = \{\lambda^* \mid \lambda \in F\}$. Then $\Phi^* = \Phi \setminus \{0^*\}$ is a group of automorphisms of the group $(V, +)$.

It is now an easy step to start with a group $(G, +)$ and a group of automorphisms Φ^* of $(G, +)$. Let $\zeta : G \to G$ be the trivial endomorphism where $x\zeta = 0$ for each $x \in G$, and then let $\Phi = \Phi^* \cup \{\zeta\}$, a semigroup of endomorphisms of $(G, +)$. Consider $f : G \to G$ where

$$\phi \circ f = f \circ \phi \qquad (14 : 4)$$

for all $\phi \in \Phi$. From $\zeta \circ f = f \circ \zeta$, we have $0(f \circ \zeta) = 0(\zeta \circ f)$, or $0f = 0$, so $f \in M_0(G)$. Define $M_\Phi(G) = \{f \in M_0(G) \mid \phi \circ f = f \circ \phi \text{ for all } \phi \in \Phi\}$. These form subnearrings of $(M_0(G), +, \circ)$ and have been studied intensely, especially when Φ^* is a regular group of automorphisms of $(G, +)$. But there is no reason to be so restrictive. Let Φ^* be a semigroup of endomorphisms of $(G, +)$, let $\Phi = \Phi^* \cup \{\zeta\}$ and again let $M_\Phi(G) = \{f \in M_0(G) \mid \phi \circ f = f \circ \phi \text{ for all } \phi \in \Phi\}$.

Even though it has been customary to study such $M_\Phi(G)$ as defined in the preceding paragraph, it has been recognized that one need not include $\zeta \in \Phi$. So if Φ is a semigroup of endomorphisms of a group $(G, +)$, define $M_\Phi(G) = \{f \in M(G) \mid \phi \circ f = f \circ \phi \text{ for all } \phi \in \Phi\}$.

(14.1) Theorem. *If $(\Phi, \circ)$ is a semigroup of endomorphisms of a group $(G, +)$, then $(M_\Phi(G), +, \circ)$ is a subnearring of $(M(G), +, \circ)$. If the trivial* *endomorphism* $\zeta \in \Phi$, *then the nearring* $(M_\Phi(G), +, \circ)$ *is a subnearring of* $(M_0(G), +, \circ)$.

Proof. Take $f, g \in M_\Phi(G)$, and $\phi \in \Phi$. Then $(f \circ g) \circ \phi = f \circ (g \circ \phi) = f \circ (\phi \circ g) = (f \circ \phi) \circ g = (\phi \circ f) \circ g = \phi \circ (f \circ g)$. Hence, $f \circ g \in M_\Phi(G)$. Also, since ϕ is an endomorphism of $(G, +)$, $(f - g) \circ \phi = [f + (-g)] \circ \phi = (f \circ \phi) + ((-g) \circ \phi) = (\phi \circ f) + (-(g \circ \phi)) = (\phi \circ f) + (-(\phi \circ g)) = (\phi \circ f) + (\phi \circ (-g)) = \phi \circ [f + (-g)] = \phi \circ (f - g)$. Hence, $f - g \in M_\Phi(G)$. So $M_\Phi(G)$ is a subnearring of $M(G)$. If $\zeta \in \Phi$, then, as noted above, $0f = 0$ if $f \in M_\Phi(G)$, hence $f \in M_0(G)$, and then $M_\Phi(G)$ is a subnearring of $M_0(G)$.

(14.2) Definition. For a semigroup Φ of endomorphisms of a group $(G, +)$, the nearring $(M_\Phi(G), +, \circ)$ is called a *centralizer* nearring.

Centralizer nearrings have assumed a very important role in nearring theory, especially in structure theory. A hint of their importance is provided by

(14.3) Theorem. *Let* $(N, +, \cdot)$ *be a nearring with identity* 1. *Then there is a semigroup* Φ *of endomorphisms of the group* $(N, +)$ *so that* $(N, +, \cdot)$ *is isomorphic to* $(M_\Phi(N), +, \circ)$.

Proof. Let $\Phi = \{n^* \mid n \in N\}$ where $n^* : N \to N$ is defined by $xn^* = nx$. Then n^* is an endomorphism of $(N, +)$. For $u \in N$, define $f_u : N \to N$ by $xf_u = xu$. Then $x(n^* \circ f_u) = (xn^*)f_u = (nx)f_u = (nx)u = n(xu) = (xu)n^* = (xf_u)n^* = x(f_u \circ n^*)$. Hence, $f_u \in M_\Phi(N)$. If $g \in M_\Phi(N)$, let $u = 1g$. Then $xg = (x1)g = (1x^*)g = 1(x^* \circ g) = 1(g \circ x^*) = (1g)x^* = ux^* = xu = xf_u$. So $g = f_u$, and $M_\Phi(N) = \{f_u \mid u \in N\}$.

This gives a surjection $\Gamma : N \to M_\Phi(N)$ defined by $\Gamma(u) = f_u$. Now $\Gamma(uv) = f_{uv}$ and $xf_{uv} = x(uv) = (xu)v = (xf_u)f_v = x(f_u \circ f_v)$. Consequently, $\Gamma(uv) = f_{uv} = f_u \circ f_v = \Gamma(u) \circ \Gamma(v)$. Also, $\Gamma(u + v) = f_{u+v}$ and $xf_{u+v} = x(u + v) = xu + xv = xf_u + xf_v = x(f_u + f_v)$. This makes $\Gamma(u + v) = f_{u+v} = f_u + f_v = \Gamma(u) + \Gamma(v)$. So Γ is a nearring epimorphism. If $u, v \in N$ and $u \neq v$, then $1u \neq 1v$, so $f_u \neq f_v$. Hence, Γ is a nearring isomorphism.

Some centralizer nearrings in $M_0(V)$

For a vector space $(V, +)$ over a field F, Let $(\mathrm{Hom}_F(V, V), +, \circ)$ be the ring of all linear transformations of V into itself, and let $(\mathrm{Hom}(V, V), +, \circ)$ be the ring of all group homomorphisms of V into itself. Then $\mathrm{Hom}_F(V, V)$ is a subring of $\mathrm{Hom}(V, V)$.

If the dimension of V over F is n, and n is finite, then, for a fixed basis, the elements $A \in \mathrm{Hom}_F(V, V)$ can be represented by $n \times n$ matrices and the elements of V can be represented by n-tuples in F^n For

$x = (x_1, x_2, \ldots, x_n) = (x_i) \in V$ and $A = (a_{ij}) \in \mathrm{Hom}_F(V, V)$, we have $xA = (\sum_{j=1}^{n} x_j a_{ji})$. One reason for calling elements in $\mathrm{Hom}_F(V, V)$ linear transformations is that the image of a line in V through 0 by $A \in \mathrm{Hom}_F(V, V)$ is either 0 or a line. Eigenvalues and eigenvectors arise when one considers the action of such an A upon all lines passing through 0. Which lines remain fixed by A?

Now straight lines are nice and they are relatively easy to work with. But the world is not completely linear and, in recent decades, it has become imperative to learn how to explain nonlinear phenomena better. A. P. J. van der Walt has developed a very nice tool to study some nonlinear phenomena with methods that are motivated by those developed for linear phenomena. His ideas lead naturally to subrings of $\mathrm{Hom}(V, V)$ other than $\mathrm{Hom}_F(V, V)$, and to centralizer nearrings in $M_0(V)$ with varying degrees of linearity. We shall present some of his ideas here.

It will be helpful for the reader to think continually of the field F as the real numbers $\mathbf{R}$ and $V = \mathbf{R} \oplus \mathbf{R}$, the euclidean plane, a vector space of dimension 2 over $F = \mathbf{R}$. In place of the straight lines passing through $(0, 0)$, it will be helpful to think frequently of all cubics of the form $y = mx^3$. One easy application of the theory developed here will be in this setting.

We shall find it suitable to enrich the structure of certain nearrings.

(14.4) Definition. Let $(N, +, \cdot)$ be a (left) nearring as is $(A, +, \cdot)$ also. Suppose $(A, +)$ is a nearring N-module where $(a \cdot b)n = a \cdot (bn)$ for all $a, b \in A$ and for all $n \in N$. We shall refer to $(A, +, \cdot)$ as an N-*nearalgebra*, or a *nearalgebra over* N.

(14.5) Examples. For a nearring $(N, +, \cdot)$, let $(G, +)$ be a nearring N-module. Now $(M(G), +, \circ)$ is a left nearring if we write xf for $x \in G$ and $f \in M(G)$. For $f \in M(G)$ and $n \in N$, define $fn \in M(G)$ by $x(fn) = (xf)n$. Then for $f \in M(G)$, $m, n \in N$, and $x \in G$, $x[f(m+n)] = (xf)(m+n) = (xf)m + (xf)n = x(fm) + x(fn) = x[(fm) + (fn)]$. This makes $(M(G), +)$ an N-module. We easily have $x[(f \circ g)n] = [x(f \circ g)]n = [(xf)g]n = (xf)(gn) = x[f \circ (gn)]$, so $(f \circ g)n = f \circ (gn)$ for all $f, g \in M(G)$ and for all $n \in N$. Hence, $(M(G), +, \circ)$ is an N-nearalgebra.

For our immediate purposes, we will take a field $(F, +, \cdot)$ for $(N, +, \cdot)$ and a vector space $(V, +)$ for $(G, +)$. Then $(M(V), +, \circ)$ is an F-nearalgebra.

(14.6) Examples. For a zero-symmetric nearring $(N, +, \cdot)$, let $(G, +)$ be an N-module. For the zero-symmetric nearring $(M_0(G), +, \circ)$, we have that $(M_0(G), +, \circ)$ is an N-nearalgebra, just as in (14.5). For our immediate purposes, we again take $(N, +, \cdot)$ to be a field $(F, +, \cdot)$ and $(G, +)$ to be a vector space $(V, +)$ over F. Then $(M_0(V), +, \circ)$ is an F-nearalgebra.

One key to understanding van der Walt's idea is the concept of a *transition matrix*. For this, we need our vector space $(V, +)$ to have dimension at least 2 and a basis indexed by some well ordered set $I = \{1 < 2 < \cdots\}$. So $|I| \geq 2$, of course. The transition matrix Φ must consist of automorphisms $\phi_{ij} \in \operatorname{Aut}F^*$, where $i, j \in I$.

(14.7) Definition. Let $\Phi = (\phi_{ij})$ be a matrix of group automorphisms of the multiplicative group F^* of a field $(F, +, \cdot)$. Then Φ is a *transition matrix for* F if and only if $\phi_{ii} = 1_{F^*} = 1$ for each $i \in I$, and $\phi_{ik}\phi_{kj} = \phi_{ij}$ for all $i, j, k \in I$.

It is easy to make transition matrices, and after we have seen what they are good for, the reader will find it entertaining to construct some and experiment with them. Take any map $\psi : I \to \operatorname{Aut}F^*$. Let

$$\phi_{ij} = \psi(i)\psi(j)^{-1}.$$

Then $\phi_{ii} = 1$, and $\phi_{ik}\phi_{kj} = \psi(i)\psi(k)^{-1}\psi(k)\psi(j)^{-1} = \psi(i)\psi(j)^{-1} = \phi_{ij}$. Hence, $\Phi = (\phi_{ij})$ is a transition matrix. From $\phi(i) = \phi_{i1} = \psi(i)\psi(1)^{-1}$, we see that we really only need a $\phi : I \setminus \{1\} \to \operatorname{Aut}F^*$.

(14.8) Theorem. *There is a bijection between the set of all transition matrices for F and all mappings $\phi : I \setminus \{1\} \to \operatorname{Aut}F^*$. Hence, there are exactly $|\operatorname{Aut}F^*|^{|I|-1}$ transition matrices for F.*

Proof. For a transition matrix Φ, we obtain $\phi : I \setminus \{1\} \to \operatorname{Aut}F^*$, defined by $\phi(i) = \phi_{i1}$. So Φ yield ϕ. Now start with $\phi : I \setminus \{1\} \to \operatorname{Aut}F^*$, set $\phi_{11} = 1$, and $\phi_{i1} = \phi(i)$. Next define $\phi_{ij} = \phi_{i1}\phi_{j1}^{-1}$. So $\phi_{ii} = 1$ for each $i \in I$, and $\phi_{ik}\phi_{kj} = \phi_{i1}\phi_{k1}^{-1}\phi_{k1}\phi_{j1}^{-1} = \phi_{i1}\phi_{j1}^{-1} = \phi_{ij}$. Now ϕ yields Φ, which in turn yields ϕ. Hence, the bijection. We should all now know that there are $|\operatorname{Aut}F^*|^{|I|-1}$ such transition matrices for F.

We have seen that we can construct all the elements of Φ from those in the first column. One can be more flexible.

(14.9) Corollary *A transition matrix Φ can be derived from any of its rows or columns.*

Proof. Consider the jth column $\Phi_{\bullet j} = (\phi_{ij})_{i \in I}$. From $\phi_{st} = \phi_{ts}^{-1}$, we know that $\phi_{i1} = \phi_{ij}\phi_{j1} = \phi_{ij}\phi_{1j}^{-1}$, and since ϕ_{ij} and ϕ_{1j} are in the jth column, we obtain the first column $\Phi_{\bullet 1} = (\phi_{i1})_{i \in I}$ from the jth column. Now proceed as in the proof of (14.8) to obtain Φ.

If we start with the ith row $\Phi_{i\bullet} = (\phi_{ij})_{j \in I}$, we obtain the ith column $\Phi_{\bullet i} = (\phi_{ji})_{j \in I} = (\phi_{ij}^{-1})_{j \in I}$. As in the preceding paragraph, we can now construct Φ.

Let $\mathcal{M}(V, F) = \{A = (a_{ij}) \mid A$ is a row-finite matrix over F and $i, j \in I\}$. This is a vector space over F; in fact, it is an F-algebra. Also,

$M_0(V)$ is a vector space over F and also a nearalgebra over F. We identify a transition matrix Φ with a mapping $\Phi : \mathcal{M}(V, F) \to M_0(V)$ by $\Phi(A) = [\Phi]A$, where $(x_i)[\Phi]A = (\sum_k (x_k \phi_{ki}) a_{ki})$. Here $(x_i) \in V$, so only finitely many of the values of the x_i are not 0 and we therefore have meaning for $\sum_k (x_k \phi_{ki}) a_{ki}$. (In truth, $0\phi_{ki}$ has no meaning, but we easily extend $\phi_{ki} \in \mathrm{Aut} F^*$ to $\phi_{ki} : F \to F$ by agreeing that $0\phi_{ki} = 0$.)

(14.10) Proposition. Φ *is a linear transformation with trivial kernel.*

Proof. Take $A, B \in \mathcal{M}(V, F)$, $(x_i) \in V$, and note that

$$(x_i)[\Phi](A + B) = \left(\sum_k (x_k \phi_{ki})(a_{ki} + b_{ki}) \right)$$

$$= \left(\sum_k (x_k \phi_{ki}) a_{ki} + \sum_k (x_k \phi_{ki}) b_{ki} \right)$$

$$= \left(\sum_k (x_k \phi_{ki}) a_{ki} \right) + \left(\sum_k (x_k \phi_{ki}) b_{ki} \right)$$

$$= (x_i)[\Phi]A + (x_i)[\Phi]B = (x_i)([\Phi]A + [\Phi]B).$$

Hence, $\Phi(A + B) = [\Phi](A + B) = [\Phi]A + [\Phi]B = \Phi(A) + \Phi(B)$.
  Also, for $b \in F$, $\Phi(Ab) = [\Phi](Ab)$, and

$$(x_i)[\Phi](Ab) = \left(\sum_k (x_k \phi_{ki})(a_{ki} b) \right) = \left(\sum_k (x_k \phi_{ki}) a_{ki} \right) b$$

$$= ((x_i)[\Phi]A)b = (x_i)([\Phi]A)b.$$

Therefore, $\Phi(Ab) = [\Phi](Ab) = ([\Phi]A)b = \Phi(A)b$. So Φ is a linear transformation.
  If $\Phi(F) = \zeta$, then $(0) = (x_i)[\Phi]A = (\sum_k (x_k \phi_{ki}) a_{ki})$ for each $(x_i) \in V$. Let $(x_i) = (\delta_{ij})$, where δ_{ij} is the Kronecker symbol. Then $(0) = (a_{ji})$, so each $a_{ji} = 0$, which makes $A = (0)$ also. Hence, $\ker \Phi = \{(0)\}$.

(14.11) Definition. Denote by $\mathcal{M}_\Phi(V, F)$ the vector subspace of $M_0(V)$ which is the image of Φ.

(14.12) Remark. Inside $M_0(V)$, which is itself a vector space over F, we have a family of vector subspaces $\mathcal{M}_\Phi(V, F)$, one for each transition matrix Φ, or, alternatively, one for each $\phi : I \setminus \{1\} \to \mathrm{Aut} F^*$.

(14.13) Definitions. For a transition matrix Φ for F, we define a family of groups, one for each row of Φ, and one for each column of Φ. For the kth row, we define

$$F^*(k, \Phi) = \{a(k, \Phi) \mid a \in F^*\},$$

where $(x_i)a(k, \Phi) = (x_i(a\phi_{ki}))$. For the kth column, we define

$$F^*(\Phi, k) = \{a(\Phi, k) \mid a \in F^*\},$$

where $(x_i)a(\Phi, k) = (x_i(a\phi_{ik}))$.

(14.14) Theorem. *Each $F^*(k, \Phi)$ and each $F^*(\Phi, k)$ is a group isomorphic to F^*, hence one extends each $F^*(k, \Phi)$ and each $F^*(\Phi, k)$ to a field isomorphic to F by simply adjoining 0.*

Proof. The map $t : F^* \to F^*(k, \Phi)$ is simply $t(a) = a(k, \Phi)$. Then $t(ab) = (ab)(k, \Phi)$, and

$$
\begin{aligned}
(x_i)(ab)(k, \Phi) &= (x_i((ab)\phi_{ki})) = (x_i(a\phi_{ki}b\phi_{ki})) \\
&= ((x_i(a\phi_{ki}))b\phi_{ki}) = (x_i(a\phi_{ki}))b(k, \Phi) \\
&= [(x_i)a(k, \Phi)]b(k, \Phi) = (x_i)[a(k, \Phi)b(k, \Phi)].
\end{aligned}
$$

Hence, $t(ab) = (ab)(k, \Phi) = a(k, \Phi)b(k, \Phi) = t(a)t(b)$.

If $t(a) = a(k, \Phi) = 1$, then $(x_i) = (x_i)a(k, \Phi) = (x_i(a\phi_{ik}))$ for all $(x_i) \in V$. Choose an $(x_i) \in V$ so that $x_j = 1$. Then $1 = 1(a\phi_{jk})$, so $a\phi_{jk} = 1$ and $a = 1$. Hence, t is an isomorphism. Replace $F^*(k, \Phi)$ by $F^*(\Phi, k)$ and $t(a) = a(\Phi, k)$ above, and we see that each $F^*(\Phi, k)$ is isomorphic to F^*.

By adjoining $0 = 0(k, \Phi)$ to $F^*(k, \Phi)$ to obtain $F(k, \Phi)$, and $0 = 0(\Phi, k)$ to $F^*(\Phi, k)$ to obtain $F(\Phi, k)$, and defining $a(k, \Phi) + b(k, \Phi) = (a+b)(k, \Phi)$ and $a(\Phi, k) + b(\Phi, k) = (a + b)(\Phi, k)$, we obtain that the $F(k, \Phi)$ and the $F(\Phi, k)$ are fields isomorphic to $(F, +, \cdot)$.

(14.15) Examples. Noticing that $(-1)\phi = -1$ for each $\phi \in \mathrm{Aut} F^*$, we see that a basis B for the vector space $(V, +)$ over F is also a basis for each $F(k, \Phi)$ and each $F(\Phi, k)$, in the sense of (13.3), if each ϕ_{ij} of Φ is an automorphism of the field F. This is because the elements of $F(k, \Phi)$ and $F(\Phi, k)$ are essentially endomorphisms of the group $(V, +)$.

(14.16) Remark. Even though the $F(k, \Phi)$ and the $F(\Phi, k)$ are isomorphic to F as fields, they define distinct actions on V. For $a \in F$, we have $(x_i)a = (x_i a)$, but for $a(k, \Phi) \in F(k, \Phi)$, we have $(x_i)a(k, \Phi) = (x_i a\phi_{ki}) \neq (x_i a)$ if some $\phi_{ki} \neq 1$.

(14.17) Theorem. *V is a vector space over each $F(k, \Phi)$ (and each $F(\Phi, k)$), if the ϕ_{ij} are automorphisms of the field F.*

Proof. Take $(x_i), (y_i) \in V$, and $a(k, \Phi), b(k, \Phi) \in F(k, \Phi)$. Then

$$
\begin{aligned}
[(x_i) + (y_i)]a(k, \Phi) &= (x_i + y_i)a(k, \Phi) = ((x_i + y_i)(a\phi_{ki})) \\
&= (x_i(a\phi_{ki}) + y_i(a\phi_{ki})) = (x_i(a\phi_{ki})) + (y_i(a\phi_{ki})) \\
&= (x_i)a(k, \Phi) + (y_i)a(k, \Phi).
\end{aligned}
$$

So each $a(k, \Phi)$ is an endomorphism of the group $(V, +)$. Also,

$$
\begin{aligned}
(x_i)[a(k, \Phi) + b(k, \Phi)] &= (x_i)(a + b)(k, \Phi) = (x_i((a + b)\phi_{ki})) \\
&= (x_i[a\phi_{ki} + b\phi_{ki}]) = (x_i(a\phi_{ki}) + x_i(b\phi_{ki})) \\
&= (x_i(a\phi_{ki})) + (x_i(b\phi_{ki})) \\
&= (x_i)a(k, \Phi) + (x_i)b(k, \Phi).
\end{aligned}
$$

So each element $(x_i) \in V$ distributes from the left over elements of the field $F(k, \Phi)$. Certainly, $(x_i)1(k, \Phi) = (x_i(1\phi_{ki})) = (x_i)$, and $(x_i)[a(k, \Phi) \circ b(k, \Phi)] = ((x_i)a(k, \Phi))b(k, \Phi)$. So V is a vector space over the field $F(k, \Phi)$. The proof for $F(\Phi, k)$ is virtually identical.

(14.18) Theorem. *The elements of $\mathcal{M}_\Phi(V, F)$ are homomorphisms of the group $(V, +)$ if and only if the ϕ_{ij} are automorphisms of the field F. In this case, $\mathcal{M}_\Phi(V, F)$ is a ring with identity.*

Proof. Suppose each ϕ_{ij} is an automorphism of the field F. Then

$$
\begin{aligned}
[(x_i) + (y_i)][\Phi]A &= (x_i + y_i)[\Phi]A = \left(\sum_k ((x_k + y_k)\phi_{ki})a_{ki} \right) \\
&= \left(\sum_k (x_k \phi_{ki} + y_k \phi_{ki})a_{ki} \right) \\
&= \left(\sum_k (x_k \phi_{ki})a_{ki} + \sum_k (y_k \phi_{ki})a_{ki} \right) \\
&= \left(\sum_k (x_k \phi_{ki})a_{ki} \right) + \left(\sum_k (y_k \phi_{ki})a_{ki} \right) \\
&= (x_i)[\Phi]A + (y_i)[\Phi]A.
\end{aligned}
$$

Hence, $[\Phi]A$ is a group homomorphism of $(V, +)$.

If each $[\Phi]A$ is a group homomorphism of V, then for a specific ϕ_{st} of Φ, we have for an A with $a_{st} = 1$,

$$
(\delta_{is}(a + b))[\Phi]A = \left(\sum_k [(\delta_{ks}(a + b))\phi_{ki}]a_{ki} \right) = ([(a + b)\phi_{si}]a_{si})
$$

and

$$
(\delta_{is}a)[\Phi]A + (\delta_{is}b)[\Phi]A = ((a\phi_{si})a_{si}) + ((b\phi_{si})a_{si}) = ((a\phi_{si} + b\phi_{si})a_{si}).
$$

Hence, for $i = t$, we obtain $(a\phi_{st} + b\phi_{st})a_{st} = [(a + b)\phi_{st}]a_{st}$, and since $a_{st} = 1$, $(a + b)\phi_{st} = a\phi_{st} + b\phi_{st}$. Since $\phi_{st} \in \text{Aut}F^*$, and since we have extended each ϕ_{st} so that $0\phi_{st} = 0$, then each ϕ_{st} is a field automorphism of F.

In this case, $\mathcal{M}_\Phi(V, F) \subseteq \mathrm{Hom}(V, V)$, which is a ring. Also, $\mathcal{M}_\Phi(V, F)$ is a vector space by (14.10). To see that $\mathcal{M}_\Phi(V, F)$ is a ring, we need only show closure for $\circ$ in $\mathcal{M}_\Phi(V, F)$, so take $[\Phi]A, [\Phi]B \in \mathcal{M}_\Phi(V, F)$. Now

$$(x_i)[\Phi]A \circ [\Phi]B = \left(\sum_j (x_j\phi_{ji})a_{ji}\right)[\Phi]B = \left(\sum_k \left(\left(\sum_j (x_j\phi_{jk})a_{jk}\right)\phi_{ki}\right)b_{ki}\right)$$

$$= \left(\sum_k \left(\sum_j [(x_j\phi_{jk})a_{jk}]\phi_{ki}\right)b_{ki}\right) = \left(\sum_k \left(\sum_j (x_j\phi_{jk}\phi_{ki})(a_{jk}\phi_{ki})\right)b_{ki}\right)$$

$$= \left(\sum_k \left(\sum_j (x_j\phi_{ji})(a_{jk}\phi_{ki})\right)b_{ki}\right) = \left(\sum_k \sum_j (x_j\phi_{ji})(a_{jk}\phi_{ki})b_{ki}\right)$$

$$= \left(\sum_j (x_j\phi_{ji})\right)\left(\sum_k (a_{jk}\phi_{ki})b_{ki}\right) = (x_i)[\Phi]C$$

where $c_{ji} = \sum_k (a_{jk}\phi_{ki})b_{ki}$. Hence, $\mathcal{M}_\Phi(V, F)$ is a ring and a vector space over F. If $A = I = (\delta_{ij})$, then $[\Phi]A \circ [\Phi]B = [\Phi]C$, where $C = (c_{ij})$ and $c_{ij} = \sum_k (a_{ik}\phi_{ik})b_{kj} = (1\phi_{ii})b_{ij} = b_{ij}$. Also, $[\Phi]B \circ [\Phi]I = [\Phi]C$, where $c_{ij} = \sum_k (b_{ik}\phi_{kj})\delta_{kj} = (b_{ij}\phi_{jj})1 = b_{ij}$. So $[\Phi]I$ is the multiplicative identity for $\mathcal{M}_\Phi(V, F)$. Since $\mathcal{M}_\Phi(V, F) \subseteq \mathrm{Hom}(V, V)$, a ring, our proof is now complete.

(14.19) Corollary. *If each ϕ_{ij} is an automorphism of the field F, and if we have $[\Phi]A, [\Phi]B \in \mathcal{M}_\Phi(V, F)$, then $[\Phi]A \circ [\Phi]B = [\Phi]C$, where $c_{ij} = \sum_k (a_{ik}\phi_{ik})b_{ki}$.*

(14.20) Theorem. *The elements of $\mathcal{M}_\Phi(V, F)$ are linear transformations over any $F(s, \Phi)$ if the ϕ_{ij} are automorphisms of the field F.*

Proof. From (14.18), each $[\Phi]A \in \mathcal{M}_\Phi(V, F)$ is a homomorphism of the group $(V, +)$. Now

$$[(x_i)a(s, \Phi)][\Phi]A = (x_i(a\phi_{si}))[\Phi]A = \left(\sum_k ([x_k(a\phi_{sk})]\phi_{ki})a_{ki}\right)$$

$$= \left(\sum_k ((x_k\phi_{ki})(a\phi_{si}))a_{ki}\right) = \left(\left(\sum_k (x_k\phi_{ki})a_{ki}\right)(a\phi_{si})\right)$$

$$= (((x_i)[\Phi]A)(a\phi_{si})) = ((x_i)[\Phi]A)a(x, \Phi).$$

So $[\Phi]A$ is a linear transformation over $F(s, \Phi)$ since the arbitrary $[\Phi]A \in \mathcal{M}_\Phi(V, F)$ commutes with any scalar $a(s, \Phi) \in F(s, \Phi)$.

(14.21) Proposition. *The elements of any $F(k, \Phi)$, or $F(\Phi, k)$, are linear transformations of V. If $a \neq 0$, then $a(k, \Phi)$ and $a(\Phi, k)$ are automorphisms of V.*

Proof. Take $(x_i), (y_i) \in V$, $b \in F$, and $a(k, \Phi) \in F(k, \Phi)$. Then

$$
\begin{aligned}
[(x_i) + (y_i)]a(k, \Phi) &= (x_i + y_i)a(k, \Phi) = ((x_i + y_i)(a\phi_{ki})) \\
&= (x_i(a\phi_{ki}) + y_i(a\phi_{ki})) = (x_i(a\phi_{ki})) + (y_i(a\phi_{ki})) \\
&= (x_i)a(k, \Phi) + (y_i)a(k, \Phi).
\end{aligned}
$$

Also,

$$
\begin{aligned}
[(x_i)b]a(k, \Phi) &= (x_i b)a(k, \Phi) = ((x_i b)(a\phi_{ki})) \\
&= ([x_i(a\phi_{ki})]b) = (x_i(a\phi_{ki}))b = [(x_i)a(k, \Phi)]b.
\end{aligned}
$$

If $a \neq 0$ and $(0) = (x_i)a(k, \Phi) = (x_i(a\phi_{ki}))$, then since each $a\phi_{ki} \neq 0$, we must have each $x_i = 0$ and $(x_i) = (0)$. So $a(k, \Phi)$ is a monomorphism. For $(y_i) \in V$, let $x_i = y_i(a\phi_{ki})^{-1}$, so $(x_i)a(k, \Phi) = (y_i(a\phi_{ki})^{-1}(a\phi_{ki})) = (y_i)$, and now $a(k, \Phi)$ is an automorphism of V.

The proof for $F(\Phi, k)$ is almost identical to that for $F(k, \Phi)$.

(14.22) Corollary. *If $V = F^n$ is a finite dimensional vector space over F, then*

$$
a(k, \Phi) = \begin{pmatrix}
a\phi_{k1} & 0 & 0 & \cdots & 0 & 0 \\
0 & a\phi_{k2} & 0 & \cdots & 0 & 0 \\
\cdots & \cdots & \cdots & \ddots & \cdots & \cdots \\
0 & 0 & 0 & \cdots & 0 & a\phi_{kn}
\end{pmatrix}
$$

and

$$
a(\Phi, k) = \begin{pmatrix}
a\phi_{1k} & 0 & 0 & \cdots & 0 & 0 \\
0 & a\phi_{2k} & 0 & \cdots & 0 & 0 \\
\cdots & \cdots & \cdots & \ddots & \cdots & \cdots \\
0 & 0 & 0 & \cdots & 0 & a\phi_{nk}
\end{pmatrix}.
$$

Proof. For each j, $1 \leq j \leq n$, $(\delta_{ij})a(k, \Phi) = (\delta_{ij}(a\phi_{ki}))$ and $(\delta_{ij})a(\Phi, k) = (\delta_{ij}(a\phi_{ik}))$ are all one needs to consider.

(14.23) Theorem. *Each $F^*(k, \Phi) = F^*(1, \Phi)$.*

Proof. Let $b = a\phi_{1k}$. Then $(x_i)b(k, \Phi) = (x_i(b\phi_{ki})) = (x_i(a\phi_{1k}\phi_{ki})) = (x_i(a\phi_{1i})) = (x_i)a(1, \Phi)$. Hence, $F^*(k, \Phi) \subseteq F^*(1, \Phi)$. Similarly, let $b = a\phi_{k1}$. Then we obtain $(x_i)b(1, \Phi) = (x_i(b\phi_{1i})) = (x_i(a\phi_{k1}\phi_{1i})) = (x_i(a\phi_{ki})) = (x_i)a(k, \Phi)$, and so $F^*(1, \Phi) \subseteq F^*(k, \Phi)$.

(14.24) Theorem. $\mathcal{M}_\Phi(V, F) \subseteq M_{F(k, \Phi)}(V)$.

Proof. Computing with $[\Phi]A \in \mathcal{M}_\Phi(V, F)$, $(x_i) \in V$, and $a(k, \Phi) \in F^*(k, \Phi)$, we obtain

$$[(x_i)a(k, \Phi)][\Phi]A = (x_i(a\phi_{ki}))[\Phi]A = \left(\sum_j \{[x_j(a\phi_{kj})]\phi_{ji}\}a_{ji}\right)$$

$$= \left(\sum_j (x_j\phi_{ji})(a\phi_{kj}\phi_{ji})a_{ji}\right) = \left(\sum_j (x_j\phi_{ji})(a\phi_{ki})a_{ji}\right)$$

$$= \left(\left(\sum_j (x_j\phi_{ji})a_{ji}\right)(a\phi_{ki})\right) = \left(\sum_j (x_j\phi_{ji})a_{ji}\right)a(k, \Phi)$$

$$= [(x_i)[\Phi]A]a(k, \Phi).$$

Hence, $a(k, \Phi)[\Phi]A = [\Phi]Aa(k, \Phi)$, so $[\Phi]A \in M_{F(k,\Phi)}(V)$.

(14.25) Proposition. *Each $M_{F(k,\Phi)}(V)$ is a nearalgebra over F, a sub-nearalgebra of $M_0(V)$.*

Proof. Take $f \in M_{F(k,\Phi)}(V)$ and $b \in F$. Then we have $[(x_i)b]a(k, \Phi) = [(x_i)a(k, \Phi)]b$ by (14.21). So

$$[(x_i)a(k, \Phi)](fb) = \{[(x_i)a(k, \Phi)]f\}b = \{[(x_i)f]a(k, \Phi)\}b$$
$$= ([(x_i)f]b)a(k, \Phi) = [(x_i)(fb)]a(k, \Phi),$$

which gives us $fb \in M_{F(k,\Phi)}(V)$. If $g \in M_{F(k,\Phi)}(V)$ also, then

$$(x_i)[(f \circ g)b] = [(x_i)(f \circ g)]b = [((x_i)f)g]b$$
$$= ((x_i)f)(gb) = (x_i)(f \circ (gb)).$$

Hence, $(f \circ g)b = f \circ (gb)$, and this completes the proof.

Each transition matrix Φ for F gives us a vector subspace $\mathcal{M}_\Phi(V, F)$ of $M_0(V)$. These are each isomorphic to $\mathrm{Hom}_F(V, V)$ as vector spaces (14.10), since $\mathrm{Hom}_F(V, V) \cong \mathcal{M}(V, F)$. If each of the ϕ_{ij} of Φ is a field automorphism of F, then $\mathcal{M}_\Phi(V, F) \subseteq \mathrm{Hom}(V, V)$, the ring of group homomorphisms of $(V, +)$ and, in fact, each $(\mathcal{M}_\Phi(V, F), +, \circ)$ is a subring of $(\mathrm{Hom}(V, V), +, \circ)$ by (14.18). From (14.24), we obtain that $\mathcal{M}_\Phi(V, F)$ is contained in the subnearring $M_{F(k,\Phi)}(V)$ even if there is a ϕ_{ij} of Φ which is not a field automorphism of F.

It may be that a $\mathcal{M}_\Phi(V, F)$ has a considerable intersection with the ring $\mathrm{Hom}(V, V)$, however. It all depends upon how many of the entries ϕ_{ij} of Φ are field automorphisms of F. Certainly, each $\phi_{ii} = 1$ is, so there will always be some nontrivial intersection. We now turn our attention to answers some basic questions in this area.

(14.26) Proposition. *Let* $\Phi = (\phi_{ij})$ *and* $\Gamma = (\gamma_{ij})$ *be transition matrices for* F *and* V. *If* $A \in \mathcal{M}(V, F)$, *then* $[\Phi]A = [\Gamma]A$ *if and only if* $a_{ij} = 0$ *whenever* $\phi_{ij} \neq \gamma_{ij}$.

Proof. Suppose $[\Phi]A = [\Gamma]A$ with $\Phi \neq \Gamma$. Then for any $(x_i) \in V$, $(x_i)[\Phi]A = (\sum_k (x_k \phi_{ki})a_{ki}) = (\sum_k (x_k \gamma_{ki})a_{ki}) = (x_i)[\Gamma]A$, so for any fixed $j \in I$, we have

$$\sum_k (x_k \phi_{kj})a_{kj} = \sum_k (x_k \gamma_{kj})a_{kj} \qquad (14:5)$$

for all $(x_i) \in V$. Let $(x_i) = (\delta_{ti}b)$. Then we have $[(\delta_{ii}b)\phi_{ij}]a_{ij} = (b\phi_{ij})a_{ij} = [(\delta_{ii}b)\gamma_{ij}]a_{ij} = (b\gamma_{ij})a_{ij}$ for each $b \in F$. So if $\phi_{ij} \neq \gamma_{ij}$, then there is a $b \in F$ such that $b\phi_{ij} \neq b\gamma_{ij}$, which implies that $a_{ij} = 0$. In summary, if $[\Phi]A = [\Gamma]A$ and $\phi_{ij} \neq \gamma_{ij}$, then $a_{ij} = 0$.

Conversely, suppose $a_{ij} = 0$ whenever $\phi_{ij} \neq \gamma_{ij}$. In (14:5), we still have $(x_k \phi_{kj})a_{kj} = (x_k \gamma_{kj})a_{kj}$ whether $x_k \phi_{kj} = x_k \gamma_{kj}$ or not. So (14:5) is valid. Hence, $(x_i)[\Phi]A = (x_i)[\Gamma]A$ for all $(x_i) \in V$, making $[\Phi]A = [\Gamma]A$.

(14.27) Proposition. *Let* Φ *and* Γ *be transition matrices for* F *and* V, *and let* $A, B \in \mathcal{M}(V, F)$. *If* $[\Phi]A = [\Gamma]B$, *then* $A = B$.

Proof. For $(x_i) = (\delta_{it})$, then $(\delta_{it})[\Phi]A = (\delta_{it})[\Gamma]B$, so $\sum_k (\delta_{kt}\phi_{ki})a_{ki} = \sum_k (\delta_{kt}\gamma_{ki})b_{ki}$ and therefore $(1\phi_{ti})a_{ti} = (1\gamma_{it})b_{ti}$, so $a_{ti} = b_{ti}$. Hence, $A = B$.

(14.28) Proposition. *For a transition matrix* Φ *and* $A \in \mathcal{M}(V, F)$, $[(x_i)a][\Phi]A = [(x_i)[\Phi]A]a$ *for all* $(x_i) \in V$ *and for all* $a \in F$ *if and only if* $a_{ij} = 0$ *whenever* $\phi_{ij} \neq 1$.

Proof. Start with $[\Phi]A$ commuting with the scalars $a \in F$, that is, we have $[(x_i)a][\Phi]A = [(x_i)[\Phi]A]a$ for all $(x_i) \in V$ and for all $a \in F$. Suppose $\phi_{st} \neq 1$ and $a_{st} \neq 0$. Then

$$[(\delta_{is})a][\Phi]A = \left(\sum_k [(\delta_{ks}a)\phi_{ki}]a_{ki}\right) = ((a\phi_{si})a_{si})$$

and

$$[(\delta_{is})[\Phi]A]a = \left(\sum_k (\delta_{ks}\phi_{ki})a_{ki}\right)a = ((1\phi_{si})a_{si})a = (a_{si})a.$$

For $i = t$, we obtain $(a\phi_{st})a_{st} = a_{st}a$ for each $a \in F$. If we insist that $\phi_{st} \neq 1$, then we are forced to accept that $a_{st} = 0$.

Conversely, suppose $\phi_{ij} \neq 1$ implies $a_{ij} = 0$. Then $(x_k \phi_{kj})a_{kj} = x_k a_{kj}$ in all cases. So $[(x_k a)\phi_{kj}]a_{kj} = (x_k a)a_{kj}$ for each $a \in F$. This makes

$$[(x_i)a][\Phi]A = (x_i a)[\Phi]A = \left(\sum_k [(x_k a)\phi_{ki}]a_{ki}\right) = \left(\sum_k (x_k a)a_{ki}\right)$$

$$= \left(\sum_k x_k a_{ki}\right)a = \left(\sum_k (x_k \phi_{ki})a_{ki}\right)a = [(x_i)[\Phi]A]a$$

for all $(x_i) \in V$ and for all $a \in F$. This completes the proof of (14.28).

(14.29) Proposition. *For a transition matrix Φ and $A \in \mathcal{M}(V, F)$, we have $[\Phi]A \in \text{Hom}(V, V)$ if and only if $a_{ij} = 0$ whenever ϕ_{ij} is not a field automorphism of F.*

Proof. Suppose $[\Phi]A \in \text{Hom}(V, V)$. Then $[(x_i) + (y_i)][\Phi]A = (x_i)[\Phi]A + (y_i)[\Phi]A$. Also, $(x_i + y_i)\phi_{ij}a_{ij} = [x_i\phi_{ij} + y_i\phi_{ij}]a_{ij}$ if $a_{ij} = 0$ or if ϕ_{ij} is a field automorphism. From

$$[(\delta_{is})a + (\delta_{is})b][\Phi]A = (\delta_{is}(a + b))[\Phi]A$$
$$= \left(\sum_k [\delta_{ks}(a + b)]\phi_{ki}a_{ki}\right) = \left((a + b)\phi_{si}a_{si}\right)$$

and

$$[(\delta_{is})a][\Phi]A + [(\delta_{is})b][\Phi]A = ((a\phi_{si})a_{si}) + ((b\phi_{si})a_{si}) = ([a\phi_{si} + b\phi_{si}]a_{si}),$$

we have $(a + b)\phi_{st}a_{st} = [a\phi_{st} + b\phi_{st}]a_{st}$ for all $a, b \in F$. If $a_{st} \neq 0$, we are forced to conclude that ϕ_{st} is a field automorphism of F. Alternatively, if we insist that ϕ_{st} is not a field automorphism, then we must accept $a_{st} = 0$.

For the converse, suppose $a_{ij} = 0$ whenever ϕ_{ij} is not a field automorphism. As noted above, we always have $(x_i + y_i)\phi_{ij}a_{ij} = [x_i\phi_{ij} + y_i\phi_{ij}]a_{ij} = (x_i\phi_{ij})a_{ij} + (y_i\phi_{ij})a_{ij}$. Therefore, for each $j \in I$, $\sum_k(x_k + y_k)\phi_{kj}a_{kj} = \sum_k(x_k\phi_{kj})a_{kj} + \sum_k(y_k\phi_{kj})a_{kj}$. But $\left(\sum_k(x_k + y_k)\phi_{kj}a_{kj}\right) = (x_j + y_j)[\Phi]A = [(x_i) + (y_i)][\Phi]A$ and $\left(\sum_k(x_k\phi_{kj})a_{kj}\right) + \left(\sum_k(y_k\phi_{kj})a_{kj}\right) = (x_j)[\Phi]A + (x_j)[\Phi]A$, putting $[\Phi]A \in \text{Hom}(V, V)$.

(14.30) Corollary. $[\Phi]A \in \text{Hom}_F(V, V)$ *if and only if $\phi_{ij} \neq 1$ implies $a_{ij} = 0$.*

Proof. Apply (14.28) and (14.29).

For each transition matrix Φ, we have that the vector space $\mathcal{M}_\Phi(V, F)$ is a subspace of each nearring $M_{F(k,\Phi)}(V)$, by (14.24). So we get $\mathcal{M}_\Phi(V, F) \subseteq \cap_{k \in I} M_{F(k,\Phi)}(V)$. From (14.23), we know that each $F(k, \Phi) = F(1, \Phi)$, so one might immediately conclude that each $M_{F(k,\Phi)}(V) = M_{F(1,\Phi)}(V)$. But doubt might occur when one realizes that even though $F(k, \Phi) = F(1, \Phi)$, their elements appear to induce different actions of V. Specifically, we have $(x_i)a(1, \Phi) = (x_i(a\phi_{1i}))$ and $(x_i)a(k, \Phi) = (x_i(a\phi_{ki}))$. We settle the matter formally once and for all.

(14.31) Theorem. $M_{F(k,\Phi)}(V) = M_{F(1,\Phi)}(V)$.

Proof. Take $f \in M_{F(k,\Phi)}(V)$, $(x_i) \in V$, and $a(1, \Phi) \in F(1, \Phi)$. When we prove that $[(x_i)a(1, \Phi)]f = [(x_i)f]a(1, \Phi)$, then we will have $f \in M_{F(1,\Phi)}(V)$

and $M_{F(k,\Phi)}(V) \subseteq M_{F(1,\Phi)}(V)$. An analogous proof will give us the reverse inclusion.

Now $a = b\phi_{k1}$ for some $b \in F$. Let $(x_i)f = (y_i)$. Then

$$[(x_i)a(1, \Phi)]f = (x_i(a\phi_{1i}))f = (x_i(b\phi_{k1}\phi_{1i}))f = (x_i(b\phi_{ki}))f$$
$$= [(x_i)b(k, \Phi)]f = [(x_i)f]b(k, \Phi) = (y_i)b(k, \Phi)$$
$$= (y_i(b\phi_{ki})) = (y_i(b\phi_{k1}\phi_{1i})) = (y_i(a\phi_{1i}))$$
$$= (y_i)a(1, \Phi) = [(x_i)f]a(1, \Phi),$$

as promised.

So each $\mathcal{M}_\Phi(V, F) \subseteq M_{F(1,\Phi)}(V)$. But if each ϕ_{ij} of Φ is a field automorphism of F, then $\mathcal{M}_\Phi(V, F)$ is a nearring, and even a ring by (14.18). But, if there is a ϕ_{ij} which is not a field automorphism, then we have no promise that $\mathcal{M}_\Phi(V, F)$ is a nearring. However, the more entries ϕ_{ij} that are field automorphisms, the more elements of $\mathcal{M}_\Phi(V, F)$ are also in $\mathrm{Hom}(V, V)$ by (14.29). This all suggests that the subnearring of $M_{F(1,\Phi)}(V)$ generated by the elements of $\mathcal{M}_\Phi(V, F)$ might be close to being a ring or quite far from being a ring, depending upon how many of the ϕ_{ij} of Φ are field automorphisms of F.

(14.32) Definition. For a transition matrix Φ for F, let $N(\Phi)$ be the subnearring of $M_{F(1,\Phi)}(V)$ generated by the elements of $\mathcal{M}_\Phi(V, F)$, that is, the intersection of all subnearrings of $M_{F(1,\Phi)}(V)$ which contain $\mathcal{M}_\Phi(V, F)$.

If each ϕ_{ij} of Φ is a field automorphism of F, then $N(\Phi) = \mathcal{M}_\Phi(V, F)$ by (14.18). Sometimes the nearring $N(\Phi)$ is simple.

(14.33) Definition. We say that $A \in \mathcal{M}(V, F)$ is a *kth row matrix* if $a_{ij} = 0$ for $i \neq k$.

(14.34) Lemma. *Let $A \in \mathcal{M}(V, F)$ be a kth row matrix, and let $f \in M_{F(1,\Phi)}(V)$. Then $[\Phi]A \circ f = [\Phi]B$ for some kth row matrix $B \in \mathcal{M}(V, F)$.*

Proof. Let $(x_i) = (a_{ki})$, the kth row of A, and let $(a_{ki})f = (b_{ki})$. Let $B \in \mathcal{M}(V, F)$ be the kth row matrix with kth row equal to (b_{ki}). For $(x_i) \in V$, we obtain

$$(x_i)[\Phi]A \circ f = \left(\sum_j (x_j\phi_{ji})a_{ji}\right)f = ((x_k\phi_{ki})a_{ki})f$$
$$= (a_{ki}(x_k\phi_{ki}))f = [(a_{ki})x_k(k, \Phi)]f$$
$$= [(a_{ki})f]x_k(k, \Phi) = (b_{ki})x_k(k, \Phi) = (b_{ki}(x_k\phi_{ki}))$$
$$= ((x_k\phi_{ki})b_{ki}) = \left(\sum_j (x_j\phi_{ji})b_{ji}\right) = (x_i)[\Phi]B.$$

Hence, $[\Phi]A \circ f = [\Phi]B$.

(14.35) Theorem. *If the dimension of V is finite, then $N(\Phi)$ is a simple nearring. In fact, if $(N', +)$ is a subgroup of $(N(\Phi), +)$ satisfying $N(\Phi)N' \cup N'N(\Phi) \subseteq N'$, then either $N' = \{\zeta\}$ or $N' = N$.*

Proof. Let $\zeta \neq f \in N'$. Then there is a $(z_i) \in V$ such that $(z_i)f = (u_i) \neq (0)$. Let $Z^{(s)}$ be the sth row matrix with sth row equal to (z_i), and, similarly, let $U^{(s)}$ be the sth row matrix with sth row equal to (u_i). By (14.34), $[\Phi]Z^{(s)} \circ f = [\Phi]U^{(s)}$. Suppose $u_k = u_{sk} \neq 0$.

Let $A^{(s)}$ have $a_{ks} = (u_k^{-1})\phi_{ks}$ and $a_{ij} = 0$ otherwise. We proceed now to prove that $[\Phi]U^{(s)} \circ [\Phi]A^{(s)} = [\Phi]E^{(s)}$, where $e_{ss} = 1$ and $e_{ij} = 0$, otherwise. For $(x_i) \in V$, we have

$$(x_i)[\Phi]U^{(s)} \circ [\Phi]A^{(k)} = \left(\sum_j (x_j\phi_{ji})u_{ji} \right)[\Phi]A^{(k)} = \left((x_s\phi_{si})u_{si} \right)[\Phi]A^{(k)}$$

$$= \left(\sum_j [(x_s\phi_{sj})u_{sj}]\phi_{ji}a_{ji} \right) = \left([(x_s\phi_{sk})u_{sk}]\phi_{ki}a_{ki} \right).$$

where

$$[(x_s\phi_{sk})u_{sk}]\phi_{ki}a_{ki} = \begin{cases} 0, & \text{if } i \neq s; \\ (x_s\phi_{ss})(u_{sk}\phi_{ks})a_{ks}, & \text{if } i = s, \end{cases}$$

$$= \begin{cases} 0, & \text{if } i \neq s; \\ x_s(u_{sk}\phi_{ks})(u_k^{-1})\phi_{ks}, & \text{if } i = s, \end{cases}$$

$$= \begin{cases} 0, & \text{if } i \neq s; \\ x_s, & \text{if } i = s. \end{cases}$$

Also, $(x_i)[\Phi]E^{(s)} = \left(\sum_j (x_j\phi_{ji})e_{ji} \right) = \left((x_s\phi_{si})e_{si} \right)$ where

$$(x_s\phi_{si})e_{si} = \begin{cases} 0, & \text{if } i \neq s; \\ (x_s\phi_{ss})e_{ss}, & \text{if } i = s, \end{cases} = \begin{cases} 0, & \text{if } i \neq s; \\ x_s, & \text{if } i = s. \end{cases}$$

Hence, $[\Phi]U^{(s)} \circ [\Phi]A^{(k)} = [\Phi]E^{(s)}$.

Since $f \in N'$, each $[\Phi]Z^{(s)} \circ f = [\Phi]U^{(s)} \in N'$, and, consequently, each $[\Phi]E^{(s)} \in N'$. But $1_V = [\Phi]E^{(1)} + [\Phi]E^{(2)} + \cdots + [\Phi]E^{(n)} \in N'$ also. This puts $N(\Phi) \subseteq N'$ making $N(\Phi) = N'$, and completes our proof.

We can find some proper ideals if we look at some subnearrings. Let $\mathcal{M}_f(V, F) = \{A \in \mathcal{M}(V, F) \mid A$ is column-finite$\}$, and let $N_f(\Phi)$ denote the subnearring of $N(\Phi)$ generated by the image $\Phi(\mathcal{M}_f(V, F))$ of $\mathcal{M}_f(V, F)$ in $\mathcal{M}_\Phi(V, F)$. So if V has finite dimension, then $\mathcal{M}_f(V, F) = \mathcal{M}(V, F)$ and $N_f(\Phi) = N(\Phi)$. We are assured that $N_f(\Phi)$ is simple when V has finite dimension. This will not be the case when V has infinite dimension.

(14.36) Definitions. For $J \subseteq I$, define $V_J = \{(x_i) \in V \mid x_j = 0$ for $j \in J\}$, a subspace of V. For $g \in N_f(\Phi)$, we say that $J = J(g) \subseteq I$ is a *threshold for g* if $V_J g \subseteq \{(0)\}$. This does not preclude g from having more than one threshold, that is, $J(g)$ need not be a function.

(14.37) Proposition. *For $g \in N_f(\Phi)$ and a finite $J \subseteq I$, there is a finite $T \subseteq I$ such that $V_T g \subseteq V_J$.*

Proof. First, consider $g = [\Phi]A$, for $A \in \mathcal{M}_f(V, F)$. For each i, define $J_i = \{k \in I \mid a_{ki} \neq 0\}$, a finite set. Let $T = \cup_{i \in J} J_i$, which is also a finite set since J is finite. We want to show that $V_T g \subseteq V_J$, so take $(x_i) \in V_T$. Then $(x_i)g = (x_i)[\Phi]A = (\sum_j (x_j \phi_{ji}) a_{ji})$. Take $i \in J$, so if

$$\Gamma = \sum_j (x_j \phi_{ji}) a_{ji} = \sum_{j \in J_i} (x_j \phi_{ji}) a_{ji}. \qquad (14:6)$$

Since $j \in J_i \subseteq T$, we have $x_j = 0$ and $x_j \phi_{ji} = 0$. Hence, $\Gamma = 0$. This gives us $(x_i)g = (x_i)[\Phi]A \in V_J$, and, consequently, $V_T g \subseteq V_J$.

If $g \in N_f(\Phi)$ has a finite $T \subseteq I$ such that $V_T g \subseteq V_J$, then $V_T(-g) = -[V_T g] \subseteq -V_J = V_J$, so $-g \in N_f(\Phi)$ has the same finite $T \subseteq I$ such that $V_T(-g) \subseteq V_J$. Suppose $g_1, g_2 \in N_f(\Phi)$ have finite T_1, T_2, respectively, with $T_1 \cup T_2 = T \subseteq I$, and where $V_{T_1} g_1 \subseteq V_J$ and $V_{T_2} g_2 \subseteq V_J$. Now T is a finite subset of I. If $(x_i) \in V_T$, then $(x_i) \in V_{T_1} \cap V_{T_2}$, for if $x_i = 0$ when $i \in T$, then $x_i = 0$ when $i \in T_v$, $v = 1, 2$. So $(x_i) \in V_{T_1} \cap V_{T_2}$. For $v \in \{1, 2\}$, $V_T g_v \subseteq V_{T_v} g_v \subseteq V_J$, so $V_T(g_1 + g_2) \subseteq V_T g_1 + V_T g_2 \subseteq V_J + V_J = V_J$. We can now conclude that $g_1 + g_2$ also has a finite $T \subseteq I$ such that $V_T(g_1 + g_2) \subseteq V_J$.

For $g_1, g_2 \in N_f(\Phi)$, suppose, for any finite set $J' \subseteq I$, that there are such finite subsets T_1', T_2' of I such that $V_{T_v'} g_v \subseteq V_{J'}$, $v = 1, 2$. So in particular, there are finite subsets T_1, T_2 of I such that $V_{T_v} g_v \subseteq V_J$, $v = 1, 2$. Now let $J' = T_2$ for g_1. Then there is a finite T_3 so that $V_{T_3} g_1 \subseteq V_{T_2}$. Hence, $V_{T_3} g_1 \circ g_2 \subseteq V_{T_2} g_2 \subseteq V_J$. This means $g_1 \circ g_2$ has a finite subset T_3 such that $T_3 g_1 \circ g_2 \subseteq V_J$.

The elements of $N_f(\Phi)$ are generated, using $+$, $-$, and $\circ$, by all the $[\Phi]A$, where $A \in \mathcal{M}_f(V, F)$. And since each $[\Phi]A$ satisfies (14.37), so do all the elements of $N_f(\Phi)$. This completes the proof of (14.37). $\quad\blacksquare$

(14.38) Corollary. *For $g \in N_f(\Phi)$ and a countable $J \subseteq I$, there is a countable $T \subseteq I$ such that $V_T g \subseteq V_J$.*

Proof. The proof of (14.37) can be easily modified. One still has the J_i finite, but now T will be countable since J is countable. $\quad\blacksquare$

(14.39) Theorem. *If V has infinite dimension over F, then $N_f(\Phi)$ is not simple. In particular,*

$$K = \{g \in N_f(\Phi) \mid g \text{ has a finite threshold } J(g)\}$$

is a proper ideal.

Proof. First, $1_V \notin K$ but $1_V \in N_f(\Phi)$, and next, $[\Phi]E^{(1)} \in K$ and $\zeta \neq [\Phi]E^{(1)}$. (We defined $[\Phi]E^{(s)}$ in the proof of (14.35).) We can take $J([\Phi]E^{(1)}) = \{1\}$. Let $J(g_1), J(g_2)$ be finite thresholds for $g_1, g_2 \in K$, respectively. Since $V_{J(g_1)}g_1 = \{(0)\}$, we have $V_{J(g_1)}(-g_1) = -[V_{J(g_1)}] = -\{(0)\} = \{(0)\}$, and so $-g_1 \in K$. Let $J = J(g_1) \cup J(g_2)$, a finite subset of I. If $(x_i) \in V_J$, then $x_i = 0$ for $i \in J$. Here $x_i = 0$ for $i \in J(g_v)$, $v = 1, 2$, and so $(x_i)g_v = (0)$, $v = 1, 2$. Consequently, $(x_i)(g_1 + g_2) = (0)$, which means $V_J(g_1 + g_2) = \{(0)\}$. Hence, $g_1 + g_2 \in K$. We now have that K is a (normal) subgroup of $(N_f(\Phi), +)$.

For $g \in K$ and $s, t \in N_f(\Phi)$, let $J(g)$ be a finite threshold for g and let $(x_i) \in V_{J(g)}$. Then $(x_i)[(s+g) \circ t - s \circ t] = ((x_i)s + (x_i)g)t - ((x_i)s)t) = (0)$, so $J(g)$ is also a finite threshold for $(s+g) \circ t - s \circ t$. Thus $(s+g) \circ t - s \circ t \in K$.

Finally, for $s \in N_f(\Phi)$ and $g \in K$ with finite threshold $J(g)$, we are assured by (14.37) that there is a finite $T \subseteq I$ such that $V_T s \subseteq V_{J(g)}$. Hence, $V_T s \circ g \subseteq V_{J(g)}g = \{(0)\}$. So T is a finite threshold for $s \circ g$, putting $s \circ g \in K$. This makes K an ideal.

(14.40) Theorem. *Suppose V has an uncountable basis. Then, in addition to the ideal K of (14.39), $N_f(\Phi)$ has a larger proper ideal*

$$L = \{g \in N_f(\Phi) \mid g \text{ has a countable threshold }\}.$$

Proof. Modify the proof of (14.39) appropriately, and use (14.38) in place of (14.37).

(14.41) Problem. What are the ideals and quotient nearrings for the $N_f(\Phi)$? For $N(\Phi)$? As a partial answer, what are the maximal ideals and corresponding quotient nearrings?

(14.42) Problem. What conditions on Φ and Γ are necessary and sufficient to make $N_f(\Phi)$ and $M_f(\Gamma)$ isomorphic?

(14.43) Problem. The more entries ϕ_{ij} of Φ that are field automorphisms, the more $\mathcal{M}_\Phi(V, F)$ intersects the ring $\mathrm{Hom}(V, V)$, hence the closer $N_f(\Phi)$ should be to a ring. Does the number of entries ϕ_{ij} that are field automorphisms affect the existence of some ideal I of $N_f(\Phi)$ such that $N_f(\Phi)/I$ is a ring? Alternatively, can one always measure in some way how close $N_f(\Phi)$ is to being a ring by looking at the entries ϕ_{ij}?

(14.44) Problem. The above three problems can also be rephrased to query $M_{F(1,\Phi)}(V)$.

The following theorem leads one to focus on the $F^*(1, \Phi)$'s in a different light.

(14.45) Theorem. *If $F \neq Z_2$, the field of two elements, then each $F^*(1, \Phi)$ is a group of fixed point free automorphisms of the group $(V, +)$, and if $a \notin \{0, 1\}$, then $-1_V + a(1, \Phi)$ is also an automorphism, hence a bijection. (We are using pointwise $+$ for functions here.) This makes each $(V, F^*(1, \Phi))$ a Ferrero pair.*

Proof. For $(x_i) \in V$ and $a \neq 1$, we have $(x_i)a(1, \Phi) = (x_i(a\phi_{1i}))$. For any $x_j \neq 0$, we have $x_j \neq x_j(a\phi_{1j})$ since $a\phi_{1j} \neq 1$. Hence, $a(1, \Phi)$ has no fixed point other than (0). For $(y_i) \in V$, let $(x_i) = (y_i(-1 + a\phi_{1i})^{-1})$. Since $a \notin \{0, 1\}$, we know $(-1 + a\phi_{1i})^{-1}$ exists. Hence, $(x_i)(-1_V + a(1, \Phi)) = (y_i)$, so $-1_V + a(1, \Phi)$ is surjective. It is certainly injective and a homomorphism, so it is an automorphism.

(14.46) Exploratory problem. Investigate the geometry, combinatorics, etc., of the various planar nearrings one obtains from the various $F^*(1, \Phi)$ as Φ varies. What conditions on Φ and Γ yield isomorphic planar nearrings, geometric structures, etc? One can also obtain a result such as (14.45) for each $F^*(\Phi, 1)$. How do the resulting Ferrero pairs $(V, F^*(\Phi, 1))$ compare to those of (14.45)?

(14.47) Remark. In §7, the results in 'The van der Walt Connection' are gleaned by looking at a Ferrero pair $(V, F^*(1, \Phi))$ where $x\phi_{21} = x\alpha$, or $\alpha(x)$ as we denoted it there.

Now we wish to explore a little the concepts of eigenvalues and eigenvectors. Essentially, what are these concepts in classical linear algebra? Each eigenvector for a linear transformation A identifies a 1-dimensional subspace of V that is left invariant by A. Each nonzero vector in this subspace is also an eigenvector. The eigenvalue for A describes A's action on the eigenvectors in this 1-dimensional subspace.

In classical linear algebra, where $\Phi = (\phi_{ij})$ and each $\phi_{ij} = 1_V$, we have $\mathrm{Hom}_F(V, V) = \mathcal{M}_\Phi(V, F) = N_f(\Phi) \subseteq M_F(V) = M_{F(1,\Phi)}(V)$. Perhaps there is no harm in extending the ideas of eigenvalues and eigenvectors to elements of $M_{F(1,\Phi)}(V)$.

(14.48) Definitions. An $f \in M_{F(1,\Phi)}(V)$ has *eigenvalue* $a(1, \Phi) \in F(1, \Phi)$ if there is a nonzero vector $(x_i) \in V$ so that $(x_i)f = (x_i)a(1, \Phi)$. Such a vector (x_i) is an *eigenvector for $a(1, \Phi)$ as well as for f.*

There are several results that should make the reader feel more at home with our extension of these concepts.

(14.49) Proposition. *If $(x_i) \in V$ is an eigenvector for $f \in M_{F(1,\Phi)}(V)$ and for $a(1, \Phi) \in F(1, \Phi)$, then so is $(x_i)b(1, \Phi)$ for each $b(1, \Phi) \in F^*(1, \Phi)$.*

Proof. If $b(1, \Phi) \in F^*(1, \Phi)$, then $(x_i)b(1, \Phi) \neq (0)$. For $f \in M_{F(1,\Phi)}(V)$, we have that $[(x_i)b(1,\Phi)]f = (x_i)[f \circ b(1,\Phi)] = [(x_i)a(1,\Phi)]b(1,\Phi) = [(x_i)b(1,\Phi)]a(1,\Phi)$.

(14.50) Proposition. *Let $p(x) = b_0 + b_1 x + \cdots + b_n x^n \in F[x]$ be a polynomial with coefficients in F. If $(x_i) \in V$ is an eigenvector for an eigenvalue $a(1,\Phi)$ and an $f \in M_{F(1,\Phi)}(V)$, then $p(a(1,\Phi)) = b_0(1,\Phi) + b_1(1,\Phi)a(1,\Phi) + b_2(1,\Phi)a(1,\Phi)^2 + \cdots + b_n(1,\Phi)a(1,\Phi)^n$ is an eigenvalue for $p(f) = b_0(1,\Phi) + b_1(1,\Phi)f + \cdots + b_n(1,\Phi)f^n$, and (x_i) is also an eigenvector for $p(f)$ and for $p(a(1,\Phi))$. (The $+$ in our expression for $p(a(1,\Phi))$ and for $p(f)$ is pointwise addition, and is not to be confused with the addition of (14.14).)*

Proof. Computing directly,

$$
\begin{aligned}
(x_i)p(f) &= (x_i)[b_0(1,\Phi) + b_1(1,\Phi)f + \cdots + b_n(1,\Phi)f^n] \\
&= (x_i)b_0(1,\Phi) + (x_i)fb_1(1,\Phi) + \cdots + (x_i)f^n b_n(1,\Phi) \\
&= (x_i)b_0(1,\Phi) + (x_i)a(1,\Phi)b_1(1,\Phi) + \cdots + (x_i)a(1,\Phi)^n b_n(1,\Phi) \\
&= (x_i)b_0(1,\Phi) + (x_i)b_1(1,\Phi)a(1,\Phi) + \cdots + (x_i)b_n(1,\Phi)a(1,\Phi)^n \\
&= (x_i)p(a(1,\Phi)).
\end{aligned}
$$

Thus $p(a(1,\Phi))$ is an eigenvalue for $p(f)$, and (x_i) is an eigenvector for $p(f)$ and for $p(a(1,\Phi))$. Now $F(1,\Phi) \subseteq M_{F(1,\Phi)}(V)$, and so $p(a(1,\Phi)), p(f) \in M_{F(1,\Phi)}(V)$.

(14.51) Corollary. *If $a(1,\Phi)$ is an eigenvalue for $f \in M_{F(1,\Phi)}(V)$ and if $p(a(1,\Phi)) = \zeta = 0(1,\Phi)$ for $p(x) \in F[x]$, then $p(f)$ is not a unit in $M_{F(1,\Phi)}(V)$.*

Proof. There is a nonzero $(x_i) \in V$ such that $(x_i)f = (x_i)a(1,\Phi)$, and so $(x_i)p(f) = (x_i)p(a(1,\Phi)) = (x_i)0(1,\Phi) = (0)$. Hence, $p(f)$ cannot be a bijection of V, so it cannot be a unit in the nearring $M_{F(1,\Phi)}(V)$.

Only a few facts have been discovered about these eigenvalues and eigenvectors in this unusual setting. That there are some interesting and amusing interpretations of these ideas is what we now wish to demonstrate.

We shall restrict our attention to $[\Phi]A \in \mathcal{M}_\Phi(V, F)$. First, let V have dimension 2 over F. From $(x_i)[\Phi]A = (x_i)\lambda(1,\Phi)$ with $(x_i) \neq (0)$, we have

$$
\begin{cases}
x_1 a_{11} + (x_2 \phi_{21})a_{21} = x_1 \lambda \\
(x_1 \phi_{12})a_{12} + x_2 a_{22} = x_2(\lambda \phi_{12}).
\end{cases}
\tag{14 : 7}
$$

To get rid of the subscripts, let $A = \begin{pmatrix} a_{11} & a_{12} \\ a_{21} & a_{22} \end{pmatrix} = \begin{pmatrix} a & b \\ c & d \end{pmatrix}$, $(x_1, x_2) = (x, y)$, $\phi_{21} = \phi^{-1}$, and $\phi_{12} = \phi$. With $t = \lambda$, we have

$$
\begin{cases}
xa + (y\phi^{-1})b = xt \\
(x\phi)c + yd = y(t\phi).
\end{cases}
\tag{14 : 8}
$$

Suppose $b = 0$. Then either $t = a$ or $x = 0$. If $x = 0$, then $y \neq 0$, and we have $dy = (t\phi)y$, so $d = t\phi$ and $t = d\phi^{-1}$. So when $b = 0$, the eigenvalues are a and $d\phi^{-1}$. Suppose $c = 0$. Then either $t = d\phi^{-1}$ or $y = 0$. If $y = 0$, then $x \neq 0$, so $t = a$. Again, when $c = 0$, the eigenvalues are a and $d\phi^{-1}$.

Suppose $0 \notin \{b, c\}$ and $a = 0$. Then from (14:8) we obtain

$$\begin{cases} -(t\phi)(x\phi) + (b\phi)y = 0 \\ c(x\phi) + (d - t\phi)y = 0, \end{cases}$$

which has a nontrivial solution $(x\phi, y)$ if and only if

$$(t\phi)^2 - d(t\phi) - c(b\phi) = 0. \tag{14 : 9}$$

Solutions for $t\phi$ of (14:9) are at most two in number, and so these yield at most two eigenvalues for $[\Phi]A$.

From (14:8) we obtain

$$\begin{cases} [-(t - a)\phi](x\phi) + (b\phi)y = 0 \\ c(x\phi) + (d - t\phi)y = 0. \end{cases} \tag{14 : 10}$$

Now (14:10) has a nontrivial solution for $(x\phi, y)$, and equivalently for (x, y), if and only if

$$[(t - a)\phi](t\phi - d) = (b\phi)c. \tag{14 : 11}$$

So (14:11) is our 'characteristic equation' for t. That is, when $\phi = 1$, then (14:11) becomes the classical characteristic equation for the eigenvalues t.

Suppose $0 \notin \{b, c\}$ and $d = 0$. Also, let $c = c'\phi$. Then (14:11) becomes

$$[(t - a)\phi](t\phi) = (b\phi)(c'\phi)$$

or

$$[(t - a)t]\phi = (bc')\phi.$$

This means $(t - a)t = bc'$, which has at most two solutions for t, and there are at most two eigenvalues for $[\Phi]A$.

In summary, if $0 \in \{a, b, c, d\}$, then $[\Phi]A$ has at most two eigenvalues.

But what happens when $0 \notin \{a, b, c, d\}$ may be a surprise! For example, let $z\phi = z^n$ define an automorphism ϕ of F^*. Specifically, let $F = \mathbf{R}$, the field of real numbers, and let n be an odd positive integer. Then from (14:11), we get

$$(t - a)^n (t^n - d) = b^n c.$$

Let $p(t) = (t - a)^n (t^n - d) - b^n c$. The roots of $p(t)$ are our eigenvalues. Now the derivative $p'(t) = n(t - a)^{n-1}(2t^n - at^{n-1} - d)$. If $g(t) = 2t^n - at^{n-1} - d$, then the derivative $g'(t)$ can have at most two roots, so $g(t)$ can have at most three roots. This means that $p'(t)$ can have at most four roots, and therefore $p(t)$ can have at most four roots since the degree of $p(t)$ is $2n$.

As a specific example, where $F = \mathbf{R}$ and $z\phi = z^3$, let $A = \begin{pmatrix} a & b \\ c & d \end{pmatrix} = \begin{pmatrix} 4 & 2 \\ -8 & -1 \end{pmatrix}$. Noting that $p(t) = (t-4)^3(t^3+1) + 64$, we compute $p(-1) = 64$, $p(-1/2) = -1007/64$, $p(0) = 0$, $p(1) = 10$, $p(2) = -8$, and $p(3) = 36$. So $p(t)$ has roots $r_1 < r_2 < r_3 < r_4$ where $-1 < r_1 < r_2 = 0 < 1 < r_3 < 2 < r_4 < 3$. In fact, approximations for r_i, $i \neq 2$, are -0.735, 1.337, and 2.318. This means that $[\Phi]A$ has *four* distinct eigenvalues.

As one might correctly guess from (14.18), the situation is even closer to the classical case when each ϕ_{ij} of Φ is an automorphism of the field F. As an example, let us let V have dimension 3 over F. From $(x_i)[\Phi]A = (x_i)\lambda(1, \Phi)$ we write

$$
\begin{cases}
(x_1\phi_{11})a_{11} + (x_2\phi_{21})a_{21} + (x_3\phi_{31})a_{31} = x_1(\lambda\phi_{11}) \\
(x_1\phi_{12})a_{12} + (x_2\phi_{22})a_{22} + (x_3\phi_{32})a_{32} = x_2(\lambda\phi_{12}) \\
(x_1\phi_{13})a_{13} + (x_2\phi_{23})a_{23} + (x_3\phi_{33})a_{33} = x_3(\lambda\phi_{13}).
\end{cases} \qquad (14:12)
$$

We would want to solve for λ and for (x_i). Since each ϕ_{ij} is a field automorphism, then (14:12) becomes

$$
\begin{cases}
x_1 a_{11} + (x_2\phi_{21})a_{21} + (x_3\phi_{31})a_{31} = x_1\lambda \\
x_1(a_{12}\phi_{21}) + (x_2\phi_{21})(a_{22}\phi_{21}) + (x_3\phi_{31})(a_{32}\phi_{21}) = (x_2\phi_{21})\lambda \\
x_1(a_{13}\phi_{31}) + (x_2\phi_{21})(a_{23}\phi_{31}) + (x_3\phi_{31})(a_{33}\phi_{31}) = (x_3\phi_{31})\lambda.
\end{cases} \qquad (14:13)
$$

One can now solve (14:13) by first solving

$$
\det \begin{vmatrix}
a_{11} - \lambda & a_{21} & a_{31} \\
a_{12}\phi_{21} & a_{22}\phi_{21} - \lambda & a_{32}\phi_{21} \\
a_{13}\phi_{31} & a_{23}\phi_{31} & a_{33}\phi_{31} - \lambda
\end{vmatrix} = 0
$$

for λ and then for $(x_1, x_2\phi_{21}, x_3\phi_{31})$. We can then compute (x_1, x_2, x_3). We see, for example, that when each ϕ_{ij} is an automorphism of F, then $[\Phi]A$ has at most three distinct eigenvalues.

For a vector $(0) \neq (a_i) \in V$, we will refer to the *orbit of* $F(1, \Phi)$ *on* (a_i) as $(a_i)F(1, \Phi) = \{(a_i)b(1, \Phi) \mid b \in F\}$. Since $(a_i) = (a_i'\phi_{1i})$ for some $(0) \neq (a_i') \in V$, then $(a_i)b(1, \Phi) = (a_i'\phi_{1i})b(1, \Phi) = ((a_i'\phi_{1i})(b\phi_{1i})) = ((a_i'b)\phi_{1i})$. If $\Phi = (\phi_{ij})$ where each $\phi_{ij} = 1$, then $F(1, \Phi) = F$, and $(a_i)F(1, \Phi)$ is just the set of all scalar multiples of (a_i). If $V = \mathbf{R} \oplus \mathbf{R}$ and $z\phi_{12} = z^3$, then $(1, b)F(1, \Phi) = \{(x, bx^3) \mid x \in F\}$, the graph of $y = bx^3$, and $(0, 1)F(1, \Phi) = \{(0, y) \mid y \in F\}$, the y-axis.

If $f \in M_{F(1,\Phi)}(V)$ and $(0) \neq (a_i) \in V$, then $(a_i)f = (b_i)$ for some $(b_i) \in V$. Since f commutes with the elements of $F(1,\Phi)$, then f maps the elements of the orbit $(a_i)F(1,\Phi)$ onto the orbit $(b_i)F(1,\Phi)$. This is true in particular for the $[\Phi]A \in \mathcal{M}_{\Phi}(V,F) \subseteq M_{F(1,\Phi)}(V)$. This is analogous to matrices or linear transformations mapping a straight line through the origin to the origin or onto a straight line through the origin.

Should $(a_i) \in V$ be an eigenvector for $f \in M_{F(1,\Phi)}(V)$ with eigenvalue $\lambda(1,\Phi)$, then $[(a_i)b(1,\Phi)]f = [(a_i)f]b(1,\Phi) = (a_i)\lambda(1,\Phi) \circ b(1,\Phi) = (a_i)b(1,\Phi) \circ \lambda(1,\Phi)$. Hence, f leaves the orbit $(a_i)F(1,\Phi)$ fixed. If $V = \mathbf{R} \oplus \mathbf{R}$ and $z\phi_{12} = z^3$, then $(1,-8)$ is an eigenvector for $[\Phi]A$ where $A = \begin{pmatrix} 4 & 2 \\ -8 & -1 \end{pmatrix}$, and for eigenvalue $0 = 0(1,\Phi)$. So the orbit $(1,-8)F(1,\Phi) = \{(x,-8x^3) \mid x \in F\}$, the graph of $y = -8x^3$, is mapped onto (0). For each of the other three eigenvalues r_1, r_3, and r_4 of $[\Phi]A$, the vector $(1,(r_i-4)/2)$ is a corresponding eigenvector, and so $[\Phi]A$ maps each of the graphs of $y = [(r_i-4)/2]x^3$ onto itself, as well as mapping any cubic $y = mx^3$ onto some cubic $y = m'x^3$.

(14.52) Exploratory problem. Investigate the theory of eigenvalues and eigenvectors for these various $M_{F(1,\Phi)}(V)$, and especially for elements of $\mathcal{M}_{\Phi}(V,F)$. Are there applications to analysis, in general, and to differential equations, in particular?

We started in §14 with a semigroup $(\Phi,\circ)$ of endomorphisms of a group $(G,+)$. With this, we constructed the nearring $(M_{\Phi}(G),+,\circ)$. We remarked that these nearrings have been studied intensely, especially when $\Phi^* = \Phi \setminus \{\zeta\}$ is a regular group of automorphisms. But some regular groups of automorphisms give us Ferrero pairs (G,Φ^*) and planar nearrings $(G,+,\cdot)$. Specifically, we noted in (14.45) that our $(V,F^*(1,\Phi))$ are Ferrero pairs.

With a Ferrero pair (G,Φ^*), we can construct a planar nearring $(G,+,\cdot)$ and a centralizer nearring $(M_{\Phi}(G),+,\circ)$ where $\Phi = \Phi^* \cup \{\zeta\}$. How are these nearrings related? We shall now explore this, but in a more general setting. The following theorem has been extremely useful in constructing examples of nearrings.

(14.53) Theorem. *Let $(G,+)$ be a group and let $(\mathrm{End}\,G,\circ)$ denote the semigroup of endomorphisms of $(G,+)$. The multiplications $\cdot$ on G which are left distributive over $+$ are in one-to-one correspondence with mappings $\phi : G \to \mathrm{End}\,G$, and are defined by $a \cdot b = a \cdot_{\phi} b = b\phi_a$, where $a\phi = \phi_a$. These left distributive multiplications are associative exactly when $\phi_{b\phi_a} = \phi_b \circ \phi_a$ for all $a,b \in G$.*

Proof. For $\phi : G \to \mathrm{End}\,G$ and $a \cdot_{\phi} b = b\phi_a$, we have $a \cdot_{\phi} (b+c) = (b+c)\phi_a = b\phi_a + c\phi_a = (a \cdot_{\phi} b) + (a \cdot_{\phi} c)$. Certainly, $\phi_1 \neq \phi_2$ implies

$\cdot_{\phi_1} \neq \cdot_{\phi_2}$. Conversely, if $\cdot$ is left distributive over $+$, then $\phi : G \to \text{End}\,G$ defined by $a\phi = \phi_a$ where $b\phi_a = a \cdot b$ makes $\phi_a \in \text{End}\,G$. If $\cdot \neq \cdot'$ and $\cdot$ defines ϕ and $\cdot'$ defines ϕ', then $\phi \neq \phi'$. Also, given $\phi : G \to \text{End}\,G$ first, we construct $\cdot_\phi$, and $\cdot_\phi$ in turn defines ϕ. Similarly, given a left distributive $\cdot$ first, we construct $\phi : G \to \text{End}\,G$, which in turn defines $\cdot$. Hence, the one-to-one correspondence.

Suppose $\cdot$ is an associative left distributive multiplication. From $a\cdot(b\cdot c) = (a \cdot b) \cdot c$ for all $a, b, c \in G$, we obtain $(c\phi_b)\phi_a = c\phi_{b\phi_a}$, so $\phi_b \circ \phi_a = \phi_{b\phi_a}$ for all $a, b \in G$. Conversely, if $\phi_b \circ \phi_a = \phi_{b\phi_a}$ for all $a, b, c \in G$, then $a\cdot(b\cdot c) = (c\phi_b)\phi_a = c(\phi_b \circ \phi_a) = c\phi_{b\phi_a} = c\phi_{a\cdot b} = (a\cdot b)\cdot c$ for all $a, b, c \in G$, completing our proof.

(14.54) Corollary. *A nearring* $(G, +, \cdot)$ *defines a semigroup of endomorphisms* $(\Phi, \circ)$ *by* $a \mapsto \phi_a$, *where* $x\phi_a = a \cdot x$.

Proof. The condition $\phi_b \circ \phi_a = \phi_{b\phi_a}$ gives us closure, which is all we need.

So starting with a nearring $(G, +, \cdot)$, we obtain a semigroup of endomorphisms $(\Phi, \circ)$ from (14.54). If G is zero-symmetric, then $\zeta \in \Phi$. In any case, we then obtain the centralizer nearring $(M_\Phi(G), +, \circ)$. Since $(G, +)$ is an $M(G)$-module, and even an $M_0(G)$-module, one sees immediately that it is also an $M_\Phi(G)$-module.

But we can also let the elements of $(G, +, \cdot)$ act on $(M_\Phi(G), +, \circ)$ if $(\Phi, \circ)$ is commutative. For $a \in G$ and $f \in M_\Phi(G)$, define $a * f = f \circ \phi_a$. Hence, $* : G \times M_\Phi(G) \to M_\Phi(G)$. Now for $b \in G$ also, $(a\cdot b)*f = (b\phi_a)*f = f \circ \phi_{b\phi_a}$ and $a * (b * f) = (f \circ \phi_b) \circ \phi_a = f \circ (\phi_b \circ \phi_a)$. Since $\phi_{b\phi_a} = \phi_b \circ \phi_a$, we can conclude that $(a \cdot b) * f = a * (b * f)$ for all $a, b \in G$ and for all $f \in M_\Phi(G)$.

Now with $g \in M_\Phi(G)$ also, $a * (f + g) = (f + g) \circ \phi_a = \phi_a \circ (f + g) = (\phi_a \circ f) + (\phi_a \circ g) = (f \circ \phi_a) + (g \circ \phi_a) = (a * f) + (a * g)$. This makes $(M_\Phi(G), +)$ a G-comodule. So $(G, +)$ is a $M_\Phi(G)$-module and $(M_\Phi(G), +)$ is a G-comodule.

In addition, $a*(f \circ g) = (f \circ g)\circ\phi_a = f\circ(g\circ\phi_a) = (f\circ\phi_a)\circ g = (a*f)\circ g$, which makes $M_\Phi(G)$ look a lot like a G-nearalgebra, but of course, it isn't. Perhaps there will be good reason to expand the idea of (14.4).

(14.55) Definition. Let $(N, +, \cdot)$ be a (left) nearring as is $(A, +, \cdot)$ also. Suppose $(A, +)$ is a nearring N-comodule where $n(a \cdot b) = (na) \cdot b$ for all $n \in N$ and for all $a, b \in A$. Then we refer to $(A, +, \cdot)$ as an *N-nearcoalgebra* or a *nearcoalgebra over* N.

We can now summarize this last discussion with

(14.56) Theorem. *Let* $(G, +, \cdot)$ *be a (left) nearring which defines the commutative semigroup* $(\Phi, \circ)$ *of endomorphisms of* $(G, +)$ *as in* (14.54). *Then the left nearring* $(M_\Phi(G), +, \circ)$ *is a G-nearcoalgebra.*

(14.57) Corollary. *Let $(N, +, \cdot)$ be a planar nearring constructed from a Ferrero pair (N, Φ) where $(\Phi, \circ)$ is commutative. Then $(M_\Phi(N), +, \circ)$ is an N-nearcoalgebra.*

(14.58) Exploratory problem. What, if any, connections are there between the algebraic structure of a nearring $(M_\Phi(N), +, \circ)$ and the various geometric structures one can obtain from a planar nearring $(N, +, \cdot)$? In particular, suppose $(N, +, \cdot)$ is constructed from a Ferrero pair $(V, F^*(a, \Phi))$ of (14.45).

15. Distributively generated nearrings

A point of frequent emphasis in Chapter 3 was that pointwise addition of endomorphisms of an algebraic structure need not result in a function which is also an endomorphism. This privilege is generally reserved for endomorphisms of ring modules.

However, for an arbitrary group $(G, +)$, if $f, g \in M(G)$, then $f + g \in M(G)$, also. This is true even in the more symmetrical setting of $M_0(G)$. We have frequently referred to $(M(G), +, \circ)$ and $(M_0(G), +, \circ)$, and certainly $M_0(G)$ is a subnearring of $M(G)$ which contains all the endomorphisms of G.

If we write $f(x)$ for $f \in M(G)$ and $x \in G$, then we have $(f + g) \circ h = (f \circ h) + (g \circ h)$ for all $f, g, h \in M(G)$, and we cannot reasonable hope that $f \circ (g + h) = (f \circ g) + (f \circ h)$. However, $\mathrm{End}\,G = \{\alpha : G \to G \mid \alpha \text{ is an endomorphism}\}$ consists of exactly those $\alpha \in M(G)$ such that

$$\alpha \circ (g + h) = (\alpha \circ g) + (\alpha \circ h) \qquad (15 : 1)$$

for all $g, h \in M(G)$. To see this, let $a, b \in G$ be arbitrary, and let $\mathbf{a}$ and $\mathbf{b}$ denote the constant functions where $\mathbf{a}(x) = a$ and $\mathbf{b}(x) = b$ for all $x \in G$. Since $\alpha \circ (\mathbf{a} + \mathbf{b}) = (\alpha \circ \mathbf{a}) + (\alpha \circ \mathbf{b})$, we obtain $\alpha(a + b) = \alpha(a) + \alpha(b)$. Hence, $\alpha \in \mathrm{End}\,G$. Conversely, if $\alpha \in \mathrm{End}\,G$, then certainly (15:1) is valid. This helps motivate

(15.1) Definition. For a left nearring $(N, +, \cdot)$, if $d \in N$ and $(x + y)d = (xd) + (yd)$ for all $x, y \in N$, then d is a *distributive* element. (If $(N, +, \cdot)$ is a right nearring, then we would require $d(x + y) = (dx) + (dy)$.)

(15.2) Proposition. *For a group $(G, +)$, the distributive elements in $(M(G), +, \circ)$ are exactly the elements in $\mathrm{End}\,G$. Likewise, in $(M_0(G), +, \circ)$, the distributive elements are exactly the elements in $\mathrm{End}\,G$.*

Let $(\Phi, \circ)$ be a subsemigroup of $\mathrm{End}\,G$, where $(G, +)$ is a group. From (2.8), we know that the intersection of all subnearrings of $M_0(G)$ which contain Φ is a nearring. Immediately after (2.8) we saw that $D(\Phi, G)$ is a nearring and that $\Phi \subseteq D(\Phi, G)$. So these two nearrings are the same.

(15.3) Definition. Let $(\Phi, \circ)$ be a subsemigroup of $(\mathrm{End}\,G, \circ)$, the endo-morphisms of a group $(G, +)$. Then $D(\Phi, G)$ is the *endomorphism nearring generated by the semigroup* $(\Phi, \circ)$.

It has been usual to take Φ to be one of the semigroups are$\mathrm{End}\,G$, $\mathrm{Aut}\,G = \{\alpha \in \mathrm{End}\,G \mid \alpha$ is an automorphism$\}$, or $\mathrm{Inn}\,G = \{\alpha \in \mathrm{Aut}\,G \mid \alpha$ is an inner automorphism$\}$. As such, they have their special names and notation. First, $D(\mathrm{End}\,G, G) = E(G)$ is *the endomorphism nearring of* G, $D(\mathrm{Aut}\,G, G) = A(G)$ is *the automorphism nearring of* G, and $D(\mathrm{Inn}\,G, G) = I(G)$ is *the inner automorphism nearring of* G. Certainly, $I(G) \subseteq A(G) \subseteq E(G) \subseteq M_0(G)$.

For an abelian group $(G, +)$, we have $\mathrm{Inn}\,G = \{1_G\}$ so we will not obtain much for $I(G)$. However, it could easily be that $A(G) = E(G) = \mathrm{End}\,G$. Fuchs [LF] has raised the question as to when $A(G) = \mathrm{End}\,G$ for an abelian group $(G, +)$. In contrast, Fröhlich [F] has shown that $I(G) = A(G) = E(G) = M_0(G)$ if G is a finite simple group. A generalization of Fuchs' question is when does $A(G) = E(G)$ for any group? In light of the above-mentioned result of Fröhlich, there are numerous variations to this question. Meldrum [M] has shown that it is also reasonable to replace 'equal' by 'isomorphic' for these variations.

Since $E(G) = \mathrm{End}\,G$ for an abelian group $(G, +)$, the idea of $E(G)$ really only takes on meaning when G is nonabelian. Dihedral groups are nonabelian groups that come about as close to being abelian as any, and $E(G)$ has been thoroughly studied for the usual dihedral groups by J. J. Malone, Jr, C. Lyons, and J. D. P. Meldrum. Let $D_n = Z_n \times_\theta Z_2$ denote a finite dihedral group of order $2n$, and $D_\infty = Z \times_\theta Z_2$ denote an analogous infinite dihedral group. Malone and Lyons studied the $E(D_n)$ in two stages, first for n odd, and then for n even. They found some dramatic contrasts between the two cases. Meldrum studied $E(D_\infty)$.

In an attempt to extend the results of Malone, Lyons, and Meldrum, G. Grainger and the author have had a rewarding experience. It was decided to look at *generalized dihedral groups*, $D_B = B \times_\theta Z_2$, where B is an abelian group, and where $(a, x) \cdot (b, y) = (a + (-1)^x b, x + y)$ for $(a, x), (b, y) \in B \times_\theta Z_2$. These groups also come as close as any nonabelian group to being abelian. It was soon discovered that some basic properties of B had a powerful influence upon the structure of $I(D_B)$, $A(D_B)$ and $E(D_B)$. Considerable success was experienced for B with the property that $a \mapsto 2a$ is an isomorphism. We called such D_B *odd* generalized dihedral groups since the development was analogous to and included the D_n, with n odd. There seem to be four major cases for B which seem relevant. First, the odd generalized dihedral groups, second, the 'infinite' generalized dihedral groups where $a \mapsto 2a$ is injective but not surjective, third, those 'even' generalized dihedral groups where $a \mapsto 2a$ has a nontrivial kernel and whose image is a proper subgroup of B, and fourth, those 'even' generalized dihedral groups where $a \mapsto 2a$ has a nontrivial kernel and is surjective.

It can be profitable to study the results we obtained for such odd generalized dihedral groups. We expect that it could also be profitable to study the other three cases mentioned above.

15.1. Endomorphism nearrings of odd generalized dihedral groups

What is done here with these odd generalized dihedral groups could easily serve as a model for future investigations. As we shall see, the results obtained here for the odd D_B can be generalized in much the same way that H. Gonshor [G] generalized abstract affine nearrings. In fact, Gonshor's example will be followed on several occasions throughout these notes, even again here in §15 when we discuss formal group laws. Let us start out with

(15.4) Definition. Let $(B, +)$ be an abelian group where $a \mapsto 2a$ defines an automorphism of B. Let $D_B = B \times_\theta Z_2$, where $(a, x) \cdot (b, y) = (a + (-1)^x b, x + y)$ is a binary operation on $B \times Z_2$. We shall refer to such a group $(D_B, \cdot)$ as an *odd generalized dihedral group*.

Consider two affine transformations A and B of a vector space V into itself. Let $A(x) = a + L(x)$ and $B(x) = b + M(x)$, where $a, b \in V$ are constants and L and M are linear transformations of V into itself. So we can identify $A = (a, L)$ and $B = (b, M)$. Since $A \circ B(x) = a + L(b) + L \circ M(x)$, we identify $A \circ B = (a + Lb, LM)$ and this motivates the binary operation $(a, L)(b, M) = (a + Lb, LM)$. This results from writing the variable x to the right of A and B. If we write xA and xB, then $xA = a + xL$ and $xB = b + xM$. If we identify $A = (a, L)$ and $B = (b, M)$, then $A \circ B = (b + aM, LM)$ and this motivates the rather awkward binary operation $(a, L)(b, M) = (b + aM, LM)$. However, if we write $xA = xL + a$ and $xB = xM + b$ and identify $A = (L, a)$ and $B = (M, b)$, then since $xA \circ B = xL \circ M + aM + b$, we have $A \circ B = (LM, aM + b)$, which motivates the binary operation $(L, a)(M, b) = (LM, aM + b)$, which is just as nice as the first one. Here we have one idea, the composition of two affine transformations. Three different ways of expressing this single idea motivate three different binary operations. Since the second one seems awkward, the first and the third will be used. The point here is that choice of notation influences the appearance or presentation of the material. This is one of the reasons that we sometimes write the variable to the left of the function, as in xA, and sometimes to the right of the function, as in $A(x)$.

For groups $(A, +)$ and $(G, +)$, we have been writing $A \times_\theta G$ for a semidirect product where $\theta : G \rightarrow \text{Aut}A$ is a group homomorphism, and where $(a, x) \cdot (b, y) = (a + \theta(x)(b), x + y)$. We could just as nicely write $G_\theta \times A$ where $(x, a) \cdot (y, b) = (x + y, a(y\theta) + b)$. It is common practice to write xb for $\theta(x)(b)$

and similarly ay for $a(y\theta)$. Then we would have $(a, x) \cdot (b, y) = (a + xb, x + y)$ or $(x, a) \cdot (y, b) = (x + y, ay + b)$. Generally, the author prefers $A \times_\theta G$ since this notation seems to flow with the notation

$$0 \longrightarrow A \longrightarrow A \times_\theta G \rightleftarrows G \longrightarrow 0$$

for a split short exact sequence.

It is easy to see that the above ideas and notation can be extended to semigroups $(A, +)$ and $(G, +)$ where $\theta : G \to \mathrm{End}\,A$ is a semigroup homomorphism. Again, $A \times_\theta G$ corresponds to $(a, x) \cdot (b, y) = (a + \theta(x)(b), x + y)$ and $G \;_\theta \times A$ corresponds to $(x, a) \cdot (y, b) = (x + y, a(y\theta) + b)$. We make another remark about notation for semigroups $(S, \cdot)$: if S has a zero ζ, where $\zeta \cdot \alpha = \alpha \cdot \zeta = \zeta$ for all $\alpha \in S$, then $S^* = S \setminus \{\zeta\}$. Of course, it may be that $(S^*, \cdot)$ is a subsemigroup and has a zero ζ^*, but since $\zeta \neq \zeta^*$, there should be no problem.

There are numerous little facts about these odd generalized dihedral groups D_B which we shall use, often without reference. Since these facts are not frequently listed in popular sources, we collect them together to make

(15.5) Proposition. *Suppose that $(B, +)$ is an abelian group and that $(D_B, \cdot)$ is an odd generalized dihedral group. Then the following are valid.*
(1) $(b, x)^{-1} = ((-1)^x(-b), x)$.
(2) For $(b, 1) \in D_B$, the order of $(b, 1)$ is $|(b, 1)| = 2$.
(3) For $b \in B \setminus \{0\}$, $|(b, 0)|$ is not even, and $|(0, 0)| = 1$ since $(0, 0)$ is the identity of D_B.
(4) D_B has a trivial centre.
(5) Each subgroup H of B is identified with a subgroup $H = \{(a, 0) \mid a \in H\}$ of D_B. Other than D_B, these are the only normal subgroups of D_B.
(6) Let $\bar{\zeta}$ denote the zero endomorphism of D_B. If $\bar{\beta} \in (\mathrm{End}\,D_B)^*$, then there is a unique $\beta \in \mathrm{End}\,B$ and there is a unique $b \in B$ such that $\bar{\beta} = [\beta, (b, 1)]$, where $[\beta, (b, 1)] : D_B \to D_B$ is defined by $(c, 0)[\beta, (b, 1)] = (c\beta, 0)$ and $(c, 1)[\beta, (b, 1)] = (c\beta + b, 1)$. Conversely, for $(\beta, b) \in (\mathrm{End}\,B) \times B$ and $\bar{\beta} = [\beta, (b, 1)]$ as defined above, we obtain $\bar{\beta} \in (\mathrm{End}\,D_B)^*$. This correspondence is bijective.
(7) We have $[\beta, (b, 1)] \in \mathrm{Aut}\,D_B$ if and only if $\beta \in \mathrm{Aut}\,B$.
(8) The inner automorphism of D_B defined by (b, x) is

$$[(-1)^x, ((-1)^x(-2b), 1)].$$

For each $b \in B$ and $x \in Z_2$, we have that $[(-1)^x, (b, 1)]$ is an inner automorphism. (Of course, $(-1)^0 = 1_B$ and $(-1)^1 = -1_B$.)
(9) $[\alpha, (a, 1)] \circ [\beta, (b, 1)] = [\alpha \circ \beta, (a\beta + b, 1)]$.

(10) $((\mathrm{End}D_B)^*, \circ)$ *is isomorphic to* $(\mathrm{End}B \,_\theta \times B, \circ)$.
(11) $\mathrm{Aut}D_B$ *is isomorphic to* $\mathrm{Aut}B \,_\theta \times B$.
(12) $|\mathrm{End}D_B| = |\mathrm{End}B| \cdot |B| + 1$.
(13) $|\mathrm{Aut}D_B| = |\mathrm{Aut}B| \cdot |B|$.
(14) $|\mathrm{Inn}D_B| = |D_B| = 2|B|$.
(15) *If* $B = Z_n$ *and* n *is odd, then* $|\mathrm{End}D_n| = n^2 + 1$, $|\mathrm{Aut}D_n| = (n\phi) \cdot n$ *where* ϕ *is the Euler* ϕ-*function, and* $|\mathrm{Inn}D_n| = 2n$
(16) *For each* $b \in B$, *the equation* $2x = b$ *has a unique solution for* $x \in B$.
(17) *If* $b = 2a$, *then* $bh = a$ *defines* $h \in \mathrm{Aut}B$.
(18) $2h = 1_B$.
(19) *If* $2h' = 1_B$ *and* $h' \in \mathrm{End}B$, *then* $h = h'$.
(20) *The map* $\alpha \mapsto 2\alpha$ *is an automorphism of* $(\mathrm{End}B, +)$.
(21) $h \circ \alpha = \alpha \circ h$ *for each* $\alpha \in \mathrm{End}B$.

Proof. Certainly, $(0,0)$ is the identity, and $(b,x)((-1)^x(-b),x) = (b + (-1)^x(-1)^x(-b), x + x) = (0,0)$, so $(b,x)^{-1} = ((-1)^x(-b), x)$. Also note $(b,1)(b,1) = (0,0)$, and $(b,0)(b,0) = (2b,0) \neq (0,0)$ if $b \neq 0$. Hence, (1), (2), and (3).

For $(a,0), (b,1) \in D_B$, we have $(a,0)(b,1) = (a+b,1)$ and $(b,1)(a,0) = (b-a,1)$. So if $a \neq 0$, then neither $(a,0)$ nor $(b,1)$ can be in the centre of D_B, hence D_B has centre $\{(0,0)\}$, and we have (4).

Let H be a normal subgroup of D_B with $(b,1) \in H$. For $a \in B$, we have $(a,0)^{-1}(b,1)(a,0) = (b-2a,1)$. Since $c \mapsto 2c$ is an automorphism and $(b,1)(b-2a,1) = (2a,0) \in H$, we have $B \subset H$ properly, which forces $H = D_B$ since $[D_B : B] = 2$. So, take H to be a subgroup of B and identify it with $H = \{(a,0) \mid a \in H\}$. Then $(b,x)^{-1}(a,0)(b,x) = ((-1)^x(-b),x)(a+b,x) = ((-1)^x a, 0) \in H$, so H is normal in D_B. We now have (5).

Turning our attention now to (6), let $\overline{\beta} \in (\mathrm{End}D_B)^*$. Then $(c,0)\overline{\beta}$ cannot have order 2, so $(c,0)\overline{\beta} = (c\beta, 0)$ for some map $\beta : B \to B$. Also, $(0,1)\overline{\beta}$ must have order 1 or 2, so $(0,1)\overline{\beta} = (0,0)$ or $(0,1)\overline{\beta} = (b,1)$ for some $b \in B$. If $(0,1)\overline{\beta} = (0,0)$, then $(0,1) \in \ker\overline{\beta}$, a normal subgroup, and so $\ker\overline{\beta} = D_B$ by 5), and so $\overline{\beta} = \zeta$. But, $\overline{\beta} \in (\mathrm{End}D_B)^*$. So we are left with $(0,1)\overline{\beta} = (b,1)$ for some $b \in B$. With $c, c' \in B$, $(c+c',0)\overline{\beta} = ((c+c')\beta, 0)$ and $(c,0)(c',0) = (c+c',0)$, so $(c+c',0)\overline{\beta} = [(c,0)(c',0)]\overline{\beta} = (c,0)\overline{\beta}(c',0)\overline{\beta} = (c\beta, 0)(c'\beta, 0) = (c\beta + c'\beta, 0)$. Hence, $\beta \in \mathrm{End}B$. Also $(c,1) = (c,0)(0,1)$, so $(c,1)\overline{\beta} = (c,0)\overline{\beta}(0,1)\overline{\beta} = (c\beta, 0)(b,1) = (c\beta + b, 1)$. Hence $\overline{\beta} = [\beta, (b,1)]$, as stated.

Conversely, take $(\beta, b) \in (\mathrm{End}B) \times B$ and define $\overline{\beta} = [\beta, (b,1)]$ accordingly. Then $[(c,x)(c',x')]\overline{\beta} = (c + (-1)^x c', x + x')\overline{\beta}$ is to be compared to $(c,x)\overline{\beta}(c',x')\overline{\beta}$. There are four cases to consider defined by $(x,x') \in \{(0,0), (0,1), (1,0), (1,1)\}$ In each of these four cases for (x,x'), one obtains that $[(c,x)(c',x')]\overline{\beta} = (c,x)\overline{\beta}(c',x')\overline{\beta}$, so $\overline{\beta} \in (\mathrm{End}D_B)^*$. Certainly, $\overline{\beta} \longleftrightarrow [\beta, (b,1)]$ is a bijection. This yields (6).

If $\overline{\beta} \in \operatorname{Aut}D_B$ and $\ker \beta \neq \{0\}$, then for $0 \neq c \in \ker \beta$, $(c,0)\overline{\beta} = (0,0)$. Hence, $\overline{\beta} \in \operatorname{Aut}D_B$ implies $\beta \in \operatorname{Aut}B$. Conversely, if $\beta \in \operatorname{Aut}B$, $b \in B$, and $\overline{\beta} = [\beta, (b,1)]$, then $(c,0)\overline{\beta} = (c\beta, 0) = (0,0)$ implies $c = 0$. So no element $(c,0) \neq (0,0)$ can be in the kernel of $\overline{\beta}$, a normal subgroup of D_B. By (5), $\ker \overline{\beta} = \{(0,0)\}$, so $\overline{\beta} \in \operatorname{Aut}D_B$. Now we have (7).

Now $(b,x)^{-1}(c,0)(b,x) = ((-1)^x c, 0)$ and $(b,x)^{-1}(c,1)(b,x) = ((-1)^x c + (-1)^x(-2b), 1)$. From (6), we see that (b,x) defines the inner automorphism $[(-1)^x, ((-1)^x(-2b), 1)]$. Since $c \mapsto 2c$ is an automorphism, each $[(-1)^x, (b,1)]$ occurs as an inner automorphism. Hence, (8).

For $[\alpha, (a,1)], [\beta, (b,1)] \in (\operatorname{End}D_B)^*$, note that $(c,0)[\alpha, (a,1)] \circ [\beta, (b,1)] = (c\alpha, 0)[\beta, (b,1)] = (c\alpha\beta, 0)$, and that $(c,1)[\alpha, (a,1)] \circ [\beta, (b,1)] = (c\alpha + a, 1)[\beta, (b,1)] = ((c\alpha + a)\beta + b, 1) = (c\alpha\beta + a\beta + b, 1)$. Hence, $[\alpha, (a,1)] \circ [\beta, (b,1)] = [\alpha\beta, (a\beta + b, 1)]$, which is (9).

Define $F : (\operatorname{End}D_B)^* \to \operatorname{End}B_\theta \times B$ by $F[\alpha, (a,1)] = (\alpha, a)$. By (6), F is a bijection, and by (9), F is a homomorphism. Hence, (10). Now restrict F to $\operatorname{Aut}D_B$. From (7) we obtain (11). From (10) we obtain (12), and from (11) we obtain (13). Apply (8) to obtain (14). Combine (12), (13), and (14) to obtain (15).

Since $c \mapsto 2c$ defines an automorphism of B, we know that each $a \in B$ has a unique $b \in B$ such that $a = 2b$. Hence, (16). As $(2c)h = c$, we have that h is the inverse of $c \mapsto 2c$, so it is an automorphism. Hence, (17). Now $(2c)(h+h) = (2c)h + (2c)h = c + c = 2c$, so $h + h = 2h = 1_B$, and we obtain (18). If $2h' = 1_B = 2h$, then $b(2h) = b(h+h) = bh + bh = b1_B = b$, as well as $b(2h') = b$. Hence, $bh + bh = bh' + bh'$. Since $c \mapsto 2c$ is an automorphism of B, we obtain $bh = bh'$ for all $b \in B$, hence $h = h'$, and this is (19).

Now $1_B = h + h$ implies that $\alpha = \alpha \circ 1_B = (\alpha \circ h) + (\alpha \circ h)$, and $\alpha = 1_B \circ \alpha = (h \circ \alpha) + (h \circ \alpha)$. If $\alpha = \beta + \beta = \beta' + \beta'$, then for $b \in B$, we obtain $b\alpha = b\beta + b\beta = b\beta' + b\beta'$, and so $2(b\beta) = 2(b\beta')$ for each $b \in B$, which forces $b\beta = b\beta'$ for each $b \in B$, and hence $\beta = \beta'$. This makes $\alpha \circ h = h \circ \alpha$, which is (21), and since $\alpha \in \operatorname{End}B$ itself is arbitrary, we have that $\alpha \mapsto 2\alpha$ is an automorphism of $(\operatorname{End}B, +)$. This is (20), and this completes the proof of (15.5).

The $[\beta, (b,1)]$ of (15.5) provide a quick way to describe and manipulate the elements of $(\operatorname{End}D_B)^*$. We will be using some or all of these to describe various endomorphism nearrings of different odd generalized dihedral groups D_B. It will be to our advantage to extend this notation to include other maps of D_B into itself. For $\alpha, \beta \in \operatorname{End}B$, $b, c \in B$, and $x, y \in Z_2$, we define $[\alpha, \beta, b, x] : D_B \to D_B$ by

$$(c,y)[\alpha, \beta, b, x] = \begin{cases} (c\alpha, 0), & \text{if } y = 0; \\ (c\beta + b, x), & \text{if } y = 1. \end{cases} \tag{15:2}$$

With this notation, we have

$$[\alpha, \alpha, b, 1] = [\alpha, (b, 1)] \tag{15 : 3}$$

and

$$[\zeta, \zeta, 0, 0] = \overline{\zeta}. \tag{15 : 4}$$

It will be convenient to have some rules to manipulate these mappings $[\alpha, \beta, b, x]$.

(15.6) Proposition. *For* $\alpha, \beta, \gamma, \delta \in \operatorname{End}B$, $a, b \in B$, *and* $x, y \in Z_2$, *we have:*

(1) $[\alpha, \beta, a, x] + [\gamma, \delta, b, y] = [\alpha + \gamma, \beta + (-1)^x \delta, a + (-1)^x b, x + y]$.

(2) $[\alpha, \beta, a, x] \circ [\gamma, \delta, b, y] = \begin{cases} [\alpha \circ \gamma, \beta \circ \gamma, a\gamma, 0], & \text{if } x = 0; \\ [\alpha \circ \gamma, \beta \circ \delta, a\delta + b, y], & \text{if } x = 1. \end{cases}$

(3) $-[\alpha, \beta, a, x] = [-\alpha, (-1)^x(-\beta), (-1)^x(-a), x]$.

(4) *for* $0 \le k \in Z$ *and* $\alpha_i, \beta_i \in \operatorname{End}B$, $i = 0, 1, \dots, k$,

$$\sum_{i=0}^{k} [\alpha_i, \beta_i, 0, 1] = \left[\sum_{i=0}^{k} \alpha_i, \sum_{i=0}^{k} (-1)^i \beta_i, \ 0, \ (k+1) \bmod 2 \right].$$

Proof. The proof will be direct. First we will take $(c, 0), (c, 1) \in D_B$. Then $(c, 0)\{[\alpha, \beta, a, x] + [\gamma, \delta, b, y]\} = (c\alpha, 0) \cdot (c\gamma, 0) = (c(\alpha + \gamma), 0)$, and similarly, $(c, 1)\{[\alpha, \beta, a, x] + [\gamma, \delta, b, y]\} = (c\beta + a, x)(c\delta + b, y) = (c\beta + a + (-1)^x(c\delta + b), x + y) = (c(\beta + (-1)^x \delta) + a + (-1)^x b, x + y)$, and we have (1). With (1) and (15:4) we see that $[\alpha, \beta, a, x] + [-\alpha, (-1)^x(-\beta), (-1)^x(-a), x] = [\alpha + (-\alpha), \beta + (-1)^x(-1)^x(-\beta), a + (-1)^x(-1)^x(-a), x + x] = [\zeta, \zeta, 0, 0]$. Now we have (3).

Continuing, $(c, 0)[\alpha, \beta, a, 0] \circ [\gamma, \delta, b, y] = (c\alpha, 0)[\gamma, \delta, b, y] = (c\alpha\gamma, 0)$, and $(c, 1)[\alpha, \beta, a, 0] \circ [\gamma, \delta, b, y] = (c\beta + a, 0)[\gamma, \delta, b, y] = ((c\beta + a)\gamma, 0) = (c\beta\gamma + a\gamma, 0)$. Hence, $[\alpha, \beta, a, 0] \circ [\gamma, \delta, b, y] = [\alpha\gamma, \beta\gamma, a\gamma, 0]$. When $x = 1$, we obtain $(c, 0)[\alpha, \beta, a, 1] \circ [\beta, \delta, b, y] = (c\alpha\gamma, 0)$ just as before, but $(c, 1)[\alpha, \beta, a, 1] \circ [\gamma, \delta, b, y] = (c\beta + a, 1)[\gamma, \delta, b, y] = ((c\beta + a)\delta + b, y) = (c\beta\delta + a\delta + b, y)$. Hence, $[\alpha, \beta, a, 1] \circ [\gamma, \delta, b, y] = [\alpha\gamma, \beta\delta, a\delta + b, y]$, and we now have (2).

We will obtain (4) with an easy induction. To begin, take $k = 0$. Then $[\sum_{i=0}^{k} \alpha_i, \sum_{i=0}^{k}(-1)^i \beta_i, 0, (k+1) \bmod 2] = [\alpha_0, \beta_0, 0, 1] = \sum_{i=0}^{k}[\alpha_i, \beta_i, 0, 1]$. Now assume $\sum_{i=0}^{k}[\alpha_i, \beta_i, 0, 1] = [\sum_{i=0}^{k} \alpha_i, \sum_{i=0}^{k}(-1)^i \beta_i, 0, (k+1) \bmod 2]$ for $k \ge 0$, and make the induction step with $k+1$, with $\sum_{i=0}^{k+1}[\alpha_i, \beta_i, 0, 1] = \sum_{i=0}^{k}[\alpha_i, \beta_i, 0, 1] + [\alpha_{k+1}, \beta_{k+1}, 0, 1] = [\sum_{i=0}^{k} \alpha_i, \sum_{i=0}^{k}(-1)^i \beta_i, 0, (k + 1) \bmod 2] + [\alpha_{k+1}, \beta_{k+1}, 0, 1] = [\sum_{i=0}^{k} \alpha_i + \alpha_{k+1}, \sum_{i=0}^{k}(-1)^i \beta_i + (-1)^{k+1}\beta_{k+1}, 0, (k + 1 + 1) \bmod 2] = [\sum_{i=0}^{k+1} \alpha_i, \sum_{i=0}^{k+1}(-1)^i \beta_i, 0, [(k + 1) + 1] \bmod 2]$, and now we are finished with (15.6).

The endomorphisms $[-1_B, (0,1)]$ and $[h, (0,1)]$ will play an important role in our study of endomorphism nearrings of an odd generalized dihedral group D_B. There will also be a lot of messy calculations. In order to focus more on the structure of the proof of (15.9) and to highlight the roles of $[-1_B, (0,1)]$ and $[h, (0,1)]$, we isolate

(15.7) Lemma. *Let D_B be an odd generalized dihedral group with h as in (15.5). Let $(S', \circ)$ be a subsemigroup of $(\mathrm{End}B, \circ)$, and let $S = S' {}_\theta \times B = \{[\alpha, (b,1)] \mid \alpha \in S', b \in B\}$. If either $h \in S'$ or $-1_B \in S'$ and $h \in D(S', B)$, then $[\zeta, (0,1)]$, $[h, (0,1)]$, $[-1_B, (0,1)]$, and $[1_B, (0,1)]$ all belong to $D(S, D_B)$. Furthermore, if either condition is satisfied, then the maps* (1) $M^\alpha = [\alpha, \alpha, 0, 0]$, (2) $\Lambda^\alpha = [\alpha, -\alpha, 0, 0]$, (3) $\Pi_1^\alpha = [\alpha, \zeta, 0, 0]$, *and* (4) $\Pi_2^\alpha = [\zeta, \alpha, 0, 0]$, *for $\alpha \in D(S', B)$, also belong to $D(S, D_B)$.*

Proof. First, note that $1_B = 1_B^2 = (h + h)^2 = h^2 + 2h^2 + h^2 = 2h^2 + 2h^2$. Consequently, $2h^2 = h$ by (19) of (15.5). For $h \in S'$, we have $[h, (0,1)] \in D(S, D_B)$, and so $2\{-[h, (0,1)] + [h, (0,1)]^2\} = f \in D(S, D_B)$. But $f = 2\{-[h, h, 0, 1] + [h \circ h, (0h + 0, 1)]\} = 2\{-[-h, (-1)^1(-h), 0, 1] + [h^2, (0,1)]\} = 2\{[-h, h, 0, 1] + [h^2, h^2, 0, 1]\} = 2[-h + h^2, h + (-1)^1 h^2, 0, 0] = [-h + h^2, h - h^2, 0, 0] + [-h + h^2, h - h^2, 0, 0] = [-2h + 2h^2, 2h - 2h^2, 0, 0] = [-h, h, 0, 1]$, hence

$$[-h, h, 0, 0] \in D(S, D_B). \qquad (15:5)$$

Now consider the sum $[h, (0,1)] + [-h, h, 0, 0] = [h, h, 0, 1] + [-h, h, 0, 0] = [\zeta, h + (-1)^1 h, 0, 1] = [\zeta, \zeta, 0, 1] = [\zeta, (0,1)]$, so we obtain

$$[\zeta, (0,1)] \in D(S, D_B). \qquad (15:6)$$

Also, consider the sum $[h, (0,1)] - [-h, h, 0, 0] = [h, h, 0, 1] + [h, -h, 0, 0] = [2h, h + (-1)^1(-h), 0, 1] = [1_B, 1_B, 0, 1] = [1_B, (0,1)]$, and so

$$[1_B, (0,1)] \in D(S, D_B). \qquad (15:7)$$

Finally, $[h, (0,1)] + 3[-h, h, 0, 0] = [h, h, 0, 1] + [-h, h, 0, 0] + [-h, h, 0, 0] + [-h, h, 0, 0] = [\zeta, \zeta, 0, 1] + [-2h, 2h, 0, 0] = [-2h, \zeta + (-1)^1(2h), 0, 1] = [-1_B, -1_B, 0, 1] = [-1_B, (0,1)]$. This makes

$$[-1_B, (0,1)] \in D(S, D_B). \qquad (15:8)$$

So, from (15:6), (15:7), and (15:8), we know that if $h \in S'$, then $[\zeta, (0,1)]$, $[h, (0,1)], [-1_B, (0,1)], [1_B, (0,1)] \in D(S, D_B)$.

Next we start with $-1_B \in S'$ and $h \in D(S', B)$. So, immediately, we obtain $[-1_B, (0,1)] \in D(S, D_B)$, and $h = \sum_{i=0}^k \alpha_i$, with $\alpha_i \in S'$, since $-1_B \in S'$. With each $\alpha_i \in S'$, we have each $[\alpha_i, (0,1)] \in D(S, D_B)$. For any

$\alpha \in S'$, we have $[\alpha, (0, 1)] \in D(S, D_B)$, and $-\{[-1_B, (0, 1)] \circ [\alpha, (0, 1)]\} =$ $-[-\alpha, (0, 1)] = -[-\alpha, -\alpha, 0, 1] = [\alpha, (-1)^1 \alpha, 0, 1] = [\alpha, -\alpha, 0, 1]$. So

$$[\alpha, -\alpha, 0, 1] \in D(S, D_B). \qquad (15:9)$$

For k an even integer, we have the sum of mappings $\sum_{i=0}^{k} [\alpha_i, (-1)^i \alpha_i, 0, 1]$ $= [\sum_{i=0}^{k} \alpha_i, \sum_{i=0}^{k} (-1)^i (-1)^i \alpha_i, 0, (k+1) \bmod 2] = [h, h, 0, 1] = [h, (0, 1)]$. For k an odd integer, the sum $-\{\sum_{i=0}^{k} [\alpha_i, (-1)^i \alpha_i, 0, 1]\} + [1_B, (0, 1)] =$ $-[\sum_{i=0}^{k} \alpha_i, \sum_{i=0}^{k} (-1)^{2i} \alpha_i, 0, (k+1) \bmod 2] + [1_B, (0, 1)] = -[h, h, 0, 0] +$ $[1_B, 1_B, 0, 1] = [-h, -h, 0, 0] + [1_B, 1_B, 0, 1] = [-h + 2h, -h + 2h, 0, 1] =$ $[h, h, 0, 1] = [h, (0, 1)]$. In any case we have

$$[h, (0, 1)] \in D(S, D_B). \qquad (15:10)$$

With $-1_B \in S'$, we surely have $1_B \in S'$, and so

$$[1_B, (0, 1)] \in D(S, D_B). \qquad (15:11)$$

We only need $[\zeta, (0, 1)] \in D(S, D_B)$. But, with $[h, (0, 1)] \in D(S, D_B)$, we proceed exactly as in the case with $h \in S'$ to obtain $f = [-h, h, 0, 1] \in D(S, D_B)$ and then to obtain $[\zeta, (0, 1)] \in D(S, D_B)$. (See the development for (15:5) and for (15:6).)

We can be assured, in either case, that $[\zeta, (0, 1)], [h, (0, 1)], [-1_B, (0, 1)]$, $[1_B, (0, 1)] \in D(S, D_B)$. We proceed now to attain the maps of (1), (2), (3), and (4) in $D(S, D_B)$.

Suppose $\alpha = \sum_{i=0}^{k} \epsilon_i \alpha_i \in D(S', B)$, where $\epsilon_i = \pm 1$ and $\alpha_i \in S'$. For $\alpha_i \in S'$, $[\alpha_i, (0, 1)] \in D(S, D_B)$, and since $[-1_B, (0, 1)] \in D(S, D_B)$, we have $[-1_B, (0, 1)] \circ [\alpha_i, (0, 1)] = [-\alpha_i, (0, 1)] \in D(S, D_B)$. Hence, $[\epsilon_i \alpha_i, (0, 1)] =$ $[\epsilon_i \alpha_i, \epsilon_i \alpha_i, 0, 1] \in D(S, D_B)$. With $[-1_B, (0, 1)] \in D(S, D_B)$, the proof of (15:9) shows that $[\epsilon_i \alpha_i, -\epsilon_i \alpha_i, 0, 1] \in D(S, D_B)$. So

$$[\epsilon_i \alpha_i, (-1)^i \epsilon_i \alpha_i, 0, 1], [\epsilon_i \alpha_i, (-1)^{i+1} \epsilon_i \alpha_i, 0, 1] \in D(S, D_B). \qquad (15:12)$$

For k odd, $\sum_{i=0}^{k} [\epsilon_i \alpha_i, (-1)^i \epsilon_i \alpha_i, 0, 1] = [\sum_{i=0}^{k} \epsilon_i \alpha_i, \sum_{i=0}^{k} (-1)^i (-1)^i \epsilon_i \alpha_i, 0,$ $(k+1) \bmod 2] = [\alpha, \alpha, 0, 0] = M^\alpha$. For k even, $\sum_{i=0}^{k} [\epsilon_i \alpha_i, (-1)^i \epsilon_i \alpha_i, 0, 1] +$ $[\zeta, (0, 1)] = [\alpha, \alpha, 0, 1] + [\zeta, \zeta, 0, 1] = [\alpha + \zeta, \alpha + (-1)^1 \zeta, 0, 0] = [\alpha, \alpha, 0, 0] =$ M^α. So, in any event, each $M^\alpha \in D(S, D_B)$.

For k odd, we take the sum of mappings $\sum_{i=0}^{k} [\epsilon_i \alpha_i, (-1)^{i+1} \epsilon_i \alpha_i, 0, 1] =$ $[\sum_{i=0}^{k} \epsilon_i \alpha_i, (-1) \sum_{i=0}^{k} (-1)^i (-1)^i \epsilon_i \alpha_i, 0, (k+1) \bmod 2] = [\alpha, -\alpha, 0, 0] =$ Λ^α, and for k even, $\sum_{i=0}^{k} [\epsilon_i \alpha_i, (-1)^{i+1} \epsilon_i \alpha_i, 0, 1] + [\zeta, (0, 1)] = [\alpha, -\alpha, 0, 1] +$ $[\zeta, \zeta, 0, 1] = [\alpha + \zeta, -\alpha + (-1)^1 \zeta, 0, 0] = [\alpha, -\alpha, 0, 0] = \Lambda^\alpha$. So we have $\Lambda^\alpha \in D(S, D_B)$.

From $(M^\alpha + \Lambda^\alpha) \circ [h, (0,1)] = \{[\alpha, \alpha, 0, 0] + [\alpha, -\alpha, 0, 0]\} \circ [h, h, 0, 1] =$
$[2\alpha, \zeta, 0, 0] \circ [h, h, 0, 1] = [(2\alpha)h, \zeta, 0, 0] = [\alpha h + \alpha h, \zeta, 0, 0] = [\alpha, \zeta, 0, 0] =$
Π_1^α, and $(M^\alpha - \Lambda^\alpha) \circ [h, (0,1)] = \{[\alpha, \alpha, 0, 0] - [\alpha, -\alpha, 0, 0]\} \circ [h, (0,1)] =$
$[\zeta, 2\alpha, 0, 0] \circ [h, h, 0, 1] = [\zeta, \alpha, 0, 0] = \Pi_2^\alpha$, so we have $\Pi_1^\alpha, \Pi_2^\alpha \in D(S, D_B)$,
and we come to an end of the proof of (15.7).

Now call (3.17) into service. For a (left) nearring $(N, +, \cdot)$, we have
$N^+ = \mathrm{Ann}(e) + eN$, a semidirect product of the normal subgroup $\mathrm{Ann}(e)$
with the subgroup eN of $N^+ = (N, +)$, where $e \in N$ and $e^2 = e$, that is, e
is an idempotent.

(15.8) Proposition. *Let $(N, +, \cdot)$ be a nearring whose additive group
$(N, +)$ is generated, using $+$, by the set of a semigroup $(S, \cdot)$ of distributive
elements. Let $e \in N$ be an idempotent. Then eN is the group generated
by $\{ed \mid d \in S\}$ using $+$, and $\mathrm{Ann}(e)$ is the normal subgroup generated by
$\{d - ed \mid d \in S\}$ using $+$.*

Proof. Certainly, $\{ed \mid d \in S\} \subseteq eN$ and $\{d - ed \mid d \in S\} \subseteq \mathrm{Ann}(e)$.
For $n = \sum_{i=0}^k \epsilon_i d_i$, where $\epsilon_i = \pm 1$, $d_i \in S$, then $e(\epsilon_i d_i) = \epsilon_i(ed_i)$, so
$en = \sum_{i=0}^k e(\epsilon_i d_i) = \sum_{i=0}^k \epsilon_i(ed_i)$. Hence, $\{ed \mid d \in S\}$ generates the
subgroup eN using $+$.

Let T be the normal subgroup of N^+ generated by $\{d - ed \mid d \in S\}$,
using $+$. Then $T \subseteq \mathrm{Ann}(e)$. Now $T + eN \subseteq \mathrm{Ann}(e) + eN = N$. Also, if
$d \in S$, then $d = (d - ed) + ed \in T + eN$, so $S \subseteq T + eN$, and $T + eN$ is
a subgroup of N^+. Since S generates N^+, using $+$, we have $N \subseteq T + eN$,
and so $N = T + eN$. Hence, $\mathrm{Ann}(e) \subseteq T + eN$. Now $\mathrm{Ann}(e) \cap eN = \{0\}$,
so for $x \in \mathrm{Ann}(e)$, $x = t + en$ for some $t \in T$ and $en \in eN$. This makes
$-t + x = en \in \mathrm{Ann}(e) \cap eN = \{0\}$, so $t = x$. Hence, $\mathrm{Ann}(e) \subseteq T \subseteq \mathrm{Ann}(e)$,
which forces $T = \mathrm{Ann}(e)$, and this is what we want to prove.

(15.9) Theorem. *Let D_B be an odd generalized dihedral group with h as
in (15.5). Let $(S', \circ)$ be a subsemigroup of $(\mathrm{End}B, \circ)$ and let $S = S'_{\ \theta} \times B =
\{[\alpha, (b, 1)] \mid \alpha \in S', b \in B\}$. If either $h \in S'$ or $(-1_B, h) \in S' \times D(S', B)$,
then the following are true.*
(1) *S is a subsemigroup of $(\mathrm{End}D_B, \circ)$.*
(2) *$[\zeta, (0,1)] \in D_B$ is an idempotent. So let $M_{[\zeta, (0,1)]} = [\zeta, (0,1)]D(S, D_B)$
and $A_{[\zeta, (0,1)]} = \mathrm{Ann}([\zeta, (0,1)])$.*
(3) *$M_{[\zeta, (0,1)]}^+$ is isomorphic to D_B as groups.*
(4) *$A_{[\zeta, (0,1)]}^+ \cong D(S', B)^+ \oplus D(S', B)^+$.*
(5) *$D(S, D_B)^+ \cong D(S', B)^+ \oplus [(D(S', B)^+ \oplus B) \times_\theta Z_2]$.*
(6) *The elements of $D(S, D_B)$ are exactly the mappings $[\alpha, \beta, b, x]$ where
$\alpha, \beta \in D(S', B)$, $b \in B$, and $x \in Z_2$.*
(7) *Addition and multiplication and additive inverses are exactly as defined
in (15.6).*

Proof. The elements of S are mappings of D_B into D_B, so associativity is immediate, and closure follows from (9) of (15.5). Also from (9) of (15.5), we see that $[\zeta, (0,1)]$ is an idempotent. Now we have (1) and (2). From (15.8), we know that $M_{[\zeta,(0,1)]}$ is generated by all $[\zeta, (0,1)] \circ [\alpha, (b,1)] = [\zeta, (b,1)] = [\zeta, \zeta, b, 1]$, where $\alpha \in S'$ and $b \in B$. So, an arbitrary element $f \in M_{[\zeta,(0,1)]}$ is $f = \sum_{i=0}^{k} \epsilon_i [\zeta, \zeta, b_i, 1]$, where each $\epsilon_i = \pm 1$. Hence $(b,0)f = (0,0)$ and $(b,1)f = \prod_{i=0}^{k}(b_i, 1)^{\epsilon_i} = (\sum_{i=0}^{k}(-1)^i b_i, (k+1) \bmod 2)$. This makes $f = [\zeta, \zeta, \sum_{i=0}^{k}(-1)^i b_i, (k+1) \bmod 2]$. So $M_{[\zeta,(0,1)]} = \{[\zeta, \zeta, b, x] \mid (b,x) \in D_B\}$. The map $[\zeta, \zeta, b, x] \mapsto (b,x)$ is easily an isomorphism by (15.6). Now we have (3).

The proof of (4) requires a fresh dose of patience. The subgroup $A_{[\zeta,(0,1)]}$ is the *normal* subgroup generated by all $[\alpha, (b,1)] - [\zeta, (0,1)] \circ [\alpha, (b,1)]$ for $\alpha \in S'$ and $b \in B$, because of (15.8). But $[\alpha, (b,1)] - [\zeta, (0,1)] \circ [\alpha, (b,1)] = [\alpha, (b,1)] - [\zeta, (b,1)] = [\alpha, \alpha, b, 1] - [\zeta, \zeta, b, 1] = [\alpha, \alpha, b, 1] + [\zeta, \zeta, b, 1] = [\alpha, \alpha, 0, 0] = M^\alpha$. So $A_{[\zeta,(0,1)]}$ is the normal subgroup generated by $\{M^\alpha \mid \alpha \in S'\}$. Since $M^\alpha \pm M^\beta = [\alpha, \alpha, 0, 0] \pm [\beta, \beta, 0, 0] = [\alpha \pm \beta, \alpha \pm \beta, 0, 0] = M^{\alpha \pm \beta}$, we have $\{M^\alpha \mid \alpha \in S'\} \subseteq \{M^\alpha \mid \alpha \in D(S', B)\} \subseteq A_{[\zeta,(0,1)]}$. Now we can say that $A_{[\zeta,(0,1)]}$ is the *normal* subgroup generated by $\{M^\alpha \mid \alpha \in D(S', B)\}$.

By normality, each $-[\beta, (b,1)] + M^\alpha + [\beta, (b,1)] \in A_{[\zeta,(0,1)]}$, where $\alpha \in D(S', B)$ and $\beta \in S'$. But $-[\beta, (b,1)] + M^\alpha + [\beta, (b,1)] = -[\beta, \beta, b, 1] + [\alpha, \alpha, 0, 0] + [\beta, \beta, b, 1] = [-\beta, (-1)^1(-\beta), (-1)^1(-b), 1] + [\alpha + \beta, \alpha + (-1)^0 \beta, 0 + (-1)^0 b, 1] = [-\beta, \beta, b, 1] + [\alpha + \beta, \alpha + \beta, b, 1] = [-\beta + \alpha + \beta, \beta + (-1)^1(\alpha + \beta), b + (-1)^1 b, 0] = [\alpha, -\alpha, 0, 0] = \Lambda^\alpha$. As the notation suggests, each Λ^α is produced independently of the $[\beta, (b,1)]$ used. Similarly, $\Lambda^\alpha = [\beta, (b,1)] + M^\alpha - [\beta, (b,1)]$. So, conjugating any M^α by elements of S or their negatives yields Λ^α.

Now let us conjugate any Λ^α by elements of S and their negatives to see what we obtain in $A_{[\zeta,(0,1)]}$. For $\alpha \in D(S', B)$ and $[\beta, (b,1)] \in S$, $-[\beta, (b,1)] + \Lambda^\alpha + [\beta, (b,1)] = [-\beta, \beta, b, 1] + [\alpha, -\alpha, 0, 0] + [\beta, \beta, b, 1] = [-\beta, \beta, b, 1] + [\alpha + \beta, -\alpha + \beta, b, 1] = [\alpha, \beta + (-1)^1(-\alpha + \beta), b + (-1)^1 b, 0] = [\alpha, \alpha, 0, 0] = M^\alpha$. Similarly, $[\beta, (b,1)] + \Lambda^\alpha - [\beta, (b,1)] = M^\alpha$. So, conjugating any Λ^α by elements of S or their negatives yields M^α again. Hence, conjugating any M^α and any Λ^α by *any* element of $D(S, D_B)$ yield only M^α and Λ^α. Hence, $A_{[\zeta,(0,1)]}$ is the group generated by $\{M^\alpha, \Lambda^\alpha \mid \alpha \in D(S', B)\}$.

Our next step is to show that $A_{[\zeta,(0,1)]}$ is the group generated by $\{\Pi_1^\alpha, \Pi_2^\alpha \mid \alpha \in D(S', B)\}$, where the $\Pi_1^\alpha, \Pi_2^\alpha$ are as defined in (15.7). From (15.7), we know that $[h, (0,1)] \in D(S, D_B)$, so $(M^\alpha + \Lambda^\alpha) \circ [h, (0,1)], (M^\alpha - \Lambda^\alpha) \circ [h, (0,1)] \in D(S, D_B)$. But $(M^\alpha + \Lambda^\alpha) \circ [h, (0,1)] = \{[\alpha, \alpha, 0, 0] + [\alpha, -\alpha, 0, 0]\} \circ [h, (0,1)] = [2\alpha, \zeta, 0, 0] \circ [h, h, 0, 1] = [2\alpha h, \zeta, 0, 0] = [\alpha h, \alpha h, 0, 0] + [\alpha h, -\alpha h, 0, 0] = M^{\alpha h} + \Lambda^{\alpha h} = \Pi_1^\alpha$, and $(M^\alpha - \Lambda^\alpha) \circ [h, (0,1)] =$

$\{[\alpha, \alpha, 0, 0] - [\alpha, -\alpha, 0, 0]\} \circ [h, (0, 1)] = [\zeta, 2\alpha, 0, 0] \circ [h, h, 0, 1] = [\zeta, 2\alpha h, 0, 0]$
$= [\alpha h, \alpha h, 0, 0] - [\alpha h, -\alpha h, 0, 0] = M^{\alpha h} - \Lambda^{\alpha h} = \Pi_2^\alpha$. Since $[\zeta, (0, 1)] \circ \Pi_1^\alpha = [\zeta, \zeta, 0, 1] \circ [\alpha, \zeta, 0, 0] = [\zeta, \zeta, 0, 0]$ and $[\zeta, (0, 1)] \circ \Pi_2^\alpha = [\zeta, \zeta, 0, 1] \circ [\zeta, \alpha, 0, 0] = [\zeta, \zeta, 0, 0]$, we have $\{\Pi_1^\alpha, \Pi_2^\alpha \mid \alpha \in D(S', B)\} \subseteq A_{[\zeta, (0, 1)]}$, so $A_{[\zeta, (0, 1)]}$ contains the group generated by $\{\Pi_1^\alpha, \Pi_2^\alpha \mid \alpha \in D(S', B)\}$. But $M^\alpha = [\alpha, \alpha, 0, 0] = [\alpha, \zeta, 0, 0] + [\zeta, \alpha, 0, 0] = \Pi_1^\alpha + \Pi_2^\alpha$, and $\Lambda^\alpha = [\alpha, -\alpha, 0, 0] = [\alpha, \zeta, 0, 0] - [\zeta, \alpha, 0, 0] = \Pi_1^\alpha - \Pi_2^\alpha$, so the set of mappings $\{M^\alpha, \Lambda^\alpha \mid \alpha \in D(S', B)\}$ is in the group generated by $\{\Pi_1^\alpha, \Pi_2^\alpha \mid \alpha \in D(S', B)\}$. Since $\{M^\alpha, \Lambda^\alpha \mid \alpha \in D(S', B)\}$ generates $A_{[\zeta, (0, 1)]}$, we have that $A_{[\zeta, (0, 1)]}$ is the group generated by $\{\Pi_1^\alpha, \Pi_2^\alpha \mid \alpha \in D(S', B)\}$.

Now let us see what $\{\Pi_1^\alpha, \Pi_2^\alpha \mid \alpha \in D(S'B)\}$ generates. First, $\Pi_1^\alpha + \Pi_1^\alpha = [\alpha, \zeta, 0, 0] + [\beta, \zeta, 0, 0] = [\alpha + \beta, \zeta, 0, 0] = \Pi_1^{\alpha+\beta}$, and $\Pi_2^\alpha + \Pi_2^\beta = [\zeta, \alpha, 0, 0] + [\zeta, \beta, 0, 0] = [\zeta, \alpha + (-1)^0\beta, 0, 0] = [\zeta, \alpha + \beta, 0, 0] = \Pi_2^{\alpha+\beta}$. Next, $-\Pi_1^\alpha = -[\alpha, \zeta, 0, 0] = [-\alpha, \zeta, 0, 0] = \Pi_1^{-\alpha}$, and $-\Pi_2^\alpha = -[\zeta, \alpha, 0, 0] = [\zeta, (-1)^0(-\alpha), 0, 0] = [\zeta, -\alpha, 0, 0] = \Pi_2^{-\alpha}$. Finally, $\Pi_1^\alpha + \Pi_2^\beta = [\alpha, \zeta, 0, 0] + [\zeta, \beta, 0, 0] = [\alpha, \beta, 0, 0] = [\zeta, \beta, 0, 0] + [\alpha, \zeta, 0, 0] = \Pi_2^\beta + \Pi_1^\alpha$. So, for $i = 1, 2$, let $P_i = \{\Pi_i^\alpha \mid \alpha \in D(S', B)\}$. Then each P_i is a subgroup of $A_{[\zeta, (0, 1)]}$ which is isomorphic to $D(S', B)$, and certainly $P_1 \cap P_2 = \{\bar\zeta\} = \{[\zeta, \zeta, 0, 0]\}$. Since $\Pi_1^\alpha + \Pi_2^\beta = \Pi_2^\beta + \Pi_1^\alpha$, we have that $A_{[\zeta, (0, 1)]}$ is abelian and that $A_{[\zeta, (0, 1)]}^+ = P_1 \oplus P_2 = \{[\alpha, \beta, 0, 0] \mid \alpha, \beta \in D(S', B)\}$. This completes the proof of 4).

Construct $G = D(S', B) \oplus [(D(S', B) \oplus B) \times_\theta Z_2]$, the group, and define $F : G \to D(S, D_B)$ by $F(\alpha, [(\beta, b), x]) = [\alpha, \beta, b, x]$. Certainly, F is a monomorphism by (1) of (15.6). By (3.17), we have $D(S, D_B) = A_{[\zeta, (0, 1)]} + M_{[\zeta, (0, 1)]}$. Now $[\alpha, \beta, 0, 0] + [\zeta, \zeta, b, x] = [\alpha+\zeta, \beta+(-1)^0\zeta, 0+(-1)^0 b, 0+x] = [\alpha, \beta, b, x]$, so F is an isomorphism. This gives us (5) and (6), and (7) is from (15.6) by virtue of (6). This completes the proof of (15.9).

Let us apply (15.9) to some specific situations.

(15.10) Corollary. *Let $(B, +)$ be an abelian group with odd exponent n, that is, the smallest positive integer n where $nb = 0$ for each $b \in B$ is an odd integer. Then:*
(1) $I(D_B)^+ \cong Z_n^+ \oplus [(Z_n^+ \oplus B) \times_\theta Z_2^+]$.
(2) $A(D_B)^+ \cong D(\text{Aut}B, B)^+ \oplus [(D(\text{Aut}B, B)^+ \oplus B) \times_\theta Z_2^+]$.
(3) $E(D_B)^+ \cong (\text{End}B)^+ \oplus [((\text{End}B)^+ \oplus B) \times_\theta Z_2^+]$.
Multiplication, addition, and additive inverses are as in (15.6).

Proof. Here, $h = [(n+1)/2] \cdot 1_B$, and by (8) of (15.5), $[-1_B, (0, 1)] \in \text{Inn}D_B$. For (1), we let $S' = \{\pm 1_B\}$, so $(-1_B, h) \in S' \times D(S', B)$. For (2), we let $S' = \text{Aut}B$, hence, by (7) of (15.5), $\text{Aut}D_B = \{[\beta, (b, 1)] \mid \beta \in \text{Aut}B, b \in B\} = \text{Aut}B \, \theta \times B = S$ of (15.9). Hence, $(-1_B, h) \in S' \times D(S', B)$. And for (3), we let $S' = \text{End}B$, so $\text{End}D_B = \{[\beta, (b, 1)] \mid \beta \in \text{End}B, b \in B\} \cup \{\bar\zeta\}$, by (6) of (15.5). Hence, $(-1_B, h) \in S' \times D(S', B)$. So, we can be assured that (15.9) is applicable.

For (1), $D(S', B) \cong Z_n^+$; for (2), $D(S', B)^+ = D(\mathrm{Aut}\,B, B)^+$; and for (3), $D(S', B)^+ = D(\mathrm{End}\,B, B)^+ = (\mathrm{End}\,B)^+$. Certainly, addition, multiplication, and additive inverses are as in (15.6). This completes the proof.

(15.11) Corollary. *Let* $(B, +) = (Z_n, +)$, *where* n *is odd. So* $D_B = D_n$, *the usual dihedral group of order* $2n$. *Then* $I(D_n) = A(D_n) = E(D_n)$, $|I(D_n)| = 2n^3$, *and* $I(D_n)^+ \cong Z_n^+ \oplus [(Z_n^+ \oplus Z_n^+) \times_\theta Z_2^+]$, *and addition, multiplication, and additive inverses are as in* (15.6).

Malone and Lyons [M&L] first proved (15.11).

For the finite simple groups G, we also have $I(G) = A(G) = E(G)$. But (15.10) could be useful to find examples where $A(D_B) = E(D_B)$. This will happen exactly when $D(\mathrm{Aut}\,B, B) = \mathrm{End}\,B$.

(15.12) Problem. For which abelian groups $(B, +)$ with odd exponent do we have $D(\mathrm{Aut}\,B, B) = \mathrm{End}\,B$?

We now follow the example of H. Gonshor.

(15.13) Theorem. *Let* R *be a ring with identity and group of units* $\mathcal{U}(R)$. *Let* M *be a unitary right* R-*module. On the set* $N = R \times R \times M \times Z_2$, *define*

$$[r, s, a, x] + [r', s', a', x'] = [r + r', s + (-1)^x s', a + (-1)^x a', x + x']$$

and

$$[r, s, a, x] \cdot [r', s', a', x'] = \begin{cases} [rr', sr', ar', 0], & \text{if } x = 0; \\ [rr', ss', as' + a', x'], & \text{if } x = 1. \end{cases}$$

Then the following are true.
(1) $(N, +, \cdot)$ *is a (left) nearring with identity* $[1, 1, 0, 1]$.
(2) $-[r, s, a, x] = [-r, (-1)^x(-s), (-1)^x(-a), x]$.
(3) *The group of units of* N *is isomorphic to the direct product* $\mathcal{U}(R) \times [\mathcal{U}(R)_\theta \times M]$; *it consists of all elements of the form* $[r, s, a, 1]$ *where* $r, s \in \mathcal{U}(R)$ *and* $a \in M$. *For a unit* $[r, s, a, 1]$, $[r, s, a, 1]^{-1} = [r^{-1}, s^{-1}, -as^{-1}, 1]$.
(4) *The distributive elements of* N *are the elements* $[r, r, a, 1]$ *for any* $r \in R$ *and for any* $a \in M$, *together with the elements* $[r, r, a, 0]$ *with* $2r = 0$ *and* $2a = 0$.
(5) *If* $D = \{d \in N \mid d \text{ is a distributive element}\}$, *then* D *generates the group* N^+ *with* $+$ *if and only if* $2 \in \mathcal{U}(R)$, *that is, if and only if* $r \mapsto 2r$ *is an automorphism of the group* R^+.
(6) *If* D *generates the group* N^+ *with* $+$, *then every element of* N *is the*

sum of three or fewer distributive elements of D, and for some elements of N, three is the minimum number required.

Proof. Taking guidance from the proof of (5) of (15.9), we construct the group $G = R^+ \oplus [(R^+ \oplus M^+) \times_\theta Z_2^+]$ and define $F : G \to N$ by $F(r, [(s, a), x])$
$= [r, s, a, x]$. One should be able to see immediately that F is an isomorphism from G onto $(N, +)$. Hence, $(N, +)$ is a group.

Associativity can often be a long exhausting process to verify. We need to show that $(N, \cdot)$ is a semigroup. Consider the products

$$[r, s, a, x]\{[r', s', a', x'][r'', s'', a'', x'']\}$$

and

$$\{[r, s, a, x][r', s', a', x']\}[r'', s'', a'', x''].$$

It should not be too difficult to verify associativity of multiplication by considering the four cases defined by $(x, x') \in \{(0, 0), (0, 1), (1, 0), (1, 1)\}$. Assuming that the curious reader has done that, we proceed similarly to see that multiplication is left distributive over $+$. We consider the expressions

$$[r, s, a, x]\{[r', s', a', x'] + [r'', s'', a'', x'']\}$$

and

$$[r, s, a, x][r', s', a', x'] + [r, s, a, x][r'', s'', a'', x''].$$

For $x = 0$,

$$[r, s, a, 0][r' + r'', s' + (-1)^{x'} s'', a' + (-1)^{x'} a'', x' + x'']$$
$$= [r(r' + r''), s(r' + r''), a(r' + r''), 0],$$

and

$$[r, s, a, 0][r', s', a', x'] + [r, s, a, 0][r'', s'', a'', x'']$$
$$= [rr', sr', ar', 0] + [rr'', sr'', ar'', 0] = [rr' + rr'', sr' + sr'', ar' + ar'', 0].$$

For $x = 1$,

$$[r, s, a, 1][r' + r'', s' + (-1)^{x'} s'', a' + (-1)^{x'} a'', x' + x'']$$
$$= [r(r' + r''), s(s' + (-1)^{x'} s''), a(s' + (-1)^{x'} s'') + a' + (-1)^{x'} a'', x' + x'']$$

and

$$[r, s, a, 1][r', s', a', x'] + [r, s, a, 1][r'', s'', a'', x'']$$
$$= [rr', ss', as' + a', x'] + [rr'', ss'', as'' + a'', x'']$$
$$= [rr' + rr'', ss' + (-1)^{x'} (ss''), as' + a' + (-1)^{x'} (as'' + a''), x' + x''].$$

Since $s(s' + (-1)^{x'}s'') = ss' + (-1)^{x'}(ss'')$, and $a(s' + (-1)^{x'}s'') + a' + (-1)^{x'}a'' = as' + a' + (-1)^{x'}(as'' + a'')$ we have that $(N, +, \cdot)$ is a left nearring.

Since $[0, 0, 0, 0]$ is the additive identity, and since

$$[r, s, a, x] + [-r, (-1)^x(-s), (-1)^x(-a), x] =$$
$$[r + (-r), s + (-1)^x(-1)^x(-s), a + (-1)^x(-1)^x(-a), x + x] = [0, 0, 0, 0],$$

we have $-[r, s, a, x] = [-r, (-1)^x(-s), (-1)^x(-a), x]$. Also, calculate that $[1, 1, 0, 1][r, s, a, x] = [r, s, a, x]$, $[r, s, a, 0][1, 1, 0, 1] = [r, s, a, 0]$, and that $[r, s, a, 1][1, 1, 0, 1] = [r, s, a, 1]$. So $[1, 1, 0, 1]$ is the multiplicative identity. We now have (1) and (2).

Take $[r, s, a, 1] \in N$ with $r, s \in \mathcal{U}(R)$. Then

$$[r, s, a, 1][r^{-1}, s^{-1}, -as^{-1}, 1] = [1, 1, as^{-1} - as^{-1}, 1] = [1, 1, 0, 1].$$

So $[r, s, a, 1]$ is a unit and $[r, s, a, 1]^{-1} = [r^{-1}, s^{-1}, -as^{-1}, 1]$. Certainly, no $[r, s, a, 0]$ can be a unit. Suppose $[r, s, a, 1]$ is a unit. There is a $[r', s', a', 1]$ such that $[1, 1, 0, 1] = [r, s, a, 1][r', s', a', 1] = [rr', ss', as' + a', 1]$. So $r, s \in \mathcal{U}(R)$, and $a' = -as'$. But $s' = s^{-1}$, so $a' = -as^{-1}$. We know now that the units of N are indeed all $[r, s, a, 1]$ with $r, s \in \mathcal{U}(R)$.

Let $(H, \cdot)$ be the multiplicative group defined by $\mathcal{U}(R) \times [\mathcal{U}(R)_\theta \times M]$, and define $F_1 : H \to \mathcal{U}(R)$ by $F_1(r, (s, a)) = [r, s, a, 1]$. Again, one should be able to see immediately that F_1 is an isomorphism. We now have (3).

To see that each $[r, r, a, 1]$ is a distributive element, we will consider

$$\{[r_1, s_1, a_1, x_1] + [r_2, s_2, a_2, x_2]\}[r, r, a, 1] \qquad (15 : 13)$$

and

$$[r_1, s_1, a_1, x_1][r, r, a, 1] + [r_2, s_2, a_2, x_2][r, r, a, 1] \qquad (15 : 14)$$

for each of the four cases $(x_1, x_2) \in \{(0, 0), (0, 1), (1, 0), (1, 1)\}$. First, with $(x_1, x_2) = (0, 0)$, (15:13) becomes

$$[r_1 + r_2, s_1 + s_2, a_1 + a_2, 0][r, r, a, 1] = [(r_1 + r_2)r, (s_1 + s_2)r, (a_1 + a_2)r, 0]$$

and (15:14) becomes

$$[r_1r, s_1r, a_1r, 0] + [r_2r, s_2r, a_2r, 0] = [r_1r + r_2r, s_1r + s_2r, a_1r + a_2r, 0].$$

Next, with $(x_1, x_2) = (0, 1)$, (15:13) becomes

$$[r_1 + r_2, s_1 + s_2, a_1 + a_2, 1][r, r, a, 1] = [(r_1 + r_2)r, (s_1 + s_2)r, (a_1 + a_2)r + a, 1]$$

and (15:14) becomes

$$[r_1r, s_1r, a_1r, 0] + [r_2r, s_2r, a_2r + a, 1] = [r_1r + r_2r, s_1r + s_2r, a_1r + a_2r + a, 1].$$

Now, with $(x_1, x_2) = (1, 0)$, (15:13) becomes

$$[r_1 + r_2, s_1 - s_2, a_1 - a_2, 1][r, r, a, 1] = [(r_1 + r_2)r, (s_1 - s_2)r, (a_1 - a_2)r + a, 1]$$

and (15:14) becomes

$$[r_1r, s_1r, a_1r + a, 1] + [r_2r, s_2r, a_2r, 0] = [r_1r + r_2r, s_1r - s_2r, a_1r + a - a_2r, 1].$$

Finally, with $(x_1, x_2) = (1, 1)$, (15:13) becomes

$$[r_1 + r_2, s_1 - s_2, a_1 - a_2, 0][r, r, a, 1] = [(r_1 + r_2)r, (s_1 - s_2)r, (a_1 - a_2)r, 0]$$

and (15:14) becomes

$$[r_1r, s_1r, a_1r + a, 1] + [r_2r, s_2r, a_2r + a, 1] = [r_1r + r_2r, s_1r - s_2r, a_1r - a_2r, 0].$$

So, each $[r, r, a, 1]$ is indeed a distributive element.

Turning our attention to elements $[r, r, a, 0]$ with $2r = 0$ and $2a = 0$, we again consider expressions such as those in (15:13) and (15:14), but with $[r, r, a, 0]$ in place of $[r, r, a, 1]$. For $(x_1, x_2) \in \{(0, 0), (0, 1)\}$, the calculations will be almost exactly like those above. But, for example, with $(x_1, x_2) = (1, 1)$, we obtain

$$[r_1 + r_2, s_1 - s_2, a_1 - a_1, 0][r, r, a, 0] = [(r_1 + r_2)r, (s_1 - s_2)r, (a_1 - a_2)r, 0]$$

and

$$[r_1r, s_1r, a_1r + a, 0] + [r_2r, s_2r, a_2r + a, 0]$$
$$= [r_1r + r_2r, s_1r + s_2r, a_1r + a + a_2r + a, 0].$$

Since $r + r = 0$, we have $(s_1 - s_2)r = s_1r - s_2r = s_1r + s_2(-r) = s_1r + s_2r$, and similarly, $(a_1 - a_2)r = a_1r + a_2r$. This, with $a + a = 0$, shows that these two expressions are equal. When $(x_1, x_2) = (1, 0)$, one needs a similar but simpler argument. Hence, these $[r, r, a, 0]$ with $2r = 0$ and $2a = 0$ are indeed distributive.

We must now show that there are no other distributive elements. Suppose $[r, s, a, x]$ is a distributive element. Then

$$\{[r_1, s_1, a_1, 1] + [r_2, s_2, a_2, 1]\}[r, s, a, x]$$
$$= [r_1, s_1, a_1, 1][r, s, a, x] + [r_2, s_2, a_2, 1][r, s, a, x]$$

for all $r_i, s_i \in R$ and all $a_i \in M$, $i = 1, 2$. Hence, $s_1 r - s_2 r = s_1 s + (-1)^x s_2 s$ and $a_1 r - a_2 r = a_1 s + a + (-1)^x (a_2 s) + (-1)^x a$ for all $s_1, s_2 \in R$ and for all $a_1, a_2 \in M$. Either $x = 0$ or $x = 1$. If $x = 1$, then setting $s_1 = 1$ and $s_2 = 0$, one obtains $r = s$. Hence, $[r, s, a, x] = [r, r, a, 1]$, which we already know to be distributive. If $s = 0$, then with $s_1 = 1$ and $s_2 = 0$, we obtain $r = s$ again. But setting $s_1 = 0$ and $s_2 = 1$ gives us $-r = s$. Hence, $r = s$ and $2r = 0$. In addition, setting $a_1 = a_2 = 0$, we obtain with $x = 0$ that $0 = 2a$. So, in this case, $[r, s, a, x] = [r, r, a, 0]$ where $2r = 0$ and $2a = 0$. Such $[r, r, a, 0]$ have already been shown to be distributive. We now have (4).

Assume $2 \in \mathcal{U}(R)$, that is, $r \mapsto 2r$ is an automorphism of the group R^+. Let h be the multiplicative inverse of $2 \in \mathcal{U}(R)$. Take any $[r, s, a, 1] \in N$. Then $[h(r - s), h(r - s), 0, 1] + [h(r - s), h(r - s), 0, 1] + [s, s, a, 1] = [2h(r - s), 0, 0, 0] + [s, s, a, 1] = [r, s, a, 1]$. So $[r, s, a, 1]$ is the sum of three distributive elements. Similarly, $[h(r + s), h(r + s), a, 1] + [h(r - s), h(r - s), 0, 1] = [2hr, 2hs, a, 0] = [r, s, a, 0]$. This shows that $[r, s, a, 0]$ is the sum of two distributive elements, hence N is generated with $+$ by the elements of D. Since $2 \in \mathcal{U}(R)$, we cannot have $2r = 0$ or $2a = 0$ nontrivially. So we will have no nontrivial distributive elements of the form $[r, r, a, 0]$. Now, notice that $[t, t, b, 1] + [u, u, c, 1] = [t + u, t - u, b - c, 0] \neq [r, s, a, 1]$, so three distributive elements must sometimes be used. This gives us (6) and one half of (5).

With the assumption that D generates the group N^+ with $+$, we proceed to show that $2 \in \mathcal{U}(R)$, and this will complete the proof of (5) as well as (15.13). We have that $[1, 0, 0, 0] = \sum_{i=1}^{k} \epsilon_i [r_i, r_i, a_i, x_i]$ for some finite sequence of distributive elements $[r_i, r_i, a_i, x_i]$ where $\epsilon_i = \pm 1$. One can easily show that an $\epsilon_i [r_i, r_i, a_i, x_i]$ has the form $[\epsilon_i' r_i', r_i', a_i', x_i']$, where $\epsilon_i' = \pm 1$. So, we may assume that $[1, 0, 0, 0] = \sum_{i=1}^{k} [\epsilon_i r_i, r_i, a_i, x_i]$. A straight-forward but slightly messy induction argument shows that $\sum_{i=1}^{k} [\epsilon_i r_i, r_i, a_i, x_i] = [\sum_{i=1}^{k} \epsilon_i r_i, \sum_{i=1}^{k} (-1)^{m_i} r_i, \sum_{i=1}^{k} (-1)^{m_i} a_i, \sum_{i=1}^{k} x_i]$ where $m_1 = 0$ and the other $m_i = \sum_{j=1}^{i-1} x_j$. So we have

$$1 = 1 + 0 = \sum_{i=1}^{k} \epsilon_i r_i + \sum_{i=1}^{k} (-1)^{m_i} r_i$$

$$= \sum_{i=1}^{k} [\epsilon_i + (-1)^{m_i}] r_i = 2 \cdot \sum_{j \in J} \eta_j r_j$$

where $J = \{i \mid \epsilon_i + (-1)^{m_i} = \pm 2\}$ and where

$$\eta_j = \begin{cases} 1, & \text{if } \epsilon_j + (-1)^{m_j} = 2; \\ -1, & \text{if } \epsilon_j + (-1)^{m_j} = -2, \end{cases}$$

since each $\epsilon_i + (-1)^{m_i} \in \{-2, 0, 2\}$. So $h = \sum_{j \in J} \eta_j r_j$, and $2 \in \mathcal{U}(R)$.

This seems to be an appropriate place to state

(15.14) Definition. A nearring $(N, +, \cdot)$ is *distributively generated* (d.g.) if there is a subsemigroup $(S, \cdot)$ of $(N, \cdot)$ of distributive elements which generate $(N, +)$ with respect to $+$ only. So each $x \in N$ can be expressed $x = \sum_{i=1}^{k} \epsilon_i d_i$, for some finite sum, where each $d_i \in S$ and each $\epsilon_i = \pm 1$.

The nearrings $D(S, D_B)$ of (5) in (15.9), and the nearrings N of (15.13) when $2 \in \mathcal{U}(R)$ are all distributively generated. If N in (15.13) is d.g., then the distributive elements of N are exactly the $[r, r, a, 1]$, with $r \in R$ and $a \in M$ chosen arbitrarily.

For N of (15.13), if we take $R = Z$ and $M = Z$, then $N = Z \times Z \times Z \times Z_2$. So M is a faithful unitary R-module, but $2 \notin \mathcal{U}(R)$. So N cannot be d.g. by (5) of (15.13).

We used the idempotent $[\zeta, (0, 1)]$ in (15.9) to compute $M_{[\zeta, (0,1)]} = [\zeta, (0, 1)]D(S, D_B)$ and $A_{[\zeta, (0,1)]} = \text{Ann}([\zeta, (0, 1)])$ and to apply (3.17). In (15.9) we noticed that $M_{[\zeta, (0,1)]}^+ \cong D_B$. So, the group D_B was recovered from the d.g. nearrings $D(S, D_B)$ using the idempotent $[\zeta, (0, 1)]$. It is natural to see if something similar can be done with the d.g. nearrings N of (15.13).

(15.15) Theorem. *Let N be a d.g. nearring of* (15.13). *Then the following are true.*
(1) *Each element $[0, 0, a, 1]$, $a \in M$, is an idempotent.*
(2) *If $M_{[0,0,a,1]} = [0, 0, a, 1]N$, then each $M_{[0,0,a,1]}^+ \cong D_M$.*
(3) $D_M \cong \cap \{M_e \mid e \in N, \ e^2 = e, \ \text{and} \ M_e^+ \ \text{is nonabelian} \}$, *if $M_e = eN$.*

Proof. For $a \in M$, $[0, 0, a, 1]^2 = [0, 0, a, 1]$, and $[0, 0, a, 1][r, s, b, x] = [0, 0, as + b, x]$. Hence $[0, 0, a, 1]N = \{[0, 0, a, x] \mid a \in M, \ x \in Z_2\}$. The map $[0, 0, a, x] \mapsto (a, x)$ defines an isomorphism from $[0, 0, a, 1]N$ onto D_M. Hence, (1) and (2).

If $[r, s, a, 0]^2 = [r, s, a, 0]$, then $[r, s, a, 0]N = \{[rr', sr', ar', 0] \mid r' \in R\}$, an abelian subgroup of N^+. If $[r, s, a, 1]^2 = [r, s, a, 1]$, then $[r, s, a, 1]N = \{[rr', ss', as' + a', x'] \mid [r', s', a', x'] \in N\}$, a nonabelian subgroup of N^+ containing $[0, 0, 0, 1]N$, since x' could be 1 and since the $a' \in M$ are arbitrary. Hence, (3). $\quad\blacksquare$

The d.g. nearrings $(N, +, \cdot)$ of (15.13) have some interesting ideals.

(15.16) Theorem. *Let $(N, +, \cdot)$ be a d.g. nearring of* (15.13). *Let*

$$L = \{[r, s, a, 0] \mid r, s \in R, \ a \in M\}.$$

Then L is an ideal of N and $N/L \cong Z_2$, the field.

Proof. Define $F : N \to Z_2$ by $F[r, s, a, x] = x$. Then by the rules for $+$ and $\cdot$ of (15.13), we see that F is a nearring epimorphism with kernel L.

(15.17) Theorem. *Let* $(N, +, \cdot)$ *be a d.g. nearring of* (15.13). *Let*

$$K = \{[0, s, a, x] \mid s \in R, \ a \in M, \ x \in Z_2\}.$$

Then K is an ideal of N, $N/K \cong R$, and $K = \mathrm{Ann}(L)$, where L is as in (15.16). *As a subnearring of N, it is not distributively generated, but it does have a left identity $[0, 1, 0, 1]$.*

Proof. Define $F : N \to R$ by $F[r, s, a, x] = r$. The rules for the operations $+$ and $\cdot$ of (15.13) assure us that F is a nearring epimorphism with kernel K. Take $[0, s, a, x] \in K$ and note that $[0, 1, 0, 1][0, s, a, x] = [0, s, a, x]$, so $[0, 1, 0, 1]$ is a left identity for K. Now take $[r', s', a', 0] \in L$ of (15.16). Then $[r', s', a', 0][0, s, a, x] = [0, 0, 0, 0]$. Hence, $K \subseteq \mathrm{Ann}(L)$. If $[r, s, a, x] \in \mathrm{Ann}(L)$, then $[0, 0, 0, 0] = [1, 0, 0, 0][r, s, a, x] = [r, 0, 0, 0]$, so $r = 0$ and then $[r, s, a, x] = [0, s, a, x] \in K$.

If $[0, s, a, x] \in K$ is a distributive element for K, then

$$\{[0, 1, a_1, 0] + [0, s_2, a_2, 1]\}[0, s, a, x]$$
$$= [0, 1, a_1, 0][0, s, a, x] + [0, s_2, a_2, 1][0, s, a, x]$$

or

$$[0, (1 + s_2)s, (a_1 + a_2)s + a, x] = [0, s_2 s, a_2 s + a, x].$$

Hence, $(1 + s_2)s = s + s_2 s = s_2 s$ forces $s = 0$. So our distributive elements of K are among those of the form $[0, 0, a, x]$, and there is no way that these will ever generate $[0, 1, 0, 0] \in K$ with $+$, for example. This means that K cannot be distributively generated.

(15.18) Proposition. *Let K and N be as in* (15.17). *Then every element of K is either nilpotent or has additive order 2. In fact, $[0, s, a, 0]^2 = [0, 0, 0, 0]$ and $[0, s, a, 1] + [0, s, a, 1] = [0, 0, 0, 0]$.*

(15.19) Theorem. *Let N and K be as in* (15.17) *and L be as in* (15.16). *Then $N/K \cap L \cong R \oplus Z_2$, and $(K \cap L, +, \cdot)$ is a zero ring.*

Proof. Define $F : N \to R \oplus Z_2$ by $F[r, s, a, x] = (r, x)$. The rules for $+$ and $\cdot$ assure us that F is a nearring epimorphism with kernel $\{[0, s, a, 0] \mid x \in R, \ a \in M\}$, which one can easily see is the same as $K \cap L$. If $[0, s, a, 0], [0, s', a', 0] \in K \cap L$, then $[0, s, a, 0][0, s', a', 0] = [0, 0, 0, 0]$.

(15.20) Problem. Suppose we know that a nearring $(N, +, \cdot)$ is an endomorphism nearring. Then N has at its source a group $(G, +)$ and a semigroup $(\Phi, \circ)$ of endomorphisms of $(G, +)$. Is there a way to reconstruct $(G, +)$ and/or $(\Phi, \circ)$ from $(N, +, \cdot)$? Parts (2) and (3) of (15.15) suggest a technique that may work sometimes. When do we have $G \cong \cap\{M_e \mid e \in N, e^2 = e, \text{ and } M_e^+ \text{ is nonabelian }\}$?

For each idempotent $[0, 0, b, 1]$ of (15.15), it is easy to see that the annihilator $\text{Ann}([0, 0, b, 1]) = \{(r, s, -bs, 0] \mid r, s \in R\}$, and in particular, $\text{Ann}([0, 0, 0, 1]) = \{[r, s, 0, 0] \mid r, s \in R\}$. And it is direct to see that each $\text{Ann}([0, 0, b, 1])$ is a subring of N. Also, L is a subring of N. A casual inspection shows that H. Gonshor's example can be followed to describe a way to make new rings from a ring R and a module M.

(15.21) Theorem. *Let R be a ring and M one of its right modules. On $R \times M$, define $+$ and $\cdot$ by $(r, m) + (r', m') = (r + r', m + m')$ and $(r, m) \cdot (r', m') = (rr', mr')$. Then $(R \times M, +, \cdot)$ is a ring with ideal $M' = \{(0, m) \mid m \in M\}$.*

Proof. The proof is as direct as can be.

But one can follow H. Gonshor's example even further.

(15.22) Corollary. *Let N be a (left) nearring and G one of its comodules. On $N \times G$ define $+$ and $*$ by $(n, g) + (n', g') = (n + n', g + g')$ and $(n, g) * (n', g') = (nn', ng')$. Then $(N \times G, +, *)$ is a left nearring with ideal $G' = \{(0, g) \mid g \in G\}$.*

Proof. Exercise.

In (9) of (15.5), one can see that multiplication $\circ$ in $(\text{End}D_B)^*$ is like that in the abstract affine nearring $B \oplus_A \text{End}B$. In fact, it is exactly the left hand version of the multiplication in $B \oplus_A \text{End}B$. This suggests that one could define a $\oplus$ on $(\text{End}D_B)^*$ by $[\alpha, (a, 1)] \oplus [\beta, (b, 1)] = [\alpha + \beta, (a + b, 1)]$ and make $((\text{End}D_B)^*, \oplus, \circ)$ into an abstract affine nearring. In fact, written this way, it is a very natural thing to do. However, it must be pointed out that $\oplus$ is *not* pointwise addition of elements of $(\text{End}D_B)^*$ as (15.10) so forcefully reminds us. But look at D_B again. Each D_B gives us a group object in $\mathbf{Ob}\mathcal{G}(Z_2)$ as (12.18) assures us.

From (12.19), we know that $(\mathbf{hom}[(D_B, \gamma), (D_B, \gamma)], \oplus, \circ)$ is isomorphic to the abstract affine nearring $Z_\theta^1(Z_2, B) \oplus \mathcal{C}_h(B, Z_2)$. One readily sees that $\mathcal{C}_h(B, Z_2) = \{\alpha \in \text{Hom}(B, B) \mid \alpha \circ \theta(x) = \theta(x) \circ \alpha \text{ for all } x \in Z_2\} = \text{Hom}(B, B)$. A $\bar{b} \in Z_\theta^1(Z_2, B)$ must satisfy $\bar{b}(0) = \bar{b}(0 + 0) = \bar{b}(0) + \theta(0)\bar{b}(0) = \bar{b}(0) + \bar{b}(0)$. Hence, $\bar{b}(0) = 0$. Let $\bar{b}(1) = b \in B$. Checking the four possibilities for $(x, x') \in Z_2 \times Z_2$, one is readily assured that $\bar{b}(x + x') = \bar{b}(x) + \theta(x)\bar{b}(x')$ in all cases. Hence, $Z_\theta^1(Z_2, B) \cong B$ with isomorphism $\Psi : B \to Z_\theta^1(Z_2, B)$ defined by $\Psi(b) = \bar{b}$, or $\Gamma : Z_\theta^1(Z_2, B) \to B$ defined by $\Gamma(l) = l(1)$, if you would rather.

This suggests that it may be more efficient to study relationships between the nearring $(\mathbf{hom}[(D_B, \gamma), (D_B, \gamma)], \oplus, \circ)$ and the group D_B than between

$(E(D_B), +, \circ)$ and the group D_B, since $\mathbf{hom}[(D_B, \gamma), (D_B, \gamma)]$ is so much smaller.

(15.23) Exploratory problem. Study the relationships among the group D_B, the endomorphism nearring $(\mathbf{hom}[(D_B, \gamma), (D_B, \gamma)], \oplus, \circ)$, and the nearring $(E(D_B), +, \circ)$. If some interesting relationships are found, try to extend them to other group objects in $\mathcal{G}(A)$ of (12.18).

Perhaps the answers to the following two Problems will be relevant to (15.23).

(15.24a) Problem. For which abelian groups B and A and homomorphism $\theta : A \to \mathrm{Aut}\, B$ do we have $\mathcal{C}_h(B, A) = \mathrm{Hom}(B, B)$ and $Z_\theta^1(A, B) \cong B$?

(15.24b) Problem. For which group objects $B \times_\theta A$ of $\mathcal{G}(A)$ do we have $\mathbf{hom}[(B \times_\theta A, \gamma), (B \times_t aA, \gamma)] = \mathrm{End}(B \times_\theta A) \setminus \{\zeta\}$?

15.2. Distributively generated nearrings from noncommutative formal group laws

We have indicated that distributively generated (d.g.) nearrings arise from a semigroup $(S, \circ)$ of endomorphisms of a nonabelian group $(G, +)$, but (15.13) shows us that we need not restrict ourselves to this source. Another source for d.g. nearrings comes from noncommutative formal group laws. This source of nearrings was introduced by Fröhlich [F, 1968], and he has suggested that they show promise of having important applications to number theory. Hazewinkel [H] has collected together lots of good news about formal group laws, together with many applications of commutative formal group laws.

We shall start with a commutative ring A with identity 1.

(15.25) Definitions. For a commutative ring A with identity, a (*one dimensional*) *formal group law over* A is a formal power series $F(X, Y) \in A[[X, Y]]$ of the form

$$F(X, Y) = X + Y + \sum_{i=1}^{\infty} \sum_{j=1}^{\infty} f_{ij} X^i Y^j$$

such that the 'associative condition'

$$F(X, F(Y, Z)) = F(F(X, Y), Z)$$

is valid. (Hopefully, the reader will understand what we will mean by the suggestive notation $F(X, F(Y, Z))$, $F(F(X, Y), Z)$, etc.) If, in addition,

$F(X,Y) = F(Y,X)$, we say that F is *commutative*. Otherwise, F is *non-commutative*.

Let $A_0[[X]]$ denote all the power series over A of the form

$$\alpha = \alpha(X) = a_1 X + a_2 X^2 + a_3 X^3 + \cdots.$$

So $A_0[[X]]$ is the family of power series over A with 0 for the constant term. For a one dimensional formal group law F over A, we define $+_F$ on $A_0[[X]]$ by

$$\alpha +_F \beta = \alpha(X) +_F \beta(X) = F(\alpha(X), \beta(X)).$$

The value of our formal group law F is a result of the next theorem.

(15.26) Theorem. *For a commutative ring A with identity, and a one dimensional formal group law F over A, we obtain that $(A_0[[X]], +_F, \circ)$ is a nearring with identity $\iota(X) = X$, where $(\alpha \circ \beta)(X) = \alpha(\beta(X)) = \sum_{i=1}^{\infty} a_i (\beta(X))^i$.*

Proof. Just before (2.10), we noted that $(A_0[[X]], +, \circ)$ is a nearring, where $+$ is defined by the formal group law $G(X,Y) = X+Y$, so we have that $\circ$ is associative and that the $(A_0[[X]], +, \circ)$ will be special cases of our theorem here. Also, $\iota(X) = X$ is the identity for $\circ$.

For $\alpha(X) = \sum_{i=1}^{\infty} a_i X^i$, $\beta(X) = \sum_{i=1}^{\infty} b_i X^i$, and $\gamma(X) = \sum_{i=1}^{\infty} g_i X^i$, we have

$$\alpha(X) +_F \beta(X) = F(\alpha(X), \beta(X)) = \alpha(X) + \beta(X) + \sum_{i=1}^{\infty}\sum_{j=1}^{\infty} f_{ij}\alpha(X)^i \beta(X)^j,$$

so $+_F$ is a binary operation on $A_0[[X]]$, and from the 'associative condition' for F, we have $\alpha(X) +_F [\beta(X) +_F \gamma(X)] = F[\alpha(X), F[\beta(x), \gamma(x)]] = F[F[\alpha(X), \beta(X)], \gamma(X)] = (\alpha(X) +_F \beta(X)) +_F \gamma(X)$, so $+_F$ is associative. The formal power series $\zeta(X) = \sum_{i=1}^{\infty} 0X^i$ is certainly the additive identity for $+_F$, and

$$[\alpha(X) +_F \beta(X)] \circ \gamma(X) = F(\alpha(X), \beta(X)) \circ \gamma(X)$$

$$= \left[\alpha(X) + \beta(X) + \sum_{i=1}^{\infty}\sum_{j=1}^{\infty} f_{ij}\alpha(X)^i \beta(X)^j\right] \circ \gamma(X)$$

$$= \alpha(\gamma(X)) + \beta(\gamma(X)) + \sum_{i=1}^{\infty}\sum_{j=1}^{\infty} f_{ij}\alpha(\gamma(X))^i \beta(\gamma(X))^j$$

$$= F[\alpha(\gamma(X)), \beta(\gamma(X))] = (\alpha \circ \gamma)(X) +_F (\beta \circ \gamma)(X),$$

so $\circ$ is right distributive over $+_F$.

It remains only to show the existence of a right inverse for $\alpha(X)$ with respect to $+_F$. If

$$\zeta(X) = \alpha(X) +_F \beta(X) = \alpha(X) + \beta(X) + \sum_{i=1}^{\infty}\sum_{j=1}^{\infty} f_{ij}\alpha(X)^i\beta(X)^j,$$

then it is necessary and sufficient to show the existence of coefficients $b_1, b_2, \ldots$ so that the coefficient of each X^n in

$$\alpha(X) + \beta(X) + \sum_{i=1}^{\infty}\sum_{j=1}^{\infty} f_{ij}\alpha(X)^i\beta(X)^j \qquad (15:15)$$

is zero.

Consider the terms, or summands, in an expansion for

$$f_{ij}\alpha(X)^i\beta(X)^j = f_{ij}\alpha(X)^{i-1}\beta(X)^{j-1}\alpha(X)\beta(X), \qquad (15:16)$$

and remember that $i \geq 1$ and $j \geq 1$. Those in $\alpha(X)\beta(X)$ are of the form

$$a_s X^s b_t X^t = a_s b_t X^{s+t},$$

and have degree $s + t \geq 2$. So, in particular, each summand in (15:16) will have degree $\geq i + j$. Also, each term, or summand, $B_m X^m$ in (15:16) will have a coefficient of the form

$$B_m = N_m a_{k_1} \cdots a_{k_u} b_{l_1} \cdots b_{l_v},$$

where $u \geq 1$, $v \geq 1$, where N_m is an integer, and where $k_1 + \cdots + k_u + l_1 + \cdots + l_v = m$. In particular, the subscripts $k_s < m$ and $l_t < m$.

So, in (15:16), the summands have degree $m \geq i + j$, and if $B_m X^m$ is a summand of degree $m \geq i + j$, the coefficient B_m has as factors an integer and some coefficients a_s and b_t, where $s < m$ and $t < m$. This means we have as the coefficient of X^m in (15:15)

$$a_m + b_m + A_m \qquad (15:17)$$

where A_m is an algebraic expression in terms of the f_{ij}, $a_1, \ldots, a_{n-1}, b_1, \ldots, b_{n-1}$, and notice that the subscripts of the a_s and b_t have $s < m$ and $t < m$.

Now let us show the existence of the required $b_1, b_2, \ldots$. First, the coefficient of X in (15:15) is $a_1 + b_1$, so $b_1 = -a_1$ is necessary and sufficient for $a_1 + b_1 = 0$, and so b_1 can be expressed in terms of the known a_1, hence in terms of the known a_1, and the known f_{ij}, but of course, the f_{ij} here really have no influence on b_1. Hence, we can assume b_1 to be known.

For $m = 2$, the coefficient of X^2 in (15:15) is $a_2 + b_2 + f_{11}a_1b_1$, so $b_2 = -a_2 - f_{11}a_1b_1$ is necessary and sufficient for $a_2 + b_2 + f_{11}a_1b_1 = 0$, and so b_2 can be expressed as an algebraic expression in terms of the known f_{ij}, a_1, a_2, and b_1. We may now assume b_2 to be known.

Assume $b_1, b_2, \ldots, b_{n-1}$ to be known. For $m = n$, we have the coefficient of X^n in (15:15) to be of the form $a_n + b_n + A_n$, where A_n is an algebraic expression in terms of the known f_{ij}, $a_1, \ldots, a_{n-1}$, $b_1, \ldots, b_{n-1}$, and so $b_n = -a_n - A_n$ is necessary and sufficient for $a_n + b_n + A_n = 0$, and so b_n can be expressed as an algebraic expression in terms of the known f_{ij}, $a_1, \ldots, a_n$, $b_1, \ldots, b_{n-1}$. So we conclude that b_n in known.

By induction, we have the existence of $b_1, b_2, \ldots$ such that $a_n + b_n + A_n = 0$ for each $n = 1, 2, \ldots$, and so the coefficient of each X^n in (15:15) is 0. Hence, $\beta(X)$ is the inverse of $\alpha(X)$ with respect to $+_F$, and the proof of (15.26) is complete.

Now we want to discover all of the distributive elements in a nearring $(A_0[[X]], +_F, \circ)$, so an element $\delta = \delta(X) \in A_0[[X]]$ is distributive if and only if $\delta \circ [\alpha +_F \beta] = (\delta \circ \alpha) +_F (\delta \circ \beta)$ for all $\alpha, \beta \in A_0[[X]]$. That is, if and only if

$$\delta\big(F\big(\alpha(X), \beta(X)\big)\big) = F\big(\delta\big(\alpha(X)\big), \delta\big(\beta(X)\big)\big) \qquad (15:18)$$

for all $\alpha(X), \beta(X) \in A_0[[X]]$. This means, for $\delta = d_1 X + d_2 X^2 + \cdots$ that

$$d_1 F(\alpha(X), \beta(X)) + d_2 F(\alpha(X), \beta(X))^2 + \cdots$$
$$= \delta(\alpha(X)) + \delta(\beta(X)) + \sum_{i=1}^{\infty} \sum_{j=1}^{\infty} f_{ij}\delta(\alpha(X))^i \delta(\beta(X))^j \qquad (15:19)$$

for all $\alpha, \beta \in A_0[[X]]$. If we let $\alpha(X) = \iota(X) = X$ and $Y = \beta(X)$, then we have

$$\delta(F(X, Y)) = F(\delta(X), \delta(Y)) \qquad (15:20)$$

and

$$d_1 F(X, Y) + d_2 F(X, Y)^2 + \cdots$$
$$= \delta(X) + \delta(Y) + \sum_{i=1}^{\infty} \sum_{j=1}^{\infty} f_{ij}\delta(X)^i \delta(Y)^j.$$

Conversely, if (15:20) is valid, then replacing X with $\alpha(X)$ and Y with $\beta(X)$, we obtain (15:18). Hence,

(15.27) Proposition. *An element* $\delta = \delta(X) \in A_0[[X]]$ *is a distributive element in the nearring* $(A_0[[X]], +_F, \circ)$ *if and only if* $\delta(F(X, Y)) = F(\delta(X), \delta(Y))$.

(15.28) Definition. For a formal group law F over A, an *endomorphism* of F is a distributive element $\delta = \delta(X) \in A_0[[X]]$.

Taking guidance from the endomorphisms of a group in $M_0(G)$, we let $\mathbf{hom}(F, F)$ denote the set of endomorphisms of the group law F (the distributive elements in $(A_0[[X]], +_F, \circ)$), and investigate the d.g. nearring $(\mathrm{Hom}(F, F), +_F, \circ)$ generated by $\mathbf{hom}(F, F)$.

Two easy classes of examples of commutative group laws are defined by $F(X, Y) = X + Y$ and $G(X, Y) = X + Y + XY$. What we need now are examples of noncommutative group laws. The following theorem provides a way to construct numerous examples.

(15.29) Theorem. *Let A be a commutative ring with identity and b a nonzero element with finite additive order which is also nilpotent. Then there is a $c \in A$ such that $c \neq 0$, $c^2 = 0$, and $pc = 0$ for some prime p. Furthermore,*

$$F(X, Y) = X + Y + cXY^p \qquad (15:21)$$

is a one dimensional noncommutative formal group law over A.

Proof. Suppose $b^n = 0$ with n the smallest positive integer with this property. If $n = 2m$ is even, then let $d = b^m$, so $d^2 = 0$. Also, let b have additive order k, so $kd = 0$ and we can, without any loss of generality, assume that k is the order for d. Let $k = pl$, where p is a prime. Take $c = ld \neq 0$. Then $pc = (pl)d = kd = 0$ and $c^2 = (ld)^2 = l^2 d^2 = 0$. If $n = 2m + 1$, let $d = b^{2m} = b^{n-1}$, so $d \neq 0$. Again, if d has additive order $k = pl$, where p is a prime, then let $c = ld \neq 0$. So $pc = (pl)d = kd = 0$ and $c^2 = (ld)^2 = l^2 d^2 = 0$. This gives the promised $c \in A$.

Since $F(Y, X) = Y + X + cYX^p$, we have $F(X, Y) \neq F(Y, X)$. Now

$$\begin{aligned}
F(X, F(Y, Z)) &= X + F(Y, Z) + cXF(Y, Z)^p \\
&= X + Y + Z + cYZ^p + cX(Y + Z + cYZ^p)^p \\
&= X + Y + Z + cYZ^p + cX\left[\sum_{k=0}^{p} \binom{p}{k}(Y + Z)^{p-k}(cYZ^p)^k\right] \\
&= X + Y + Z + cYZ^p + cX(Y + Z)^p \\
&= X + Y + Z + cYZ^p + cXY^p + cXZ^p,
\end{aligned}$$

since $pc = 0$, $c^2 = 0$, and since p divides $\binom{p}{k}$ for $k = 1, 2, \ldots, p-1$. Also,

$$\begin{aligned}
F(F(X, Y), Z) &= F(X, Y) + Z + cF(X, Y)Z^p \\
&= X + Y + cXY^p + Z + c[X + Y + cXY^p]Z^p \\
&= X + Y + Z + cXY^p + cXZ^p + cYZ^p
\end{aligned}$$

Hence, F is a one dimensional noncommutative formal group law over A, and we have completed the proof of (15.29).

(15.30) Corollary. *Let $A = Z_{p^2}$, the ring of integers modulo p^2, where p is a prime. Let $c = p \in A$. Then*

$$F(X,Y) = X + Y + pXY^p \qquad (15:22)$$

is a one dimensional noncommutative formal group law over A.

(15.31) Corollary. *Let K be a field of characteristic $p > 0$, and let t be an indeterminate. Suppose (t^2) is the principal ideal generated by the polynomial t^2 in the ring of all polynomials $K[t]$ over K with indeterminate t. For $A = K[t]/(t^2)$ and $c = t$, then*

$$F(X,Y) = X + Y + tXY^p \qquad (15:23)$$

is a one dimensional noncommutative formal group law over A.

Our next goal is to compute $\mathbf{hom}(F, F)$ and then $\mathrm{Hom}(F, F)$ for each of the families of group laws defined by (15:22) and (15:23). What we suggest is a parallel development, since the results and proof for (15:22) are analogous to those for (15:23). In each case, there is a considerable amount of computation required, and we invite the reader to fill in these computations since we shall only provide an outline here. As indicated below in Problems, it is felt that there is considerably more to the story than outlined here, and the reader will be invited to discover much of what is left untold, and so filling in the computational details may give the reader the insight required to discover some interesting mathematics.

(15.32) Theorem. *For F as in (15:22), we have*

$$\mathbf{hom}(F, F) = \left\{ aX + p\sum_{k=0}^{\infty} a_k X^{p^k} \mid a \in \{0, 1\}, \ a_k \in \{0, 1, \ldots, p-1\} \right\}.$$

For F as in (15:23), we have

$$\mathbf{hom}(F, F) = \left\{ aX + t\sum_{k=0}^{\infty} a_k X^{p^k} \mid a \in \{0, 1\}, \ a_k \in K \right\}.$$

Proof. The first step is to show that if $\alpha \in \mathbf{hom}(F, F)$, then

$$\alpha(X) = a_1 X + \sum_{k=1}^{\infty} a_{kp} X^{kp} \qquad (15:24)$$

where $a_i \in A$. This will be for (15:22) and for (15:23). When we consider $\alpha \in \mathbf{hom}(F, F)$ and F as in (15:22), we shall write

$$\alpha(X) = D_1 X + D_2 X^2 + \cdots.$$

When $\alpha \in \mathbf{hom}(F, F)$ and F is as in (15:23), we shall write

$$\alpha(X) = B_1 X + B_2 X^2 + \cdots.$$

In each case, we consider the implications of the condition

$$\alpha(F(X, Y)) = F(\alpha(X), \alpha(Y)), \qquad (15 : 25)$$

which is just (15:20) with the endomorphism (distributive element) α in place of the endomorphism δ.

Now $\alpha(F(X, Y)) = \sum_{k=1}^{\infty} D_k F(X, Y)^k$ for (15:22) and for (15:23) we get $\alpha(F(X, Y)) = \sum_{k=1}^{\infty} B_k F(X, Y)^k$. For $k \geq 2$, we have

$$F(X, Y)^k = \begin{cases} (X + Y)^k + k(X + Y)^{k-1}(pXY^p), & \text{for (15:22);} \\ (X + Y)^k + k(X + Y)^{k-1}(tXY^p), & \text{for (15:23).} \end{cases}$$

So we obtain

$$\alpha(F(X, Y)) = \sum_{k=1}^{\infty} D_k(X + Y)^k + (pXY^p) \sum_{k=1}^{\infty} kD_k(X + Y)^{k-1}$$

and

$$\alpha(F(X, Y)) = \sum_{k=1}^{\infty} B_k(X + Y)^k + (tXY^p) \sum_{k=1}^{\infty} kB_k(X + Y)^{k-1}.$$

The right hand side of (15:25) gives us

$$F(\alpha(X), \alpha(Y)) = \sum_{k=1}^{\infty} D_k(X^k + Y^k) + p \sum_{i=1}^{\infty} \sum_{j=1}^{\infty} D_i D_j^p X^i Y^{jp}$$

and

$$F(\alpha(X), \alpha(Y)) = \sum_{k=1}^{\infty} B_k(X^k + Y^k) + t \sum_{i=1}^{\infty} \sum_{j=1}^{\infty} B_i B_j^p X^i Y^{jp}.$$

Suppose $1 < i < p$. Collecting terms of degree i, we can conclude that $D_i = 0$ and $B_i = 0$. Collecting terms of degree $p + i$, we conclude that

$D_{p+i} = 0$ and $B_{p+i} = 0$. Also, $pD_p = 0$ and $tB_p = t(B_{p,1} + tB_{p,2}) = tB_{p,1}$, where $B_k = B_{k,1} + tB_{k,2}$.

By a rather messy induction argument, one shows that $D_{kp+i} = 0$ and $B_{kp+i} = 0$ if $1 < i < p$, and that $pD_{kp} = 0$ and $tB_{kp} = tB_{kp,1}$. We can then conclude that α has the form (15:24). To obtain α, as promised, by the theorem requires a considerable amount of calculation. We shall outline the steps for (15:22). The procedure for (15:23) is parallel, but more involved. For that case, we shall point out a few adjustments that must be made.

Continuing, with $\alpha(X) = D_1 X + \sum_{k=1}^{\infty} D_{kp} X^{kp}$, where each $pD_{kp} = 0$, we obtain, upon insisting that (15:25) be valid, the equation

$$pD_1 XY^p + \sum_{k=1}^{\infty} D_{kp}(X+Y)^{kp} = pD_1^{p+1} XY^p + \sum_{k=1}^{\infty} D_{kp}(X^{kp} + Y^{kp}).$$

$$(15:26)$$

Using $pD_{kp} = 0$, we obtain

$$D_{kp}(X^{kp} + Y^{kp}) = D_{kp} \sum_{j=0}^{k} \binom{k}{j} (X^p)^{k-j} (Y^p)^j,$$

which forces $kD_{kp} = 0$. So, if $D_{kp} \neq 0$, then p divides k. Setting $k = k_0 p^{s-1}$ with $(k_0, p) = 1$, one shows that $k_0 = 1$ and so $k = p^{s-1}$. This in turn makes

$$\alpha(X) = D_1 X + \sum_{k=1}^{\infty} D_{p^k} X^{p^k}.$$

Turning our attention again to (15:26), we now consider the consequence of $pD_1 = pD_1^{p+1}$. Certainly, this can be true if $pD_1 = 0$, and so $\alpha(X) = \sum_{k=0}^{\infty} D_{p^k} X^{p^k}$ with $pD_{p^k} = 0$. Alternatively, if $pD_1 \neq 0$, then $D_1 = mp + i$ with $(i, p) = 1$, $1 \leq i \leq p - 1$. So $pD_1 = pi$ and $D_1^p \equiv (mp + i)^p \equiv i^p \pmod{p^2}$. Hence, $pD_1^{p+1} = (pi)i^p = pi^{p+1}$. This means $pi = pi^{p+1}$ and, consequently, $p = pi^p$, which in turn makes $i^p = 1$ and so $i = np + 1$ for some n. Thus $D_1 = (m+n)p + 1$, so $\alpha(X) = X + \sum_{k=0}^{\infty} D_{p^k} X^{p^k}$, where $pD_{p^k} = 0$ for each k. So, in this case, $\alpha(X)$ must be as promised in (15.32).

Conversely, if $\alpha(X) = aX + p\sum_{k=0}^{\infty} a_k X^{p^k}$, with $a \in \{0, 1\}$ and $a_k \in \{0, 1, \ldots, p-1\}$, then a direct proof assures us that $\alpha(X) \in \mathrm{hom}(F, F)$.

As indicated above, the procedure for (15:23) is parallel to that for (15:22), with a few modifications. Where we proved that $D_1 = (m+n)p+1$ if $pD_a \neq 0$, we also obtain from the analogous condition $B_1 t = B_1^{p+1} t$, that $B_1 = 1 + B_{1,2} t$ or $B_1 = B_{1,2} t$. Consider each of these two cases separately.

Suppose $B_1 = B_{1,2} t$, and that (15:25) is valid. We are forced to have

$$\sum_{k=1}^{\infty} B_{kp}(X+Y)^{kp} = \sum_{k=1}^{\infty} B_{kp}(X^{kp} + Y^{kp}) + t \left(\sum_{k=1}^{\infty} B_{kp} X^{kp} \right) \left(\sum_{k=1}^{\infty} B_{kp}^p X^{kp^2} \right).$$

For $1 < i < p$, we can conclude that $B_{ip} = 0$ and that $B_p = B_{p,2}t$. We now set up the following condition for an induction argument:

first segment; (a) $p^0 < i < p^1$ implies $B_i = 0$,
                (b) $B_{p^0} = B_{p^0,2}t$,

second segment; (a) $p^1 < i < p^2$ implies $B_i = 0$,
                 (b) $B_{p^1} = B_{p^1,2}t$.

Assume the

kth segment; (a) $p^{k-1} < i < p^k$ implies $B_i = 0$,
              (b) $B_{p^{k-1}} = B_{p^{k-1},2}t$.

Then prove the

$(k+1)$th segment; (a) $p^k < i < p^{k+1}$ implies $B_i = 0$,
                    (b) $B_{p^k} = B_{p^k,2}t$.

Next, suppose $B_1 = 1 + B_{1,2}t$. The proof is now closer to that for (15:22). We can then conclude that $\alpha(X)$ must be as promised in (15.32).

Conversely, if $\alpha(X) = aX + t\sum_{k=0}^{\infty} a_k X^{p^k}$, with $a \in \{0,1\}$ and $a_k \in K$, a direct proof assures one that $\alpha(X) \in \mathbf{hom}(F,F)$. If one does all that is asked above, one can be assured that the proof of (15:32) is now complete.

With (15.32), we know the structure of the elements of $\mathbf{hom}(F,F)$ for F either for the family defined by (15:22) or for the family defined by (15:23). Our next job is to describe the corresponding d.g. nearrings $(\mathrm{Hom}(F,F), +_F, \circ)$. Perhaps the reader will be as surprised as the author to realize that within the corresponding nearrings $(A_0[[X]], +_F, \circ)$ with nonabelian groups $(A_0[[X]], +_F)$, these d.g. nearrings for F as in (15:22) or in (15:23) are actually commutative rings!

(15.33) Theorem. *For F as defined in (15:22) or (15:23), the d.g. nearring* $(\mathrm{Hom}(F,F), +_F, \circ)$ *is actually a commutative ring with identity* $\iota(X) = X$.

Proof. For $\alpha, \beta \in \mathbf{hom}(F,F)$, we set $\alpha = aX + \lambda \sum_{k=0}^{\infty} a_k X^{p^k}$ and $\beta = bX + \lambda \sum_{k=0}^{\infty} b_k X^{p^k}$, where λ is either p or t, depending upon whether F is as in (15:22) or (15:23). In each case $\lambda^2 = 0$ and we obtain

$$\alpha +_F \beta = \alpha + \beta + \lambda ab X^{p+1}. \qquad (15:27)$$

In deriving this, we use the fact that $b \in \{0,1\}$ and so $b^p = b$. Hence,

$$\beta +_F \alpha = \beta + \alpha + \lambda ba X^{p+1}$$

for the same reasoning with $a \in \{0, 1\}$. So $\alpha +_F \beta = \beta +_F \alpha$ and now we know that $(\mathrm{Hom}(F, F), +_F, \circ)$ is a ring with identity $\iota(X) = X$.

Similarly,

$$\alpha \circ \beta = abX + \lambda \sum_{k=0}^{\infty} (ab_k + ba_k) X^{p^k},$$

and so

$$\beta \circ \alpha = baX + \lambda \sum_{k=0}^{\infty} (ba_k + ab_k) X^{p^k},$$

from which we see that the theorem is true.

From (15:27), we see that $\mathbf{hom}(F, F)$ is properly contained in $\mathrm{Hom}(F, F)$, so there is more to do. In the following theorems we will obtain the structure of each element in $\mathrm{Hom}(F, F)$.

(15.34) Theorem. *For F as defined in (15:23), the elements of the commutative ring $(\mathrm{Hom}(F, F), +_F, \circ)$ are the elements*

$$mX + t \sum_{k=0}^{\infty} a_k X^{p^k} + \frac{m(m-1)}{2} t X^{p+1}$$

where $m \in \{0, 1, \ldots, p-1\}$ and $a_k \in K$.

Proof. For $\alpha \in \mathbf{hom}(F, F)$, let $m\alpha$ be the sum of m copies of α with respect to the usual $+$, and let $m \cdot \alpha = [(m-1) \cdot \alpha] +_F \alpha$. One easily obtains that

$$m \cdot \alpha = m\alpha + \frac{m(m-1)}{2} at X^{p+1} \qquad (15:28)$$

for any $\alpha \in \mathbf{hom}(F, F)$. So a nonzero $\alpha \in \mathbf{hom}(F, F)$ has order p.

From $(m \cdot \alpha) +_F (n \cdot \alpha) = (n \cdot \alpha) +_F (m \cdot \alpha) = (m + n) \cdot \alpha$, we obtain

$$\left[\frac{m(m-1)}{2} + \frac{n(n-1)}{2} + mn^p\right] t = \left[\frac{n(n-1)}{2} + \frac{m(m-1)}{2} + nm^p\right] t$$

$$= \frac{(m+n)(m+n-1)}{2} t. \qquad (15:29)$$

Looking at the four possibilities for $a, b \in \{0, 1\}$, one can be assured that

$$\left[\frac{m(m-1)}{2} a + \frac{n(n-1)}{2} b + mnab\right] t = \frac{(m+n)(m+n-1)}{2} t.$$

Now

$$(m \cdot \alpha) +_F (n \cdot \beta) = (ma + nb)X$$
$$+ t \sum_{k=0}^{\infty} (ma_k + nb_k)X^{p^k} + \left[\frac{m(m-1)}{2}a + \frac{n(n-1)}{2}b + mnab\right]tX^{p+1},$$

which completes the proof of (15.34).

(15.35) Proposition. *The rules for $+_F$ and $\circ$ in $\mathrm{Hom}(F, F)$ of (15.34)
are: if*

$$\alpha = mX + t\sum_{k=0}^{\infty} a_k X^{p^k} + \left[\frac{m(m-1)}{2}\right]tX^{p+1}$$

and

$$\beta = nX + t\sum_{k=0}^{\infty} b_k X^{p^k} + \left[\frac{n(n-1)}{2}\right]tX^{p+1},$$

then

$$\alpha +_F \beta = (m+n)X + t\sum_{k=0}^{\infty}(a_k + b_k)X^{p^k} + \left[\frac{(m+n)(m+n-1)}{2}\right]tX^{p+1}$$

and

$$\alpha \circ \beta = mnX + t\sum_{k=0}^{\infty}(na_k + mb_k)X^{p^k} + \left[\frac{mn(mn-1)}{2}\right]tX^{p+1}.$$

Proof. The proof offers no complications.

For a commutative ring R with identity, and a unitary R-module M, one
constructs the *idealization of M* by defining the following binary operations
on the cartesian product $R \times M$:

$$(r, m) + (r', m') = (r + r', m + m') \tag{15 : 30}$$

and

$$(r, m) \cdot (r', m') = (rr', rm' + r'm). \tag{15 : 31}$$

(Notice how this is like working with fractions.) With direct verification,
one obtains

(15.36) Theorem. *For a commutative ring R with identity, and a unitary
R-module M, the idealization of M $(R \times M, +, \cdot)$ is a commutatative ring*

with identity $(1,0)$. *The subring* $M' = \{(0,m) \mid m \in M\}$ *is an ideal in* $R \times M$ *isomorphic to* M *as an* R-*module*.

The group of units U *of* $R \times M$ *is*

$$U = \{(r,m) \mid r \text{ is a unit of } R,\ m \in M\},$$

and has normal subgroups $A = \{(r,0) \mid r \text{ is a unit of } R\}$, *and* $B = \{(1,m) \mid m \in M\}$, *where* A *is isomorphic to the group of units of the ring* R, $(B,\cdot)$ *is isomorphic to the group* $(M,+)$, $A \cap B = \{(1,0)\}$, *and* $AB = U$. *An element* $(r,m) \in U$ *has* $(r,m)^{-1} = (r^{-1}, -r^{-2}m)$ *and factorization* $(r,0)(1,r^{-1}m)$.

The nilpotent elements of $R \times M$ *are the elements of* $J(R) \times M$, *where* $J(R)$ *consists of the nilpotent elements of* R, *since* $(r,m)^k = (r^k, kr^{k-1}m)$.

The ideals of $R \times M$ *are the* $R' \times M'$ *where* R' *is an ideal of* R *and* M' *is a submodule of* M *with the additional property that* $R'M \subseteq M'$.

For F as in (15:23), our rings $(\mathrm{Hom}(F,F), +_F, \circ)$ have this structure. For $\alpha = mX + t\sum_{k=0}^{\infty} a_k X^{p^k} + [m(m-1)/2]tX^{p+1}$, define $\Psi(\alpha) = \bigl(m, (a_i)\bigr) \in Z_p \times M$, where $M = K^N$ is the Z_p-module of all sequences from $N = \{0,1,2,\ldots\}$ into K. So M is a unitary Z_p-module. It is easy to see that Ψ is an isomorphism. So, we have

(15.37) Theorem. *For* F *as in (15:23), the ring* $(\mathrm{Hom}(F,F), +_F, \circ)$ *is isomorphic to the idealization of the unitary* Z_p-*module* M, *where* $M = K^N$. *As such, (15.36) applies to determine the group of units of* $\mathrm{Hom}(F,F)$, *the nilpotent elements, and the ideals.*

In studying the elements of $\mathrm{Hom}(F,F)$ for F as in (15:22), let us note that $\alpha \in \mathbf{hom}(F,F)$ can be written as in (15.32) or as

$$\alpha = (a + pa_0)X + \sum_{k=1}^{\infty} a_k X^{p^k}.$$

For convenience here, we shall write

$$\alpha = aX + p\sum_{k=1}^{\infty} a_k X^{p^k}$$

and realize that $a \equiv 1 \bmod p$. Again, one obtains that

$$m \cdot \alpha = m\alpha + \frac{m(m-1)}{2} pa^{p+1} X^{p+1}.$$

Also, $(m \cdot \alpha) +_F (n \cdot \alpha) = (n \cdot \alpha) +_F (m \cdot \alpha) = (m+n) \cdot \alpha$, so

$$p\left[\frac{m(m-1)}{2} + \frac{n(n-1)}{2} + mn^p\right]a^{p+1}$$
$$= p\left[\frac{n(n-1)}{2} + \frac{m(m-1)}{2} + nm^p\right]a^{p+1} = p\left[\frac{(m+n)(m+n-1)}{2}\right]a^{p+1}.$$

Alternatively, if $m \in Z_{p^2}$, then $pm^p \equiv pm \bmod p^2$. So $m \cdot \alpha = ma + [m(m-1)/2]paX^{p+1}$. For $\beta = bX + p\sum_{k=1}^{\infty} b_k X^{p+1}$, where $b \equiv 1 \bmod p$, we have

$$n \cdot \beta = n\beta + \frac{n(n-1)}{2} pb^{p+1} X^{p+1} = n\beta + \frac{n(n-1)}{2} pb X^{p+1}$$

and so

$$(m \cdot \alpha) +_F (n \cdot \beta) = (ma + nb)X + p\sum_{k=1}^{\infty}(ma_k + nb_k)X^{p^k}$$
$$+ p\left[\frac{m(m-1)}{2}a + \frac{n(n-1)}{2}b + mnab\right]X^{p+1}$$
$$= (ma + nb)X + p\sum_{k=1}^{\infty}(ma_k + nb_k)X^{p^k}$$
$$+ p\left[\frac{(ma+nb)(ma+nb-1)}{2}\right]X^{p+1}.$$

We summarize with

(15.38) Theorem. *For F as defined in* (15:22), *the elements of the commutative ring* $(\mathrm{Hom}(F,F), +_F, \circ)$ *are the elements*

$$mX + p\sum_{k=1}^{\infty}a_k X^{p^k} + \frac{m(m-1)}{2}pX^{p+1}$$

where $m \in Z_{p^2}$ and $a_k \in \{0, 1, \dots, p-1\}$.

Note that the summation in (15.34) starts from $k = 0$.

(15.39) Proposition. *The rules for $+_F$ and $\circ$ in* $\mathrm{Hom}(F,F)$ *of* (15.38) *are: if*

$$\alpha = mX + p\sum_{k=1}^{\infty}a_k X^{p^k} + \left[\frac{m(m-1)}{2}\right]pX^{p+1}$$

and

$$\beta = nX + p\sum_{k=1}^{\infty}b_k X^{p^k} + \left[\frac{n(n-1)}{2}\right]pX^{p+1},$$

then

$$\alpha +_F \beta = (m+n)X + p\sum_{k=1}^{\infty}(a_k + b_k)X^{p^k} + \left[\frac{(m+n)(m+n-1)}{2}\right]pX^{p+1}$$

and

$$\alpha \circ \beta = mnX + p\sum_{k=1}^{\infty}(na_k + mb_k)X^{p^k} + \left[\frac{mn(mn-1)}{2}\right]pX^{p+1}.$$

Define M as the set of all sequences from $\{1, 2, \ldots\}$ into Z_p. Then M is a unitary Z_{p^2}-module, and so we have the idealization of M, namely, $Z_{p^2} \times M$. Define $\Psi : \text{Hom}(F, F) \to Z_{p^2} \times M$ by $\Psi(\alpha) = (m, (a_k))$ for $\alpha = mX + p\sum_{k=1}^{\infty}x_k X^{p^k} + [m(m-1)/2]pX^{p+1}$ and F as in (15:22). Then Ψ is a ring isomorphism. So we have

(15.40) Theorem. *For F as in (15.22), the ring $(\text{Hom}(F, F), +_F, \circ)$ is isomorphic to the idealization of the unitary Z_{p^2}-module M, where M is the set of sequences from $\{1, 2, \ldots\}$ into Z_p. As such, (15.36) applies to determine the group of units of $\text{Hom}(F, F)$, the nilpotent elements, and the ideals.*

If the idealization of a module had not already been discovered, we could have discovered the process here. But it has been discovered, and one can find a lot of interesting theory in [N, FGR], where it is called the *principle of idealization*. The reader is invited to see how this theory applies to these idealizations developed here.

There seem to be only nuances between our development for an F as in (15:22) and for an F as in (15:23). Perhaps the results for these two families of formal group laws F are part of a stronger theory.

(15.41) Problem. Develop a unifying theory that includes any F as in (15:22) and any F as in (15:23).

Using (15.24), one should now be able to construct numerous families of formal group laws as defined by (15:21).

(15.42) Problems. Develop a unifying theory for the F of (15:21). In particular, when is $(\text{Hom}(F, F), +_F, \circ)$ a commutative ring? When does $(\text{Hom}(F, F), +_F, \circ)$ have the structure of the idealization of an R-module?

(15.43) Problem. Are there one dimensional noncommutative formal group laws for which $(\text{Hom}(F, F), +_F, \circ)$ is not a (commutative) ring?

(15.44) Exploratory problem. Find one dimensional noncommutative formal group laws F other than those of the form (15:21). Study their $(\text{Hom}(F, F), +_F, \circ)$. Are there some where infinitely many $f_{ij} \neq 0$?

(15.45) Exercise. For a commutative ring R with identity, consider all R-modules M and hence the idealizations of any such M. Make a category

$\mathcal{I}(R)$ from these objects. Define an appropriate category so that the theory of Chapter 3 applies. Can you ever add the ring endomorphisms of an $R \times M$? What do you get?

(15.46) Exploratory problem. For each noncommutative group law, we obtain the nearring $(A_0[[X]], +_F, \circ)$. Study the structure and invariants of some of these nearrings.

(15.47) Problem. These group laws are beginning to suggest group or cogroup objects of an appropriate category. Does the theory of Chapter 3 apply here in some nice way?

If by chance the reader should find all this too trivial, then he is invited to extend these ideas to noncommutative n-dimensional formal group laws. See [H] for definitions.

(15.48) Exploratory problem. Find some interesting noncommutative n-dimensional formal group laws F and explore the resulting nearrings $(A_0[[X]], +_F, \circ)$ and $(\mathrm{Hom}(F, F), +_F, \circ)$.

16. The ideals of abstract affine nearrings

In §2, we defined an abstract affine nearring as follows. We took a ring R and a left R-module M. On the cartesian product $M \times R$ we defined $+$ and $\cdot$ by $(m, r) + (m', r') = (m + m', r + r')$ and $(m, r) \cdot (m', r') = (m + rm', rr')$. Then $(M \times R, +, \cdot)$ is a right nearring which we denote by $M \oplus_A R$, and call an *abstract affine nearring*. This came about from considering the affine transformations of a vector space into itself. For a vector space V over a field F, let $\mathrm{Hom}_F(V, V)$ be the ring of linear transformations of V into itself. For an $L \in \mathrm{Hom}_F(V, V)$ and $\alpha \in V$, define $A : V \to V$ by $Ax = \alpha + Lx$. Then A is an *affine transformation* of V. If $\mathcal{A}(V, F)$ denotes all the affine transformations of V, then $(\mathcal{A}(V, F), +, \circ)$ is a right nearring isomorphic to the abstract affine nearring $V \oplus_A \mathrm{Hom}_F(V, V)$.

The abstract affine nearrings given in Chapter 3 were all right distributive also. That is partly because some came from endomorphisms of cogroup objects. With the cogroup objects, the variable was placed to the left of the function. Above, for the affine transformations of a vector space, the variable was placed to the right of the affine transformations. Again, we see that the placement of the variable in relationship to the function does not in itself decide which distributive law will be valid.

Since the abstract affine nearrings that we have seen up to now are right distributive, we will continue to keep our abstract affine nearrings right distributive. But we should be aware that they could all be left distributive.

For a vector space V over F, let a linear transformation $L \in \mathrm{Hom}_F(V, V)$ act on the right of a vector x. For $\alpha \in V$, define $xA = xL + \alpha$. Then

$(\mathcal{A}(V, F), +, \cdot)$ is a left nearring. If we identify A with (L, α), then $\circ$ will motivate $(L, \alpha)(M, \beta) = (LM, \alpha M + \beta)$. So, for a ring R, let M be a right R-module. On $R \times M$, define $+$ and $\cdot$ by $(r, m) + (r', m') = (r + r', m + m')$ and $(r, m) \cdot (r', m') = (rr', mr' + m')$. Then $(R \times M, +, \cdot)$ is a left nearring. So we see that our abstract affine nearrings could just as easily be left distributive, so the reader should not be concerned if a left distributive abstract affine nearring arises naturally in some context.

We proceed now to calculate the ideals of an abstract affine nearring $M \oplus_A R$, and remember, we are assuming that the right distributive law is valid. Actually, we will do a little more. We can almost concurrently calculate the bimodules in $M \oplus_A R$. These results are entirely due to H. Gonshor [G].

(16.1) Theorem. *Let* $M \oplus_A R$ *be an abstract affine nearring. The ideals of* $M \oplus_A R$ *are exactly the* $S \oplus I$ *where* S *is a submodule of* M *containing* IM, *and* I *is an ideal of* R. *The bimodules in* $M \oplus_A R$ *are exactly the* $M \oplus I$ *where* I *is an ideal of* R.

Proof. Since $IM \subseteq M$ for every ideal I of R, we see that each bimodule in $M \oplus_A R$ will be an ideal of $M \oplus_A R$. Let us review what an ideal J of $M \oplus_A R$ must satisfy. An ideal J must be a normal subgroup, it must be a comodule in $M \oplus_A R$, that is, $J(M \oplus_A R) \subseteq J$, and for any $x, y \in M \oplus_A R$ and any $a \in J$, we must have $x(y + a) - xy \in J$. Since $+$ in $M \oplus_A R$ is commutative, we need not worry about J being a normal subgroup.

Let J be an ideal of $M \oplus_A R$ or a bimodule in $M \oplus_A R$. For $(a, t) \in J$, we must have $(a, t)(0, 0) = (a, 0) \in J$. Hence, $(a, 0), (0, t) \in J$, and so $J = S \oplus I$ where $(S, +)$ is a subgroup of $(M, +)$ and $(I, +)$ is a subgroup of $(R, +)$. If $t \in I$, then $(0, t) \in J$, and so $(0, t)(m, r) = (tm, tr) \in J$, putting $tm \in S$ and $tr \in I$. Hence, $IM \subseteq S$ and I is a right ideal of R. When J is a bimodule, then $(m, r)(0, t) = (m, rt) \in J$, and so $rt \in I$. Hence, I is an ideal of R. Also, $m \in S$, so $S = M$. Thus a bimodule J has the structure $J = M \oplus I$ where I is an ideal of R. Let $J' = M \oplus I$ where I is an ideal of R. For $(a, t) \in J' = M \oplus I$ and $(m, r) \in M \oplus_A R$, we have $(m, r)(a, t) = (m + ra, rt) \in M \oplus I$ and $(a, t)(m, r) = (a + tm, tr) \in M \oplus I$, so $J' = M \oplus I$ is a bimodule in $M \oplus_A R$.

Continuing with $J = S \oplus I$ an ideal of $M \oplus_A R$, we have from above that I is a right ideal of R, that $(S, +)$ is a subgroup of $(M, +)$, and that $IM \subseteq S$. We need to show that S is a submodule of M and that I is a left ideal of R. For $(a, t) \in J = S \oplus I$, and for $(m_1, r_1), (m_2, r_2) \in M \oplus_A R$, we have $(m_1, r_1)[(m_2, r_2) + (a, t)] - (m_1, r_1)(m_2, r_2) = (r_1 a, r_1 t) \in J$. Hence, $r_1 a \in S$ and $r_1 t \in I$. This makes S a submodule of M and I a left ideal of R, hence an ideal of R.

Let S be a submodule of M, I an ideal of R, and suppose that $IM \subseteq S$. For $(a, t) \in S \oplus I$ and $(m, r) \in M \oplus_A R$, we have $(a, t)(m, r) = (a + $

tm, tr). Since $IM \subseteq S$, we have $a + tm \in S$, and since I is an ideal of R, $tr \in I$. Hence, $S \oplus I$ is a comodule in $M \oplus_A R$. As in the preceding paragraph, $(m_1, r_1)[(m_2, r_2) + (a, t)] - (m_1, r_1)(m_2, r_2) = (r_1 a, r_1 t)$. Since S is a submodule, $r_1 a \in S$, and since I is an ideal, $r_1 t \in I$. Hence, $S \oplus I$ is an ideal of $M \oplus_A R$. This completes the proof of (16.1).

With an ideal I of R, we have the quotient ring R/I. With a submodule S of the left R-module M, we have the quotient R-module M/S. But M/S is also a left R/I-module if $IM \subseteq S$. One defines $(r + I)(m + S) = rm + S$, and so we need to show that this is well defined. If $r + I = r' + I$ and $m + S = m' + S$, then $r - r' \in I$, so $(r - r')m, (r - r')m' \in S$, and $rm + S = r'm + S$ and $rm' + S = r'm' + S$. From $m - m' \in S$, we have $r(m - m') \in S$, so $rm + S = rm' + S$. Hence, $rm + S = rm' + S = r'm' + S$. Thus $(r + I)(m + S) = rm + S = r'm' + S = (r' + S)(m' + S)$. Now that the action of R/I on M/S is well defined, one readily obtains that M/S is a left R/I-module. So one can construct the abstract affine nearring $(M/S) \oplus_A (R/I)$. In fact, one obtains

(16.2) Theorem. *Let R be a ring. For a left R-module M, let $S \oplus I$ be an ideal of $M \oplus_A R$. Then*

$$\frac{M \oplus_A R}{S \oplus I} \cong \frac{M}{S} \oplus_A \frac{R}{I}.$$

Proof. Define $\Psi : (M \oplus_A R)/(S \oplus I) \to (M/S) \oplus_A (R/I)$ by $\Psi[(m, r) + (S \oplus I)] = (m + S, r + I)$. It is easy to see that Ψ is well defined and surjective. Also, it is easy to see that Ψ is a group homomorphism, and in fact, a group monomorphism. As you may have guessed, it is also easy to see that Ψ is a semigroup homomorphism, and so, one can summarize by saying that the proof of the theorem is direct.

Theorems (16.1) and (16.2) together show us that the ideals and the homomorphic images of abstract affine nearrings are again abstract affine nearrings.

Exercise. Take one of your favourite rings R with one of its left modules M. Using (16.1), calculate all the ideals of $M \oplus_A R$. Perhaps, with R and M chosen carefully, it might be interesting to find M, R, S, and I so that $IM \subseteq S$.

17. Polynomial nearrings

In §2, we introduced polynomial nearrings $(R[x], +, \circ)$, where R is a commutative ring with identity. We saw in §12 that $(R[x], +, \circ)$ describes the algebra endomorphisms of the R-algebra $(R[x], +, \cdot)$. From (3.51), we get some ideals of $(F[x], +, \circ)$ where F is a field, and from (3.54), we learned that these ideals are maximal as long as $|F| > 2$.

Actually, polynomial nearrings have been studied intensively. Over the centuries, man has found polynomials to be extremely interesting as well as valuable, but Nöbauer has shown that there is still much to be learned about these curiously simple mathematical objects. He demonstrated that it is suitable to talk about polynomials over structures other than rings. His ideas are best explained in the book [L & N]. Also, in [P] one finds many fascinating facts about polynomials. Since this subject has been so well studied, we shall not include much about polynomials in this work, but we wish to emphasize that it is a very rich and interesting field of study, and that there are many more interesting problems of all degrees of difficulty awaiting a solution.

In this section, we shall study some of the ideals of $(R[x], +, \circ)$, their quotient nearrings, and, in a later section, we will look at the groups of units of some of these quotient nearrings.

If $f(x) = f = \sum_{i=0}^{m} a_i x^i$, $g(x) = g = \sum_{j=0}^{n} b_j x^j \in R[x]$, then

$$f \circ g = f(x) \circ g(x) = f \circ g(x) = f(g(x)) = \sum_{i=0}^{m} a_i \Big(\sum_{j=0}^{n} b_j x^j \Big)^i,$$

so $\circ$ is dependent upon regular polynomial multiplication $\cdot$, where $f(x) \cdot g(x) = \sum_{k=0}^{m+n} \big(\sum_{i+j=k} a_i b_j \big) x^k$. In fact, it is convenient and customary to study the structure $(R[x], +, \cdot, \circ)$ concurrently. Since

$$(f(x) \cdot g(x)) \circ h(x) = \sum_{k=0}^{m+n} \Big(\sum_{i+j=k} a_i b_j \Big) (h(x))^k$$

$$= \Big(\sum_{i=0}^{m} a_i h(x)^i \Big) \cdot \Big(\sum_{j=0}^{n} b_j h(x)^j \Big) = (f \circ h(x)) \cdot (g \circ h(x)),$$

we see that $\circ$ is right distributive over $\cdot$ as well as over $+$. (Note: we also write $f \circ g(x)$ for $f(x) \circ g(x) = f \circ g$.) Such structures are called *composition rings*.

(17.1) Definition. Let $(R, +, \cdot)$ be a ring, and suppose that $\circ$ is an associative binary operation on R which is right distributive over $\cdot$ and over $+$. Then $(R, +, \cdot, \circ)$ is a *composition ring*.

(17.2) Theorem. *Let R and S be commutative rings with identity, and let $\phi : R \to S$ be a ring epimorphism. Define $\Phi : R[x] \to S[x]$ by*

$$\Phi \Big(\sum_{i=0}^{m} a_i x^i \Big) = \sum_{i=0}^{m} \phi(a_i) x^i.$$

Then Φ is a composition ring epimorphism.

Proof. It is well known that Φ is a ring epimorphism. From this,

$$\Phi(f \circ g) = \Phi\left(\sum_{i=0}^{m} a_i \left(\sum_{j=0}^{n} b_j x^j\right)^i\right) = \sum_{i=0}^{m} \phi(a_i)\left(\sum_{j=0}^{n} \phi(b_j)x^j\right)^i$$

$$= \left(\sum_{i=0}^{m} \phi(a_i)x^i\right) \circ \left(\sum_{j=0}^{n} \phi(b_j)x^j\right) = \Phi(f) \circ \Phi(g).$$

Hence, Φ is a composition ring epimorphism.

(17.3) Corollary. *The Φ of (17.2) is a nearring epimorphism.*

Composition ring homomorphisms have kernels which are ring ideals as well as nearring ideals.

(17.4) Definition. The kernel of a composition ring homomorphism is a *full ideal*.

(17.5) Theorem. *Let J be an ideal of a commutative ring R with identity 1. Then*

$$(J) = \left\{f = \sum_{i=0}^{m} a_i x^i \in R[x] \mid each\ a_i \in J\right\}$$

is a full ideal of $(R[x], +, \cdot, \circ)$.

Proof. Let $\phi : R \to R/J$ be the natural map $\phi(r) = r + J$. Now apply (17.2) and note that $\ker \Phi = (J)$.

(17.6) Theorem. *Let J be an ideal of a commutative ring R with identity 1. Then*

$$\langle J \rangle = \{f \in R[x] \mid \overline{f}(r) \in J \text{ for all } r \in R\}$$

is a full ideal of $(R[x], +, \cdot, \circ)$, and $(J) \subseteq \langle J \rangle$. (Recall that $\overline{f}$ is the polynomial function $\overline{f} : R \to R$ defined by $f \in R[x]$.)

Proof. Take $f, g \in R[x]$, $a, b \in \langle J \rangle$, and $r \in R$. Then $\overline{(a + b)}r = (\overline{a} + \overline{b})r = \overline{a}r + \overline{b}r \in J$, and $\overline{(-a)}r = -(\overline{a}r) \in J$, so $\langle J \rangle$ is a (normal) subgroup of $(R[x], +)$. Also $\overline{(a \cdot f)}r = \overline{a}r \cdot \overline{f}r \in J$, and $\overline{(a \circ f)}r = \overline{a}(\overline{f}(r)) \in J$, so $a \cdot f$ and $a \circ f$ are each in $\langle J \rangle$. This makes $\langle J \rangle$ a ring ideal of $(R[x], +, \cdot)$, and also gives us $\langle J \rangle \circ R[x] \subseteq \langle J \rangle$.

Finally, $\overline{(f \circ (g + a) - f \circ g)}(r) = \overline{f}(\overline{g}(r) + \overline{a}(r)) - \overline{f}(\overline{g}(r))$. Letting $\overline{g}(r) = t \in R$ and $\overline{a}(r) = j \in J$ and taking $f = \sum_{i=0}^{m} a_i x^i$, we have

$$\overline{f}(t + j) - \overline{f}(t) = \sum_{i=0}^{m} a_i \left[(t + j)^i - t^i\right]$$

$$= \sum_{i=0}^{m} a_i \left[t^i + \sum_{s=1}^{i} \binom{i}{s} t^{i-s} j^s - t^i\right] = \sum_{i=0}^{m} a_i \left(\sum_{s=1}^{i} \binom{i}{s} t^{i-s} j^s\right)$$

Since each of the $\sum_{s=1}^{i}\binom{i}{s}t^{i-s}j^{s} \in J$, $1 \leq i \leq m$, we have all the values $(f \circ (g+a) - f \circ g)(r) \in J$ for all $r \in R$. Hence, $f \circ (g+a) - f \circ g \in \langle J \rangle$, and so $\langle J \rangle$ is a nearring ideal, and also a full ideal.

If we have $f = \sum_{i=0}^{m} a_i x^i$ with each $a_i \in J$, then, certainly, $\overline{f}(r) \in J$ for each $r \in R$. Hence, $(J) \subseteq \langle J \rangle$.

The following theorem shows how the (J) and $\langle J \rangle$ capture the full ideals.

(17.7) Theorem. *For a commutative ring R with 1, if I is a full ideal of $(R[x], +, \cdot, \circ)$, then there is a unique ideal J of R such that $(J) \subseteq I \subseteq \langle J \rangle$, namely $J = I \cap R$.*

Proof. Take $J = I \cap R$ as promised. Then J is an ideal of the subring R of the ring $(R[x], +, \cdot)$. So $(J) = J[x] \subseteq I$. If $a \in I$ and $r \in R$, then $\overline{a}(r) = a \circ r \in I \cap R = J$, so $I \subseteq \langle J \rangle$. Hence, $(J) \subseteq I \subseteq \langle J \rangle$.

Suppose we have an ideal K of R such that $(K) \subseteq I \subseteq \langle K \rangle$. Then $(K) \subseteq I \subseteq \langle J \rangle$, and since $K \subseteq (K)$, we get $K \subseteq \langle J \rangle$. The elements of K are constant polynomials when we think of $K \subseteq \langle J \rangle$, and the constant polynomials of $\langle J \rangle$ are the elements of J, hence $K \subseteq J$. Reversing the roles of K and J, we obtain that $J \subseteq K$. Hence, $J = K$.

(17.8) Definition. Let I and J be as in (17.7). Then J is called the *enclosing ideal* of I.

The enclosing ideals are particularly powerful for $R = F$, a field.

(17.9) Theorem. *For a field F, the composition ring $(F[x], +, \cdot, \circ)$ has a unique maximal ideal $\langle (0) \rangle$, and so $(F[x], +, \cdot, \circ)$ is simple if F is infinite.*

Proof. The only two ideals in F are (0) and F. Now $(F) = \langle F \rangle = F[x]$, so all other ideals I must satisfy $((0)) \subseteq I \subseteq \langle (0) \rangle$. Hence, $\langle (0) \rangle$ is the unique maximal ideal. If $f \in \langle (0) \rangle$, then $\overline{f}(r) = 0$ for all $r \in F$, so if F is infinite, we must have $f = 0$, and so any ideal of $(F[x], +, \cdot, \circ)$ other than $F[x]$ itself must be (0).

At this point, we have a nice opportunity to discuss the question as to whether mathematicians are scientists or not. Now I do not particularly care whether the answer is 'yes' or 'no'; it is enough to be a mathematician. I do remember, however, a departmental faculty meeting where a colleague emphatically said that mathematicians are *not* scientists, and to support his claim, he said that mathematicians do not conduct experiments. This colleague was a physicist by background and training, but had found a home in a mathematics department. I am still surprised that I was the only one who challenged his argument.

Whether or not we have just reported the results of an experiment here in §17, we can apply the scientific method. Let us make some observations about the composition rings $(F[x], +, \cdot, \circ)$ where F is a field.

(1) $(F) = \langle F \rangle$.

(2) There are neither full ideals, ring ideals, nor nearring ideals strictly between (F) and $\langle F \rangle$.

(3) $\langle (0) \rangle$ is the unique maximal ideal.

(4) $\langle (0) \rangle$ is a maximal ideal.

(5) If $|F| < \infty$, then $F[x]/\langle (0) \rangle \cong M(F)$. (See (3.53).)

(6) If $|F| = \infty$, then $\langle (0) \rangle = \big((0) \big) = \{0\}$.

(7) If $|F| = \infty$, then there are neither full ideals, ring ideals, nor nearring ideals strictly between $\big((0) \big)$ and $\langle (0) \rangle$.

(8) If $|F| = \infty$, then $(F[x], +, \cdot, \circ)$ is simple.

We also notice that up to now:

(A) all our full ideals have been of the form (I) or $\langle I \rangle$;

(B) all our nearring ideals in $(R[x], +, \circ)$ have been ideals of the ring $(R[x], +, \cdot)$;

(C) all our maximal ideals have been unique.

To what extent do some of these observations illustrate a broader idea? For example, we could take (1) and (6), and wonder if there are proper ideals I of a commutative ring R with identity 1 for which $(I) = \langle I \rangle$. Likewise, we could take (2) and (7) and wonder if there are proper ideals I of R for which there are neither full ideals, ring ideals, nor nearring ideals strictly between (I) and $\langle I \rangle$ if $(I) \neq \langle I \rangle$. As we proceed through §17, the reader should observe how these observations have influenced what we do. Then he should answer for himself if the scientific method is valid for mathematics, if mathematicians conduct experiments, and even if mathematicians are scientists. Better than that, he should notice which observations listed above have not been adequately addressed, and equally valid, he should make further observations. These are important techniques for successful research in all areas.

It is easy enough to conclude that $(R) = \langle R \rangle = R[x]$ for every commutative ring R with identity 1. It is also quite easy to find any number of maximal full ideals. Because of (17.7), we know that maximal full ideals will be of the form $\langle J \rangle$ where J is a maximal ideal of R. We can use various $(Z_n, +, \cdot)$ to obtain rings R with the desired number of maximal full ideals if this desired number is finite. We can use $(Z, +, \cdot)$ if we want a countably infinite number of maximal full ideals. If X is any topological space and $(\mathcal{C}(X, \mathbf{R}), +, \cdot)$ is the ring of continuous functions $f : X \to \mathbf{R}$, then for $x \in X$, $M_x = \{f \in \mathcal{C}(X, \mathbf{R}) \mid f(x) = 0\}$ is a maximal ideal of $\mathcal{C}(X, \mathbf{R})$, and so $\mathcal{C}(X, \mathbf{R})$ has at least $|X|$ maximal ideals. The $\langle M_x \rangle$ are then maximal full ideals of $(\mathcal{C}(X, \mathbf{R})[x], +, \cdot, \circ)$. Likewise, if R has a unique maximal ideal, then so does $(R[x], +, \cdot, \circ)$ have a unique maximal ideal, and conversely.

Note. To make life easier for all of us, we will assume throughout §17 that R is always a commutative ring with identity 1. Also, if $p \in R[x]$ and $r \in R$, we will write $p(r)$ for $\bar{p}(r)$.

(17.10) Theorem. $R[x]/\langle(0)\rangle \cong M(R)$ *if and only if* R *is a finite field.*

Proof. Suppose $R[x]/\langle(0)\rangle \cong M(R)$. If R is infinite, then $|R| = |R[x]| < |M(R)|$, and since $|R[x]/\langle(0)\rangle| \leq |R[x]|$, we cannot have $R[x]/\langle(0)\rangle \cong M(R)$. So our R must be finite. If R has a proper ideal J, then $\langle(0)\rangle \subseteq \langle J\rangle$, and so $\langle J\rangle/\langle(0)\rangle$ is isomorphic to a nearring ideal in $M(R)$. Hence, $R = Z_2$ by (3.46). Otherwise, R must be simple and so R is a finite field. The converse is just (5).

The proof of (17.10) helps us to find a nearring ideal of $(Z_2[x], +, \circ)$ that is maximal and is not a full ideal. We just look to (3.45) and the fact that $Z_2[x]/\langle(0)\rangle \cong M(Z_2)$. Note that our ideals $\langle(0)\rangle$ in the nearrings $F[x]$ are exactly the $A(p^n)$ of (3.51). Let B denote the polynomials of $Z_2[x]$ which define constant functions on Z_2. From $Z_2[x]/\langle(0)\rangle \cong M(Z_2)$, B corresponds to $M(Z_2)_c$, and B is a nearring ideal of $(Z_2[x], +, \circ)$, and B is certainly maximal since $|Z_2[x]/B| = 2$. It is not a full ideal since there is no ideal J of Z_2 for which $(J) \subseteq B \subseteq \langle J\rangle$. In fact, $B = \{(x^2 - x) \cdot f + j \mid f \in Z_2[x], j \in Z_2\}$. Summarizing, we have

(17.11) Theorem. *The nearring* $(Z_2[x], +, \circ)$ *has* $B = \{(x^2 - x) \cdot f + j \mid f \in Z_2[x], j \in Z_2\}$ *as a maximal ideal, and* $(Z_2[x]/B, +, \circ) \cong (Z_2, +, \cdot)$. *Since* $\langle(0)\rangle$ *is properly contained in* B, *we obtain that* B *is not a full ideal of* $(Z_2[x], +, \cdot, \circ)$.

Now 2 is really an *odd* prime. The ideal B of (17.11) is not the only maximal ideal.

(17.12) Theorem. *Let c be a root of the irreducible polynomial $1 + x + x^2 \in Z_2[x]$, and let $F = Z_2(c)$ be the quadratic extension of Z_2. Define $\Gamma : Z_2[x] \to Z_2$ by $\Gamma(p) = p(c)^2 + p(c)$. Then Γ is a nearring epimorphism, and so $T = \ker \Gamma$ is a maximal ideal of $(Z_2[x], +, \circ)$ which is not a full ideal of $(Z_2[x], +, \cdot, \circ)$, and is distinct from B of (17.11), and consists of all the $p \in Z_2[x]$ such that $p(c)^2 + p(c) = 0$.*

Proof. For $p \in F[x]$, $p(c) \in F = \{0, 1, c, 1 + c\}$. If $p(c) \in \{0, 1\}$, then $p(c)^2 + p(c) = 0$. If $p(c) \in \{c, 1 + c\}$, then $p(c)^2 + p(c) = 1$. Hence, $p(c)^2 + p(c) \in Z_2$ for all $p \in F[x]$, and in particular, for $p \in Z_2[x]$. Hence, $\Gamma(p) = p(c)^2 + p(c)$ defines $\Gamma : Z_2[x] \to Z_2$.

Since $c^2 + c + 1 = 0$, we have $\Gamma(1 + x^2) = 1$ and $\Gamma(1 + x + x^2) = 0$, so Γ is surjective. It is easy to see that $\Gamma(p + q) = \Gamma(p) + \Gamma(q)$, so Γ is a group epimorphism with kernel T. It remains to show that $\Gamma(p \circ q) = \Gamma(p) \cdot \Gamma(q)$.

Take $p = p_0 + p_1 x + \cdots + p_m x^m \in Z_2[x]$. Now

$$\Gamma(p \circ q) = [p(q(c))]^2 + p(q(c)) \qquad (17:1)$$

and

$$\Gamma(p) \cdot \Gamma(q) = [p(c)^2 + p(c)] \cdot [q(c)^2 + q(c)]. \qquad (17:2)$$

If $q(c) = 0$, then $\Gamma(p \circ q) = p(0)^2 + p(0) = 0$ and $\Gamma(q) = 0$, so $\Gamma(p \circ q) = \Gamma(p) \cdot \Gamma(q)$. If $q(c) = 1$, then $\Gamma(p \circ q) = p(1)^2 + p(1) = (p_0 + \cdots + p_m)^2 + (p_0 + \cdots + p_m) = 0$ and $\Gamma(q) = 0$, so $\Gamma(p \circ q) = \Gamma(p) \cdot \Gamma(q)$. Also, if $q(c) = c$, then $\Gamma(p \circ q) = p(c)^2 + p(c)$, and $\Gamma(q) = 1$, so $\Gamma(p \circ q) = \Gamma(p) \cdot \Gamma(q)$.

Finally, if $q(c) = 1+c$, then $\Gamma(p \circ q) = p(1+c)^2 + p(1+c)$ and $\Gamma(p) \cdot \Gamma(q) = p(c)^2 + p(c)$. It remains to show that $p(c)^2 + p(c) = p(1+c)^2 + p(1+c)$. From $p(x)^2 = (p_0 + p_1 x + \cdots + p_m x^m)^2 = p_0 + p_1 x^2 + \cdots + p_m x^{2m}$, we obtain $p(x)^2 + p(x) = p_1(x^2 + x) + p_2(x^4 + x^2) + \cdots + p_m(x^{2m} + x^m)$. Hence,

$$p(1+c)^2 + p(1+c) = p_1\left[(1+c)^2 + (1+c)\right] + \cdots + p_m\left[(1+c)^{2m} + (1+c)^m\right]$$

and

$$p(c)^2 + p(c) = p_1[c^2 + c] + \cdots + p_m[c^{2m} + c^m].$$

We will be through upon showing that $(1+c)^{2k} + (1+c)^k = c^{2k} + c^k$. From $1 + c + c^2 = 0$, we get $1 + c^2 = c$ and $1 + c = c^2$. Hence, $(1+c)^{2k} + (1+c)^k = [(1+c)^2]^k + (c^2)^k = (1+c^2)^k + c^{2k} = c^k + c^{2k} = c^{2k} + c^k$, so we are through. This means that $\Gamma(p \circ q) = \Gamma(p) \cdot \Gamma(q)$ when $q(c) = 1 + c$. Since we have considered all possibilities for $q(c)$, we have that Γ is a nearring epimorphism from $Z_2[x]$ onto the field Z_2, a nearring without proper ideals, hence $T = \ker \Gamma = \{p \in Z_2[x] \mid p(c)^2 + p(c) = 0\}$ is a maximal ideal. Since $x^3 \in T$ and $x^3 \notin B$, we have $B \neq T$. Since $1 \in T$, we have that T is not a full ideal of $(Z_2[x], +, \cdot, \circ)$.

What are the elements of T, really? We know $p \in T$ if and only if $p(c)^2 + p(c) = 0$. This next theorem defines the elements of T more explicitly.

(17.13) Theorem. *The ideal T of (17.12) is the subspace of the vector space $(Z_2[x], +)$ over Z_2 generated by*

$$S = \{1, x + x^2, x^3, x + x^4, x + x^5, x^6, \ldots, x + x^{3k+1}, x + x^{3k+2}, x^{3k+3}, \ldots\}.$$

Proof. If $p \in S$, then $p(c)^2 + p(c) = 0$, so if q is a linear combination of elements in S, we have $q(c)^2 + q(c) = 0$, also. Hence, if T' is the subgroup (hence subspace) of $Z_2[x]$ generated by the elements of S, then $T' \subseteq T$. As soon as we show that the index $[Z_2[x] : T'] = 2$, we will have $T' = T$. We will do this by showing that any $p \in Z_2[x] \setminus T'$ has the form $p = x + d$, where $d \in T'$, and this will be proven by induction on the degree of p. With $p \notin T'$, we have $p \notin \{0, 1\}$. If p has degree 1, then $p = x$ or $p = x + 1$. Hence, p has the required form. If p has degree 2, then $p = x^2$ or $p = x^2 + 1$ since $p \notin T'$. If $p = x^2$, then $p = x + (x + x^2)$, and if $p = x^2 + 1$, then $p = x + (x + x^2 + 1)$. Certainly, $x + x^2, x + x^2 + 1 \in T'$.

Assume that the polynomials of degree $< s$ which are not in T' have the desired form. Take $p = p_0 + p_1 x + \cdots + p_s x^s$ with $p_s = 1$ and $r = p_0 + p_1 x + \cdots + p_{s-1} x^{s-1}$. If $r \notin T'$, then $r = x + r'$ with $r' \in T'$. Hence, $p = x + r' + x^s$. If $s = 3k$, then $p = x + r' + x^{3k}$ and $r' + x^{3k} \in T'$. If $s \in \{3k + 1, 3k + 2\}$, then $p = x + r' + x^s = r' + x + x^s \in T'$, a contradiction. Next, suppose $r \in T'$, and we start with $p = r + x^s$. If $s = 3k$, then $p \in T'$. If $s \in \{3k + 1, 3k + 2\}$, then $p = r + x + x + x^s = x + (r + x + x^s)$ with $r + x + x^s \in T'$. By induction, we have our desired result for p. So T' has exactly two cosets, T' and $x + T'$, so $[Z_2[x] : T'] = 2$, which forces $T' = T$.

We know that $B \cap T = I$ is an ideal of $(Z_2[x], +, \circ)$, but exactly what is $Z_2[x]/I$? For

$$
\begin{aligned}
p &= p_0 + p_1 x + p_2 x^2 + p_3 x^3 + (x^{3k_1} + \cdots + x^{3k_s}) \\
&\quad + (x^{3l_1+1} + \cdots + x^{3l_t+1}) + (x^{3m_1+2} + \cdots + x^{3m_u+2}) \\
&= p_0 + (p_1 + t + u)x + p_2 x^2 + (p_3 + s)x^3 \\
&\quad + [(x^3 + x^{3k_1}) + \cdots + (x^3 + x^{3k_s})] \\
&\quad + [(x + x^{3l_1+1}) + \cdots + (x + x^{3l_t+1})] \\
&\quad + [(x + x^{3m_1+2}) + \cdots + (x + x^{3m_u+2})]
\end{aligned}
\tag{17:3}
$$

in $Z_2[x]$, we have each $p_0, x^3 + x^{3k_i}, x + x^{3l_i+1}, x + x^{3m_i+2} \in B \cap T = I$, so $p + I = (p_1 + t + u)x + p_2 x^2 + (p_3 + s)x^3 + I$, and the possible coset representatives of I are $0, x^3, x^2, x^2 + x^3, x, x + x^3, x + x^2$, and $x + x^2 + x^3$. But $I = x + x^2 + I$, $x^3 + I = x + x^3 + I$, $x + I = x^2 + I$, and $x^2 + x^3 + I = x + x^3 + I$. So $|Z_2[x]/I| \leq 4$. With $|Z_2[x]/B| = |Z_2[x]/T| = 2$ and $B \neq T$, we must have $Z_2[x]/I = \{I, x + I, x^3 + I, x + x^3 + I\}$. It is easy to see that each element in $Z_2[x]/I$ is an idempotent for $\circ$ and that $\circ$ is commutative for $Z_2[x]/I$, and so $Z_2[x]/I \cong Z_2 \oplus Z_2$, the direct sum of the field Z_2 with itself. Hence,

(17.14) Theorem. *For $I = B \cap T$, the nearring $Z_2[x]/I \cong Z_2 \oplus Z_2$, the direct sum of the field Z_2 with itself.*

Note that $1 \in B \cap T$. This must be for each maximal ideal of $(Z_2[x], +, \circ)$.

(17.15) Lemma. *Every maximal ideal of $(Z_2[x], +, \circ)$ contains 1.*

Proof. If M is a maximal ideal of $(Z_2[x], +, \circ)$ and $M \neq B$, then there is a $p \in M$ such that $p(0) \neq p(1)$. Now $p \circ 0, p \circ 1 \in M$ for $0, 1 \in Z_2[x]$, and one of these is 1.

With 1 in every maximal ideal M of $(Z_2[x], +, \circ)$, it is in the intersection of all such maximal ideals. But what is the smallest ideal K of $(Z_2[x], +, \circ)$ with $1 \in K$? We want to show that $K = I = B \cap T$.

If K is an ideal of $(Z_2[x], +, \circ)$ and $1 \in K$, what else is in K? Since $x^3 \circ (x+1) - x^3 \circ x = x^2 + x + 1 \in K$ and $1 \in K$, we obtain the useful fact that $x^2 + x \in K$. Also, note that $0 = x + x = x + x^1 \in K$, so $x + x^1, x + x^2 \in K$. We want to show that each $B_k = \{x + x^{3k+1}, x + x^{3k+2}\} \subseteq K$, $k = 0, 1, 2, \ldots$. Toward this end, $(x + x^2) \circ x^k = x^k + x^{2k}$, so each

$$x^k + x^{2k} \in K, \quad k = 0, 1, 2, \ldots. \qquad (17:4)$$

If $t \in K$, then $x^k \circ (t + 0) - x^k \circ 0 = t^k \in K$, so

$$t \in K \implies t^k \in K, \quad k = 0, 1, 2, \ldots. \qquad (17:5)$$

Let $P_3 = \{3k + 1, 3k + 2 \mid k = 0, 1, 2, \ldots\}$, so P_3 is the disjoint union of the pairs $A_k = \{3k + 1, 3k + 2\}$, $k = 0, 1, 2, \ldots$. For each pair A_k, we will show $B_k \subseteq K$ and we already have $B_0 \subseteq K$.

Each A_k has an even number $2s_k \in A_k$. If $x + x^s \in K$, then $x + x^s + x^s + x^{2s} = x + x^{2s} \in K$, by (17:4). Hence,

$$x + x^s \in K \implies x + x^{2s} \in K. \qquad (17:6)$$

So, if we have $B_0 \cup B_1 \cup \cdots \cup B_{k-1} \subseteq K$, then $x + x^{2s_k} \in K$. Since $x + x^2 \in K$, we obtain $x + x^4, x + x^8, \ldots \in K$. Also, $(x + x^2)^3 - (x + x^2) \circ x^3 = x^4 + x^5 \in K$. With $x + x^4 \in K$, we now have $x + x^5 \in K$, and so $B_1 \subseteq K$. We have only to get $x + x^7 \in K$ in order to get $B_2 \subseteq K$, and this requires some computation.

To obtain $x + x^7 \in K$, first note that $(x + x^5)^3 - (x + x^5) \circ x^3 = x^7 + x^{11} \in K$, and that $(x + x^4)^5 - (x + x^4) \circ x^5 = x^8 + x^{17} \in K$. Hence, $x + x^{17}, x^4 + x^{17} \in K$. Next, $(x^4 + x^{17})^3 - (x^4 + x^{17}) \circ x^3 = x^{25} + x^{38} \in K$, and $x + x^5 \in K$ implies $(x + x^5) \circ x^5 = x^5 + x^{25} \in K$. Then we obtain $x^5 + x^{38} \in K$, and so $x + x^{38} \in K$. But (17:4) assures us that $x^{19} + x^{38} \in K$, so $x + x^{19} \in K$. With $x + x^{17} \in K$ and $x + x^5 \in K$, we have $(x + x^{17})^3 - (x + x^{17}) \circ x^3 - (x + x^5) \circ x^4 = x^7 + x^{19} \in K$. From $x + x^{19} \in K$, we finally obtain $x + x^7 \in K$. Now we have $B_2 \subseteq K$.

If $B_0 \cup B_1 \cup \cdots \cup B_{2u} \subseteq K$, we obtain $B_{2u+1} \subseteq K$. We will now prove this. Certainly, $A_{2u+1} = \{6u + 4, 6u + 5\}$. With $x + x^2, x + x^{3(2u+1)-2} \in K$, we obtain $(x^2 + x^{6u+1})^3 - (x^2 + x^{6u+1}) \circ x^3 = x^{6u+5} + x^{12u+4} \in K$. Now $6u + 2 \in P_3$ and $6u + 2 < 6u + 4$, so $x + x^{6u+2} \in K$. Hence, $x + x^{12u+4} \in K$ by (17:6). Adding this to $x^{6u+5} + x^{12u+4}$, we obtain $x + x^{6u+5} \in K$. Also, $3u + 2 < 6u + 4$, so $x + x^{3u+2} \in K$, and so $x + x^{6u+4} \in K$. We now have $B_{2u+1} \subseteq K$.

Suppose m and b are positive integers satisfying (1) 3 does not divide b, (2) $1 \leq b \leq 3m - 1$, (3) m is odd, and (4) $3m + b = 6s + 1$. We have $B_{2s+1} \subseteq K$ if $B_0 \cup B_1 \cup \cdots \cup B_{2s} \subseteq K$, but how do we obtain $B_{2s} \subseteq K$ if $B_0 \cup B_1 \cup \cdots \cup B_{2s-1} \subseteq K$? Now $A_{2s} = \{6s + 1, 6s + 2\}$, and we have $x + x^{6s+2} \in K$ by induction hypothesis and (17:6). So we need to obtain $x + x^{6s+1} \in K$. With an m and b satisfying (1)–(4), we have $x + x^b, x + x^{3m-b} \in K$, so $x^b + x^{3m-b} \in K$. This assures us that $(x^b + x^{3m-b})^3 - (x^b + x^{3m-b}) \circ x^3 = x^{3m+b} + x^{6m-b} \in K$. With m odd and $3m + b = 6s + 1$, we have b even, so $6m - b$ is even also. From $-1 < 3m$, we have $3m - 1 < 6m$, so $b \leq 3m - 1 < 6m$ makes $0 < 6m - b$. Since $6m$ and b are even, we have that $3m - b/2$ is a positive integer which is not divisible by 3 since 3 does not divide b. So $3m - b/2 < 2m + b = 6s + 1$, and so $x + x^{3m-b/2} \in K$, and then $x + x^{6m-b} \in K$ by (17:6). With $x^{3m+b} + x^{6m-b} \in K$ already, we have $x + x^{3m+b} = x + x^{6s+1} \in K$.

As long as we have (m, b) as above and $B_0 \cup \cdots \cup B_{2s-1} \subseteq K$, we also obtain $B_{2s} \subseteq K$. One can quickly see that for $s = 1$, there is no such (m, b), and this is why we had to treat $x + x^7$ separately. But, for $s \geq 2$, $3(2s - 1) + 4 = 6s + 1$, so $(m, b) = (2s - 1, 4)$ is a candidate. Certainly, 3 does not divide 4 and $2s - 1 = m$ is odd. By definition, $3m + b = 6s + 1$, so we only need to show $1 \leq b \leq 3m - 1$, that is, $1 \leq 4 \leq 6s - 4$. But $2 \leq s$, so $8 \leq 6s - 4$, hence $1 \leq 4 < 8 \leq 6s - 4$ assures us that conditions (1)–(4) are met. So we have $x + x^{6s+1} \in K$, and with $x + x^{6s+2} \in K$, we conclude that $B_{2s} \subseteq K$.

By induction, we have each $B_k \subseteq K$. This means $x + x^u \in K$ for each $u \in P_3$. Our next task will be to show that each $x^{3m} + x^{3n} \in K$, where m and n are positive integers.

With $x + x^2, x + x^4, x + x^5 \in K$, we have $(x^2 + x^5)^3 - (x^2 + x^5) \circ x^3 - (x + x^4) \circ x^3 = x^3 + x^9 \in K$. We already have $x^3 + x^6 \in K$ by (17:4), and so $x^6 + x^9 \in K$. Let $b = cd$, where $c = 3^a$ and 3 does not divide d. We have seen that $x + x^d \in K$, so $(x + x^d) \circ x^c = x^c + x^{cd} \in K$. For $c \geq 9$, $(x^3 + x^9) \circ x^{c/9} = x^{c/3} + x^c \in K$. So we have $x^{3^k} + x^{3^{k+1}} \in K$ for $k = 1, 2, \ldots$, and in particular, $x^9 + x^{27} \in K$. We easily obtain $x^3 + x^{3^k} \in K$ for each $k = 1, 2, \ldots$. Since $x^3 + x^{cd} = (x^c + x^{cd}) + (x^3 + x^c) \in K$, and $x^3 + x^{3d} = (x + x^d) \circ x^3 \in K$, and $x^9 + x^{9d} = (x + x^d) \circ x^9 \in K$, we obtain $x^3 + x^{9d} \in K$, and so $x^3 + x^{cd} \in K$ for each $c = 3^k$, $k = 0, 1, 2, \ldots$. This puts $x^3 + x^{3m} \in K$ for each $m = 1, 2, \ldots$, and so $(x^3 + x^{3m}) + (x^3 + x^{3n}) = x^{3m} + x^{3n} \in K$.

In the proof of (17.14), we started by asking what $Z_2[x]/I$ was. Following that same argument for $Z_2[x]/K$, we arrive at exactly the same conclusion, since all $x + x^d \in K$, when 3 does not divide d, since all $x^{3m} + x^{3n} \in K$, and since $1 \in K$. Hence, $K = I$. So we have

(17.16) Theorem. *The intersection of all maximal ideals M of the near-*

ring $(Z_2[x], +, \circ)$ *is* $I = B \cap T$.

Exercise. Show that B and T are the only maximal ideals of the nearring $(Z_2[x], +, \circ)$.

For which fields F does the nearring $(F[x], +, \circ)$ have an ideal which is not an ideal of $(F[x], +, \cdot)$, that is, which is not a full ideal? The next result shows that they will not be among the infinite fields.

(17.17) Theorem. *For an infinite field F, the nearring $(F[x], +, \circ)$ is simple.*

Proof. Suppose $I \neq \{0\}$ is an ideal of the nearring $(F[x], +, \circ)$. For $0 \neq f \in I$ and $\beta \in F \subset F[x]$, we have $f \circ \beta = \alpha \in I$ for some $\alpha \in F$. Since $|F| = \infty$, there is a nonzero $\alpha \in I \cap F$. In fact, this puts all but finitely many of F's elements in I, since f has only finitely many roots in F. So we have $F \subseteq I$. Since $x, x^2 \in F[x]$, we have $x^2 \circ (x + \alpha) - x^2 \circ x = 2\alpha x + \alpha^2 \in I$ for each $\alpha \in F$.

Assume first that the characteristic of F is $\neq 2$. Then with $\alpha^2 \in I$, we have $2\alpha x \in I$. With $\alpha = 2^{-1} \in I$, we have $x \in I$, so $I = F[x]$.

Now we turn to the case where the characteristic of F is 2. With $\alpha^{-1}x^3 \circ (x + \alpha) - (\alpha^{-1}x^3) \circ x = x^2 + \alpha x + \alpha^2 \in I$ and with $\alpha^2 \in I$, we have all $x^2 + \alpha x \in I$. In particular, $x^2 + x \in I$, so all $(\alpha + 1)x \in I$, for $\alpha \neq 0$. Take $\alpha \in F \setminus \{0, 1\}$. So $(\alpha + 1)x \circ (\alpha + 1)^{-1}x = x \in I$, and again we have $I = F[x]$.

So, are there any finite fields F, other than Z_2, for which $(F[x], +, \circ)$ has an ideal which is not a full ideal? The next result shows that they will not have characteristic > 2.

(17.18) Theorem. *If F is a finite field with characteristic > 2, then every ideal of $(F[x], +, \circ)$ is a full ideal.*

Proof. Let K be an ideal of $(F[x], +, \circ)$. When we get $f \cdot p \in K$ for each $f \in F[x]$ and each $p \in K$, we will know that K is an ideal of $(F[x], +, \cdot)$, and hence a full ideal. Now $x^2 \circ (f + p) - x^2 \circ f = 2fp + p^2 \in K$. Also, $x^2 \circ (0 + p) - x^2 \circ 0 = p^2 \in K$. Hence, $2fp \in K$. Since F has characteristic > 2, we have $fp = f \cdot p \in K$, as was to be shown.

We finally get a positive answer with all finite fields of characteristic 2.

(17.19) Theorem. *Let F be a finite field of order q and characteristic 2. For each n, $n = 1, 2, \ldots$, define*

$$K_n = (x^{q^n} + x)^2 F[x^2] + (x^{q^n} + x)^4 F[x].$$

Then each K_n is an ideal of $(F[x], +, \circ)$ which is not a full ideal.

Proof. It is easy to see that each K_n is a subgroup of $(F[x], +)$. We will let $k = k(x) = (x^{q^n} + x)^2 h(x^2) + (x^{q^n} + x)^4 l(x)$ represent an arbitrary element of K_n.

The terms of $(x^{q^n} + x)^2 h(x^2)$ all have even degree, and the terms of $(x^{q^n} + x)^4 l(x)$ have degree ≥ 4. So k cannot have a term of degree 3. Now $(x^{q^n} + x)^2 \in K_n$ and $(x^{q^n} + x)^2 \cdot x$ has a term of degree 3, so $(x^{q^n} + x)^2 \cdot x \notin K_n$. This shows that K_n cannot be a full ideal. We proceed now to show that K_n is an ideal of $(F[x], +, \circ)$. Now $w = x^{q^n} + x$ is the monic polynomial which has every element of a field of order q^n as a root of multiplicity 1. If $f \in F[x]$, then $w \circ f$ has every element of this field of order q^n as a root, so w divides $w \circ f$, that is, $(f(x)^{q^n} + f(x)) = (x^{q^n} + x)m(x)$ for some $m \in F[x]$. Hence, $(f(x)^{q^n} + f(x))^2 = (x^{q^n} + x)^2 m(x)^2 \in (x^{q^n} + x)^2 F[x^2]$ since $m(x)^2 \in F[x^2]$. Also, $(f(x)^{q^n} + f(x))^4 = (x^{q^n} + x)^4 m(x)^4$, so for $k \in K_n$, we have $k \circ f = (f(x)^{q^n} + f(x))^2 h(f(x)^2) + (f(x)^{q^n} + f(x))^4 l(f(x)) = (x^{q^n} + x)^2 m(x)^2 h(f(x)^2) + (x^{q^n} + x)^4 m(x)^4 l(f(x)) \in K_n$ since $f(x)^2 \in F[x^2]$ also.

It remains to show that $f \circ (g + k) - f \circ g \in K_n$ for arbitrary $f, g \in F[x]$. Let $f = \sum_{i=0}^n a_i x^i$. Then

$$f \circ (g + k) - f \circ g = \sum_{i=1}^n a_i \left[(g + k)^i - g^i \right]$$

$$= a_1 k + a_2 k^2 + \sum_{i=3}^n a_i \left[(g + k)^i - g^i \right]$$

$$= a_1 k + a_2 k^2 + \sum_{i=3}^n a_i \left[\sum_{j=1}^i \binom{i}{j} g^{i-j} k^j \right]$$

$$= a_1 k + a_2 k^2 + \sum_{i=3}^n a_i \left[ig^{i-1} k + \sum_{j=2}^i \binom{i}{j} g^{i-j} k^j \right]$$

$$= a_1 k + \left(\sum_{i=3}^n i a_i g^{i-1} \right) k + a_2 k^2 + \left[\sum_{i=3}^n \left(\sum_{j=2}^i \binom{i}{j} g^{i-j} k^{j-2} \right) \right] k^2$$

$$= Gk + Hk^2,$$

for $G = a_1 + a_3 g^2 + \cdots + a_{2s+1} g^{2s}$ for some s, and

$$H = a_2 + \sum_{i=3}^n \left(\sum_{j=2}^i \binom{i}{j} g^{i-j} k^{j-2} \right).$$

Now $Hk^2 \in (x^{q^n} + x)^4 F[x]$ and $G(x)k(x) = (x^{q^n} + x)^2 h(x^2)G(x) + (x^{q^n} + x)^4 l(x)G(x)$. Since each $g^{2i} \in F[x^2]$, so is $h(x^2)G(x)$, hence $Gk \in K_n$.

Now we have $f \circ (g + k) - f \circ g = Gk + Hk^2 \in K_n$, and so each K_n is an ideal.

Let us recollect some of what we have seen. For a finite field F, $F[x]/\langle(0)\rangle$ $\cong M(F)$, and since $M(F)$ is simple, except for $F = Z_2$, we have that $\langle(0)\rangle$ is a maximal ideal of $(F[x], +, \circ)$ for all finite fields F except Z_2. With Z_2, there are two maximal ideals of $(Z_2[x], +, \circ)$, namely B and T. If F is infinite, then $(F[x], +, \circ)$ is simple. If $|F| < \infty$ and the characteristic of F is > 2, then each ideal of $(F[x], +, \circ)$ is a full ideal. If $|F| = q = 2^t$, then $(F[x], +, \circ)$ has ideals K_n which are not full ideals.

For Z_2, $I = B \cap T$ is the intersection of all maximal ideals of $(Z_2[x], +, \circ)$, with B and T being the only maximal ideals. In fact I is the ideal generated by 1. The nearrings $Z_2[x]/B$ and $Z_2[x]/T$ are each isomorphic to the field Z_2; $Z_2[x]/\langle(0)\rangle \cong M(Z_2)$, and $Z_2[x]/I \cong Z_2 \oplus Z_2$. Of course, $A = \langle(0)\rangle \cap I$ is also an ideal of $(Z_2[x], +, \circ)$. It will be rewarding to study A and $Z_2[x]/A$ further. It is easy to see that $T \cap B = I = \{(x^2 + x)f + j \mid f \in T, \ j \in Z_2\}$, and that $T \cap \langle(0)\rangle = \{(x^2 + x)f \mid f \in T\}$. Recall that $f \in T$ if and only if $f(c)^2 + f(c) = 0$.

Now I is generated by 1 and all $x^3 + x^{3a}$, and all $x + x^b$, where 3 and b are relatively prime. All of these generators are in $\langle(0)\rangle$ except 1. So A is generated by all $x^3 + x^{3a}$ and all $x + x^b$ with $(3, b) = 1$. As we did for (17.14), we consider the representatives for $p + A \in Z_2[x]/A$. Consider p as in (17:3). Then $p + A = p_0 + (p_1 + t + u)x + p_2 x^2 + (p_3 + s)x^3 + A$. Of these sixteen possible elements of $Z_2[x]/A$, we note that $A = x + x^2 + A$, so $|Z_2[x]/A| = 8$ and the elements of $Z_2[x]/A$ have representatives $0, 1, x, 1 + x, x^3, 1 + x^3, x + x^3$, and $1 + x + x^3$.

If we identify $[a, (b, c)] \equiv a + bx + cx^3 + A$, then from $(a + bx + cx^3) \circ (d + ex + fx^3) + A = (a + bd) + bex + (bf + cf + ce)x^3 + A \equiv [a + (b + c)d, (be, bf + cf + ce)]$, we are motivated to start with a ring R, an R-module A, and an abelian group M which is a left R-module and a left A-module. Define $+$ on $N = M \times (R \times A)$ by $[m, (r, a)] + [n, (s, b)] = [m + n, (r + s, a + b)]$, and define $\cdot$ on N by

$$[m, (r, a)] \cdot [n, (s, b)] = [m + rn + an, (rs, rb + sa + ab)]. \qquad (17:7)$$

This in turn motivates defining $+$ on $S = R \times A$ by $(r, a) + (s, b) = (r + s, a + b)$, defining $\cdot$ on S by

$$(r, a) \cdot (s, b) = (rs, rb + sa + ab), \qquad (17:8)$$

and letting S act on M by

$$(r, a)n = rn + an. \qquad (17:9)$$

It is now routine to verify

(17.20) Theorem. *For a ring R, an R-algebra A, and an abelian group M which is both a left R-module and a left A-module, let $(N, +, \cdot)$ and $(S, +, \cdot)$ be as defined above. Then $(S, +, \cdot)$ is a ring, M is an S-module, and N is the abstract affine nearring $M \oplus_A S$. Furthermore, if R is commutative so is $(S, +, \cdot)$. If $1 \in R$ is an identity, then $(1, 0)$ is the identity of S. If M is unitary for R, then M is unitary for S.*

Note. With $R = Z$, the ring of integers, we have that $(S, +, \cdot)$ is the standard way to embed the ring A into a ring with identity.

(17.21) Corollary. *The nearring $Z_2[x]/A$ is isomorphic to the abstract affine nearring $M \oplus_A S$ where S is obtained by letting $R = Z_2$, and $M = Z_2$, also.*

There are lots of full ideals between $((0)) = \{0\}$ and $\langle (0) \rangle$ for $(F[x], +, \cdot, \circ)$ when F is a finite field.

(17.22) Theorem. *Let F be a finite field with $q = |F|$. Suppose that $n_1, n_2, \ldots, n_k$ are positive integers with $1 \leq n_1 < n_2 < \cdots < n_k$, and that each m_i is a positive integer, $1 \leq i \leq k$. Let p be the least common multiple*

$$p = \text{l.c.m.}\{(x^{q^{n_i}} - x)^{m_i} \mid 1 \leq i \leq k\}.$$

Then the principal ideal (p) of $(F[x], +, \cdot)$ is a full ideal of $(F[x], +, \cdot, \circ)$.

Proof. For $f, g \in F[x]$, we have $(p \cdot f) \circ g = (p \circ g) \cdot (f \circ g)$. So, to get $a \circ g \in (p)$ for all $a \in (p)$ and all $g \in F[x]$, we need only show $p \circ g \in (p)$. To do that, we need only show that $x^{q^n} - x$ divided $(x^{q^n} - x) \circ g$ in $(F[x], \cdot)$.

Let $g = g_0 + g_1 x + \cdots + g_k x^k$. Then $(x^{q^n} - x) \circ g = g^{q^n} - g = (g_0 + g_1 x + \cdots + g_k x^k)^{q^n} - (g_0 + g_1 x + \cdots + g_k x^k) = (g_0 + g_1 x^{q^n} + g_2 x^{2q^n} + \cdots + g_k x^{kq^n}) - (g_0 + g_1 x + g_2 x^2 + \cdots + g_k x^k) = g_1(x^{q^n} - x) + g_2(x^{2q^n} - x^2) + \cdots + g_k(x^{kq^n} - x^k)$. Upon being satisfied that $x^{q^n} - x$ divides each $x^{iq^n} - x^i$, $1 \leq i \leq k$, we should be satisfied that $x^{q^n} - x$ divides $(x^{q^n} - x) \circ g$ in $(F[x], \cdot)$. If $c \in K$ and K is a field of order q^n, then c is a root of $x^{iq^n} - x^i$, hence $x^{q^n} - x$ divides $x^{iq^n} - x^i$.

Now take $h = h_0 + h_1 x + \cdots + h_m x^m \in F[x]$. We must show that $h \circ (g + (p \cdot f)) - h \circ g \in (p)$. But

$$h \circ (g + (p \cdot f)) - h \circ g = h_1(p \cdot f) + \sum_{i=2}^{m} h_i \left[(g + (p \cdot f))^i - g^i \right]$$

$$= h_1(p \cdot f) + (p \cdot f) \left\{ \sum_{i=2}^{m} h_i \left[\sum_{j=1}^{i} \binom{i}{j} g^{i-j} (p \cdot f)^{j-1} \right] \right\}$$

$$= (p \cdot f) \cdot \left\{ h_1 + \sum_{i=2}^{m} h_i \left[\sum_{j=1}^{i} \binom{i}{j} g^{i-j} (p \cdot f)^{j-1} \right] \right\}.$$

Hence, (p) is a full ideal.

One very nice thing about (17.22) is that the converse is also true.

(17.23) Theorem. *For a finite field F with $q = |F|$, suppose V is a full ideal of $(F[x], +, \cdot, \circ)$ which is properly contained in $F[x]$. Then there are positive integers $1 \leq n_1 < n_2 < \cdots < n_k$ and positive integers m_i, $1 \leq i \leq k$, so that*

$$p = \text{l.c.m.}\{(x^{q^{n_i}} - x)^{m_i} \mid 1 \leq i \leq k\}$$

generates V, that is, $V = (p)$.

Proof. The full ideal V is a ring ideal of the principal ideal domain $(F[x], +, \cdot)$, so there is a nonconstant monic polynomial $p \in F[x]$ such that $V = (p)$.

Let C be the algebraic closure of F, and let c_1 be any root of p. If $a \in F[x]$, then $p \circ a \in V$, so $p \circ a = p \cdot r$ for some $r \in F[x]$. Hence, $p \circ a(c_1) = p(a(c_1)) = p(c_1) \cdot r(c_1) = 0$.

Consider any $\alpha = a_0 + a_1 c_1 + \cdots + a_t c_1^t$ in the extension field $F(c_1)$ and $a = a_0 + a_1 x + \cdots + a_t x^t \in F[x]$. Since $p \circ a = f \cdot r$ for some $r \in F[x]$, and $p(\alpha) = p(a(c_1)) = p(c_1) \cdot r(c_1) = 0$, we see that every element $\alpha \in F(c_1)$ is also a root of p.

Let $c_1, c_2, \ldots, c_s$ be the distinct roots of p in C, and let b_j be the multiplicity of the root c_j of p, $1 \leq j \leq s$. Then $p = (x - c_1)^{b_1} (x - c_2)^{b_2} \cdots (x - c_s)^{b_s}$ where $s > 1$, and each $b_i > 0$. For $c_2 \in F(c_1)$, we have $c_2 \neq c_1$, and we take $g \in F[x]$ so that $g(c_1) = c_2$. Define

$$h = \begin{cases} g, & \text{if } g'(c_1) \neq 0; \\ g + (x^{q^{e_1}} - x), & \text{if } g'(c_1) = 0, \end{cases}$$

where $e_1 = [F(c_1) : F]$, the dimension of the vector space $F(c_1)$ over the field F, and where g' represents, as usual, the derivative of $g \in F[x]$. Then $h(c_1) = c_2$, and $h'(c_1) \neq 0$. So

$$p \circ h = (h - c_1)^{b_1} (h - c_2)^{b_2} \cdots (h - c_2)^{b_s},$$

so $p \circ h(c_1) = (c_2 - c_1)^{b_1} (c_2 - c_2)^{b_2} \cdots (c_2 - c_s)^{b_s} = 0$. Taking the derivative of $p \circ h$, one obtains that c_1 is a root of $p \circ h$ with multiplicity b_2. But $p \circ h$ is a multiple of p, so $b_2 \geq b_1$. Since $c_2 \in F(c_1)$ was almost arbitrary, only restricting $c_2 \neq c_1$, we have that $(x^{q^{e_1}} - x)^{b_1}$ divides p. But c_1 was chosen as an arbitrary root of p, so we could use any c_i in this argument and conclude that $(x^{q^{e_i}} - x)^{b_i}$ divides p, where $e_i = [F(c_i) : F]$. Let

$$f = \text{l.c.m.}\{(x^{q^{e_i}} - x)^{b_i} \mid 1 \leq i \leq s\},$$

so f divides p.

Suppose $f \neq p$. Then $p = f \cdot g$, where g is a nonconstant polynomial in $F[x]$. But g has a root $c_i \in C$, so $x - c_i$ divides $x^{q^{e_i}} - x$, and $(x - c_i)^{b_i+1}$ divides p, a contradiction. Hence, $f = p$.

There is no loss of generality in assuming $e_1 \leq e_2 \leq \cdots \leq e_s$. In this case, we let $n_1 = e_1 = \cdots = e_{s_1} < n_2 = e_{s_1+1} = \cdots = e_{s_2} < \cdots < n_k = e_{s_{k-1}+1} = \cdots = e_s$ and we have $1 \leq n_1 < n_2 < \cdots < n_k$. Also, let $m_i = \max\{b_{s_{i-1}+1}, \ldots, b_{s_i}\}$. Then certainly,

$$f = \text{l.c.m.}\{(x^{q^{n_i}} - x)^{m_i} \mid 1 \leq i \leq k\},$$

and this concludes the proof of the theorem, since $f = p$.

18. Power series nearrings

It has been nearly 30 years since H. Cartan reminded us in [Ca] that the entire functions $(\mathcal{D}(\mathbf{C}), +, \circ)$ provide us with a nearring. In fact, $(\mathcal{D}(\mathbf{C}), +, \cdot, \circ)$ is a composition ring. What are some of the interesting properties of this nearring? To my knowledge, they are still awaiting discovery.

We noticed in §2, that if we consider the formal power series $R[[x]]$ over a commutative ring R, and take $f = \sum_{n=0}^{\infty} f_n x^n$, $g = \sum_{n=0}^{\infty} g_n x^n \in R[[x]]$, then $f \circ g \in R[[x]]$ if $\sum_{n=0}^{\infty} f_n g_0^n \in R[[x]]$ has meaning. Of course, the easiest way for this is for $g_0 = 0$. Before we take a little look at $(\mathcal{D}(\mathbf{C}), +, \cdot, \circ)$, we will present some nice results about these composition rings $(R_0[[x]], +, \cdot, \circ)$ where $R_0[[x]]$ denotes all power series $\sum_{n=1}^{\infty} a_n x^n$, that is, all power series in $R[[x]]$ with constant term equal to 0.

It was Hermann Kautschitsch [HK] who led the way in opening up this field for harvesting. Other students of Nöbauer, specifically Winfried Müller and Rainer Mlitz, have joined Kautschitsch's pioneering effort. See [HK & Mü] and [HK & Ml] for interesting extensions of some of the ideas we present here. After we present some of the main results of [HK, 1977], we shall include some nice consequences which H. Kautschitsch and I were fortunate enough to find. The influence of Nöbauer, as one can see in §17, will also be obvious here.

We begin by noticing that a composition ring $(R_0[[x]], +, \cdot, \circ)$ obtains ideals from the ideals A of the ring R. It will be convenient throughout §18 to always have R to be a commutative ring with identity 1.

(18.1) Theorem. *Let A be an ideal of the commutative ring R. Define*

$$(A) = \{f \mid f \in A_0[[x]]\}$$

and

$$\langle A \rangle = \{f \mid f'(0) \in A\}.$$

Then (A) and $\langle A \rangle$ are full ideals of the composition ring $(R_0[[x]], +, \cdot, \circ)$.

Note. If $f = \sum_{i=1}^{\infty} f_i x^i$, then $f' = \sum_{i=1}^{\infty} i f_i x^{i-1}$ and $f'(0) = f_1$.

Proof. For $a, b \in (A) = A_0[[x]]$, we certainly have $a - b \in A_0[[x]]$, also. For $a = \sum_{i=1}^{\infty} a_i x^i \in A_0[[x]]$ and $f = \sum_{i=1}^{\infty} f_i x^i \in R_0[[x]]$, we have $a \circ f = \sum_{i=1}^{\infty} a_i f^i = \sum_{i=1}^{\infty} a_i [\sum_{j=1}^{\infty} f_j x^j]^i$, so all the coefficients of the various powers of x are in A, since A is an ideal of R. Hence, $a \circ f \in A_0[[x]] = (A)$. Now take $g = \sum_{i=1}^{\infty} g_i x^i \in R_0[[x]]$ also. Since $f \circ [g+a] - f \circ g = \sum_{i=1}^{\infty} f_i (g + a)^i - \sum_{i=1}^{\infty} f_i g^i = \sum_{i=1}^{\infty} f_i [(g + a)^i - g^i] = f_1 a + \sum_{i=2}^{\infty} f_i [\sum_{j=1}^{i} \binom{i}{j} g^{i-j} a^j]$, we will have $f \circ (g + a) - f \circ g \in A_0[[x]] = (A)$ once we show $a \cdot g \in (A)$. But $a \cdot g = (\sum_{i=1}^{\infty} a_i x^i) \cdot (\sum_{i=1}^{\infty} g_i x^i) = \sum_{k=2}^{\infty} (\sum_{i+j=k} a_i g_j) x^k$, and since A is an ideal of R, each $\sum_{i+j=k} a_i g_j \in A$, hence $a \cdot g \in (A)$. So $(A) = A_0[[x]]$ is a full ideal of $(R_0[[x]], +, \cdot, \circ)$.

With $\langle A \rangle = \{a = \sum_{i=1}^{\infty} a_i x^i \mid a_1 \in A\}$, if $a, b \in \langle A \rangle$, then $a - b \in \langle A \rangle$. For $f = \sum_{i=1}^{\infty} f_i x^i \in R_0[[x]]$, we have $a \circ f = \sum_{i=1}^{\infty} a_i (\sum_{j=1}^{\infty} f_j x^j)^i = a_1 f_1 x + \sum_{k=2}^{\infty} c_k x^k$ for some $c_k \in R$, but $a_1 f_1 \in A$, so $a \circ f \in \langle A \rangle$. We easily have $a \cdot f = \sum_{k=2}^{\infty} (\sum_{i+j=k} a_i f_j) x^k \in \langle A \rangle$. It remains to show that $f \circ (g + a) - f \circ g \in \langle A \rangle$ for arbitrary $a \in \langle A \rangle$ and arbitrary $f, g \in R_0[[x]]$. Continuing the calculations for this corresponding step in the preceding paragraph, we have $f \circ (g + a) - f \circ g = f_1 \sum_{i=1}^{\infty} a_i x^i + \sum_{i=2}^{\infty} f_i [\sum_{j=1}^{i} \binom{i}{j} g^{i-j} a^j]$, and since $f_1 a_1 \in A$, we are assured that $f \circ (g + a) - f \circ g \in \langle A \rangle$. This makes $\langle A \rangle$ a full ideal of $(R_0[[x]], +, \cdot, \circ)$.

The next theorem shows, as our notation suggests, that the ideals (A) and $\langle A \rangle$ are *enclosing ideals*. Compare (17.5)–(17.8) with what we have here.

(18.2) Theorem. *Let I be a nearring ideal of $(R_0[[x]], +, \circ)$. Then there is an ideal A of the ring R such that $(A) \subseteq I \subseteq \langle A \rangle$.*

Note. Here, we take I to be a nearring ideal, whereas in (17.7) we took I to be a full ideal.

Proof. Let $A = \{a'(0) \mid a \in I\}$, so A is the set of coefficients of x for the various $a \in I$. For $a'(0), b'(0) \in A$, with $a, b \in I$, we have $a'(0) - b'(0) = (a' - b')(0) = (a - b)'(0)$. Since $a - b \in I$, we obtain that $(A, +)$ is a subgroup of $(R, +)$. For $a'(0) \in A$, $a \in I$, and $r \in R$, we have $rx \in R_0[[x]]$, and so $a \circ (rx) \in I$. So $[a \circ (rx)]'(0) = [[a' \circ (rx)] \cdot r](0) = a'(0) \cdot r \in A$, hence A is an ideal of R, since all our rings R are commutative.

If $A = \{0\}$, then $\langle A \rangle$ consists of all the power series of the form $f = \sum_{i=2}^{\infty} f_i x^i$. This suggests the following theorem.

(18.3) Theorem. *For each positive integer k, let*

$$I_k = \{x^k \cdot f \mid f \in R_0[[x]]\}. \tag{18:1}$$

Then each I_k is a full ideal of $(R_0[[x]], +, \cdot, \circ)$.

Proof. For $f, g \in R_0[[x]]$, $x^k \cdot f - x^k \cdot g = x^k \cdot (f - g) \in I_k$, so each $(I_k, +)$ is a subgroup. Also, $(x^k \cdot f) \cdot g = x^k \cdot (f \cdot g) \in I_k$ since $f \cdot g \in R_0[[x]]$. Now $f \circ (g + x^k \cdot a) - f \circ g = \sum_{i=1}^{\infty} f_i[g + x^k \cdot a]^i - \sum_{i=1}^{\infty} f_i g^i = \sum_{i=1}^{\infty} f_i\{(g + x^k \cdot a)^i - g^i\} = f_1(x^k \cdot a) + \sum_{i=2}^{\infty} f_i\{(g + x^k \cdot a)^i - g^i\} = x^k \cdot (f_1 a) + \sum_{i=2}^{\infty} f_i[\sum_{j=1}^{i} \binom{i}{j} g^{i-j}(x^k \cdot a)^j] = x^k \cdot (f_1 a) + x^k \cdot h = x^k \cdot (f_1 a + h)$ for some $h \in R_0[[x]]$. This puts $f \circ (g + x^k \cdot a) - f \circ g \in I_k$. Finally, $(x^k \cdot a) \circ f = f^k \cdot (a \circ f)$. Certainly, $a \circ f \in R_0[[x]]$ and $f^k = (\sum_{i=1}^{\infty} f_i x^i)^k = x^k \cdot (\sum_{i=1}^{\infty} f_i x^{i-1})^k = x^k \cdot g$ with $g \in R[[x]]$, so $(x^k \cdot a) \circ f = (x^k \cdot g) \cdot (a \circ f) = x^k \cdot [g \cdot (a \circ f)]$. But $a \circ f \in R_0[[x]]$ puts $g \cdot (a \circ f) \in R_0[[x]]$, so all is well. This completes the proof that each I_k is indeed a full ideal of $(R_0[[x]], +, \cdot, \circ)$.

If F is a field of characteristic $\neq 2$, then the ideals of $(F_0[[x]], +, \circ)$ are precisely those I_k described in (18.3). Towards the goal of proving this, we first show that for such fields, all ideals of $(F_0[[x]], +, \circ)$ are full ideals of $(F_0[[x]], +, \cdot, \circ)$.

(18.4) Theorem. *Let F be a field of characteristic $\neq 2$. Then every ideal I of $(F_0[[x]], +, \circ)$ is a full ideal of $(F_0[[x]], +, \cdot, \circ)$.*

Proof. For any $a \in I$ and $f \in F_0[[x]]$, we need only to show that $a \cdot f \in I$. Since I is a nearring ideal, we have $x^2 \circ (p + a) - x^2 \circ p \in I$ for any $p \in F_0[[x]]$, but $x^2 \circ (p + a) - x^2 \circ p = (p + a)^2 - p^2 = 2pa + a^2$, so with $p = 0$, we have $a^2 \in I$, hence $2pa \in I$. With $p = f/2$, we obtain $f \cdot a \in I$, completing our proof.

It will be useful to know (1) the units of the ring $(F[[x]], +, \cdot)$, and (2) the units of the nearring $(F_0[[x]], +, \circ)$.

(18.5) Proposition. *For a commutative ring R with identity 1, $f = \sum_{i=0}^{\infty} f_i x^i$ is a unit in the ring $(R[[x]], +, \cdot)$ if and only if f_0 is a unit in R.*

Proof. Suppose $f = \sum_{i=0}^{\infty} f_i x^i$ is a unit in the ring $(R[[x]], +, \cdot)$. Then there is a $g = \sum_{i=0}^{\infty} g_i x^i \in R[[x]]$ with $1 = f \cdot g = \sum_{k=0}^{\infty}(\sum_{i+j=k} f_i g_j)x^k$. Hence, $f_0 g_0 = 1$ and f_0 is a unit in R.

Conversely, suppose f_0 is a unit in R. We need the existence of a $g = \sum_{i=0}^{\infty} g_i x^i \in R[[x]]$ such that $1 = f \cdot g = \sum_{k=0}^{\infty}(\sum_{i+j=k} f_i g_j)x^k$. Let $g_0 = f_0^{-1}$, so $f_0 g_0 = 1$. For $k \geq 1$, we need $\sum_{i+j=k} f_i g_j = 0$. For $k = 1$, this means $f_0 g_1 + f_1 g_0 = 0$, or $g_1 = f_0^{-1}(-f_1 g_0)$. We now have g_0 and g_1. Suppose we have $g_0, g_1, \ldots, g_{k-1}$. From $0 = \sum_{i+j=k} f_i g_j$, we obtain $g_k = f_0^{-1}[-\sum_{i=1}^{k} f_i g_{k-i}]$ in terms of the f_i and $g_0, g_1, \ldots, g_{k-1}$. With these values for g_i, we then have $g = \sum_{i=0}^{\infty} g_i x^i$ where $f \cdot g = 1$.

(18.6) Proposition. *For a commutative ring R with identity 1, $f =$*

$\sum_{i=1}^{\infty} f_i x^i$ is a unit in the nearring $(R_0[[x]], +, \circ)$ if and only if f_1 is a unit in R.

Proof. Throughout the proof, we will make reference to $g = \sum_{i=1}^{\infty} g_i x^i \in R_0[[x]]$ where $f \circ g = x = f_1 g + f_2 g^2 + f_3 g^3 + \cdots$.

If there is a g such that $f \circ g = x$, then since the powers of x in $f_i g^i$ are all greater than or equal to i, we have $f_1 g_1 = 1$, so f_1 is a unit in R, and we can focus on the converse.

Start with f_1 being a unit of the ring R. Do there exist $g_1, g_2, \ldots \in R$ such that $f \circ g = x$? First, set $g_1 = f_1^{-1}$, so for any $g_2, g_3, \ldots \in R$, we have for $g = g_1 x + g_2 x^2 + \cdots$ that $f \circ g = x + b_2 x^2 + b_3 x^3 + \cdots$ for some $b_2, b_3, \ldots \in R$. We need to choose $g_2, g_3, \ldots \in R$ so that these $b_i = 0$, $i = 2, 3, \ldots$.

In $f_i g^i$, there are only terms of x with exponent $\geq i$, so to find the coefficient of x^k we need only look within the $f_i g^i$, $1 \leq i \leq k$. For x, we have already seen that $f_1 g_1 = 1$ if and only if $g_1 = f_1^{-1}$. To have the coefficient of x^2 to be 0, we take $f_1 g_2 x^2$ from $f_1 g$ and $f_2 g_1^2 x^2$ from $f_2 g^2$. So we have

$$(f_1 g_2 + f_2 g_1^2) x^2 = 0,$$

or $g_2 = f_1^{-1}(-f_2 g_1^2)$, and so $f \circ g = x + \sum_{i=3}^{\infty} b_i x^i$ if and only if $g_1 = f_1^{-1}$ and $g_2 = f_1^{-1}(-f_2 g_1^2)$. From $f_1 g_2 + f_2 g_1^2 = 0$, we have $f_1 g_2 + H_2 = 0$, where H_2 is a (trivial) sum of products from $\{f_2, g_1\}$. In general, to obtain the coefficient of x^k in $f \circ g$, we have $f_1 g_k x^k$ from $f_1 g$, and from each $f_i g^i$, $2 \leq i \leq k$, we obtain coefficients of x^k in terms of products from $\{f_2, \ldots, f_k, g_1, \ldots, g_{k-1}\}$. That is, we have

$$(f_1 g_k + H_k) x^k = 0,$$

where H_k is a sum of products from $\{f_2, \ldots, f_k, g_1, \ldots, g_{k-1}\}$. From $f_1 g_k + H_k = 0$, we obtain that $f \circ g = x$ if and only if $g_1 = f_1^{-1}$, $g_2 = f_1^{-1}(-H_2)$, $\ldots$, $g_k = f_1^{-1}(-H_k)$, $\ldots$. This completes the proof of our proposition.

(18.7) Corollary. *Let I be an ideal of $(F_0[[x]], +, \circ)$ where F is a field. Suppose I contains an element $l = \sum_{i=1}^{\infty} l_i x^i$ with $l_1 \neq 0$. Then $I = F_0[[x]]$.*

Proof. With l a unit in $(F_0[[x]], +, \circ)$, we obtain $x \in I$ and hence $x \circ f = f \in I$ for each $f \in F_0[[x]]$.

(18.8) Theorem. *If F is a field of characteristic $\neq 2$, then every nontrivial ideal of $(F_0[[x]], +, \circ)$, hence each full ideal of $(F_0[[x]], +, \cdot, \circ)$, is of the form $I_k = \{x^k \cdot g \mid g \in F_0[[x]]\}$, for some integer $k > 0$.*

Proof. We know that each I_k is an ideal of $(F_0[[x]], +, \circ)$ and that each ideal of $(F_0[[x]], +, \circ)$ is a full ideal of $(F_0[[x]], +, \cdot, \circ)$.

Start with a nontrivial ideal I of $(F_0[[x]], +, \circ)$, so it is a full ideal of $(F_0[[x]], +, \cdot, \circ)$. From (18.2), we have the existence of an ideal A of F such that $(A) \subseteq I \subseteq \langle A \rangle$. Since F is a field, we do not have many choices for A. If $A = F$, then $(A) = F_0[[x]] \subseteq I$, so I is no longer nontrivial. So we may proceed with $A = \{0\}$, and so $I \subseteq \langle A \rangle = I_1$. Let k be the minimum positive integer with the property that there is an $f = \sum_{i=k+1}^{\infty} f_i x^i \in I$ with $f_{k+1} \neq 0$. So each $g \in I$ has the form $g = x^k \cdot h$ for some $h \in F_0[[x]]$. This means that $I \subseteq I_k$. Since I is a full ideal, we have $f \cdot x = x^{k+2} \cdot (f_{k+1} + f_{k+2}x + \cdots) = x^{k+2} \cdot l$, where $l = f_{k+1} + f_{k+2}x + \cdots \in F[[x]]$, and $f_{k+1} \neq 0$. From (18.5) we have the existence of an $n \in F[[x]]$ such that $l \cdot n = 1$. Now $(f \cdot x) \cdot n = x^{k+2} \cdot (l \cdot n) = x^{k+2} = (x^{k+1} \cdot l) \cdot (x \cdot n)$. Since $x \cdot n \in F_0[[x]]$ and $x^{k+1} \cdot l = f \in I$, we obtain $x^{k+2} \in I$ since I is a full ideal. Hence, $I_{k+2} \subseteq I$.

Continuing, we note that $f_{k+3}x^{k+3} + \cdots = x^{k+2} \cdot (f_{k+3}x + \cdots) \in I_{k+2}$, so $f_{k+1}x^{k+1} + f_{k+2}x^{k+2} \in I$, with $f_{k+1} \neq 0$. Take $ax^{k+1} + bx^{k+2} \in I$ with $a \neq 0$. Then $(ax^{k+1} + bx^{k+2}) \cdot ba^{-1}x = bx^{k+2} + b^2a^{-1}x^{k+3} \in I$. But $b^2a^{-1}x^{k+3} \in I$, since $b^2a^{-1}x^{k+3} \in I_{k+2} \subseteq I$. Hence, $bx^{k+2} \in I$ and so $ax^{k+1} \in I$ also. Now $a^{-1}x \circ (0 + ax^{k+1}) - a^{-1}x \circ 0 = x^{k+1} \in I$, so $bx \circ (0 + x^{k+1}) - bx \circ 0 = bx^{k+1} \in I$ for each $b \in F$. Also, all $x^{k+1} \cdot h$, $h \in F_0[[x]]$, are in I, that is, $I_{k+1} \subseteq I$. So we have $I_{k+1} \subseteq I \subseteq I_k$. With $f = \sum_{i=k+1}^{\infty} f_i x^i \in I$, we see that $f \notin I_{k+1}$, so $I_{k+1} \subset I \subseteq I_k$. We would like to obtain $I = I_k$, of course. Take $g = x^k \cdot h = \sum_{i=k+1}^{\infty} g_i x^i \in I_k$. Now $\sum_{i=k+2}^{\infty} g_i x^i \in I_{k+1} \subset I$, and $g_{k+1}x^{k+1} \in I$, so $g \in I$. Thus $I_k \subseteq I$ and we have $I \subseteq I_k \subseteq I$, giving us $I = I_k$ and the end of our proof.

Exercise. For each positive integer n, show the existence of nearrings $(N, +, \cdot)$ with exactly n proper ideals $I_1, \ldots, I_n$. Can the ideals of your example form a chain $\{0\} = I_0 \subset I_1 \subset \cdots \subset I_n \subset I_{n+1} = N$?

(18.9) Problem and Exploratory problem. Of course, the reader is curious to know what happens when the field F has characteristic 2. Others would also like to know if $(F_0[[x]], +, \circ)$ has any ideals other than the I_k. But what curious things happen with the ideals in $(R_0[[x]], +, \circ)$ when R is a commutative ring with identity?

We have suggested that our study of $(R_0[[x]], +, \circ)$ and $(R_0[[x]], +, \cdot, \circ)$ is motivated by the composition ring of entire functions $(\mathcal{D}(\mathbf{C}), +, \cdot, \circ)$. We have also indicated that this composition ring, or nearring $(\mathcal{D}(\mathbf{C}), +, \circ)$, has not been seriously studied. Certainly, $\mathcal{D}(\mathbf{C}) \subset \mathbf{C}[[x]]$, but since $f \circ g \in \mathcal{D}(\mathbf{C})$ if $f, g \in \mathcal{D}(\mathbf{C})$, one need not consider $\mathcal{D}(\mathbf{C}) \cap \mathbf{C}_0[[x]]$. One quite easily obtains that $(\mathcal{D}(\mathbf{C}), +, \circ)$ is simple.

(18.10) Theorem. *The nearring* $(\mathcal{D}(\mathbf{C}), +, \circ)$ *is simple.*

Proof. Suppose $I \neq \{0\}$ is an ideal of $(\mathcal{D}(\mathbf{C}), +, \circ)$. For $0 \neq f \in I$ and

$\beta \in \mathbf{C}$, we have $f \circ \beta = \alpha \in I$ for some $\alpha \in \mathbf{C}$. Since $f \neq 0$, there is a $\beta \in \mathbf{C}$ such that $\alpha = f \circ \beta \neq 0$. Since $x, x^2 \in \mathcal{D}(\mathbf{C})$, we have $x^2 \circ (x + \alpha) - x^2 \circ x = 2\alpha x + \alpha^2 \in I$ for each $\alpha \in I \cap \mathbf{C}$. With $\alpha \in I \cap \mathbf{C}$, $\alpha \neq 0$, we have $(2\alpha x + \alpha^2) \circ \beta = 2\alpha\beta + \alpha^2 \in I$ for each $\beta \in \mathbf{C}$. This puts $\mathbf{C} \subseteq I$, and hence $\alpha^2 \in I$. Since $2\alpha x + \alpha^2 \in I$, we also have $2\alpha x \in I$, and so $2\alpha x \circ (2\alpha)^{-1}x = x \in I$. Now we have $\mathcal{D}(\mathbf{C}) \subseteq I$.

In the ring $(\mathcal{D}(\mathbf{C}), +, \cdot)$, there are lots of maximal ideals. If $a \in \mathbf{C}$, we define $F_a : \mathcal{D}(\mathbf{C}) \to \mathbf{C}$ by $F_a(f) = f(a)$. Then $F_a(f + g) = F_a(f) + F_a(g)$ and $F_a(f \cdot g) = F_a(f) \cdot F_a(g)$, so F_a is a ring epimorphism onto the field $\mathbf{C}$, so the kernel of F_a is a maximal ideal, and $\ker F_a = \{f \in \mathcal{D}(\mathbf{C}) \mid f(a) = 0\}$. This is in dramatic contrast to the nearring $(\mathcal{D}(\mathbf{C}), +, \circ)$, which, as we have just seen, has no nontrivial ideals

This approach does come close to giving us some ideals in $(\mathcal{D}(\mathbf{C}), +, \circ)$, however. But, to quote Golden M. Wood of Burley High School, 'close doesn't count except in horseshoes and dancing'. Note that $F_a(f \circ g) = (f \circ g)(a) = f(g(a)) = f \bullet g(a) = f \bullet F_a(g)$, where $\mathcal{D}(\mathbf{C})$ acts on $\mathbf{C}$ by $f \bullet b = f(b)$. The (right) nearring $(\mathcal{D}(\mathbf{C}), +, \circ)$ acts on the groups $(\mathcal{D}(\mathbf{C}), +)$ and $(\mathbf{C}, +)$ on the left. For $f \in (\mathcal{D}(\mathbf{C}), +, \circ)$ and $g \in (\mathcal{D}(\mathbf{C}), +)$, we have $f \circ g$. For $b \in \mathbf{C}$, we have $f \bullet b = f(b)$. With these actions, $(\mathcal{D}(\mathbf{C}), +)$ and $(\mathbf{C}, +)$ are $\mathcal{D}(\mathbf{C})$-modules and each F_a is a $\mathcal{D}(\mathbf{C})$-homomorphism. We have already noted that $\ker F_a$ is an ideal of the ring $(\mathcal{D}(\mathbf{C}), +, \cdot)$, but as the kernel of a $\mathcal{D}(\mathbf{C})$-homomorphism, $\ker F_a$ is a *left ideal* of the right nearring $(\mathcal{D}(\mathbf{C}), +, \circ)$.

(18.11) Definitions. For a (left) nearring $(N, +, \cdot)$ let $(R, +)$ be a normal subgroup of $(N, +)$. Suppose $(x + r)y - xy \in R$ for each $r \in R$ and for all $x, y \in N$. Then R is a *right ideal* for the nearring $(N, +, \cdot)$. (If we had $xr \in R$, instead, then R would be a *left ideal* of $(N, +, \cdot)$.)

Exercise. Formulate the definition of a left ideal (and a right ideal) for a right nearring $(N, +, \cdot)$.

Exercise. For the right nearring $(\mathcal{D}(\mathbf{C}), +, \circ)$, prove that each $\ker F_a$ is a left ideal but not an ideal.

(18.12) Theorem. *Let R be a right ideal for a left nearring $(N, +, \cdot)$. Let N act on the quotient group $(N/R, +)$ by $(n + R) \cdot x = nx + R$. Then N/R is an N-module.*

Proof. The only point that should cause a little problem is in showing that $(n + R) \cdot x = nx + R$ is well-defined. Suppose $n + R = n' + R$. Can we be assured that $nx + R = n'x + R$ for each $x \in N$? Since $n' = n + r$ for some $r \in R$, then $n'x = (n + r)x$ for each $x \in N$. Hence, $0 = (n + r)x - n'x = (n + r)x - nx + nx - n'x$. But $(n + r)x - nx \in R$, so $nx - nx' \in R$

for each $x \in R$. In particular, $n(-x) - n'(-x) = -nx + n'x \in R$, hence $nx + R = n'x + R$. The other details of the proof are direct.

Even though one does not get any nontrivial ideals in $(\mathcal{D}(\mathbf{C}), +, \circ)$, one does get lots of ideals in $(\mathcal{D}(\mathbf{C})_0, +, \cdot, \circ)$, however. If we let each I_k be an ideal of $(\mathbf{C}_0[[x]], +, \cdot, \circ)$, as described in (18.3), then each $J_k = I_k \cap \mathcal{D}(\mathbf{C})_0$ is a full ideal. This is a consequence of an important result, whose proof is left as an exercise, and the corresponding result for ring theory.

(18.13) Theorem. *Let $(N, +, \cdot)$ be a nearring with subnearring $(S, +, \cdot)$. If I is an ideal of N, then $S \cap I$ is an ideal of S.*

One could have a field day with (18.13) and the various subnearrings of the $(R_0[[x]], +, \circ)$. In particular, one could take $R_0[x]$ to be all the polynomials of the form $r_1x + r_2x^2 + \cdots + r_nx^n$, $n > 0$, that is, all the polynomials with constant term 0.

With the full ideals I_k of $(R_0[[x]], +, \cdot, \circ)$ and the full ideals $J_k = I_k \cap R_0[x]$ of $(R_0[x], +, \cdot, \circ)$, one can construct the quotient composition rings $R_0[[x]]/I_k$ and $R_0[x]/J_k$. Understanding the quotient structure of all these is most likely to be very difficult, but one can get a lot just by looking at $k \in \{1, 2, 3\}$. For $k = 1$, the quotient structure is easy to understand, but interesting, nevertheless.

(18.14) Theorem. *For a commutative ring R with identity, let I_1 be as in (18.3) and define $J_1 = I_1 \cap R_0[x]$. Then the quotient composition rings $R_0[[x]]/I_1$ and $R_0[x]/J_1$ are such that:*
(1) $(R_0[[x]]/I_1, +, \cdot)$ and $(R_0[x], +, \cdot)$ are trivial rings.
(2) $(R_0[[x]]/I_1, +, \circ)$ and $(R_0[x], +, \circ)$ are rings isomorphic to $(R, +, \cdot)$.

Proof. Let K_1 denote either I_1 or J_1. Then $(r_1x + K_1) \cdot (r_2x + K_1) = r_1r_2x^2 + K_1 = K_1$, hence (1). If N denotes either $R_0[[x]]$ or $R_0[x]$, respectively, then $\Psi : N/K_1 \to R$ defined by $\Psi(rx + K_1) = r$ is a well defined nearring isomorphism. So we have (2).

We now describe a way to construct composition rings from a commutative ring R with identity 1 and a unitary R-module M. This construction was developed from looking at the quotient composition ring $(R_0[[x]]/I_2, +, \cdot, \circ)$, as (18.16) suggests.

(18.15) Theorem. *For a commutative ring R with identity 1, let M be a unitary R-module. On the set $M \times R$ define $+$ componentwise, define $\circ$ by*

$$(a, x) \circ (b, y) = (xb + y^2a, xy), \qquad (18:2)$$

and for an arbitrary but fixed $m \in M$, define $\cdot_m$ by

$$(a, x) \cdot_m (b, y) = (a, x)(b, y) = (xym, 0). \qquad (18:3)$$

Then $(M \times R, +, \cdot_m, \circ)$ is a right composition ring, $(M \times R, +, \cdot_m)$ is a commutative ring in which the product of any three elements is 0, $(M \times R, +, \circ)$ is a zero-symmetric (right) nearring with identity $(0, 1)$, and if $r + r = 0$ for each $r \in R$, then $(M \times R, +, \cdot_m, \circ)$ is a 'double ring', that is, $(M \times R, +, \circ)$ is also a ring. Furthermore, if $r^2 = r$ for each $r \in R$, then $\circ$ is also left distributive over $\cdot_m$.

Proof. Take any three elements (a, x), (b, y), $(c, z) \in M \times R$. Their product $(a, x)[(b, y)(c, z)] = [(a, x)(b, y)](c, z) = (0, 0)$. Also $(a, x)(b, y) = (xym, 0) = (yxm, 0) = (b, y)(a, x)$. From $(a, x)[(b, y) + (c, z)] = (a, x)(b + c, y + z) = (x(y + z)m, 0) = (xym, 0) + (xzm, 0) = (a, x)(b, y) + (a, x)(c, z)$, we have that $(M \times R, +, \cdot_m)$ is a commutative ring in which the product of any three elements is 0.

Next, consider $\circ$. For (a, x), (b, y), $(c, z) \in M \times R$, $(a, x) \circ [(b, y) \circ (c, z)] = (a, x) \circ (yc + z^2 b, yz) = (xyc + xz^2 b + (yz)^2 a, x(yz))$ and $[(a, x)(b, y)](c, z) = (xb + y^2 a, xy) \circ (c, z) = (xyc + z^2 xb + z^2 y^2 a, (xy)z)$. Since R is a commutative ring, we have that $\circ$ is associative. Also, $[(a, x) + (b, y)] \circ (c, z) = (a + b, x + y) \circ (c, z) = ((x + y)c + z^2(a + b), (x + y)z)$ and $(a, x) \circ (c, z) + (b, y) \circ (c, z) = (xc + z^2 a, xz) + (yc + z^2 b, yz) = (xc + z^2 a + yc + z^2 b, xz + yz)$ and we have that $(M \times R, +, \circ)$ is a right nearring. Certainly, $(0, 1)$ is an identity for $\circ$.

To tie $\cdot_m$ and $\circ$ together, note that $[(a, x) \cdot_m (b, y)] \circ (c, z) = (xym, 0) \circ (c, z) = (z^2 xym, 0)$ and $[(a, x) \circ (c, z)] \cdot_m [(b, y) \circ (c, z)] = (xc + z^2 a, xz) \cdot_m (yc + z^2 b, yz) = (xzyzm, 0)$. Since R is a commutative ring, we get that $\circ$ is right distributive over $\cdot_m$, and so $(M \times R, +, \cdot_m, \circ)$ is a right composition ring.

To finish the remaining details, first note that $(0, 0) \circ (a, x) = (0, 0) = (a, x) \circ (0, 0)$, so $(M \times R, +, \circ)$ is zero-symmetric. When $r + r = 0$ for each $r \in R$, we have $(a, x) \circ [(b, y) + (c, z)] = (a, x) \circ (b + c, y + z) = (x(b + c) + (y + z)^2 a, x(y + z))$ and $(a, x) \circ (b, y) + (a, x) \circ (c, z) = (xb + y^2 a, xy) + (xc + z^2 a, xz) = (xb + y^2 a + xc + z^2 a, xy + xz)$. But $(y + z)^2 a = (y^2 + 2yz + z^2)a = (y^2 + z^2)a$, so $\circ$ is also left distributive over $+$, making $(M \times R, +, \circ)$ into a ring. We now assume $r^2 = r$ for each $r \in R$. Then $(a, x) \circ [(b, y) \cdot_m (c, z)] = (a, x) \circ (yzm, 0) = (xyzm, 0)$ and $[(a, x) \circ (b, y)] \cdot_m [(a, x) \circ (c, z)] = (xb + y^2 a, xy) \cdot_m (xc + z^2 a, xz) = (xyxzm, 0) = (xyzm, 0)$, and so $\circ$ is left distributive over $\cdot_m$. This completes the proof of (18.15).

(18.16) Theorem. *Let R be a commutative ring with identity 1, and let I_2 of $(18:1)$ be the full ideal of the composition ring $(R_0[[x]], +, \cdot, \circ)$. Then $(R_0[[x]]/I_2, +, \cdot, \circ) \cong (R \times R, +, \cdot_1, \circ)$ where the latter is as in (18.15). If $r + r = 0$ for each $r \in R$, then $(R_0[[x]]/I_2, +, \cdot, \circ)$ is a 'double ring' where neither $\cdot$ nor $\circ$ is trivial.*

Proof. This is really a corollary of (18.15). For $ax + bx^2 + I_2 \in R_0[[x]]/I_2$, associate $(b, a) \in R \times R$. It is easy to see that this is a well defined isomorphism.

(18.17) Exploratory problem. In (18.15), let R, M, and m vary. Discover something interesting, amusing, and/or significant.

The ideals I_3 and J_3 add considerably to the complexity. By looking at $(R_0[[x]]/I_3, +, \cdot, \circ)$ or $(R_0[x]/J_3, +, \cdot, \circ)$, one is led to

(18.18) Theorem. *For a commutative ring R with identity, let A be a commutative R-algebra with identity, and let M be a unitary A-module. With $S = M \times A \times R$, choose $(m, x) \in M \times A$ and fix it. Define $+$ componentwise, define $\cdot_{(m,x)}$ by*

$$(c, b, a) \cdot_{(m,x)} (f, e, d) = (c, b, a) \cdot (f, e, d) = ((ae + db)m, adx, 0), \quad (18:4)$$

and define $\circ_m$ by

$$
\begin{aligned}
(c, b, a) \circ_m (f, e, d) &= (c, b, a) \circ (f, e, d) \\
&= (af + 2bdem + d^3c, ae + d^2b, ad).
\end{aligned}
\quad (18:5)
$$

Then:
(1) $(S, +, \cdot_{(m,x)})$ is a commutative ring in which the product of any four elements is 0.
(2) $(S, +, \circ_m)$ is a right nearring.
(3) If $x = 1$, then $(S, +, \cdot_{(m,1)}, \circ_m)$ is a right composition ring.
(4) If R is a boolean ring, then $(S, +, \cdot_{(m,1)}, \circ_m)$ is a 'double ring', that is, $(S, +, \circ_m)$ is also a ring.

Proof. The proof is straight-forward, so we will only put up some milestones. Take arbitrary $(c, b, a), (f, e, d), (i, h, g) \in S$. Then $(c, b, a) \cdot [(f, e, d) \cdot (i, h, g)] = (adgxm, 0, 0) = [(c, b, a) \cdot (f, e, d)] \cdot (i, h, g)$. Now it is easy to see that the product with respect to $\cdot_{(m,x)}$ of any four elements is 0, and even without this, it is easy to see that $\cdot_{(m,x)}$ is commutative. We also have $(c, b, a) \cdot [(f, e, d) + (i, h, g)] = ((ae + ah + bd + bg)m, (ad + ag)x, 0) = (c, b, a) \cdot (f, e, d) + (c, b, a) \cdot (i, h, g)$, and so $(S, +, \cdot_{(m,x)})$ is a commutative ring. Associativity of $\circ_m$ follows from $[(c, b, a) \circ (f, e, d)] \circ (i, h, g) = (adi + 2(ae + d^2b)ghm + g^3(af + 2bdem + d^3c), adh + g^2(ae + d^2b), adg) = (a(di + 2eghm + g^3f) + 2bdg(dh + g^2e)m + (dg)^3c, a(dh + g^2e) + (dg)^2b, adg) = (c, b, a) \circ [(f, e, d) \circ (i, h, g)]$.

From $[(c, b, a) + (f, e, d)] \circ (i, h, g) = ((a + d)i + 2(b + e)ghm + g^3(c + f), (a + d)h + g^2(b + e), (a + d)g) = (ai + 2bghm + g^3c + di + 2eghm + g^3f, ah + g^3b + dh + g^2e, ag + dg) = (c, b, a) \circ (i, h, g) + (f, e, d) \circ (i, h, g)$, we obtain that $(S, +, \circ_m)$ is a right nearring. Similarly, $[(c, b, a) \cdot (f, e, d)] \circ (i, h, g) = (2adxghm + g^3(ae + db)m, g^2adx, 0)$ and $[(c, b, a) \circ (i, h, g)] \cdot [(f, e, d) \circ (i, h, g)] = (2adghm + g^3(ae + db)m, g^2adx, 0)$. So if $x = 1$, then $\circ_m$ is right distributive over $\cdot_{(m,1)}$ making $(S, +, \cdot_{(m,1)}, \circ_m)$ a right composition ring.

Letting R be a boolean ring, we have $(f, e, d) \circ_m (c, b, a) = (dc + af, db + ae, ad)$, and so $\circ_m$ is commutative. This makes $(S, +, \circ_m)$ a ring also.

(18.19) Theorem. *Let R be a commutative ring with identity 1, and let I_3 be the full ideal of the composition ring $(R_0[[x]], +, \cdot, \circ)$. Then the quotient composition ring $(R_0[[x]]/I_3, +, \cdot, \circ) \cong (R \times R \times R, +, \cdot_{(1,1)}, \circ_1)$, where the latter is as in (18.18). If R is a boolean ring, then $(R \times R \times R, +, \cdot_{(1,1)}, \circ_1)$ is a 'double ring'.*

Proof. This is really a corollary of (18.18). To $ax + bx^2 + cx^3 + I_3 \in R_0[[x]]$, associate $(c, b, a) \in R \times R \times R$. It is easy to see that this is a well defined isomorphism.

The structures $(M \times R, +, \cdot_1, \circ)$ of (18.15) have a rather nice theory, and it is hoped that the reader can see that this is what follows.

(18.20) Definition. For a commutative ring R with identity 1, let M be a unitary R-module with submodule N. Let $I = (N : M) = \{r \in R \mid rM \subseteq N\}$ and call it the *quotient of N by M*. (Note: $rm = mr$ for $r \in R$ and $m \in M$.)

(18.21) Proposition. *The quotient of N by M, $(N : M)$, is an ideal of R.*

Proof. Certainly, $0 \in (N : M)$, and if $r, r' \in (N : M)$ and $m \in M$, then $(r - r')m = rm - r'm \in N$. Hence, $(N : M)$ is a subgroup of $(R, +)$. If $x \in R$, then $(xr)m = x(rm) \in N$ since $rm \in N$ and N is a submodule. Since R is a commutative ring, I is an ideal.

Perhaps the reader is tired of being reminded that R is a commutative ring with identity 1. With this assumption, we will take R to denote such a ring throughout the remainder of this section, and we will also let M be a unitary R-module.

(18.22) Theorem. *Let N be a submodule of M and let J be an ideal of R with $J \subseteq I = (N : M)$. Then $N \times J$ is an ideal in $(M \times R, +, \circ)$. If $u \in N$, then $N \times J$ is a full ideal in $(M \times R, +, \cdot_u, \circ)$.*

Proof. Certainly, $N \times J$ is a (normal) subgroup of $M \times R$. For $(n, j) \in N \times J$ and $(m, x) \in M \times R$, we have $(n, j) \circ (m, x) = (jm + x^2 n, jx) \in N \times J$, since $j \in J \subseteq I$. This makes $N \times J$ a right ideal of the right nearring $(M \times R, +, \circ)$. If $(a, y) \in M \times R$, also, then $(m, x) \circ [(a, y) + (n, j)] - (m, x) \circ (a, y) = (xn + 2yjm + j^2 m, xj) \in N \times J$ since $j \in J \subseteq I$. Now we have that $N \times J$ is a nearring ideal. Since $(M \times R, +, \cdot)$ is a commutative ring, and since $(n, j) \cdot_u (m, x) = (jxu, 0)$, we have that $N \times J$ is a full ideal.

Exercise. Find rings R for which $(R_0[[x]], +, \circ)$ has proper ideals other than the I_k of (18:1). Contrast your examples with the results of (18.8).

The ideals or full ideals $N \times J$ are developed from a submodule N of M. With this submodule N, we constructed the ideal $(N : M)$ of R and then took J to be any ideal of R contained in this quotient $(N : M)$. Let us now take an ideal I of R and construct a submodule $N = IM$, the submodule generated by all im, where $i \in I$ and $m \in M$. Since $I \subseteq (N : M)$, we have

(18.23) Corollary. *For an ideal I of R and the submodule $N = IM$ of M, we have the ideal $N \times I$ of $(M \times R, +, \circ)$; and if $u \in N$, then $N \times I$ is a full ideal of $(M \times R, +, \cdot_u, \circ)$.*

Note. Compare (18.23) with (15.36) and (16.1).

Every nearring ideal K of $M \times R$ is contained in some $N \times J$.

(18.24) Theorem. *Let K be a nearring ideal of $(M \times R, +, \circ)$. Define $I(K) = \{i \in R \mid (m, i) \in K \text{ for some } m \in M\}$ and $N(K) = \{n \in M \mid (n, r) \in K \text{ for some } r \in R\}$. Then $I(K)$ is an ideal of R, $N(K)$ is a submodule of M, $I(K) \subseteq (N(K) : M)$, and so $N(K) \times I(K)$ is an ideal of $(M \times R, +, \circ)$ with $K \subseteq N(K) \times I(K) \subseteq N(K) \times (N(K) : M)$.*

Proof. For $i, j \in I(K)$, and $(m, i), (n, j) \in K$, we obtain $(m, i) - (n, j) = (m - n, i - j) \in K$, so $i - j \in I(K)$. The same type of argument puts $m - n \in N(K)$. For $r \in R$, we obtain $(0, r) \in M \times R$, and $(m, i) \circ (0, r) = (r^2 m, ir) \in K$, and this puts $ir \in I(K)$. Now $I(K)$ is an ideal. Also, $(0, r) \circ (m, i) = (rm, ri) \in K$, so $rm \in N(K)$, and so $N(K)$ is a submodule. (Recall that the nearring $M \times R$ is zero-symmetric, so (3.36) applies. This is why we casually stated that $(0, r) \circ (m, i) \in K$.)

If $(m, i) \in K$, then $m \in N(K)$ and $i \in I(K)$, so $(m, i) \in N(K) \times I(K)$. Now we have $K \subseteq N(K) \times I(K)$. Once we obtain that $I(K) \subseteq (N(K) : M)$, we can apply (18.22) to see that $N(K) \times I(K)$ is an ideal, and we will certainly have $N(K) \times I(K) \subseteq N(K) \times (N(K) : M)$. So take $i \in I(K)$ and $m \in M$. We will show $im \in N(K)$. Now there is an $(x, i) \in K$, and certainly $(m, 1) \in M \times R$, so $(x, i) \circ (m, 1) = (im + x, i) \in K$. With $x, im + x \in N(K)$, we obtain $im \in N(K)$, and so $I(K) \subseteq (N(K) : M)$.

With an ideal K of $M \times R$ and any $(n, i) \in K$, we have two sets. Define $N_i(K) = \{m \in M \mid (m, i) \in K\}$ and $I_n(K) = \{j \in R \mid (n, j) \in K\}$. We will eventually describe K in terms of these $N_i(K)$ and $I_n(K)$. The following proposition gives us a hint of this description.

(18.25) Proposition. *For an ideal K of $M \times R$ and $(n, i) \in K$, we obtain that $N_0(K)$ is a submodule of M, $I_0(K)$ is an ideal of R, $N_i(K) = n + N_0(K)$, and $I_n(K) = i + I_0(K)$.*

Proof. For $(m, 0), (m', 0) \in K$ and $(0, r) \in M \times R$, we have $(m - m', 0) \in K$ and $(0, r) \circ (m, 0) = (rm, 0) \in K$. Hence, $N_0(K)$ is a submodule of M. Similarly, if $(0, k), (0, k') \in K$, then $(0, k - k') \in K$ and $(0, r) \circ (0, k) = (0, rk) \in K$, so $I_0(K)$ is an ideal of R.

If $(n', 0) \in K$, then $(n + n', i) \in K$, and so $n + N_0(K) \subseteq N_i(K)$. If $(m, i) \in K$, then $m \in N_i(K)$. Hence, $(m - n, 0) \in K$ and we obtain $m - n \in N_0(K)$. Thus $m + N_0(K) = n + N_0(K)$, and so $m \in n + N_0(k)$. Now we have $N_i(K) \subseteq n + N_0(K)$, and so $N_i(K) = n + N_0(K)$.

For $j \in I_n(K)$ we have $(n, j) \in K$. For $r \in I_0(K)$, we have $(0, r) \in K$. Hence, $(n, j + r) \in K$. This means $j + r \in I_n(K)$, $j + I_0(K) \subseteq I_n(K)$, and $i + I_0(K) \subseteq I_n(K)$ since $i \in I_n(K)$. With $i, j \in I_n(K)$, we have $(n, i), (n, j) \in K$, so $(0, i - j) \in K$ and $i - j \in I_0(K)$. Hence, $j \in i + I_0(K)$. Now we have $I_n(K) \subseteq i + I_0(K)$ and therefore $I_n(K) = i + I_0(K)$.

The ideal K gives us $I_0(K)$ and $N_0(K)$. But each of these can give us the other. Define $\mathcal{I}(N_0(K)) = \{i \in R \mid N_0(K) \times \{i\} \subseteq K\}$ and $\mathcal{N}(I_0(K)) = \{m \in M \mid \{m\} \times I_0(K) \subseteq K\}$.

(18.26) Proposition. *For an ideal K of $M \times R$, $\mathcal{I}(N_0(K))$ is an ideal of R and $\mathcal{N}(I_0(K))$ is a submodule of M. Furthermore, $\mathcal{I}(N_0(K)) = I_0(K)$ and $\mathcal{N}(I_0(K)) = N_0(K)$.*

Proof. Take $r \in I_0(K)$ and $m \in M$. Now $(0, r) \in K$ and $(0, r) \circ (m, 0) = (rm, 0) \in K$. Hence, $rm \in N_0(K)$. This puts $r \in (N_0(K) : M)$.

(18.27) Corollary. *We have $\mathcal{N}(\mathcal{I}(N_0(K))) = N_0(K)$ and $\mathcal{I}(\mathcal{N}(I_0(k)))$, making $\mathcal{N}$ and $\mathcal{I}$ inverse operators of one another.*

Proof. Apply (18.26).

(18.28) Proposition. *We have $I_0(K) \subseteq (N_0(K) : M)$.*

Proof. The proof offers no difficulty.

(18.29) Proposition. *For an ideal K of $M \times R$, $N_0(K) \times I_0(K)$ is an ideal of $M \times R$.*

Proof. With $I_0(K)$ being an ideal of R, by (18.25), and with $I_0(K) \subseteq (N_0(K) : M)$ by (18.28), we obtain our desired result by applying (18.22).

We met it first in (17.8), we have seen it again here in (18.1), and, yet again, we have the idea of *enclosing ideals*.

(18.30) Theorem. *If K is an ideal of $M \times R$, then $N_0(K) \times I_0(K) \subseteq K \subseteq N(K) \times I(K)$.*

Proof. From (18.24) we obtain $K \subseteq N(K) \times I(K)$. Let $(n, i) \in N_0(K) \times I_0(K)$. Then $(n, 0), (0, i) \in K$, so $(n, i) \in K$ also. Now we have $N_0(K) \times I_0(K) \subseteq K$.

(18.31) Proposition. *For an ideal K of $M \times R$, if $N_0(K) = N(K)$, then $I_0(K) = I(K)$, and conversely.*

Proof. Start with $N_0(K) = N(K)$. If $i \in I_0(K)$, then $(0, i) \in K$, so $i \in I(K)$. So it is easy to obtain $I_0(K) \subseteq I(K)$. If $i \in I(K)$, then there is an $m \in M$ such that $(m, i) \in K$. Thus $i \in I_m(K) = i + I_0(K)$. Also, $m \in N_i(K) = m + N_0(K)$. But $m \in N(K) = N_0(K)$ and now we have $(m, 0), (m, i) \in K$, so $(0, i) \in K$ and $i \in I_0(K)$. With $I(K) \subseteq I_0(K) \subseteq I(K)$, we have equality.

Next, take $I_0(K) = I(K)$. As above, it is easy to obtain $N_0(K) \subseteq N(K)$. So take $m \in N(K)$. With i such that $(m, i) \in K$, we obtain $m \in N_i(K)$ and $i \in I_m(K) = i + I_0(K) = i + I(K) = I(K) = I_0(K)$. We have $(0, i), (m, i) \in K$, so $(m, 0) \in K$. Hence, $m \in N_0(K)$ and we see that $N(K) \subseteq N_0(K) \subseteq N(K)$, which yields the desired equality.

(18.32) Theorem. *For an ideal K of $M \times R$:*
(1) $K = \cup_{(m,r) \in K} \left[N_r(K) \times I_m(K) \right]$.
(2) *If $(m, r) \in K$, then $(m, r) + N_0(K) \times I_0(K) = N_r(K) \times I_m(K) = (m + N_0(K)) \times (r + I_m(K))$.*

Proof. For $(m, r) \in K$, we have $m \in N_r(K)$ and $r \in I_m(K)$, so $(m, r) \in N_r(K) \times I_m(K)$. Now we have $K \subseteq \cup_{(m,r) \in K} [N_r(K) \times I_m(K)]$. Next take $(m', r') \in N_r(K) \times I_m(K)$ with $(m, r) \in K$. So $(m', r), (m, r') \in K$ and, consequently, $(m' - m, 0) \in K$. Now we have $(m' - m, 0) + (m, r') = (m', r') \in K$ and (1).

For (2), we already have $N_r(K) = m + N_0(K)$ and $I_m(K) = r + I_0(K)$ from (18.25). Take $(m + n, r + j) \in N_r(K) \times I_m(K)$. Then $(m, r) + (n, j) \in (m, r) + N_0(K) \times I_0(K)$. If $(m, r) + (n, j) \in (m, r) + N_0(K) \times I_0(K)$, then $(m + n, r + j) \in (m + N_0(K)) \times (r + I_m(K)) = N_r(K) \times I_m(K)$. Now we have (2).

We have lots of R-modules.

(18.33) Proposition. *For an ideal K of $M \times R$, let $A = N_0(K) \times I_0(K)$. Then $M \times R$, K, K/A, $N(K)/N_0(K)$ and $I(K)/I_0(K)$ are all R-modules where $r(m, x) = (rm, rx)$, $r[(m, x) + A] = (rm, rx) + A$, $r(m + N_0(K)) = rm + N_0(K)$, and $r(x + I_0(K)) = rx + I_0(K)$.*

Proof. As the reader may have guessed, the proofs are very routine. One needs only take a little care to be certain that the action of r on K/A is well defined. We will leave it at that.

Perhaps the reader has noticed already that the ring R is embedded into the nearring $M \times R$.

(18.34) Proposition. *The ring R is isomorphic to a subring of the nearring $M \times R$.*

Proof. $R' = \{(0, r) \mid r \in R\}$ is a subnearring of $M \times R$.

(18.35) Theorem. *For an ideal K of the nearring $M \times R$, let $A = N_0(K) \times I_0(K)$, an ideal of $M \times R$. Suppose $A \subset K \subset N(K) \times I(K)$. Define*

$$f_M : \frac{K}{A} \to \frac{N(K)}{N_0(K)} \quad and \quad f_R : \frac{K}{A} \to \frac{I(K)}{I_0(K)}$$

by $f_M[(m, j) + A] = m + N_0(K)$ and $f_R[(m, j) + A] = j + I_0(K)$. Then f_M and f_R are R-isomorphisms.

Proof. If $(m, j) + A = (n, k) + A$, then $(m - n, j - k) \in A = N_0(K) \times I_0(K)$. With $m - n \in N_0(K)$, we have that f_M is well defined. With $j - k \in I_0(K)$, we have that f_R is well defined. The remainder of the proof is even easier, as the reader is invited to discover.

(18.36) Lemma. *Take an ideal K of $M \times R$. If $I(K) = R$ or if $I_0(K) = R$, then $K = M \times R$.*

Proof. Since $1 \in R = I(K)$, there is an $m \in M$ such that $(m, 1) \in K$. But $(m, 1) \circ (-m, 1) = (0, 1)$, the identity of $M \times R$. With the identity $(0, 1)$ in K, we obtain $M \times R \subseteq K$.

One obtains the maximal ideals of $M \times R$ from the maximal ideals of R.

(18.37) Theorem. *If K is a maximal ideal of $M \times R$, then $K = M \times I_0(K) = M \times I(K)$ and $I_0(K) = I(K)$ is a maximal ideal of R. Conversely, if J is a maximal ideal of R, and $K = M \times J$, then K is a maximal ideal of $M \times R$ and $I_0(K) = I(K) = J$.*

Proof. Let us start with J, a maximal ideal of R. We want to see that $K = M \times J$ is a maximal ideal of $M \times R$. Certainly, $I_0(K) = I(K) = J$. Let us consider the possibility of an ideal L of $M \times R$ where $K = M \times J \subset L \subseteq M \times R$. Then $N_0(L) = N(L) = M$, and, from (18.31), $J \subseteq I_0(L) = I(L) \subseteq R$. With J being maximal, we consider $J \subset I_0(L)$. In this case $I_0(L) = R$, so $L = M \times R$ by (18.36). We are left to consider $J = I_0(L) = I(L) \subset R$, and so by (18.30), $N_0(L) \times I_0(L) = L = N(L) \times I(L)$, or $M \times I_0(L) = L = M \times I(L) = M \times J = K$. This makes K maximal.

Now we take K to be a maximal ideal of $M \times R$. Then $N_0(K) \times I_0(K) \subseteq K \subseteq N(K) \times I(K) \subseteq M \times R$. Either $N(K) \times I(K) = M \times R$ or $K = N(K) \times I(K)$. The former implies $I(K) = R$ and $K = M \times R$, so we must have $K = N(K) \times I(K)$. Consequently, $N_0(K) = N(K)$ and $I_0(K) = I(K)$, which forces $I(K) \subset R$. Thus $K = N(K) \times I(K) \subseteq M \times I(K) \subset M \times R$. But $M \times I(K)$ is an ideal, and K is maximal, so $N(K) = M$ and $K = M \times I(K)$. It remains to show that $I(K)$ is a maximal ideal of R. Suppose J is an ideal of R and that $I_0(K) \subset J \subset R$. Then $M \times J$ is an ideal of

$M \times R$ and $K = M \times I(K) \subset M \times J \subset M \times R$, contrary to K being a maximal ideal. Hence, $I_0(K) = I(K)$ is a maximal ideal.

(18.38) Corollary. *Let K be a maximal ideal of the nearring $M \times R$. Then the quotient nearring $(M \times R)/K$ is a field.*

Proof. The map $(m, r) + M \times I(K) = (m, r) + K \mapsto r + I(K)$ is an isomorphism from $(M \times R)/K$ onto $R/I(K)$. Since $I(K)$ is a maximal ideal of the commutative ring R with identity, $R/I(K)$ is a field.

The nilpotent elements of $M \times R$ are also obtained from the nilpotent elements of R.

(18.39) Theorem. *If $x \in R$ is nilpotent, and $a \in M$, then (a, x) is nilpotent in the nearring $(M \times R, +, \circ)$. Conversely, if $(a, x) \in M \times R$ is nilpotent, then $x \in R$ is nilpotent.*

Proof. Take $(a, x) \in M \times R$. Observe that $(a, x)^n = (x^{n-1}a + x^n a + \cdots + x^{2(n-1)}a, x^n)$. So if $x^n = 0$, then $(a, x)^{n+1} = (0, 0)$, and if $(a, x)^{n+1} = (0, 0)$, then $x^{n+1} = 0$.

Let us take the idea of a nilpotent ideal from ring theory. So, an ideal I of a nearring N is *nilpotent* provided $I^n = \{0\}$ for some positive integer n. Here, I^n is the ideal generated by all products $a_1 a_2 \cdots a_n$, where each $a_i \in I$.

The nilpotent ideals of any $M \times R$ are obtained from the nilpotent ideals of R.

(18.40) Theorem. *Let J be a nilpotent ideal of R. Then any ideal K of $M \times R$ with $K \subseteq M \times J$ is nilpotent. Conversely, if K is a nilpotent ideal of $M \times R$, then $I(K)$ is a nilpotent ideal of R and $K \subseteq M \times I(K)$, which is a nilpotent ideal of $M \times R$.*

Proof. Suppose $J^n = \{0\}$. If $(m_1, r_1), (m_2, r_2), \ldots, (m_{2n}, r_{2n}) \in M \times R$, then $[(m_1, r_1) \circ \cdots \circ (m_n, r_n)] \circ [(m_{n+1}, r_{n+1} \circ \cdots \circ (m_{2n}, r_{2n})] = (a, 0) \circ (b, 0) = (0, 0)$ for some $a, b \in M$. Hence, $K^{2n} \subseteq (M \times J)^{2n} = \{(0, 0)\}$. Now let $K^n = \{(0, 0)\}$, where K is a nilpotent ideal of $M \times R$. Then $K \subseteq N(K) \times I(K) \subseteq M \times I(K)$ by (18.24). For $r_1, \ldots, r_n \in I(K)$, there are $m_1, \ldots, m_n \in M$ so that each $(m_i, r_i) \in K$, $1 \leq i \leq n$. Now $(0, 0) = (m_1, r_1) \circ \cdots \circ (m_n, r_n) = (0, r_1 \cdots r_n)$, so $I(K)^n = \{0\}$. By the first part of our theorem, $M \times I(K)$ is nilpotent, and we have $K \subseteq M \times I(K)$.

Finally, let us extend the idea of a prime ideal from ring theory to nearring theory. If I and J are ideals of a nearring N, then IJ denotes the set of all products ij where $i \in I$ and $j \in J$. An ideal P of N is a *prime ideal* if $IJ \subseteq P$ implies $I \subseteq P$ or $J \subseteq P$ for all ideals I and J of N.

Once again, the prime ideals of R yield prime ideals of $M \times R$.

(18.41) Theorem. *Let P be a prime ideal of R. Then $M \times P$ is a prime ideal of $M \times R$.*

Proof. With the prime ideal P of R, let K and L be ideals of $M \times R$ with $KL \subseteq M \times P$. For $k \in I(K)$ and $l \in I(L)$, we have $m, m' \in M$ where $(m, k) \in K$ and $(m', l) \in L$. Hence, $(m, k) \circ (m', l) \in KL \subseteq M \times P$. So $kl \in P$. This allows us to conclude that $I(K)I(L) \subseteq P$. We may assume that $I(K) \subseteq P$. Now $K \subseteq M \times I(K) \subseteq M \times P$, so $M \times P$ is indeed a prime ideal of $M \times R$.

(18.42) Exploratory problems. We have been studying power series nearrings $(R_0[[x]], +, \circ)$ here in §18 and polynomial nearrings $(R[x], +, \circ)$ in §17. From (15.26), we see that one could also study the power series nearrings $(R_0[[x]], +_F, \circ)$ for a one dimensional formal group law F, and if F is a polynomial in $R[x, y]$, then one could also study the nearrings $(R_0[x], +_F, \circ)$. The author is confident that studying these $(R_0[[x]], +_F, \circ)$ and $(R_0[x], +_F, \circ)$ would lead to a rewarding theory. (This is really a reinforcement of (15.46).)

Exercise. If a one dimensional formal group law F over R is a polynomial in $R[x, y]$, can one conclude that $(R[x], +_F, \circ)$ is a nearring?

The appreciation of mathematical beauty is not like the enjoyment of literature, music, and other art forms. It requires serious effort and hard study. It is much more difficult for a mathematician to explain his triumphs and masterpieces than for any other kind of artist or scientist. Consequently, most mathematicians do not try to interpret their work to the general public, but only communicate with colleagues having similar interests. For this reason, a mathematician is often considered to be a rather aloof person who lives partly in this world and partly in some mysterious realm. This is in fact a fairly accurate conception. However, the door to the world of mathematics is never locked, and anyone who will make the effort can enjoy the beauties of an intellectual domain which comes closer to aesthetic perfection than any other science.

Ross A. Beaumont and Richard S. Pierce, 1963

CHAPTER 5

Some structure of groups of units

The intimate connection between the geneses of nearring theory and group theory provides fertile ground for harvesting interesting groups. Here, groups arise naturally in at least two ways within the *field* of nearrings with identity. First, there is the additive group of the nearring, and then there is the group of units of the nearring.

As one studies any mathematical structure, one should always try to become familiar with the invariants of that structure. One such invariant for a nearring with identity is the group of units of that nearring. The groups of units are often particularly interesting for nearrings; it is a nice source for groups whose structure can be nicely described in classical group theoretic terms, for example, direct products, semidirect products, wreath products, and factor sets. The groups themselves are often rather complicated in structure, but yet not too complicated. This is a nice place to find the beauty of nearring theory. As well, it is a nice place to show how some rather classical group theoretic constructions arise in a natural setting.

In [R], Rotman raises the question, 'When does one "know" a group G?' His first answer is when '... all the elements of G are known and all possible products can be computed'. His second answer is when '... the isomorphism class of G can be characterized'. His second answer is particularly relevant to the study of these invariants of a nearring. In so doing, we get, as a byproduct, his first answer.

The investigations of this chapter take a family of nearrings with identity and focus on their groups of units. For such a nearring $(N, +, \cdot)$, if $\mathcal{U}(N) = \mathcal{U}(N, \cdot)$ denotes the group of units, then we shall describe the structure of $\mathcal{U}(N)$ in terms of a normal subgroup $K \triangleleft \mathcal{U}(N)$ and its quotient group $Q = \mathcal{U}(N)/K$. Often, K and $Q = \mathcal{U}(N)/K$ are isomorphic to better known groups, and so $\mathcal{U}(N)$ will be isomorphic to an extension of these two better known groups. This determines the isomorphism class of $\mathcal{U}(N)$, and gives us a rule for computing the product of all elements as well.

19. Preliminaries

Perhaps we should review here the approach and notation we will frequently use in describing the structure of various groups of units, since the author has experienced that not everyone views these constructions in the same way.

As one develops understanding of the structure of groups, one passes through three popular construction stages in describing the short exact sequence

$$1 \longrightarrow K \longrightarrow G \longrightarrow Q \longrightarrow 1.$$

That is, if a group G has a normal subgroup K whose quotient group G/K is isomorphic to Q, how can one possibly reconstruct G, or an isomorphic copy of G, from isomorphic copies of K and Q?

The first such construction is the *direct product*. With groups K and Q, one defines, on the cartesian product $K \times Q$, the binary operation

$$(k, q) \cdot (k', q') = (kk', qq'), \qquad (19:1)$$

and then observes that $(K \times Q, \cdot)$ is a group, that K and Q are each isomorphic to *normal* subgroups K' and Q' of $K \times Q$, that $K' \cap Q' = \{(1, 1)\}$, and that $K \times Q = K'Q'$. And, in fact, if G has normal subgroups K and Q where $KQ = G$ and $K \cap Q = \{1\}$, then G is isomorphic to this *direct product* $K \times Q$. With a handful of groups, it is easy to construct examples. For example, if Z_n denotes the ring of integers modulo n, then the group Z_6 is isomorphic to $Z_3 \times Z_2$.

The next construction is the *semidirect product*. If one has subgroups K and Q of G, where K is normal, where $KQ = G$, and where $K \cap Q = \{1\}$, then for $x = kq$ and $x' = k'q'$, we obtain $xx' = [k(qk'q^{-1})][qq']$, and $\theta(q)(k') = qk'q^{-1}$ not only defines an automorphism $\theta(q)$ of K for each $q \in Q$, but a homomorphism $\theta : Q \to \operatorname{Aut}K$ of Q into the automorphism group $\operatorname{Aut}K$ of K. With this θ, one can define, on the cartesian product $K \times Q$, the binary operation

$$(k, q) \cdot (k', q') = (k\theta(q)(k'), qq'), \qquad (19:2)$$

and with this, $(K \times Q, \cdot)$ is a group, which we shall denote by $K \times_\theta Q$ to emphasize θ's role. One often writes the operation in K with $+$, and then (19:2) becomes

$$(k, q) \cdot (k', q') = (k + \theta(q)(k'), qq'). \qquad (19:3)$$

The group $(K \times_\theta Q, \cdot)$ is isomorphic to G, and in fact, given two groups $(K, +)$ and $(Q, \cdot)$, and a group homomorphism $\theta : Q \to \operatorname{Aut}K$, one defines $\cdot$ as in (19:3) and then $(K \times Q, \cdot)$ is a group, which we also denote by $K \times_\theta Q$, since $K' = \{(k, 1) \mid k \in K\}$ is a normal subgroup of $K \times_\theta Q$, $Q' = \{(0, q) \mid q \in Q\}$ is a subgroup of $K \times_\theta Q$, $K' \cap Q' = \{(0, 1)\}$, and $K'Q' = K \times_\theta Q$. Of course, it is somewhat more difficult to find examples, but one really doesn't need to look far. If S_n denotes the group of permutations on n letters, then S_3 is isomorphic to $Z_3 \times_\theta Z_2$ for a nontrivial θ.

Going from a direct product to a semidirect product, one relaxed the requirement that Q be a normal subgroup to that of just having Q to be a subgroup. Our third construction relaxes the requirement of Q again, and to make the construction reasonable, one also relaxes conditions on K. We still want K to be a normal subgroup of G, but we will take $(K, +)$ to be abelian. All we require from Q is that it is isomorphic to the quotient G/K.

Let $\pi : G \to Q$ be an epimorphism with kernel K, that is,

$$1 \longrightarrow K \longrightarrow G \overset{\pi}{\longrightarrow} Q \longrightarrow 1$$

is a short exact sequence, or, simply, G is an extension of K by Q. Since G/K is isomorphic to Q, we can and do think of π as the natural epimorphism $\pi(g) = Kg$. For each $x = Kg \in Q = G/K$, choose $l(x) \in G$ so that $\pi(l(x)) = x$, and choose $l(1) = l(K) = 0$, the identity in G. (So, we are taking the operation in G to be $+$, but, of course, G need not be abelian.) Then $\theta(x)(k) = l(x) + k - l(x)$ defines a homomorphism $\theta : Q \to \mathrm{Aut}\,K$, and $f : Q \times Q \to K$, defined by $f(x, y) = l(x) + l(y) - l(xy)$, is a *factor set*. So a factor set measures how far l is from being a homomorphism. That is, if $(K, +)$ is an abelian group, $(Q, \cdot)$ is a group, $\theta : Q \to \mathrm{Aut}\,K$ is a homomorphism, and $f : Q \times Q \to K$ satisfies (a) $f(1, y) = 0 = f(x, 1)$ for every $x, y \in Q$, and (b) $\theta(x)[f(y, z)] - f(xy, z) + f(x, yz) - f(x, y) = 0$ for all $x, y, z \in Q$, then f is a *factor set*. With a factor set f for an abelian group $(K, +)$, a group $(Q, \cdot)$, and a homomorphism $\theta : Q \to \mathrm{Aut}\,K$, one defines $+$ on $K \times Q$ by

$$(k, q) + (k', q') = (k + \theta(q)(k') + f(q, q'), qq'), \qquad (19:4)$$

and obtains that $(K \times Q, +)$ is a group, which we denote by $K \times_\theta^f Q$. The group $K \times_\theta^f Q$ has a normal subgroup $K' = \{(k, 1) \mid k \in K\}$ which is isomorphic to K, and the map $\pi : K \times_\theta^f Q \to Q$ defined by $\pi(k, q) = q$ is a group epimorphism with kernel K'. So $K \times_\theta^f Q$ is an extension of K by Q.

Certainly, $K = \{0, 2\}$ is a normal subgroup of Z_4 and $Z_4/K \cong Z_2 \cong K$. So one has that Z_4 is an extension of Z_2 by Z_2, or, alternatively described, there is a short exact sequence

$$0 \longrightarrow Z_2 \longrightarrow Z_4 \longrightarrow Z_2 \longrightarrow 0.$$

One can take $\theta : Z_2 \to \mathrm{Aut}\,Z_2$ as $\theta(x) = 1_{Z_2}$, the identity automorphism of Z_2, for each $x \in Z_2$, and with $f(x, y) = xy$, one obtains that f is a factor set for Z_2, Z_2, and θ, and $Z_4 \cong Z_2 \times_\theta^f Z_2$. Here, (19:4) becomes

$$(a, x) + (b, y) = (a + b + xy, x + y). \qquad (19:5)$$

In fact, for any Z_n, Z_n, and θ defined by $\theta(x) = 1_{Z_n}$ for each $x \in Z_n$, the map $f : Z_n \times Z_n \to Z_n$ defined by $f(x,y) = xy$ is a factor set. Hence, $Z_n \times_\theta^f Z_n$ is a group. It is natural to suspect that $Z_3 \times_\theta^f Z_3$ would be isomorphic to Z_9, but the author was very surprised to discover that it is not. Being a group of order 3^2, one must then accept that $Z_3 \times_\theta^f Z_3$ is isomorphic to $Z_3 \times Z_3$.

Accepting that $Z_3 \times_\theta^f Z_3 \cong Z_3 \times Z_3$, and noticing that the corresponding operations defined in (19:5) and (19:1), respectively, are quite different, one naturally wonders what an explicit description of the isomorphism would be. Also, if R is any commutative ring, if we take its additive group for $(K, +)$ and $(Q, +)$, and if we take $\theta : (R, +) \to \mathrm{Aut}(R, +)$ as $\theta(x) = 1_R$ for each $x \in R$, then $f(x, y) = xy$ defines a factor set $f : R \times R \to R$, and so the operation defined by (19:5) makes a group $R \times_\theta^f R$. This led to the discovery of the following theorem.

(19.1) Theorem *Suppose S is a subring of a commutative ring R with identity 1. Suppose also that $1/2 \in R$ and that there is an element $a \in S$ with the property that $x(x - a)/2 \in S$ for each $x \in S$. Let N be the group constructed from $S \times S$ with binary operation (19:5). Then*

$$F_a(b, x) = (b + x(x - a)/2, x)$$

defines an isomorphism from the direct sum (product) $S^+ \oplus S^+$ onto N.

Proof. The proof offers no surprises, but it does show that the 'parabola'

$$T_a = \{(x(x - a)/2, x) \mid x \in S\}$$

is a subgroup of N isomorphic to S^+. If $N_1 = \{(b, 0) \mid b \in S\}$, then N_1 is also a subgroup isomorphic to S^+, $N_1 \cap T_a = \{(0, 0)\}$, and $N_1 + T_a = N$, so $N = N_1 \oplus T_a$.

To see that each F_a is an isomorphism, we first show that it is a homomorphism, then show that it is a monomorphism, and finally show that it is an epimorphism.

For (b, x), $(c, y) \in S^+ \oplus S^+$,

$$\begin{aligned}
F_a[(b, x) + (c, y)] &= F_a(b + c, \ x + y) \\
&= (b + c + (x + y)(x + y - a)/2, \ x + y) \\
&= (b + c + [(x^2 + 2xy + y^2) - ax - ay]/2, \ x + y) \\
&= (b + (x^2 - ax)/2 + c + (y^2 - ay)/2 + xy, \ x + y) \\
&= (b + x(x - a)/2, \ x) + (c + y(y - a)/2, \ y) \\
&= F_a(b, x) + F_a(c, y).
\end{aligned}$$

Thus, F_a is a homomorphism.

If $(0,0) = F_a(b,x) = (b+x(x-a)/2, x)$, then $x = 0$, and $b+x(x-a)/2 = b+0 = b = 0$. This makes the kernel of F_a trivial, and, consequently, F_a is a monomorphism.

For an arbitrary $(c,x) \in N$, note that $F_a(c - x(x-a)/2, x) = (c - x(x-a)/2 + x(x-a)/2, x) = (c,x)$, so F_a is surjective, making F_a an epimorphism. Altogether, we have that F_a is an isomorphism.

Since $F_a(0,x) = (x(x-a)/2, x)$ we have that T_a is a subgroup of N isomorphic to S^+. Since $F_a(b,0) = (b,0)$, we have that N_1 is a subgroup of N isomorphic to S^+. Certainly, $N_1 \cap T_a = \{(0,0)\}$, and since $(b,0) + (x(x-a)/2, x) = (b+x(x-a)/2, x) = F_a(b,x)$, we have that $N = N_1 + T_a$, so $N = N_1 \oplus T_a$, a direct sum of N_1 and T_a. This completes the proof of the theorem and the remarks at the beginning of our proof.

Our discussion here about semidirect products and group extensions is highly influenced by the development in [R]. One obtains from (19.1) that $F^+ \times_\theta^f F^+ \cong F^+ \oplus F^+$ for any field of characteristic $\neq 2$, and, in particular, $Z_p^+ \times_\theta^f Z_p \cong Z_p \oplus Z_p$. In fact, $Z_n \times_\theta^f Z_n \cong Z_n \oplus Z_n$ if and only if n is odd. For other surprising isomorphic groups see [R] and [Du].

In describing our group structures, we make heavy use of the semidirect product. It will be to our advantage to go over the semidirect product once again from a slightly different point of view, and then add an unorthodox twist to the presentation.

If a group G has subgroups A and B, where A is normal in G, denoted by $A \lhd B$, and B is simply a subgroup of G, denoted by $B < G$, where $A \cap B = \{1\}$, and $G = AB$, then G is a *semidirect product* of A with B. We often denote this by simply writing $G \cong A \times_\theta B$, and we shall now explain why.

First of all, any $g \in G$ has a unique factorization $g = ab$ with $a \in A$ and $b \in B$. If $g' = a'b'$ with $a' \in A$ and $b' \in B$, then $gg' = aba'b' = a(ba'b^{-1})bb'$, where $ba'b^{-1} \in A$ since $bAb^{-1} = A$ for each $b \in B$. So there is an automorphism $\theta(b)$ of A, where $\theta(b)(a') = ba'b^{-1}$, and the map $\theta : B \to \mathrm{Aut}A$ from B to the group of automorphisms of A, $\mathrm{Aut}A$, is a homomorphism.

On the cartesian product $A \times B$, we define a binary operation $\cdot$ by

$$(a,b) \cdot (a',b') = (a[\theta(b)(a')], bb') = (a[ba'b^{-1}], bb').$$

With this, $(A \times B, \cdot)$ is a group, which we denote by $A \times_\theta B$ to emphasize the role of θ. It is direct to see the $\psi : G \to A \times_\theta B$ defined by $\psi(g) = \psi(ab) = (a,b)$ is a group isomorphism.

If we can describe A and B as being isomorphic to friendly, more familiar groups, say A' and B', respectively, then with the appropriate modification of θ, say to $\theta' : B' \to \mathrm{Aut}A'$, we have $G \cong A' \times_{\theta'} B'$.

Sometimes the factorization $g = ab$ is not so easy, or not so elementary. Since $AB = BA$, it may be easier to factor $g = b'a'$ with $b' \in B$ and $a' \in A$. In this case, suppose $g = ba$ and $g' = b'a'$. Then $g'g = b'a'ba = b'b(b^{-1}a'b)a$ where $b^{-1}a'b \in A$. So for each $b \in B$, there is an automorphism $b\phi$ of A, where $a(b\phi) = b^{-1}ab$, and the map $\phi : B \to \text{Aut}A$ is a homomorphism. (If we had written $\phi(b)(a)$ for $a(b\phi)$, then we would have been required to have ϕ to be an anti-homomorphism.)

On the cartesian product $B \times A$, we define $\cdot$ by

$$(b', a') \cdot (b, a) = (b'b, [a'(b\phi)]a) = (b'b, (b^{-1}a'b)a).$$

Then $(B \times A, \cdot)$ is a group, which we denote by $B \mathbin{_\phi\times} A$, and we have $\Gamma : G \to B \mathbin{_\phi\times} A$ defined by $\Gamma(g) = \Gamma(ba) = (b, a)$, a group isomorphism.

We have three occasions for semidirect products $A \times_\theta B$ where the θ will be understood from context. The first is when $(A, +, \cdot)$ is a nearring with identity and B is a subgroup of the group of units $\mathcal{U}(A, \cdot)$. Then we have for $(a, x), (b, y) \in A \times B$, that $(a, x) \cdot (b, y) = (a + xb, xy)$, and so $\theta(x)(b) = xb$ defines θ as well as each $\theta(x)$. The second is when $(A, +)$ is a group and B is a group of automorphisms of $(A, +)$. Again, for $(a, x), (b, y) \in A \times B$, we have $(a, x) \cdot (b, y) = (a + xb, x \circ y)$, and we have $\theta(x)(b) = xb$ defining θ as well as each $\theta(x)$. For the third, we have for $(A, +)$ a unitary R-module M, and for $(B, \cdot)$ we have a group of units of R. As before, $(a, x) \cdot (b, y) = (a + xb, xy)$ gives, similarly, that $\theta(x)(b) = xb$, which defines θ and each $\theta(x)$.

There is another special class of semidirect products that we will have occasion to use. These are the *wreath products*. Wreath products arise very naturally in describing groups of units of nearrings. For a semidirect product $A \times_\theta B$, we need (1) a group A, (2) a group B, and (3) a group homomorphism $\theta : B \to \text{Aut}A$. For a wreath product, we take B to be a group of permutations Π on a nonempty set I. We take A to be the group K^I of all mappings $f : I \to K$ from I to a group $(K, \cdot)$, with $if \in K$ for all $i \in I$ and all $f \in K^I$. For $f, g \in K^I$ we define $f \cdot g$ by $i(f \cdot g) = (if) \cdot (ig)$. So K^I is the complete direct product of K with itself $|I|$ times.

For $\pi \in \Pi$, we define $\theta(\pi)$ as follows. For $f \in K^I$, $\theta(\pi)(f) = \pi \circ f$, where $i(\pi \circ f) = (i\pi)f$. We must show (a) each $\pi \circ f \in K^I$, (b) each $\theta(\pi)$ is an automorphism of K^I, and (c) θ is a group homomorphism. Certainly, $\pi \circ f \in K^I$, hence $\theta(\pi) : K^I \to K^I$. If $\pi_1, \pi_2 \in \Pi$, then $\theta(\pi_1 \circ \pi_2)(f) = (\pi_1 \circ \pi_2) \circ f = \pi_1 \circ (\pi_2 \circ f) = \theta(\pi_1)(\theta(\pi_2)(f)) = \theta(\pi_1) \circ \theta(\pi_2)(f)$ for any $f \in K^I$. Hence, $\theta(\pi_1 \circ \pi_2) = \theta(\pi_1) \circ \theta(\pi_2)$ and so θ is a homomorphism. Just to be tidy, we will show that each $\theta(\pi)$ is an automorphism of K^I.

Take $f, g \in K^I$. Then $\theta(\pi)(f \cdot g) = \pi \circ (f \cdot g)$, and if $i \in I$, we have $i[\pi \circ (f \cdot g)] = (i\pi)(f \cdot g) = (i\pi)f \cdot (i\pi)g = [i(\pi \circ f)][i(\pi \circ g)] = i[(\pi \circ f) \cdot (\pi \circ g)]$. Hence, $\pi \circ (f \cdot g) = (\pi \circ f) \cdot (\pi \circ g)$, or $\theta(\pi)(f \cdot g) = \theta(\pi)(f) \cdot \theta(\pi)(g)$, and so $\theta(\pi)$ is a homomorphism. Let $ie = 1$ for each $i \in I$. Then $e \in K^I$ is the identity. If $\theta(\pi)(f) = e$, then $i(\pi \circ f) = ie = 1$, so $(i\pi)f = 1$ for every $i \in I$. Since π is a permutation of I then $jf = 1$ for each $j \in I$. Hence, $f = e$, and so $\theta(\pi)$ is a monomorphism. For $\pi \in \Pi$, we have $\pi^{-1} \in \Pi$ also, and $\pi^{-1} \circ f \in K^I$. But $\theta(\pi)(\pi^{-1} \circ f) = f$, so $\theta(\pi)$ is an epimorphism. Now we have seen directly that $\theta(\pi)$ is an automorphism. With $A = K^I$, $B = \Pi$ and $\theta : \Pi \to \operatorname{Aut}K^I$, we can construct $K^I \times_\theta \Pi$, and we denote it by $K \wr \Pi$, and call it the *wreath product* of K by Π. The binary operation in $K \wr \Pi = K^I \times_\theta \Pi$ is

$$(f, \pi) \cdot (g, \gamma) = (f \cdot (\pi \circ g), \pi \circ \gamma). \qquad (19:6)$$

Now every group B is a permutation group on its underlying set B, where $b \in B$ defines $b^\bullet : B \to B$ by $ab^\bullet = ab$. So if $A = K^B$, we have the wreath product $K \wr B = K^B \times_\theta B$. The binary operation in $K \wr B = K^B \times_\theta B$ is

$$(f, a) \cdot (g, b) = (f \cdot (a^\bullet \circ g), ab). \qquad (19:7)$$

Let us now outline the approach we shall take throughout this chapter and the next, when we want to understand the structure of a group G, for example, a group of units of a nearring. In short, we wish to find an epimorphism $\pi : G \to Q$ with kernel $\ker \pi = K$. What amounts to the same thing is to find a normal subgroup K of G.

If we can find another normal subgroup Q of G such that $K \cap Q = \{1\}$, and $KQ = G$, then we will have G as the direct product KQ. If K and Q are each isomorphic to more familiar groups K' and Q', then we will have $G \cong K' \times Q'$. If we cannot find such a Q, perhaps we can find a subgroup Q that is not normal, but yet $K \cap Q = \{1\}$ and $KQ = G$. Then $G/K \cong Q$, and for each $q \in Q$, there is an $l(q) \in G$ such that $l(q) = kq$, with $k \in K$. (We could take $l(q) = q$ under these circumstances.) The map $l : Q \to G$ is called a *lifting*, and is used to define a homomorphism $\theta : G \to \operatorname{Aut}K$ by $\theta(q)(k) = l(q)kl(q)^{-1}$, or if $(K, +)$ is considered, $\theta(q)(k) = l(q) + k - l(q)$. We then have $G \cong K \times_\theta Q$, so if K and Q are isomorphic to more familiar groups K' and Q', we can define $\theta' : Q' \to \operatorname{Aut}K'$ so that $G \cong K' \times_{\theta'} Q'$.

If we cannot find a subgroup Q as above, we can still be lucky if K is abelian. Now G/K plays the role of Q and we tend to identify these two. We choose for each $q \in Q$ an element $l(q) \in G$ such that $\pi(l(q)) = q$ where $\pi : G \to Q$ is an epimorphism with kernel $\ker \pi = K$. Again, $l : Q \to G$ is called a *lifting*. Our lifting defines a group homomorphism $\theta : Q \to \operatorname{Aut}K$ by $\theta(q)(k) = l(q)kl(q)^{-1}$ or $\theta(q)(k) = l(q) + k - l(q)$ if we consider $(K, +)$. (We should make sure to choose $l(1) = 1$ or $l(1) = 0$, as appropriate. That is, l lifts the identity of Q to the identity of G.) Our lifting also defines a factor set which is related to θ. Our factor set $f : Q \times Q \to K$ is defined by $f(x, y) = l(x) \cdot l(y) \cdot l(xy)^{-1}$ or $f(x, y) = l(x) + l(y) - l(xy)$ if appropriate.

This factor set f satisfies (a) $f(1, y) = 0 = f(x, 1)$ for all $x, y \in Q$, and (b) $\theta(x)[f(y, z)] - f(xy, z) + f(x, yz) - f(x, y) = 0 \in K$ for all $x, y, z \in Q$. (Here, we have lapsed into thinking of the operation in K as $+$, although it need not be. Hopefully, we will be able to adjust when the operation is not $+$.) Conversely, for an abelian group $(K, +)$, a group $(Q, \cdot)$, a homomorphism $\theta : Q \to \mathrm{Aut}K$, any map $f : Q \times Q \to K$ satisfying (a) and (b) is a *factor set*. With this factor set, one obtains a group $K \times_\theta^f Q$ from $K \times Q$ by defining

$$(a, x) \cdot (b, y) = (a + \theta(x)(b) + f(x, y), xy), \qquad (19 : 8)$$

or $(a, x) \cdot (b, y) = (a \cdot \theta(x)(b) \cdot f(x, y), xy)$ if we consider $(K, \cdot)$. Returning now to our investigation, we have a group G, and abelian normal subgroup $(K, +)$ with quotient group equal or isomorphic to Q, a lifting $l : Q \to G$ such that $l(1) = 0$, a homomorphism $\theta : Q \to \mathrm{Aut}K$ defined using the lifting by $\theta(q)(k) = l(q) + k - l(q)$, and a factor set $f : Q \times Q \to K$, defined, also from the lifting, by $f(x, y) = l(x) + l(y) - l(xy)$. We then have $G \cong K \times_\theta^f Q$, so if K and Q are more familiar groups, then we may now 'know' the group G by characterizing its isomorphism class, and also how to multiply its elements.

Having cut enough bait, it is time to fish.

20. Direct products in groups of units

Direct products play an important role in describing the structure of groups of units in polynomial nearrings and in quotient nearrings of polynomial nearrings.

Let R be a commutative ring with identity. Let I be an ideal of R. Then $(I) = I[x]$ and $\langle I \rangle$ are ideals of $(R[x], +, \circ)$. See (17.5) and (17.6). It is interesting to study the structure of the group of units of $(R[x]/(I), +, \circ)$ and of $(R[x]/\langle I \rangle, +, \circ)$ for various R and I. This next theorem shows that their study can be reduced to the study of the group of units of $(R[x], +, \circ)$ and $(R[x]/\langle (0) \rangle, +, \circ)$.

(20.1) Theorem.   *For commutative rings with identity R and S, let $\phi : R \to S$ be a ring epimorphism with kernel I. If Φ is the nearring epimorphism from $R[x]$ to $S[x]$, as provided by (17.2), then $R[x]/(I) \cong S[x]$ and $R[x]/\langle I \rangle \cong S[x]/\langle (0) \rangle$.*

Proof. Since $(I) = I[x]$, we have from (17.2) that $R[x]/(I) = R[x]/I[x] \cong S[x]$. Certainly, $(I) \subseteq \langle I \rangle \subseteq R[x]$. Take $f \in \langle (0) \rangle \subseteq S[x]$, and $g = \sum g_i x^i \in R[x]$ such that $\Phi(g) = \sum \phi(g_i) x^i = f$. If $a \in S$ and $\phi(b) = a$, then $0 = \sum \phi(g_i) a^i = \sum \phi(g_i) \phi(b)^i = \phi(\sum g_i b^i)$. Hence, $\sum g_i b^i \in \ker \phi = I$ for each $b \in R$, which puts $\sum g_i x^i = g \in \langle I \rangle$. This means $\Phi^{-1} (\langle (0) \rangle) \subseteq \langle I \rangle$. Conversely, take $g \in \langle I \rangle$. So $g = \sum g_i x^i$, and if $b \in R$, then $\sum g_i b^i \in I$, which means, $\phi(\sum g_i b^i) = \sum \phi(g_i) g(b)^i = 0$. For $c \in S$, there is a

$b \in R$ such that $\phi(b) = c$. Hence, $\sum \phi(g_i)c^i = 0$ for each $c \in S$. This means $\Phi(g) = \sum \phi(g_i)x^i \in \langle(0)\rangle$, so $\Phi(\langle I \rangle) \subseteq \langle(0)\rangle$. We already have $\Phi^{-1}\langle(0)\rangle \subseteq \langle I \rangle$, so $\langle(0)\rangle = \Phi\Phi^{-1}\langle(0)\rangle \subseteq \Phi(\langle I \rangle) \subseteq \langle(0)\rangle$. This makes $\Phi(\langle I \rangle) = \langle(0)\rangle$. From the correspondence theorem (3.55) we obtain the ideal $\langle I \rangle$ in $R[x]$ corresponding to the ideal $\langle(0)\rangle$ in $S[x]$ and so $S[x]/\langle(0)\rangle \cong (R[x]/\langle I \rangle)/(\langle I \rangle/\langle I \rangle) \cong R[x]/\langle I \rangle$, and this completes the proof.

This next theorem is a key to describing our groups of units as direct products of simpler groups of units.

(20.2) Theorem. *Let R be a commutative ring with identity, and let A and B be full ideals of $(R[x], +, \cdot, \circ)$. Define $\Gamma : R[x]/A \cap B \to R[x]/A \oplus R[x]/B$ by $\Gamma(f + A \cap B) = (f + A, f + B)$. Then Γ is a composition ring monomorphism. If $A + B = R[x]$, then Γ is an isomorphism.*

Proof. Let us be assured that Γ is well defined. If $f + A \cap B = f' + A \cap B$, then $f - f' \in A \cap B$, so $f - f' \in A$ and $f - f' \in B$. Hence, $(f + A, f + B) = (f' + A, f' + B)$. It should now be easy to see that Γ is a monomorphism.

With $A + B = R[x]$, we take $(f + A, g + B) \in R[x]/A \oplus R[x]/B$, and begin our search for an h so that $\Gamma(h + A \cap B) = (f + A, g + B)$. Now $f = f_A + f_B$, and $g = g_A + g_B$, where $f_A, g_A \in A$ and $f_B, g_B \in B$. So $f + A = f_B + A$ and $g + B = g_A + B$. Let $h = f_B + g_A$. Then $\Gamma(h + A \cap B) = (f + A, g + B)$, and we are finished with our proof.

(20.3) Proposition. *With R a commutative ring with identity, let A and B be ideals of R. Then $(A \cap B) = (A) \cap (B)$ and $\langle A \cap B \rangle = \langle A \rangle \cap \langle B \rangle$.*

Proof. For $f \in (A \cap B)$, we have $f \in A \cap B[x]$, so $f = \sum f_i x^i$ with each $f_i \in A \cap B$. Hence, $f \in A[x] \cap B[x] = (A) \cap (B)$. Conversely, if $f \in (A) \cap (B)$, then $f \in A[x] \cap B[x]$ and $f = \sum f_i x^i$ with each $f_i \in A \cap B$. Hence $f \in A \cap B[x] = (A \cap B)$. Now we have $(A \cap B) = (A) \cap (B)$.

If $f \in \langle A \cap B \rangle$, then $f(c) \in A \cap B$ for each $c \in R$. Thus $f(c) \in A$ for each $c \in R$ and $f(c) \in B$ for each $c \in R$. Now we have $f \in \langle A \rangle \cap \langle B \rangle$. If $f \in \langle A \rangle \cap \langle B \rangle$, then $f(c) \in A \cap B$ for each $c \in R$, so $f \in \langle A \cap B \rangle$. This gives us $\langle A \cap B \rangle = \langle A \rangle \cap \langle B \rangle$.

(20.4) Theorem. *For a commutative ring R with identity, let $A_1, \ldots, A_n$ be ideals of R having $A_1 \cap \cdots \cap A_n = (0)$. Then $R[x] \cong R[x]/(A_1) \oplus \cdots \oplus R[x]/(A_n)$ and $R[x]/\langle(0)\rangle \cong R[x]/\langle A_1 \rangle \oplus \cdots \oplus R[x]/\langle A_n \rangle$.*

Proof. Applications of (20.3) give us $(A_1 \cap \cdots \cap A_n) = ((0)) = (A_1) \cap \cdots \cap (A_n)$ and $\langle A_1 \cap \cdots \cap A_n \rangle = \langle(0)\rangle = \langle A_1 \rangle \cap \cdots \cap \langle A_n \rangle$. Then applications of (20.2) give us $R[x]/((0)) \cong R[x] \cong R[x]/(A_1) \oplus \cdots \oplus R[x]/(A_n)$ and $R[x]/\langle(0)\rangle \cong R[x]/\langle A_1 \rangle \oplus \cdots \oplus R[x]/\langle A_n \rangle$.

(20.5) Corollary. *Let a positive integer n have prime factorization $n = p_1^{t_1} p_2^{t_2} \cdots p_k^{t_k}$. The principal ideal (n) of Z gives us $Z[x]/((n)) \cong Z_n[x] \cong$*

$Z_{p_1^{t_1}}[x] \oplus \cdots \oplus Z_{p_k^{t_k}}[x]$ and $Z[x]/\langle(n)\rangle \cong Z_n[x]/\langle(0)\rangle \cong Z_{p_1^{t_1}}[x]/\langle(0)\rangle \oplus \cdots \oplus Z_{p_k^{t_k}}[x]/\langle(0)\rangle$.

Proof. First, use (20.1) to obtain $Z[x]/((n)) \cong Z_n[x]$ and $Z[x]/\langle(n)\rangle \cong Z_n[x]/\langle(0)\rangle$. Since $(n) = (p_1^{t_1}) \cap \cdots \cap (p_k^{t_k})$ as ideals in Z, we have, using (20.2), that $Z[x]/((n)) \cong Z[x]/((p_1^{t_1})) \oplus \cdots \oplus Z[x]/((p_k^{t_k}))$ and $Z[x]/\langle(n)\rangle \cong Z[x]/\langle(p_1^{t_1})\rangle \oplus \cdots \oplus Z[x]/\langle(p_k^{t_k})\rangle$. Using (20.1) again, we obtain that each $Z[x]/((p_j^{t_j})) \cong Z_{p_j^{t_j}}[x]$ and that each $Z[x]/\langle(p_j^{t_j})\rangle \cong Z_{p_j^{t_j}}[x]/\langle(0)\rangle$. Combining the above gives us our desired results.

If $(N, +, \cdot)$ is a nearring with identity, we let $\mathcal{U}(N) = \mathcal{U}(N, \cdot)$ denote its group of units. The following theorem and corollaries have very easy proofs, which we will not give.

(20.6) Theorem. *If M and N are nearrings with identity and $M \oplus N$ is their direct sum, then $\mathcal{U}(M \oplus N) \cong \mathcal{U}(M) \times \mathcal{U}(N)$.*

(20.7) Corollary. *For R and all the A_i as in (20.4), we have $\mathcal{U}(R[x]) \cong \mathcal{U}(R[x]/(A_1)) \times \cdots \times \mathcal{U}(R[x]/(A_n))$ and $\mathcal{U}(R[x]/\langle(0)\rangle) \cong \mathcal{U}(R[x]/\langle A_1\rangle) \times \cdots \times \mathcal{U}(R[x]/\langle A_n\rangle)$.*

(20.8) Corollary. *For n as in (20.5), $\mathcal{U}(Z_n[x]) \cong \mathcal{U}(Z_{p_1^{t_1}}[x]) \times \cdots \times \mathcal{U}(Z_{p_k^{t_k}}[x])$ and $\mathcal{U}(Z_n[x]/\langle(0)\rangle) \cong \mathcal{U}(Z_{p_1^{t_1}}[x]/\langle(0)\rangle) \times \cdots \times \mathcal{U}(Z_{p_k^{t_k}}[x]/\langle(0)\rangle)$.*

For a commutative ring R with identity 1, and a unitary R-module M, we discussed the idealization of M in (15.36) and computed its group of units to be $U = \{(r, m) \mid r \in \mathcal{U}(R), m \in M\}$. We also saw that U is a direct product of $A = \{(r, 0) \mid r \in \mathcal{U}(R)\}$ and $B = \{(1, m) \mid m \in M\}$, that $A \cong \mathcal{U}(R)$, and $(B, \cdot) \cong (M, +)$. Thus $U \cong \mathcal{U}(R) \times M$, but $(r, m) \mapsto (r, r^{-1}m)$ is the required isomorphism, which is not what one might expect. In §15, some d.g. nearrings actually were rings isomorphic to an idealization of an appropriate R-module M, so here, the direct product is used to describe the group of units of some nearrings.

Earlier in §15, we studied some d.g. nearrings of odd generalized dihedral groups. In (15.13), these results were shown to be part of a larger class of nearrings which were sometimes also distributively generated. Also in (15.13), the group of units was shown to be isomorphic to a direct product, with one of the subgroups itself being isomorphic to a semidirect product. We'll return to this in §23.

21. Semidirect products and wreath products

Perhaps the easiest place to see a semidirect product is in some polynomial nearring $(R[x], +, \circ)$, but it is just about as easy to see it in an abstract affine nearring $M \oplus_A R$.

(21.1) Theorem. *Let D be a commutative integral domain with identity. Then the nearring $(D[x], +, \circ)$ has its group of units $\mathcal{U}(D[x])$ isomorphic to $D^+ \times_\theta \mathcal{U}(D)$, where $D^+ = (D, +)$ and $\mathcal{U}(D) = \mathcal{U}(D, \cdot)$.*

Proof. Now $f \in \mathcal{U}(D[x])$ if and only if there is a $g \in D[x]$ such that $f \circ g = x$. But $f = \sum f_i x^i$ and $g = \sum g_i x^i$ with $x = f \circ g = \sum f_i (\sum g_j x^j)^i$ forces $f_2 = f_3 = \cdots = f_n = 0$, so $f = f_0 + f_1 x$, and, similarly, $g = g_0 + g_1 x$. From $x = f \circ g = (f_0 + f_1 g_0) + f_1 g_1 x$ we see that $f_1, g_1 \in \mathcal{U}(D)$. Conversely, if $f_1 \in \mathcal{U}(D)$ and $f_0 \in D$, then $g = -f_1^{-1} f_0 + f_1^{-1} x$ has the property that $f \circ g = x$. This means $\mathcal{U}(D[x]) = \{f_0 + f_1 x \mid f_0 \in D, f_1 \in \mathcal{U}(D)\}$. If $A = \{a + x \mid a \in D\}$ and $B = \{ux \mid u \in \mathcal{U}(D)\}$, then one easily sees that A is a normal subgroup of $\mathcal{U}(D[x])$ isomorphic to D^+, that B is a subgroup of $\mathcal{U}(D[x])$ isomorphic to $\mathcal{U}(D)$, that $A \cap B = \{x\}$, the identity of $\mathcal{U}(D[x])$, and that $AB = \mathcal{U}(D[x])$. This shows that $\mathcal{U}(D[x])$ is a semidirect product of A with B. From $(f_0 + f_1 x) \circ (g_0 + g_1 x) = (f_0 + f_1 g_0) + (f_1 g_1)x$, we see that the map $F : \mathcal{U}(D[x]) \to D^+ \times_\theta \mathcal{U}(D)$, defined by $F(f_0 + f_1 x) = (f_0, f_1)$ is a group isomorphism.

(21.2) Theorem. *Let $M \oplus_A R$ be an abstract affine nearring. Then $\mathcal{U}(M \oplus_A R) \cong M^+ \times_\theta \mathcal{U}(R)$.*

Proof. As in the proof of (21.1), we see that $(m, r) \cdot (m', r') = (m + rm', rr') = (0, 1)$ if and only if $r' = r^{-1}$ and $m' = -r^{-1}m$, so $r \in \mathcal{U}(R)$.

If F is a field, then the nearring $(F[x], +, \circ)$ has its group of units isomorphic to $F^+ \times_\theta F^*$. In particular, each prime field Z_p has its polynomial nearring $(Z_p[x], +, \circ)$ with its group of units isomorphic to $Z_p \times_\theta Z_p^*$.

Semidirect products are also used to describe groups of units of the quotient nearrings $(R_0[[x]]/I_2, +, \circ)$. In (18.16) we noted that these nearrings are special cases of those described in (18.15), so let us look at those of (18.15). For a commutative ring R with identity 1 and unitary R-module M, we have the nearring $(M \times R, +, \circ)$, where $\circ$ is defined by (18:2). This nearring is a zero-symmetric right nearring and has the identity $(0, 1)$.

Let us now study the structure of $\mathcal{U}(M \times R)$. Using (18:2), we have $(0, 1) = (a, x) \circ (b, y) = (xb + y^2 a, xy)$, so if $(a, x) \in \mathcal{U}(M \times R)$, then $x \in \mathcal{U}(R)$ and $(a, x)^{-1} = (b, x^{-1})$ for some b. In fact, if $b = -x^{-3}a$, then $(a, x) \circ (-x^{-3}a, x^{-1}) = (0, 1)$, so $(a, x)^{-1} = (-x^{-3}a, x^{-1})$. Hence, $\mathcal{U}(M \times R) = \{(m, u) \mid m \in M, u \in \mathcal{U}(R)\}$. Let $K = \{(m, 1) \mid m \in M\}$ and $Q = \{(0, u) \mid u \in \mathcal{U}(R)\}$. Then $(m, 1) \circ (m', 1) = (m + m', 1)$, and $(m, 1)^{-1} = (-m, 1)$. Also, $(a, x) \circ (m, 1) \circ (a, x)^{-1} = (x^{-1}m, 1)$, so $K \lhd \mathcal{U}(M \times R)$ and $(K, \circ) \cong (M, +)$. As well, $(0, u) \circ (0, v) = (0, uv)$ and $(0, u)^{-1} = (0, u^{-1})$, so $Q < \mathcal{U}(M \times R)$ and $Q \cong \mathcal{U}(R)$. But since $(a, x) \circ (0, u) \circ (a, x)^{-1} = (-ux^{-2}a + u^2 x^{-1}a, u)$, we see that, in general, Q is not normal in $\mathcal{U}(M \times R)$. This means that $\mathcal{U}(M \times R) \cong M^+ \times_\theta \mathcal{U}(R)$ for some $\theta : \mathcal{U}(R) \to \text{Aut}\, M^+$. The map $f : \mathcal{U}(M \times R) \to \mathcal{U}(R)$ defined by

$f(a, u) = u$ has ker $f = K$, and $l(u) = (0, u)$ is an appropriate lifting. Thus $\theta(u)(m, 1) = l(u) \circ (m, 1) \circ l(u)^{-1} = (0, u) \circ (m, 1) \circ (0, u^{-1}) = (u^{-1}m, 1)$ defines θ and each $\theta(u)$. Hence, $\mathcal{U}(M \times R) \cong M^+ \times_\theta \mathcal{U}(R)$ where $(m, u) \cdot (m', u') = (m + \theta(u)(m'), uu') = (m + u^{-1}m', uu')$. (Compare this to (21.2).) In summary, we have

(21.3) Theorem. *Let $(M \times R, +, \circ)$ be a right nearring as described in (18.15). Then $\mathcal{U}(M \times R) \cong M^+ \times_\theta \mathcal{U}(R)$ where $(m, u) \circ (n, v) = (m + u^{-1}n, uv)$.*

(21.4) Corollary. *Consider a right nearring $(R_0[[x]]/I_2, +, \circ)$ as in (18.16). Then $\mathcal{U}(R_0[[x]]/I_2) \cong R^+ \times_\theta \mathcal{U}(R)$ where $(r, u) \cdot (r', u') = (r + u^{-1}r', uu')$, and $\mathcal{U}(R_0[[x]]/I_2) = \{ux + rx^2 + I_2 \mid r \in R, \ u \in \mathcal{U}(R)\}$.*

The group of units of $(R_0[[x]]/I_3, +, \circ)$ is also a semidirect product. These nearrings are special cases of those described in (18.18), so we will focus on the nearrings of (18.18).

(21.5) Theorem. *Let R, A, M, and S be as in (18.18). For any fixed $m \in M$, consider the right nearring $(S, +, \circ_m)$ with identity $(0, 0, 1)$. Then $\mathcal{U}(S) = \{(c, b, u) \mid (c, b) \in M \times A, \ u \in \mathcal{U}(R)\}$, and $\mathcal{U}(S)$ has normal subgroup $K = \{(c, b, 1) \mid (c, b) \in M \times A\}$ and subgroup $Q = \{(0, 0, u) \mid u \in \mathcal{U}(R)\}$, which is isomorphic to $\mathcal{U}(R)$, and $\mathcal{U}(S)$ is a semidirect product of K by Q, and is isomorphic to $K \times_\theta \mathcal{U}(R)$ where $\theta(u)(c, b, 1) = (u^{-2}c, u^{-1}b, 1)$.*

Proof. For $(c, b, u), (f, e, v) \in S$, with $u, v \in \mathcal{U}(R)$, we have $(c, b, u) \circ_m (f, e, v) = (uf + 2bvem + v^3c, ue + v^2b, uv)$, so $(c, b, u)^{-1} = (-u^{-4}c + 2b^2u^{-5}m, -u^{-3}b, u^{-1})$. Define $F : \mathcal{U}(S) \to \mathcal{U}(R)$ by $F(c, b, u) = u$. Then F is an epimorphism with kernel K, making K a normal subgroup. We easily obtain that $Q < \mathcal{U}(S)$, and that $K \cap Q = \{(0, 0, 1)\}$. From $(u^{-3}c, u^{-2}b, 1) \circ_m (0, 0, u) = (c, b, u)$, we have that $\mathcal{U}(S) = KQ$. Hence, $\mathcal{U}(S)$ is a semidirect product of K and Q and is isomorphic to a semidirect product $K \times_\theta \mathcal{U}(R)$ for some $\theta : \mathcal{U}(R) \to \mathrm{Aut}K$. To obtain θ and $\theta(u)$, we need a lifting $l : \mathcal{U}(R) \to \mathcal{U}(S)$. The easiest thing to do is to take $l(u) = (0, 0, u)$. Then $\theta(u)(c, b, 1) = l(u) \circ_m (c, b, 1) \circ_m l(u)^{-1} = (u^{-2}a, u^{-1}b, 1)$. This completes the proof of (21.5).

If we take $(c, b, 1), (f, e, 1) \in K$ of (21.5), then $(c, b, 1) \circ_m (f, e, 1) = (c + f + 2bem, b + e, 1)$, so we should not conclude that $K \cong M^+ \oplus A^+$. We will return to this in §22 and §23.

(21.6) Corollary. *The group of units $\mathcal{U}(R_0[[x]]/I_3, \circ)$ is a semidirect product of the normal subgroup $K = \{x + ax^2 + bx^3 + I_3 \mid a, b \in R\}$ by the subgroup $Q = \{ux + I_3 \mid u \in \mathcal{U}(R)\}$.*

Exercises. Let $R = K$, a commutative integral domain with identity. Let $G = V = \{1, a, b, c\}$ be Klein's four group, and refer to (12.16) and

(12.17). Here, one can have a rewarding experience in studying the near-rings $(\mathbf{hom}(KV, KV), \oplus, \circ) \cong (\mathbf{M}(r, K), \oplus_\sigma, \cdot)$ of (12.16).

(a) Determine the elements of order 2 in $\mathcal{U}(KV)$. (Hint: consider two classifications, each with two subclassifications. There are the integral domains K with characteristic 2, and those with characteristic $\neq 2$. Of those with characteristic 2, there is $K = Z_2$ and those $K \neq Z_2$. Of those with characteristic $\neq 2$, there are those with $1/2 \in K$ and there are those with $1/2 \notin K$.)

(b) If $K = Z_2$, show that $\mathbf{hom}(Z_2V, Z_2V)$ has exactly 64 elements.

(c) If $K \neq Z_2$ but has characteristic 2, show that $\mathbf{hom}(KV, KV)$ has $|K|^6$ elements.

(d) If $1/2 \notin K$ and the characteristic of K is $\neq 2$, then $\mathbf{hom}(KV, KV)$ has exactly 64 elements.

(e) If $1/2 \in K$, then $\mathbf{hom}(KV, KV)$ has exactly 256 elements.

(f) If $K = Z_2$ or K has characteristic $\neq 2$, then $\mathcal{U}(\mathbf{hom}(KV, KV)) \cong \mathrm{Hol}V = V \times_\theta \mathrm{Aut}V$, a group of order 24, where $\mathrm{Hol}V$ is the *holomorph* of V.

(g) If $K \neq Z_2$ and K has characteristic 2, then $\mathcal{U}(\mathbf{hom}(KV, KV))$ has a proper subgroup isomorphic to $\mathrm{Hol}V = V \times_\theta \mathrm{Aut}V$, but if K is infinite, then so is $\mathcal{U}(\mathbf{hom}(KV, KV))$.

(21.7) Problem. Let K be a commutative integral domain with characteristic 2 and $K \neq Z_2$. Determine the structure of $\mathbf{hom}(KV, KV)$ and $\mathcal{U}(\mathbf{hom}(KV, KV))$.

(21.8) Exploratory problem. Let R and/or G vary from those defined in the *exercises* above. What happens to $(\mathbf{hom}(RG, RG), \oplus, \circ)$ and to $\mathcal{U}(\mathbf{hom}(RG, RG))$?

Have you ever seen a wreath product in its natural habitat? I have had trouble with most introductions of the wreath product, both in conveying the basic ideas to others and in motivation. Hopefully, the reader can see from the introduction of the wreath product in §19 that it is really quite a natural application of the semidirect product construction. And, hopefully, the reader can see from our presentation below, that the wreath product is exactly what is needed to describe the structure of some groups of units of nearrings.

In order to obtain our next group of units to work out nicely as a wreath product, we must take care in our choice of notation. We start with a nontrivial group of fixed point free automorphisms Γ of a group $(G, +)$. Let $\Phi = \Gamma \cup \{\zeta\}$ as we did just before (14.1). Then $M_\Phi(G)$ is a subnearring of $M_0(G)$. The nontrivial orbits of Γ on G are the $\Gamma(a) = \{\gamma a \mid \gamma \in \Gamma\}$, with $a \neq 0$. For $f \in M_\Phi(G)$, we have $(\gamma a)f = \gamma(af)$ for all $\gamma \in \Gamma$ and for all $a \in G$. Hence, $\gamma \in \Gamma$ acts on the left of an $a \in G$, and $f \in M_\Phi(G)$

acts on the right of an $a \in G$. This is a notational convention which makes everything work out nicely.

Choose a representative e_i of each nontrivial orbit $\Gamma(e_i)$, and let I be the set of representatives of the family of nontrivial orbits of Γ acting on G. We start with an $f \in \mathcal{U}(M_\Phi(G))$ and obtain (1) a permutation π_f of I, and (2) a mapping $F_f : I \to \Gamma$. The permutation π_f is defined by $e_i \pi_f = e_j$ where $e_i f \in \Gamma(e_j)$. Since f is a unit, it is a bijection of G into itself, and since $f \in M_\Phi(G)$, then $(\Gamma(e_i))f = \Gamma(e_j f)$, and so f permutes the nontrivial orbits of Γ. This is essentially what π_f does. So, $e_i \pi_f = e_j$ if and only if $(\Gamma(e_i))f = \Gamma(e_j)$.

The map $F_f : I \to \Gamma$ is as follows. Again, we have $e_i f \in \Gamma(e_j)$, so there is a unique $\gamma_i \in \Gamma$ such that $\gamma_i(e_j) = e_i f$. So we define $e_i F_f = \gamma_i$. With $f \in \mathcal{U}(M_\Phi(G))$, we associate $(F_f, \pi_f) \in \Gamma^I \times S_I$, where S_I is the group of all permutations on I, and Γ^I is the group of all mappings from I into Γ, as we described in §19 in our development of the wreath product there. We can say, for $0 \neq a \in \Gamma(e_i)$, that π_f tells us which orbit contains af, and F_f tells us how to find af in that orbit. For $a = \lambda e_i$, $\lambda \in \Gamma$, then $af = (\lambda e_i)f = \lambda(e_i f) = \lambda(\gamma_i(e_i \pi_f)) = (\lambda \gamma_i)(e_j)$, where $e_i \pi_f = e_j$.

Certainly, $f \mapsto (F_f, \pi_f)$ defines a mapping from $\mathcal{U}(M_\Phi(G))$ into $\Gamma^I \times S_I$. We want to see that this mapping is a bijection, so we will address injectivity first. Suppose $g \in \mathcal{U}(M_\Phi(G))$ and $f \neq g$. Then $af \neq ag$ for some $a \neq 0$. If $af \in \Gamma(e_j)$ and $ag \in \Gamma(e_k) \neq \Gamma(e_j)$, then for $a \in \Gamma(e_i)$, we have $e_i \pi_f = e_j$ and $e_i \pi_g = e_k$, so $\pi_f \neq \pi_g$. If $af, ag \in e_j \Gamma$, then there are $\gamma_f, \gamma_g \in \Gamma$ such that $\gamma_f e_j = af$ and $\gamma_g e_j = ag$. But $e_i F_f = \gamma_f$ and $e_i F_g = \gamma_g$ and $\gamma_f \neq \gamma_g$. So, from $f \neq g$, we infer $(F_f, \pi_f) \neq (F_g, \pi_g)$, giving us injectivity.

Next comes surjectivity. Start with $(F, \pi) \in \Gamma^I \times S_I$. We want to construct $f \in M_\Phi(G)$ so that $f \mapsto (F_f, \pi_f) = (F, \pi)$. First, define $0f = 0$, and then, since $e_i f$ must be an element in $\Gamma(e_i \pi) = \Gamma(e_j)$, where $e_i \pi = e_j$, and since F will locate $e_i f$ in $\Gamma(e_j)$ by $\gamma_i = e_i F$, we define $e_i f = \gamma_i(e_j) = \gamma_i(e_i \pi) = (e_i F)(e_i \pi)$. Having each $e_i f$ defined, it is easy to define af for any $a \neq 0$. Let $a \in \Gamma(e_i)$ and suppose $a = \mu e_i$. Then $af = (\mu e_i)f = \mu(e_i f)$. In summary,

$$af = \begin{cases} 0, & \text{if } a = 0; \\ (e_i F)(e_i \pi), & \text{if } a = e_i; \\ [\mu \circ (e_i F)](e_i \pi), & \text{if } a = \mu e_i. \end{cases} \qquad (21:1)$$

Take $a \in G$ and $\phi \in \Phi$. If $a = 0$ or $\phi = \zeta$, then $(\phi a)f = 0f = 0 = \phi(af)$. If $a \neq 0$ and $\phi \neq \zeta$, then $a = \mu e_i$ for some unique $e_i \in I$ and some unique $\mu \in \Gamma$. Then $(\phi a)f = [(\phi \circ \mu)e_i]f = [(\phi \circ \mu) \circ (e_i F)](e_i \pi) = \phi[\mu \circ (e_i F)(e_i \pi)] = \phi(af)$. Hence, $f \in M_\Phi(G)$. So, an $(F, \pi) \in \Gamma^I \times S_I$ defines an $f \in M_\Phi(G)$. We need to obtain $f \in \mathcal{U}(M_\Phi(G))$, and to do this, it is enough to get f to be a bijection, since it is already in $M_\Phi(G)$.

From (21:1), we see that $af \neq 0$ if $a \neq 0$. If $a \neq b$, then (21:1) also assures us that $af \neq bf$. So f is injective. It remains to show that f is surjective. Take $b \in \Gamma(e_j)$, so $b \neq 0$. There is a unique $e_i \in I$ such that $e_i\pi = e_j$. Also, $(e_iF)(e_i\pi) = \gamma_i(e_j)$, where $e_iF = \gamma_i$. This means that $e_if = \gamma_i(e_j) \in \Gamma(e_j)$. So, there is a unique $\mu \in \Gamma$ such that $\mu(e_if) = (\mu \circ \gamma_i)(e_i) = b$. Take $a = \mu(e_i)$. Then $af = (\mu e_i)f = \mu(e_if) = b$. With f being surjective, we now have that $f \in M_\Phi(G)$ is a bijection, so $f \in \mathcal{U}(M_\Phi(G))$.

Does $(F_\phi, \pi_f) = (F, \pi)$? From $e_if = (e_iF)(e_i\pi)$, we have $e_i\pi_f = e_i\pi$ for each $e_i \in I$. Hence, $\pi_f = \pi$. Also, $e_iF_f = \gamma_i$, where $\gamma_i(e_j) = e_if$, so $F_f = F$, and we do have $(F_f, \pi_f) = (F, \pi)$, and the map $f \mapsto (F_f, \pi_F)$ from $\mathcal{U}(M_\Phi(G))$ to $\Gamma^I \times S_I$ is a bijection.

What binary operation does $f \circ g$ induce on $\Gamma^I \times S_I$ using this bijection? Take $g \in \mathcal{U}(M_\Phi(G))$ with $g \mapsto (F_g, \pi_g)$. Then $e_i(f \circ g) = [(e_iF_f)(e_i\pi_f)]g = (e_iF_f)[(e_i\pi_f)g] = (e_iF_f)[(e_i\pi_f)F_g((e_i\pi_f)\pi_g)] = [(e_iF_f)\circ((e_i\pi_f)F_g)](e_i(\pi_f \circ \pi_g)) = [e_i(F_f \circ (\pi_f \circ F_g))](e_i(\pi_f \circ \pi_g))$. Hence, $f \circ g \mapsto (F_{f\circ g}, \pi_{f\circ g}) = (F_f \cdot (\pi_f \circ F_g), \pi_f \circ \pi_g) = (F_f, \pi_f) \cdot (F_g, \pi_g)$. This is exactly the binary operation as defined in (19:6), hence $\mathcal{U}(M_\Phi(G)) \cong \Gamma \wr S_I$. So, here are wreath products in their natural habitat! We should summarize this.

(21.9) Theorem. *Let Γ be a nontrivial group of fixed point free automorphisms of a group $(G, +)$. Let $\Phi = \Gamma \cup \{\zeta\}$, where $g\zeta = 0$ for each $g \in G$. Then the zero-symmetric nearring $M_\Phi(G)$ has its group of units $\mathcal{U}(M_\Phi(G))$ isomorphic to the wreath product $\Gamma \wr S_I$, where I is a set of representatives e_i of the nontrivial orbits $\Gamma(e_i)$ of Γ acting on G, and S_I is the group of all permutations on I.*

C. J. Maxson and J. D. P. Meldrum have extended (21.9) to a wider class of nearrings $M_\Phi(G)$. See [CMM]. Again, wreath products are used to describe the structure of the groups of units. Also in [CMM], one learns how to define a function D on $M_\Phi(G)$ that behaves analogously to a determinant function. This reinforces the kinship of the $M_\Phi(G)$ with the ring of linear transformations of a vector space. Also, in [R], one learns how the wreath product is just what is needed to describe Sylow subgroups of the finite symmetric groups S_n.

22. Group extensions with factor sets

The easiest place for us to display the use of a factor set in describing the structure of a group of units is with the subgroup K of (21.5), which generalizes the structure of a subgroup of units of $R_0[[x]]/I_3$. See (21.6). As we remarked just before (21.6), $(c, b, 1) \circ (f, e, 1) = (c + f + 2bem, b + e, 1)$.

(22.1) Theorem. *Let M be an R-module, where R is a commutative ring with identity. Suppose $m \in M$, and define $f : R \times R \to M$ by $f(x, y) = mxy$. Then f is a factor set for $\theta : R^+ \to M$ with $\theta(x) = 1_M$ for* *each $x \in R$. Hence,*

$$(a, x) +_m (b, y) = (a + b + mxy, x + y)$$

makes $(M \times R, +_m)$ a group, which we denote by $M \times^f R^+$. If there is a $d \in R$ with the property that $mx(x - d)/2 \in M$ for each $x \in R$, then $F_d : M \oplus R^+ \to M \times^f R^+$ is a group isomorphism, where

$$F_d(a, x) = (a + mx(x - d)/2, x).$$

Proof. This is really just a generalization of (19.1), and the proof of (19.1) can easily be modified to apply here.

(22.2) Corollary. *The subgroup K of (21.5) is isomorphic to $R^+ \times^f R^+$ where $f(b, e) = 2bem$.*

Proof. This should be obvious from the equation $(c, b, 1) \circ (f, e, 1) = (c + f + 2bem, b + e)$.

(22.3) Corollary. *The subgroup K of (21.5) is isomorphic to $R^+ \oplus R^+$ with the isomorphism $F_0 : R^+ \oplus R^+ \to R^+ \times^f R^+$ defined by $F_0(a, y) = (a + y^2, y)$.*

Proof. Use (22.2) and (22.1). In using (22.1), we take $m = 2$ and $d = 0$.

Perhaps the reader is a little surprised at the isomorphisms of (22.1), and, in particular, that in (22.3). If so, then perhaps the reader will appreciate the isomorphism of (22.3) even more when $R = F$, a field.

When $R = F$ is a field, then $F \oplus F$ is the classical setting for an affine plane $(F \oplus F, \mathcal{L}, \in)$. The points, of course, are the elements of $F \oplus F$. The lines $\mathcal{L}$ consists of the 'vertical' lines $l_a = \{(a, y) \mid y \in F\}$, the 'horizontal' lines $l(b) = \{(x, b) \mid x \in F\}$, and the lines with nonzero 'slope' m and y-intercept b, namely, $l(m, b) = \{(x, mx + b) \mid x \in F\}$. In $\mathcal{U}(F_0[[x]]/I_3, \circ)$, we have the subgroup $K = \{x + bx^2 + ax^3 + I_3 \mid a, b \in F\}$. By (22.3), we know that $f : F \oplus F \to K$, defined by $f(a, b) = x + bx^2 + (a + b^2)x^3 + I_3$, is an isomorphism. Hence, f induces on K an affine plane $(K, \mathbf{L}, \in)$. A 'vertical' line l_a in $\mathcal{L}$ corresponds to $f(l_a) = L_a = \{x + tx^2 + (a + t^2)x^3 + I_3 \mid t \in F\} \in \mathbf{L}$, a 'horizontal' line $l(b)$ in $\mathcal{L}$ corresponds to $f(l(b)) = L(b) = \{x + bx^2 + (s + b^2)x^3 + I_3 \mid s \in F\} \in \mathbf{L}$, and a line $l(m, b)$ with nonzero 'slope' m and y-intercept b in $\mathcal{L}$ corresponds to $f(l(m, b)) = L(m, b) = \{f(u, mu + b) \mid u \in F\} = \{x + (mu + b)x^2 + (u + (mu + b)^2)x^3 + I_3 \mid u \in F\}$.

Whereas K may be an unusual setting for an affine plane, the points being a set of cosets of a nearring ideal, it may be even more interesting to bring the setting back to $F \times F$. The isomorphism $F_0 : F^+ \oplus F^+ \to F^+ \times^f F^+$, defined by $F_0(a, y) = (a + y^2, y)$, induces an amusing affine plane

$(F \times F, \Lambda, \in)$ back on the set $F \times F$. A 'horizontal' line $l(b)$ corresponds to itself, the same set of points $l(b)$ in Λ, that is, $F_0(l(b)) = \{(x + b^2, b) \mid x \in F\} = l(b) = \lambda(b)$, but a 'vertical' line l_a corresponds to the 'parabola' $\lambda_a = F_0(l_a) = \{(a + y^2, y) \mid y \in F\}$, and a line $l(m, b)$, $m \neq 0$, corresponds to the 'parabola' $\lambda(m, b) = F_0(l(m, b)) = \{(x + (mx + b)^2, mx + b) \mid x \in F\} = \{((y + 1/2m)^2 - b/m, y) \mid y \in F\}$. So Λ, consisting of all $\lambda(b)$, all λ_a, and all $\lambda(m, b)$, is mostly a family of 'parabolas', but yet $(F \times F, \Lambda, \in)$ necessarily satisfies the requirements to be an affine plane.

So far, our use of factor sets occurred to describe the structure of groups of units, which could also have been described simply with direct products. That will not be the case for our next use.

We turn our attention to the structure of $\mathcal{U}(Z_{p^2}[x])$, where p is a prime. We already have $\mathcal{U}(Z_p[x]) \cong Z_p^+ \times_\theta Z_p^*$, and we next consider the groups of units $\mathcal{U}(Z_{p^2}[x])$. Let $\phi : Z_{p^2} \to Z_p$ be the natural map, where $\phi(a) \equiv a \bmod p$ for each $a \in Z_{p^2}$, and let $\Phi : Z_{p^2}[x] \to Z_p[x]$ be its extension as in (17.2). If $f, g \in \mathcal{U}(Z_{p^2}[x])$ and $f \circ g = x$, then $x = \Phi(x) = \Phi(f \circ g) = \Phi(f) \circ \Phi(g)$. Hence, $\mathcal{U}(Z_{p^2}[x]) \subseteq \Phi^{-1}(\mathcal{U}(Z_p[x]))$. Now $\Phi^{-1}(\mathcal{U}(Z_p[x])) = \{a + bx + ps(x) \mid a, b \in Z_p, b \neq 0, s(x) \in Z_p[x]\}$. (Here, we are asking the reader to begin to take special care, since we are taking liberties with notation, for example, $Z_p = \{0, 1, \ldots, p-1\} \subset Z_{p^2} = \{0, 1, \ldots, p^2 - 1\}$, and $Z_p^* = \{1, \ldots, p-1\} \subset \mathcal{U}(Z_{p^2})$.) We wish to show that $\Phi^{-1}(\mathcal{U}(Z_p[x])) = \mathcal{U}(Z_{p^2}[x])$.

Take $x + ps(x) \in Z_{p^2}[x]$, with $s(x) \in Z_p[x]$. Then $(x + ps(x)) \circ (x - ps(x)) = x$, so $x + ps(x) \in \mathcal{U}(Z_{p^2}[x])$. In fact, if $T = \{x + ps(x) \in Z_{p^2}[x] \mid s(x) \in Z_p[x]\}$, then $T < \mathcal{U}(Z_{p^2}[x])$ and $T \cong Z_p[x]^+$. This follows from $(x + ps(x)) \circ (x + pt(x)) = x + p[s(x) + t(x)]$. Certainly, $a + bx \in \mathcal{U}(Z_{p^2}[x])$ if $a \in Z_p$ and $b \in Z_p^*$. Given $a + bx + ps(x) \in \Phi^{-1}(\mathcal{U}(Z_p[x]))$, we have $(a + bx) \circ (x + pb^{-1}s(x)) = a + bx + ps(x)$, so $\Phi^{-1}(\mathcal{U}(Z_p[x])) \subseteq \mathcal{U}(Z_{p^2}[x])$. Now we have $\Phi^{-1}(\mathcal{U}(Z_p[x])) \subseteq \mathcal{U}(Z_{p^2}[x])$.

Let $\Gamma : \mathcal{U}(Z_{p^2}[x]) \to \mathcal{U}(Z_p[x])$ be the restriction of Φ to $\mathcal{U}(Z_{p^2}[x])$, that is, $\Gamma = \Phi \mid \mathcal{U}(Z_{p^2}[x])$. Then Γ is an epimorphism and $\ker \Gamma = T$, so $T \triangleleft \mathcal{U}(Z_{p^2}[x])$. We have the short exact sequence

$$1 \to T \to \mathcal{U}(Z_{p^2}[x]) \overset{\Gamma}{\longrightarrow} \mathcal{U}(Z_p[x]) \to 1,$$

or, equivalently,

$$0 \to Z_p[x]^+ \to \mathcal{U}(Z_{p^2}[x]) \to Z_p^+ \times_\theta Z_p^* \to 1. \qquad (22 : 1)$$

We need a lifting $l : \mathcal{U}(Z_p[x]) \to \mathcal{U}(Z_{p^2}[x])$ to define our $\gamma : \mathcal{U}(Z_p[x]) \to \mathrm{Aut}\, T$. The natural choice for l is $l(a + bx) = a + bx$. (The $a + bx$ in $l(a + bx)$ is in $Z_p[x]$, but the $a + bx$ to the right of the equality sign is in $Z_{p^2}[x]$.)

With this lifting, $\gamma(a+bx)(x+ps(x)) = l(a+bx)\circ(x+ps(x))\circ l(a+bx)^{-1} = (a+bx)\circ(x+ps(x))\circ(a+bx)^{-1} = a+b(-ab^{-1}+b^{-1}x)+pbs(-ab^{-1}+b^{-1}x) = x + pbs(-ab^{-1}+b^{-1}x)$. (Here $(a+bx)^{-1} = -ab^{-1}+b^{-1}x \in Z_{p^2}[x]$, with $b^{-1}, -ab^{-1} \in Z_{p^2}$.)

With the lifting, we obtain the required factor set $f(a+bx, c+dx) = l(a+bx)\circ l(c+dx)\circ l[(a+bx)\circ(c+dx)]^{-1} = (a+bx)\circ(c+dx)\circ((a+bc)_p + (bd)_px)^{-1} = [(a+bc)+(bd)x]\circ[(a+bc)_p + (bd)_px]^{-1}$. (Here, $(a+bx)\circ(c+dx) = (a+bc)+(bd)x$ is in $Z_{p^2}[x]$, but $(a+bx)\circ(c+dx) = (a+bc)_p + (bd)_px \in Z_p[x]$, and $[(a+bc)_p + (bd)_px]^{-1} \in Z_{p^2}[x]$.) This means it could very well be that $f(a+bx, c+dx) = \alpha + \beta x + pA(x)$ for some $A(x) \in Z_p[x]$, $A(x) \neq 0$. For example, with $Z_9[x]$ and $Z_3[x]$, $f(1+x, 2+x) = (1+x)\circ(2+x)\circ l[(1+x)\circ(2+x)]^{-1} = (3+x)\circ l(x)^{-1} = (3+x)\circ x = 3 + x = x + 3\cdot 1$, so $A(x) = 1$. Let us summarize this first step with

(22.4) Proposition. *For a prime p, $\mathcal{U}(Z_{p^2}[x]) = \{a+bx+ps(x) \mid a, b \in Z_p, b \neq 0, s(x) \in Z_p[x]\}$ and has a normal subgroup $T = \{x+ps(x) \mid s(x) \in Z_p[x]\}$ isomorphic to $Z_p[x]^+$. There is the short exact sequence*

$$1 \to T \to \mathcal{U}(Z_{p^2}[x]) \xrightarrow{\Gamma} \mathcal{U}(Z_p[x]) \to 1.$$

The group $\mathcal{U}(Z_{p^2}[x])$ is isomorphic to $T \times_\gamma^f \mathcal{U}(Z_p[x])$ where $\gamma(a+bx)(x+ps(x)) = x + pbs(-ab^{-1}+b^{-1}x)$, and where $f(a+bx, c+dx) = [(a+bc)+(bd)x]\circ[(a+bc)_p + (bd)_px]^{-1}$.

Now, we wish to extend (22.4) to $\mathcal{U}(Z_{p^n}[x])$. Let $\phi_n : Z_{p^n} \to Z_{p^{n-1}}$ be the natural map where $\phi_n(a) \equiv a \bmod p^{n-1}$ for each $a \in Z_{p^n}$, and let $\Phi_n : Z_{p^n}[x] \to Z_{p^{n-1}}[x]$ be its extension as in (17.2). Let $\Gamma_n : \mathcal{U}(Z_{p^n}[x]) \to \mathcal{U}(Z_{p^{n-1}}[x])$ be the restriction of Φ_n to $\mathcal{U}(Z_{p^n}[x])$, so $\Gamma_n = \Phi|\mathcal{U}(Z_{p^n}[x])$.

(22.5) Proposition. *The kernel of Γ_n is $T_n = \{x+p^{n-1}s(x) \mid s(x) \in Z_p[x]\}$, a normal subgroup of $\mathcal{U}(Z_{p^n}[x])$. Also, $T_n \cong Z_p[x]^+$.*

Proof. The monoid epimorphism $\Phi_n : (Z_{p^n}[x], \circ) \to (Z_{p^{n-1}}[x], \circ)$ has $\Phi_n^{-1}(x) = \{x+p^{n-1}s(x) \mid s(x) \in Z_p[x]\} = T_n$, hence $\ker \Gamma_n \subseteq T_n$. But $(x+p^{n-1}s(x))\circ(x+p^{n-1}t(x)) = x + p^{n-1}[s(x)+t(x)]$ not only shows that $T_n \cong Z_p[x]^+$ but that $x - p^{n-1}s(x)$ is the inverse of $x + p^{n-1}s(x)$. Hence, $T_n \subseteq \mathcal{U}(Z_{p^n}[x])$ and $\Phi_n(T_n) = \{x\}$. Now we can conclude that $T_n = \ker \Gamma_n$.

(22.6) Proposition. $\mathcal{U}(Z_{p^n}[x]) = \Phi_n^{-1}(\mathcal{U}(Z_{p^{n-1}}[x]))$.

Proof. Certainly, $\mathcal{U}(Z_{p^n}[x]) \subseteq \Phi_n^{-1}(\mathcal{U}(Z_{p^{n-1}}[x]))$. Let us start with $f \in \Phi_n^{-1}(\mathcal{U}(Z_{p^{n-1}}[x]))$, and $\Phi_n(f) = p$. Then there are a $q \in \mathcal{U}(Z_{p^{n-1}}[x])$

and a $g \in \Phi_n^{-1}(\mathcal{U}(Z_{p^{n-1}}[x]))$ such that $x = p \circ q$ and $\Phi_n(g) = q$. But $\Phi_n(f \circ g) = \Phi_n(f) \circ \Phi_n(g) = p \circ q = x$, hence $f \circ g \in \Phi_n^{-1}(x) = T_n$ of (22.5). This puts $f \circ g, g \circ f \in \mathcal{U}(Z_{p^n}[x])$. So there are $r, s \in \mathcal{U}(Z_{p^n}[x])$ such that $(f \circ g) \circ r = r \circ (f \circ g) = x = (g \circ f) \circ s = s \circ (g \circ f)$. Hence, $g \circ f \circ g \circ r = g = s \circ g \circ f \circ g$, so $g \circ r^{-1} = g \circ f \circ g = s^{-1} \circ g$, or $s \circ g = g \circ r$. But $x = s \circ (g \circ f) = (s \circ g) \circ f = (g \circ r) \circ f$ and $x = (f \circ g) \circ r$ imply $f \circ (g \circ r) = (g \circ r) \circ f = x$. Hence, $f \in \mathcal{U}(Z_{p^n}[x])$. Now we have $\Phi_n^{-1}(\mathcal{U}(Z_{p^{n-1}}[x])) = \mathcal{U}(Z_{p^n}[x])$.

(22.7) Theorem. *We have the extension defined by the short exact sequence*

$$1 \to T_n \to \mathcal{U}(Z_{p^n}[x]) \xrightarrow{\Gamma_n} \mathcal{U}(Z_{p^{n-1}}[x]) \to 1.$$

Hence $\mathcal{U}(Z_{p^n}[x]) \cong T_n \times_{\gamma_n}^{f_n} \mathcal{U}(Z_{p^{n-1}}[x])$ for appropriate γ_n and f_n.

Proof. This is another way of stating that $\Gamma_n : \mathcal{U}(Z_{p^n}[x]) \to \mathcal{U}(Z_{p^{n-1}}[x])$ is a group epimorphism with kernel $\ker \Gamma_n = T_n$.

There are various ways to describe the elements of $\mathcal{U}(Z_{p^n}[x])$.

(22.8) Theorem. *We have $L = M = N = \mathcal{U}(Z_{p^n}[x])$ where*

$$L = \{a + bx + ps_1(x) + \cdots + p^{n-1}s_{n-1}(x) \mid$$
$$a, b \in Z_p, b \neq 0, s_1(x), \ldots, s_{n-1}(x) \in Z_p[x]\};$$
$$M = \{a + bx + px^2 s(x) \mid a \in Z_{p^n}, b \in \mathcal{U}(Z_{p^n}), s(x) \in Z_{p^n}[x]\};$$
$$N = \{a + bx + ps(x) \mid a, b \in Z_p, b \neq 0, s(x) \in Z_{p^{n-1}}[x]\}.$$

Proof. We have $\mathcal{U}(Z_p[x]) = \Gamma_2 \circ \cdots \circ \Gamma_n(\mathcal{U}(Z_{p^n}[x])) = \{a + bx \mid a, b \in Z_p, b \neq 0\}$. We also have $\mathcal{U}(Z_{p^n}[x]) = \Phi_n^{-1}(\Phi_{n-1}^{-1}(\cdots(\Phi_2^{-1}(\mathcal{U}(Z_p[x]))) \cdots))$. Hence,

$$\mathcal{U}(Z_p[x]) = \{a + bx \mid a, b \in Z_p, b \neq 0\},$$

$$\mathcal{U}(Z_{p^2}[x]) = \{a + bx + ps(x) \mid a, b \in Z_p, b \neq 0, s(x) \in Z_p[x]\},$$

$$\cdots \cdots \cdots \cdots \cdots \cdots \cdots \cdots$$

$$\mathcal{U}(Z_{p^n}[x]) = \{a + bx + ps_1(x) + \cdots + p^{n-1}s_{n-1}(x) \mid$$
$$a, b \in Z_p, b \neq 0, s_1(x), \ldots, s_{n-1}(x) \in Z_p[x]\},$$

from which we infer $\mathcal{U}(Z_{p^n}[x]) = L$.

We will next show that $L \subseteq M$. Take $f(x) = a + bx + ps_1(x) + \cdots + p^{n-1}s_{n-1}(x) \in L$. Each $s_i(x) = s_{i0} + s_{i1}x + \cdots + s_{im}x^m = s_{i0} + s_{i1}x +$

$x^2 t_i(x)$. Hence, $p^i s_i(x) = p^i(s_{i0} + s_{i1}x) + p^i x^2 t_i(x)$, and so $f(x) = (a + ps_{10} + \cdots + p^{n-1}s_{n-1,0}) + (b + ps_{11} + \cdots + p^{n-1}s_{n-1,1})x + px^2 t_1(x) + p^2 x^2 t_2(x) + \cdots + p^{n-1}x^2 t_{n-1}(x) = A + Bx + px^2 F(x)$ with $A = a + ps_{10} + \cdots + p^{n-1}s_{n-1,0}$, $B = b + ps_{11} + \cdots + p^{n-1}s_{n-1,1}$, and $F(x) = t_1(x) + pt_2(x) + \cdots + p^{n-2}t_{n-1}(x)$. Since p and B are relatively prime, $B \in \mathcal{U}(Z_{p^n})$, and we have $f(x) \in M$, so $L \subseteq M$.

Next, take $g(x) = a + bx + px^2 s(x) \in M$, so $a \in Z_{p^n}$, $b \in \mathcal{U}(Z_{p^n})$, and $s(x) \in Z_{p^n}[x]$. Now $a = pA + a'$, $0 \le a' < p$, $b = pB + b'$, $0 < b' < p$, and $s(x) = s_0 + s_1 x + \cdots + s_m x^m$, where each $s_i = p^{n-1}S_i + s_i'$, $0 \le s_i' < p^{n-1}$. Hence, $g(x) = a' + b'x + p(A + Bx) + px^2[(p^{n-1}S_0 + s_0') + (p^{n-1}S_1 + s_1')x + \cdots + (p^{n-1}S_m + s_m')x^m] = a' + b'x + p(A + Bx) + px^2(s_0' + s_1'x + \cdots + s_m'x^m) = a' + b'x + pt(x)$, where $t(x) = A + Bx + s_0'x^2 + \cdots + s_m'x^{m+2}$. So $g(x) \in N$, and, consequently, $M \subseteq N$.

Finally, take $h(x) \in N$, so $h(x) = a + bx + ps(x)$, where $a, b \in Z_p$, $b \ne 0$. and $s(x) = s_0 + s_1 x + \cdots + s_m x^m \in Z_{p^{n-1}}[x]$, so each $s_i \in \{0, 1, \ldots, p^{n-1} - 1\}$. Writing each s_i in base p, we have $s_i = s_{i0} + s_{i1}p + \cdots + s_{i,n-1}p^{n-1}$ with each $0 \le s_{ij} < p$. Hence, $s_i x^i$ contributes a term $p^j s_{ij}x^i$. This means $ps(x) = p\sum_{i=0}^{m} s_i x^i = p\sum_{i=0}^{m}\left(\sum_{j=0}^{n-1} s_{ij}p^j\right)x^i = p\sum_{j=0}^{n-1} p^j \sum_{i=0}^{m} s_{ij}x^i = p\sum_{j=0}^{n-1} p^j s_{j+1}(x) = ps_1(x) + p^2 s_2(x) + \cdots + p^{n-1}s_{n-1}(x)$ where $s_{j+1}(x) = \sum_{i=0}^{m} s_{ij}x^i$. Hence, $h(x) \in L$ and so $N \subseteq L$. With $L \subseteq M \subseteq N \subseteq L$, we have equality and the completion of the proof of (22.8)

With (22.8), we can say something nice about appropriate γ_n and f_n of (22.7). For each of these, we shall need a lifting $l_n : \mathcal{U}(Z_{p^{n-1}}[x]) \to \mathcal{U}(Z_{p^n}[x])$. We will worry about which lifting to take in a few moments, but right now, using M of (22.8), we are assured that $l_n(f) = a + bx + px^2 s(x)$ for some a, b, and $s(x)$. With

$$
\begin{aligned}
\gamma_n(f)(x + p^{n-1}t(x)) &= l_n(f) \circ (x + p^{n-1}t(x)) \circ l_n(f)^{-1} \\
&= l_n(f) \circ (l_n(f)^{-1} + p^{n-1}t(l_n(f)^{-1})) \\
&= a + b[l_n(f)^{-1} + p^{n-1}t(l_n(f)^{-1})] \\
&\quad + p[l_n(f)^{-1} + p^{n-1}t(l_n(f)^{-1})]^2 s[l_n(f)^{-1} + p^{n-1}t(l_n(f)^{-1})] \\
&= a + bl_n(f)^{-1} + p[l_n(f)^{-1}]^2 s[l_n(f)^{-1}] + p^{n-1}bt(l_n(f)^{-1}) \\
&= l_n(f) \circ l_n(f)^{-1} + p^{n-1}bt(l_n(f)^{-1}) \\
&= x + p^{n-1}bt(l_n(f)^{-1}),
\end{aligned}
$$

we get the rather nice formula

$$
\gamma_n(f)(x + p^{n-1}t(x)) = x + p^{n-1}bt(l_n(f)^{-1}),
$$

which defines not only γ_n but each $\gamma_n(f)$, in terms of our lifting $l_n(f) = a + bx + px^2 s(x)$ and $x + p^{n-1}t(x)$. In summary, we have the follosing lemma.

(22.9) Lemma. *For a lifting $l_n : \mathcal{U}(Z_{p^{n-1}}[x]) \to \mathcal{U}(Z_{p^n}[x])$ for Γ_n of (22.7), there is, for each $f \in \mathcal{U}(Z_{p^{n-1}}[x])$, an $l_n(f) = a + bx + px^2 s(x)$, and the γ_n of (22.7) is defined by the equation $\gamma_n(f)(x + p^{n-1} t(x)) = x + p^{n-1} bt(l_n(f)^{-1})$.*

Of course, (22.9) is valid for any lifting l_n for Γ_n, as will be both (22.10) and (22.11). But we may as well define l_n by the equation $l_n(a + bx + px^2 s(x)) = a + bx + px^2 s(x)$, since we can and do think of $Z_{p^{n-1}}[x] \subset Z_{p^n}[x]$ and $Z_{p^{n-1}} \subset Z_{p^n}$.

To get a better understanding of the factor set f_n of (22.7), we will make use of

(22.10) Lemma. *For l_n as in (22.9), there is an $A(x) \in Z_p[x]$ such that $l_n(f \circ g)^{-1} = l_n(g)^{-1} \circ l_n(f)^{-1} + p^{n-1} A(x)$.*

Proof. With Γ_n from (22.7), we have $x = \Gamma_n(x) = \Gamma_n[l_n(f) \circ l_n(f)^{-1}] = \Gamma_n(l_n(f)) \circ \Gamma_n(l_n(f)^{-1}) = f \circ \Gamma_n(l_n(f)^{-1})$. Hence, $\Gamma_n(l_n(f)^{-1}) = f^{-1}$. Consequently,

$$\Gamma_n[l_n(f \circ g)^{-1}] = (f \circ g)^{-1} = g^{-1} \circ f^{-1}$$
$$= \Gamma_n(l_n(g))^{-1} \circ \Gamma_n(l_n(f))^{-1}$$
$$= \Gamma_n(l_n(g)^{-1}) \circ \Gamma_n(l_n(f)^{-1})$$
$$= \Gamma_n[l_n(g)^{-1} \circ l_n(f)^{-1}],$$

or simply $\Gamma_n[l_n(f \circ g)^{-1}] = \Gamma_n[l_n(g)^{-1} \circ l_n(f)^{-1}]$. But remember that Γ_n is a restriction of Φ_n, a nearring epimorphism. Hence, $0 = \Phi_n[l_n(f \circ g)^{-1}] - \Phi_n[l_n(g)^{-1} \circ l_n(f)^{-1}] = \Phi_n[l_n(f \circ g)^{-1} - l_n(g)^{-1} \circ l_n(f)^{-1}]$. This means that $l_n(f \circ g)^{-1} - l_n(g)^{-1} \circ l_n(f)^{-1}$ is a polynomial which Φ_n sends to 0, that is, $l_n(f \circ g)^{-1} - l_n(g)^{-1} \circ l_n(f)^{-1} = p^{n-1} A(x)$ for some polynomial $A(x) \in Z_p[x]$.

Now we are ready to compute the factor set f_n of (22.7).

(22.11) Lemma. *For l_n as in (22.9), the f_n of (22.7) is defined by the equation*

$$f_n(f, g) = x + p^{n-1} bd A(x),$$

where $l_n(f) = a + bx + px^2 s(x)$, $l_n(g) = c + dx + px^2 t(x)$, and $l_n(f \circ g)^{-1} = l_n(g)^{-1} \circ l_n(f)^{-1} + p^{n-1} A(x)$ as in (22.10).

Proof. Computing directly,

$$
\begin{aligned}
f_n(f,g) &= l_n(f) \circ l_n(g) \circ l_n(f \circ g)^{-1} \\
&= l_n(f) \circ l_n(g) \circ [l_n(g)^{-1} \circ l_n(f)^{-1} + p^{n-1}A(x)] \\
&= l_n(f) \circ \{c + d[l_n(g)^{-1} \circ l_n(f)^{-1} + p^{n-1}A(x)] \\
&\quad + p[l_n(g)^{-1} \circ l_n(f)^{-1} + p^{n-1}A(x)]^2 t[l_n(g)^{-1} \circ l_n(f)^{-1} + p^{n-1}A(x)]\} \\
&= l_n(f) \circ \{c + d[l_n(g)^{-1} \circ l_n(f)^{-1} + p^{n-1}A(x)] \\
&\quad + p[l_n(g)^{-1} \circ l_n(f)]^2 t[l_n(g)^{-1} \circ l_n(f)^{-1}]\} \\
&= l_n(f) \circ \{c + dl_n(g)^{-1} \circ l_n(f)^{-1} + p^{n-1}dA(x) \\
&\quad + p[l_n(g)^{-1} \circ l_n(f)^{-1}]^2 t[l_n(g)^{-1} \circ l_n(f)^{-1}]\} \\
&= l_n(f) \circ \{c + d[l_n(g)^{-1} \circ l_n(f)^{-1}] \\
&\quad + p[l_n(g)^{-1} \circ l_n(f)^{-1}]^2 t[l_n(g)^{-1} \circ l_n(f)^{-1}] + p^{n-1}dA(x)\} \\
&= l_n(f) \circ \{l_n(g) \circ [l_n(g)^{-1} \circ l_n(f)^{-1}] + p^{n-1}dA(x)\} \\
&= l_n(f) \circ [l_n(f)^{-1} + p^{n-1}dA(x)] \\
&= a + b[l_n(f)^{-1} + p^{n-1}dA(x)] + p[l_n(f)^{-1} \\
&\quad + p^{n-1}dA(x)]^2 s[l_n(f)^{-1} + p^{n-1}dA(x)] \\
&= a + bl_n(f)^{-1} + p[l_n(f)^{-1}]^2 s[l_n(f)^{-1}] + p^{n-1}bdA(x) \\
&= l_n(f) \circ l_n(f)^{-1} + p^{n-1}bdA(x) = x + p^{n-1}bdA(x),
\end{aligned}
$$

as promised.

With the lifting $l_n(a + bx + px^2 s(x)) = a + bx + px^2 s(x)$, we have from (22.7), (22.9), and (22.10) a nice description of $\mathcal{U}(Z_{p^n}[x])$ in terms of T_n, or $Z_p[x]^+$, and $\mathcal{U}(Z_{p^{n-1}})$. For now let us let $\mathcal{U}_n = \mathcal{U}(Z_{p^n}[x])$. With what we have done here, we have $\mathcal{U}_1 \cong Z_p^+ \times_\theta Z_p^*$, $\mathcal{U}_2 \cong Z_p[x]^+ \times_{\gamma_2}^{f_2} \mathcal{U}_1 = Z_p[x]^+ \times_{\gamma_2}^{f_2} [Z_p^+ \times_\theta Z_p^*]$, and, in general, $\mathcal{U}_n \cong Z_p^+[x] \times_{\gamma_n}^{f_n} \mathcal{U}_{n-1}$, or

$$
\mathcal{U}_n \cong Z_p[x]^+ \times_{\gamma_n}^{f_n} [Z_p[x]^+ \times_{\gamma_{n-1}}^{f_{n-1}} [\cdots [Z_p[x]^+ \times_{\gamma_2}^{f_2} (Z_p^+ \times_\theta Z_p^*)] \cdots]].
$$

In particular, $\mathcal{U}_n$ is infinite if and only if $n \geq 2$.

23. A mixture of the above

In this section, we discuss groups of units of nearrings where, to describe the structure of the group of units, we need at least two of the three constructions described in §19 and used in §20, §21, and §22.

Theorem (20.6) and its corollary (20.8) set us up for our first examples. From (21.1), we obtain that $\mathcal{U}(Z_p[x]) \cong Z_p^+ \times_\theta Z_p^*$. Hence,

(23.1) Theorem. *Let the positive integer n have prime factorization $n = p_1 p_2 \ldots p_k$, where the primes $p_1, p_2, \ldots, p_k$ are all distinct. Then*

$$\mathcal{U}(Z_n[x]) \cong (Z_{p_1}^+ \times_\theta Z_{p_1}^*) \times \cdots \times (Z_{p_k}^+ \times_\theta Z_{p_k}^*).$$

From (3.52), we have $F[x]/\langle(0)\rangle \cong M(F)$, for any finite field F, hence $\mathcal{U}(F[x]/\langle(0)\rangle) \cong S_F$, the group of all permutations on F. So, from this observation and (20.8), we get

(23.2) Theorem. *Let a positive integer n have prime factorization $n = p_1 p_2 \ldots p_k$ as in (23.1). Then*

$$\mathcal{U}(Z_n[x]/\langle(0)\rangle) \cong S_{p_1} \times \cdots \times S_{p_k},$$

where S_{p_i} is the group of all permutations on p_i distinct elements.

In (21.5), we obtained $\mathcal{U}(S) \cong K \times_\theta \mathcal{U}(R)$. Then from (22.1), we obtained $K \cong M \times^f A^+$, hence

(23.3) Theorem. *Let R, A, M, and S be as in (18.18). For any fixed $m \in M$, the right nearring $(S, +, \circ_m)$ has its group of units*

$$\mathcal{U}(S) \cong (M \times^f A^+) \times_\theta \mathcal{U}(R),$$

where f is as in (22.1) and θ is as in (21.5).

In §20, we noted that (15.13) gave us nearrings whose groups of units were isomorphic to a direct product with one of the factors itself being a semidirect product. In fact, the group of units of a nearring considered in (15.13) is isomorphic to $\mathcal{U}(R) \times [\mathcal{U}(R)_\theta \times M]$ where R is a ring with identity and M is a unitary right R-module. Note that $u(R)_\theta \times M$ is the alternative semidirect product construction discussed in §19.

In §22, we had the short exact sequence (22:1). Hence,

(23.4) Theorem. *The group of units $\mathcal{U}(Z_{p^2}[x]) \cong Z_p[x]^+ \times_\gamma^f (Z_p^+ \times_\theta Z_p^*)$, for some homomorphism $\gamma : Z_p^+ \times_\theta Z_p^* \to \mathrm{Aut}\, Z_p[x]^+$ and some factor set $f : [Z_p^+ \times_\theta Z_p^*] \times [Z_p^+ \times_\theta Z_p^*] \to Z_p[x]$.*

We are able to give an amusing characterization of square free positive integers.

(23.5) Theorem. *Let the positive integer n have prime factorization $n = p_1^{t_1} \cdots p_k^{t_k}$ as in (20.5). then $\mathcal{U}(Z_n[x])$ is finite if and only if each $t_i = 1$.*

Proof. From (20.8), the group $\mathcal{U}(Z_n[x]) \cong \mathcal{U}(Z_{p_1^{t_1}}[x]) \times \cdots \times \mathcal{U}(Z_{p_k^{t_k}}[x])$, so $|\mathcal{U}(Z_n[x])| = \prod_{i=1}^k |\mathcal{U}(Z_{p_i^{t_i}}[x])|$ is finite if and only if each $|\mathcal{U}(Z_{p_i^{t_i}}[x])|$ is finite, which is so if and only if each $t_i = 1$ by (22.7) and (21.1).

In (23.5), for each $t_i \geq 2$, we have $\mathcal{U}(Z_{p_i^{t_i}}[x]) \cong Z_{p_i^{t_i}}[x]^+ \times_\gamma^f \mathcal{U}(Z_{p_i^{t_i-1}}[x])$ for some nontrivial f and γ, so when $k \geq 2$ and at least one $t_i \geq 2$, then the structure of $\mathcal{U}(Z_n[x])$ is given in terms of direct products, semidirect products, and nontrivial factor sets.

A good stack of examples, as large as possible, is indispensable for a thorough understanding of any concept, and when I want to learn something new, I make it my first job to build one.

Paul R. Halmos

CHAPTER 6

Avant-guarde families of nearrings

One can be confident that others will feel that certain families of nearrings should have been included in Chapter 4 among the 'first families'. (For example, nearfields would certainly be among the first families, but they really do not have many ideals.) Also, among those families included in Chapter 4, one can assume with probability 0.9813 that others will feel that the best stories concerning these families were left untold. Decisions as to what to include were influenced by (1) what was already told in [P], [M], and [W], and (2) the directions and limitations agreed upon between the author and his publisher.

Here in this final chapter, we touch only lightly upon various families of nearrings. These are but a small sampling of topics of current interest to rerearchers in nearring theory, but each of these topics shows promise of being a source for very interesting new mathematics and a strong potential for applications. As the name of the chapter suggests, topics included here have not yet been developed sufficiently to be among the 'first families'. It is hoped that others will be encouraged and even motivated to develop various families into prominence. With such development, we can only expect considerable enrichment of the theory.

24. Sandwich and laminated nearrings

This topic should have been in Chapter 4. It has been around for a quarter of a century, but, for reasons beyond the comprehension of the author, has not attracted enough researchers for adequate development. The author can think of but few underdeveloped ideas which offer as much promise of reward as do the basic ideas which we present here.

The whole idea behind the sandwich nearring is both simple and powerful. For a topological space $(X, \mathcal{T})$ and a topological group $(G, +, \mathcal{T}')$, let $\mathcal{C}(X, G) = \{f : X \to G \mid f \text{ is continuous }\}$. Then $+$ for G induces the obvious $\oplus$ for $\mathcal{C}(X, G)$, where $(f \oplus g)(x) = f(x) + g(x)$. Now $(\mathcal{C}(X, G), \oplus)$ is a group, and when $(X, \mathcal{T}) = (G, \mathcal{T}')$, one easily sees that $(\mathcal{C}(X, G), \oplus, \circ)$ is a nearring. But the requirement that $(X, \mathcal{T}) = (G, \mathcal{T}')$ was too restrictive for K. D. Magill, Jr. He wanted to determine the influence of an $(X, \mathcal{T}) \neq (G, \mathcal{T}')$. With a continuous $\phi : G \to X$, he defined $*_\phi$ on $\mathcal{C}(X, G)$ by $f *_\phi g = f \circ \phi \circ g$, and with such a ϕ, $(\mathcal{C}(X, G), \oplus, *_\phi)$ is a nearring.

Hence the term 'sandwich'. The f and the g provide the bread, and ϕ is the answer to 'where's the beef?' The $(\mathcal{C}(X,G), \oplus, *_\phi)$ are the *sandwich nearrings*, but if the word 'topology' brings about intimidation, all one has to do is to let $\mathcal{T}$ and $\mathcal{T}'$ be discrete topologies, and then everybody is continuous. Now $\mathcal{C}(X,G)$ is simply the set of all mappings $f : X \to G$, i.e., $\mathcal{C}(X,G) = G^X$.

From the idea of a sandwich nearring, one is led very gently to the idea of a *laminated nearring*. If one starts with a nearring $(N, +, \cdot)$, then one need only take any $e \in N$ to define $*_e$ on N by $x *_e y = x \cdot e \cdot y$. Then $(N, +, *_e)$ is a *laminated nearring*. The element e is the *laminating element*, or the *laminator*.

The games K. D. Magill, Jr, has played with sandwich and laminated nearrings often include two classical games. He adapts each of these games to interesting settings, and has invited others to play the games with him or to play their own favourite game. The first is to consider two sandwich nearrings $(\mathcal{C}(X,G), \oplus, *_\phi)$ and $(\mathcal{C}(Y,H), \oplus, *_\lambda)$, or two laminated nearrings $(N, +, *_e)$ and $(M, +, *_f)$, and ask when they are isomorphic. The second is to compute the automorphism group $\mathrm{Aut}(\mathcal{C}(X,G), \oplus, *_\phi)$ of a sandwich nearring or the automorphism group $\mathrm{Aut}(N, +, *_\phi)$ of a laminated nearring.

Of course, the ideas of a sandwich nearring and a laminated nearring blend into one when $(X, \mathcal{T}) = (G, \mathcal{T}')$. In this case, the sandwich nearring $(\mathcal{C}(X,G), \oplus, *_\phi)$ is then a laminated nearring of $(\mathcal{C}(X,G), \oplus, \circ)$ with laminator ϕ.

Another variation on the idea of a sandwich nearring is to take abelian groups $(A, +)$ and $(B, +)$ and consider the abelian group $(\mathrm{Hom}(A,B), \oplus)$ of all homomorphisms from A to B. With $\phi \in \mathrm{Hom}(B, A)$, one defines $*_\phi$ on $\mathrm{Hom}(A,B)$ by $f *_\phi g = f \circ \phi \circ g$. Then $(\mathrm{Hom}(A,B), \oplus, *_\phi)$ is a nearring. Of course, we could extend this to R-modules A and B and do the same type of thing for $(\mathrm{Hom}_R(A,B), \oplus, *_\phi)$. The reader is invited, yes, even encouraged, to find his own variation of this theme.

Basic to Magill's games is the following theorem.

(24.1) Theorem. *Let $(\mathcal{C}(X,G), \oplus, *_\phi)$ and $(\mathcal{C}(Y,H), \oplus, *_\lambda)$ be two sandwich nearrings. Suppose $h : X \to Y$ is a homeomorphism, and that $t : G \to H$ is a group isomorphism as well as a homeomorphism such that $h \circ \phi = \lambda \circ t$. Then the map $F : \mathcal{C}(X,G) \to \mathcal{C}(Y,H)$ defined by $F(f) = t \circ f \circ h^{-1}$ is a nearring isomorphism.*

Proof. With t, f, and h^{-1} continuous, we certainly have $F(f) \in \mathcal{C}(Y,H)$ for each $f \in \mathcal{C}(X,G)$. Now $F(f \oplus g) = t \circ (f \oplus g) \circ h^{-1} = t \circ [(f \circ h^{-1}) \oplus (g \circ h^{-1})] = t \circ (f \circ h^{-1}) \oplus t \circ (g \circ h^{-1}) = F(f) \oplus F(g)$, and $F(f *_\phi g) = t \circ (f *_\phi g) \circ h^{-1} = t \circ f \circ \phi \circ g \circ h^{-1} = t \circ f \circ h^{-1} \circ h \circ \phi \circ g \circ h^{-1} = F(f) \circ \lambda \circ t \circ g \circ h^{-1} = F(f) \circ \lambda \circ F(g) = F(f) *_\lambda F(g)$. Now we know that F is a nearring homomorphism.

Let ζ_G and ζ_H be the additive identities of $\mathcal{C}(X,G)$ and $\mathcal{C}(Y,H)$ respectively. If $\zeta_H = F(f) = t \circ g \circ h^{-1}$, then $\zeta_H \circ h = t \circ f$. From $0 = \zeta_H \circ h(x) = t(f(x))$ for each $x \in G$ and the fact that t is a group isomorphism, we are confident and secure with the decision that $f = \zeta_G$. Now this makes F a monomorphism. If $g \in \mathcal{C}(Y,H)$, then $t^{-1} \circ g \circ h \in \mathcal{C}(X,G)$ and $F(t^{-1} \circ g \circ h) = g$, so F is an epimorphism. Putting the pieces together allows us to conclude that F is an isomorphism, which completes the proof of (24.1).

The isomorphism F of (24.1) depends upon the homeomorphism h and the topological isomorphism t, and since this construction for F from h and t is such a natural approach, we call such an isomorphism *natural*. A central theme of Magill's efforts has been to find 'reasonable' conditions on X, Y, G, and H to ensure that every isomorphism from $\mathcal{C}(X,G)$ onto $\mathcal{C}(Y,H)$ is natural. We will present a simplified version of some of his results. See [Mag, 1967, 1967a, 1980].

Let X and Y be nonempty sets with a surjective map $\phi : Y \to X$. If $\mathrm{Map}(X,Y)$ denotes all functions $f : X \to Y$, we define the binary operation $*_\phi$ on $\mathrm{Map}(X,Y)$ by $f *_\phi g = f \circ \phi \circ g$. Then $(\mathrm{Map}(X,Y), *_\phi)$ is a semigroup. Let $\mathcal{S}(X,Y)$ denote a subsemigroup of $\mathrm{Map}(X,Y)$. We say that $\mathcal{S}(X,Y)$ *has all the constant functions*, if for each $y \in Y$, the constant function $T_y \in \mathcal{S}(X,Y)$, where $T_y(x) = y$ for all $x \in X$. We also say that $\mathcal{S}(X,Y)$ *separates points* if for any two distinct points $x, x' \in X$ there is an $f \in \mathcal{S}(X,Y)$ such that $f(x) \neq f(x')$. (Note that T_y can be ambiguous. If $y \in Y \cap V$, then is $T_y \in \mathcal{S}(X,Y)$ or is $T_y \in \mathcal{S}(U,V)$ for some other U? We will rely upon context and the mental dexterity of the reader to keep everything in place.)

(24.2) Theorem. *Suppose $X, Y, U,$ and V are nonempty sets with surjective maps $\phi : Y \to X$ and $\lambda : V \to U$. Suppose $\mathcal{S}(X,Y)$ and $\mathcal{S}(U,V)$ are subsemigroups of $(\mathrm{Map}(X,Y), *_\phi)$ and $(\mathrm{Map}(U,V), *_\lambda)$, respectively, each of which has all the constant functions, and each of which separates points. Also, suppose $F : \mathcal{S}(X,Y) \to \mathcal{S}(U,V)$ is a semigroup isomorphism. Let $\mathcal{K}(X,Y)$ and $\mathcal{K}(U,V)$ denote, respectively, the constant functions of $\mathcal{S}(X,Y)$ and $\mathcal{S}(U,V)$. Then $\mathcal{K}(X,Y)$ and $\mathcal{K}(U,V)$ are subsemigroups of $\mathcal{S}(X,Y)$ and $\mathcal{S}(U,V)$, respectively, and F restricted to $\mathcal{K}(X,Y)$ is an isomorphism onto $\mathcal{K}(U,V)$. In addition, there are unique bijections $t : Y \to V$ and $h : X \to U$ such that $\lambda \circ t = h \circ \phi$, and such that $t \circ f = F(f) \circ h$ for arbitrary $f \in \mathcal{S}(X,Y)$.*

Proof. For $T_y \in \mathcal{K}(X,Y)$ and $f \in \mathcal{S}(X,Y)$, we have $T_y *_\phi f = T_y \circ \phi \circ f$, and for $x \in X$, $T_y \circ \phi \circ f(x) = y$. Hence, $T_y *_\phi f = T_y$. In particular, $*_\phi$ defines a binary operation on $\mathcal{K}(X,Y)$. Hence, $\mathcal{K}(X,Y)$ and $\mathcal{K}(U,V)$ are subsemigroups.

From $T_y = T_y *_\phi f = T_y \circ \phi \circ f$ for each $f \in \mathcal{S}(X, Y)$, we have $F(T_y) = F(T_y) *_\lambda F(f) = F(T_y) \circ \lambda \circ g$ for each $g \in \mathcal{S}(U, V)$. Take $g = T_v$ for some $v \in V$, and let $u, u' \in U$. Then $T_v(u) = v = T_v(u')$, so $F(T_y)(u) = F(T_y)(u')$. Hence, $F(T_y)$ is a constant function, and $F(\mathcal{K}(X, Y)) \subseteq \mathcal{K}(U, V)$. This result also applies to F^{-1}, so $F^{-1}(\mathcal{K}(U, V)) \subseteq \mathcal{K}(X, Y)$. Hence $F(\mathcal{K}(X, Y)) = \mathcal{K}(U, V)$, and so F restricted to $\mathcal{K}(X, Y)$ defines an isomorphism onto $\mathcal{K}(U, V)$.

The restriction of F to $\mathcal{K}(X, Y)$ helps us to define a bijection $t : Y \to V$. In fact, we define

$$t(y) = v \text{ if and only if } F(T_y) = T_v. \tag{24:1}$$

Hence,

$$F(T_y) = T_{t(y)}. \tag{24:2}$$

Certainly, t is a bijection.

The bijection t has an important relationship with ϕ and λ. In fact

$$\phi(y) = \phi(y') \implies \lambda(t(y)) = \lambda(t(y')). \tag{24:3}$$

To see this, we begin with any $g \in \mathcal{S}(U, V)$. Now there is a unique $f \in \mathcal{S}(X, Y)$ such that $F(f) = g$. From $f(\phi(y)) = f(\phi(y'))$ we obtain $f *_\phi T_y = f *_\phi T_{y'}$. Suppose $F(T_y) = T_v$ and $F(T'_y) = T_{v'}$. Then $F(f) *_\lambda T_v = F(f) *_\lambda T_{v'}$, so $g(\lambda(v)) = g(\lambda(v'))$. But $F(T_y) = T_v$ and $F(T_{y'}) = T_{v'}$ imply $t(y) = v$ and $t(y') = v'$. Hence, $g(\lambda(t(y))) = g(\lambda(t(y')))$. Remember that g is arbitrary in $\mathcal{S}(U, V)$ and that $\mathcal{S}(U, V)$ separates points. Hence, $\lambda(t(y)) = \lambda(t(y'))$, and we have (24:3).

We are now ready to define $h : X \to U$. Define

$$h(x) = \lambda(t(y)) \text{ where } \phi(y) = x. \tag{24:4}$$

If $\phi(y) = \phi(y') = x$, then $\lambda(t(y)) = \lambda(t(y'))$ by (24:3), so h is well defined.

For any $y \in Y$, $\phi(y) = x$ for some $x \in X$. Since $h(x) = \lambda(t(y))$, we also have $h(\phi(y)) = \lambda(t(y))$, which means

$$h \circ \phi = \lambda \circ t, \tag{24:5}$$

which is part of the conclusion of our theorem.

To obtain the other equation, start with $x \in X$ and $y \in Y$ such that $\phi(y) = x$. Certainly, $T_{f(x)} \in \mathcal{K}(X, Y)$ for any $f \in \mathcal{S}(X, Y)$, and if $x, x' \in X$, then $T_{f(x)}(x') = f(x)$. Also, $f \circ \phi \circ T_y(x') = f(\phi(y)) = f(x)$. Hence, $T_{f(x)} = f *_\phi T_y$. Using (24:1), we have $F(T_{f(x)}) = T_v$ if and only if $t(f(x)) = v$. Hence, $T_{t(f(x))} = T_v = F(T_{f(x)}) = F(f *_\phi T_y) = F(f) *_\lambda F(T_y) = F(f) \circ \lambda \circ T_{t(y)}$. So, if $u \in U$, use (24:5) to obtain $(t \circ f)(x) = T_{t(f(x))}(u) =$

$F(f) \circ \lambda \circ T_{t(y)}(u) = F(f) \circ \lambda \circ t(y) = F(f) \circ h \circ \phi(y) = F(f) \circ h(x)$. But $(t \circ f)(x) = (F(f) \circ h)(x)$ for arbitrary $x \in X$ is exactly what is meant by

$$t \circ f = F(f) \circ h, \qquad (24:6)$$

and since $f \in \mathcal{S}(X, Y)$ is arbitrary, then (24:6) is valid for arbitrary $f \in \mathcal{S}(X, Y)$.

We already have that t is a bijection. We turn our attention now to showing that h is also a bijection. Suppose $x, x' \in X$ and $x \neq x'$, and suppose $\phi(y) = x$ and $\phi(y') = x'$, so $\phi(y) \neq \phi(y')$. Choose $f \in \mathcal{S}(X, Y)$ so that $f(x) \neq f(x')$. Now for any $a \in X$, $(f \circ \phi \circ T_y)(a) = f(\phi(y)) = f(x)$ and $(f \circ \phi \circ T_{y'})(a) = f(x')$, so we know that $f *_\phi T_y \neq f *_\phi T_{y'}$ and $F(f) \circ \lambda \circ F(T_y) \neq F(f) \circ \lambda \circ F(T_{y'})$. But $F(T_y) = T_{t(y)}$ and $F(T_{y'}) = T_{t(y')}$, so $F(f)\big(\lambda(t(y))\big) \neq F(f)\big(\lambda(t(y'))\big)$, and, consequently, $\lambda \circ t(y) \neq \lambda \circ t(y')$. With (24:5) we conclude that $h(\phi(y)) \neq h(\phi(y'))$. Since $\phi(y) = x$ and $\phi(y') = x'$, we finally obtain that $h(x) \neq h(x')$, and so h is injective. Take $u \in U$ and $v \in V$ such that $\lambda(v) = u$. Then take $y \in Y$ such that $t(y) = v$. But $\phi(y) \in X$ and $h(\phi(y)) = \lambda(t(y)) = \lambda(v) = u$. Hence, h is surjective, making h a bijection.

Finally, we turn our attention to uniqueness of t and h. Suppose $t' : Y \to V$ and $h' : X \to U$ are such that $\lambda \circ t' = h' \circ \phi$ and such that $t' \circ f = F(f) \circ h'$ for any $f \in \mathcal{S}(X, Y)$. Choose any $y \in Y$ and $x \in X$. Then $t(y) = t(T_y(x)) = t \circ T_y(x) = F(T_y)(h(x))$. Similarly, $t'(y) = F(T_y)(h'(x))$. But $F(T_y)$ is a constant function, which means $t(y) = t'(y)$ for any $y \in Y$. Of course, this means $t = t'$.

For any $x \in X$, there is a $y \in Y$ such that $\phi(y) = x$. Now $h(x) = h(\phi(y)) = \lambda(t(y))$. Similarly, $h'(x) = \lambda(t'(y))$. We already have $t(y) = t'(y)$, hence $h'(x) = h(x)$. This means $h = h'$, and completes the proof of our theorem.

So (24.2) provides a big step towards making every isomorphism from $\mathcal{C}(X, G)$ onto $\mathcal{C}(Y, H)$ natural. We need to get the t and h of (24.2) to be homeomorphisms. Our next step will be to make h a homeomorphism.

Start with nonempty topological spaces X and Y, a continuous function $\phi : Y \to X$ which is surjective, and an $\mathcal{S}(X, Y)$ as in (24.2) with the additional requirement that each $f \in \mathcal{S}(X, Y)$ is continuous. In addition, we want X and Y to be topological spaces with two particularly important properties. First, single point subsets must be closed, that is, if $u \in X \cup Y$, then $\{u\}$ is a closed set. We say that 'points are closed'. (Topologists say such spaces are T_1.) The next property is a little more complicated since it involves both X and Y. If $A \subseteq X$ is closed, and $x \in X \setminus A$, we need an $f_x \in \mathcal{S}(X, Y)$ and a point $p_x \in X$ such that $(\phi \circ f_x)(a) = p_x$ for each $a \in A$ and such that $(\phi \circ f_x)(x) \neq p_x$. If we have all this, then we shall say that the semigroup $(\mathcal{S}(X, Y), *_\phi)$ is *admissible*. This next result provides an abundance of admissible semigroups, but let us remind ourselves that a topological space X is *completely regular* if every closed set $A \subseteq X$ and point $x \in X \setminus A$ has a continuous function $f_{A,x} : X \to [0,1]$, the closed unit interval, such that $f_{A,x}(a) = 0$ for all $a \in A$ and $f_{A,x}(x) = 1$.

(24.3) Proposition. *Let X be a completely regular topological space in which points are closed. Let Y be a topological space in which points are closed, and which has two points $a, b \in Y$, $a \neq b$, and a continuous function $\gamma : [0,1] \to Y$ such that $\gamma(0) = a$ and $\gamma(1) = b$. Let $\mathcal{S}(X,Y) = \mathcal{C}(X,Y)$, the set of all continuous functions from X into Y. Then $(\mathcal{S}(X,Y), *_\phi)$ is admissible for any continuous surjective map $\phi : Y \to X$.*

Proof. Take $y \in Y$ and consider T_y. If $U \subseteq Y$ and U is open, then either $y \in U$ or $y \notin U$. If $y \notin U$, then $T_y^{-1}(U) = \varnothing$, an open set. If $y \in U$, then $T_y^{-1}(U) = X$, an open set. Hence, each $T_y \in \mathcal{S}(X,Y)$, and so $\mathcal{S}(X,Y)$ has all the constant functions. If $x, x' \in X$ and $x \neq x'$, there is a continuous function $f_{\{x\},x'} : X \to [0,1]$ such that $\gamma \circ f_{\{x\},x'}(x) = a$ and $\gamma \circ f_{\{x\},x'}(x') = b$. Since $\gamma \circ f_{\{x\},x'} \in \mathcal{S}(X,Y)$, we know that $\mathcal{S}(X,Y)$ separates points.

Finally, take $A \subseteq X$ to be a closed set, and let $c \in X \setminus A$. We need an $f_c \in \mathcal{S}(X,Y)$ and a point $p_c \in X$ such that $(\phi \circ f_c)(d) = p_c$ for each $d \in A$ and such that $(\phi \circ f_c)(c) \neq p_c$. We have $f_{A,c} : X \to [0,1]$ which is continuous and has $f_{A,c}(d) = 0$ for each $d \in A$ and has $f_{A,c}(c) = 1$. Hence $\gamma \circ f_{A,c}(d) = a$ for each $d \in A$ and $\gamma \circ f_{A,c}(c) = b$. Therefore, $\phi \circ \gamma \circ f_{A,c}(d) = \phi(a)$ for each $d \in A$, and $\phi \circ \gamma \circ f_{A,c}(c) = \phi(b)$. Let $f_c = \gamma \circ f_{A,c}$ and $p_c = \phi(a)$. Then $(\phi \circ f_c)(d) = p_c$ for each $d \in A$, and $(\phi \circ f_c)(c) = \phi(b) \neq p_c$. Hence, $\mathcal{S}(X,Y)$ is admissible.

(24.4) Theorem. *Let X, Y, U, and V be nonempty topological spaces in which points are closed. Suppose $\phi : Y \to X$ and $\lambda : V \to U$ are continuous surjections. Also suppose that $\mathcal{S}(X,Y)$ and $\mathcal{S}(U,V)$ are admissible semigroups with an isomorphism $F : \mathcal{S}(X,Y) \to \mathcal{S}(U,V)$. Then the h of (24.2) is a homeomorphism.*

Proof. For $x \in X$ and $f \in \mathcal{S}(X,Y)$, define

$$H(x, f) = f^{-1}(\phi^{-1}(x)) = \{x' \in X \mid \phi \circ f(x') = x\},$$

a closed subset of X since $\{x\}$ is closed and $\phi \circ f$ is continuous. Our first step is to show that the following three statements are equivalent.
(1) $u \in h[H(x, f)]$, $u = h(x')$, and $(\phi \circ f)(x') = x$.
(2) $h(x) = (h \circ \phi \circ f)(x') = (\lambda \circ F(f) \circ h)(x') = (\lambda \circ F(f))(u)$, where $u = h(x')$.
(3) $u \in H(h(x), F(f))$.
(We take the liberty of letting $H(u, g) = g^{-1}(\lambda^{-1}(u))$ for arbitrary $u \in U$ and arbitrary $g \in \mathcal{S}(U,V)$.)

So assume $u = h(x')$, $(\phi \circ f)(x') = x$, so $x' \in H(x, f)$, and $h(x') = u \in h[H(x, f)]$. (More simply stated, we take $x' \in H(x, f)$.) Hence, using (24.2), $h(x) = h[(\phi \circ f)(x')] = (h \circ \phi \circ f)(x') = \lambda \circ t \circ f(x') = (\lambda \circ F(f) \circ h)(x') = (\lambda \circ F(f))(u)$. Hence, (1) implies (2). Now take $u = h(x')$ and assume $h(x) = (h \circ \phi \circ f)(x') = (\lambda \circ F(f) \circ h)(x') = (\lambda \circ F(f))(u)$. Since h is injective, we get $\phi \circ f(x') = x$, so $x' \in H(x, f)$ and $u = h(x') \in h[H(x, f)]$. Thus, (2) implies (1).

Again, with $u = h(x')$ and $h(x) = (h \circ \phi \circ f)(x') = (\lambda \circ F(f))(u)$, we obtain $u \in H(h(x), F(f))$, and so (2) implies (3). Conversely, if $u \in H(h(x), F(f))$, then $\lambda \circ F(f)(u) = h(x)$. But there is a unique x' such that $h(x') = u$, so $h(x) = (\lambda \circ F(f) \circ h)(x') = (\lambda \circ F(f))(u)$, and $(\lambda \circ F(f) \circ h)(x') = (\lambda \circ t \circ f)(x') = (h \circ \phi \circ f)(x')$, and we conclude that (3) implies (2).

As a consequence of (1) and (3), we have that $h[H(x, f)] \subseteq H(h(x), F(f))$. Certainly, if $u \in H(h(x), F(f))$, then (3) implies $u \in h[H(x, f)]$, and so we get $H(h(x), F(f)) \subseteq h[H(x, f)]$. We can now state

$$h[H(x, f)] = H(h(x), F(f)). \qquad (24:7)$$

Starting with our isomorphism F and the h and t from (24.2), we have $\lambda \circ t = h \circ \phi$, and for any $f \in \mathcal{S}(X, Y)$, we obtain $t \circ f = F(f) \circ h$. We also have the isomorphism $F^{-1} : \mathcal{S}(U, V) \to \mathcal{S}(X, Y)$. So, by (24.2), there are unique bijections $t' : V \to Y$ and $h' : U \to X$ such that $h' \circ \lambda = \phi \circ t'$, and for any $g \in \mathcal{S}(U, V)$, we obtain $t' \circ g = F^{-1}(g) \circ h'$. We proceed to show that $h' = h^{-1}$ and that $t' = t^{-1}$. From $\lambda \circ t = h \circ \phi$, we obtain $h^{-1} \circ \lambda = \phi \circ t^{-1}$. Any $g \in \mathcal{S}(U, V)$ has a unique $f \in \mathcal{S}(X, Y)$ so that $F(f) = g$. From $F^{-1}(g) = f$ we have $t \circ f = F(f) \circ h$, so $f \circ h^{-1} = t^{-1} \circ F(f)$, or $t^{-1} \circ g = F^{-1}(g) \circ h^{-1}$. With t^{-1} and h^{-1} doing what t' and h' do, we let uniqueness force $t' = t^{-1}$ and $h' = h^{-1}$. Hence, applying (24:7), we get

$$h^{-1}[H(u, g)] = H(h^{-1}(u), F^{-1}(g)) \qquad (24:8)$$

for any $u \in U$ and any $g \in \mathcal{S}(U, V)$.

Our next step is to show that the family of all $H(x, f)$, $x \in X$ and $f \in \mathcal{S}(X, Y)$, forms a basis for the closed sets of X. That is, each closed set $A \subseteq X$ is the intersection of a set of $H(x, f)$. Let $A \subseteq X$ be any closed set, and let $x \in X \setminus A$. Then there are a $p_x \in X$ and an $f_x \in \mathcal{S}(X, Y)$ such that $(\phi \circ f_x)(a) = p_x$ for each $a \in A$ and such that $(\phi \circ f_x)(x) \neq p_x$. We wish to demonstrate that $A = \cap \{H(p_x, f_x) \mid x \in X \setminus A\}$.

Take $b \in \cap \{H(p_x, f_x) \mid x \in X \setminus A\}$. Then $b \in H(p_x, f_x)$ for each $x \in X \setminus A$, so $\phi \circ f_x(b) = p_x$. If $b \notin A$, then $b \in H(p_b, f_b)$, so $= \phi \circ f_b(b) = p_b$. But $(\phi \circ f_b)(b) \neq p_b$. Hence, $b \in A$ and so $\cap \{H(p_x, f_x) \mid x \in X \setminus A\} \subseteq A$. Now take $b \in A$. We have for any $x \in X \setminus A$, $(\phi \circ f_x)(a) = p_x$ for all $a \in A$.

Hence, $(\phi \circ f_x)(b) = p_x$ and so $b \in H(p_x, f_x)$ for each $x \in X \setminus A$. This means $A \subseteq \cap\{H(p_x, f_x) \mid x \in X \setminus A\}$, and, consequently, $A = \cap\{H(p_x, f_x) \mid x \in X \setminus A\}$.

Now we are ready to show that h and h^{-1} are continuous. A function is continuous if and only if the inverse image of every closed set is a closed set. If $A = \cap\{H(p_x, f_x) \mid x \in X \setminus A\}$ is a closed set in X, then $(h^{-1})^{-1}(A) = h(A) = h[\cap\{H(p_x, f_x) \mid x \in X \setminus A\}] = \cap\{(h^{-1})^{-1}(H(p_x, f_x)) \mid x \in X \setminus A\} = \cap\{h(H(p_x, f_x)) \mid x \in X \setminus A\} = \cap\{H(h(p_x), F(f_x)) \mid x \in X \setminus A\}$, a closed set. This makes h^{-1} continuous. Reversing the roles of $\mathcal{S}(X, Y)$ and $\mathcal{S}(U, V)$, F and F^{-1}, and h and h^{-1}, we conclude that h is continuous. Hence, h is a homeomorphism.

Since (24.4) provides a way to get h of (24.2) to be a homeomorphism, we need to focus on getting t to be a homeomorphism. For our purposes here, there is a very easy way to do this.

(24.5) Theorem. *Let $\mathcal{S}(X, Y)$ and $\mathcal{S}(U, V)$ be as in (24.4), but with the property that $\phi : Y \to X$ and $\lambda : V \to U$ are homeomorphisms. Then the t of (24.2) is a homeomorphism.*

Proof. From (24.2) we have $\lambda \circ t = h \circ \phi$, and so $t = \lambda^{-1} \circ h \circ \phi$. Since λ^{-1}, h, and ϕ are all homeomorphisms, then so is t.

The following result is motivated by results of Magill. See [Mag, 1980] and (24.10) below.

(24.6) Theorem. *Let $(N, +, \cdot)$ be a (left) nearring with identity 1 and a group of units $\mathcal{U}(N)$. For $u \in N$, we have the laminated nearring $(N, +, *_u)$. Then the following are equivalent.*
(1) $u \in \mathcal{U}(N)$.
(2) $(N, +, \cdot) \cong (N, +, *_u)$.
(3) $(N, +, *_u)$ *has identity.*

Proof. If $u \in \mathcal{U}(N)$, then $u^{-1} *_u x = u^{-1} \cdot u \cdot x = x = x \cdot u \cdot u^{-1} = x *_u u^{-1}$. So u^{-1} is the identity for $*_u$. With u' being the identity for $(N, +, *_u)$, we have $u' *_u 1 = u' \cdot u \cdot 1 = u' \cdot u = 1$. Hence, $u' = u^{-1}$, and so $u \in \mathcal{U}(N)$. We now have(1) and (3) equivalent.

Define $F : (N, +, \cdot) \to (N, +, *_u)$ by $F(x) = u^{-1} \cdot x$. Then $F(x + y) = u^{-1}(x + y) = (u^{-1}x) + (u^{-1}y) = F(x) + F(y)$. From $0 = F(x) = u^{-1}x$, we get $0 = u \cdot 0 = x$. Hence, F is injective. Since $F(u \cdot y) = u^{-1} \cdot (u \cdot y) = y$, we conclude that F is a group isomorphism. Finally, $F(x \cdot y) = u^{-1} \cdot (x \cdot y) = (u^{-1} \cdot x) \cdot u \cdot (u^{-1} \cdot y) = F(x) *_u F(y)$, and so F is a nearring isomorphism. So either (1) or (3) implies (2). If we start with an isomorphism $F : (N, +, \cdot) \to (N, +, *_u)$, and let $F(1) = e$ and $F(v) = 1$, then $1 = F(v) = F(v \cdot 1) = F(v) *_u F(1) = 1 \cdot u \cdot e = u \cdot e$, so $e = u^{-1} \in \mathcal{U}(N)$. Hence (2) and (1) are equivalent, completing our proof of (24.6).

(24.7) Theorem. *Suppose we have the hypotheses of* (24.2). *In addition, suppose Y and V are topological groups with* $F : \mathcal{S}(X, Y) \to \mathcal{S}(U, V)$ *a nearring isomorphism. Then the* $t : Y \to V$ *of* (24.2) *is also a group isomorphism.*

Proof. We easily obtain $T_y + T_{y'} = T_{y+y'}$ for all $y, y' \in Y$, and $T_v + T_{v'} = T_{v+v'}$ for all $v, v' \in V$. From (24:2), we have $F(T_y) = T_{t(y)}$. Hence, $T_{t(y+y')} = F(T_{y+y'}) = F(T_y + T_{y'}) = F(T_y) + F(T_{y'}) = T_{t(y)} + T_{t(y')} = T_{t(y)+t(y')}$. Since F is an isomorphism, we have $t(y + y') = t(y) + t(y')$. This is all we needed to get t to be an isomorphism, since (24.2) assures us that t is a bijection.

(24.8) Lemma. *Let* $\mathcal{HA}(\mathbf{R}^+)$ *denote the set of homeomorphisms* $f : \mathbf{R} \to \mathbf{R}$ *such that* f *is also a group automorphism of* $(\mathbf{R}, +)$. *Then* $\mathcal{HA}(\mathbf{R}^+)$ *is a group isomorphic to* $\mathbf{R}^*$, *the multiplicative group of the field* $(\mathbf{R}, +, \cdot)$. *In fact, each element* $f \in \mathcal{HA}(\mathbf{R}^+)$ *is defined by* $f(x) = ax$ *for some* $a \in \mathbf{R}^*$.

Proof. If $f \in \mathcal{HA}(\mathbf{R}^+)$, then there is an $a \in \mathbf{R}^*$ such that $f(1) = a$. From $a = f(1) = f(n \cdot (1/n))$ we obtain $f(1/n) = a/n$ for each positive integer n, and then we readily get $f(m/n) = a(m/n)$ for each rational number m/n. If the sequence $\{m_k/n_k\}$ converges to $r \in \mathbf{R}$, then continuity of f assures us that $f(r) = ar$. If $f(x) = ax$, $g(x) = bx$, and $a \neq b$, then $f \neq g$ and $(f \circ g)(x) = abx$. Conversely, for any $a \in \mathbf{R}^*$, $f(x) = ax$ defines $f \in \mathcal{HA}(\mathbf{R}^+)$. The isomorphism between $\mathcal{HA}(\mathbf{R}^+)$ and $\mathbf{R}^*$ is obviously $f \leftrightarrow a$ when $f(x) = ax$.

(24.9) Proposition. *If* $\phi : \mathbf{R} \to \mathbf{R}$ *is a homeomorphism, then the groups* $\mathrm{Aut}(\mathcal{C}(\mathbf{R}, \mathbf{R}), +, *_\phi)$ *and* $\mathbf{R}^*$ *are isomorphic.*

Proof. Since $\phi \in \mathcal{U}(\mathcal{C}(\mathbf{R}, \mathbf{R}), +, \circ)$, we have by (24.6) that the nearring $(\mathcal{C}(\mathbf{R}, \mathbf{R}), +, *_\phi)$ is isomorphic to the nearring $(\mathcal{C}(\mathbf{R}, \mathbf{R}), +, \circ)$. We need only show that $\mathrm{Aut}(\mathcal{C}(\mathbf{R}, \mathbf{R}), +, *_\phi) \cong \mathbf{R}^*$.

For $F \in \mathrm{Aut}(\mathcal{C}(\mathbf{R}, \mathbf{R}), +, *_\phi)$, we have homeomorphisms $h : \mathbf{R} \to \mathbf{R}$ and $t : \mathbf{R} \to \mathbf{R}$ such that $t \in \mathrm{Aut}(\mathbf{R}, +)$, $\iota \circ t = h \circ \iota$ for $\iota(x) = x$, the identity of $\mathcal{C}(\mathbf{R}, \mathbf{R})$, and such that $t \circ f = F(f) \circ h$ for all $f \in \mathcal{C}(\mathbf{R}, \mathbf{R})$. Since $h = t$, we have $F(f) = t \circ f \circ t^{-1}$. We will be finished when we show that distinct $t_1, t_2 \in \mathcal{HA}(\mathbf{R}^+)$ define distinct $F_i(f) = t_i \circ f \circ t_i^{-1}$. But this follows from the uniqueness guaranteed by (24.2).

We can extend (24.6) for $(\mathcal{C}(\mathbf{R}, \mathbf{R}), +, \circ)$. For our purposes here we say that a surjective $\phi \in \mathcal{C}(\mathbf{R}, \mathbf{R})$ is *nearly increasing* if there is an interval $(-r, r)$ so that ϕ is increasing on $\mathbf{R} \setminus (-r, r) = (-\infty, -r] \cup [r, +\infty)$, that is, if $x, x' \in \mathbf{R} \setminus (-r, r)$ and $x < x'$, then $\phi(x) < \phi(x')$. Similarly, a surjective $\phi \in \mathcal{C}(\mathbf{R}, \mathbf{R})$ is *nearly decreasing* if there is an interval $(-r, r)$ so that ϕ is decreasing on $\mathbf{R} \setminus (-r, r)$, that is, if $x, x' \in \mathbf{R} \setminus (-r, r)$ and $x < x'$, then $\phi(x) > \phi(x')$. A surjective $\phi \in C(\mathbf{R}, \mathbf{R})$ is *nearly monotonic* if it is either nearly increasing or nearly decreasing. Polynomials of odd degree define nearly monotonic functions, as does the hyperbolic sine function $f(x) = \sinh x = (e^x - e^{-x})/2$.

(24.10) Theorem. *Let $\phi \in C(\mathbf{R}, \mathbf{R})$ be surjective and nearly monotonic. Then the following are equivalent.*
(1) *The mapping ϕ is a homeomorphism.*
(2) $(C(\mathbf{R}, \mathbf{R}), +, \circ) \cong (C(\mathbf{R}, \mathbf{R}), +, *_\phi)$.
(3) $(C(\mathbf{R}, \mathbf{R}), +, *_\phi)$ *has an identity.*
(4) $\mathrm{Aut}(C(\mathbf{R}, \mathbf{R}), +, *_\phi) \cong \mathbf{R}^*$.
(5) $\mathrm{Aut}(C(\mathbf{R}, \mathbf{R}), +, *_\phi)$ *has more than two elements.*

Proof. The equivalence of (1), (2), and (3) follow from (24.6). From (24.9), we have that (1) implies (4), and certainly (4) implies (5). We will be finished when we show that (5) implies (1).

Let F_1, F_2, and F_3 be three distinct automorphisms of $(C(\mathbf{R}, \mathbf{R}), +, *_\phi)$. Each F_i has its h_i and t_i of (24.2). The mappings t_i are distinct, for if $t_i = t_j$, then $h_i \circ \phi = \phi \circ t_i = \phi \circ t_j = h_j \circ \phi$. For arbitrary $x \in \mathbf{R}$, there is a $y \in \mathbf{R}$ such that $\phi(y) = x$. Hence, $h_i(x) = h_i(\phi(y)) = \phi \circ t_i(y) = \phi \circ t_j(y) = h_j \circ \phi(y) = h_j(x)$. This makes $h_i = h_j$, and with $F_i(f) = t_i \circ f \circ h_i^{-1} = t_j \circ f \circ h_j^{-1} = F_j(f)$ for each $f \in C(\mathbf{R}, \mathbf{R})$, we obtain $F_i = F_j$. So the mappings t_i are distinct, and by (24.8), there are $a_i \in \mathbf{R}^*$ such that $t_i(x) = a_i x$, from which it follows that one of the $a_i \notin \{\pm 1\}$.

So, there is an automorphism F of $(C(\mathbf{R}, \mathbf{R}), +, *_\phi)$ which has an h and t of (24.2) where $t(x) = ax$ and $a \neq \pm 1$. Our next step is to show that we may, without loss of generality, assume $a > 1$. First, if $a < 0$, then F^2 has t^2 and $t^2(x) = t(t(x)) = a^2 x$ and $a^2 > 0$ and $a^2 \neq 1$. So we may assume that $0 < a$. If $0 < a < 1$, then F^{-1} has t^{-1} and $t^{-1}(x) = a^{-1}x$, with $1 < a^{-1}$. So, without loss of generality, we take $a > 1$.

In fact, without loss of generality, we may assume a to be very large. For with F having t with $t(x) = ax$ and $1 < a$, then F^n has t^n with $t^n(x) = a^n x$, and $a^n \to +\infty$ as $n \to +\infty$. In particular, if ϕ is monotonic on $\mathbf{R} \setminus (-r, r)$, and if $x, y \in \mathbf{R}^*$ are distinct, then there is an n so that $a^n x, a^n y \in \mathbf{R} \setminus (-r, r)$ and $a^n x \neq a^n y$.

We are ready to show that ϕ is injective, hence ϕ is a homeomorphism of $\mathbf{R}$. If $\phi(x) = \phi(y)$ and $x \neq y$, then $\phi(a^n x) = \phi \circ t^n(x) = h^n \circ \phi(x) = h^n \circ \phi(y) = \phi \circ t^n(y) = \phi(a^n y)$. If $x = 0$, then $\phi(0) = \phi(a^n y)$ for each positive integer n, and this contradicts ϕ being nearly monotonic. If $0 \notin \{x, y\}$, then there is an n so that $a^n x, a^n y \in \mathbf{R} \setminus (-r, r)$ and $a^n x \neq a^n y$. Since ϕ is nearly monotonic, we also have $\phi(a^n x) \neq \phi(a^n y)$. Since the assumption $\phi(x) = \phi(y)$ with $x \neq y$ gets us into this unacceptable situation, we must have $\phi(x) \neq \phi(y)$ if $x \neq y$. Hence, ϕ is injective, and, consequently, a homeomorphism. Having shown that (5) implies (1), we have finished the proof of (24.10).

(24.11) Corollary. *Let* $\phi \in C(\mathbf{R}, \mathbf{R})$ *be surjective. Also, suppose that* ϕ *is not a homeomorphism. If* $(C(\mathbf{R}, \mathbf{R}), +, *_\phi)$ *has more than two automorphisms, then it has infinitely many automorphisms.*

Proof. From the proof of (24.10), there is an automorphism F which has a t where $t(x) = ax$ and $a \neq \pm 1$. Then for every integer n, F^n has t^n where $t^n(x) = a^n x$, and F^n is not the identity. If $m \neq n$, then $F^m \neq F^n$. Hence, each F^n is a distinct automorphism of $(C(\mathbf{R}, \mathbf{R}), +, *_\phi)$.

A $\phi \in C(\mathbf{R}, \mathbf{R})$ which is surjective and nearly monotonic has an important property, which was communicated to the author by K. D. Magill, Jr.

(24.12) Proposition. *Suppose* $\phi \in C(\mathbf{R}, \mathbf{R})$ *is surjective and nearly monotonic. Then* $A \subseteq \mathbf{R}$ *is a closed subset of* $\mathbf{R}$ *whenever* $\phi^{-1}(A)$ *is closed.*

Proof. If ϕ is nearly decreasing, then $-\phi$ is nearly increasing, so we may assume that ϕ is nearly increasing. Since ϕ is nearly increasing and since ϕ is surjective, there is a positive number r so that ϕ is increasing on $(-\infty, -r]$, increasing on $[r, +\infty)$, and satisfies $\phi(-r) \leq \phi(x) \leq \phi(r)$ for $-r \leq x \leq r$. Hence, ϕ restricted to $(-\infty, -r]$ defines a homeomorphism from $(-\infty, -r]$ onto $(-\infty, \phi(-r)]$, and ϕ restricted to $[r, +\infty)$ defines a homeomorphism from $[r, +\infty)$ onto $[\phi(r), +\infty)$.

For $A \subseteq \mathbf{R}$, suppose $\phi^{-1}(A)$ is closed. Then $\phi^{-1}(A) = B \cup C \cup D$ where $B = \phi^{-1}(A) \cap (-\infty, -r]$, $C = \phi^{-1}(A) \cap [-r, r]$, and $D = \phi^{-1}(A) \cap [r, +\infty)$. With ϕ being surjective, we are assured that $\phi(\phi^{-1}(A)) = A$, and, as a consequence, $A = \phi(B) \cup \phi(C) \cup \phi(D)$. Certainly, B, C, and D are closed subsets of $\mathbf{R}$, and since ϕ restricted to $(-\infty, -r]$ and ϕ restricted to $[r, +\infty)$ are homeomorphisms, then $\phi(B)$ and $\phi(D)$ are closed subsets of $(-\infty, \phi(-r)]$ and $[\phi(r), +\infty)$, respectively, and so are closed subsets of $\mathbf{R}$. Now C is a closed subset of $[-r, r]$, and so C is compact. This makes $\phi(C)$ closed in $\mathbf{R}$. With $\phi(B)$, $\phi(C)$, and $\phi(D)$ closed in $\mathbf{R}$, we have $A = \phi(B) \cup \phi(C) \cup \phi(D)$ closed also.

For a nearly monotonic surjective $\phi \in C(\mathbf{R}, \mathbf{R})$, if ϕ is not a homeomorphism, then by (24.10), there can be at most one nontrivial automorphism F of $(C(\mathbf{R}, \mathbf{R}), +, *_\phi)$. What must such a ϕ look like? With F, there are the h and t from (24.2) and $t \in \mathcal{HA}(\mathbf{R}^+)$, so by (24.8), $t(x) = ax$ for some $a \in \mathbf{R}^*$. If $a \neq \pm 1$, then we'd get ϕ to be a homeomorphism, as we did in the proof of (24.10), so we have $a = \pm 1$. If $a = 1$, then $t = \iota$, the identity map, and since $\phi \circ t = h \circ \phi$, we have $\phi = h \circ \phi$, or $\phi(x) = h(\phi(x))$ for all $x \in \mathbf{R}$. However, ϕ is surjective, so $h(y) = y$ for each $y \in \mathbf{R}$, which makes $h = \iota$, also. But $F(f) = t \circ f \circ h^{-1}$ for each $f \in \mathcal{C}(\mathbf{R}, \mathbf{R})$, so F is the identity automorphism. We must insist that $a = -1$, that is, $t(x) = -x$ for each $x \in \mathbf{R}$. With ϕ not being a homeomorphism, if $\phi(x) = \phi(y)$, then $\phi(-x) = \phi \circ t(x) = h \circ \phi(x) = h \circ \phi(y) = \phi \circ t(y) = \phi(-y)$. That is, $\phi(x) = \phi(y)$ implies $\phi(-x) = \phi(-y)$. Let us call such $\phi \in \mathcal{C}(\mathbf{R}, \mathbf{R})$ a *pseudosymmetric* mapping.

So we start with a surjective nearly monotonic pseudosymmetric $\phi \in \mathcal{C}(\mathbf{R}, \mathbf{R})$ which is not a homeomorphism. Define $h : \mathbf{R} \to \mathbf{R}$ as follows. For $y \in \mathbf{R}$, choose x such that $\phi(x) = y$. Let $h(y) = \phi(-x)$. For if $\phi(x) = \phi(x')$, then $\phi(-x) = \phi(-x')$, so $h(y)$ is well defined. If $h(y) = h(y')$ and $\phi(x) = y$ and $\phi(x') = y'$, then $h(y) = \phi(-x)$ and $h(y') = \phi(-x')$. From $\phi(-x) = \phi(-x')$, we obtain $\phi(x) = \phi(x')$, or $y = y'$. This makes h injective. If $u \in \mathbf{R}$, then there is $-x \in \mathbf{R}$ such that $\phi(-x) = u$. Now $\phi(x) = y$ for some $y \in \mathbf{R}$, and $h(y) = \phi(-x) = u$. This makes h surjective and bijective.

To obtain h continuous, we take a closed set A. Now $(h \circ \phi)^{-1}(A) = \phi^{-1}(h^{-1}(A))$. But $h \circ \phi(x) = h(\phi(x)) = \phi(-x)$, so $h \circ \phi$ is continuous. This makes $\phi^{-1}(h^{-1}(A))$ a closed set. From (24.12), we obtain that $h^{-1}(A)$ is a closed set. Now we have $h \in \mathcal{C}(\mathbf{R}, \mathbf{R})$. Since h is a bijection, we have that h is a homeomorphism.

For $t(x) = -x$, we have $t \in \mathcal{HA}(\mathbf{R}^+)$, and so $h \circ \phi = \phi \circ t$. Define $F(f) = t \circ f \circ h^{-1}$, and with the aid of (24.1), conclude that F is an automorphism of $(\mathcal{C}(\mathbf{R}, \mathbf{R}), +, *_\phi)$. Do we have a nontrivial automorphism F? If F is trivial, then $T_a = F(T_a) = t \circ T_a \circ h^{-1}$ for each constant function T_a. But $t \circ T_a \circ h^{-1}(x) = t(a)$, and $T_a(x) = a$. This implies $t(a) = a$ for each $a \in \mathbf{R}$, which is clearly not the case. So we have a nontrivial F, and we can now state

(24.13) Theorem. *Let $\phi \in \mathcal{C}(\mathbf{R}, \mathbf{R})$ be surjective and nearly monotonic. If ϕ is not a homeomorphism, and if $\mathrm{Aut}(\mathcal{C}(\mathbf{R}, \mathbf{R}), +, *_\phi) \cong Z_2$, then ϕ is pseudosymmetric. Conversely, if ϕ is pseudosymmetric, then $\mathrm{Aut}(\mathcal{C}(\mathbf{R}, \mathbf{R}), +, *_\phi) \cong Z_2$.*

(24.14) Corollary. *Let $\phi \in \mathcal{C}(\mathbf{R}, \mathbf{R})$ be surjective and nearly monotonic. Suppose ϕ is not a homeomorphism and is not pseudosymmetric. Then $(\mathcal{C}(\mathbf{R}, \mathbf{R}), +, *_\phi)$ has only the trivial automorphism.*

Proof. If $(\mathcal{C}(\mathbf{R}, \mathbf{R}), +, *_\phi)$ has exactly two automorphisms, then ϕ is pseudosymmetric. If $(\mathcal{C}(\mathbf{R}, \mathbf{R}), +, *_\phi)$ has more than two automorphisms, then, as in the proof of (24.10), ϕ is injective, and so ϕ is a homeomorphism.

To obtain a $(\mathcal{C}(\mathbf{R}, \mathbf{R}), +, *_\phi)$ with exactly two automorphisms, we need a $\phi \in \mathcal{C}(\mathbf{R}, \mathbf{R})$ which (1) is surjective, (2) is pseudosymmetric, (3) is not a homeomorphism, and (4) is nearly monotonic. The following theorem helps in constructing such functions.

(24.15) Theorem. *Let* $\lambda \in C(\mathbf{R}, \mathbf{R})$ *be a homeomorphism. Let* $f \in C(\mathbf{R}, \mathbf{R})$ *be surjective and odd. (Recall that* f *is odd if and only if* $f(-x) = -f(x)$ *for all* $x \in \mathbf{R}$.) *Define* $\phi = \lambda \circ f$. *Then* ϕ *is* (1) *surjective and* (2) *pseudosymmetric. If* f *is not a homeomorphism, then* (3) ϕ *is not a homeomorphism. If* f *is nearly monotonic, then* (4) ϕ *is nearly monotonic.*

Proof. Since λ and f are surjective, so is $\phi = \lambda \circ f$. If $\phi(x) = \phi(y)$, then $\lambda(f(x)) = \lambda(f(y))$. Since λ is injective, then $f(x) = f(y)$. Since f is odd, we then obtain $-f(x) = -f(y)$ and so $f(-x) = f(-y)$. We now have $\lambda(f(-x)) = \lambda(f(-y))$, or $\phi(-x) = \phi(-y)$. This is exactly what we need in order to be assured that ϕ is pseudosymmetric. If f is not a homeomorphism, then there are distinct $x, y \in \mathbf{R}$ such that $f(x) = f(y)$. But then $\phi(x) = \lambda(f(x)) = \lambda(f(y)) = \phi(y)$, which assures us that ϕ is also not a homeomorphism.

Now suppose that f is nearly monotonic. Then there is an $r > 0$ such that f is monotonic on $\mathbf{R} \backslash (-r, r)$. Since λ is a homeomorphism, it is either strictly increasing or strictly decreasing on all of $\mathbf{R}$. There are actually four cases to consider. We shall examine one of them, and rely upon the serious reader to complete the other three cases. Let us suppose that f is decreasing on $\mathbf{R} \backslash (-r, r)$ and that λ is strictly increasing on $\mathbf{R}$. If $x, y \in \mathbf{R} \backslash (-r, r)$ and $x < y$, then $f(x) > f(y)$, so $\lambda(f(x)) > \lambda(f(y))$, that is, $\phi(x) > \phi(y)$. Hence, ϕ is decreasing on $\mathbf{R} \backslash (-r, r)$. The remaining three cases are analogous. So we conclude that ϕ is nearly monotonic.

We now give $\phi_1, \phi_2 \in C(\mathbf{R}, \mathbf{R})$ which (1) are surjective, (2) are pseudosymmetric, (3) are not homeomorphisms, and (4) are nearly monotonic. The first is $\phi_1(x) = x(x^2 - 1)$, and the second is

$$\phi_2(x) = \begin{cases} x + 2, & \text{if } x \leq -1; \\ -x & \text{if } -1 < x \leq 1; \\ x - 2, & \text{if } 1 < x. \end{cases}$$

It is easy to see that each ϕ_i has the stated properties.

The theory we have developed so far applies to nearly monotonic functions $\phi \in C(\mathbf{R}, \mathbf{R})$, and in particular, to all polynomials $p(x) \in \mathbf{R}[x]$ of odd degree. Magill has investigated what happens when a polynomial $p(x) \in \mathbf{R}[x]$ has even degree in [Mag, 1986]. The same types of groups $\text{Aut}(C(\mathbf{R}, \mathbf{R}), +, *_\phi)$ are obtained. The situation becomes very different, however, for $(C(\mathbf{C}, \mathbf{C}), +, *_p)$ with $p(x) \in \mathbf{C}[x]$ where $\mathbf{C}$ is the field of complex numbers. See [MMT, 1983, 1983a] and [Mag *et al.*].

(24.16) Problems. Can $\text{Aut}(C(\mathbf{R}, \mathbf{R}), +, *_\phi)$ be anything but $\{1\}$, Z_2, or $\mathbf{R}^*$ if $\phi \in C(\mathbf{R}, \mathbf{R})$? If $\phi \in C(\mathbf{R}, \mathbf{R})$ and ϕ is surjective? If $\phi \in C(\mathbf{R}, \mathbf{R})$, ϕ is surjective, and ϕ is pseudosymmetric?

(24.17) Exploratory problem. We began this section by considering semigroups which contain all the constant functions and which separate points. Certainly $\mathcal{C}(\mathbf{R}, \mathbf{R})$ and $\mathcal{C}(\mathbf{C}, \mathbf{C})$ are such semigroups. Are there any other such semigroups, for example, $\mathbf{R}[x]$ and $\mathbf{C}[x]$, which yield interesting results for the automorphisms?

(24.18) Exploratory problem. The automorphism groups, when $\phi \in \mathbf{C}[x]$, are explored in [MMT] and [Mag *et al.*]. What happens when one only requires $\phi \in \mathcal{C}(\mathbf{C}, \mathbf{C})$?

(24.19) Exploratory problem. Just before (24.1), we noticed that the ideas of sandwich nearrings and laminated nearrings easily extend to $(\mathrm{Hom}_R(A, B), \oplus, *_\phi)$. Doing so is likely to lead to some very interesting results.

What we have done so far in this section is to show what the principal player, K. D. Magill, Jr, has chosen to look at. We have seen only a small part of the interesting results he has obtained. But, finding automorphism groups is not the only game to play. Peter Fuchs has recently shown that it is fun to play another traditional game, that of finding ideals. See [Fu].

We close this section with a couple of results, in the spirit of (24.6), which were motivated by the work of Magill, and another exploratory problem.

(24.20) Proposition. *Let* $(N, +, \cdot)$ *be a nearring with identity. Let* D *be the set of distributive elements of* N. *Then* $\mathcal{U}(N) \cap D$ *is a subgroup of the group of units* $\mathcal{U}(N)$ *of* N.

Proof. For $u \in \mathcal{U}(N) \cap D$, define $f_u : N \to N$ by $f_u(x) = xu$. Then f_u is an automorphism of $(N, +)$. Consider $(a + b)u^{-1}$. Now $a = xu$ and $b = yu$ for unique $x, y \in N$. So $(a + b)u^{-1} = (xu + yu)u^{-1} = x + y = au^{-1} + bu^{-1}$, so $u^{-1} \in D$. With $u^{-1} \in \mathcal{U}(N)$, we have $u^{-1} \in \mathcal{U}(N) \cap D$. Certainly, $1 \in \mathcal{U}(N) \cap D$, and so is uv if $u, v \in \mathcal{U}(N) \cap D$.

(24.21) Theorem. *Let* $(N, +, \cdot)$ *be a nearring with identity and with* $\mathcal{U}(N)$ *and* D *as in* (24.20). *For* $u \in \mathcal{U}(N) \cap D$, *define* $F_u : N \to N$ *by* $F_u(x) = uxu^{-1}$. *Then* F_u *is an automorphism of the nearring* $(N, +, \cdot)$.

Proof. It is direct and easy.

(24.22) Exploratory problem. The groups $(\mathbf{R}, +)$ and $(\mathbf{C}, +)$ are certainly interesting topological groups, but the fields $\mathbf{R}$ and $\mathbf{C}$ have other nice topological groups. First, there are the multiplicative groups $\mathbf{R}^*$ and $\mathbf{C}^*$, and let us not forget the group $\mathbf{S}^1 = \{z \in \mathbf{C}^* \mid |z| = 1\}$, the unit circle. Some still find it surprising and/or amusing that, as groups, $\mathbf{C}^*$ and $\mathbf{S}^1$ are actually isomorphic. See [Du]. But $\mathbf{S}^1$ is compact and $\mathbf{C}^*$ is not, so they are not isomorphic as topological groups. This fact makes the study of sandwich nearrings $(\mathcal{C}(\mathbf{S}^1, \mathbf{C}^*), \oplus, *_\phi)$ and $(\mathcal{C}(\mathbf{C}^*, \mathbf{S}^1), \oplus, *_\phi)$ tempting. Actually, what we are suggesting here is to take some topological groups, in addition to $(\mathbf{R}, +)$ and $(\mathbf{C}, +)$, and see what can happen with the various sandwich and laminated nearrings. The beautiful work of Magill *et al.* with $(\mathbf{R}, +)$ and $(\mathbf{C}, +)$ strongly suggests that such activity would be rewarding indeed!

25. Syntactic nearrings

Techniques of obtaining nearrings from automata were first explained in the interesting work by M. Holcombe [Ho, 1983]. Then Ambassador Pilz focused attention on semiautomata [HP, P 1987]. In keeping with his calling, Pilz took the message to Yuen Fong, who in turn took the news to Wen-Fong Ke and the author. We present here some especially satisfying results about semiautomata, satisfying in the sense that such elementary ideas naturally lead to some rather complicated nearrings, nearrings which have, nevertheless, a strong harmony in their internal structure. The influence of Yuen Fong is significant.

The basic idea is really elementary. From a group $(G, +)$, any subset S of the transformation nearring $(M(G), +, \circ)$ will generate a subnearring $(N(S), +, \circ)$ of $(M(G), +, \circ)$. There is no end to what one might discover by taking arbitrary S, but the trick is to find an S which leads to useful or interesting $(N(S), +, \circ)$. It is hoped that the reader will also find the $(N(S), +, \circ)$ which we present here interesting or useful.

Automata theory has given rise to a lot of interesting mathematics, as well as applications. One source for interesting mathematics is the *semiautomaton*. For certain semiautomata, one obtains nearrings in a natural way, and these nearrings can be related and analogous to distributively generated nearrings. At least for a dihedral group $D_n = Z_n \times_\theta Z_2$, n even, one obtains nearrings with very interesting structure.

(25.1) Definitions. A *semiautomaton* is an ordered triple $\mathbf{S} = (G, I, \delta)$, where G is the set of *states*, I is the set of *inputs*, and $\delta : G \times I \to G$ is the *transition function*. If $(G, +)$ is a group, then $\mathbf{S}$ is called a *group-semiautomaton*. For a group-semiautomaton $\mathbf{S} = (G, I, \delta)$, the transition function δ defines a family of functions $M_\delta = \{f_x \mid x \in I\}$, where $af_x = \delta(a, x)$ defines $f_x : G \to G$. The group $(G, +)$ has a left nearring $(M(G), +, \circ)$, where $M(G)$ denotes the set of all mappings of G into itself. Certainly, $M_\delta \subseteq M(G)$, so M_δ, together with the identity map 1_G of G, generates a subnearring of $M(G)$, denoted by $N(\mathbf{S})$, and is called the *syntactic nearring of* $\mathbf{S}$.

It has been productive and interesting to study a group-semiautomaton $\mathbf{S} = (G, G, \delta)$, where the inputs $I = G$, the set of states, and to take the transition function δ to be some function intimately connected with the group G. For example, if G is an abelian group of exponent n, then G is a Z_n-module, where $(Z_n, +, \cdot)$ denotes the ring of integers modulo n. Let $\delta(x, y) = x + y$. In this case $N(\mathbf{S}) \cong Z_n \oplus_A G$, where $\oplus_A$ denotes that $Z_n \oplus_A G$ is the abstract affine nearring of the ring Z_n and the Z_n-module G. This is (25.8). Also, see [Fo & C]. Other interesting applications have come from $\delta(x, y) = x\phi + y\mu$, where $\phi, \mu \in \text{End}G$, the endomorphisms of an abelian group G. See [Fo].

We continue to look at traditional algebraic structures here, but one could also consider traditional topics from automata theory. For example, in [HP], Hofer and Pilz have a very nice theorem concerning reachability in $\mathbf{S}$. In that theorem, reachability in $\mathbf{S}$ is equivalent to an $N(\mathbf{S})$-module Q being strictly monogenic with $0N(\mathbf{S}) = Q$.

We first take $I = G = D_n$, n even, where $D_n = Z_n \times_\theta Z_2$, the dihedral group of order $2n$, and take $\delta(x, y) = x + y$ for all $x, y \in D_n$. In some sense, the dihedral groups are the nonabelian groups closest to being abelian, but we shall see that the resulting $N(\mathbf{S})$ will be considerably more complicated, and, at the same time, more interesting. Of basic interest for any new algebraic structure is to determine the algebraic invariants. Towards that end, we shall examine the structure of the additive group of $N(\mathbf{S})$ carefully, as well as the group of units of the nearring $(N(\mathbf{S}), +, \circ)$. Each will have structure described in terms of their elements, which will be represented by 5-tuples $[2t, 2u : m \mid s, \epsilon] \in 2Z_n \times 2Z_n \times Z_n \times Z_n \times Z_2$, which in turn enables us to express the group structures as semidirect products. Finally, results that are related to the ideal structure of syntactic nearrings are presented. These concepts are applied to give a better picture of the properties of these syntactic nearrings of $(D_n, +)$, n even.

Preliminaries

In describing our group structures, we make use of the semidirect product. The reader can refer to the remarks made early in the chapter on groups of units. Let $(G, +)$ denote a finite group of order n, and let $\mathbf{G} = (G, G, \delta)$ with $\delta(x, y) = x + y$. For $x f_y = x + y$, we have $f_y = 1_G + \phi_y = 1 + \phi_y$ where $1_G = 1$ is the identity map on G and ϕ_y is the constant map $x\phi_y = y$ for each $x \in G$. Now $f_0 = 1 + \phi_0 = 1$, so $1 \in N(\mathbf{G})$. Thus $-1 + f_y = \phi_y$, so each $\phi_y \in N(\mathbf{G})$. If T' is the subnearring of $M(G)$ generated by 1 and all the mappings ϕ_y, then $T' \subseteq N(\mathbf{G})$. But $f_y = 1 + \phi_y$ shows that $N(\mathbf{G}) = T'$. Hence,

(25.2) Proposition. *For a finite group G, the nearring $N(\mathbf{G})$, with $\delta(x, y) = x + y$, is generated by 1_G and all the constant functions ϕ_y, $y \in G$.*

The next two propositions are special cases of the more general results which were first presented by Fong and Ke at the General Algebra conference in Krems, 1988. Their proofs are included for their own interest, as well as for completeness.

(25.3) Proposition. *If $(G, +)$ is a group with finite exponent $n > 0$, then*

$$N(\mathbf{G}) = \left\{ \sum_{i=1}^{m} f_{a_i} \mid a_i \in G, \ m \geq 1 \right\}.$$

Proof. Since G has exponent n, each f_a has additive order some $m \leq n$, and, in particular, $-f_a$ is then a sum of $m - 1$ of the f_a. For $f_a, f_b \in N(\mathbf{G})$, we have $x(f_a \circ f_b) = (x + a) + b = x f_{a+b}$. For $f_{a_1}, f_{a_2}, \ldots, f_{a_m} \in N(\mathbf{G})$, we have $x(f_{a_1} + \cdots + f_{a_m}) \circ f_b = (x + a_1) + \cdots + (x + a_m) + b = x(f_{a_1} + \cdots + f_{a_{m-1}} + f_{a_m + b})$. Hence,

$$\left(\sum_{i=1}^{m} f_{a_i} \right) \circ \left(\sum_{j=1}^{k} f_{b_j} \right) = \sum_{j=1}^{k} \left(\sum_{i=1}^{m} f_{a_i} \right) \circ f_{b_j} = \sum_{j=1}^{k} \left[\sum_{i=1}^{m-1} f_{a_i} + f_{a_m + b_j} \right].$$

So the group generated by the mappings f_a under $+$ is closed with respect to $\circ$, and is in this way related to a distributively generated (d.g.) nearring. Since $(M(G), +, \circ)$ is a nearring,

$$N(\mathbf{G}) = \left\{ \sum_{i=1}^{m} f_{a_i} \mid a_i \in G, \ m \geq 1 \right\},$$

which is the subgroup of $(M(G), +)$ generated by all the mappings f_{a_i}, $a_i \in G$.

(25.4) Proposition. *Let $\mathbf{G} = (G, G, \delta)$ where $(G, +)$ is a group of finite exponent $n > 0$ and $\delta(x, y) = x + y$. Then $N(\mathbf{G}) = I(G) + M_c(G)$, where $I(G)$ is the d.g. nearring generated by the inner automorphisms of G, and $M_c(G)$ is the subnearring of $M(G)$ consisting of the constant maps ϕ_y, $y \in G$. Thus $N(\mathbf{G})^+ \cong I(G)^+ \times_\theta M_c(G)^+$.*

Proof. Certainly, each $\phi_y \in N(\mathbf{G})$, and an inner automorphism of G has the structure $\phi_x + 1 - \phi_x = \phi_x + f_0 - \phi_x$, so all inner automorphisms of G are in $N(\mathbf{G})$. This means that $I(G)$ and $M_c(G)$ are subnearrings of $N(\mathbf{G})$, so $I(G) + M_c(G) \subseteq N(\mathbf{G})$, and $I(G) \cap M_c(G) = \{0\}$. We shall take an arbitrary $f \in N(\mathbf{G})$ and show that it is in $I(G) + M_c(G)$. If $f = \sum_{i=1}^{m} f_{a_i}$, then

$$f = \sum_{i=1}^{m}(f_0 + \phi_{a_i}) = f_0 + (\phi_{a_1} + f_0 - \phi_{a_1})$$

$$+ [(\phi_{a_1} + \phi_{a_2}) + f_0 - (\phi_{a_1} + \phi_{a_2})] + \cdots$$
$$+ [(\phi_{a_1} + \cdots + \phi_{a_{m-1}}) + f_0 - (\phi_{a_1} + \cdots + \phi_{a_{m-1}})]$$
$$+ (\phi_{a_1} + \cdots + \phi_{a_{m-1}}) + \phi_{a_m},$$

which is an element of $I(G) + M_c(G)$ since $(\phi_{a_1} + \cdots + \phi_{a_{m-1}}) + \phi_{a_m} = \phi_{a_1+a_2+\cdots+a_m}$.

It is easy to see the $I(G)^+ \triangleleft N(\mathbf{G})^+$, so, as groups, $N(\mathbf{G})^+ \cong I(G)^+ \times_\theta M_c(G)^+$.

From (25.4), if we know $I(G)$ and G, then we know $N(\mathbf{G})$, and if we know the sizes of $I(G)$ and G, we know the size of $N(\mathbf{G})$. We shall find it helpful to do this for $G = D_n$, with n even. For a contrasting proof, see [M & L, 1973].

It will be useful sometimes to identify elements of $D_n = Z_n \times_\theta Z_2$ by $ia + \epsilon b \equiv (i, \epsilon)$. So $a = (1,0)$ and $b = (0,1)$ would represent the traditional generators of D_n.

An inner automorphism is of the form $\phi_x + 1 - \phi_x$ for $x \in D_n$. Since $I(D_n)$ is finite, each $g \in I(D_n)$ is of the form $g = \sum_{j=1}^{m}(\phi_{k_j a+\epsilon_j b} + 1 - \phi_{k_j a+\epsilon_j b})$ where $k_j \in Z_n$ and $\epsilon_j \in Z_2$. From $(ia)g = \sum_{j=1}^{m}[k_j a+\epsilon_j b+ia-\epsilon_j b-k_j a] = \left(\sum_{j=1}^{m}(-1)^{\epsilon_j}\right)(ia)$, and $(ia + b)g = \sum_{j=1}^{m}[k_j a + \epsilon_j b + ia + b - \epsilon_j b - k_j a] = \sum_{j=1}^{m}[(2k_j + (-1)^{\epsilon_j}i)a + b] = \left[\sum_{j=1}^{m}(-1)^{j+1}(2k_j + (-1)^{\epsilon_j}i)\right]a + mb = \left[\sum_{j=1}^{m}(-1)^{j+1}(2k_j)\right]a + \left[\sum_{j=1}^{m}(-1)^{j+1}(-1)^{\epsilon_j}\right](ia) + mb$, we get

$$\begin{cases} (ia)g = A(ia); \\ (ia + b)g = 2Ba + C(ia) + Mb, \end{cases} \qquad (25:1)$$

where $A \equiv \sum_{j=1}^{m}(-1)^{\epsilon_j} \bmod n$, $C \equiv \sum_{j=1}^{m}(-1)^{j+1}(-1)^{\epsilon_i} \bmod n$, $2B \equiv 2\sum_{j=1}^{m}(-1)^{j+1}k_j \bmod n$, and $M \equiv m \bmod 2$. In fact, g is defined by this $(2B, A, C, M) \in 2Z_n \times Z_n \times Z_n \times Z_2$, and distinct elements g in $I(D_n)$ have distinct values $(2B, A, C, M)$ in $2Z_n \times Z_n \times Z_n \times Z_2$. But perhaps some values $(2B, A, C, M)$ do not define a $g \in I(D_n)$ by (25:1). It makes sense to ask which of the $(2B, A, C, M) \in 2Z_n \times Z_n \times Z_n \times Z_2$ define a $g \in I(D_n)$ by (25:1). Before we answer that question, we must notice a few facts.

First, note that we have $\sum_{j=1}^{m}(-1)^{\epsilon_j} + \sum_{j=1}^{m}(-1)^{j+1}(-1)^{\epsilon_j} = \sum_{j=1}^{m}[1 + (-1)^{j+1}](-1)^{\epsilon_j}$, and each $[1 + (-1)^{j+1}] \in \{0,2\}$. So $A + C \in 2Z_n$ if $(2B, A, C, M)$ represents a $g \in I(D_n)$ by (25:1). Hence, $A \in 2Z_n$ if and only if $C \in 2Z_n$.

Next, notice that $\sum_{j=1}^{m}(-1)^{\epsilon_j} = s - t$, where s is the number of indices j such that $\epsilon_j = 0$, and t is the number of indices j such that $\epsilon_j = 1$. This makes $s + t = m$. If $s - t$ is even, then either s and t are both even, or s and t are both odd, and conversely. In either case, m is even and so $M = 0$. Otherwise, m is odd, one of s or t is odd, and $M = 1$. In summary, $M = 0$ implies m is even and $A, C \in 2Z_n$; and $M = 1$ implies m is odd and $A - 1$, $C - 1 \in 2Z_n$.

Clearly, there are g in $I(D_n)$ whose $M = 0$; also, there are $g \in I(D_n)$ where $M = 1$. First, we will take any M and consider which $(2B, A, C, M)$ define a $g \in I(D_n)$ by (25:1). From $2B \equiv 2\sum_{j=1}^{m}(-1)^{j+1}k_j$, and our right to choose the values of each k_j as we please, we realize that any value of $2B \in 2Z_n$ can be used with our arbitrary M. Take $g \in I(D_n)$ and consider its $(2B, A, C, M)$.

Let $h = (\phi_b + 1 - \phi_b) + (\phi_b + 1 - \phi_b)$, so $(ia)h = -2(ia)$, and $(ia+b)h = 0$. If $f = g + h$, we have $(ia)f = A(ia) - 2(ia) = (A - 2)(ia)$, and $(ia + b)f = 2Ba + C(ia) + Mb$, and so f is defined by $(2B, A - 2, C, M)$. So, with any $2B$, C, and M, all $A - 2j$ can be used in the A-position.

Take $h = (\phi_0 + 1 - \phi_0) + (\phi_b + 1 - \phi_b)$. So $(ia)h = 0$ and $(ia+b)h = 2(ia)$. For $f = g + h$, we obtain $(ia)f = (ia)g = A(ia)$ and $(ia + b)f = 2Ba + C(ia) + Mb + 2(ia)$. Hence, f is defined by $(2B, A, C \pm 2, M)$, with $C + 2$ for $M = 0$ and $C - 2$ for $M = 1$.

Summarizing, for $M = 0$, the $2B \in 2Z_n$ can be arbitrary, as can $A, C \in 2Z_n$. For $M = 1$, $2B \in 2Z_n$ can be arbitrary, as can $A - 1$, $C - 1 \in 2Z_n$. Hence, $|I(D_n)| = n^3/8 + n^3/8 = n^3/4$, and we obtain

(25.5) Theorem. *If n is even, the elements of $I(D_n)$ are defined by* (25 : 1) *where* $(2B, A, C, M) \in 2Z_n \times Z_n \times Z_n \times Z_2$, *where $M = 0$ implies* $A, C \in 2Z_n$, *and $M = 1$ implies $A - 1$, $C - 1 \in 2Z_n$. Distinct $(2B, A, C, M)$ define distinct elements of $I(D_n)$. Hence, $|I(D_n)| = n^3/4$.*

(25.6) Corollary. *If $G = D_n$, with n even, then $N(\mathbf{D}_n)$ has order $n^4/2$.*

Proof. $M_c(D_n) \cong D_n$, and has order $2n$. From (25.5) we know that $I(D_n)$ has order $n^3/4$. Since $|N(\mathbf{D}_n)| = |I(D_n)| \cdot |D_n|$, we have our desired result.

In §15, the structure of $I(D_B)$ was studied in detail when D_B is an odd generalized dihedral group. That is, $D_B = B \times_\theta Z_2$, where B is an abelian group with the property that $b \mapsto 2b$ is an automorphism of B, and where $(b, x)(b', x') = (b + (-1)^x b', x + x')$. In particular, (15.11) tells us that $I(D_B)^+ \cong Z_n \oplus [(Z_n \oplus B) \times_{\theta_1} Z_2]$. If we restrict B to having finite exponent $n > 0$, then a direct application of (25.4) gives us

(25.7) Corollary. *Let $(B, +)$ be an abelian group with finite exponent $n > 0$, and so that D_B is an odd generalized dihedral group. For $\mathbf{D}_B =$*

(D_B, D_B, δ) *as in* (25.4), *we have*

$$N(\mathbf{D}_B)^+ \cong \left\{ Z_n \oplus \left[(Z_n \oplus B) \times_{\theta_1} Z_2 \right] \right\} \times_\theta (B \times_{\theta_2} Z_2)$$

where each $\theta_i(\epsilon) = (-1)^\epsilon$, *and* θ *is as for* (25.4).

(25.8) Corollary. *Let* $(B, +)$ *be an abelian group with exponent* n, *let* $\delta(x, y) = x + y$, *and* $\mathbf{B} = (B, B, \delta)$. *Then* $N(\mathbf{B}) \cong Z_n \oplus_A B$, *the abstract affine nearring constructed from the ring* Z_n *and the* Z_n-*module* B.

Proof. From (25.4), we have $N(\mathbf{B})^+ \cong I(B)^+ \times_\theta M_c(B)^+$, and we have $I(B) \cong Z_n$ since B is abelian and has exponent n. Certainly, $M_c(B)^+ \cong B^+$, and B is a Z_n-module. What happens to the multiplication in $N(\mathbf{B})$? Identify $k \equiv k \cdot 1$ and $b \equiv \phi_b$, where $k \in Z_n$, $b \in B$, and $1 = 1_{Z_n}$, the identity map of Z_n. If we identify $k \cdot 1 + \phi_a \equiv (k, a)$, and if $k \cdot 1 + \phi_a \in N(\mathbf{B})$, then $k \cdot 1 + \phi_a \in \mathrm{End} Z_n + M_c(B)$. So, if $k' \cdot 1 + \phi_{a'} \in N(\mathbf{B})$ also, we have $(k \cdot 1 + \phi_a) \circ (k' \cdot 1 + \phi_{a'}) = (k \cdot 1 + \phi_a) \circ (k' \cdot 1) + \phi_{a'} = kk' \cdot 1 + \phi_{ak' + a'}$, hence $(k \cdot 1 + \phi_a) \circ (k' \cdot 1 + \phi_{a'}) \equiv (k, a)(k', a') = (kk', ak' + a') \equiv kk' \cdot 1 + \phi_{ak' + a'}$. This makes $N(\mathbf{B}) \cong Z_n \oplus_A B$, an abstract affine nearring.

The elements of $N(\mathbf{D}_n)$, when n is even

We wish to study in detail $N(\mathbf{D}_n)$, where n is even, and where $D_n = Z_n \times_\theta Z_2$, the dihedral group of order $2n$. It will be useful for us to describe each element of $N(\mathbf{D}_n)$ as a 5-tuple. Define

$$T = \{ [2t, 2u : m \mid s, \epsilon] \mid 2t, 2u \in 2Z_n, \ m, s \in Z_n, \ \text{and} \ \epsilon \in Z_2 \},$$

where

$$(ia)[2t, 2u : m \mid s, \epsilon] = mia + sa + \epsilon b \qquad (25 : 2a)$$

and $(ia + b)[2t, 2u : m \mid s, \epsilon] =$

$$\begin{cases} 2ta + 2uia + sa + \epsilon b, & \text{if } m \text{ is even;} \\ 2ta + 2uia - sa + ia + b + \epsilon b, & \text{if } m \text{ is odd,} \end{cases} \qquad (25 : 2b)$$

so each $[2t, 2u : m \mid s, \epsilon] : D_n \to D_n$ is in $M(D_n)$. We have $|T| = n^4/2$, so by (25.6), $T = N(\mathbf{D}_n)$ if we can show that $T \subseteq N(\mathbf{D}_n)$.

(25.9) Theorem. $T = N(\mathbf{D}_n)$.

Proof. First, $1 = [0, 0 : 1 \mid 0, 0]$ and $\phi_{sa + \epsilon b} = [0, 0 : 0 \mid s, \epsilon]$, so all of $N(\mathbf{D}_n)$'s generators are in T. Let $f = [2t, 2u : m \mid s, \epsilon]$ and $g = [2t, 2u : m \mid 0, 0] + \phi_{sa + \epsilon b}$. Using (25:2a) and (25:2b), one obtains that $f = g$. Since

$\phi_{sa+eb} \in T \cap N(\mathbf{D}_n)$, we need only show that $h = [2t, 2u : m \mid 0,0] \in N(\mathbf{D}_n)$.

Similarly, let $h = [2t, 2u : m \mid 0,0]$ and $k = [2t, 2u : 0 \mid 0,0] + m \cdot 1$. Using (25:2a) and (25:2b), we obtain $h = k$. Since $m \cdot 1 \in T \cap N(\mathbf{D}_n)$, it is sufficient to show that $l = [2t, 2u : 0 \mid 0,0] \in N(\mathbf{D}_n)$ in order to conclude that $h \in T \cap N(\mathbf{D}_n)$.

Before showing that $l = [2t, 2u : 0 \mid 0,0] \in N(\mathbf{D}_n)$, we examine carefully what l does.

$$xl = \begin{cases} 0, & \text{if } x = ia; \\ 2ta + 2uia, & \text{if } x = ia + b. \end{cases} \qquad (25:3)$$

We shall construct an element of $N(\mathbf{D}_n)$ that does exactly this.

For an odd k, $\phi_{sa} + k \cdot 1 - \phi_{sa}$ and $\phi_{sa+b} + k \cdot 1 - \phi_{sa+b}$ are elements of $N(\mathbf{D}_n)$. Now

$$x[\phi_{sa} + k \cdot 1 - \phi_{sa}] = \begin{cases} kia, & \text{if } x = ia; \\ 2sa + ia + b, & \text{if } x = ia + b, \end{cases} \qquad (25:4)$$

and

$$x[\phi_{sa+b} + k \cdot 1 - \phi_{sa+b}] = \begin{cases} -kia, & \text{if } x = ia; \\ 2sa - ia + b, & \text{if } x = ia + b. \end{cases} \qquad (25:5)$$

Let $f = [\phi_{sa} + k \cdot 1 - \phi_{sa}] + [\phi_{s'a+b} + k' \cdot 1 - \phi_{s'a+b}]$ for odd k and k'. Then $f \in N(\mathbf{D}_n)$ and

$$xf = \begin{cases} (k - k')ia, & \text{if } x = ia; \\ 2(s - s')a + 2ia; & \text{if } x = ia + b. \end{cases} \qquad (25:6)$$

The $k - k', 2(s - s') \in 2Z_n = \{0, 2, \ldots, 2n - 2\}$ and the k, k', s, s' are arbitrary enough so that we can conclude that all $\langle 2m, 2t, 1 \rangle \in N(\mathbf{D}_n)$ where

$$x\langle 2m, 2t, 1 \rangle = \begin{cases} 2mia, & \text{if } x = ia; \\ 2ta + 2ia, & \text{if } x = ia + b. \end{cases} \qquad (25:7)$$

Sums $\sum_{j=1}^{u} \langle 2m_j, 2t_j, 1 \rangle \in N(\mathbf{D}_n)$ also. If $\langle \cdot \rangle = \sum_{j=1}^{u} \langle 2m_j, 2t_j, 1 \rangle$, then one computes, using (25:7),

$$x\langle \cdot \rangle = \begin{cases} 2(\sum_{j=1}^{u} m_j)ia, & \text{if } x = ia; \\ 2(\sum_{j=1}^{u} t_j)a + 2uia, & \text{if } x = ia + b. \end{cases} \qquad (25:8)$$

The values $2(\sum_{j=1}^{u} m_j)$, $2(\sum_{j=1}^{u} t_j)$, and $2u$ can be any value in $2Z_n$, so each $\langle 2m, 2t, 2u \rangle \in N(\mathbf{D}_n)$, where

$$x\langle 2m, 2t, 2u \rangle = \begin{cases} 2mia, & \text{if } x = ia; \\ 2ta + 2uia, & \text{if } x = ia + b. \end{cases} \qquad (25:9)$$

Since $(ia)\langle 0, 2t, 2u \rangle = 0$ and $(ia + b)\langle 0, 2t, 2u \rangle = 2ta + 2uia$, we have $\langle 0, 2t, 2u \rangle = l \in N(\mathbf{D}_n)$, from (25:3).

Now $[2t, 2u : 0 \mid 0, 0]$, $m \cdot 1$, $\phi_{sa+b} \in N(\mathbf{D}_n)$. Thus $T = N(\mathbf{D}_n)$, completing the proof of (25.9).

(25.10) Corollary. $[2t, 2u : m \mid s, \epsilon] = [2t, 2u : 0 \mid 0, 0] + m \cdot 1 + \phi_{sa+\epsilon b}.$

(25.11) Corollary. *The zero-symmetric part of* $N(\mathbf{D}_n)$ *is*

$$N(\mathbf{D}_n)_0 = \{[2t, 2u : m \mid 0, 0] \mid [2t, 2u : m \mid 0, 0] \in N(\mathbf{D}_n)\},$$

and the constant part of $N(\mathbf{D}_n)$ *is*

$$N(\mathbf{D}_n)_c = \{[0, 0 : 0 \mid s, \epsilon] \mid [0, 0 : 0 \mid s, \epsilon] \in N(\mathbf{D}_n)\},$$

and $N(\mathbf{D}_n)_c^+ \cong D_n.$

(25.12) Corollary. $N(\mathbf{D}_n)_0 = I(D_n)$ *and* $N(\mathbf{D}_n)_c = M_c(D_n).$

It will be very useful for the reader to have a rule to add and multiply two elements of $N(\mathbf{D}_n)$ when expressed as elements of T. The addition rule is summarized below in four parts, or one could write one rather complex rule:

$$[2t, 2u : m \mid s, \epsilon] + [2t', 2u' : m' \mid s', \epsilon']$$
$$= \Big[2t + (-1)^{m+\epsilon}(2t') + (-1)^m[1 - (-1)^{m'}]s,$$
$$2u + (-1)^{m+\epsilon}(2u') + \left[\frac{1 - (-1)^{m'}}{2}\right][1 - (-1)^\epsilon](-1)^{m+m'} :$$
$$m + (-1)^\epsilon m' \mid s + (-1)^\epsilon s', \epsilon + \epsilon'\Big].$$

Alternatively, $[2t, 2u : m \mid s, \epsilon] + [2t', 2u' : m' \mid s', \epsilon'] =$

$$[2t + (-1)^\epsilon(2t'), 2u + (-1)^\epsilon(2u') : m + (-1)^\epsilon m' \mid s + (-1)^\epsilon s', \epsilon + \epsilon'] \quad (25 : 10a)$$

if m and m' are both even;

$$[2t + [(-1)^\epsilon(2t') + 2s], 2u + [(-1)^\epsilon(2u') + (-1)^\epsilon - 1] :$$
$$m + (-1)^\epsilon m' \mid s + (-1)^\epsilon s', \epsilon + \epsilon'] \quad (25 : 10b)$$

if m is even and m' is odd;

$$[2t - (-1)^\epsilon(2t'), 2u - (-1)^\epsilon(2u') : m + (-1)^\epsilon m' \mid s + (-1)^\epsilon s', \epsilon + \epsilon'] \quad (25 : 10c)$$

if m is odd and m' is even;

$$[2t - [(-1)^\epsilon (2t') + 2s], 2u - [(-1)^\epsilon (2u') + (-1)^\epsilon - 1] :$$
$$m + (-1)^\epsilon m' \mid s + (-1)^\epsilon s', \epsilon + \epsilon'] \qquad (25 : 10d)$$

if m and m' are both odd. Collectively, equations (25:10a) through (25:10d) will be referred to as (25:10) and as

The Addition Rule of $T = N(\mathbf{D}_n)$.

Now we need to tell the reader how to multiply: $[2t, 2u : m \mid s, \epsilon] \circ [2t', 2u' : m' \mid s', \epsilon'] =$

$$[2tm', 2um' : mm' \mid sm' + s', \epsilon'] \qquad (25 : 11a)$$

if m and m' are both even and $\epsilon = 0$;

$$[(2t)(2u'), (2u)(2u') : m(2u') \mid s(2u') + 2t' + s', \epsilon'] \qquad (25 : 11b)$$

if m and m' are both even and $\epsilon = 1$;

$$[2tm', 2um' : mm' \mid sm' + s', \epsilon'] \qquad (25 : 11c)$$

if m is even, m' is odd, and $\epsilon = 0$;

$$[(2t)(2u'+1), (2u)(2u'+1) : m(2u'+1) \mid s(2u'+1)+2t'-s', \epsilon+\epsilon'] \quad (25 : 11d)$$

if m is even, m' is odd, and $\epsilon = 1$;

$$[2t' + (2t)(2u') - s(2u') - sm', (2u+1)(2u') : mm' \mid sm' + s', \epsilon'] \quad (25 : 11e)$$

if m is odd, m' is even, and $\epsilon = 0$;

$$[2tm' - sm' - 2t' - (2u')s, (2u+1)m' : m(2u') \mid s(2u')+2t'+s', \epsilon'] \quad (25 : 11f)$$

if m is odd, m' is even, and $\epsilon = 1$;

$$[2t' + (2u'+1)(2t-s) + sm', (2u)(2u') + (2u) + (2u') :$$
$$mm' \mid sm' + s', \epsilon'] \qquad (25 : 11g)$$

if m and m' are both odd, and $\epsilon = 0$;

$$[(2t)m' + 2t' + (2u'+1)s - sm', (2u+1)m' - 1 :$$
$$(2u'+1)m \mid s(2u'+1) + 2t' - s', \epsilon + \epsilon'] \qquad (25 : 11h)$$

if m and m' are both odd, and $\epsilon = 1$. The equations (25:11a) through (25:11h) will be referred to collectively as (25:11) and as

The Multiplication Rule of $T = N(\mathbf{D}_n)$.

It will also be very helpful to the reader to have handy the formulae for $x = -[2t, 2u : m \mid s, \epsilon]$. One readily obtains from the addition rule that

$$x = [(-1)^{\epsilon}(-2t), (-1)^{\epsilon}(-2u) : (-1)^{\epsilon}(-m) \mid (-1)^{\epsilon}(-s), \epsilon] \qquad (25:12a)$$

if m is even, and

$$x = [(-1)^{\epsilon}(2t - 2s), (-1)^{\epsilon}(2u + 1 - (-1)^{\epsilon}) : \\ (-1)^{\epsilon}(-m) \mid (-1)^{\epsilon}(-s), \epsilon] \qquad (25:12b)$$

if m is odd.

The thorough reader may want to verify the addition and multiplication rules.

The structure of $N(\mathbf{D}_n)_0$

In discussing T above, we considered the maps $\langle 0, 2t, 2u \rangle = [2t, 2u : 0 \mid 0, 0]$. These maps also play a role in describing the structure of $N(\mathbf{D}_n)_0$.

From (25.12), we have $N(\mathbf{D}_n)_0 = I(D_n)$, so in discussing the structure of $N(\mathbf{D}_n)_0$, we are also discussing the structure of the d.g. nearring generated by the inner automorphisms of the group $(D_n, +)$. The reader may wish to contrast our development here with that in [M & L, 1973].

Define

$$A = \{\langle 0, 2t, 2u \rangle \mid 2t, 2u \in 2Z_n\}$$

and

$$B = \{k \cdot 1 \mid k \in Z_n\}.$$

Then $A \cup B \subseteq N(\mathbf{D}_n)_0$, $B < N(\mathbf{D}_n)_0^+$, and $A \cap B = \{0\}$.

(25.13) Lemma. $A \lhd N(\mathbf{D}_n)_0$ *and* $A \cong 2Z_n \oplus 2Z_n$.

Proof. It is direct and easy to show $A \cong 2Z_n \oplus 2Z_n$. For normality, let $f = [2t, 2u : m \mid 0, 0] + [2t', 2u' : 0 \mid 0, 0] - [2t, 2u : m \mid 0, 0]$ and note that $(ia)f = 0$.

(25.14) Theorem. *The group* $N(\mathbf{D}_n)_0^+ \cong (2Z_n \oplus 2Z_n) \times_\theta Z_n$ *where* $\theta : Z_n \to \mathrm{Aut}(2Z_n \oplus 2Z_n)$ *is defined by*

$$\theta(m)[2t, 2u] = [(-1)^m(2t), (-1)^m(2u)].$$

Proof. For $\langle 0, 2t, 2u \rangle \in A$, $\langle 0, 2t, 2u \rangle = [2t, 2u : 0 \mid 0, 0]$, so $\langle 0, 2t, 2u \rangle + m \cdot 1 = [2t, 2u : m \mid 0, 0]$. Hence, the isomorphism $\Psi : N(\mathbf{D}_n)_0^+ \to (2Z_n \oplus 2Z_n) \times_\theta Z_n$ is $\Psi[2t, 2u : m \mid 0, 0] = ([2t, 2u], m)$. We see directly that the equation $\theta(m)[2t, 2u] = [(-1)^m(2t), (-1)^m(2u)]$ is a consequence of the equation $[0, 0 : m \mid 0, 0] + [2t, 2u : 0 \mid 0, 0] - [0, 0 : m \mid 0, 0] = [(-1)^m(2t), (-1)^m(2u) : 0 \mid 0, 0]$, as described in §19.

We now turn our attention to the structure of the group of units of $N(\mathbf{D}_n)_0$, denoted by $\mathcal{U}(N(\mathbf{D}_n)_0)$. To do this satisfactorily, we make use of Jacobson's circle operation $\circ$ for a ring $(R, +, \cdot)$, where

$$a \circ b = a + b + ab. \tag{25 : 13}$$

All we really need is that $0 \in R$ is the identity for $\circ$, and that if R has an identity 1, then R's group of units $(\mathcal{U}(R), \cdot)$ is isomorphic to R's group of units with respect to $\circ$, namely, $(\mathcal{U}(R), \circ)$, and the isomorphism is $\Lambda : (\mathcal{U}(R), \circ) \to (\mathcal{U}(R), \cdot)$ defined by $\Lambda(a) = 1 + a$. We use this for $(R, +, \cdot) = (Z_n, +, \cdot)$. Since n is even, the elements of $(\mathcal{U}(Z_n), \cdot)$ are odd, so the elements of $(\mathcal{U}(Z_n), \circ)$ are even. Since Λ is a group isomorphism, $(1 + (2u)) \circ (2u')) = (1 + 2u) \cdot (1 + 2u')$. If $(2u) \circ (2u') = 0$, then we denote $(2u)$'s inverse with respect to $\circ$ by $(2u)^*$. That is, if $(2u) \in \mathcal{U}(Z_n, \circ)$, then $(2u)^*$ is the element so that $(2u) \circ (2u)^* = 0$.

(25.15) Theorem. *The group of units of* $(N(\mathbf{D}_n)_0, +, \cdot)$ *is*

$$\mathcal{U}(N(\mathbf{D}_n)_0) = \{[2t, 2u : m \mid 0, 0] \mid$$
$$2t \in 2Z_n,\ 2u \in \mathcal{U}(Z_n, \circ),\ \text{and } m \in \mathcal{U}(Z_n, \cdot)\}.$$

For such a unit,

$$[2t, 2u : m \mid 0, 0]^{-1} = [-((2u)^* + 1)(2t), (2u)^* : m^{-1} \mid 0, 0]. \tag{25 : 14}$$

Also, $|\mathcal{U}(N(\mathbf{D}_n))_0| = n|\mathcal{U}(Z_n, \cdot)|^2/2$.

Proof. Since $1 \in Z_n$ is odd, we see from the multiplication rule (25:11) that our units will be as described for the case where m and m' are odd, and $\epsilon = 0$, that is, $[2t, 2u : m \mid 0, 0]$ where m is odd; in fact, where $m \in \mathcal{U}(Z_n, \cdot)$. If m and m' are odd, and

$$f = [2t, 2u : m \mid 0, 0] \circ [2t', 2u', m' \mid 0, 0], \tag{25 : 15}$$

then

$$xf = mm'ia, \tag{25 : 16a}$$

if $x = ia$, and

$$xf = [2t' + (2u' + 1)(2t)]a + [(2t) \circ (2u')]ia + ia + b, \tag{25 : 16b}$$

if $x = ia + b$. Therefore,

$$[2t, 2u :m \mid 0, 0] \circ [2t', 2u' : m' \mid 0, 0]$$
$$= [2t' + (2u' + 1)(2t), (2u) \circ (2u') : mm' \mid 0, 0]. \qquad (25:17)$$

From this, if $[2t', 2u' : m' \mid 0, 0]$ is to be the inverse of $[2t, 2u : m \mid 0, 0]$, we see that it is necessary and sufficient for

$$[2t, 2u : m \mid 0, 0]^{-1} = [-((2u)^* + 1)(2t), (2u)^* : m^{-1} \mid 0, 0].$$

There are $n/2$ elements in $2Z_n$ for the $2t$. Since $\mathcal{U}(Z_n, \circ) \cong \mathcal{U}(Z_n, \cdot)$, we have now that

$$|\mathcal{U}(N(\mathbf{D}_n)_0)| = \frac{n|\mathcal{U}(Z_n, \cdot)|^2}{2}.$$

This completes the proof of (25.15).

Define
$$C = \{[2t, 0 : 1 \mid 0, 0] \mid 2t \in 2Z_n\},$$

and
$$D = \{[0, 2u : m \mid 0, 0] \mid 2u \in \mathcal{U}(Z_n, \circ), \ m \in \mathcal{U}(Z_n, \cdot)\}.$$

We will prove

(25.16) Theorem. *With* $\mathcal{U}(N(\mathbf{D}_n)_0) \cong C \times_\theta D$, $C \cong (2Z_n)^+$, *and* $D \cong \mathcal{U}(Z_n, \circ) \times \mathcal{U}(Z_n, \cdot)$, *we have*

$$\mathcal{U}(N(\mathbf{D}_n)_0) \cong (2Z_n)^+ \times_\theta \left[\mathcal{U}(Z_n, \circ) \times \mathcal{U}(Z_n, \cdot)\right],$$

with isomorphism $\Psi[2t, 2u : m \mid 0, 0] = \big((2t)(2u + 1)^{-1}, [2u, m]\big)$ *and where* $\theta[2u, m](2t) = (2u + 1)^{-1}(2t)$.

Proof. Take $[2t, 0 : 1 \mid 0, 0]$, $[2t', 0 : 1 \mid 0, 0] \in C$. Then by (25:17), their product is $[2t + 2t', 0 : 1 \mid 0, 0]$, so $C \cong (2Z_n)^+$. From

$$[2t, 2u : m \mid 0, 0] \circ [2t', 0 : 1 \mid 0, 0] \circ [2t, 2u : m \mid 0, 0]^{-1}$$
$$= [((2u)^* + 1)(2t'), 0 : 1 \mid 0, 0] \in C,$$

we obtain $C \triangleleft \mathcal{U}(N(\mathbf{D}_n)_0)$.

For $[0, 2u : m \mid 0, 0]$, $[0, 2u' : m' \mid 0, 0] \in D$, we obtain, using (25:17), that their product is $[0, (2u) \circ (2u') : mm' \mid 0, 0]$, so $D \cong \mathcal{U}(Z_n, \circ) \times \mathcal{U}(Z_n, \cdot)$, the direct product of these two isomorphic groups. Certainly, $C \cap D = \{1\}$, and

$$[2t, 2u : m \mid 0, 0] = [(2t)(2u + 1)^{-1}, 0 : 1 \mid 0, 0] \circ [0, 2u : m \mid 0, 0],$$

hence $\mathcal{U}(N(\mathbf{D}_n)_0) \cong C \times_\theta D$ for some θ. Again, from the discussion in §19, if we define $\Psi[2t, 2u : m \mid 0, 0] = ((2t)(2u + 1)^{-1}, [2u, m])$, we see that θ must be defined by $\theta[2u, m](2t) = (2u + 1)^{-1}(2t)$.

(25.17) Corollary. *We have*

$$\mathcal{U}(N(\mathbf{D}_n)_0) \cong D \,_\phi \times C \cong \left(\mathcal{U}(Z_n, \circ) \times \mathcal{U}(Z_n, \cdot)\right) \,_\phi \times (2Z_n)^+$$

where $(2t)[2u, m]\phi = (2u + 1)(2t)$.

Proof. This is the alternative description of the semidirect product, as discussed in §19. In this case, it is only slightly easier, since $[0, 2u : m \mid 0, 0] \circ [2t, 0 : 1 \mid 0, 0] = [2t, 2u : m \mid 0, 0]$. Then $\Gamma[2t, 2u : m \mid 0, 0] = ([2u, m], 2t)$ and $(2t)[2u, m]\phi = (2t)(2u + 1)$ are the consequences.

The structure of $N(\mathbf{D}_n)$

The structure of $N(\mathbf{D}_n)^+$ and the group of units $\mathcal{U}(N(\mathbf{D}_n))$ is considerably more challenging and interesting. In both cases, the semidirect product must be used twice to obtain a description of these groups in terms of relatively elementary building blocks. Let us first investigate $\mathcal{U}(N(\mathbf{D}_n))$.

For a $[2t, 2u : m \mid s, \epsilon]$, we wonder if there is a $[2t', 2u' : m' \mid s', \epsilon']$ where

$$[2t, 2u : m \mid s, \epsilon] \circ [2t', 2u' : m' \mid s', \epsilon'] = [0, 0 : 1 \mid 0, 0].$$

If so, then from the multiplication rule given in (25.11), $[2t, 2u : m \mid s, \epsilon]$ must satisfy cases where m and m' are both odd and $m \in \mathcal{U}(Z_n, \cdot)$. One can verify that

$$\begin{aligned}
[2t, 2u : m \mid s, 0]^{-1} =&[-\{(2u)^*(2t) - (2u)^*s + (2t) - s + sm^{-1}\}, \\
&(2u)^* : m^{-1} \mid -sm^{-1}, 0],
\end{aligned}$$

$$(25 : 18)$$

and

$$\begin{aligned}
[2t, 2u : m \mid s, 1]^{-1} =&[(s - 2t)(2u + 1)^{-1} - m^{-1}s, \; m^{-1} - 1 : \\
&(2u + 1)^{-1} \mid (2u + 1)^{-1}(s - 2t), 1].
\end{aligned}$$

$$(25 : 19)$$

Hence, one needs $2u \in \mathcal{U}(Z_n, \circ)$ also.

The multiplication rule of (25:11) also gives us

$$\begin{aligned}
[2t, 2u : m \mid s, 0] \circ [2t', 2u' : m' \mid s', \epsilon'] =& \\
[2t' + (1 + 2u')(2t) - (2u')s - s + sm', & \\
(2u) \circ (2u') : mm' \mid sm' + s', \epsilon'],
\end{aligned}$$

$$(25 : 20a)$$

and

$$[2t, 2u : m \mid s, 1] \circ [2t', 2u' : m' \mid s', \epsilon'] =$$
$$[(2t)m' + 2t' + (2u')s - (m' - 1)s, (2u + 1)m' - 1 : (2u' + 1)m$$
$$\mid (2u' + 1)s + 2t' - s', 1 + \epsilon'].$$

$$(25 : 20b)$$

We have, therefore,

(25.18) Theorem. *The units of $N(\mathbf{D}_n)$ are given by*

$$\mathcal{U}(N(\mathbf{D}_n)) = \{[2t, 2u : m \mid s, \epsilon] \mid$$
$$2t \in 2Z_n, \ 2u \in \mathcal{U}(Z_n, \circ), \ m \in \mathcal{U}(Z_n, \cdot), \ s \in Z_n, \ \epsilon \in Z_2\}.$$

The inverse of a unit $[2t, 2u : m \mid s, 0]$ is given by $(25 : 18)$. The inverse of a unit $[2t, 2u : m \mid s, 1]$ is given by $(25 : 19)$. Multiplication of units is given by equations $(25 : 20a)$ and $(25 : 20b)$.

(25.19) Corollary. *The additive group $N(\mathbf{D}_n)^+$ and the multiplicative group $\mathcal{U}(N(\mathbf{D}_n))$ each has a subgroup isomorphic to our source group D_n. For $N(\mathbf{D}_n)^+$, it is*

$$N(\mathbf{D}_n)_c = \{[0, 0 : 0 \mid s, \epsilon] \mid s \in Z_n, \epsilon \in Z_2\},$$

and for $\mathcal{U}(N(\mathbf{D}_n))$, it is

$$D_n^* = \{[0, 0 : 1 \mid s, \epsilon] \mid s \in Z_n, \epsilon \in Z_2\}.$$

Proof. For $N(\mathbf{D}_n)_c$, the isomorphism is $sa + \epsilon b \mapsto [0, 0 : 0 \mid s, \epsilon]$. For D_n^*, the isomorphism is $sa + \epsilon b \mapsto [0, 0 : 1 \mid s, \epsilon]$.

Using the multiplication formulae (25:20a) and (25:20b), one sees that

$$\mathcal{U}_1 = \{[2t, 2u : m \mid s, 0] \mid [2t, 2u : m \mid s, 0] \in \mathcal{U}(N(\mathbf{D}_n))\}$$

is a normal subgroup of index 2 in $\mathcal{U}(N(\mathbf{D}_n))$, and that

$$B_1 = \{[0, 0 : 1 \mid 0, \epsilon] \mid \epsilon \in Z_2\}$$

is its complement, that is, $B_1 < \mathcal{U}(N(\mathbf{D}_n))$, $\mathcal{U}_1 \cap B_1 = \{1\}$, and $\mathcal{U}_1 B_1 = \mathcal{U}(N(\mathbf{D}_n))$. This means we obtain

(25.20) Theorem. *We have isomorphic groups*

$$\mathcal{U}(N(\mathbf{D}_n)) \cong \mathcal{U}_1 \times_\theta B_1 \cong \mathcal{U}_1 \times_\theta Z_2,$$

with isomorphism $\Psi[2t, 2u : m \mid s, \epsilon] = \langle[2t, 2u : m \mid s, 0], \epsilon\rangle$ and with $\theta : Z_2 \to \text{Aut}\,\mathcal{U}_1$ given by $\theta(0) = 1$ and $\theta(1)[2t, 2u : m \mid s, 0] = [2t, m - 1 : 2u + 1 \mid 2t - s, 0]$.

We can do more with $\mathcal{U}_1$. Let

$$\mathcal{U}_2 = \{[2t, 0 : 1 \mid s, 0] \mid 2t \in 2Z_n, s \in Z_n\}$$

and

$$B_2 = \{[0, 2u : m \mid 0, 0] \mid 2u \in \mathcal{U}(Z_n, \circ) \text{ and } m \in \mathcal{U}(Z_n, \cdot)\}.$$

We will prove

(25.21) Theorem. *We have* $\mathcal{U}_2 \cong 2Z_n \oplus Z_n$, $B_2 \cong \mathcal{U}(Z_n, \circ) \times \mathcal{U}(Z_n, \cdot)$, *and*

$$\mathcal{U}_1 \cong \Big(\mathcal{U}(Z_n, \circ) \times \mathcal{U}(Z_n, \cdot)\Big)_\mu \times \big(2Z_n \oplus Z_n\big)$$

with isomorphism $\Gamma[2t, 2u : m \mid s, 0] = \big([2u, m], [2t, s]\big)$, *where*

$$[2t, s]\big([2u, m]\mu\big) = [(1 + 2u)(2t - s) + sm, sm].$$

Proof. Certainly, $\mathcal{U}_2 \cap B_2 = \{1\}$. That $B_2 \cong \mathcal{U}(Z_n, \circ) \times \mathcal{U}(Z_n, \cdot)$ follows directly from multiplying two elements of B_2 using (25:20a). Again, using (25:20a), one obtains that $\mathcal{U}_2 \cong 2Z_n \oplus Z_n$. To see that $\mathcal{U}_2 \lhd \mathcal{U}_1$, compute the $2u$-component and the m-component of $[2t, 2u : m \mid s, 0] \circ [2t', 0 : 1 \mid s', 0] \circ [2t, 2u : m \mid s, 0]^{-1}$ using (25:20a) and (25.18).

It is rather messy to show that $\mathcal{U}_1 = \mathcal{U}_2 B_2$, but rather easy to see that $\mathcal{U}_1 = B_2 \mathcal{U}_2$. Just note that $[2t, 2u : m \mid s, 0] = [0, 2u : m \mid 0, 0] \circ [2t, 0 : 1 \mid s, 0]$. This means that $\Gamma[2t, 2u : m \mid s, 0] = \big([2u, m], [2t, s]\big)$ provides the isomorphism $\Gamma : \mathcal{U}_1 \to \Big(\mathcal{U}(Z_n, \circ) \times \mathcal{U}(Z_n, \cdot)\Big)_\mu \times \big(2Z_n \oplus Z_n\big)$, where $\mu : \mathcal{U}(Z_n, \circ) \times \mathcal{U}(Z_n, \cdot) \to \text{Aut}(2Z_n \oplus Z_n)$ is defined by $[2t, s]\big([2u, m]\mu\big) = [(1 + 2u)(2t - s) + sm, sm]$.

(25.22) Corollary.

$$\mathcal{U}(N(\mathbf{D}_n)) \cong \Big\{ [\mathcal{U}(Z_n, \circ) \times \mathcal{U}(Z_n, \cdot)]_\mu \times [2Z_n \oplus Z_n] \Big\} \times_\theta Z_2.$$

Proof. Apply (25.20), and then (25.21).

This being a rather satisfactory description of the structure of $\mathcal{U}(N(\mathbf{D}_n))$ in terms of quite elementary components, we turn our attention to the structure of $N(\mathbf{D}_n)^+$, and we start with an unorthodox approach. The obvious place to start is with the decomposition $N(\mathbf{D}_n)^+ = N(\mathbf{D}_n)_0^+ + N(\mathbf{D}_n)_c^+$, but we will save that for last.

We shall need the addition rule of (25:10) frequently. Let

$$S_1 = \{[2t, 2u : m \mid s, 0] \mid [2t, 2u : m \mid s, 0] \in N(\mathbf{D}_n)\},$$

and

$$T_1 = \{[0, 0 : 0 \mid 0, \epsilon] \mid \epsilon \in Z_2\}.$$

Obviously, $T_1 \cong Z_2$, $S_1 \triangleleft N(\mathbf{D}_n)$ since it is of index 2, $S_1 + T_1 = N(\mathbf{D}_n)$, and one defines an isomorphism $\Psi : N(\mathbf{D}_n)^+ \to S_1 \times_\sigma Z_2$ where $\Psi[2t, 2u : m \mid s, \epsilon] = ([2t, 2u : m \mid s, 0], \epsilon)$ and we obtain

(25.23) Theorem. *We have* $N(\mathbf{D}_n)^+ \cong S_1 \times_\sigma Z_2$ *where*

$$\sigma(1)[2t, 2u : m \mid s, 0] = \begin{cases} [-2t, -2u : -m \mid -s, 0], & \textit{if } m \textit{ is even;} \\ [-2t, -2u - 2 : -m \mid -s, 0], & \textit{if } m \textit{ is odd.} \end{cases}$$

We cannot do much to simplify $T_1 \cong Z_2$, but we can break S_1 down further. Let us define

$$S_2 = \{[2t, 2u : 0 \mid s, 0] \mid [2t, 2u : 0 \mid s, 0] \in S_1\}$$

and

$$T_2 = \{[0, 0 : m \mid 0, 0] = m \cdot 1 \mid m \in Z_n\}.$$

Then $T_2 \cong Z_n$ and $T_2 < S_1$. With the m-component set to 0 in the elements of S_2, we readily obtain $[2t, 2u : 0 \mid s, 0] + [2t', 2u' : 0 \mid s', 0] = [2t + 2t', 2u + 2u' : 0 \mid s + s', 0]$ and so $2Z_n \oplus 2Z_n \oplus Z_n \cong S_2 < S_1$. To see $S_2 \triangleleft S_1$, use the definition of normality, and realize that the m-component of $[2t, 2u : m \mid s, 0] + [2t', 2u' : 0 \mid s', 0] - [2t, 2u : m \mid s, 0]$ is 0, as is the ϵ-component.

It is elementary to see that $[2t, 2u : 0 \mid s, 0] + [0, 0 : m \mid 0, 0] = [2t, 2u : m \mid s, 0]$ if m is even, and $[2t - 2s, 2u : 0 \mid s, 0] + [0, 0 : m \mid 0, 0] = [2t, 2u : m \mid s, 0]$ if m is odd, so $S_1 \cong S_2 \times_\lambda T_2$, and we obtain

(25.24) Theorem. $S_1 \cong [2Z_n \oplus 2Z_n \oplus Z_n] \times_\lambda Z_n$, *where*

$$\lambda(m)[2t, 2u, s] = \begin{cases} [2t, 2u, s], & \textit{if } m \textit{ is even;} \\ [-2t - 2s, -2u, s], & \textit{if } m \textit{ is odd.} \end{cases}$$

(25.25) Theorem. $N(\mathbf{D}_n)^+ \cong \left\{ [2Z_n \oplus 2Z_n \oplus Z_n] \times_\lambda Z_n \right\} \times_\sigma Z_2$.

Proof. Combine the consequences of (25.23) and (25.24).

As promised, we will study the decomposition $N(\mathbf{D}_n)^+ = N(\mathbf{D}_n)_0^+ + N(\mathbf{D}_n)_c^+$ further. Theorem (25.25) gives us at least one decomposition in terms of elementary components.

Of course, we have $N(\mathbf{D}_n)_c^+ \cong D_n \cong Z_n \times_\mu Z_2$, where $(a, x) + (b, y) = (a + (-1)^x b, x + y)$. It is well known that the decomposition $N(\mathbf{D}_n)^+ = N(\mathbf{D}_n)_0^+ + N(\mathbf{D}_n)_c^+$, as a group, is a semidirect product, so we have

$$N(\mathbf{D}_n)^+ \cong N(\mathbf{D}_n)_0^+ \times_\gamma N(\mathbf{D}_n)_c^+ \cong N(\mathbf{D}_n)_0^+ \times_\gamma D_n.$$

Before we define $\gamma : D_n \to \mathrm{Aut} N(\mathbf{D}_n)^+$, we shall recall our description of $N(\mathbf{D}_n)_0$.

$$N(\mathbf{D}_n)_0 = \{[2t, 2u : m \mid 0, 0] \mid [2t, 2u : m \mid 0, 0] \in N(\mathbf{D}_n)\}$$

and we will see that it will be convenient to identify, or define

$$([2t, 2u], m) \equiv [2t, 2u : m \mid 0, 0].$$

Let $f = [2t, 2u : m \mid 0, 0] + [2t', 2u' : m' \mid 0, 0] = ([2t, 2u], m) + ([2t', 2u'], m')$. From (25.10), one easily obtains

$$N(\mathbf{D}_n)^+ \cong (2Z_n \oplus 2Z_n) \times_\theta Z_n,$$

where $\theta(m)[2t, 2u] = [(-1)^m (2t), (-1)^m (2u)]$. We will soon have

(25.26) Theorem.

$$N(\mathbf{D}_n) \cong \left\{(2Z_n \oplus 2Z_n) \times_\theta Z_n\right\} \times_\gamma D_n \cong \left\{(2Z_n \oplus 2Z_n) \times_\theta Z_n\right\} \times_\gamma (Z_n \times_\mu Z_2),$$

where $\gamma(sa + \epsilon b) = \gamma(s, \epsilon)$ *is defined by*

$$\gamma(s, \epsilon)\Big([2t, 2u], m\Big) = \Big([(-1)^\epsilon (2t), (-1)^\epsilon (2u)], (-1)^\epsilon m\Big),$$

if m is even, and

$$\gamma(s, \epsilon)\Big([2t, 2u], m\Big) = \Big([(-1)^\epsilon (2t) + 2s, (-1)^\epsilon (2u) + (-1)^\epsilon - 1], (-1)^\epsilon m\Big),$$

if m is odd, where $\theta(m)[2t, 2u] = [(-1)^m (2t), (-1)^m (2u)]$, *with* $\mu(\epsilon) = (-1)^\epsilon$.

Proof. The map $\Psi[2t, 2u : m \mid s, \epsilon] = \Big[([2t, 2u], m), (s, \epsilon)\Big]$ defines the isomorphism, which in turn, defines γ by $\gamma(s, \epsilon)([2t, 2u], m) = \phi_{sa+\epsilon b} + ([2t, 2u], m) - \phi_{sa+\epsilon b}$.

Some ideals in $N(\mathbf{D}_n)$

One can obtain a maximal ideal of $N(\mathbf{D}_n)$ rather quickly. For the moment, we say that $[2t, 2u : m \mid s, \epsilon]$ is *odd* if $m \in \{1, 3, \ldots, n-1\}$, and *even* if $m \in \{0, 2, \ldots, n-2\}$. We will abbreviate 'odd' with 'o' and 'even' with 'e'.

(25.27) Theorem. *Let*

$$M_1 = \{[2t, 2u : m \mid s, \epsilon] \in N(\mathbf{D}_n) \mid m \in \{0, 2, \ldots, n-2\}\}.$$

Then M_1 is a maximal ideal of the nearring $N(\mathbf{D}_n)$ and the quotient nearring $N(\mathbf{D}_n)/M_1$ is isomorphic to the prime field Z_2.

Proof. So, M_1 consists of the 'even' elements of $N(\mathbf{D}_n)$. One must show (1) M_1 is a normal subgroup of $N(\mathbf{D}_n)$, (2) that $N(\mathbf{D}_n)M_1 \subseteq M_1$, and (3) that $(x + a)y - xy \in M_1$ for all $x, y \in N(\mathbf{D}_n)$ and for all $a \in M_1$. Keep a lazy eye on the addition rule (25:10), an active eye on the multiplication rule (25:11), and an active eye on the table below to be assured that M_1 is indeed an ideal. Since M_1 has index 2, M_1 will be a normal subgroup and a maximal ideal, and the quotient nearring $N(\mathbf{D}_n)/M_1$ will have order 2 and an identity, hence will be isomorphic to the prime field Z_2.

x	a	xa	$x + a$	y	$(x+a)y$	$-xy$	$(x+a)y - xy$
e	e	e	e	e	e	e	e
e	e	e	e	o	e	e	e
o	e	e	o	e	e	e	e
o	e	e	o	o	o	o	e

Table. Proof that M_1 is an ideal.

(25.28) Theorem. *Let*

$$I = \{[2t, 0 : 0 \mid s, 0] \mid 2t \in 2Z_n, \ s \in Z_n\}$$

and

$$T = \{[0, 2u : m \mid 0, \epsilon] \mid 2u \in 2Z_n, \ m \in Z_n, \ \epsilon \in Z_2\}.$$

Then T is a subnearring of $N(\mathbf{D}_n)$ and the map $\Psi : N(\mathbf{D}_n) \to T$ defined by

$$\Psi[2t, 2u : m \mid s, \epsilon] = [0, 2u : m \mid 0, \epsilon]$$

is a nearring epimorphism with kernel I, hence I is an ideal of $N(\mathbf{D}_n)$ and $N(\mathbf{D}_n)/I \cong T$.

Proof. To obtain that T is a subnearring of $N(\mathbf{D}_n)$, construct an addition rule and a multiplication rule for T using those corresponding addition and multiplication rules (25:10) and (25:11) for $N(\mathbf{D}_n)$. Notice that the $2u$-position, the m-position, and the s-position of a resulting $[2t, 2u : m \mid s, \epsilon]$ is only influenced by what comes from these $2u$-positions, these m-positions, and these s-positions of the summands and the factors. This not only shows that T is a subnearring of $N(\mathbf{D}_n)$ but also that Ψ is a nearring epimorphism. Obviously, the kernel of Ψ is I, so I is an ideal and $N(\mathbf{D}_n)/I \cong T$.

In describing our next ideal, it will be helpful to make use of the nearring of mappings $(M(Z_2), +, \circ)$, where Z_2 is the group of integers modulo 2, and where we write xf for $x \in Z_2$ and $f \in M(Z_2)$. Then $M(Z_2) = \{\phi_0, \phi_1, 1_{Z_2}, \phi_1 - 1_{Z_2}\}$ where ϕ_0 and ϕ_1 are the constant functions $x\phi_0 = 0$ and $x\phi_1 = 1$, and where 1_{Z_2} is the identity function $x1_{Z_2} = x$.

(25.29) Theorem. *Define* $\Gamma : N(\mathbf{D}_n) \rightarrow M(Z_2)$ *by*

$$\Gamma[2t, 2u : m \mid s, \epsilon] = \begin{cases} \phi_0, & \text{if } m \in 2Z_n \text{ and } \epsilon = 0; \\ 1_{Z_2}, & \text{if } m \notin 2Z_n \text{ and } \epsilon = 0; \\ \phi_1, & \text{if } m \in 2Z_n \text{ and } \epsilon = 1; \\ \phi_1 - 1_{Z_2}, & \text{if } m \notin 2Z_n \text{ and } \epsilon = 1. \end{cases}$$

Then Γ is a nearring epimorphism with kernel

$$J = \{[2t, 2u : 2m \mid s, 0] \mid 2t, 2u, 2m \in 2Z_n, s \in Z_n\},$$

so J is an ideal of $N(\mathbf{D}_n)$ and $N(\mathbf{D}_n)/J \cong M(Z_2)$.

Proof. To see that Γ is a nearring epimorphism is easy if one goes through an intermediate step. Let $M = Z_2 \oplus Z_2$ be Klein's four-group. Define $\Gamma' : N(\mathbf{D}_n) \rightarrow M$ by

$$\Gamma'[2t, 2u : m \mid s, \epsilon] = (m \bmod 2, \epsilon).$$

With the aid of the addition rule (25:10), one can be quickly assured that Γ' is a group epimorphism. Let the multiplication rule (25:11) induce, via Γ', a binary operation $*$ on M so that Γ' will be a nearring epimorphism. That is, define $*$ by $\Gamma'(x) * \Gamma'(y) = \Gamma'(x \circ y)$ for all $x, y \in N(\mathbf{D}_n)$. Then verify that $*$ is defined by $(a, x) * (b, y) = (ab, xb + y)$. Define $\phi : M \rightarrow M(Z_2)$ by $\phi[0, 0] = \phi_0$, $\phi[0, 1] = \phi_1$, $\phi[1, 0] = 1_{Z_2}$, and, therefore, $\phi[1, 1] = 1_{Z_2} + \phi_1$. Then ϕ is a nearring isomorphism and $\Gamma = \phi \circ \Gamma'$ gives us our desired result.

Since it is well known by (3.45) that $M(Z_2)$ has a proper ideal, one can use the correspondence theorem to obtain the ideal M_1 of (25.27). However, the understanding and proof of (25.27) led to the discovery and understanding of the proofs of (25.28) and (25.29).

(25.30) Theorem. *Let*

$$K = \{[2t, 0 : 0 \mid 2s, 0] \mid 2t, 2s \in 2Z_n\}.$$

Then K is an ideal of $N(\mathbf{D}_n)$.

Proof. Notice that K is a subset of the ideal I. This allows us to focus on the entry in the s-position as we check the details for K being an ideal. It is easy to see that K is a subgroup of $N(\mathbf{D}_n)$. From $[2t, 2u : m \mid s, \epsilon] + [2t', 0 : 0 \mid 2s', 0] - [2t, 2u : m \mid s, \epsilon]$, we note that the s-position is $(-1)^\epsilon(2s')$, so K is a normal subgroup of $N(\mathbf{D}_n)$.

From $[2t, 2u : m \mid s, \epsilon] \circ [2t', 0 : 0 \mid 2s', 0]$ and the multiplication rule (25:11), one sees that the s-position in this product is either $2s'$ or $2t' + 2s'$, so we have $N(\mathbf{D}_n)K \subseteq K$.

It remains to show that $(x + a)y - xy \in K$ for all $x, y \in N(\mathbf{D}_n)$ and all $a \in K$. Consider $\{[2t, 2u : m \mid s, \epsilon] + [2\bar{t}, 0 : 0 \mid 2\bar{s}, 0]\} \circ [2t', 2u' : m' \mid s', \epsilon'] - [2t, 2u : m \mid s, \epsilon] \circ [2t', 2u' : m' \mid s', \epsilon']$ separately with the eight cases defined in the multiplication rule (25:11). Remember, we need only focus on what results in the s-position. For cases a, c, e, and g, we obtain $(-1)^\epsilon(2\bar{s})m'$ in the s-position. For cases b and f, we obtain $(-1)^\epsilon(2\bar{s})(2u')$ in the s-position. And for cases d and h, we get $(-1)^\epsilon(2\bar{s})(2u' + 1)$ in the s-position. Hence, K is truly an ideal of $N(\mathbf{D}_n)$.

Even though it takes a lot of computation to see that K is an ideal, it is relatively easy compared to the computation required for our next result, which is much stronger, however, and has K as a special case. The proof of our next theorem is greatly simplified by the knowledge that K is an ideal.

(25.31) Theorem. *Let H be a subgroup of $2Z_n$, and define*

$$\mathcal{K}(H) = \{[a, 0 : 0 \mid b, 0] \mid a, b \in H\}.$$

Then $\mathcal{K}(H)$ is an ideal of $N(\mathbf{D}_n)$. Conversely, if L is an ideal of $N(\mathbf{D}_n)$ and $L \subseteq K = \mathcal{K}(2Z_n)$, then there is a subgroup H of $2Z_n$ such that $L = \mathcal{K}(H)$. So, the lattice of ideals of $N(\mathbf{D}_n)$ between (0) and K is isomorphic to the lattice of subgroups of $2Z_n$.

Proof. Consider $[2t, 2u : m \mid s, \epsilon] \circ [a, 0 : 0 \mid b, 0]$. Cases a, b, e, and f of the multiplication (25:11) rule apply, and one obtains, respectively,

$$[0, 0 : 0 \mid b, 0], \tag{25:21a}$$

$$[0, 0 : 0 \mid a + b, 0], \tag{25:21b}$$

$$[a, 0 : 0 \mid b, 0], \tag{25:21c}$$

$$[-a, 0 : 0 \mid a + b, 0].$$ (25 : 21d)

Hence, $N(\mathbf{D}_n)\mathcal{K}(H) \subseteq \mathcal{K}(H)$. Next, consider

$$[2t, 2u : m \mid s, \epsilon] + [a, 0 : 0 \mid b, 0] - [2t, 2u : m \mid s, \epsilon].$$

There are two cases to consider: (i) m is even, and (ii) m is odd. One obtains, respectively,

$$[(-1)^\epsilon a, 0 : 0 \mid (-1)^\epsilon b, 0],$$ (25 : 22a)

and

$$[(-1)^\epsilon(-a - 2b), 0 : 0 \mid (-1)^\epsilon b, 0].$$ (25 : 22b)

Hence, $\mathcal{K}(H)$ is a left ideal of $N(\mathbf{D}_n)$. Finally, consider $(x + i)y - xy$ for $x, y \in N(\mathbf{D}_n)$ and $i \in \mathcal{K}(H)$, that is, all possible

$$\begin{aligned}
\{[2t, 2u : m \mid s, \epsilon] &+ [a, 0 : 0 \mid b, 0]\} \circ [2t', 2u' : m' \mid s', \epsilon'] \\
&- [2t, 2u : m \mid s, \epsilon] \circ [2t', 2u' : m' \mid s', \epsilon'] \\
= [2t \pm (-1)^\epsilon a, 2u : m \mid s &+ (-1)^\epsilon b, \epsilon] \circ [2t', 2u' : m' \mid s', \epsilon'] \\
&- [2t, 2u : m \mid s, \epsilon] \circ [2t', 2u' : m' \mid s', \epsilon'] = (*).
\end{aligned}$$

Since $K = \mathcal{K}(2Z_n)$ is an ideal already, and $\mathcal{K}(H) \subseteq K$, we know that the $2u$-, m-, and ϵ-components of $(*)$ are each 0. So it will help considerably to focus only on the $2t$- and s-components of $(*)$. The eight cases of the multiplication rule (25:11) yield, for $(*)$, respectively,

$$[\pm(-1)^\epsilon am', 0 : 0 \mid (-1)^\epsilon bm', 0],$$ (25 : 23a)

$$[\pm(-1)^\epsilon a(2u'), 0 : 0 \mid (-1)^\epsilon b(2u'), 0],$$ (25 : 23b)

$$[\pm(-1)^\epsilon am', 0 : 0 \mid (-1)^\epsilon bm', 0],$$ (25 : 23c)

$$[\pm(-1)^\epsilon a(2u' + 1), 0 : 0 \mid (-1)^\epsilon b(2u' + 1), 0],$$ (25 : 23d)

$$[(-1)^\epsilon\{(\pm a - b)(2u') - bm'\}, 0 : 0 \mid (-1)^\epsilon bm', 0],$$ (25 : 23e)

$$[(-1)^\epsilon\{(\pm a - b)m' - b(2u')\}, 0 : 0 \mid (-1)^\epsilon b(2u'), 0],$$ (25 : 23f)

$$[(-1)^\epsilon\{(\pm a - b)(2u' + 1) - bm'\}, 0 : 0 \mid (-1)^\epsilon bm', 0],$$ (25 : 23g)

$$[(-1)^\epsilon\{(\pm a - b)m' - b(2u' + 1)\}, 0 : 0 \mid (-1)^\epsilon b(2u' + 1), 0].$$ (25 : 23h)

Hence, $\mathcal{K}(H)$ is an ideal of $N(\mathbf{D}_n)$.

Conversely, suppose $(0) \neq L \subseteq K = \mathcal{K}(2Z_n)$ and that L is an ideal of $N(\mathbf{D}_n)$ with nonzero $[a, 0 : 0 \mid b, 0] \in L$. So $a \neq 0$ or $b \neq 0$. With $2u' = 0$ and $m' = 1$, use (25:23e) to conclude that $[(-1)^\epsilon(-b), 0 : 0 \mid$

$(-1)^\epsilon b, 0] \in L$. Hence, since (25:21a) yields $[0, 0 : 0 \mid b, 0] \in L$, we obtain $[(-1)^\epsilon(-b), 0 : 0 \mid 0, 0]$, $[b, 0 : 0 \mid 0, 0] \in L$. From (25:21a), we obtain $[0, 0 : 0 \mid b, 0]$, $[0, 0 : 0 \mid -b, 0] \in L$, so $[a, 0 : 0 \mid 0, 0] \in L$. With (25 : 21b), we obtain $[0, 0 : 0 \mid a, 0] \in L$. Summarizing, we can be assured from $[a, 0 : 0 \mid b, 0] \in L$, that $[a, 0 : 0 \mid 0, 0]$, $[0, 0 : 0 \mid a, 0]$, $[0, 0 : 0 \mid b, 0]$, $[b, 0 : 0 \mid 0, 0] \in L$. Certainly, $\{a, b \mid [a, 0 : 0 \mid b, 0] \in L\} = H$ is a subgroup of $2Z_n$, and then $\mathcal{K}(H) = L$.

The next theorem gives us two ideals whose quotient nearrings are particularly nice, being isomorphic to abstract affine nearrings. It also provides the motivation to apply the ideas of (25.31) to see that the lattice of subgroups of $2Z_n$ is again represented in the lattice of ideals of $N(\mathbf{D}_n)$.

(25.32) Theorem. *Let P and Q be the Z_2-modules $P = Z_2$ and $Q = Z_2 \oplus Z_2$, and consider the abstract affine nearrings $Z_2 \oplus_A P$ and $Z_2 \oplus_A Q$ where addition is $(a, x) + (b, y) = (a+b, x+y)$, of course, and multiplication is $(a, x)(b, y) = (ab, xb + y)$. Define $F : N(\mathbf{D}_n) \to Z_2 \oplus_A P$ and $G : N(\mathbf{D}_n) \to Z_2 \oplus_A Q$ by*

$$F[2t, 2u : m \mid s, \epsilon] = (m \bmod 2, \ s \bmod 2)$$

and

$$G[2t, 2u : m \mid s, \epsilon] = (m \bmod 2, \ s \bmod 2, \ \epsilon).$$

Then F and G are nearring epimorphisms with

$$\ker F = N = \{[2t, 2u : 2m \mid 2s, \epsilon] \mid 2t, 2u, 2m, 2s \in 2Z_n, \ \epsilon \in Z_2\}$$

and

$$\ker G = L = \{[2t, 2u : 2m \mid 2s, 0] \mid 2t, 2u, 2m, 2s \in 2Z_n\}$$

Proof. First show that F is a nearring epimorphism using the addition rule (25:10) and multiplication rule (25:11). Then it will be very easy to see that G is also a nearring epimorphism.

(25.33) Theorem. *Let k be be an even divisor of n. The map $1 \mapsto 1$ defines an epimorphism of Z_n onto Z_k with kernel H, and if $x \in Z_n$, let x' be the image of x in Z_k. Define $f : N(\mathbf{D}_n) \to N(\mathbf{D}_k)$ by*

$$f[2t, 2u : m \mid s, \epsilon] = [2t', 2u' : m' \mid s', \epsilon].$$

Then f is a nearring epimorphism with

$$\ker f = \mathcal{L}(H) = \{[a, b : c \mid d, 0] \mid a, b, c, d \in H\}.$$

Proof. Again, one needs only to realize from (25.26) that f is surjective, and then study the addition rule and multiplication rule.

(25.34) Corollary. *For a subgroup* $H \triangleleft 2Z_n$,

$$\mathcal{L}(H) = \{[a, b : c \mid d, 0] \mid a, b, c, d \in H\}$$

is an ideal of $N(\mathbf{D}_n)$.

Now (25.33) and its corollary, together with (25.31), show that the lattice of subgroups H of $2Z_n$ is represented at least twice in $N(\mathbf{D}_n)$. We can summarize the ideals we have discovered here in the following diagram.

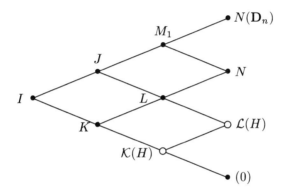

Our Ideals of $N(\mathbf{D}_n)$.

Here, K—$\mathcal{K}(H)$—(0) and L—$\mathcal{L}(H)$—(0) represent, respectively, the sublattice of ideals defined by the lattice of subgroups $H \triangleleft 2Z_n$ and (25.31) for $\mathcal{K}(H)$ and (25.33) for $\mathcal{L}(H)$.

The reader should notice that *none* of our nontrivial ideals are entirely in $N(\mathbf{D}_n)_0$, that is, each involves some constant functions. Now $N(\mathbf{D}_n)_0 \cong I(D_n)$, as we have seen in (25.4). So the intersection of any of our ideals of $N(\mathbf{D}_n)$ with $N(\mathbf{D}_n)_0$ will yield an ideal of $I(D_n)$. Malone and Lyons did not present any results on the ideals of $I(D_n)$, n even, in [M & L, 1973], except for radicals, whereas they did discuss the ideals of $I(D_n)$, n odd, thoroughly in [M & L, 1972]. In §15, the ideas of Malone and Lyons were extended to the odd generalized dihedral groups, and numerous ideals were provided. Preliminary investigations indicate that extension of Malone and Lyons' work to even generalized dihedral groups will be considerably more complex.

26. The cornucopia

In this section, we introduce, without much detail, several topics which have received attention during recent years. Each of these topics has already provided the world with some interesting mathematics, and it is believed that each shows considerable promise for further research. What we present here is only a hint of the interesting results which have appeared in research journals, and only a sample of some of the families of nearrings receiving attention.

26.1. Distributive nearrings

Suppose one dearly wants to make matrix nearrings from an arbitrary nearring $(N, +, \cdot)$. Let

$$\mathcal{M}(2, N) = \left\{ \begin{pmatrix} a & b \\ c & d \end{pmatrix} \mid a, b, c, d \in N \right\}$$

and let $+$ and $\cdot$ be the usual matrix addition and multiplication, respectively. Let's insist that $(\mathcal{M}(2, N), +, \cdot)$ be a nearring. In what we do now, we need only look at the $(1, 1)$-position of the resulting matrices. With

$$\begin{pmatrix} a & b \\ c & d \end{pmatrix} \left[\begin{pmatrix} e & f \\ g & h \end{pmatrix} + \begin{pmatrix} w & x \\ y & z \end{pmatrix} \right] = \begin{pmatrix} a & b \\ c & d \end{pmatrix} \begin{pmatrix} e & f \\ g & h \end{pmatrix} + \begin{pmatrix} a & b \\ c & d \end{pmatrix} \begin{pmatrix} w & x \\ y & z \end{pmatrix},$$

we see that

$$aw + bg = bg + aw \qquad (26:1)$$

for all $a, b, g, w \in N$. If N has a multiplicative identity, then with $w = g = 1$, we obtain

$$a + b = b + a \qquad (26:2)$$

for all $a, b \in N$.

From

$$\left[\begin{pmatrix} a & b \\ c & d \end{pmatrix} \begin{pmatrix} e & f \\ g & h \end{pmatrix} \right] \begin{pmatrix} w & x \\ y & z \end{pmatrix} = \begin{pmatrix} a & b \\ c & d \end{pmatrix} \left[\begin{pmatrix} e & f \\ g & h \end{pmatrix} \begin{pmatrix} w & x \\ y & z \end{pmatrix} \right],$$

we see that

$$(ae + bg)w + (af + bh)y = aew + afy + bgw + bhy \qquad (26:3)$$

for all $a, b, e, f, g, h, w, y \in N$. With $f = h = y = 0$, we obtain

$$(ae + bg)w = aew + bgw \qquad (26:4)$$

for all $a, b, e, g, w \in N$. Again, if $1 \in N$, then with $e = g = 1$, equation (26:4) becomes

$$(a + b)w = (aw) + (bw) \qquad (26:5)$$

for all $a, b, w \in N$. This forces $(N, +, \cdot)$ to be a ring. So nearrings with identity which are not rings are not useful in making matrix nearrings in the obvious way. We have

(26.1) Theorem. *Let $(N, +, \cdot)$ be a nearring with identity. Suppose $(\mathcal{M}(2, N), +, \cdot)$ is also a (left) nearring. Then $(N, +, \cdot)$ is a ring.*

Without an identity $1 \in N$, and with multiplication in N satisfying (26:1), (26:3) and (26:4), we might get $(\mathcal{M}(2, N), +, \cdot)$ to be a nearring, but this rather motivates us to use a distributive nearring for $(N, +, \cdot)$. We would obtain (26:3) and (26:4) with no problem, but we would also obtain (26:1).

(26.2) Lemma. *If $(N, +, \cdot)$ is a distributive nearring and $a, b, g, w \in N$, then $aw + bg = bg + aw$.*

Proof. First, $(a + b)(g + w) = (a + b)g + (a + b)w = ag + bg + aw + bw$. Next, $(a + b)(g + w) = a(g + w) + b(g + w) = ag + aw + bg + bw$. From $ag + aw + bg + bw = ag + bg + aw + bw$ we conclude that $aw + bg = bg + aw$.

These observations lead to

(26.3) Theorem. *Let $(N, +, \cdot)$ be a distributive nearring. Let $\mathcal{M}(n, N)$ be the set of all n by n matrices with entries from N. Then $(\mathcal{M}(n, N), +, \cdot)$ is a distributive nearring.*

Proof. Certainly, each $(\mathcal{M}(n, N), +)$ is a group. For $(a_{ij}), (b_{ij}), (c_{ij}) \in \mathcal{M}(n, N)$, we have the equation $(a_{ij})[(b_{ij})(c_{ij})] = (a_{ij})(\sum_{k=1}^{n} b_{ik}c_{kj}) = (\sum_{m=1}^{n} a_{im}(\sum_{k=1}^{n} b_{mk}c_{kj})) = (\sum_{m=1}^{n} \sum_{k=1}^{n} a_{im}b_{mk}c_{kj})$, and the equation $[(a_{ij})(b_{ij})](c_{ij}) = (\sum_{m=1}^{n} a_{im}b_{mj})(c_{ij}) = (\sum_{k=1}^{n} [\sum_{m=1}^{n} a_{im}b_{mk}]c_{kj}) = (\sum_{k=1}^{n} \sum_{m=1}^{n} a_{im}b_{mk}c_{kj})$. But (26.2) gives us $\sum_{m=1}^{n} \sum_{k=1}^{n} a_{im}b_{mk}c_{kj} = \sum_{k=1}^{n} \sum_{m=1}^{n} a_{im}b_{mk}c_{kj}$, so $\cdot$ is associative in $\mathcal{M}(n, N)$.

Also $(a_{ij})[(b_{ij}) + (c_{ij})] = (a_{ij})[(b_{ij} + c_{ij})] = (\sum_{k=1}^{n} a_{ik}(b_{kj} + c_{kj})) = (\sum_{k=1}^{n} [a_{ik}b_{kj} + a_{ik}c_{kj}]) = (\sum_{k=1}^{n} a_{ik}b_{kj}) + (\sum_{k=1}^{n} a_{ik}c_{kj})) = (a_{ij})(b_{ij}) + (a_{ij})(c_{ij})$ since (26.2) is valid. But $[(a_{ij}) + (b_{ij})](c_{ij}) = (a_{ij} + b_{ij})(c_{ij}) = (\sum_{k=1}^{n} (a_{ik} + b_{ik})c_{kj}) = (\sum_{k=1}^{n} (a_{ik}c_{kj} + b_{ik}c_{kj})) = (\sum_{k=1}^{n} a_{ik}c_{kj}) + (\sum_{k=1}^{n} b_{ik}c_{kj}) = (a_{ij})(c_{ij}) + (b_{ij})(c_{ij})$ since N is distributive, and also since (26.2) is valid. This completes the proof of (26.3).

With (26.3) we have a way of making lots of distributive nearrings from a given distributive nearring $(N, +, \cdot)$. In §2, we (1) saw the simplest honest distributive nearring, (2) saw how to make a noncommutative distributive nearring from a nonabelian group $(G, +)$ and a noncommutative ring

$(R, +, \cdot)$, and (3) learned how to make distributive nearrings from nilpotent groups $(N, +)$ of class 2. We will expand upon these as well as introduce more construction methods. Henry Heatherly [HH] was one of the discoverers of the following construction, and his work with distributive nearrings is reflected in nearly everything we present here on that subject.

(26.4) Theorem. *Let $(N, +, \cdot)$ be a nonabelian group with a proper normal subgroup H such that N/H is cyclic. We also require that N/H be infinite, or that each element of N has finite order. In either event, $(N, +)$ is the additive group of a nontrivial commutative distributive nearring $(N, +, \cdot)$.*

Proof. Suppose $N/H = \langle x + H \rangle$. For $u, u' \in N$, there are integers n, n' such that $u \in nx + H$ and $u' \in n'x + H$. If N/H is infinite, then n and n' are unique, and we define $uu' = (nn')x$. One sees directly that $(N, +, \cdot)$ is a commutative and distributive nearring. If $N/H \cong Z_k$, then we use the fact that x has finite order jk, a multiple of k. So we define $uu' = (nn'j)x$. Again, it is easy to verify that $(N, +, \cdot)$ is a commutative distributive nearring.

The symmetric groups S_n have normal subgroups A_n and $S_n/A_n \cong Z_2$. Hence, (26.4) applies to each S_n, and in particular to the S_3 we used in §2. So the construction for the simplest honest nearring is a model for (26.4), but (26.4) applies to any nonabelian group $(N, +)$ whose elements have finite order and whose derived group $N' \neq N$.

An ideal I of a nearring $(N, +, \cdot)$ must satisfy (1) $(I, +)$ is a normal subgroup of $(N, +)$, (2) $NI \subseteq I$, and (3) $(x + i)y - xy \in I$ for all $x, y \in N$ and for all $i \in I$. If N is distributive, then (3) reduces to $iy \in I$, or $IN \subseteq I$ because of (26.2). Also in distributive nearrings, each $d \in N$ induces two group homomorphisms $x \mapsto xd$ and $x \mapsto dx$. Hence, if d has finite order, then each xd and each dx has order dividing the order of d. So, if d has order p^k for some prime p, then each xd and each dx has order a power of p. With these observations, we can easily prove

(26.5) Theorem. *Suppose $(N, +, \cdot)$ is a finite distributive nearring whose additive group $(N, +)$ is nilpotent. Then each Sylow p-subgroup S_p of N is an ideal of N, and N is the direct sum of all these subgroups S_p.*

Proof. Since $(N, +)$ is finite and nilpotent, we have $N^+ \cong S_{p_1}^+ \oplus \cdots \oplus S_{p_k}^+$ if $|N| = p_1^{s_1} \cdots p_k^{s_k}$ is the prime factorization of $|N|$. Certainly, each $S_{p_i}^+$ is normal in N^+, and if $d \in S_{p_i}$, we have $xd, dx \in S_{p_i}$, hence $NS_{p_i} \subseteq S_{p_i}$ and $S_{p_i}N \subseteq S_{p_i}$. This makes each S_{p_i} an ideal, and finishes our proof.

Let us return now to the *commutator nearrings* $(N, +, \cdot)$ that we made in §2 from a nilpotent group $(N, +)$ of class 2, where $a \cdot b = [a, b] = -a - b + a + b$. For an integer k, define $*_k$ by $a *_k b = k(a \cdot b) = k[a, b]$. Then $a *_k (b + c) =$

$k(a \cdot (b+c)) = k((a \cdot b)+(a \cdot c)) = k(a \cdot b)+k(a \cdot c) = (a *_k b)+(a *_k c)$, and note that we have used (26.2). Also, $(a+b)*_k c = k((a+b) \cdot c) = k((a \cdot c)+(b \cdot c)) = k(a \cdot c) + k(b \cdot c) = (a *_k c) + (b *_k c)$. So each $*_k$ is distributive. Also, $a *_k (b *_k c) = a *_k (k(b \cdot c)) = k(a \cdot [k(b \cdot c)]) = k(k(a \cdot (b \cdot c))) = k^2(a \cdot (b \cdot c)) = k^2((a \cdot b) \cdot c) = k[k((a \cdot b) \cdot c)] = k[(k(a \cdot b)) \cdot c] = (k(a \cdot b)) *_k c = (a *_k b) *_k c$. Hence,

(26.6) Theorem. *For a commutator nearring* $(N, +, \cdot)$ *and an integer* k, *define* $*_k$ *by* $a *_k b = k(a \cdot b)$. *Then* $(N, +, *_k)$ *is a distributive nearring.*

This next result makes several observations about commutator nearrings.

(26.7) Theorem. *Let* $(N, +, \cdot)$ *be a commutator nearring. Then:*
(1) $a \cdot a = 0$ *for each* $a \in N$.
(2) $a \cdot b = -(b \cdot a)$ *for all* $a, b \in N$.
(3) $a \cdot b \cdot c = 0$ *for all* $a, b, c \in N$.
(4) *The derived subgroup* N' *is a subset of the annihilator of* N.
(5) *Every normal subgroup of* $(N, +)$ *is an ideal.*
(6) *Every* N-*module in* N *is an ideal.*
(7) *Every abelian subgroup of* $(N, +)$ *is a trivial ring.*

Proof. First, $a \cdot a = [a, a] = -a - a + a + a = 0$. Next, $b \cdot a = -b - a + b + a$, so $-(b \cdot a) = -a - b + a + b = a \cdot b$. Then $a \cdot (b \cdot c) = a \cdot (-b - c + b + c) = a(-b) + a(-c) + ab + ac = -(ab) + ab + [-(ac)] + ac = 0$. We now have (1), (2), and (3).

The annihilator of N is $\text{Ann}(N) = \{n \in N \mid xn = 0 \text{ for all } x \in N\}$. If $y \in N'$, then $y = \sum \epsilon_i [a_i, b_i]$, where $\epsilon_i = \pm 1$. But $-[a_i, b_i] = -(-a_i - b_i + a_i + b_i) = -b_i - a_i + b_i + a_i = [b_i, a_i]$, so y has the form $y = \sum [c_i, d_i] = \sum c_i d_i$. If $x \in N$, then $xy = \sum x c_i d_i = 0$, by (3). Hence, $N' \subseteq \text{Ann}(N)$, and so (4).

Take a normal subgroup K^+ of N^+. For $x \in N$ and $a \in K$, we have $a \cdot x = -a + (-x + a + x) \in K$ and $x \cdot a = (-x + (-a) + x) + a \in K$. Hence, K is an ideal of N, and thus (5).

Let M be an N-module in N. So M^+ is a subgroup of N^+, and $MN \subseteq M$. For $x \in N$, $a \in M$, we have $a \cdot x = -a - x + a + x \in M$, hence $a + a \cdot x = -x + a + x \in M$. This makes M^+ normal in N^+. By (5), we have (6).

If A^+ is an abelian subgroup of N^+ and $a, b \in A$, then $a \cdot b = -a - b + a + b = 0$. Hence, (7).

The *anticommutative* property $a \cdot b = -(b \cdot a)$ in 2) of (26.7) makes a nearring $(N, +, \cdot)$ distributive. For $(a + b)(-x) = -[(a + b)x] = x(a + b) = (xa) + (xb) = -(ax) + [-(bx)] = a(-x) + b(-x)$. Since $-x \in N$ is arbitrary, as are a and b, we have that $(N, +, \cdot)$ is distributive.

The commutator nearrings are anticommutative. Those provided by (26.4) are commutative. And nearrings which are commutative, or are anticommutative, are distributive. Their symmetry provides them, as would be expected, with nice properties, some of which are explained in

(26.8) Theorem. *Suppose the nearring* $(N, +, \cdot)$ *is commutative or anti-commutative. Then:*
(1) *Every normal subgroup* M^+ *of* N^+ *with* $MN \subseteq M$ *is an ideal.*
(2) *The set of nilpotent elements* $\mathcal{N}(N)$ *is an ideal.*
(3) *We have* $N' \subseteq Ann(N) \subseteq \mathcal{N}(N)$.
(4) *The quotient nearring* $N/\mathcal{N}(N)$ *is a ring without nilpotent elements.*

Proof. To begin with, let us remember that $(N, +, \cdot)$ is distributive. To obtain (1), we need only show $nm \in M$ for all $n \in N$ and for all $m \in M$. But $nm = -(mn)$ and $mn \in M$, so $-(mn) \in M$. We have $\mathcal{N}(N) = \{x \in N \mid x^k = 0 \text{ for some positive integer } k\}$. Before we start to show that $\mathcal{N}(N)$ is an ideal, we should note that in a distributive nearring, $0a = a0 = 0$ and $a(-b) = (-a)b = -(ab)$ for all $a, b \in N$. If N is commutative or anticommutative, then $(ab)^k = a^k b^k$ for all positive integers k. This is certainly true for $k = 1$ and when N is commutative. If it is true for k, then $(ab)^{k+1} = (ab)^k(ab) = a^k b^k ab = a^k[b^k a]b = \{a^k[-(ab^k)]\}b = \{-[a^k(ab^k)]\}b = \{-(a^{k+1}b^k)\}b = (b^k a^{k+1})b = b^k(a^{k+1}b) = b^k[-(ba^{k+1})] = -[b^k(ba^{k+1})] = -[b^{k+1}a^{k+1}] = a^{k+1}b^{k+1}$. So for $n \in N$ and $x, y \in \mathcal{N}(N)$, there are positive integers s and t such that $x^s = y^t = 0$. Hence $(xn)^s = x^s n^s = 0n^s = 0$, so $xn \in \mathcal{N}(N)$, and then $nx = -(xn) \in \mathcal{N}(N)$. Also, $(-x)^s = x^s$ is s is even, and $(-x)^s = -x^s$ is s is odd. In either case, $(-x)^n = 0$.

If N is commutative, then $(x + y)^{s+t} = 0$ follows from (26.2), since the proof is just as in commutative ring theory. If N is anticommutative, then $(x + y)^2 = x^2 + y^2$, and so $(x + y)^{2^k} = x^{2^k} + y^{2^k}$. Just choose k so that $s < 2^k$ and $t < 2^k$. Then $(x + y)^{2^k} = 0$. We could conclude that $\mathcal{N}(N)$ is an ideal if we but had that the subgroup $\mathcal{N}(N)^+$ is normal.

For normality, $(-n + x + n)^2 = (-n + x + n)(-n) + (-n + x + n)x + (-n + x + n)n = (n^2 - xn - n^2) + (-nx + x^2 + nx) + (-n^2 + xn + n^2) = -xn + x^2 + xn = x^2$. Hence, $(-n + x + n)^{2m} = (x^2)^m = x^{2m} = 0$ if $s \leq 2m$. This puts $-n + x + n \in \mathcal{N}(N)$, and so $\mathcal{N}(N)^+$ is normal in N^+, and we are assured of $\mathcal{N}(N)$ being an ideal.

If $x \in N$, and $y = \sum[a_i, b_i] \in N'$, then $xy = \sum x[a_i, b_i]$. But $x[a, b] = x(-a - b + a + b) = x(-a) + x(-b) + xa + xb = x(-a) + xa + x(-b) + xb = 0$. Hence, $N' \subseteq Ann(N)$. If $a \in Ann(N)$, then $a \cdot a = 0$, so $a \in \mathcal{N}(N)$. This gives us (3).

With $N' \subseteq \mathcal{N}(N)$, we know that $(N/\mathcal{N}(N), +)$ is abelian. Distributivity in N carries over to distributivity in $(N/\mathcal{N}(N), +, \cdot)$ Hence, $(N/\mathcal{N}(N), +, \cdot)$ is a ring. Suppose $(x + \mathcal{N}(N))^s = \mathcal{N}(N)$. Then $x^s \in \mathcal{N}(N)$, and so $x \in \mathcal{N}(N)$. Hence, $N/\mathcal{N}(N)$ has no nilpotent elements, and we are finished with the proof of (26.8).

There is a subideal of $\text{Ann}(N)$, if N is a distributive nearring, which is helpful in constructing distributive nearrings.

(26.9) Theorem. *Let* $(N, +, \cdot)$ *be a distributive nearring. Let* $A = \{a \in N \mid an = na = 0 \text{ for every } n \in N\}$. *Then* A *is an ideal of* N. *If* $(N/A, +)$ *is cyclic, then* N *is commutative. If* $m \in x + A$ *and* $n \in y + A$, *then* $mn = xy$.

Proof. Suppose $N/A = \langle x + A \rangle$. For $m, n \in N$, there are integers s and t and elements $a, b \in A$ such that $m = sx + a$ and $n = tx + b$. Hence, $mn = (sx + a)(tx + b) = (sx + a)(tx) = (sx)(tx) = stx^2 = tsx^2 = (tx + b)(sx + a) = nm$. So we will have N commutative if A is an ideal and $(N/A, +)$ is cyclic. Similarly, $m = x + a$ and $n = y + b$ for some $a, b \in A$ imply $mn = (x + a)(y + b) = (x + a)y = xy$. So let us see that A really is an ideal.

For $a, b \in A$, and $n \in N$, we have $(a + b)n = n(a + b) = 0$, and $n(-a) = -(na) = 0 = -(an) = (-a)n$. Since $0 \in A$, we have that $(A, +)$ is a subgroup. If $x \in N$ also, then $(an)x = 0x = 0 = x0 = x(na)$. Hence, $na, an \in A$. Finally, $(-n + a + n)x = (-n)x + ax + nx = -(nx) + nx = 0$, and $x(-n + a + n) = x(-n) + xa + xn = -(xn) + xn = 0$. This puts $-n + a + n \in A$, so $(A, +)$ is a normal subgroup. Sure enough, A is an ideal!

To apply (26.9) for the construction of distributive nearrings, one starts with a normal subgroup A of a group $(N, +)$. If it has a ring $(N/A, +, \cdot)$, then $\cdot$ can be 'lifted' to N as illustrated by the following example.

For the quaternion group Q_8 of order 8, let $Q_8 = \langle a, b \mid 4a = 0, 2a = 2b, -b + a + b = -a \rangle$, and let $A = \{0, b, 2a, 2a + b\} = \langle b \rangle$. If $\{x, y\} \cap A \neq \varnothing$, then define $x \cdot y = 0$. If $\{x, y\} \cap A = \varnothing$, then define $x \cdot y = 2a$. Then one can easily see that $(Q_8, +, \cdot)$ is a nontrivial distributive nearring.

The method we used in §2 to make a nontrivial distributive nearring from a nonabelian group $(G, +)$ and a noncommutative ring $(R, +, \cdot)$ can easily be generalized. Just construct any semidirect product $G \times_\theta R^+$ and define $(g, r) \cdot (g', r') = (0, rr')$. It is easy to see that $(G \times_\theta R^+, +, \cdot)$ is a distributive nearring.

The next result shows that a nearring with a certain curious mapping must be a distributive nearring.

(26.10) Theorem. *If a nearring* $(N, +, \cdot)$ *has a map* $\phi : N \to N$ *such that* (1) ϕ *is a group automorphism of* $(N, +)$, (2) $\phi^2 = 1_N$, *and* (3) $\phi(a \cdot b) = \phi(b) \cdot \phi(a)$ *for all* $a, b \in N$, *then* $(N, +, \cdot)$ *is distributive.*

Proof. For $a, b, c \in N$, $(a + b) \cdot c = \phi \circ \phi[(a + b) \cdot c] = \phi(\phi(c) \cdot \phi(a + b)) =$

$\phi(\phi(c) \cdot [\phi(a) + \phi(b)]) = \phi[\phi(c) \cdot \phi(a) + \phi(c) \cdot \phi(b)] = \phi[\phi(a \cdot c) + \phi(b \cdot c)] = \phi^2(a \cdot c) + \phi^2(b \cdot c) = (a \cdot c) + (b \cdot c)$. That's it!

The complex number field guides us to another construction method for distributive nearrings. If we start with a distributive nearring $(N, +, \cdot)$, construct the direct sum of the groups $N^+ \oplus N^+$, and then define $*$ on $N^+ \oplus N^+$ by $(a, b) * (c, d) = (ac - bd, ad + bc)$, then $(N \oplus N, +, *)$ is also a distributive nearring.

Gary F. Birkenmeier [Bi] provides us with an amusing way to construct distributive nearrings. In §25, we used Jacobson's circle operation ∘ to describe the structure of groups of units. (See (25:13).) Birkenmeier reminds us that we should always be willing to look at things in an alternative way.

(26.11) Theorem. *Let $(R, +, \cdot)$ be a ring in which $xyz = 0$ for all $x, y, z \in R$. If $a \circ b = a + b + ab$, then $(R, \circ, \cdot)$ is a distributive nearring.*

Proof. Since $0 \circ a = 0 + a + 0a = a$ and $a \circ 0 = a + 0 + a0 = a$, we easily obtain that 0 is the identity for ∘. Also, $a \circ (-a + a^2) = a + (-a + a^2) + a(-a + a^2) = a^3 = 0$, and $(-a + a^2) \circ a = -a + a^2 + a + (-a + a^2)a = a^2 - a^2 + a^3 = 0$, so $-a + a^2$ is the inverse of a with respect to ∘. From $a \circ (b \circ c) = a + (b \circ c) + a(b \circ c) = a + b + c + bc + a(b + c + bc) = a + b + c + bc + ab + ac + abc$ and $(a \circ b) \circ c = (a \circ b) + c + (a \circ b)c = a + b + ab + c + (a + b + ab)c = a + b + ab + c + ac + bc + abc$, we easily see that ∘ is associative. Hence, $(R, \circ)$ is a group. Since $a \circ b = a + b + ab$ and $b \circ a = b + a + ba$, we see that ∘ is not commutative if $\cdot$ is not commutative.

We already know that $\cdot$ is associative, so it remains only to see that $\cdot$ is distributive over ∘. Now $a \cdot (b \circ c) = a \cdot (b + c + bc) = (a \cdot b) + (a \cdot c) + a \cdot (b \cdot c) = (a \cdot b) + (a \cdot c)$, and $(a \cdot b) \circ (a \cdot c) = (a \cdot b) + (a \cdot c) + (a \cdot b) \cdot (a \cdot c) = (a \cdot b) + (a \cdot c)$. Hence, $\cdot$ is left distributive over ∘. Similarly, $(a \circ b) \cdot c = (a + b + ab) \cdot c = (a \cdot c) + (b \cdot c)$ and $(a \cdot c) \circ (b \cdot c) = (a \cdot c) + (b \cdot c) + (a \cdot c) \cdot (b \cdot c) = (a \cdot c) + (b \cdot c)$, and so $\cdot$ is right distributive over ∘. This completes our proof.

We wish to provide some examples illustrating (26.11). In the ring of integers modulo 8, $(Z_8, +, \cdot)$, let $R = \{0, 2, 4, 6\}$, a subring of Z_8 in which $xyz = 0$ for arbitrary $x, y, z \in R$. Here, $(R, \circ)$ is an abelian group. In fact, $(R, \circ) \cong Z_2 \oplus Z_2$, so ∘ changes the nature of the group since $R^+ \cong Z_4$. Next, let S be any ring, and let

$$R = \left\{ \begin{pmatrix} 0 & a & b \\ 0 & 0 & c \\ 0 & 0 & 0 \end{pmatrix} \;\middle|\; a, b, c \in S \right\}.$$

Then $(R, +, \cdot)$ is a noncommutative ring for which $xyz = 0$ for arbitrary $x, y, z \in R$. Hence, (26.11) will provide us with nontrivial distributive nearrings.

Exercises. (1) Take some friendly rings for $(S, +, \cdot)$ above, and compute the structure of the groups $(R, \circ)$ obtained from the corresponding R as described above. Can you discover a theorem from your observations?

(2) If $(N, +, \cdot)$ is a distributive nearring with identity, then $(N, +, \cdot)$ is a ring.

26.2. Matrix nearrings

We have already seen that distributive nearrings provide an appropriate medium to construct matrix nearrings in a traditional way, and that nearrings in general are not appropriate for such activity. With a distributive nearring, the resulting matrix nearring was also distributive. But Birkenmeier also taught us to look at things in an alternative way. This is exactly what J. D. P. Meldrum and A. P. J. van der Walt did to develop an expanded point of view for matrix nearrings. In J. H. Meyer's dissertation [Me], one can obtain a better and more thorough introduction to the theory of matrix nearrings, as introduced by Meldrum and van der Walt.

Let $\mathcal{M}(n, R)$ be the n by n matrices over a ring R. Let $[r; i, j]$ be the matrix with an r in the (i, j)-position, and with 0 elsewhere. Certainly, all the $[r; i, j]$ generate all of $\mathcal{M}(n, R)$. Another way to look at $[r; i, j]$ is to consider it as the mapping $\pi_i \circ r^* \circ e_j$, where $\pi_i : R^n \to R$ is defined as the ith projection $(x_1, \ldots, x_n)\pi_i = x_i$, r^* is right multiplication $r^* : R \to R$ defined by $ar^* = ar$, and e_j is the jth insertion $e_j : R \to R^n$ defined by $ye_j = (y\delta_{1j}, y\delta_{2j}, \ldots, y\delta_{nj})$, with $y\delta_{tj} = \begin{cases} 0, & \text{if } t \neq j; \\ y, & \text{if } t = j. \end{cases}$ In this setting, $[r; i, j]$ makes sense for a nearring, that is, $[r; i, j] = \pi_i \circ r^* \circ \epsilon_j$ where $\pi_i : N^n \to N$, $r^* : N \to N$, and $e_j : N \to N^n$ are defined as before. So each $[r; i, j] \in M(N^n)$, the nearring of all mappings of the group N^n into itself. So we let $\mathbf{M}_n(N)$ denote the subnearring of $M(N^n)$ generated by all the $[r; i, j]$, with $r \in N$, and $i, j \in \{1, 2, \ldots, n\}$. If $N = R$, a ring, then $\mathbf{M}_n(N)$ is essentially the traditional ring of all n by n matrices over R. But if N is not a ring, we have no such assurances. An element of $\mathbf{M}_n(N)$ may not even be a matrix, but nevertheless, we call $\mathbf{M}_n(N)$ the *nearring of n by n matrices over N*. This should not trouble the world too much. After all, many things are not what they are called. Leberkäse is neither liver (Leber) nor cheese (Käse), and a rose of Sharon is not a rose.

(26.12) Theorem. *For a nearring $(N, +, \cdot)$. the following are equivalent:*
(1) *N is zero-symmetric.*
(2) *$\mathbf{M}_n(N)$ is zero-symmetric for each positive integer n.*
(3) *$\mathbf{M}_k(N)$ is zero-symmetric for some positive integer k.*

Proof. First, note that $-[r; i, j] = [-r; i, j]$. Hence, the group generated by all the $[r; i, j]$ consists of all finite sums $\sum [r_t; i_t, j_t]$. But we have $(0, \ldots, 0)[r; i, j] = (0, \ldots, 0)$, so we get the equation $(0, \ldots, 0) \sum [r_t; i_t, j_t] =$

$(0, \ldots, 0) \left[\sum [r_s; i_s, j_s] \right] \left[\sum [r'_t; u_t, v_t] \right] = (0, \ldots, 0)$. This means that (1) implies (2), and certainly (2) implies (3). Suppose there is a k such that $\mathbf{M}_k(N)$ is zero-symmetric. If $0r \neq 0$ for some $r \neq 0$, then $(0, \ldots, 0)[r; i, j] = ((0r)\delta_{1j}, \ldots, (0r)\delta_{kj})$ and $(0r)\delta_{jj} = 0r \neq 0$, contrary to $\mathbf{M}_k(N)$ being zero-symmetric. Hence, (3) implies (1).

(26.13) Lemma. *For a nearring* $(N, +, \cdot)$ *with identity 1, the following are equivalent:*

(1) $r \in N$ *is distributive.*

(2) *Each* $[r; i, j]$ *is distibutive, for* $i, j \in \{1, 2, \ldots, n\}$.

(3) *There exist* $s, t \in \{1, 2, \ldots, n\}$ *so that* $[r; s, t]$ *is distributive.*

Proof. Suppose r is distributive and that $i, j \in \{1, 2, \ldots, n\}$. Then

$$
\begin{aligned}
[(x_1, \ldots, x_n) + (y_1, \ldots, y_n)][r; i, j] &= (x_1 + y_1, \ldots, x_n + y_n)[r; i, j] \\
&= ((x_1 + y_1)r\delta_{1j}, \ldots, (x_1 + y_1)r\delta_{nj}) \\
&= (x_1 r\delta_{1j} + y_1 r\delta_{1j}, \ldots, x_1 r\delta_{nj} + y_1 r\delta_{nj}) \\
&= (x_1 r\delta_{1j}, \ldots, x_1 r\delta_{nj}) + (y_1 r\delta_{1j}, \ldots, y_1 r\delta_{nj}) \\
&= (x_1, \ldots, x_n)[r; i, j] + (y_1, \ldots, y_n)[r; i, j].
\end{aligned}
$$

Since $[r; i, j]$ defines an endomorphism of $(N^n, +)$, we must conclude that $[r; i, j]$ is a distibutive element. Since $i, j \in \{1, 2, \ldots, n\}$ are arbitrary, we have (1) implies (2), and certainly (2) implies (3).

Now, suppose $[r; s, t]$ is distributive in $\mathbf{M}_n(N)$. Then $([1; 1, s] + [1; 2, s]) \circ [r; s, t] = [1; 1, s] \circ [r; s, t] + [1; 2, s] \circ [r; s, t] = [r; 1, t] + [r; 2, t]$, and so

$$
\begin{aligned}
(a_1, a_2, \ldots, a_n) &\left([r; 1, t] + [r; 2, t] \right) \\
&= ((a_1 r)\delta_{1t}, \ldots, (a_1 r)\delta_{nt}) + ((a_2 r)\delta_{1t}, \ldots, (a_2 r)\delta_{nt}) \\
&= ((a_1 r)\delta_{1t} + (a_2 r)\delta_{1t}, \ldots, (a_1 r)\delta_{nt} + (a_2 r)\delta_{nt})
\end{aligned}
$$

is equal to

$$
\begin{aligned}
(a_1, a_2, \ldots, a_n) &\left([1; 1, s] + [1; 2, s] \right) \circ [r; s, t] \\
&= [(a_1 \delta_{1s}, \ldots, a_1 \delta_{ns}) + (a_2 \delta_{1s}, \ldots, a_2 \delta_{ns})][r; s, t] \\
&= (a_1 \delta_{1s} + a_2 \delta_{1s}, \ldots, a_1 \delta_{ns} + a_2 \delta_{ns})[r; s, t] \\
&= ([a_1 \delta_{ss} + a_2 \delta_{ss}]r\delta_{1t}, \ldots, [a_1 \delta_{ss} + a_2 \delta_{ss}]r\delta_{nt}) \\
&= ((a_1 + a_2)r\delta_{1t}, \ldots, (a_1 + a_2)r\delta_{nt}),
\end{aligned}
$$

from which follows $(a_1 r)\delta_{tt} + (a_2 r)\delta_{tt} = (a_1 + a_2)r\delta_{tt}$, or $(a_1 r) + (a_2 r) = (a_1 + a_2)r$. Since $a_1, a_2 \in N$ are arbitrary, we can conclude that r is distributive in N.

Exercises. (1) For a nearring $(N, +, \cdot)$ with identity, $\mathbf{M}_n(N)$ is distributively generated if and only if N is distributively generated.

(2) Choose some ot(her properties for a nearring N and see if N having this property is equivalent to $\mathbf{M}_n(N)$ having the same property, for example, suppose N is an abstract affine nearring.

Let us refer to the generators $[a; i, j]$ as *first generation formulae for elements of* $\mathbf{M}_n(N)$, and assign them *length* $l[a; i, j] = 1$. Let F_1 be all first generation formulae for elements of $\mathbf{M}_n(N)$. Having the $(k-1)$th generation formulae for elements of $\mathbf{M}_n(N)$, we construct the set F_k of kth *generation formulae for elements of* $\mathbf{M}_n(N)$ as all finite sums $\sum_{i=1}^{m} \epsilon_i X_i$ and all finite products $X_1 X_2 \cdots X_m$, where each $\epsilon_i = \pm 1$, each $X_i \in F_{k-1}$, and $m \in \{1, 2, \ldots, \}$. We assign *lengths* $l(\sum_{i=1}^{m} \epsilon_i X_i) = \sum_{i=1}^{m} l(X_i)$ and $l(X_1 \cdots X_m) = \sum_{i=1}^{m} l(X_i)$ to elements in F_k. Let $\mathcal{F} = \cup_{k=1}^{\infty} F_k$. The set $\mathcal{F}$ is the set of *formulae* for elements in $\mathbf{M}_n(N)$. So each $X \in \mathcal{F}$ is really a formula for some $A(X) \in \mathbf{M}_n(N)$, and each $A \in \mathbf{M}_n(N)$ is defined by some formula $F(A) \in \mathcal{F}$. (This is analogous to $(x+1)^2$ and $x^2 + 2x + 1$ being two different formulae for the same function $f : \mathbf{R} \to \mathbf{R}$.)

With each $A \in \mathbf{M}_n(N)$, there is a formula $F(A) \in \mathcal{F}$, and with each $F(A)$ there is a length $l(F(A))$. Define the *weight of* A as

$$w(A) = min\{l(F(A)) \mid F(A) \text{ is a formula for } A\}.$$

These concepts will permit an induction proof of

(26.14) Theorem. *If R is a right ideal of a nearring N, then R^n is an* $\mathbf{M}_n(N)$*-ideal of the* $\mathbf{M}_n(N)$*-module N^n.*

Proof. Clearly, R^n is a right ideal of N^n, and N^n is an $\mathbf{M}_n(N)$-module. For $r \in R^n$, $x \in N^n$, and $A \in \mathbf{M}_n(N)$, we need to show that $(x+r)A - xA \in R^n$. For $A = [a; i, j]$, that is, a formula of length 1, we have $(x+r)A - xA = r'$ where $r'_k = 0$ if $k \neq j$, and $r'_j = (x_i + r_i)a - x_i r \in R$, so $(x+r)A - xA \in R^n$.

Suppose $(x+r)A' - xA' \in \mathbf{M}_n(N)$ for all $A' \in \mathbf{M}_n(N)$ with weight $w(A')$ satisfying $1 \leq w(A') < k$, and suppose $w(A) = k$. Then $A = B + C$ or $A = BC$ with $w(B), w(C) \in \{1, 2, \ldots, k-1\}$. First consider $A = B + C$. Then $(x+r)(B+C) - x(B+C) = (x+r)B + (x+r)C - xC - xB = (x+r)B - xB + r' \in R^n$ since $(x+r)C - xC$, $(x+r)B - xB \in R^n$ by induction hypothesis, and since R^n is a right ideal of N^n.

Next consider $A = BC$. From $(x+r)B - xB \in R^n$, we have $(x+r)B = r'_B + xB = xB + r_B$ for some $r'_B, r_B \in R^n$. Hence, $(x+r)BC - xBC = (xB + r_B)C - (xB)C \in R^n$. This completes the proof of (26.14).

(26.15) Corollary. *For R as in (26.14), $(R^n : N^n)$ is an ideal of* $\mathbf{M}_n(N)$.

Proof. Apply (13.26).

So, for any ideal A of a nearring N, we have two ideals in any $\mathbf{M}_n(N)$, namely, $\mathbf{A}^* = (A^n : N^n)$ and the ideal $\mathbf{A}^\circ$ generated by the set of all $[a; i, j]$ where $a \in A$, and $i, j \in \{1, 2, \ldots, n\}$. We first make a few observations.

(26.16) Proposition. *For an ideal A of a nearring N with identity:*
(1) $\mathbf{A}^\circ \subseteq \mathbf{A}^*$.
(2) *The map* $A \mapsto \mathbf{A}^\circ$ *is injective.*
(3) *The map* $A \mapsto \mathbf{A}^*$ *is injective.*

Proof. For $x \in N^n$ and $a \in A$, $x[a; i, j] = (x_i a \delta_{1j}, \ldots, x_i a \delta_{nj})$, and each $x_i a \delta_{kj}$ is either 0 or $x_i a$. Since $a \in A$ and A is an ideal, we have $x_i a \in A$. Hence, $[a; i, j] \in \mathbf{A}^*$, forcing the conclusion that $\mathbf{A}^\circ \subseteq \mathbf{A}^*$.

Suppose we have two distinct ideals A and B of N, with $b \in B \setminus A$. With $[b; 1, 1] \in \mathbf{B}^\circ \subseteq \mathbf{B}^*$, and $(1, 0, \ldots, 0)[b; 1, 1] = (b, 0, \ldots, 0) \in B^n \setminus A^n$, we have $[b; 1, 1] \in \mathbf{B}^\circ \setminus \mathbf{A}^\circ$ and $[b; 1, 1] \in \mathbf{B}^* \setminus \mathbf{A}^*$, and so our proof is complete.

But can we have $\mathbf{A}^\circ \subset \mathbf{A}^*$? An affirmative answer comes from a trivial nearring $(Z_4, +, \cdot)$ with ideal $A = \{0, 2\}$, where $1 \cdot x = 3 \cdot x = x$ and $0 \cdot x = 2 \cdot x = 0$ for all $x \in Z_4$. Consider $X = [3; 1, 1] + [3; 2, 1] + ([1; 1, 1] + [3; 2, 1])[1; 1, 1]$. So $(a, b)X = (a3 + b3 + (a1 + b3)1, 0)$, and $a3 + b3 + (a1 + b3)1 \in \{0, 2\}$ for all $a, b \in Z_4$. Hence, $X \in \mathbf{A}^*$, and we would like to show that $X \notin \mathbf{A}^\circ$. First notice that $(a, b)X = (a', 0)$, that is, the second coordinate of any $(a, b)X$ is 0, and the first coordinate is an element $a' \in A$. Also note that $(1, 0)X = (0, 1)X = (0, 0)$, and $(1, 1)X = (2, 0)$. We shall endeavour to show that $\mathbf{A}^\circ$ has no such element T with the properties that (1) $(a, b)T = (a', 0)$ with $a' \in A$, (2) $(1, 0)T = (0, 1)T = (0, 0)$, and (3) $(1, 1)T = (2, 0)$.

Suppose $T \in \mathbf{A}^\circ$ and has the properties of X referred to above. But $(1, 0)[2; 1, 1] = (2, 0)$, $(0, 1)[2; 2, 1] = (2, 0)$, and $(1, 1)\{[2; 1, 1] + [2; 2, 1]\} = (0, 0)$. Hence, $T \notin \{[2; 1, 1], [2; 2, 1], [2; 1, 1] + [2; 2, 1], [0; 1, 1]\} = \mathbf{S}$. Our efforts now will be to show that if $(a, b)Y = (a', 0)$ and $Y \in \mathbf{A}^\circ$, then $Y \in \mathbf{S}$. Let $\mathbf{A} = \{[a_{11}; 1, 1] + [a_{12}; 1, 2] + [a_{21}; 2, 1] + [a_{22}; 2, 2] \mid$ each $a_{ij} \in \{0, 2\} = A\}$. If we show that $\mathbf{A} = \mathbf{A}^\circ$, then for $Y \in \mathbf{A}$, $(a, b)Y = (aa_{11}, 0) + (0, aa_{12}) + (ba_{21}, 0) + (0, ba_{22}) = (aa_{11} + ba_{21}, aa_{12} + ba_{22})$, so if $(a, b)Y = (a', 0)$, with $a' \in A$, and for all $(a, b) \in Z_4^2$, then $a_{12} = a_{22} = 0$, and so $Y \in \mathbf{S}$. With $\mathbf{A} \subseteq \mathbf{A}^\circ$, we need to show that $\mathbf{A}$ is an ideal, and then we will have $\mathbf{A} = \mathbf{A}^\circ$.

Our first step will be to demonstrate that $\mathbf{M}_2(Z_4)\mathbf{A} \subseteq \mathbf{A}$, but the reader should note immediately that $\mathbf{A}$ is a (normal) subgroup of $\mathbf{M}_2(Z_4)$. Our demonstration will be analogous to the proof of (26.14), that is, by induction on the weight of $Y \in \mathbf{M}_2(Z_4)$. For $Y = [r; i, j]$ of weight 1, note that
$$[r; i, j][a; s, t] = \begin{cases} [0; 1, 1], & \text{if } s \neq j; \\ [ra; i, t], & \text{if } s = j. \end{cases}$$ Hence, with $ra \in A$ for $r \in Z_4$ and $a \in A$, we have $Y\mathbf{A} \subseteq \mathbf{A}$. Assume $Y'\mathbf{A} \subseteq \mathbf{A}$ for each Y' with weight

$w(Y') < k$. Suppose $w(Y) = k$. Then $Y = Y_1 + Y_2$ or $Y = Y_1 Y_2$ with each $w(Y_i) < k$. If $Y = Y_1 Y_2$, then $Y\mathbf{A} = Y_1 Y_2 \mathbf{A} \subseteq Y_1 \mathbf{A} \subseteq \mathbf{A}$. If $Y = Y_1 + Y_2$ and $W = [a_{11}; 1, 1] + [a_{12}; 1, 2] + [a_{21}; 2, 1] + [a_{22}, 2, 2] \in \mathbf{A}$, then since each $[a_{ij}; i, j]$ is distibutive (since $a \in A$ is distributive, so (26.13) applies), $YW = Y[a_{11}; 1, 1] + Y[a_{12}; 1, 2] + Y[a_{21}; 2, 1] + Y[a_{22}; 2, 2]$. But $Y[a_{ij}; i, j] = (Y_1 + Y_2)[a_{ij}; i, j] = Y_1[a_{ij}; i, j] + Y_2[a_{ij}; i, j] \in \mathbf{A}$ for each (i, j). Hence, $YW \in \mathbf{A}$. By induction, we are assured that $\mathbf{M}_2(Z_4)\mathbf{A} \subseteq \mathbf{A}$.

Do we have $(Y + B)Z - YZ \in \mathbf{A}$ for all $Y, Z \in \mathbf{M}_2(Z_4)$ and for all $B \in \mathbf{A}$? First, let us make an induction proof on $w(Y)$ to show that

$$(x + a)Y = xY \qquad (26:6)$$

for all $x \in Z_4^2$, for all $a \in A^2$, and for all $Y \in \mathbf{M}_2(Z_4)$. If $w(Y) = 1$, so $Y = [y; i, j]$, then $(x+a)Y = (x+a)[y; i, j] = ((x_i + a_i)y\delta_{1j}, (x_i + a_i)y\delta_{2j}) = (x_i y \delta_{1j}, x_i y \delta_{2j}) = x[y; i, j] = xY$, since $(x_i + a_i)y = x_i y$ for all $x_i, y \in Z_4$ and all $a_i \in A = \{0, 2\}$. Assume (26:6) for Y' with $w(Y') < k$, and suppose $w(Y) = k$. Then $Y = Y_1 + Y_2$ or $Y = Y_1 Y_2$ with each $w(Y_i) < k$. If $Y = Y_1 Y_2$, then $(x + a)Y = (x + a)Y_1 Y_2 = xY_1 Y_2 = xY$. If $Y = Y_1 + Y_2$, then $(x + a)Y = (x + a)Y_1 + (x + a)Y_2 = xY_1 + xY_2 = x(Y_1 + Y_2) = xY$. Hence, (26:6) is valid for all $x \in Z_4^2$, all $a \in A^2$, and for all $Y \in \mathbf{M}_2(Z_4)$.

Returning to consideration of $(Y + B)Z - YZ$, we take $x \in Z_4^2$, and note that $x(Y+B)Z - xYZ = (xY + xB)Z - (xY)Z = (xY)Z - (xY)Z = 0$, since $xB \in A^2$ for all $x \in Z_4^2$ and for all $B \in \mathbf{A}$. This means that $(Y + B)Z - YZ$ is the additive identity of $\mathbf{M}_2(Z_4)$, hence is an element of $\mathbf{A}$. Now we have that $\mathbf{A}$ is an ideal, and so $\mathbf{A} = \mathbf{A}^\diamond$.

We have noted that any element $Y \in \mathbf{A} = \mathbf{A}^\diamond$ which satisfies $(a, b)Y = (a', 0)$, with $a' \in A$, and for all $(a, b) \in Z_4^2$, must also satisfy $Y \in \mathbf{S}$, and no element $T \in \mathbf{S}$ satisfies (2) and (3) above, so we have no element $T \in \mathbf{A}^\diamond = \mathbf{A}$ which satisfies (1), (2), and (3). But yet X does satisfy (1), (2), and (3), and $X \in \mathbf{A}^*$. What can we do but conclude that $\mathbf{A}^\diamond \subset \mathbf{A}^*$?

It is natural to wonder if $\mathbf{A}^\diamond = \mathbf{A}^*$ if our nearring N has an identity. In Meyer's dissertation [Me], he shows that if $(N, +, \cdot) = (Z_0[x], +, \circ)$, the nearring of polynomials over the integers Z with constant term 0, then with A being all the polynomials in $Z_0[x]$ having even coefficients, we have that A is an ideal of N, N has an identity x, but yet $\mathbf{A}^\diamond \subset \mathbf{A}^*$. The proof is somewhat analogous to our proof here, but more complex in detail.

Let us review some of what we have just seen. For an ideal A of a nearring N, we have, for any positive integer n, two ways to obtain an ideal in $\mathbf{M}_n(N)$, namely, $\mathbf{A}^\diamond$ and $\mathbf{A}^*$, and we have $\mathbf{A}^\diamond \subseteq \mathbf{A}^*$, and it may be that $\mathbf{A}^\diamond \subset \mathbf{A}^*$. This time, let us start with an ideal $\mathbf{A}$ of some $\mathbf{M}_n(N)$. Define

$$\mathbf{A}_* = \{xU\pi_j \mid x \in N^n, \ U \in \mathbf{A}, \ 1 \leq j \leq n\}.$$

Our next milestone will be to show that $\mathbf{A}_*$ is an ideal in N, and then to prove that $(\mathbf{A}_*)^\diamond \subseteq \mathbf{A} \subseteq (\mathbf{A}_*)^*$. We will need a few tools to do this.

(26.17) Definition. For $1 \leq k \leq n$, let

$$\mathbf{R}_k = \{[r_1; k, 1] + \cdots + [r_n; k, n] \mid r_1, \ldots, r_n \in N\},$$

and think of $\mathbf{R}_k$ as the set of all *kth row matrices of* $\mathbf{M}_n(N)$.

(26.18) Theorem. *For* $1 \leq k \leq n$, $\mathbf{R}_k$ *is a right* $\mathbf{M}_n(N)$-*submodule of the* $\mathbf{M}_n(N)$-*module* $\mathbf{M}_n(N)$.

Proof. To begin, let $U = [u_1; k, 1] + \cdots + [u_n; k, n]$ and $V = [v_1; k, 1] + \cdots + [v_n; k, n]$. For $x = (x_1, \ldots, x_n) \in N^n$, $x(U - V) = xU - xV = (x_k u_1, \ldots, x_k u_n) - (x_k v_1, \ldots, x_k v_n) = (x_k(u_1 - v_1), \ldots, x_k(u_n - v_n)) = x\{[u_1 - v_1; k, 1] + \cdots + [u_n - v_n; k, n]\}$. Hence, $U - V = [u_1 - v_1; k, 1] + \cdots + [u_n - v_n; k, n]$, and so $U - V \in \mathbf{R}_k$, which makes $\mathbf{R}_k$ a subgroup.

Our next step is to show that $\mathbf{R}_k \mathbf{M}_n(N) \subseteq \mathbf{R}_k$, or $UX \in \mathbf{R}_k$ for arbitrary $U \in \mathbf{R}_k$ and arbitrary $X \in \mathbf{M}_n(N)$. Our proof will be by induction on $w(X)$. If $X = [r; i, j]$, that is, $w(X) = 1$, then $UX = [u_i r; k, j]$. To see this, calculate $(x_1, \ldots, x_n)U[r; i, j] = (x_k u_1, \ldots, x_k u_n)[r; i, j] = ((x_k u_i r)\delta_{1j}, \ldots, (x_k u_i r)\delta_{nj}) = (x_1, \ldots, x_n)[u_i r; k, j]$. With $U[r; i, j] \in \mathbf{R}_k$, we have $\mathbf{R}_k X \subseteq \mathbf{R}_k$ for $X \in \mathbf{M}_n(N)$ with $w(X) = 1$. Assume $\mathbf{R}_k X' \subseteq \mathbf{R}_k$ for $X' \in \mathbf{M}_n(N)$ with $w(X') < k$ and take $X \in \mathbf{M}_n(N)$ with $w(X) = k$.

Now $X = X_1 + X_2$ or $X = X_1 X_2$ with each $w(X_i) < k$. So if $X = X_1 + X_2$, then $UX = U(X_1 + X_2) = UX_1 + UX_2$, and since each $UX_i \in \mathbf{R}_k$, we have $UX \in \mathbf{R}_k$. If $X = X_1 X_2$, then $UX = (UX_1)X_2$. By induction hypothesis, $UX_1 \in \mathbf{R}_k$ and so $(UX_1)X_2 \in \mathbf{R}_k$. Hence, $UX \in \mathbf{R}_k$. We now have $\mathbf{R}_k X \subseteq \mathbf{R}_k$ for each $X \in \mathbf{M}_n(N)$, and so $\mathbf{R}_k$ is a right $\mathbf{M}_n(N)$-submodule of the $\mathbf{M}_n(N)$-module $\mathbf{M}_n(N)$.

(26.19) Lemma. *Suppose N is a nearring with identity. Let* $\mathbf{L}$ *be an* $\mathbf{M}_n(N)$-*comodule in* $\mathbf{M}_n(N)$. *Then* $N^n \mathbf{L} = (1, 0, \ldots, 0)\mathbf{L}$.

Proof. Clearly, $(1, 0, \ldots, 0)\mathbf{L} \subseteq N^n \mathbf{L}$. Let $a = (a_1, \ldots, a_n) \in N^n$. Then $a\mathbf{L} = (1, 0, \ldots, 0)\{[a_1; 1, 1] + \cdots + [a_n; 1, n]\}\mathbf{L} \subseteq (1, 0, \ldots, 0)\mathbf{L}$ since $\mathbf{L}$ is an $\mathbf{M}_n(N)$-comodule in $\mathbf{M}_n(N)$, hence $\{[a_1; 1, 1] + \cdots + [a_n; 1, n]\}\mathbf{L} \subseteq \mathbf{L}$.

(26.20) Proposition. *Let N be a nearring with identity, and let* $\mathbf{A}$ *be an ideal of* $\mathbf{M}_n(N)$. *Then* $a \in \mathbf{A}_*$ *if and only if* $[a; 1, 1] \in \mathbf{A}$.

Proof. If $[a; 1, 1] \in \mathbf{A}$, then $(1, 0, \ldots, 0)[a; 1, 1]\pi_1 = a$, so $a \in \mathbf{A}_*$. The converse is not so easy.

Suppose $a \in \mathbf{A}_*$. Then there are a $U \in \mathbf{A}$, $x \in N^n$, and $j \in \{1, \ldots, n\}$ such that $a = xU\pi_j$. But $xU \in N^n \mathbf{A} = (1, 0, \ldots, 0)\mathbf{A}$ by (26.19). So, there is a $V \in \mathbf{A}$ such that $(1, 0, \ldots, 0)V = xU$. Hence, we get the equation $xU = (1, 0, \ldots, 0)\{[1; 1, 1] + [0; 1, 2] + \cdots + [0; 1, n]\}V$. Since $\mathbf{R}_1$ is a right $\mathbf{M}_n(N)$-submodule of $\mathbf{M}_n(N)$, we apply (26.18) to obtain $a_1, \ldots, a_n \in N$ such that $\{[1; 1, 1] + [0; 1, 2] + \cdots + [0; 1, n]\}V = [a_1; 1, 1] + \cdots + [a_n; 1, n]$. Hence,

$(1, 0, \ldots, 0)\{[1; 1, 1] + [0; 1, 2] + \cdots + [0; 1, n]\}V = (1, 0, \ldots, 0)\{[a_1; 1, 1] + \cdots + [a_n; 1, n]\} = (a_1, a_2, \ldots, a_n)$. Since $(a_1, a_2, \ldots, a_n)\pi_j = a_j$ and $xU\pi_j = a$, it follows that $a = a_j$.

Our next step is to prove that

$$\{[1; 1, 1] + [0; 1, 2] + \cdots + [0; 1, n]\}V[1; j, 1] = [a; 1, 1]. \tag{26:7}$$

Certainly, $(x_1, \ldots, x_n)[a; 1, 1] = (x_1 a, 0, \ldots, 0)$. Also,

$$
\begin{aligned}
(x_1, \ldots, x_n)&\{[1; 1, 1] + [0; 1, 2] + \cdots + [0; 1, n]\}V[1; j, 1] \\
&= (x_1, \ldots, x_n)\{[a_1; 1, 1] + \cdots + [a_n; 1, n]\}[1; j, 1] \\
&= (x_1 a_1, x_1 a_2, \ldots, x_1 a_n)[1; j, 1] = (x_1 a_j, 0, \ldots, 0) \\
&= (x_1 a, 0, \ldots, 0).
\end{aligned}
$$

With $V \in \mathbf{A}$, and $\mathbf{A}$ being an ideal, we have $[a_1; 1, 1] + [a_2; 1, 2] + \cdots + [a_n; 1, n] = \{[1; 1, 1] + [0; 1, 2] + \cdots + [0; 1, n]\}V \in \mathbf{A}$. By (26.13), $[1; j, 1]$ is a distributive element of $\mathbf{M}_n(N)$. If we show that $UD \in \mathbf{A}$ for any distributive element $D \in \mathbf{M}_n(N)$ and any $U \in \mathbf{A}$, then we will have the function of (26:7) in $\mathbf{A}$. Since $\mathbf{A}$ is an ideal, we have $(X + U)D - XD = XD + UD - XD \in \mathbf{A}$ for all $X \in \mathbf{M}_n(N)$, for all $U \in \mathbf{A}$, and for each distributive element D of $\mathbf{M}_n(N)$. Since $(\mathbf{A}, +)$ is a normal subgroup, we have $-XD + (XD + UD - XD) + XD = UD \in \mathbf{A}$. Hence, $[a; 1, 1] \in \mathbf{A}$, and our proof of (26.20) is complete.

(26.21) Proposition. *Suppose* $1 \in N$, *and let* $\mathbf{M}_{11} = \{[a; 1, 1] \mid a \in N\}$. *Then* $\mathbf{M}_{11}$ *is a subnearring of* $\mathbf{M}_n(N)$ *which is isomorphic to* N.

Proof. The map $f : N \to \mathbf{M}_{11} \subseteq \mathbf{M}_n(N)$ defined by $f(a) = [a; 1, 1]$ is easily seen to be an isomorphism, hence an embedding of N into $\mathbf{M}_n(N)$.

(26.22) Corollary. *Every nearring* N *can be embedded into each matrix nearring* $\mathbf{M}_n(N)$.

(26.23) Theorem. *For a nearring* N *with identity and for any ideal* $\mathbf{A}$ *of* $\mathbf{M}_n(N)$, *we have that* $\mathbf{A}_*$ *is an ideal of* N.

Proof. With an ideal $\mathbf{A}$ of $\mathbf{M}_n(N)$ and a subnearring $\mathbf{M}_{11}$ of $\mathbf{M}_n(N)$ as in (26.21), we have that $\mathbf{A} \cap \mathbf{M}_{11}$ is an ideal of $\mathbf{M}_{11}$. By (26.21), $\mathbf{A} \cap \mathbf{M}_{11}$ is isomorphic to an ideal I of N, and, by (26.20) and (26.21), we obtain $I = \mathbf{A}_*$.

(26.24) Theorem. *For a nearring* N *with identity, and an ideal* $\mathbf{A}$ *of* $\mathbf{M}_n(N)$, *we have*

$$(\mathbf{A}_*)^\circ \subseteq \mathbf{A} \subseteq (\mathbf{A}_*)^*.$$

Proof. Since $(\mathbf{A}_*)^{\circ}$ is the ideal generated by all $[a; i, j]$, $a \in \mathbf{A}_*$, and since $[1; i, 1][a; 1, 1][1; 1, j] = [a; i, j]$, we have that $(\mathbf{A}_*)^{\circ}$ is the ideal generated by all $[a; 1, 1]$, $a \in \mathbf{A}_*$. (Recall that $[1; 1, j]$ is a distributive element, so not only is $[1; i, 1][a; 1, 1]$ in the ideal generated by all $[a; 1, 1]$, $a \in \mathbf{A}_*$, but so is $[1; i, 1][a; 1, 1][1; 1, j]$.) By (26.20), $a \in \mathbf{A}_*$ if and only if $[a; 1, 1] \in \mathbf{A}$. Hence, $(\mathbf{A}_*)^{\circ} \subseteq \mathbf{A}$.

Suppose $U \in \mathbf{A}$ and $U \notin (\mathbf{A}_*)^*$. Then there is an $a \in N^n$ such that $aU \notin \mathbf{A}_*^n$, and so $aU\pi_j \notin \mathbf{A}_*$ for some $j \in \{1, \ldots, n\}$. But then $U \notin \mathbf{A}$.

If the reader did not experience déjà vu with this last theorem, then he should refer to (17.8), (18.2), and (18.30).

Perhaps the major contribution of this development of matrix nearrings so far is the fresh approach to the subject. Heretofore, it was the general opinion that the idea of a matrix nearring was necessarily restricted to distributive nearrings. But now, with the approach developed by Meldrum and van der Walt, any nearring is suitable. There has been a major application of these ideas, however. In [B, 1971], Gerhard Betsch develops a beautiful theory of 2-primitive nearrings. One of his key theorems is about nearrings with identity and suitable idempotent. If the nearring is not a ring, then, theoretically, one of two conditions must theoretically occur. Until this development of matrix nearrings, it was unknown if each theoretical condition could actually occur. In J. H. Meyer's dissertation [Me], one finds examples, constructed from matrix nearrings, showing that each alternative is indeed possible. Another application of this development of matrix nearrings is the expansion of the idea of a group nearring, but first, we need to take a fresh look at the classical idea of a group ring.

26.3. Group nearrings

Let R be a ring with identity 1. The ring $\mathcal{M}(n, R)$ of n by n matrices over R is generated by all the $[r; i, j]$, and each $[r; i, j]$ defines a mapping of R^n into R^n. These ideas can readily be extended to the group $R^{(I)}$ of all mappings with finite support of an index set I into the ring R. Here $(x_i) \in R^{(I)}$ defines a function $\mathbf{x} : I \to R$ where $\mathbf{x}(i) = x_i$, and all but finitely many of the values x_i are zero. If $s, t \in I$, we have $(x_i)[r; s, t] = (x_s r \delta_{it})$ where $x_s r \delta_{it} = \begin{cases} x_s r, & \text{if } i = t; \\ 0, & \text{if } i \neq t. \end{cases}$ Let $\mathcal{M}(I, R)$ be the nearring generated by all the $[r; s, t]$ in the zero-symmetric nearring $M_0(R^{(I)})$. Of course, as long as R is a ring, then $\mathcal{M}(I, R)$ will also be a ring, but that will all change soon. We also call $\mathcal{M}(I, R)$ a *matrix ring*, the *matrix ring of all matrices over R for I*.

Along with our ring R with identity 1, suppose we have a group $(G, \cdot)$ with identity e. We wish to show how the group ring $R[G]$ can be thought of as a subring of a matrix ring $\mathcal{M}(G, R)$. For bookkeeping reasons, we index the elements of G with an index set I, so $G = \{g_i \mid i \in I\}$ and if $i, j \in I$ and $i \neq j$, then $g_i \neq g_j$. For $\mathbf{x} \in R^{(G)}$, we write $\mathbf{x} = (x_i)$ where $\mathbf{x}(g_i) = x_i \in R$.

One often thinks of the elements of a group ring $R[G]$ as the formal sums $\sum_i x_i g_i$, with $x_i \in R$ and $g_i \in G$, where it is understood that all but finitely many of the values x_i are zero. So we can readily identify $\sum_i x_i g_i \in R[G]$ with $(x_i) \in R^{(G)}$.

Each $xg \in R[G]$ defines a map $\langle x, g \rangle \in M_0(R^{(G)})$. Specifically, from $(\sum_i x_i g_i)(xg) = \sum_i (x_i x)(g_i g) = \sum_i (x_{\gamma(i)} x) g_i$, where $g_i g = g_{\gamma^{-1}(i)}$ defines a permutation γ^{-1} on I, we are motivated to define

$$(x_i)\langle x, g \rangle = (x_{\gamma(i)} x), \qquad (26:8)$$

where $g_{\gamma(i)} g = g_i$ defines a permutation γ of the index set I. With all but finitely many of the x_i being zero, the same is true for the values $x_{\gamma(i)} x$. Certainly, xg, and $x'g'$ represent the same element of $R[G]$ if and only if $x = x'$ and $g = g'$.

Now, with $\sum_i x_i g_i \in R[G]$, we define the map $\sum_i \langle x_i, g_i \rangle \in M_0(R^{(G)})$, and one quickly sees that $\sum_i x_i g_i \leftrightarrow \sum_i \langle x_i, g_i \rangle$ defines an isomorphism between $R[G]$ and a subring of the nearring $M_0(R^{(G)})$.

Our next step is to show that each $\langle x, g \rangle \in \mathcal{M}(G, R)$, and so it will follow that all $\sum_i \langle x_i, g_i \rangle \in \mathcal{M}(G, R)$. In this way, we will think of the group ring $R[G]$ as being embedded in the matrix ring $\mathcal{M}(G, R)$. First, we have $(x_i)\langle x, g \rangle = (x_{\gamma(i)} x) = \mathbf{y}$ where $\mathbf{y}(i) = x_{\gamma(i)} x$. Also, $(x_i) \sum_k [x; \gamma(k), k] = \sum_k (x_i)[x; \gamma(k), k] = \sum_k (x_{\gamma(k)} x \delta_{ik}) = (\sum_k x_{\gamma(k)} x \delta_{ik}) = \mathbf{z}$ where $\mathbf{z}(i) = \sum_k x_{\gamma(k)} x \delta_{ik} = x_{\gamma(i)} x$. Since $\mathbf{y} = \mathbf{z}$, we conclude that

$$\langle x, g \rangle = \sum_k [x; \gamma(k), k]. \qquad (26:9)$$

But does $\sum_k [x; \gamma(k), k]$ really have any meaning if G, and, consequently, I are infinite? As above, we define $(x_i) \sum_k [x; \gamma(k), k] = (\sum_k x_{\gamma(k)} x \delta_{ik}) = \mathbf{z}$ where $\mathbf{z}(i) = \sum_k x_{\gamma(k)} x \delta_{ik} = x_{\gamma(i)} x$, and where γ is the permutation of I defined by $g_{\gamma(i)} g = g_i$. The big question is whether all but finitely many of the values $x_{\gamma(i)} x$ are zero? Since all but finitely many of the values x_i are zero, and since γ is a permutation of I, we have that all but finitely many of the values $x_{\gamma(i)} x$ are zero. Hence, $\sum_k [x; \gamma(k), k]$ defines a mapping of $R^{(G)}$ into $R^{(G)}$.

Exercise. Let G be a finite cyclic group of order n and let R be any ring with identity. Compute the isomorphic copy of $R[G]$ inside the matrix ring $\mathcal{M}(n, R)$ as we have explained here. This singles out a particularly nice subring of $\mathcal{M}(n, R)$.

We now replace our ring R with a zero-symmetric nearring N with identity. Our group $(G, \cdot)$ with identity e stays the same, as does our index set I, so $G = \{g_i \mid i \in I\}$ as before. Any $g \in G$ defines a permutation γ of I by $g_{\gamma(i)}g = g_i$. We still have that $N^{(G)}$ is a group, and so we have the zero-symmetric nearring $M_0(N^{(G)})$. From our discussion of matrix nearrings, we also have each $[x; i, j]$ if $i, j \in I$ and I is finite. But, as we have done here, an $[x; s, t]$ still has meaning if $s, t \in I$ and I is infinite. All the $[x; s, t]$ will generate a subnearring of $M_0(N^{(G)})$, which we denote by $\mathbf{M}_I(N)$, and call it the *nearring of matrices over N for I*. For $x \in N$ and $g \in G$, we define $\langle x, g \rangle$ exactly as in (26:9), and let $N[G]$ be the subnearring of $\mathbf{M}_I(N)$ generated by all the $\langle x, g \rangle$. Of course, we call $N[G]$ the *group nearring* constructed from N and G.

This approach to the group nearring has evolved. First J. D. P. Meldrum used distributively generated nearrings, but after he and A. P. J. van der Walt succeeded in their matrix nearring development, it became clear what to do for a group nearring. Meldrum and van der Walt joined with L. R. le Riche to bring about the idea as we present it here [lR *et al.*]. Actually, their development is for nearrings in general, whereas we restrict ourselves here to zero-symmetric nearrings.

We have $N[G] \subseteq \mathcal{M}(G, R) \subseteq M_0(R^{(G)})$, and if G is finite of order n, $N[G] \subseteq \mathcal{M}(n, N) \subseteq M_0(R^n)$. As we continue, we will consider only finite groups $(G, \cdot)$, but one should not have any trouble extending what we have done for $\mathbf{M}_n(N)$ to $\mathbf{M}_I(N)$, and then one can extend what we do for $N[G]$ with G finite to $N[G]$ with G infinite.

For an ideal A of N, and a finite group $(G, \cdot)$ of order n, we have that N^n is an $\mathbf{M}_n(N)$-module. So it is also an $N[G]$-module. See (26.14). As a corollary of (26.14) and (26.15) we have

(26.25) Theorem. *Let N be a zero-symmetric nearring with identity 1, and let $(G, \cdot)$ be a finite group of order n. If R is a right ideal of N, then R^n is an $N[G]$-ideal of the $N[G]$-module N^n. Also, $(R^n : N^n)$ is an ideal of $N[G]$. (Here, $(R^n : N^n)$ takes its meaning within $N[G]$, whereas in (26.15), it takes its meaning within $\mathbf{M}_n(N)$.)*

So, for an ideal A of N, we have the ideal $A[G]^* = (A^n : N^n)$, and we also have the ideal $A[G]^\circ$ generated by all the $\langle a, e \rangle$ for $a \in A$ and the identity $e \in G$. Since $\langle a, e \rangle = \sum_k [a; k, k]$, we have each $\langle a, e \rangle \in \mathbf{A}^\circ$, and so $A[G]^\circ \subseteq \mathbf{A}^\circ$, and certainly $A[G]^* \subseteq \mathbf{A}^*$.

For an ideal A of N, we have $A[G]^*$ and $A[G]^\circ$ as ideals of $N[G]$, and these are constructed similarly to the ideals $\mathbf{A}^*$ and $\mathbf{A}^\circ$ in the matrix nearring $\mathbf{M}_n(N)$. In fact, we have that $\mathbf{A}^* \cap N[G]$ and $\mathbf{A}^\circ \cap N[G]$ are also ideals of $N[G]$ determined by the ideal A of N. Are the ideals $A[G]^*$, $\mathbf{A}^* \cap N[G]$, $A[G]^\circ$, and $\mathbf{A}^\circ \cap N[G]$ distinct? In the preceding paragraph, we noted that

$A[G]^* \subseteq \mathbf{A}^* \cap N[G]$, and that $A[G]^\circ \subseteq \mathbf{A}^\circ \cap N[G]$. We can say more.

(26.26) Proposition. *For a zero-symmetric nearring N with identity, let A be an ideal of N. Then $A[G]^* = \mathbf{A}^* \cap N[G]$.*

Proof. We need only show that $\mathbf{A}^* \cap N[G] \subseteq A[G]^*$. Take $\mathbf{x} \in \mathbf{A}^* \cap N[G]$. Then $\mathbf{x} \in N[G]$, and $\mathbf{x} \in \mathbf{A}^* = (A^n : N^n)$ for $\mathbf{M}_n(G)$. Hence, $\mathbf{x} \in A[G]^* = (A^n : N^n)$ for $N[G]$, and so we have $A[G]^* = \mathbf{A}^* \cap N[G]$.

With each $\langle a, e \rangle = \sum_k [a; k, k] \in \mathbf{A}^\circ \cap N[G]$, we also have $A[G]^\circ \subseteq \mathbf{A}^\circ \cap N[G]$, and since $\mathbf{A}^\circ \subseteq \mathbf{A}^*$, we have $A[G]^\circ \subseteq \mathbf{A}^\circ \cap N[G] \subseteq \mathbf{A}^* \cap N[G] = A[G]^*$. Hence,

(26.27) Proposition. *For a zero-symmetric nearring N with an identity and an ideal A, we have $A[G]^\circ \subseteq A[G]^*$.*

With matrix nearrings $\mathbf{M}_n(N)$, we also took an ideal $\mathbf{A}$ of $\mathbf{M}_n(N)$ and constructed an ideal $\mathbf{A}_*$ of N. We wish to do something similar for group nearrings $N[G]$, but first we need a concept analogous to the weight of an $A \in \mathbf{M}_n(N)$.

Since $\mathbf{a} \in N[G]$ is generated by the $[x; s, t]$, we say that a *generating sequence* for $\mathbf{a}$ is a finite sequence $\mathbf{a}_1, \dots, \mathbf{a}_k$ of elements in $N[G]$ such that $\mathbf{a} = \mathbf{a}_k$, and for each $i \in \{1, \dots, k\}$, one of the following three cases is valid: (1) $\mathbf{a}_i = \langle x, g \rangle$ for some $x \in N$ and some $g \in G$; (2) $\mathbf{a}_i = \mathbf{a}_s + \mathbf{a}_t$ for some $s, t \in \{1, \dots, i-1\}$; (3) $\mathbf{a}_i = \mathbf{a}_s \mathbf{a}_t$ for some $s, t \in \{1, \dots, i-1\}$. Of all the generating sequences for $\mathbf{a}$, there is one with least k, and for this least k we give it a name, the *complexity of $\mathbf{a}$*, and denote it by $c(\mathbf{a})$. So the complexity $c(\mathbf{a})$ of $\mathbf{a}$ is the smallest value of k for which there is a generating sequence $\mathbf{a}_1, \dots, \mathbf{a}_k$ for $\mathbf{a}$.

We have $c(\mathbf{a}) \geq 1$ for all $\mathbf{a} \in N[G]$, and $c(\mathbf{a}) = 1$ if and only if $\mathbf{a} = \langle a, g \rangle$ for some $a \in N$ and some $g \in G$. If $c(\mathbf{a}) > 1$, then $\mathbf{a} = \mathbf{b} + \mathbf{d}$ or $\mathbf{a} = \mathbf{bd}$ for some $\mathbf{b}, \mathbf{d} \in N[G]$ with $c(\mathbf{b}), c(\mathbf{d}) \in \{1, \dots, c(\mathbf{a}) - 1\}$. For if $\mathbf{a}_1, \dots, \mathbf{a}_k$ is a generating sequence for $\mathbf{a}$, then $\mathbf{a}_1, \dots, \mathbf{a}_i$ is a generating sequence for $\mathbf{a}_i$, where $i \in \{1, \dots, k\}$.

It is interesting, as well as useful, to note that $N^{(G)}$ is a unitary N-comodule, and that each $\mathbf{a} \in N[G]$ is a scalar mapping. That is,

(26.28) Theorem. *For a zero-symmetric nearring N with identity and a finite group $(G, \cdot)$, we let N act on $N^{(G)}$ by $x(u_i) = (xu_i)$ for each $x \in N$ and each $(u_i) \in N^{(G)}$. With this scalar multiplication, $N^{(G)}$ is a unitary N-comodule, and each $\mathbf{a} \in N[G]$ has the property that $[x(u_i)]\mathbf{a} = x[(u_i)\mathbf{a}]$ for all $x \in N$ and for all $(u_i) \in N[G]$.*

Proof. It is easy to see that $N^{(G)}$ is a unitary N-comodule. It is also direct to see that $[x(u_i)]\mathbf{a} = x[(u_i)\mathbf{a}]$ for arbitrary $x \in N$ and arbitrary $(u_i) \in N^{(G)}$. Nevertheless, it may be helpful to go through the details of this part of the proof. The proof will be by induction on $c(\mathbf{a})$, the complexity of $\mathbf{a} \in N[G]$.

Suppose $c(\mathbf{a}) = 1$, so $\mathbf{a} = \langle a, g \rangle = \sum_k [a; \gamma(k), k]$. Hence, $[x(u_i)]\mathbf{a} = \sum_k (xu_i)[a; \gamma(k), k] = \sum_k (xu_{\gamma(k)} a \delta_{ik}) = (\sum_k xu_{\gamma(k)} a \delta_{ik}) = (xu_{\gamma(i)} a) = x(u_{\gamma(i)} a) = x(\sum_k u_{\gamma(k)} a \delta_{ik}) = x[\sum_k (u_{\gamma(k)} a \delta_{ik})] = x[\sum_k (u_i)[a; \gamma(k), k]] = x[(u_i)\langle a, g \rangle] = x[(u_i)\mathbf{a}]$. So our conclusion is true for $c(\mathbf{a}) = 1$. Assume $\mathbf{a} = \mathbf{b} + \mathbf{d}$ where $c(\mathbf{b}), c(\mathbf{d}) \in \{1, \ldots, m-1\}$ and $c(\mathbf{a}) = m$. Then $[x(u_i)]\mathbf{a} = [x(u_i)](\mathbf{b} + \mathbf{d}) = [x(u_i)]\mathbf{b} + [x(u_i)]\mathbf{d} = x[(u_i)\mathbf{b}] + x[(u_i)\mathbf{d}] = x[(u_i)\mathbf{b} + (u_i)\mathbf{d}] = x[(u_i)(\mathbf{b} + \mathbf{d})] = x[(u_i)\mathbf{a}]$. Similarly, assume $\mathbf{a} = \mathbf{bd}$. Then $[x(u_i)]\mathbf{a} = ([x(u_i)]\mathbf{b})\mathbf{d} = (x[(u_i)\mathbf{b}])\mathbf{d} = x([(u_i)\mathbf{b}]\mathbf{d}) = x[(u_i)\mathbf{a}]$. This completes our proof.

We have that $N^{(G)}$ is a unitary N-comodule. But $N^{(G)}$ is also an $N[G]$-module. Note that (26.28) assures us that $N^{(G)}$ is an $(N[G], N)$-bimodule. Also, for $x \in N$, we define $x^\bullet : N^{(G)} \to N^{(G)}$ by $x^\bullet(u_i) = (xu_i)$. Then (26.28) assures us that $x^\bullet$ is an $N[G]$-endomorphism of $N^{(G)}$, so we have

(26.29) Corollary. *For N and G as in (26.28), and for $x \in N$, we have that $x^\bullet(u_i) = (xu_i)$ defines $x^\bullet$ as an $N[G]$-endomorphism of $N^{(G)}$, and that $N^{(G)}$ is an $(N[G], N)$-bimodule.*

With (26.28) handy for use, we are prepared to prove

(26.30) Theorem. *For a zero-symmetric nearring N with identity 1 and a finite group $(G, \cdot)$, we take an ideal $\mathbf{A}$ of $N[G]$ and define*

$$\mathbf{A}[G]_* = \{p_1(\mathbf{a}) \mid (\delta_{1i})\mathbf{a} = (p_i(\mathbf{a})), \, \mathbf{a} \in \mathbf{A}\},$$

where $(\delta_{1i}) \in N^{(G)}$ is the element defined by $\delta_{1i} = \begin{cases} 1 \in N, & \text{if } i = 1; \\ 0 \in N, & \text{if } i \neq 1, \end{cases}$ and where $G = \{g_1, \ldots, g_n\}$. We obtain that $\mathbf{A}[G]_*$ is an ideal of N.

Proof. Let $\mathbf{z}$ be the additive identity of $\mathbf{A}$. So $(\delta_{1i})\mathbf{z} = (p_i(\mathbf{z}))$, where each $p_i(\mathbf{z}) = 0 \in N$. Hence, $p_1(\mathbf{z}) = 0$, and so $0 \in \mathbf{A}[G]_*$. Suppose $a, b \in \mathbf{A}[G]_*$. Then there are $\mathbf{a}, \mathbf{b} \in \mathbf{A}$ such that $p_1(\mathbf{a}) = a$ and $p_1(\mathbf{b}) = b$. Hence, $a - b = p_1(\mathbf{a}) - p_1(\mathbf{b})$ where $(p_i(\mathbf{a}) - p_i(\mathbf{b})) = (p_i(\mathbf{a})) - (p_i(\mathbf{b})) = (\delta_{1i})\mathbf{a} - (\delta_{1i})\mathbf{b} = (\delta_{1i})[\mathbf{a} - \mathbf{b}]$. Since $\mathbf{a} - \mathbf{b} \in \mathbf{A}$, we obtain $a - b \in \mathbf{A}[G]_*$.

If $e \in G$ is the identity and $\mathbf{a} \in \mathbf{A}$, then $-\langle x, e \rangle + \mathbf{a} + \langle x, e \rangle \in \mathbf{A}$ for each $x \in N$. Hence, $(\delta_{1i})[-\langle x, e \rangle + \mathbf{a} + \langle x, e \rangle] = -[(\delta_{1i})\langle x, e \rangle] + (\delta_{1i})\mathbf{a} + (\delta_{1i})\langle x, e \rangle = (-\delta_{1i}x) + (p_i(\mathbf{a})) + (\delta_{1i}x) = (-\delta_{1i}x + p_i(\mathbf{a}) + \delta_{1i}x)$, and $-\delta_{11}x + p_1(\mathbf{a}) + \delta_{11}x = -x + a + x$. So if $a \in \mathbf{A}[G]_*$, we obtain $-x + a + x \in \mathbf{A}[G]_*$ for every $x \in N$. This makes $\mathbf{A}[G]_*$ a normal subgroup of $(N, +)$.

For $a \in \mathbf{A}[G]_*$ and $x, y \in N$, we need $(x + a)y - xy \in \mathbf{A}[G]_*$. With $\mathbf{a} \in \mathbf{A}$ yielding $p_1(\mathbf{a}) = a$, we have $[\langle x, e \rangle + \mathbf{a}]\langle y, e \rangle - \langle x, e \rangle \langle y, e \rangle \in \mathbf{A}$, and $(\delta_{1i})\{[\langle x, e \rangle + \mathbf{a}]\langle y, e \rangle - \langle x, e \rangle \langle y, e \rangle\} = [(\delta_{1i}x) + (p_i(\mathbf{a}))]\langle y, e \rangle - (\delta_{1i}xy) =$

$((\delta_{1i}x + p_i(\mathbf{a}))y) - (\delta_{1i}xy) = ((\delta_{1i}x + p_i(\mathbf{a}))y - (\delta_{1i}xy))$, and for $i = 1$, we have $(x + a)y - xy$, so $(x + a)y - xy \in \mathbf{A}[G]_*$.

It remains to show that $xa \in \mathbf{A}[G]_*$ if $a \in \mathbf{A}[G]_*$ and $x \in N$. Since $\mathbf{A}$ is an ideal of $N[G]$, we know that $\langle x, e \rangle \mathbf{a} \in \mathbf{A}$ for each $\mathbf{a} \in \mathbf{A}$ and each $x \in N$. But, using (26.28), we obtain $(\delta_{1i})[\langle x, e \rangle \mathbf{a}] = (\delta_{1i}x)\mathbf{a} = [x(\delta_{1i})]\mathbf{a} = x[(\delta_{1i})\mathbf{a}] = x(p_i(\mathbf{a})) = (xp_i(\mathbf{a}))$, and since $p_1(\mathbf{a}) = a$, we have $xa \in \mathbf{A}[G]_*$. This completes our proof of (26.30).

As one might guess, we obtain $\mathbf{A} \subseteq (\mathbf{A}[G]_*)^*$.

(26.31) Theorem. *Let N, G, $\mathbf{A}$, and $\mathbf{A}[G]_*$ be as in (26.29). Then $\mathbf{A} \subseteq (\mathbf{A}[G]_*)^*$.*

Proof. For $\mathbf{a} \in \mathbf{A}$, we want to show that $\mathbf{a} \in (\mathbf{A}[G]_*)^*$, and so we will obtain what we want when we have $(u_i)\mathbf{a} \in \mathbf{A}[G]_*^{(G)}$ for each $(u_i) \in N^{(G)}$. For an arbitrary but fixed $(u_i) \in N^{(G)}$, denote the image of (u_i) by $\mathbf{a}$ with $(u_i)\mathbf{a} = (\alpha_i(\mathbf{a}))$. We want to show that each $\alpha_i(\mathbf{a}) \in \mathbf{A}[G]_*$, $i \in \{1, \ldots, n\}$.

For each u_j, we may choose $m_j \in G$ so that $\mu_j^{-1}(1) = j$, where $\langle u_j, m_j \rangle = \sum_k[u_j; \mu_j(k), k]$. Then $(u_i) = (\delta_{1i})[\langle u_1, m_1 \rangle + \cdots + \langle u_n, m_n \rangle]$. For this u_j, we may also choose $t_j \in G$ so that $\tau_j(1) = j$, where $\langle 1, t_j \rangle = \sum_k[1; \tau_j(k), k]$. For $(v_i) \in N^{(G)}$, we have $(v_i)\langle 1, t_j \rangle = (v_{\tau_j(i)})$ where $\tau_j(1) = j$.

Since $N[G] \subseteq M_0(N^{(G)})$, we know that $N[G]$ is zero-symmetric. Hence, $\mathbf{b} = [\langle u_1, m_1 \rangle + \cdots + \langle u_n, m_n \rangle]\mathbf{a}\langle 1, t_j \rangle \in \mathbf{A}$. With $(\delta_{1i})\mathbf{b} = (p_i(\mathbf{b}))$, we have that $p_1(\mathbf{b}) \in \mathbf{A}[G]_*$. But $(\delta_{1i})\mathbf{b} = (\delta_{1i})[\langle u_1, m_1 \rangle + \cdots + \langle u_n, m_n \rangle]\mathbf{a}\langle 1, t_j \rangle = (u_i)\mathbf{a}\langle 1, t_j \rangle = (\alpha_i(\mathbf{a}))\langle 1, t_j \rangle = (\alpha_{\tau_j(i)}(\mathbf{a}))$. Hence, $p_1(\mathbf{b}) = \alpha_{\tau_j(1)}(\mathbf{a}) = \alpha_j(\mathbf{a}) \in \mathbf{a}[G]_*$, which is exactly what is wanted. This completes our proof of (26.31).

Exercises. Throughout these notes one finds numerous **Problems** and **Exploratory problems**, but it is important for a mathematician to formulate his own **Problems** and **Exploratory problems**. Our discussions here on matrix nearrings and group nearrings were designed to encourage the serious reader to do exactly that. So the goal of this exercise is to use our discussions of matrix nearrings and group nearrings to formulate some specific **Problems**. After having done so, formulate some **Exploratory problems.** (You then may want to try your hand at solving your own problems.)

26.4. Meromorphic products

Let $J \neq \varnothing$ be an index set and let $(G, +)$ be a group. Then G^J is a group, the group of all mappings (g_j) of J into G, and G^J is a unitary $M(G)$-module, where for $(g_j) \in G^J$ and $\phi \in M(G)$, we have $(g_j) \cdot \phi = (g_j\phi)$. So $M(G)$ acts on G^J, and if H is any subgroup of G^J, we define

$$M(G, J, H) = \{\phi \in M(G) \mid (h_j) \cdot \phi \in H \text{ for each } (h_j) \in H\}.$$

Then $M(G, J, H)$ is a subnearring of $M(G)$. If $J = \{1, 2, \ldots, k\}$, a finite set, we also write $M(G, k, H)$ for $M(G, J, H)$.

One could focus on ways to construct subgroups H so that $M(G, J, H)$ is interesting. For our purpose here, we want to make $M(G, J, H)$ a finite nearfield. We have constructed nearfields from fields using the Dickson process, and we have noted that all but seven of the finite nearfields can be constructed in this way. We have also noted that each finite nearfield with order greater than two can be constructed by the Ferrero Planar Nearring Factory. We shall see that each finite zero-symmetric nearfield is isomorphic to some $M(G, J, H)$.

Let us first turn our attention to a method of constructing subgroups H of G^J. For each $j \in J$, let A_j be a subgroup of G, and suppose each A_j has a normal subgroup B_j of A_j such that the quotient groups A_j/B_j are all isomorphic. For example, we could take each $A_j = G$ and each $B_j = \{0\}$. For $s, t \in J$, let $\epsilon_{st} : A_s/B_s \to A_t/B_t$ be an isomorphism with each ϵ_{ss} being the identity map.

Let us assume that our index set J has $1 \in J$, and suppose I is another index set with $0 \in I$. With I we wish to index the elements of each A_j/B_j. Let $R_j = \{r_{ij} \mid i \in I\}$ be a complete set of coset representatives of B_j in A_j with each $r_{0j} = 0$ and where $\epsilon_{1j}(r_{i1} + B_1) = r_{ij} + B_j$. For $i \in I$, define

$$H_i = \times_{j \in J}[r_{ij} + B_j] = \times_{j \in J}\epsilon_{1j}(r_{i1} + B_1),$$

a cartesian product of the cosets $r_{ij} + B_j = \epsilon_{1j}(r_{i1} + B_1)$, with $j \in J$.

(26.32) Lemma. *If $s, t \in I$ and $s \neq t$, then $H_s \cap H_t = \varnothing$.*

Proof. If $x \in H_s \cap H_t$, then x can be expressed as $x = (r_{sj} + c_j)$ and as $x = (r_{tj} + d_j)$ where each $c_j, d_j \in B_j$. So, for each $j \in J$, $r_{sj} + c_j = r_{tj} + d_j \in [r_{sj} + B_j] \cap [r_{tj} + B_j] = \varnothing$, since $s \neq t$ and $r_{sj}, r_{tj} \in R_j$.

(26.33) Lemma. *If $x \in H_s$ and $y \in H_t$, then there is a unique u such that $x + y \in H_u$.*

Proof. For $x = (r_{sj} + c_j) \in H_s$ and $y = (r_{tj} + d_j) \in H_t$, we have, for each $j \in J$, $r_{sj} + c_j + r_{tj} + d_j \in [r_{sj} + B_j] + [r_{tj} + B_j] = r_{sj} + r_{tj} + B_j = r_{u_j, j} + B_j$ for some u_j. But $r_{u_j, j} + B_j = [r_{sj} + B_j] + [r_{tj} + B_j] = \epsilon_{1j}(r_{s1} + B_1) + \epsilon_{1j}(r_{t1} + B_1) = \epsilon_{1j}(r_{s1} + r_{t1} + B_1) = \epsilon_{1j}(r_{u_1, 1} + B_1) = r_{u_1, j} + B_j$. Hence, each $u_j = u_1$, and so each u_j is independent of j. Set $u = u_1$, and so $r_{u_j, j} + B_j = r_{uj} + B_j$, and we have $r_{sj} + c_j + r_{tj} + d_j \in r_{uj} + B_j$ for each $j \in J$. This puts $x + y \in H_u$ as desired.

(26.34) Lemma. *Given H_s, there is a unique H_u such that if $x \in H_s$, then $-x \in H_u$.*

Proof. Now $-x = -(r_{sj} + c_j) = (-(r_{sj} + c_j))$. For each $j \in J$, $r_{sj} + c_j \in r_{sj} + B_j$, and so $-(r_{sj} + c_j) \in -r_{sj} + B_j = r_{u_j, j} + B_j = -[r_{sj} + B_j] = $

$-\epsilon_{1j}[r_{s1} + B_1] = \epsilon_{1j}[-r_{s1} + B_1] = \epsilon_{1j}[r_{u_1,1} + B_1] = r_{u_1,j} + B_j$. Hence, each $u_j = u_1 = u$, and so $-(r_{sj} + c_j) = r_{uj} + e_j$ for some $e_j \in B_j$. This puts $-x \in H_u$. This argument is independent of c_j, and as c_j varies through B_j, the x varies through H_s, and so our proof is complete.

(26.35) Theorem. *If $H = \cup_{i \in I} H_i$, then H is a subgroup of G^J. Also H_0 is a subgroup of H and each H_i is a coset of H_0 in H.*

Proof. Since each $H_i \neq \varnothing$, we have from (26.33) and (26.34) that H is a subgroup of G^J.

Take $x = (r_{0j} + c_j)$, $y = (r_{0j} + d_j) \in H_0$. Then $x + y = (r_{0j} + c_j + d_j)$ since each $r_{0j} = 0$. Also, $-x = (-c_j) \in H_0$. Hence, H_0 is a subgroup of H.

With $x = (r_{sj} + c_j) = (r_{sj}) + (c_j) \in (r_{sj}) + H_0$, we obtain $H_s \subseteq (r_{sj}) + H_0$. Also, $(r_{sj}) + (c_j) = (r_{sj} + c_j) \in H_s$, so $(r_{sj}) + H_0 \subseteq H_s$. Thus $H_s = (r_{sj}) + H_0$, and this completes our proof.

(26.36) Definition. The group H of (26.35) is a J-*meromorphic product* of the quotient groups A_j/B_j, $j \in J$, and we denote this by $H = \overset{\times}{\underset{\sim}{}}_{j \in J} A_j/B_j$. If $J = \{1, 2, \ldots, k\}$, then $\overset{\times}{\underset{\sim}{}}_{j \in J} A_j/B_j = A_1/B_1 \overset{\times}{\underset{\sim}{}} \cdots \overset{\times}{\underset{\sim}{}} A_k/B_k$.

If each $B_j = \{0\}$, then $A_j = R_j = \{r_{ij} \mid i \in I\}$ where $|I| = |A_j|$ for each $j \in J$. Also, $H_i = \times_{j \in J}(r_{ij} + B_j) = \times_{j \in J}(r_{ij} + \{0\}) \equiv (r_{ij}) \in G^J$, a single element of G^J.

Take N to be any finite zero-symmetric nearfield, and since N is isomorphic to a subnearring of some $M(G)$, with $|G| \geq 4$ (3.27), we may take $N \subseteq M_0(G) \subseteq M(G)$ with $G^* = \{x_1, x_2, \ldots, x_k\}$. One easily obtains

(26.37) Lemma. *Each $x_i N$ is a unitary N-module and is an N-submodule of G.*

Proof. With $x_i m, x_i n \in x_i N$, we have $x_i m - x_i n = x_i(m - n) \in x_i N$. Also $(x_i m) \cdot n = x_i(mn) \in x_i N$. For $1 \in N$, $(x_i n) \cdot 1 = x_i(n \cdot 1) = x_i n$. Now G is a unitary $M_0(G)$-module, so we have that G is a unitary N-module, and so $x_i N$ is a unitary N-module, an N-submodule of G.

(26.38) Lemma. *If $\epsilon_{1j} : x_1 N \to x_j N$ is defined by $\epsilon_{1j}(x_1 n) = x_j n$, then ϵ_{1j} is an N-isomorphism.*

Proof. Suppose $x_1 m = x_1 n$. Then $0 = x_1(m - n)$. If $m \neq n$, then $(m - n)^{-1}$ exists, since N is a nearfield, and so $0(m - n)^{-1} = 0 = x_1$. So, for each $a \in x_1 N$, there is a unique $n_a \in N$ such that $a = x_1 n_a$. Hence, $\epsilon_{1j}(a) = \epsilon_{1j}(x_1 n_a) = x_j n_a$, and so each ϵ_{1j} is well defined. If $0 = \epsilon_{1j}(x_1 n) = x_j n$, then $0 = 0n^{-1} = (x_j n)n^{-1} = x_j$ if $x_1 n \neq 0$. Hence, ϵ_{1j} is a monomorphism if it is a homomorphism. Similarly, for $x_j m \in x_j N$, we have $\epsilon_{1j}(x_1 m) = x_j m$, so ϵ_{1j} is surjective. It remains to show that each ϵ_{1j} is an N-homomorphism.

First, $\epsilon_{1j}[(x_1 m)n] = \epsilon_{1j}[x_1(mn)] = x_j(mn) = (x_j m)n = \epsilon_{1j}(x_1 m)n$. Finally, $\epsilon_{1j}(x_1 m + x_1 n) = \epsilon_{1j}(x_1(m+n)) = x_j(m+n) = x_j m + x_j n = \epsilon_{1j}(x_1 m) + \epsilon_{1j}(x_1 n)$.

We now have isomorphic subgroups $x_j N$. Let $A_j = x_j N$ and $B_j = \{0\}$, so each $A_j/B_j \cong A_j \cong A_1 = x_1 N$. Let $H = A_1/\{0\} \overset{\times}{\sim} \cdots \overset{\times}{\sim} A_k/\{0\}$. We are now set to prove

(26.39) Theorem. *Let N be a zero-symmetric nearfield contained in $M(G)$ for some G with $|G^*| = k$. Then $N = M(G, k, H)$ for some subgroup H of G^k.*

Proof. Certainly, $H_i \equiv (r_{i1}, \ldots, r_{ik})$, and if $n \in N$, then $(r_{i1}, \ldots, r_{ik}) \cdot n = (r_{i1}n, \ldots, r_{ik}n)$. And if $\phi \in M(G, k, H)$, then $(r_{i1}, \ldots, r_{ik}) \cdot \phi = (r_{i1}\phi, \ldots, r_{ik}\phi)$, so $r_{i1}\phi \in x_1 N$.

Now $\epsilon_{1j}(x_1 1) = x_j 1 = x_j$, so $(x_1, \ldots, x_k) \in H$. For $\phi \in M(G, k, H)$, we have $x_1\phi = x_1 n$ for some $n \in N$. Also, $(x_1\phi, \ldots, x_k\phi) \in H$ and $x_j\phi = \epsilon_{1j}(x_1\phi) = \epsilon_{1j}(x_1 n) = \epsilon_{1j}(x_1)n = x_j n$. Hence, $x_j\phi = x_j n$ for $j = 1, 2, \ldots, k$. So ϕ and n agree on $G^* = G \setminus \{0\}$. With $(0, \ldots, 0) \in H$, and for $\phi \in M(G, k, H)$, we have $(0\phi, \ldots, 0\phi) \in H$. If $0\phi \neq 0$, then $0\phi = x_1 m$ for some $m \in N \setminus \{0\}$. Hence, $(0\phi, \ldots, 0\phi) = (x_1 m, \epsilon_{12}(x_1 m), \ldots, \epsilon_{1k}(x_1 m)) = (x_1 m, x_2 m, \ldots, x_k m)$. This means that each $x_i m = 0\phi$. But $m^{-1} \in N$, so $(x_i m)m^{-1} = (x_1 m)m^{-1}$, or $x_i = x_1$. Hence, $x_1 = x_2 = \cdots = x_k$, and so $k = 1$. Since $k > 1$, we obtain $0\phi = 0$ for each $\phi \in M(G, k, H)$. We also have $0n = 0$. Hence, $n = \phi$, so $\phi \in N$. Finally, $M(G, k, H) \subseteq N$.

With $n \in N \subseteq M_0(G)$, we have

$$(r_{i1}, \ldots, r_{ik}) \cdot n = (r_{i1}n, \ldots, r_{ik}n) = (r_{i1}n, \epsilon_{12}(r_{i1})n, \ldots, \epsilon_{1k}(r_{i1})n)$$
$$= (r_{i1}n, \epsilon_{12}(r_{i1}n), \ldots, \epsilon_{1k}(r_{i1}n))$$
$$= (r_{u1}, \epsilon_{12}(r_{u1}), \ldots, \epsilon_{1k}(r_{u1})) = (r_{u1}, r_{u2}, \ldots, r_{uk})$$

for some u. Hence, $n \in M(G, k, H)$, and so $N \subseteq M(G, k, H)$. Now we have $N = M(G, k, H)$.

In the proof of (26.39), we have utilized the matrix nature of H. We have

$$
\begin{array}{ccccccc}
r_{01} & r_{02} & \cdots & r_{0k} & \longleftarrow & H_0 \\
r_{11} & r_{12} & \cdots & r_{1k} & \longleftarrow & H_1 \\
\cdots & \cdots & \cdots & \cdots & \cdots & \cdots \\
r_{n1} & r_{n2} & \cdots & r_{nk} & \longleftarrow & H_n \\
\uparrow & \uparrow & \cdots & \uparrow & & \\
A_1 & A_2 & \cdots & A_n & &
\end{array}
\qquad (26:10)
$$

for $H = A_1/\{0\} \overset{\times}{\sim} \cdots \overset{\times}{\sim} A_k/\{0\}$ with $|A_1| = n+1$. Let $L_i = \{r_{i1}, r_{i2}, \ldots, r_{ik}\}$ and so $L_0 = \{0\}$. The L_i correspond to the rows; they consist of the entries of $H_i = (r_{i1}, r_{i2}, \ldots, r_{ik})$. The A_j correspond to the columns; they consist of the elements of the group $A_j = \{r_{0j}, r_{1j}, \ldots, r_{nj}\}$. Let $\mathcal{L} = \{L_1, L_2, \ldots, L_n\}$. We will say that $\mathcal{L}$ is *connected* if, for some $m_1 \in \{1, 2, \ldots, n\}$, we can find an $m_2 \in \{1, 2, \ldots, n\} \setminus \{m_1\}$ such that $L_{m_1} \cap L_{m_2} \neq \varnothing$, and then find an $m_3 \in \{1, 2, \ldots, n\} \setminus \{m_1, m_2\}$ such that $L_{m_2} \cap L_{m_3} \neq \varnothing$, and continuing, having found $m_s \in \{1, 2, \ldots, n\} \setminus \{m_1, \ldots, m_{s-1}\}$ such that $L_{m_{s-1}} \cap L_{m_s} \neq \varnothing$, we can find a next m_{s+1} in the smaller set $\{1, 2, \ldots, n\} \setminus \{m_1, \ldots, m_s\}$ such that $L_s \cap L_{s+1} \neq \varnothing$, until we have $\{1, 2, \ldots, n\} \setminus \{m_1, \ldots, m_{n-1}\}$, and then m_n is the remaining element in $\{1, 2, \ldots, n\} \setminus \{m_1, \ldots, m_{n-1}\}$ and we have $L_{m_{n-1}} \cap L_{m_n} \neq \varnothing$.

Before we prove our last theorem, we will state and prove a needed theorem which is interesting all by itself.

(26.40) Theorem. *Let $(N, +, \cdot)$ be a finite nearring with identity 1. Also, suppose N has order greater than 1. Suppose $a, b \in N$ and $ab = 0$ implies $0 \in \{a, b\}$, that is, $(N, +, \cdot)$ is an integral nearring. Then $(N, +, \cdot)$ is a nearfield.*

Proof. If $a \in N$ and $a \neq 0$, then $\{ax \mid x \in N\} = N$, and so there is an $a' \in N$ such that $aa' = 1$. So, for $N^* = N \setminus \{0\}$, there is an identity 1 and for $a \in N^*$, there is a right inverse. This is enough to make $(N^*, \cdot)$ a group. Hence, $(N, +, \cdot)$ is a nearfield.

(26.41) Theorem. *For a finite group $(G, +)$, suppose there are isomorphic subgroups $A_1, \ldots, A_k$, not necessarily distinct, so we can construct $H = A_1/\{0\} \overset{\times}{\sim} \cdots \overset{\times}{\sim} A_k/\{0\}$. Then $N = M(G, k, H)$ is a nearfield if and only if (1) N is zero-symmetric, (2) $\cup_{i=1}^k A_i = G$, and (3) $\mathcal{L}$ is connected.*

Proof. Note that the elements of H can be constructed from the rows of (26:10). So they are the $(r_{i1}, \ldots, r_{ik})$ for $i \in \{0, 1, \ldots, n\}$.

Let us start with $N = M(G, k, H)$ being a nearfield. If N is not zero-symmetric, then $(N, +, \cdot) \cong (Z_2, +, *)$, where $a * b = b$ for all $a, b \in Z_2$. But N has an identity and $(Z_2, +, *)$ does not. Hence, $(N, +, \cdot)$ is not isomorphic to $(Z_2, +, *)$. All other nearfields have an additive identity 0 where $0 \cdot x = x \cdot 0 = 0$ for all x. Hence, N is zero-symmetric.

Suppose $\cup_{i=1}^k A_i \subset G$. Recall that $\cup_{i=1}^k A_i$ consists of all r_{ij} as displayed in (26:10). Define $g : G \to G$ by $xg = x$ if $x \in \cup_{i=1}^k A_k$, and $xg = 0$, otherwise. Looking at (25:10), one can quickly see that $g \in M(G, k, H) = N$. But g is not bijective, so g cannot have an inverse, contrary to $g \neq \zeta$, the additive identity of $N \subset M(G)$, and the fact that N is a nearfield. So, we had better conclude that $\cup_{i=1}^k A_i = G$.

If $\mathcal{L}$ is not connected, then we have a maximal subset $\{m_1, \ldots, m_s\} \subset \{1, 2, \ldots, n\}$ such that $L_{m_i} \cap L_{m_{i+1}} \neq \varnothing$, for $i \in \{1, \ldots, s-1\}$. Define $h : G \to G$ by $xh = 0$ if $x \in L_0 \cup L_{m_1} \cup \cdots \cup L_{m_s}$, and $xh = x$, otherwise. Now $L_0 \cup L_{m_1} \cup \cdots \cup L_{m_s} \neq G$, for otherwise one could find $m_{s+1} \in$

$\{1, 2, \ldots, n\} \setminus \{m_1, \ldots, m_s\}$ where $L_{m_s} \cap L_{m_{s+1}} \neq \varnothing$. Certainly, with the help of (26:10), we can conclude that $h \neq \zeta$, $h \in N = M(G, k, H)$, and that h is not a bijection, and so h^{-1} does not exist. But N is a nearfield, and with $h \neq \zeta$, h^{-1} not only exists but it is in N. So, we are forced to conclude that $\mathcal{L}$ is connected.

Now we are ready to prove the converse. Certainly, $N = M(G, k, H)$ is zero-symmetric and has an identity. It is also finite. If $xf \neq 0$ for all $x \in G \setminus \{0\} = G^*$, and if $f \in N$, then we can never have $f \circ g = \zeta$ or $g \circ f = \zeta$ for $g \neq \zeta$. So we will show that if $f \in N$, and $xf = 0$ for some $x \in G^*$, then $f = \zeta$. Hence, either $xf = 0$ for all $x \in G^*$, or $xf \neq 0$ for all $x \in G^*$.

We have $\cup_{i=1}^{k} A_i = G$, so $G^* = \{r_{ij} \mid i \in \{1, \ldots, n\}, j \in \{1, \ldots, k\}\}$. See (26:10). For $f \in N$, if $r_{ij}f = 0$ for some r_{ij}, $i \neq 0$, then $r_{is}f = 0$ for $s \in \{1, \ldots, k\}$. By connectedness of $\mathcal{L}$, there is an element of L_i in common with some other L_j, and so $x_{js}f = 0$ for $s \in \{1, 2, \ldots, k\}$. With connectedness of $\mathcal{L}$, we continue in this way to obtain that $r_{us}f = 0$ for all $r_{us} \in G^*$. Certainly, $0f = 0$, so $f = \zeta$. Hence, $f, g \in N$ and $f \circ g = \zeta$ implies $\zeta \in \{f, g\}$. We apply (26.40) to conclude that $N = M(G, k, H)$ is a nearfield.

The work we have presented here is a simplified version of the nice work done by Peter Fuchs and Carl Maxson in [Fu & Ma]. In that work, they develop a criterion to decide when $M(G, k, H)$ is a finite nearfield for finite G and $H = A_1/B_1 \overset{\times}{\sim} \cdots \overset{\times}{\sim} A_k/B_k$.

Mathematicians are like Frenchmen. Whenever you say something to them, they translate it into their own language, and at once it is something entirely different.

Goethe, (Maxims and Reflections, 1829)

Appendix

In this appendix, we provide various diagrams which are related to the material of the text, but which do not flow nicely with our presentation. There are four parts. The first part is a description of all the 10 point graphs of all the E_c^1 of §6 for the circular planar nearrings $(Z_p, \mathcal{B}_{10}^*, \in)$ of the table in §5. The second part consists of a pair of interesting graphs which we have found to have an interesting alternative display using the shape of prisms. The third part consists of an interesting way to display the points of various $(Z_n, \mathcal{B}_6^*, \in)$ where $n = 1 + 3u(u+1)$, which suggests that Z_n is hexagonal in shape and that the points of Z_n can be obtained from translations of the 'basic hexagon' from Φ, where (Z_n, Φ) is an appropriate Ferrero pair, with $|\Phi| = 6$, and where the elements of Φ provide direction for the translations. The fourth part is a matrix from which one can see (a) the BIBD structure, (b) the row code $\mathcal{C}^A(v)$, and (c) the column code $\mathcal{C}_A(b)$ for a specific circular planar nearring.

We are indebted to Wen-Fong Ke not only for the discovery of all the graphs here but also for his mastery of display using TEX.

A1. The graphs of E_c^1 of the $(Z_p, \mathcal{B}_{10}^*, \in)$ from the table in §5

Below is a pictoral description of all the 10 point graphs of the various E_c^1 for the circular planar nearrings $(Z_p, \mathcal{B}_{10}^*, \in)$ of the table in §5. The pictures are followed by a table, and the table is in matrix form which tells where the various graphs represented by the pictures can be found. The rows are labelled by the primes of the table in §5. The columns are labelled by the names of the graphs whose pictures are below. The entries are the values of c. For example, the row labelled 421 and the column labelled I_1 provide

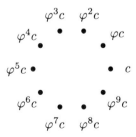

Figure T. The graph of various $T(1; c)$.

$c = 366$, and so E^1_{366} has graph I_1 found among the pictures of the various graphs. All the missing representative values of c give E^1_c whose graphs are as in Figure T, which represents a torus $T(1; c)$. See (6.6).

The following pictoral display of the graphs of the various E^1_c provided in the table below are built upon the display of Figure T, where the vertices are exactly as in Figure T, and where the solid edges and dashed edges are as defined in §6.

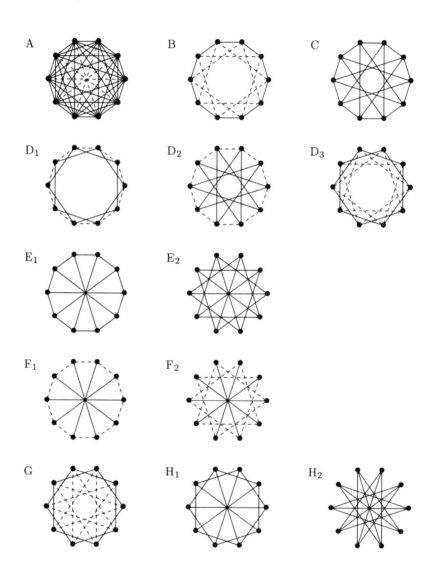

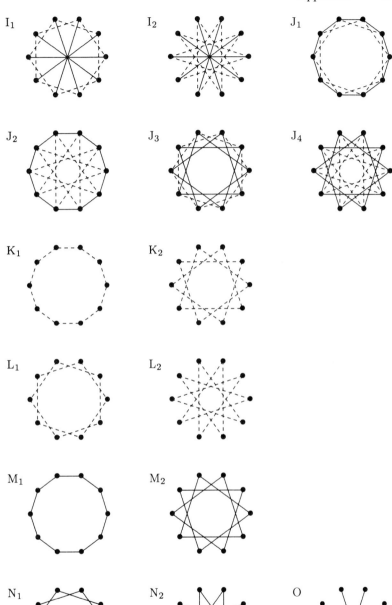

p	A	B	C	D$_1$	D$_2$	D$_3$	E$_1$	E$_2$	F$_1$	F$_2$	G	H$_1$	H$_2$	I$_1$	I$_2$
131	1	32		107			8	125				4	64		
151	1	133						107			65	43	9		
181	1		150			8		7							
191	1											19	54		
211	1			64		41	16	63							
241	1								102						213
251	1	23						216					6		
271	1			49		234									
281	1								210						204
311	1				237	185									
331	1												227		
401	1									379					
421	1								282					366	
431	1														
461	1														
491	1														
521	1														
541	1													497	
571	1														
601	1														
631	1														
641	1									423					602
661	1														
691	1														
701	1														
751	1														
761	1														648
811	1														
821	1														
881	1								300					66	
911	1														
941	1													767	267
971	1														
991	1														

p	J_1	J_2	J_3	J_4	K_1	K_2	L_1	L_2	M_1	M_2
131			128	83					35	
151	148				54				38	6, 88
181					28		64	75	57, 94	16, 47
191	135			143	178	12			174, 52	130, 37
211							4	82	180, 90	117, 149
241						137	49		113, 41, 7	111, 188, 237
251		161			236		175		129	3, 36
271						166	6	44	15, 229	114, 118, 90
281						27	116		264, 39, 67	117, 169, 50
311						297	284	189	146, 260	17, 35, 66
331		280			201	81	243		27, 9	113, 182, 270
401					255		211	85	116, 3, 68	18, 208, 54
421						201		128	364, 376, 94	143, 256, 4
431					370	7	317	399	156, 176, 316	187, 303, 423
461					8	404	355	204	104, 16, 26	237, 281, 405
491					343	16	355	385	233, 359, 391	182, 474, 64
521					248	200	488	349	103, 223, 293	441, 79, 96
541					378	530		242	16, 319, 64	436, 453, 484
571					178	415	103	243	290, 299, 372	325, 326, 91
601					508	309	598	435	138, 31, 40	113, 498, 9
631					413	458	416	90	366, 384, 617	270, 562, 589
641					156		140		170, 326, 606	264, 89, 90
661					229	16	618	317	256, 32, 634	520, 64, 65
691					129	476	3	28	414, 43, 500	303, 389, 88
701					472	124	95	282	343, 383, 64	339, 463, 666
751					589	547	157	243	500, 553, 683	257, 574, 721
761					18	184	74		149, 249, 535	160, 311, 433
811					480	198	565	26	27, 453, 626	197, 402, 684
821					221	324	807	195	211, 512, 748	23, 442, 793
881						227		438	405, 436, 57	281, 5, 791
911					149	724	504	248	360, 401, 462	47, 616, 66
941					115	487			300, 593, 64	171, 534, 715
971					623	159	298	930	269, 64, 694	172, 327, 631
991					595	609	305	839	133, 436, 798	36, 824, 980

p	N_1	N_2	O
131	119	16, 2	
151	22	75, 77	36
181	128, 14	32, 4	114, 119, 2
191	33, 59	43, 53	166, 170
211	32	126, 2, 87	164, 8
241	178, 74, 85	206, 232, 236	45, 46
251	147, 150, 46	21, 213	18, 41
271	110, 212	188, 19, 269	14, 199, 216
281	150, 220, 243	41, 70, 9	88, 98
311	113, 222, 55	142, 42	153, 173, 267
331	148, 215, 3	171, 57	224, 30, 90
401	118, 243, 328	290, 295, 306	145, 27, 34
421	122, 141, 64	117, 183, 311	188, 281
431	16, 230, 246	375, 406, 57	122, 228, 273, 45
461	102, 249, 4	202, 256, 408	101, 13, 2, 349
491	237, 417, 438	423, 67, 8	4, 441, 454, 91
521	190, 422, 430	21, 243, 343	208, 3, 508, 63
541	121, 189, 4	118, 265, 519	365, 403, 8
571	273, 3, 407	173, 337, 493	137, 145, 331, 422
601	137, 251, 82	130, 456, 480	173, 211, 317, 355
631	243, 467, 552	521, 532, 98	139, 334, 377, 551
641	231, 248, 309	160, 453, 9	375, 601
661	445, 489, 617	309, 607, 8	128, 363, 4, 553
691	358, 555, 558	122, 27, 470	185, 186, 354, 580
701	141, 236, 323	398, 461, 62	118, 31, 32, 582
751	417, 44, 81	478, 540, 733	139, 180, 336, 481
761	324, 528, 9	235, 36, 37	498, 727, 80
811	228, 459, 551	336, 463, 501	276, 35, 641, 667
821	451, 508, 521	162, 677, 814	256, 422, 671, 81
881	594, 661, 832	301, 327, 848	584, 9
911	358, 505, 531	176, 428, 610	480, 516, 696, 706
941	528, 604, 75	278, 714, 854	264, 357
971	466, 530, 869	757, 90, 905	120, 282, 537, 570
991	465, 634, 884	349, 6, 808	314, 591, 893, 99

(A1.1) Problem. In the table above, notice that as the prime p gets larger, there seem to be exactly 21 of the E_c^1 represented. This means that all the other E_c^1 have a graph as that in Figure T, so these E_c^1 each form a torus. Exactly what conditions give us a torus; that is, exactly when does an E_c^1 have a graph without edges?

(A1.2) Problem. In the table above, notice that as the prime p gets larger, one seems to obtain the same 10 graphs plus the graph of Figure T. In addition, these same 10 graphs seem to always occur the same number of times. This suggests that graphs B—J_4 are an anomaly. This phenomenon occurs with circles of all small sizes, so there must be some nice theorems waiting to be discovered. What are they?

Exercise. Find isomorphisms among the graphs pictured here.

A2. An alternate display of some graphs

There are alternative ways to display the graphs of the various E_c^r. We have had some ways displayed already in §6. Here, we present an alternative way for the graph E_{49}^1 in $(Z_{313}, \mathcal{B}_{12}^*, \in)$ and the graph of E_{98}^1 in $(Z_{229}, \mathcal{B}_{12}^*, \in)$.

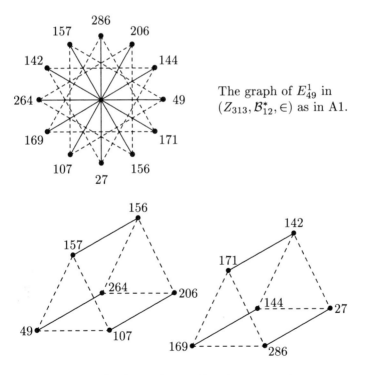

The graph of E_{49}^1 in $(Z_{313}, \mathcal{B}_{12}^*, \in)$ as in A1.

Graph of E_{49}^1 in $(\mathbf{Z}_{313}, \mathcal{B}_{12}^*, \in)$ in prism form.

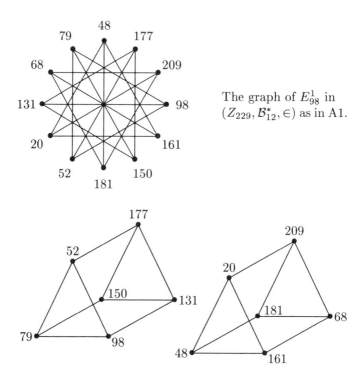

The graph of E_{98}^1 in $(Z_{229}, \mathcal{B}_{12}^*, \in)$ as in A1.

Graph of E_{98}^1 in $(\mathbf{Z}_{229}, \mathcal{B}_{12}^*, \in)$ in prism form.

(A2.1) Problems. This alternative display of some graphs in prism form is not restricted to the two examples here. For example, one obtains analogous pentagon prisms with E_{22}^1 in $(Z_{2141}, \mathcal{B}_{20}^*, \in)$ and with E_{22}^1 in $(Z_{1801}, \mathcal{B}_{20}^*, \in)$. How many prisms can one have in a single E_c^1? Which polygons, in addition to triangles and pentagons, occur in graphs with prism form? Exactly when does one obtain graphs with prism form?

(A2.2) Exploratory problem. These graphs having prism form seem to be generalizations of the simple closed 2–link chains of (6.11). Can our more complex graphs of the E_c^1 be described in some way in terms of simpler graphs? What are the relationships between the structure of the graphs of an E_c^1 and the structure of the group Φ of the Ferrero pair (N, Φ) of the circular planar nearring $(N, +, \cdot)$?

Exercise. (Wen-Fong Ke and Hubert Kiechle) Find a $(Z_p, \mathcal{B}_{15}^*, \in)$ having an E_c^1 with a 'simple closed chain' of five triangles and a 'simple closed chain' of three pentagons.

A3. A representative display of points of $(Z_n, \mathcal{B}_6^*, \in)$ when $n = 1 + 3u(u + 1)$

Suppose $n = 1 + 3u(u + 1)$. In the ring $(Z_n, +, \cdot)$, take $\phi = 3u + 2$. Then ϕ is a unit. In fact, one can see from elementary computation that $\Phi = \langle \phi \rangle = \{3u + 2, 3u + 1, -1, -(3u + 2), -(3u + 1), 1\}$, a cyclic group of order 6. To see that (Z_n, Φ) will give us a ring generated planar nearring, it is enough to show that $3u$, -2, and $-3u - 3$ are units. See (4.16). From $0 = 1 + 3u(u + 1)$, we obtain $3u(-u - 1) = 1$, and from $0 = 1 + u(3u + 3)$, we obtain $u(-3u - 3) = 1$. Now -2 is a unit if and only if 2 is a unit, and since 2 and $1 + 3u(u + 1)$ are relatively prime, we are assured that 2 is a unit. Wen-Fong Ke* first observed that one always obtains a Ferrero pair (Z_n, Φ) in this way. If $n > 19$ is a prime, then any resulting planar nearring $(Z_n, +, *)$ will be circular. This follows from recalling from §5 that $\mathcal{P}_6^* = \{2, 3, 7, 13, 19\}$. A nice and curious consequence of the above is that one can display the elements of such a Z_n in the shape of a 'hexagonal plane'. As we proceed with our discussion, the reader is invited to follow using the diagram given below for $n = 1 + 3 \cdot 4 \cdot (4 + 1) = 61$.

Put 0 in the centre of a circle and then put the six elements of $\Phi = \langle \phi \rangle$, with $\phi = 3u + 2$, on this circle in a similar was as was done in Figure T. These six points of Φ, connected by line segments between each ϕ^i and ϕ^{i+1}, form our 'basic hexagon', and each ϕ^i also defines a 'direction' for a 'basic geometric translation'. (These are the points 14, 13, 60, 47, 48, and 1 for our 'hexagonal plane' displayed below.) Using these 'basic geometric translations', one quickly obtains another layer added to the 'basic hexagon'. Continue with these 'basic geometric translations' to add additional layers to the 'basic hexagon', and every additional layer results again in a hexagonal shape. Upon adding the $(u-1)$th layer, one obtains all the points of Z_n exactly once! (In the example given below, our $(u - 1)$th $= (4 - 1)$th layer has corners 56, 52, 57, 5, 9, and 4.) Continuing, one obtains in the uth layer only points which have already been obtained. (These are the points circled in the diagram below.) But which points are they? These are natural 'neighbours' of points on the outer hexagon of the $(u - 1)$th layer, and one sees in the diagram below that each side of this outer hexagon should bave the opposite side of this hexagon as a 'neighbouring side'. For example, the 'neighbours' of 24 should include 23 and 37, and 23 and 37 are found on the opposite side. Hence, ths side consisting of $\{57, 10, 24, 38, 52\}$ of the outer hexagon in the third layer has the side consisting of $\{9, 23, 37, 51, 4\}$ as a 'neighbouring side'.

---

* Hubert Kiechle extended these observations to construct Ferrero pairs (R, Φ), where R is a commutative ring with identity and $\Phi = \langle \phi \rangle$, where ϕ is the root of the nth cyclotomic polynomial and n is a unit in R.

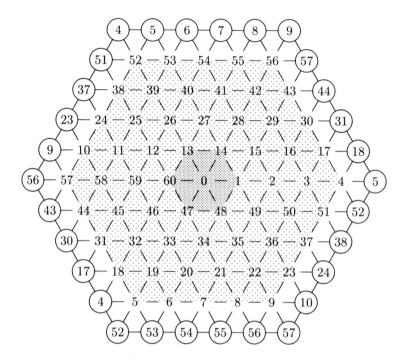

The plane $(Z_{61}, \mathcal{B}_6^*, \in)$ is hexagonal.

Exercise. Take various primes p where $p - 1$ is divisible by 6. Let $(Z_p, +, *)$ be the field generated nearring from (Z_p, Φ), the Ferrero pair with Φ cyclic of order 6. If $\Phi = \langle \phi \rangle$, arrange the elements of Φ on a circle with centre 0 as we have discussed here. Using this 'basic hexagon', continue with the 'basic geometric translations' defined by the elements of Φ until you obtain all the elements of Z_p. What is the shape of your plane $(Z_p, \mathcal{B}_6^*, \in)$? Watch one hexagonal plane grow into another as you increase from one prime $p = 1 + 3u(u + 1)$ to the next prime $q = 1 + 3v(v + 1)$.

(A3.1) Exploratory problem. Are there other interesting ways to illustrate the points of a planar nearring $(N, +, \cdot)$?

Exercise. If $n = 1 + 3 \cdot 5 \cdot (5 + 1) = 91$, show that the ring $(Z_{91}, +, \cdot)$ has two cyclic subgroups of order 6 of the group of units. Call them Φ_1 and Φ_2. One, say Φ_1, is as we have presented here. What happens when you use Φ_2 in place of Φ_1?

A4. An incidence matrix

In §8 we constructed an incidence matrix A from a BIBD $(N, \mathcal{B}, \in)$. From A we constructed a row code $\mathcal{C}^A(v)$ and a column code $\mathcal{C}_A(b)$. The column code $\mathcal{C}_A(b)$ had the particularly nice property that $d(\mathbf{a}, \mathbf{b}) = 2(r - \lambda)$ if $\mathbf{a} \neq \mathbf{b}$. See (8.3). Also, Johnson's bound was reached in $\mathcal{C}_A(b)$. See (8.9). When $(N, \mathcal{B}, \in)$ was circular, the error correcting capability of $\mathcal{C}^A(v)$ was particularly nice. See (8.6). One has all this from $(N, \mathcal{B}^*, \in)$ if $(N, +, \cdot)$ is a finite circular planar nearring, and these concepts are illustrated here with $(Z_7, \mathcal{B}_3^*, \in)$.

$$
A = \begin{bmatrix}
0 & 1 & 1 & 0 & 1 & 0 & 0 \\
0 & 0 & 1 & 1 & 0 & 1 & 0 \\
0 & 0 & 0 & 1 & 1 & 0 & 1 \\
1 & 0 & 0 & 0 & 1 & 1 & 0 \\
0 & 1 & 0 & 0 & 0 & 1 & 1 \\
1 & 0 & 1 & 0 & 0 & 0 & 1 \\
1 & 1 & 0 & 1 & 0 & 0 & 0 \\
0 & 0 & 0 & 1 & 0 & 1 & 1 \\
1 & 0 & 0 & 0 & 1 & 0 & 1 \\
1 & 1 & 0 & 0 & 0 & 1 & 0 \\
0 & 1 & 1 & 0 & 0 & 0 & 1 \\
1 & 0 & 1 & 1 & 0 & 0 & 0 \\
0 & 1 & 0 & 1 & 1 & 0 & 0 \\
0 & 0 & 1 & 0 & 1 & 1 & 0
\end{bmatrix}
$$

The incidence matrix A for $(Z_7, \mathcal{B}_3^*, \in)$.

The rows represent a BIBD with parameters $(v, b, r, k, \lambda) = (7, 14, 6, 3, 2)$. The row code $\mathcal{C}^A(7)$ is a $(7, 14, 2)$-code with constant weight 3. The column code $\mathcal{C}_A(14)$ is a $(14, 7, 8)$-code with constant weight 6. For any two distinct columns $\mathbf{a}$ and $\mathbf{b}$ of A, we have $d(\mathbf{a}, \mathbf{b}) = 2(r - \lambda) = 8$. (See §7 and §8.)

Symmetry is a vast subject, significant in art and nature. Mathematics lies at its root, and it would be hard to find a better one on which to demonstrate the working of the mathematical intellect.

Hermann Weyl, *Symmetry*

List of symbols

The following is a list of symbols whose meanings are more or less global throughout this book. Most of the notation not listed here has a rather local meaning, so one can probably find its meaning in the section of its use.

$f \circ g$	function f composed with function g; also, polynomial f composed with polynomial g	1
$\mathrm{End}_R M$	set of all R-endomorphisms of an R-module M over a ring R	2
$f + g$	pointwise sum of functions f and g	3
$f \mid X$	function f restricted to set X	3
$\mathbf{R}$	set of real numbers	3
$\mathbf{R}[x]$	set of all polynomials with coefficients in $\mathbf{R}$	3
$\mathcal{F}(X)$	elements of free group $(\mathcal{F}(X), +)$ on set X	3
$\lvert X \rvert$	cardinality of set X	3
$\mathrm{Hom}(A, B)$	set of all homomorphisms of structure A into structure B	4
ζ	often identity function with respect to a binary operation $+$ or $\oplus$	4
$\mathcal{C}(G, H)$	set of continuous functions from topological space G into topological space H	6
$\mathcal{T}$	a topology for a topological space	6
$f : A \to B$	a function with domain A and codomain B	7
$x \mapsto y$	there is a function, perhaps unnamed, which maps x to y	6
$\mathbf{C}$	set of all complex numbers	6
A^*	a set A without some of its elements, usually 0 or some elements which behave something like 0	7
$M(G)$	set of all functions from G into G	7
$M_\Phi(G)$	set of all functions in $M(G)$ which commute with elements of Φ with respect to $\circ$	8
$A \subseteq B$	set A is a subset of set B	8
$M < N$	a nearring M is a subnearring of a nearring N, or perhaps a group M is a subgroup of a group N	8
$M_0(G)$	subset of $M(G)$, each of whose elements maps 0 onto 0	9

$R[[x]]$	set of power series with indeterminate x and coefficients from ring R	10
$A \times B$	cartesian product of sets A and B	11
$\mathcal{A}_n(R)$	affine transformations of an n-dimensional free R-module	12
$\mathcal{M}_n(R)$	n by n matrices over R	12
$R[x, y]$	polynomials with indeterminates x and y over R	12
$\mathcal{U}(R)$	group of units of R	12
RG	group algebra for a ring R and group G	13
$\text{End}_R A$	R-endomorphisms of A	14
Z_n	integers modulo n	15
$\mathbf{Q}$	rational numbers	16
S_n	permutations on n symbols	17
$[a, b]$	$-a - b + a + b$, the commutator	18
N_0	zero-symmetric part of nearring N	20
N_c	constant part of nearring N	21
$A \subset B$	set A is a proper subset of set B	22
$\varnothing$	empty set	22
$\text{Ann}(x)$, $\text{Ann}(Y)$	annihilators	22
$f^{-1}(A)$	pre-image of set A	23
M/N	quotient structure of M by N	25
$a + N$	element of M/N containing a for $(M, +)$	26
ST	set of all st with $s \in S$ and $t \in T$	27
$\bar{p}$	function defined by polynomial p	31
$\overline{F[x]}$	functions defined by polynomials in $F[x]$	31
$X/\equiv$	set of equivalence classes defined on X by equivalence relation $\equiv$	39
$A \setminus B$	elements of A but not of B	40
$\text{Aut}N$	automorphisms of N	43
$\cup \mathcal{C}$	union of members of $\mathcal{C}$	44
$A \lhd B$	A is so that the quotient structure B/A is possible	66
F_{p^n}	field with p^n elements	68
Z	set of integers	70
$\mathcal{N}(\Phi)$	normalizer of Φ in $\text{Aut}(N, +)$	74
$\langle g \rangle$	subgroup generated by g	75
$\cap \mathcal{C}$	intersection of members of $\mathcal{C}$	92
$GF(p^n)$	field with p^n elements	157
$[x]$	greatest integer function	167
A^X	all mappings from X to A	230
$R[X]$	free R-algebra on X; all polynomials with indeterminates in X over R	232
$\otimes$	symbol used with tensor products	232

$R[T, T^{-1}]$	polynomials with indeterminates in $T \cup T^{-1}$	232
$\sum_{|X|}^{*} A$	complete direct sum of $|X|$ copies of A	235
$\oplus \sum_n R$	direct sum of n copies of R	237
$\prod_{|T|}^{*} \mathcal{U}(A)$	complete direct product of $|T|$ copies of $\mathcal{U}(A)$	237
$\prod_{t \in T} \phi(t)$	product of all $\phi(t)$, with $t \in T$	239
$M \oplus_A R$	abstract affine nearrings of R-module M	241
$K \times_\theta A$	semidirect product	243
$\mathrm{Map}(X, G)$	all mappings from X into G	263
$(A : B)$	$\{n \in N \mid bn \in A \text{ for each } b \in B\}$	267
det	determinant	287
D_B	generalized dihedral group for abelian group B	291
$\mathrm{Inn}G$	inner automorphisms of G	291
$A(G)$	nearring generated by the automorphisms of G	291
$E(G)$	nearring generated by the endomorphisms of G	291
$I(G)$	nearring generated by $\mathrm{Inn}(G)$	291
$D(S, G)$	nearring generated by the semigroup of endomorphisms S of G	291
$A_0[[x]]$	power series over A with zero constant term	310
$\mathcal{A}(V, F)$	all affine transformations of the F-vector space V	324
l.c.m.	least common multiple	339
$(N : M)$	the quotient of N by M	350
$K \times_\theta^f Q$	extension of K by Q	359
$B \,_\theta \times A$	also a semidirect product	362
$K \wr A$	wreath product	363
$\mathrm{Hol}(V)$	holomorph of V	369
$(X, \mathcal{T})$	a topological space	381
F^n	$F \circ F^{n-1}$	390
$a \circ b$	Jacobson's circle operation	404
$R[G]$	group algebra for a ring R and group G	432
$R^{(I)}$	functions from I into R with finite support	432

The best notation is no notation; whenever it is possible to avoid the use of a complicated alphabetic apparatus, avoid it. A good attitude to the preparation of written mathematical exposition is to pretend that you are explaining the subject to a friend on a long walk in the woods, with no paper available; fall back on symbolism only when it is really necessary.

Paul R. Halmos, 1970

Bibliography

For an extensive bibliography on nearrings, see [P, 1983] below. In [W], one can find an excellent bibliography on nearfields.

[B]   Betsch, G.
(1959) Ein Radikal für Fastringe, *Math. Z.*, **78**, 86–90.
(1963) *Struktursätze für Fastringe*, Dissertation, Univ. Tübingen.
(1971) Some structure theorems on 2-primitive near-rings, *Colloq. Math. Soc. Janus Bolyai 6, Rings, Modules, and Radicals*, Keszthely, Hungary, 73–102.

[B et al.]   Brillhart, J., Lehmer, D., Selfridge, J., Tuckerman, B. and Wagstaff, S.
(1988) Factorization of $b^n \pm 1$, *Contemporary Math.*, **22**.

[B & H]   Blackburn, N. and Huppert, B.
(1982) *Finite Groups*, Springer-Verlag, Berlin.

[B & P]   Beaumont, R. A. and Pierce, R. S.
(1963) *The Algebraic Foundations of Mathematics*, Addison-Wesley, Reading.

[B & S]   Berman, G. and Silverman, R. J.
(1959) Near-rings, *Am. Math. Monthly,* **66**, 23–34.

[Be]   Beutelspacher, A.
(1988) Enciphered geometry. Some applications of geometry to cryptography, *Annals of Discrete Math.*, **37**, 59–68.
(1991) Applications of finite geometry to cryptography, *Geometry, Codes, and Cryptography*, (G. Longo, M. Marchi, A. Sgarro, eds.) CISM Courses and Lectures, No. 313, Springer-Verlag, 161–186.

[Be & R]   Beutelspacher, A. and Rosenbaum, U.
(1990) Geometric authentication system, *Ratio Math.*, **1**, 39–50.

[Be & V]   Beutelspacher, A. and Vedder, K.
(1989) Geometric structures as threshold schemes, *Cryptography and Coding (Cirencester)*, Oxford Univ. Press, New York, 255–268

460   *Bibliography*

[Bi]	Birkenmeier, G.
	(1989) Seminearrings and nearrings induced by the circle operation, *Riv. Mat. Pura et Apl.*, **5**, 59–68.
[Bi & He]	Birkenmeier, G. and Heatherly, H.
	(1989) Medial near-rings, *Monatsh. Math.*, **107**, No. 2, 89–110.
[Bl]	Blackett, D. W.
	(1953) Simple and semi-simple near-rings, *Proc. Am. Math. Soc.*, **4**, 772–785.
[Ca]	Cartan, H.
	(1963) *Theory of Analytic Functions*, Addison-Wesley, Reading.
[CMM]	Clay, J. R., Maxson, C. J. and Meldrum, J. D. P.
	(1984) The group of units of centralizer near-rings, *Comm. in Algebra*, **12** (21), 2591–2618.
[C & M]	Clay, J. R. and Meldrum, J. D. P.
	(1983) Amalgamated product near-rings, *Contributions to General Algebra 2, Proceedings of the Klagenfurt conference, June 10–13, 1982,* Verlag Hölder-Pichler-Tempsky, Wien, 43–70.
[De]	Dembowski, P.
	(1968) *Finite Geometries*, Springer-Verlag, Berlin.
[Di]	Dickson, L. E.
	(1905) Definitions of a group and a field by independent postulates, *Trans. Am. Math. Soc.*, **6**, 198–204.
	(1905a) On finite algebras, *Nachr. Akad. Wiss. Göttingen*, 358–393.
[Du]	Duffy, L. R.
	(1983) An elementary proof of the isomorphism $\mathbf{C}^* \approx \mathbf{S}^1$, *Am. Math. Monthly*, **90**, 201–202.
[EH]	Eckmann, B. and Hilton, P.
	(1962) Group-like structures in general categories I: multiplications and comultiplications, *Math. Ann.*, **145**, 227–255.
	(1963) Group-like structures in general categories II: equalizers, limits, lengths, *Math. Ann.*, **151**, 150–186.
	(1963a) Group-like structures in general categories III: primitive categories, *Math. Ann.*, **150**, 165–187.
[Fe]	Ferrero, G.
	(1970) Stems planari e BIB-disegni, *Riv. Mat. Univ. Parma* (2), **11**, 79–96.

(1972) Su certe geometrie gruppali naturali, *Riv. Mat. Univ. Parma* (3), **1**, 97–111.

[F-C] Ferrero-Cotti, C.
(1972) Sugli stems il cui produtto è distributivo respetto a se stesso, *Riv. Mat. Univ. Parma* (3), **1**, 203–220.

[F] Fröhlich, A.
(1958) Distributively generated near-rings I. Ideal Theory, *Proc. London Math. Soc.*, **8**, 76–94.

(1958a) The near-ring generated by the inner automorphisms of a finite simple group, *J. London Math. Soc.*, **33**, 95–107.

(1968) *Formal Groups*, Lecture Notes in Math., 74, Springer-Verlag, Berlin.

[FHP] Fuchs, P., Hofer, G. and Pilz, G.
(1990) Codes from planar near-rings, *IEEE Trans. on Information Theory*, **36**, 647–651.

[FGR] Fossum, R., Griffith, P. and Reiten, I.
(1975) *Trivial Extensions of Abelian Categories*, Lecture Notes in Math., 456, Springer-Verlag, Berlin.

[FL] Fuchs, L.
(1963) Some results and problems on Abelian groups, *Topics in Abelian Groups*, Scott, Foresman, Glenview, IL.

[Fo] Fong, Y.
(1990) On the structure of abelian syntactic near-rings, *First Inter. Symposium on Algebraic Structures and Number Theory, Hong Kong, August 8–13, 1988*, World Scientific (1990), 114–123.

[Fo & C] Fong, Y. and Clay, J. R.
(1988) Computer programs for investigating syntactic near-rings of finite group-automata, *Bull. Math. Acad. Sinica*, **16**, No. 4, 295–304.

[Fou] Foulser, D. A.
(1962) *On Finite Affine Planes and their Collineation Groups*, Dissertation, U. of Michigan, Ann Arbor.

[Fu] Fuchs, P.
(1990) On the structure of ideals in sandwich near-rings, *Results in Math.*, **17**, 256–271.

[Fu & Ma] Fuchs, P. and Maxson, C. J.
(1989) Meromorphic products determining near-fields, *J. Austral. Math. Soc. (Series A)*, **46**, 365–370.

[G] Gonshor, H.

(1964) On abstract affine near-rings, *Pacific J. Math.*, **14**,
1237–1240.

[GG]　　　Grainger, G.

(1988) *Left Modules for Left Nearrings*, Dissertation, Univ.
of Arizona, Tucson.

[H]　　　Hazewinkel, M.

(1978) *Formal Groups and Applications*, Academic Press,
New York.

[Ha]　　　Hall, M.

(1959) *The Theory of Groups*, Macmillan, New York.

(1971) Designs with transitive automorphism groups, *Proc.
of Symposia in Pure Math.*, AMS, **19**, T. L. Motzkin,
ed., 109–113.

[Hi]　　　Higman, D. G.

(1967) Intersection matrices for finite permutation groups,
J. of Algebra, **6**, 22–42.

[HH]　　　Heatherly, H.

(1973) Distributive near-rings, *Quart. J. Math. Oxford Ser.*
(2), **24**, 63–70.

[HK]　　　Kautschitsch, H.

(1976) Über Vollideale in Potenzreihenringen, *Period Math.
Hungar.*, **7**, 141–152.

(1977) Connections between the ring-, near-ring- and compo-
sition-ideals in the algebra of formal power series,
Colloquia Mathematica Societatis János Bolyai, 29.
Universal Algebra, Esztergom (Hungary), 453–458.

[HK & Mü]　Kautschitsch, H. and Müller, W.

(1980) Ideale in Kompositionsringen formaler Potenzreihen
mit nilpotenten Anfangskoeffizienten, *Arch. der
Math.*, **34**, 517–525.

[HK & Ml]　Kautschitsch, H. and Mlitz, R.

(1989) Maximal ideals in composition-rings of formal power
series, *Contributions to General Algebra 6*, Verlag
Hölder-Pichler-Tempsky, Wien, 131–140.

[Ho]　　　Holcombe, W. M.

(1977) Categorical representations of endomorphism nearrings, *J. London Math. Soc.*, (2) **16**, 21–37.

(1983) The syntactic near-ring of a linear sequential machine, *Proc. Edinb. Math. Soc.*, **26**, 15–24.

[HP]　　　Hofer, G. and Pilz, G.

(1983) Near-rings and automata, *Contributions to General Algebra 2, Proceedings of the Klagenfurt Conference, June 10–13, 1982*, Verlag Hölder-Pichler-Tempsky, Wien, 153–162.

[Hu]   Huppert, B.
(1967) *Endliche Gruppen I*, Springer-Verlag, Berlin.

[HT]   Hungerford, T.
(1974) *Algebra*, Holt, Rinehart, and Winston, New York.

[I-H]   Chen, I.-Hsing.
(1991) *Some Combinatorial Structures Arising from Finite Planar Near-rings*, Thesis, National Chiao Tung University, Hsinchu.

[J]   Jacobson, N.
(1974) *Basic Algebra I*, Freeman and Co., San Francisco.

[K]   Karzel, H.
(1965) *Inzidenzgruppen*, Vorlesungsausarbeitung von I. Pieper und K. Sörensen, Universität Hamburg.

[K & O]   Karzel, H. and Oswald, A.
(1990) Near-rings (MSD)- and Laguerre Codes, *J. of Geometry*, **37**, 105–117.

[Ke]   Ke, W. F.
(1992) *Structures of Circular Planar Nearrings*, Dissertation, University of Arizona, Tucson.

[L & N]   Lausch, H. and Nöbauer, W.
(1973) *Algebra of Polynomials*, North Holland/American Elsivier, Amsterdam.

[lR *et al.*]   le Riche, L. R., Meldrum, J. D. P. and van der Walt, A. P. J.
(1989) On group near-rings, *Arch. Math.*, **52**, 132–139.

[M]   Meldrum, J. D. P.
(1985) *Near-rings and their Links with Groups*, Pitman Publ. Co. (Research Note Series No. 134.)

[M & B]   MacLane, S. and Birkhoff, G.
(1967) *Algebra*, Macmillan, New York.

[M & L]   Malone, J. J. and Lyons, C.
(1972) Finite dihedral groups and d.g. near-rings I, *Comp. Math.*, **24**, 305–312.
(1973) Finite dihedral groups and d.g. near-rings II, *Comp. Math.*, **26**, 249–259.

[Ma]   Maxson, C. J.
(1970) Dickson near-rings, *J. Algebra*, **14**, 152–169.
(1971) On morphisms of Dickson near-rings, *J. Algebra*, **17**, 404–411.

464   *Bibliography*

[Mag]   Magill, K. D. Jr
               (1967) Semigroup structures for families of functions, I; some
                   homomorphism theorems, *J. Aust. Math. Soc.*, **7**,
                   81–94.
               (1967a) Semigroup structures for families of functions, II;
                   continuous functions, *J. Aust. Math. Soc.*, **7**, 95–
                   107.
               (1980) Automorphism groups of laminated near-rings, *Proc.
                   Edin. Math. Soc.*, **23**, 97–102.
               (1986) Isomorphisms of sandwich near-rings of continuous
                   functions, *Bolletino U. M. I.*, (6) **5**-B, 209–222.
[Mag et al.]   Magill, K. D. Jr, Blevens, D. K., Misra, P. R., Parnami, J.
                   C. and Tewari, U. B.
               (1988) More on automorphism groups of laminated near-
                   rings, *Proc. Edinburgh Math. Soc.*, **31**, 185–195.
[M & S]   McWilliams, F. J. and Sloane, N. J. A.
               (1977) *The Theory of Error-correcting Codes*, North-Holland,
                   Amsterdam.
[MM]   Modisett, M. C.
               (1988) *A Characterization of the Circularity of Certain Bal-
                   anced Incomplete Block Designs*, Dissertation, Univ.
                   of Arizona, Tucson.
               (1989) A characterization of the circularity of balanced in-
                   complete block designs, *Utilitas Math.*, **35**, 83–94.
[MMT]   Magill, K. D. Jr, Misra, P. R. and Tewari, U. B.
               (1983) Automorphism groups of laminated near-rings de-
                   termined by complex polynomials, *Proc. Edinburgh
                   Math. Soc.*, **26**, 73–84.
               (1983a) Finite automorphism groups of laminated near-rings,
                   Proc. Edinburgh Math. Soc., **26**, 297–306.
[Me]   Meyer, J. H.
               (1986) *Matrix Near-rings*, Dissertation, University of Stellen-
                   bosch.
[N]   Nagata, M.
               (1962) *Local Rings*, Wiley and Sons, New York.
[NH]   Neumann, H.
               (1956) On varieties of groups and their associated near-rings,
                   Math. Z., **65**, 36–69.
               (1967) *Varieties of Groups*, Springer-Verlag, Berlin.
[P]   Pilz, G.
               (1983) *Near-rings*, North-Holland/American Elsevier,
                   Amsterdam.

(1987) Near-rings, 5 lectures, *Sem. Alg. non Commutativa*, Siena, 1–35.

[R] Rotman, J. J.

(1984) *An Introduction to the Theory of Groups*, Allyn and Bacon, Boston.

[St] Stevenson, F. W.

(1972) *Projective Planes*, Freeman, San Francisco.

[S] Suvak, J. A.

(1971) *Full Ideals and their Ring Groups for Commutative Rings with Identity*, Dissertation, Univ. of Arizona, Tucson.

[VW] Veblen, O. and MacLagan-Wedderburn, J. H.

(1907) Non-desarguesian and non-pascalian geometries, *Trans. Am. Math. Soc.*, **26**, 379–388.

[W] Wähling, H.

(1987) *Theorie der Fastkörper*, Thales Verlag, Essen.

[Wi] Wielandt, Helmut

(1964) *Finite Permutation Groups*, Academic Press, New York.

[Z] Zassenhaus, H.

(1936) Über endliche Fastkörper, *Abh. Math. Sem. Hamburg*, **11**, 187–220.

It is tribute to the genius of Galois that he recognized that those subgroups for which the left and right cosets coincide are distinguished ones. Very often in mathematics the crucial problem is to recognize and to discover what are the relevant concepts; once this is accomplished the job may be more than half done.

I. N. Herstein, *Topics in Algebra*

Index

After each entry will appear the page number of the text on which the entry is defined.

Abstract affine transformation 213
Acts 213
Action 213
Admissible semigroup 387
Affine configuration 136
  with one pencil 136
  with two pencils 136
Affine plane 120
Affine space 120
  isomorphic 120
Affine transformation 326
Annihilator 23, 263
Association scheme 161

Balanced incomplete block design 58
  automorphism 66
  efficiency 122
  geometrical 124
  isomorphism 66
  statistical 123
Basis 262
BIBD 57
Bimodule 259, 267
Binary code 168
  column 168
  constant weight 168
  row 168
Blocks 58

Category 212
  concrete 214
  isomorphism 214
  morphism 211

  objects 211
Chain 90
  compound closed 113
  simple closed 90, 92
Characteristic 36
Circle 57
  centre 57
  radius 57
Coboundary 249
Code 168
  column 168
  constant weight 168
  row 168
Codeword 168
  length 168
  weight 168
Collinear 119
Comodule 261
  unitary 261
Completely regular 388
Complexity 437
Composition ring 329
Connected 443
Constant function 21
  punctured 29
Coproduct 214
Crossed homomorphism 247

Difference set 96
Dilatation 132
Distributive element 292

Edge 99

even 99
odd 100
Eigenvalue 286
Eigenvector 286
Eliminant 71
Embedded 25
Embedding 25
Entire 11
Equivalent multipliers 40
Euclid's axiom of parallelism 130
Exploratory problem 8
Exterior point 165

Ferrero pair 45
FHP-procedure 177
Fibered product 217
Field generated 48
Field generated design 68
Fixed point free 44
Formal group law 312
  commutative 313
  endomorphism of 316
  noncommutative 313
  one dimensional 312
Formulae 429
  1st generation 429
  kth generation 429
  length 429
Free addition 262, 265
Frobenius group 49
  complement 49
  kernel 49
Functor 218
  contravariant 218
  covariant 218

Generalized dihedral group 293
  odd 294
Generating sequence 437
Group-semiautomaton 10

Homomorphism 24
  automorphism 24
  endomorphism 24

epimorphism 24
isomorphism 24
monomorphism 24
Hull 119

Ideal 28
  enclosing 331, 344, 354
  full 330
  left 348
  nilpotent 357
  prime 357
  quotient 352
  right 348
  trivial 29
Idealization 322
Idempotent 23
Incidence matrix 168
Incidence relation 57
Incidence space 119
Incidence structure 57
Incident 57
Inputs 397
Interior point 165
Inversive plane 141

J-meromorphic product 441
Johnson's bound 171

Kernel 27
kth-row matrix 282, 432

Laminator 384
Linear fractional transformation 142
Lines 58

Matrix ring 434
Meromorphic product 441
Möbius plane 141, 175
Möbius transformation 142
Module 261
  faithful 258
  generator 263
  homomorphism etc. 263
  idealization of 322
  left 258

monogenic 263
nearring 261
right 259
unitary 261
Morphism function 218

Natural isomorphism 385
Nearalgebra 272
Nearcoalgebra 291
Neardomain 193
planar 201
Nearfield 16
characteristic 36
Dickson 205
planar 40
Nearfield generated 48
Nearly decreasing 391
Nearly increasing 391
Nearly monotonic 392
Nearring 4, 6
abelian 6
abstract affine 13, 326
anticommutative 423
centralizer 9, 271
circular 57
commutator 422
constant 22, 24
constant part 22
Dickson 204
distributive 6
distributively generated 9, 309
double planar 164
endomorphism etc. 293
field generated 48
group 436
integral 41
laminated 384
left 6, 21
matrix 427
nearfield generated 48
nonassociative 203
of matrices 427, 436
planar 40

right 6
ring generated 49
sandwich 384
simple 29
subnearring 9
syntactic 10
trivial 15
zero-symmetric 21
zero-symmetric part 22
Nil group 15

Object 214
cogroup 221
equivalent 214
group 220
initial 214
isomorphic 214
terminal 214
Object function 218
Odd function 395
Orbit 45, 289
trivial 45
Orbital design 162

Parallel 120
Partial plane 130
with parallelism 130
Partially balanced incomplete block design
161
PBIBD 161
Peirce decomposition 24
Plane 120
Polynomial function 32
Principle of idealization 325
Problem 8
Product 214
Pseudosymmetric 384

Regular 44
Representation 258
Resultant 70
Right identity 41
Right inverse 42
Ring 6

Ring generated 49
Ring generated design 75

Semiautomaton 397
Semidirect product 22
    complement 22
Scalar multiplication 258
States 397
Subspace 119
Support system 265
Sylvester's determinant 71

Tactical configuration 58
Threshold 284
Transition function 397

Transition matrix 273
Transitive 143, 192
    sharply 143, 192
Trivial multiplication 15
Trivial orbit 45
Topological group 7
Torus 86
    major radius 86
    minor radius 86

Underlying set 214

Varieties 58
Vertices 99

Weight 429

A mathematician, like a painter or a poet, is a maker of patterns. If his patterns are more permanent than theirs, it is because they are made with ideas. The mathematician's patterns, like the painter's or the poet's, must be beautiful; the ideas, like the colours or the words, must fit together in a harmonious way. Beauty is the first test: there is no permanent place in the world for ugly mathematics.

G. H. Hardy (A Mathematician's Apology)

MAR 1 0 1993